SERENGETI IV

Sustaining Biodiversity in a Coupled Human-Natural System

Edited by
Anthony R. E. Sinclair, Kristine L. Metzger, Simon A. R. Mduma, and John M. Fryxell

The University of Chicago Press
Chicago and London

Anthony R. E. Sinclair is professor emeritus of zoology at the University of British Columbia and coeditor of *Serengeti I, II*, and *III*. **Kristine L. Metzger** is a landscape ecologist working for the US Fish and Wildlife Service in Albuquerque, New Mexico. **Simon A. R. Mduma** is director of the Tanzania Wildlife Research Institute and coeditor of *Serengeti III*. **John M. Fryxell** is professor of integrative biology at the University of Guelph and coeditor of *Serengeti III*.

The University of Chicago Press, Chicago 60637
The University of Chicago Press, Ltd., London

Printed in the United States of America

24 23 22 21 20 19 18 17 16 15   1 2 3 4 5

ISBN-13: 978-0-226-19583-4 (cloth)
ISBN-13: 978-0-226-19616-9 (paper)
ISBN-13: 978-0-226-19633-6 (e-book)
DOI: 10.7208/chicago/9780226196336.001.0001

Library of Congress Cataloging-in-Publication Data

Serengeti IV : sustaining biodiversity in a coupled human-natural system / edited by Anthony R. E. Sinclair, Kristine L. Metzger, Simon A. R. Mduma, and John M. Fryxell.
pages ; cm
Includes bibliographical references and index.
ISBN 978-0-226-19583-4 (cloth : alk. paper) — ISBN 978-0-226-19616-9 (pbk. : alk. paper) — ISBN 978-0-226-19633-6 (e-book) 1. Animal ecology—Tanzania—Serengeti National Park Region. 2. Biodiversity conservation—Tanzania—Serengeti National Park Region. 3. Ecosystem management—Tanzania—Serengeti National Park Region. I. Sinclair, A. R. E. (Anthony Ronald Entrican), editor. II. Metzger, Kristine L., editor. III. Mduma, Simon A. R., editor. IV. Fryxell, John M., 1954– editor.
QL337.T3S425 2015
591.709678—dc23

2014030650

♾ This paper meets the requirements of ANSI/NISO Z39.48-1992 (Permanence of Paper).

CONTENTS

Preface and Acknowledgments *ix*

1 Conservation in a Human-Dominated World *1*
Anthony R. E. Sinclair and Andy Dobson

2 Shaping the Serengeti Ecosystem *11*
Anthony R. E. Sinclair, Andy Dobson, Simon A. R. Mduma, and Kristine L. Metzger

I. Natural Sources of Heterogeneity and Disturbance

3 Scales of Change in the Greater Serengeti Ecosystem *33*
Kristine L. Metzger, Anthony R. E. Sinclair, Sandy Macfarlane, Michael Coughenour, and Junyan Ding

4 Fire in the Serengeti Ecosystem: History, Drivers, and Consequences *73*
Stephanie Eby, Jan Dempewolf, Ricardo M. Holdo, and Kristine L. Metzger

5 Spatial and Temporal Drivers of Plant Structure and Diversity in Serengeti Savannas *105*
T. Michael Anderson, John Bukombe, and Kristine L. Metzger

6 Why Are Wildebeest the Most Abundant Herbivore in the Serengeti Ecosystem? *125*
J. Grant C. Hopcraft, Ricardo M. Holdo, Ephraim Mwangomo, Simon A. R. Mduma, Simon J. Thirgood, Markus Borner, John M. Fryxell, Han Olff, and Anthony R .E. Sinclair

7 Climate-Induced Effects on the Serengeti Mammalian Food Web *175*
John M. Fryxell, Kristine L. Metzger, Craig Packer, Anthony R. E. Sinclair, and Simon A. R. Mduma

II. Response of Biodiversity to Disturbance

8 From Bacteria to Elephants: Effects of Land-Use Legacies on Biodiversity and Ecosystem Processes in the Serengeti-Mara Ecosystem *195*
Louis V. Verchot, Naomi L. Ward, Jayne Belnap, Deborah Bossio, Michael Coughenour, John Gibson, Olivier Hanotte, Andrew N. Muchiru, Susan L. Phillips, Blaire Steven, Diana H. Wall, and Robin S. Reid

9 Biodiversity and the Dynamics of Riverine Forests in Serengeti *235*
Roy Turkington, Gregory Sharam, and Anthony R. E. Sinclair

10 Invertebrates of the Serengeti: Disturbance Effects on Arthropod Diversity and Abundance *265*
Sara N. de Visser, Bernd P. Freymann, Robert F. Foster, Ally K. Nkwabi, Kristine L. Metzger, Andrew W. Harvey, and Anthony R. E. Sinclair

11 The Butterflies of Serengeti: Impact of Environmental Disturbance on Biodiversity *301*
Anthony R. E. Sinclair, Ally K. Nkwabi, and Kristine L. Metzger

12 Small Mammal Diversity and Population Dynamics in the Greater Serengeti Ecosystem *323*
Andrea E. Byrom, Wendy A. Ruscoe, Ally K. Nkwabi, Kristine L. Metzger, Guy J. Forrester, Meggan E. Craft, Sarah M. Durant, Stephen Makacha, John Bukombe, John Mchetto, Simon A. R. Mduma, Denne N. Reed, Katie Hampson, and Anthony R. E. Sinclair

13 Bird Diversity of the Greater Serengeti Ecosystem: Spatial Patterns of Taxonomic and Functional Richness and Turnover *359*
Jill E. Jankowski, Anthony R. E. Sinclair, and Kristine L. Metzger

14 The Effect of Natural Disturbances on the Avian Community of the Serengeti Woodlands *395*
Ally K. Nkwabi, Anthony R. E. Sinclair, Kristine L. Metzger, and Simon A. R. Mduma

15 Carnivore Communities in the Greater Serengeti Ecosystem *419*
Meggan E. Craft, Katie Hampson, Joseph O. Ogutu, and Sarah M. Durant

III. The Human Ecosystem and Its Response to Disturbance

16 The Plight of the People: Understanding the Social-Ecological Context of People Living on the Western Edge of Serengeti National Park *451*
Eli J. Knapp, Dennis Rentsch, Jennifer Schmitt, and Linda M. Knapp

17 Transitions in the Ngorongoro Conservation Area: The Story of Land Use, Human Well-Being, and Conservation *483*
Kathleen A. Galvin, Randall B. Boone, J. Terrence McCabe, Ann L. Magennis, and Tyler A. Beeton

18 Agricultural Expansion and Human Population Trends in the Greater Serengeti Ecosystem from 1984 to 2003 *513*
Anna B. Estes, Tobias Kuemmerle, Hadas Kushnir, V. C. Radeloff, and H. H. Shugart

19 Infectious Diseases in the Serengeti: What We Know and How We Know It *533*
Tiziana Lembo, Harriet Auty, Katie Hampson, Meggan E. Craft, Andy Dobson, Robert Fyumagwa, Eblate Ernest, Dan Haydon, Richard Hoare, Magai Kaare, Felix Lankester, Titus Mlengeya, Dominic Travis, and Sarah Cleaveland

IV. Coupled Human-Natural Interactions

20 Socioecological Dynamics and Feedbacks in the Greater Serengeti Ecosystem *585*
Ricardo M. Holdo and Robert D. Holt

21 Living in the Greater Serengeti Ecosystem: Human-Wildlife Conflict and Coexistence *607*
Katie Hampson, J. Terrence McCabe, Anna B. Estes, Joseph O. Ogutu, Dennis Rentsch, Meggan E. Craft, Cuthbert B. Hemed, Eblate Ernest, Richard Hoare, Bernard Kissui, Lucas Malugu, Emmanuel Masenga, and Sarah Cleaveland

V. Consequences of Disturbance for Policy, Management, and Conservation

22 Bushmeat Hunting in the Serengeti Ecosystem: An Assessment of Drivers and Impact on Migratory and Nonmigratory Wildlife *649*
Dennis Rentsch, Ray Hilborn, Eli J. Knapp, Kristine L. Metzger, and Martin Loibooki

23 Human Health in the Greater Serengeti Ecosystem *679*
Linda M. Knapp, Eli J. Knapp, Kristine L. Metzger, Dennis Rentsch, Rene Beyers, Katie Hampson, Jennifer Schmitt, Sarah Cleaveland, and Kathleen A. Galvin

24 Multiple Functions and Institutions: Management Complexity in the Serengeti Ecosystem *701*
Deborah Randall, Anke Fischer, Alastair Nelson, Maurus Msuha, Asanterabi Lowassa, and Camilla Sandström

25 Sustainability of the Serengeti-Mara Ecosystem for Wildlife and People *737*
Robin S. Reid, Kathleen A. Galvin, Eli J. Knapp, Joseph O. Ogutu, and Dickson S. Kaelo

VI. Synthesis

26 The Role of Research in Conservation and the Future of the Serengeti *775*
Anthony R. E. Sinclair, Julius D. Keyyu, Simon A. R. Mduma, Mtango Mtahiko, Emily Kisamo, J. Grant C. Hopcraft, John M. Fryxell, Kristine L. Metzger, and Markus Borner

27 The Future of Conservation: Lessons from the Serengeti *797*
Anthony R. E. Sinclair, Andy Dobson, Kristine L. Metzger, John M. Fryxell, and Simon A. R. Mduma

Contributors *813*

Index *819*

PREFACE AND ACKNOWLEDGMENTS

The motivation for this series of volumes is to provide a synthesis of the scientific research conducted in the greater Serengeti ecosystem and also to show how a major protected area can be of benefit to society and the world as a whole. The most recent of these volumes was *Serengeti III: Human Impacts on Ecosystem Dynamics* published in 2008. That volume was based on a series of workshops that modeled the system. It laid the theoretical foundation for the present book which puts together the field data that have been collected to test these ideas.

As we outline in chapter 1, the Serengeti National Park is under threat from development projects surrounding the ecosystem. Development from proposed roads through the protected area and airports adjacent to it will have negative impacts. There is also a groundswell of opinion in some quarters that protected area conservation has failed in meeting its objectives of conserving biota for posterity and that we must change direction and focus on conservation in human-dominated systems, which is community-based conservation. Whereas conservation in human landscapes is essential, this cannot be the sole solution to preserving biota. There are many species that cannot live there and these must find refuge in protected areas. However, even the largest protected areas cannot exist alone; they are dependent on the modified ecosystems surrounding them. It is therefore important to create sustainable ecosystems in the surrounding modified landscapes. This in turn requires that human societies are sustainable in terms of wealth, health, and education. We, therefore, in this volume ask three important questions: Do protected areas play a role in conservation that is not achieved in human ecosystems? Can human-dominated systems contribute to conservation objectives? Do these two—protected areas and human-dominated areas—support each other? The natural ecosystem is changing from environmental disturbances within and human impacts outside the system. We ask how the greater Serengeti ecosystem, including both the human and natural components, can be made sustainable in the face of these changes. The chapters, therefore, address both the function of the protected ecosystem and the contribution of the human system that surrounds it.

We have held a series of workshops first at Santa Barbara in 2001–2003 at the National Center for Ecosystem Analysis and Synthesis (NCEAS), then at the Serengeti National Park headquarters at Seronera (2004), and finally at the University of British Columbia (2007), funded by the Peter Wall Institute for Advanced Studies (PWIAS). The most recent workshop in 2007 addressed research that was to test these models. Thus the present book, which is the outcome of that workshop and the subsequent research, first documents more of the biodiversity in the system, this time including microbes, plants, insects, rodents, and birds as well as the larger mammals. Second, it explores both the environmental factors that impinge on this biodiversity and the human impacts from outside the protected area, the human component in the greater Serengeti ecosystem. This aspect explores how human livelihoods are impacted by living next to a famous national park such as the Serengeti, and throws light on what is needed for these peoples to value the area. Finally we summarize the contributions that science has made to both the Serengeti National Park and to Tanzania as a whole over the past fifty years to highlight its value for conservation and the well-being of Tanzanians.

A large number of people have been involved with this latest volume. Some fifty scientists attended the 2007 workshop. By the time the research was completed and the chapters written, we had 91 authors contributing to the book. Most encouraging is the considerable increase in the number of Tanzanians and Kenyans (22), a quarter of the contributors. We hope this trend will continue. As before, many disciplines are represented—ecologists, molecular biologists, geologists, economists, social scientists, mathematicians, and disease specialists—to name some. Members have come from many different countries. Apart from Tanzania and Kenya, they come from Ethiopia, Holland, Germany, Sweden, New Zealand, Australia, Britain, Canada, and the United States, and members are also working in Afghanistan and Indonesia.

We are very grateful to the Tanzania Wildlife Research Institute, which has administered our work and provided the facilities at the Serengeti Wildlife Research Centre in Seronera. The Tanzania National Parks have always been generous in allowing us to conduct our work and we hope this provides some return for them. The several chief park wardens of Serengeti have been very helpful over the years as were the other wardens and rangers, too many to mention by name.

Reviewers, who are experts in their field, commented on each of the chapters so as to provide an outside perspective. They include R. Bengis, R. Boone, M. Boyce, N. Bunnefeld, C. Burton, N. Carter, C. Chapman, L. Coppock, J. Cory, N. DeCrappeo, J. Detling, A. Ford, A. Gaylard, S. Gergel, B. Godley, J. Gross, K. Hodges, S. Huckett, P. Hudson, J. Jankowski, B. Klinkenberg, C. Krebs, W. Laurance, P. Lundberg, J. Luzar, A. MacDougall, A. Marin, R. McCulley, A. Middleton, R. Naidoo, A. Nuno, J. Rist, K. Rogers, N. Stronach, S. van Rensburg, B. van Wilgen, and K. Wilson. We also thank the anonymous reviewers who looked at the whole manuscript. We thank them all for their time and consideration. At the University of British Columbia (UBC) Eric Leinberger kindly drew the geology map, while Andy Leblanc and Alistair Blachford helped with computing. Dianne Newell, as director of the PWIAS, facilitated the workshop and funding in numerous different ways. In Arusha we thank Jo Driessen and Judith Jackson for looking after us.

We thank the Natural Sciences and Engineering Research Council of Canada and the Frankfurt Zoological Society (FZS) for supporting the editors. Markus Borner of FZS has consistently supported our work over many years. He retired in 2012 and Rob Muir has taken over and continues to help us. A. R. E. Sinclair was also supported by a Canadian Senior Killam Research Fellowship for two years and by the Peter Wall Institute for Advanced Studies at the University of British Columbia to produce this book.

Anne C. Sinclair helped with collating, editing, and formatting the chapters. Her help was invaluable. Finally we remember our two colleagues, Magai Kaare and Simon Thirgood, who died from freak accidents during the production of this book. Magai Kaare championed canine rabies control in communities adjacent to the Serengeti National Park. His passion for infectious disease research contributed greatly to the health of the ecosystem. He died too early to appreciate the remarkable impacts of his work, but his legacy lives on through the work of his team.

Simon Thirgood was passionate about ecology, conservation, and Africa. He saw the Serengeti ecosystem as the ultimate natural research laboratory and knew he was privileged to have spent some of the happiest years of his life there. He was a gifted teacher and trainer, always fair and always honest, and was using this talent to inspire a new generation of African biologists in the years before he died.

Both made a significant contribution to the work in Serengeti and their names appear on the chapters.

The editors

ONE

Conservation in a Human-Dominated World

Anthony R. E. Sinclair and Andy Dobson

The world's few remaining protected ecosystems are becoming progressively threatened from human exploitation. Serengeti is one outstanding example of a protected area that retains an almost complete biota that has existed for millennia (Peters et al. 2008). It also generates substantial revenue from tourism that goes toward supporting the remaining protected areas in Tanzania. The system is therefore important as a globally significant biodiversity site and as a mainstay of the Tanzanian economy. Yet the Serengeti National Park is under threat both from within and outside of Tanzania. Development from proposed roads through the protected area and airports adjacent to it will have negative impacts possibly causing the decline of the wildebeest migration and alter the whole ecosystem (Dobson et al. 2010; Sinclair 2010; Holdo et al. 2010). Rhetoric against protected area conservation in general has lead to proposals for a focus on conservation in human-dominated ecosystems at the expense of protected areas (Shellenberger and Nordhaus 2011; Kareiva and Marvier 2012). These statements beg three important questions: Do protected areas play a role in conservation that is not achieved in human ecosystems? Can human dominated systems contribute to conservation objectives? Do these two—protected areas and human-dominated areas—support each other? We examine these questions in this volume. The natural ecosystem is changing from environmental disturbances within and human impacts outside the system. We ask how the greater Serengeti ecosystem, including both the human and natural components, can be made sustainable in the face of these changes.

Many of the chapters in this volume are central to a larger argument in conservation biology that urges society to move away from the protected area paradigm and to focus on altered landscapes outside parks, a world that is being taken over and modified by humans. We recognize that new approaches to conservation are needed in a world of burgeoning human numbers, but this does not mean that protected areas have no useful function. The chapters, therefore, address both the function of the protected ecosystem and the contribution of the human system that surrounds it.

To place this volume in a broader context we review briefly the background to the debate on what was originally called community-based conservation in contrast to protected area conservation. The original, historical policy of conservation has been to secure areas such as national parks and reserves to maintain a suite of biota that was disappearing in the face of human exploitation. Yellowstone National Park in the United States was an early example, a response to the combination of illegal settlement, vandalism, and wildlife slaughter which put the US Army in charge of the park in 1886 (D. Houston pers. comm.; see also Olliff et al. 2013). Kruger National Park in South Africa was the first such protected area in Africa (proclaimed by the Transvaal Republic in 1898 as the Sabi Game Reserve), a reaction to encroaching agriculture and extermination of wildlife from grazing lands (Carruthers 1995; N. Owen-Smith, pers. com.). Since then a large number of protected areas have been set up around the world specifically to provide protection for biota (Wright 1996; Nelson and Serafin 1997; Terborgh et al. 2002; Stolton and Dudley 2010) and clearly larger ones do provide protection (Cantú-Salazar and Gaston 2010). A selected set of such areas has been designated as World Heritage Sites since 1972.

However, the past two decades have seen a movement toward community-based conservation (CBC), the attempt to maintain sustainable biological communities in the presence of human exploitation, largely in agriculture and forestry-dominated systems (Bhagwat et al. 2008; Harvey et al. 2008; Chazdon et al. 2009). The motive for CBC was threefold. First, a considerable proportion of the world's biota, some 50%, fall outside of protected areas (Sinclair 2008) and it is necessary to find some way of preserving this. Second, protected areas are failing to meet their objectives; many are experiencing attrition of territory (Sinclair et al. 1995) and losing biota (Craigie et al. 2010). For example, in an analysis covering 20–30 years of 60 tropical forest reserves around the globe, Laurance et al. (2012) found that half experienced loss of biodiversity over a wide array of animal groups. Habitat modification, exploitation, and hunting were the main disturbances. Third, no protected area is a self-sustaining system in isolation, not even the largest area (Lindenmayer, Franklin, and Fischer 2006). The analysis of forest reserves showed that environmental changes outside were as important as those inside in determining the course of ecological change. All protected areas rely on processes that emanate from outside, whether these be water flows from rivers, recolonization of plant communities, or dispersal of animals; in short, no park is an island (Janzen 1983). Thus, to maintain a protected area indefinitely we must maintain the greater ecosystem within which it is embedded. The combination of reserves and agriculture can maintain a majority of the biota in some cases; Daily et al. (2003) demonstrate this for forests and coffee estates in Costa Rica. So in this volume we examine whether the Serengeti is maintaining its biota, and how it relies on outside influences.

The problems with protected area conservation provide the rationale for renewed conservation efforts in human ecosystems. Much of community-based conservation still focuses on preserving the biota rather than the needs of the people (Harvey et al. 2008), but unless the aspirations of humans are placed as a top priority there will be no conservation (Wells and McShane 2004; Garcia et al. 2009). Several of the chapters in this volume document the livelihoods, health, and welfare of the peoples and explore how the Serengeti benefits and impedes them.

The new debate argues that a world already at 7 billion and heading toward 10 billion people will simply overrun protected areas by 2100 in the scramble for new resources, and we may as well recognize that eventuality. To some extent this is already happening (Scholte 2003; Scholte and de Groot 2009). These peoples, largely in the developing world, emulating present- day western societies, will demand and achieve energy-expensive lifestyles. Thus, if we are to save the world's natural heritage we must embrace technology and "garden" our environments—we create novel ecosystems (Rosenzwieg 2003; Shellenberger and Nordhaus 2011; Kareiva and Marvier 2012; Doak et al. 2013, 2014; Marvier and Kareiva 2014). Humans live not in natural ecosystems but in human ecosystems modified by agriculture, forestry, urbanization, and industry. Ellis (2011), Sagoff (2011), and Kareiva, Lalasz, and Marvier (2011) argue that these areas have been extremely resilient to human population and climate change over the past centuries. This view contrasts with what they claim was a prevailing concept of nature as fragile, ready to collapse with any disturbance. Thus, conservation in the twenty-first century must embrace human-made systems. If it is to be relevant it must move away from protected areas as the old model, and use technology to conserve the new biota (Kareiva, Lalasz, and Marvier 2011). This, then, is the argument of the new "gardeners" of the world.

Despite the certainty with which these statements are made, evidence is not entirely consistent with them. The authors point to the meta-analysis of Jones and Schmitz (2009), which showed that many ecosystems can recover from single human perturbations, a pulse; these could be "accidents" from inadvertent human impacts. But this is not the issue. The question is whether human ecosystems under persistent, chronic abuse can maintain both viable processes and a majority of the biota—a press perturbation—and the data would suggest that the systems would not recover under such persistent exploitation (Rosenzweig 2003). Thus, one major set of problems with community-based conservation relates to its lack of sustainability. Problems arise with sustainable harvesting of wildlife populations, increasing expectation of livelihoods, shrinking land areas, and increasing human numbers (Sinclair 2008). The claim that agricultural systems have been sustainable and resilient over historical time is questionable. The progressive loss of both biodiversity and ecosystem processes in British and European agricultural systems, due to intensification of the agribusiness, is now well documented (Donald, Green, and Heath 2001; Gregory, Noble, and Custance 2004; Gaston and Fuller 2008; Gaston 2010; Cardinale et al. 2012). There is a similar decline in the avifauna of the eastern forests of North America probably due to changes in both the Neotropical forests and temperate woodlands (Terborgh 1992). Soil loss destroyed 40,000 km^2 of cropland in the United States by 1979 due to the continued application of fertilizers; productivity has declined (Jackson 1980 in Rosenzweig 2003; Myers and Kent 1998) and caused distortions of the ecosystem (Jefferies, Rockwell, and Abraham 2004). The development of agriculture across Australia, through the removal of eucalypt woodlands, has resulted in progressive salinization of soils and continuing loss of agricultural production (Grieve 1987; McFarlane, George, and Farrington 1993). Rangelands that have been shared by humans and abundant wildlife for thousands of years now show signs of collapse (Harris et al. 2009)—the great herds of migrating Tibetan antelope (*Panthalops hodgsoni*), gazelle (*Procapra picticaudata*), and wild ass (*Equus kiang*) on the Tibetan Chang Tang Reserve have almost gone (Schaller 1998); the migrations of Saiga antelope (*Saiga tatarica*) in Kazakhstan have collapsed (Milner-Gulland et al. 2001), and the wildebeest migrations of Botswana have disappeared (Williamson, Williamson, and Ngwamotsoko 1988). The staggering loss of biota in New Zealand with the arrival of humans in 1300 resulted in a change in species composition and loss of ecosystem function (Atkinson and Cameron 1993; Cooper et al. 1993; Campbell and Atkinson 1999; Worthy and Holdaway 2002); there is no sense of sustainability and species continue to decline toward extinction today (Sinclair and Byrom 2006).

In 1700 most of the world's terrestrial biomes were without humans (over 50%) or seminatural (45%), with only minor use for agriculture and settlement. By 2000 only 25% was wild and 20% seminatural, the rest now under human modification. For the future most terrestrial ecosystems will be under human modification (Ellis et al. 2010). These authors conclude that conservation must focus on the remaining remnants, or recovering ecosystems embedded within human-modified systems. It has also been suggested that the progressive deforestation of Europe from the Middle Ages (1250) to the 1800s resulted in the concomitant increase in albedo and hence environmental cooling, seen as the "little ice-age." This temperature trend was only reversed by a similarly unsustainable increase in temperature starting with the Industrial Revolution in the early 1800s (Britannica 2008). In short, although one can point to some examples where biota have returned to human-modified landscapes, there is a far greater array of cases showing that the historical impacts of humans were unsustainable and continue to be so into the future.

But the debate is not really about that dichotomy. Rather, it concerns whether we can create sustainable human landscapes and whether protected areas are a necessary component of this, as Daily et al. (2003) have illustrated. The present volume addresses this issue.

The assertions made by the new "gardeners" of human ecosystems that they are sustainable and resilient need to be tested by comparison with areas that have less human impact. These are de facto the protected areas; they are the controls for human impacts. This is one fundamental reason why protected areas must not be lost; their presence is the best way of judging whether community-based conservation is sustainable, and that human ecosystems are robust. What protected areas are not—and in the modern context not intended to be—are "pristine, prehuman landscapes" as suggested by Kareiva, Lalasz, and Marvier (2011). Historically some may have thought that way (Adams 2003), but it is not the prevailing concept of protected areas today (Jope and Dunstan 1996; Wright and Mattson 1996).

In this book we examine the greater Serengeti ecosystem with two sets of questions in mind. First, *what is the biodiversity inside the protected area, and the nature of the heterogeneity that affects it, and is the system sustainable?* To address these questions we describe the spatial and temporal variation in the environment, the responses of the biota to this variation, and the changes in the system as a whole.

Second, *what are the dynamics of the human ecosystem, and what would be required to make the human-wildlife interaction sustainable?* We analyze aspects of human livelihoods to identify what is required to further the development of local communities. We further document changes in the biota and suggest ways to mitigate human impacts while advancing the aspirations of the people.

We divide the chapters into five sections. The first section addresses the issue of what determines biodiversity in the Serengeti. We address the effects of abiotic environmental heterogeneity (Metzger et al.), disturbance from fire (Eby et al.), and climate (Fryxell et al.) on biodiversity, then go on to describe plant diversity (Anderson et al.) and the role of the huge migrating herds in seasonal and spatial diversity (Hopcraft et al.).

In the second section, disturbance effects on biodiversity are documented for microbes (Verchot et al.), plant dynamics in forests (Turkington et al.), insects in general (de Visser et al.) and butterflies in particular (Sinclair et al.), rodents (Byrom et al.), birds (Jankowski et al., Nkwabi et al.), and carnivores (Craft et al.). These provide the data to examine how disturbance within the protected area and human influences outside alter biodiversity.

The third section documents the human social and ecological systems in the agricultural west (Eli Knapp et al.) and pastoral east (Galvin et al.), the changes in agriculture (Estes et al.), and the threats from disease (Lembo et al.). The consequences of the conflicts between human and wildlife systems are described by Hampson et al. and modeled by Holdo and Holt in the fourth section.

The fifth section analyzes the consequences of human-wildlife interactions for policy, management, and conservation. Rentsch et al. document the degree of bushmeat hunting and the implications for wildlife population. Linda Knapp et al. analyze health concerns in local communities and what is needed to improve welfare. Randall et al. examine the institutions that manage the human-wildlife interaction—the various forms of legal protection from parks to wildlife management areas. Reid et al. look at the sustainability of community-based conservation in the greater Serengeti ecosystem. We conclude with an overview of the management problems over the past half century and how research has helped to solve those problems, and make suggestions on how to develop a sustainable protected area within a human-dominated larger region.

REFERENCES

Adams, W. M. 2003. When nature won't stay still: Conservation, equilibrium and control. In *Decolonizing nature. Strategies for conservation in a post-colonial era*, ed. W. M. Adams and M. Mulligan, 221–46. London: Earthscan.

Atkinson, I. A. E., and E. K. Cameron. 1993. Human influence on the terrestrial biota and biotic communities of New Zealand. *Trends in Ecology and Evolution* 8:447–51.

Bhagwat, S. A., K. J. Willis, H. J. B. Birks, and R. J. Whittaker. 2008. Agroforestry: A refuge for tropical biodiversity? *Trends in Ecology and Evolution* 23:261–67.

Britannica. 2008. *The britannica guide to climate change.* London: Constable and Robinson, Ltd.

Campbell, D. J., and I. A. E. Atkinson. 1999. Effects of kiore (*Rattus exulans* Peale) on recruitment of indigenous coastal trees on northern offshore islands of New Zealand. *Journal of the Royal Society of New Zealand* 29:265–90.

Cantú-Salazar, L., and K. J. Gaston. 2010. Very large protected areas and their contribution to terrestrial biological conservation. *BioScience* 60:808–18.

Cardinale, B. J, J. E. Duffy, A. Gonzalez, D. U. Hooper, C. Perrings, P. Venail, A. Narwani, G. M. Mace, D. Tilman, and D. A. Wardle, et al. 2012. Biodiversity loss and its impact on humanity. *Nature* 486:59–67.

Carruthers, J. 1995. *The Kruger National Park. A social and political history.* Pietermaritzburg: University of Natal Press.

Chazdon, R. L., C. A. Harvey, O. Komar, D. M. Griffith, B. G. Ferguson, M. Martínez-Ramos, H. Morales, R. Nigh, L. Soto-Pinto, and M. Van Breugel, et al. 2009. Beyond reserves: A research agenda for conserving biodiversity in human-modified tropical landscapes. *Biotropica* 41:142–53.

Cooper, A., I. A. E. Atkinson, W. G. Lee, and T. H. Worthy. 1993. Evolution of the Moa and their effect on the New Zealand flora. *Trends in Ecology and Evolution* 8:433–42.

Craigie, D., J. M. B. Baillie, A. Balmford, C. Carbone, B. Collen, R. E. Green, and J. M. Hutton. 2010. Large mammal population declines in Africa's protected areas. *Biological Conservation* 143:2221–28.

Daily, G. C., G. Ceballos, J. Pacheco, G. Suzan, and A. Sanchez-Azofeifa. 2003. Countryside biogeography of Neotropical mammals: Conservation opportunities in agricultural landscapes of Costa Rica. *Conservation Biology* 17:1814–26.

Doak, D. F., V. J. Bakker, B. E. Goldstein, and B. Hale. 2013. What is the future of conservation? *Trends in Ecology and Evolution* 29:77–81.

———. 2014. Moving forward with effective goals and methods for conservation: A reply to Marvier and Kareiva. *Trends in Ecology and Evolution* 29:132–33.

Dobson, A. P., M. Borner, A. R. E. Sinclair, P. J. Hudson, T. M. Anderson, G. Bigurube, T. B. B. Davenport, J. Deutsch, S. M. Durant, and R. D. Estes, et al. 2010. Road will ruin Serengeti. *Nature* 467:272–74.

Donald, P. F., R. E. Green, and M. F. Heath. 2001. Agricultural intensification and the collapse of Europe's farmland bird populations. *Proceedings of the Royal Society, B.* 268:25–9.

Ellis, E. 2011. The planet of no return. Human resilience on an artificial earth. In *Love your monsters. Post-environmentalism and the Anthropocene*, ed. M. Shellenberger and T. Nordhaus, 37–46. Washington, DC: Breakthrough Institute.

Ellis, E. C., K. K. Goldewijk, S. Siebert, D. Lightman, and N. Ramankutty. 2010. Anthropogenic transformation of the biomes, 1700 to 2000..*Global Ecology and Biogeography* 19:589–606.

Garcia, C. A., S. A. Bhagwat, J. Ghazoul, C. D. Nath, K. M. Nanaya, C. G. Kushalappa, Y. Raghuramulu, R. Nasi, and P. Vaast. 2009. Biodiversity conservation in agricultural landscapes: Challenges and opportunities of coffee agroforests in the Western Ghats, India. *Conservation Biology* 24:479–88.

Gaston, K. J. 2010. Valuing common species. *Science* 327:154–55.

Gaston, K. J., and R. A. Fuller. 2008. Commonness, population depletion and conservation biology. *Trends in Ecology and Evolution* 25:372–80.

Gregory, R. D., D. G. Noble, and J. Custance. 2004. The state of play of farmland birds: Population trends and conservation status of lowland farmland birds in the United Kingdom. *Ibis* 146 (Supplement 2): 1–13.

Grieve, A. M. 1987. Salinity and waterlogging in the Murray-Darling basin. *Search* 18: 72–74.

Harris, G., S. Thirgood, J. G. C. Hopcraft, J. P. G. M. Cromsigt, and J. Berger. 2009. Global decline in aggregated migrations of large terrestrial mammals. *Endangered Species Research* 7:55–76.

Harvey, C. A., O. Komar, R. Chazdon, B. G. Ferguson, B. Finegan, D. M. Griffith, M. Martínez-Ramos, H. Morales, R. Nigh, and L. Soto-Pinto, et al. 2008. Integrating agricultural landscapes with biodiversity conservation in the Mesoamerican hotspot. *Conservation Biology* 22:8–15.

Holdo, R. M., J. M. Fryxell, A. R. E. Sinclair, A. Dobson, and R. D. Holt. 2010. Predicted impact of barriers to migration on the Serengeti wildebeest migration. *PLoS ONE* 6 (1): 1–7, e16370. doi:10.1371/journal.pone.0016370.

Janzen, D. H. 1983. No park is an island: Increase in interference from outside as park size decreases. *Oikos* 41:402–10.

Jefferies, R. L., R. F. Rockwell, and K. E. Abraham. 2004. Agricultural food subsidies, migratory connectivity and large-scale disturbance in arctic ecosystems. *Integrative and Comparative Biology* 44:130–39.

Jones, H. P., and O. J. Schmitz. 2009. Rapid recovery of damaged ecosystems. *PLoS ONE* 4 (5): e5653. doi:10.1371/journal.pone.0005653.

Jope, K. L., and J. C. Dunstan. 1996. Ecosystem-based management: Natural processes and systems theory. In *National parks and protected areas. Their role in environmental protection*, ed. R. G. Wright, 45–62. Oxford: Blackwell Science.

Kareiva, P., R. Lalasz, and M. Marvier. 2011. Conservation in the Anthropocene. Beyond solitude and fragility. In *Love your monsters. Post-environmentalism and the Anthropocene*, ed. M. Shellenberger and T. Nordhaus, 26–36. Washington, DC: Breakthrough Institute.

Kareiva, P., and M. Marvier. 2012. What is conservation science? *BioScience* 62:962–69.

Laurance, W. F., D. C. Useche, J. Rendeiro, M. Kalka, C. J. A. Bradshaw, S. P. Sloan, S. G. Laurance, M. Campbell, K. Abernethy, and P. Alvarez, et al. 2012. Averting biodiversity collapse in tropical forest protected areas. *Nature* 489:290–94. doi:10.1038/nature11318.

Lindenmayer, D., B. J. F. Franklin, and J. Fischer. 2006. General management principles and a checklist of strategies to guide forest biodiversity conservation. *Biological Conservation* 131:433–45.

Marvier, M., and P. Kareiva. 2014. The evidence and values underlying 'new conservation.' *Trends in Ecology and Evolution* 29:131–32.

McFarlane, D. J., R. J. George, and P. Farrington. 1993. Changes in the hydrologic cycle. In *Reintegrating fragmented landscapes*, ed. R. J. Hobbs and D. A. Saunders, 147–86. New York: Springer-Verlag.

Milner-Gulland, E. J., M. V. Kholodova, A. Bekenov, O. M. Bukreeva, Iu. A. Grachev,

L. Amgalan, and A. A. Lushchekina. 2001. Dramatic decline in Saiga antelope populations. *Oryx* 35:340–45.

Myers, N., and J. Kent. 1998. *Perverse subsidies*. Institute for Sustainable Development, Winnipeg, Saskatchewan.

Nelson, J. G., and R. Serafin, eds. 1997. *National parks and protected areas. Keystones to conservation and sustainable development*. New York: Springer-Verlag.

Olliff, S. T., P. Schullery, G. E. Plumb, and L. H. Whittlesey. 2013. Understanding the past: The history of wildlife and resource management in the greater Yellowstone area. In *Yellowstone's wildlife in transition*, ed. P. J. White, R. A. Garrott, and G. E. Plumb, 10–28. Cambridge, MA: Harvard University Press.

Peters, C. R., R. J. Blumenschine, R. L. Hay, D. A. Livingstone, C. W. Marean, T. Harrison, M. Armour-Chelu, P. Andrews, R. L. Bernor, and R. Bonnefille, et al. 2008. Paleoecology of the Serengeti-Mara ecosystem. In *Serengeti III: Human impacts on ecosystem dynamics*, eds. A. R. E. Sinclair, C. Packer, S. A. R. Mduma, and J. M. Fryxell, 47–94. Chicago: University of Chicago Press.

Rosenzweig, M. 2003. *Win-win ecology. How Earth's species can survive in the midst of human enterprise*. Oxford: Oxford University Press.

Sagoff, M. 2011. The rise and fall of ecological economics. A cautionary tale. In *Love your monsters. Post-environmentalism and the Anthropocene*, ed. M. Shellenberger and T. Nordhaus, 47–65. Washington, DC: Breakthrough Institute.

Schaller, G. B. 1998. *Wildlife of the Tibetan steppe*. Chicago: University of Chicago Press.

Scholte, P. 2003. Immigration, a potential time-bomb under the integration of conservation and development. *Ambio* 32:58–64.

Scholte, P., and W. T. de Groot. 2009. From debate to insight: Three models of immigration to protected areas. *Conservation Biology* 24:630–32.

Shellenberger, M., and T. Nordhaus. 2011. Introduction. In *Love your monsters. Post-environmentalism and the Anthropocene*, ed. M. Shellenberger and T. Nordhaus, 5–7. Washington, DC: Breakthrough Institute.

Sinclair, A. R. E. 2008. Integrating conservation in human and natural systems. In *Serengeti III: Human impacts on ecosystem dynamics*, ed. A. R. E. Sinclair, C. Packer, S. A. R. Mduma, and J. M. Fryxell, 471–95. Chicago: University of Chicago Press.

———. 2010. Road proposal threatens existence of Serengeti. *Oryx* 44:478–79.

Sinclair, A. R. E., and A. Byrom. 2006. Understanding ecosystems for the conservation of biota. *Journal of Animal Ecology* 75:64–79.

Sinclair, A. R. E., D. S. Hik, O. J. Schmitz, G. G. E. Scudder, D. H. Turpin, and N. C. Larter. 1995. Biodiversity and the need for habitat renewal. *Ecological Applications* 5:579–87.

Stolton, S., and N. Dudley, eds. 2010. *Arguments for protected areas. Multiple benefits for conservation and use*. Washington, DC: Earthscan.

Terborgh, J. 1992. Why American songbirds are vanishing. *Scientific American* 264:98–104.

Terborgh, J., C. v. Schaik, L. Davenport, and M. Rao, eds. 2002. *Making parks work: Strategies for preserving tropical nature*. Washington, DC: Island Press.

Wells, M. P., and T. O. McShane. 2004. Integrating protected area management with local needs and aspirations. *Ambio* 33:513–19.

Williamson D., J. Williamson, and K. T. Ngwamotsoko. 1988. Wildebeest migration in the Kalahari. *African Journal of Ecology* 26:269–80.

Worthy, T. H. and R. N. Holdaway. 2002. *The lost world of the Moa*. Bloomington: Indiana University Press.

Wright, R. G., ed. 1996. *National parks and protected areas. Their role in environmental protection*. Oxford: Blackwell Science

Wright, R. G. and D. J. Mattson. 1996. The origin and purpose of national parks and protected areas. In *National parks and protected areas. Their role in environmental protection*, ed. R. G. Wright, 3–14. Oxford: Blackwell Science.

TWO

Shaping the Serengeti Ecosystem

Anthony R. E. Sinclair, Andy Dobson, Simon A. R. Mduma, and Kristine L. Metzger

The future of the Serengeti as a self-sustaining natural ecosystem has come into question as a result of political, economic, and social pressures developing in Tanzania in the late 2000s. In order to understand what might happen to the ecosystem we need to know what lives there and the processes that keep it going. In addition, we must identify the processes and pressures that are liable to change it. Such processes can be natural such as the effects of El Niño and the occasional eruption of Ol Doinyo Lengai, which destroys vegetation locally, but enriches the soil as volcanic dust clouds settle across the southern Serengeti, or they can be human induced. Humans have had impacts through hunting, agriculture, and tourism. We must ask to what extent these influences can be absorbed and tolerated by the system and at what point they tip the balance and become unsustainable.

We outline in this chapter the historical events that have shaped the nature of the ecosystem and the surrounding human populations. First we describe the Serengeti of today to provide the baseline for other chapters to come. Then we explain the events of the nineteenth century that underlie the Serengeti as it was first found. Finally we document the formation of the protected area. Policies have changed over time and these have altered the course of history and the human impacts on the ecosystem. We conclude with a summary of the implications of these historical events on the human and natural ecosystems of today.

THE SERENGETI ECOSYSTEM

The Serengeti-Mara ecosystem (fig. 2.1) is an area of some 25,000 km^2 on the border of Tanzania and Kenya, East Africa (34° to 36° E, 1° to 3°o 30′ S) defined by the movements of the migratory wildebeest (see Hopcraft et al., chapter 6). Outside the boundaries lie the agricultural and pastoralist tribes. North of the Mara Reserve there occur the Loita plains and Loita hills that have been the lands of the Kenya Maasai since the nineteenth century; they are now managed as ranches with fences. In the northwest, above the Isuria escarpment, original highland forest with grassy glades from a century ago has been transformed by pastoralists and more recently, agriculturalists, into grassland. Wildebeest used to climb the escarpment in Kenya to graze these grasslands in the 1960s and early 1970s (pers. obs.) but they are now excluded. On the Tanzania side, dense settlement has precluded use by wildlife for at least the past 100 years and probably much longer. These peoples belong to the Wakuria, a tribe that stretches from inside Kenya, along the western boundary in Tanzania as far south as Mugumu. The Waikoma tribe takes over further south, and they stretch to the Western Corridor boundary and west to Lake Victoria. South of the corridor, along the southwest boundary, through Maswa to Makao and then east to Lake Eyasi are the Wasukuma, the largest tribe in Tanzania. These tribes are all agricultural with smallholdings and small herds of livestock. The Maasai inhabit the eastern side of the ecosystem from the Narok district in Kenya south through the Loliondo area, across Ngorongoro Conservation Area to the southern edge of the plains where they meet the Wasukuma.

The eastern boundary of the ecosystem is formed by the Crater Highlands and the Rift Valley. An arm, called the "Western Corridor" stretches west almost to Speke Gulf of Lake Victoria. The remaining western boundary is formed by dense cultivation that has intensified from recent immigration in the past few decades. The northern boundary is formed by the Isuria escarpment and the Loita plains in Kenya. The southern and southwestern boundary in the Maswa area is formed by cultivation, kopjes, and baobab (*Adansonia*) woodland into which wildebeest rarely travel.

Apart from the Serengeti National Park (14,763 km^2) the ecosystem includes several other conservation administrations. The Ngorongoro Conservation Area (NCA, 8,288 km^2) southeast of the park includes half of the short grass plains, the Olduvai Gorge, and Gol Mountains. North of the NCA is the Loliondo Game Controlled Area. This comprises savanna toward the north and plains in the south. In the far east there are the Salai plains with sand dunes, used by wildebeest if wet, and they end at the Gregory Rift escarpment.

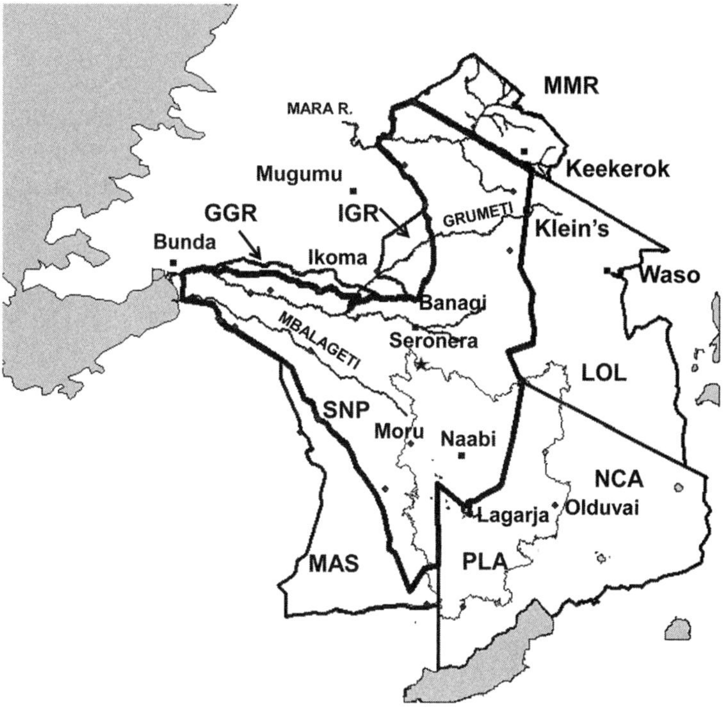

Fig. 2.1 Boundaries of the administrative areas covering the Serengeti ecosystem as they have been since 1967. SNP, Serengeti National Park; NCA, Ngorongoro Conservation Area; MAS, Maswa Game Reserve; GGR, Grumeti Game Reserve; IGR, Ikorongo Game Reserve; LOL, Loliondo Game Control Area; MMR, Maasai Mara Reserve; PLA, Serengeti Plains. Small diamond markers are guard posts. Squares are place names. Seronera (star) is the park headquarters.

Maswa Game Reserve (2,200 km^2) lies to the southwest of the park and comprises dense *Acacia* woodland on flat alluvial soils. This is the refuge for wildebeest during the December–April period if the rains fail. Maswa is the closest woodland area to the southern plains. North of the Western Corridor lie the Grumeti and Ikorongo Game Reserves. Adjacent to Serengeti in Kenya, the Mara Reserve (1,672 km^2) and adjoining group ranches lie at the foot of the Isuria escarpment and eastward to include the Loita hills. The Mara Reserve is largely grassland and relict *A. gerrardii* savanna, but it includes the permanently flowing Mara River that flows south into Serengeti and west to Lake Victoria at Musoma.

THE EVENTS OF THE 19TH CENTURY (1840–1890)

The Ivory Trade

In the nineteenth century, Zanzibar Island was an Arab state under the Sultan of Oman. It was the base from which Arab caravans set out for the interior of Africa. The central staging post was Tabora in the center of what is now Tanzania. From Tabora these caravans went west into Congo and north along both the western and eastern sides of Lake Victoria. One such route went across the Serengeti plains, through the central woodlands and then west to Lake Victoria (Wakefield 1870, 1882; Farler 1882). Other routes started from the Kenya and Mozambique coasts, and again from Khartoum traveling west and south. These routes effectively covered east, central, and southern Africa.

These caravans developed to serve the ivory and slave trades; these had operated on a small scale for a thousand years or more to supply India and Arabia. However, starting around 1840 the trade expanded rapidly as the demand for ivory in Europe and North America suddenly increased. The fundamental motivation for the increase in demand was the Industrial Revolution lead by Britain in the 1780s. This led to the invention of iron piano frames which were relatively cheap and did not warp or crack under the tension of tuned strings. By 1840 the increase in wealth in Britain created a demand for luxuries such as pianos and snooker tables—ivory was used for piano keys, billiard balls, and even knife handles. Imports of ivory to Britain were steady before 1840, but thereafter increased linearly until about 1880 (Sheriff 1987) (fig. 2.2). However, by 1890 export records from Zanzibar and Khartoum showed that the flow of ivory had dropped to near zero (Spinage 1973).

Arab caravans purchased ivory from African tribes that hunted elephants within their territories. Demand was so great that after 50 years elephants were almost eradicated. This extirpation started near the east coast in the 1850s and then extended westward, and by the 1890s elephants were so scarce throughout East Africa that exports collapsed (Sheriff 1987), despite the attempted introduction of a quota system (Simon 1962). Tsavo Park, Kenya, which is now famous as an elephant park and in the early nineteenth century was the domain of the Wakamba elephant hunters, had no elephants in the 1890s (Patterson 1907). The central Serengeti woodlands had no resident tribes in the 1860s but nomadic Wandorobo hunted elephant there (Farler 1882). At a place near what we now call Nyaraboro plateau, close to Moru kopjes, Farler comments there is a "tribe of elephant-hunters, who neither cultivate nor keep cattle, but live entirely upon the flesh of the animals they kill in hunting. They supply the caravans with a great deal of

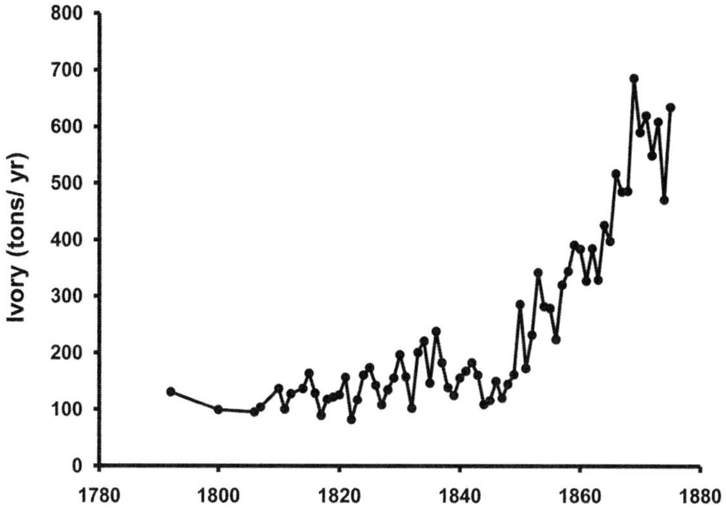

Fig. 2.2 Imports of elephant ivory from Africa to England in the mid-nineteenth century, which caused the collapse of elephant populations in Africa. The record stops in 1875 (data from Sheriff 1987).

ivory. Their country is full of elephants and other big game. They do not mix at all with other tribes." This is clear evidence that elephants were abundant in the Serengeti in the 1860s and '70s. Then elephants disappeared and were effectively absent in Serengeti throughout the period 1890–1938 (Baumann 1894; White 1915; Johnson 1929; Moore 1937).

It was not until the 1950s that elephants were seen commonly both in and out of protected areas. Elders of the Waikoma tribe remembered seeing two elephants in the west of the Serengeti in the 1930s, the first they had ever seen (J. Hando pers. comm.). In 1958 there were some 800 elephants in Serengeti, in 1961 over 1,000, and by 1965 numbers had increased to 2,000; a decade later they had reached 3,000 (Sinclair et al. 2008).

THE GREAT RINDERPEST

Rinderpest, a viral disease of cattle that occurs naturally in Asia, was introduced into Ethiopia in 1887 by cattle brought from India by Italian invaders. There is no evidence that it was ever in Africa prior to this event. It spread to West Africa, then south through East Africa, reaching the Cape by 1896. The resulting panzootic killed over 95% of cattle throughout Africa; in many cases complete herds died off, and famines decimated the human population of Africa. This ranks as one of the greatest socioecological disasters in human history, on a par with the plague in Europe and smallpox in the New World (Plowright 1982). In 1891 starving Maasai pastoralists abandoned the Serengeti plains as their cattle died (Baumann 1894). By 1892 widespread famine had occurred in Ethiopia, Somalia, southern Sudan, and eastern Africa (Pankhurst 1966, Waller 1988). Buffalo and many species of antelope, particularly wildebeest, were also decimated Spinage (2003).

The repercussions from this panzootic have had a profound influence on the ecology of the Serengeti over the last century. The Serengeti ecosystem had very few human inhabitants in the nineteenth century. Early travelers in the 1860s reported only nomadic elephant hunters in the central woodland areas and no farmers or pastoralists. Indeed, Maasai arrived on the Serengeti plains only in the 1850s, and even then reached no further west than halfway across the plains as shown on Farler's map (see boundary in fig. 2.3) which was similar to the distribution found by Baumann (Farler 1882; Baumann 1894). Baumann's map of human settlements in 1891 showed none within the present ecosystem, and human boundaries were similar to or even outside those of the present (fig. 2.3). As at present, there were agriculturalists in the west and pastoralists in the east. Both groups suffered severe declines from the rinderpest, famine, and secondary outbreaks of smallpox.

These events in the 1890–1900 period, which resulted in a collapse of the wildebeest numbers (Baumann records many skeletons due to the epidemic), would have resulted in a broad change in ecological processes. Over most of the Serengeti there was a severe reduction of burning caused by humans (because humans had left), which resulted in a major regeneration of young trees. The outbreak of trees, which has been documented empirically, occurred both inside and outside the current protected area (Sinclair et al. 2008). The dense vegetation in turn resulted in another secondary epidemic, namely typanosomiasis (sleeping sickness), which is spread by tsetse flies (*Glossina* spp.), because tsetses thrive in dense vegetation. Once tsetses invade an area cattle cannot live there, so humans are unable to return. Indeed, humans and cattle did not return until the vegetation surrounding the Serengeti had been cleared by mechanical means and by deliberate burning in the 1930–1950 period (Ford 1971).

Meanwhile, rinderpest epizootics occurred every 10–20 years over the period 1900 to 1963 (Talbot and Talbot 1963; Lowe 1942). Hence, for over half a century, cattle and wild bovines remained at low density throughout the Serengeti. Only ruminants are affected by rinderpest, the greater morbidity being in species more closely related to cattle. Thus, buffalo were most affected, followed by wildebeest. Infections were also reported in giraffe (*Giraffa camelopardalis*) and warthog (*Phacochoerus aethiopicus*), but other

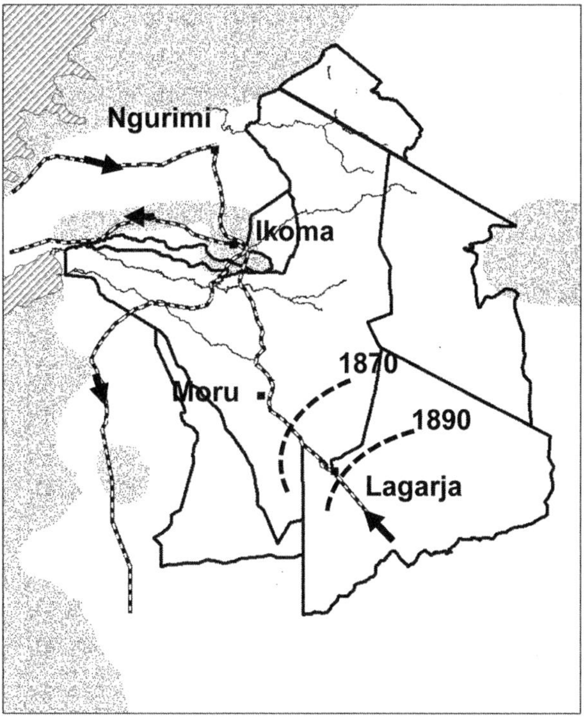

Fig. 2.3 The distribution of human agriculturalists and some pastoralists in 1891 and the route of Oskar Baumann (from the map in Baumann 1894) in relation to present protected area boundaries. The approximate limit of the Maasai distribution on the Serengeti plains (broken line) during the 1870s from the map in Farler (1882) is consistent with the more accurate limit indicated by Baumann for 1890 east of Lake Lagarja.

ruminants appear to have been less influenced by the disease (Rossiter et al. 1987).

The virus then disappeared from wildlife populations as a result of a cattle vaccination campaign. The rinderpest vaccine was developed by Walter Plowright who initially came to East Africa to work on malignant catarrhal fever for his PhD thesis (Plowright, Ferris, and Scott 1960; Plowright 1968). He set up a team that monitored the efficacy of the vaccine in cattle and gathered evidence for the disappearance of the pathogen from wildlife. This took the form of antibody titers in blood sera collected from animals of known age—some 80–100% of wildebeest and buffalo born in the 1950s had rinderpest antibodies, but only 50% of those born during 1960–62 had antibodies, and none born after 1963 suffered from the disease. By eliminating the disease from the domestic reservoir, the vaccination program effectively protected wildlife from infectious yearling cattle, and consequently the disease died out rapidly in wild populations (Sinclair 1977; Plowright 1982; Dobson 1995).

Removal of rinderpest was in many ways a remarkable large scale "experiment" for examining the impact of viral diseases on wildlife. As a consequence of the removal of rinderpest in 1963, juvenile survival in wildebeest and buffalo doubled. Both populations increased exponentially, with buffalo increasing from an estimated low of 15,000 and leveling out after 1973 near 75,000, and wildebeest growing from a low of 200,000 to 1.3 million after 1977. Most significantly, zebra (*Equus burchelli*), which as nonruminants are not affected by rinderpest, have remained at constant numbers, around 200,000, for the 45 years 1958 to 2003. This species has provided the critical test for the rinderpest theory because it confirmed the prediction that only the ruminants should have increased in numbers. Ironically, it also illustrated the considerable power of pathogens to regulate host abundance, relative to other natural enemies such as predators. When wildebeest and buffalo numbers increased, the numbers of lion and hyena also increased, suggesting that while pathogens might have a strong, top-down effect on population regulation, the abundance of predators in the system may be driven by bottom-up effects.

THE GERMAN ERA (1890–1920)

The expedition of Oskar Baumann in 1891 took place only a few months after rinderpest had hit the Serengeti region, and so we have an account of what the ecosystem looked like before rinderpest since little would have changed in the vegetation by the time of his journey. What then happened to the ecology of this area for the 30 years following rinderpest during the German era to 1920? We have a few reports that provide clues to the changes that were taking place following rinderpest.

In January 1904 G. E. Smith carried out the survey of the Anglo-German boundary from Lake Victoria to Kilimanjaro, and therefore he crossed the northern Serengeti and Mara Reserve (Smith 1907). Traveling east from the lake he mentions that for 20 miles west of the Isuria escarpment on the British side the landscape consisted of highland grassland with patches of forest, uninhabited except for wandering bands of Wandorobo hunters, and providing good grazing with abundant wildlife. This fits the scene at least until the 1980s. South of the border human settlement occurred further east than on the British side but was confined to the area above the escarpment. Below the Isuria escarpment there were no inhabitants. A photo of the Mara River at the boundary shows it to be much narrower than at present but with more substantial gallery forest. On either side of the river were wide plains with sand rivers. No people were encountered between the Isuria escarpment and Kuka hill.

The hunter S. E. White made the first recorded expedition on the west side of the Rift Valley across the northern Serengeti to Lake Victoria (fig. 2.4). From July to September 1913 he and his guide, the professional hunter R. J. Cunninghame (who also acted as Roosevelt's guide a couple of years later), and a dozen or so porters started in the Nguruman forests of Kenya and traveled south along the escarpment until they reached the land of the Sonjo tribe opposite Lake Natron. Then they turned west, crossing plains with wildebeest and zebra, and reached Waso. From there they continued west to the hill called Longossa (now the eastern boundary of Serengeti National Park) where they turned north past Lobo hill. Cunninghame had to make a side trip to Fort Ikoma at this point, walking south to Togoro kopjes then west to Ikoma. He returned to Lobo hill following the Grumeti River. The expedition walked north to the Bologonja River and then downstream some 20 miles before crossing west to Wogakuria hill. From there they traveled north to the Mara River reaching it at Rhino plain, crossed it a few miles east of Kogatende, and conducted a loop around the Lamai wedge before returning to Rhino plain. They then headed west staying south of the Mara River until they reached the Ikorongo hills. They then spent several weeks traveling south toward Ikoma and Isenye before turning northwest and ending up at Musoma, a new port on Lake Victoria. His descriptions of geography, vegetation, and animals are so detailed that we can trace his track within a kilometer or so (fig. 2.4).

From the reports of Baumann, Smith, and White we get a picture of the ecology of the Serengeti in the 1890–1920 period. Following the rinderpest, young trees regenerated in the woodland areas so that the vegetation was relatively dense in the decade before the First World War, as can be seen in White's photos. More importantly White notes on his map that many areas west of the current protected area south of the Mara River had dense vegetation and heavy tsetse fly infestations. These prevented human occupation and allowed wildlife to live there. Baumann, Smith, and White all show that human occupation was limited to areas above the Isuria escarpment, west of the Ikorongo hills far outside the present ecosystem boundaries, and north of the alluvial plains of the Ruwana River in the west (figs. 2.3, 2.4). In the east, Maasai were limited to areas east of the Kuka to the Grumechen line of hills that form the present national park boundary; they were restricted by the presence of the tsetse fly. Even the Wandorobo that Cunninghame met in 1913 had their base in the Ikorongo area west of the park and not in the park.

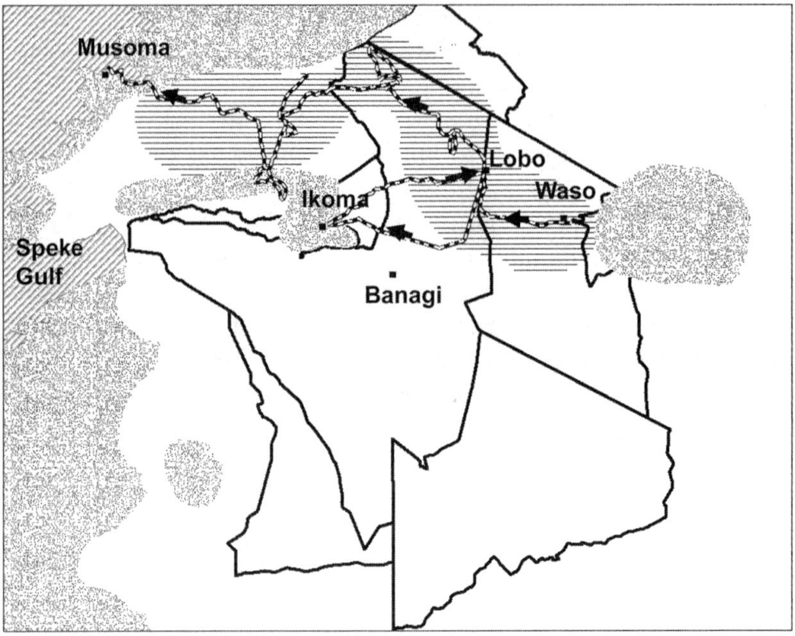

Fig. 2.4 The distribution of the migratory wildebeest and zebra in 1913 (horizontal hatching) and the route of S. E. White and R. J. Cunninghame in relation to present protected area boundaries. The villages of Waso, Ikoma, and Musoma were present but Lobo and Banagi did not exist at that time (from the map in White 1915). Distribution of peoples (stippling) as for fig. 2.3.

In July 1912 the Kenya game warden A. B. Percival (1928) noted that wildebeest and topi were very concentrated on the Mara River due to the great drought of that year. In August and September 1913, White and Cunninghame record large herds of wildebeest, zebra, and topi from near Waso in the east almost continuously as far west as the Masirori swamp, indeed almost as far as Musoma. These animals were part of the migration using lands along the Mara River at that time of year. Wildebeest, zebra, and topi were very abundant throughout, whereas kongoni were abundant in the east. White also described steinbuck, dikdik, both gazelle species, impala, giraffe, and both species of reedbuck. Chanler's reedbuck was seen only on the highest hills in the east, where they are still found today. Rhino were everywhere and seen frequently, as were lions and leopard. Wild dogs and cheetah were also recorded across the north. These distributions are similar to those recorded at least until the early 1970s within the park boundary.

However, there are differences from the distributions of today. White's records are the first clear evidence that wildlife lived far to the west of the park in the absence of human settlement. Dense vegetation and the tsetse fly contributed to this situation which lasted until the 1950s because humans were not in this area when the northern extension was added in the late 1950s (see below).

Roan antelope were clearly far more widespread; they were seen at Lobo, Bologonja, and west to the Ikorongo hills. In contrast, in the 1960s a small group of roan were recorded at Lobo but otherwise they were confined to the Lamai and Mara triangle and the *Terminalia* woodlands of the northwest. They went extinct in northern Serengeti by the early 1990s. Waterbuck was more abundant in 1913, occurring in large groups of 50, which we do not see today. Buffalo were recorded in 1913 across the north but only as lone males or very small herds; in the 1970s they occurred in herds of thousands. Elephant were unknown to the German officer and local peoples at Ikoma and they were only recorded at the Masirori swamp on the Mara River near Musoma. This swamp would have been a dry season refuge for elephants, a refuge that is no longer available for them. Ostrich were more abundant on the plains judging from Baumann's report. Greater kudu is indicated on the map at Kuka hill. In 1911 White shot a male on the Nguruman escarpment and so they were probably in the Loita forests. Kuka hill had montane forest at that time and may have supported greater kudu.

THE BEGINNING OF CONSERVATION

The 1920s was the era of uncontrolled hunting by foreign tourists, expeditions killing many lion, leopard, and rhino in the Seronera area on a single trip. Such wanton slaughter so infuriated one of the professional hunters, Denys Finch-Hatton, that he wrote a letter in 1928 to *The Times of London* denouncing both these indiscriminate practices and the irresponsible Tanganyika administration. He pushed for some form of protection (Sinclair 2012a). The British government also sent Julian Huxley, the famous evolutionist and ecologist, on a fact-finding tour of East Africa in 1929, and his report recommended that the Serengeti plains be set aside as a national park (along with other areas in Kenya and Uganda) (Huxley 1936). The Tanganyika government declared the area a closed reserve in 1929 and dispatched a game warden to the Serengeti in 1930. Denys Finch-Hatton must be credited with the creation of Serengeti as a wildlife conservation area (Wheeler 2006).

The posting of Major Monty Moore (Moore 1937; Turner 1987) as the first game warden signaled the beginning of Serengeti as a protected area for conservation and the end of the free-for-all massacres of lions. His headquarters was at Banagi hill overlooking the Mugungu River. It was the headquarters of western Serengeti until Seronera was developed in 1959.

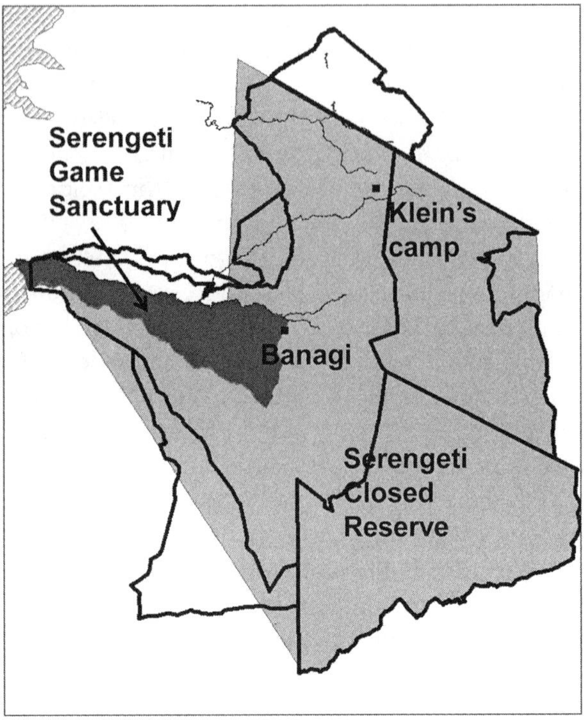

Fig. 2.5 The Serengeti Closed Reserve and Serengeti Game Sanctuary in the period 1930–50 (from the map in Moore 1937). After 1950 the headquarters were at Ngorongoro on the crater rim, while Banagi was the western warden's station from 1930–59.

The 1930 "Serengeti Closed Reserve" was a large area which included the present Serengeti, the Ngorongoro Conservation Area, and the Loliondo district as far north as the Kenya border and east to the edge of the Gregory Rift Valley (Moore 1937) (fig. 2.5). The "closed reserve" in the 1930s provided nominal control of hunting, though in practice the wardens had little ability to supervise, and abuses of hunting continued. Nevertheless, Moore worked at getting protection for lions and in 1935 a 2,200 km^2 area where lion hunting was prohibited was declared the "Serengeti Game Sanctuary" (fig. 2.5). The boundaries were between the Mbalageti and Grumeti Rivers as far as Speke Gulf in the west—what we now call the "corridor"—and along the road in the east from Banagi to about Lake Magadi. In 1937 the colonial government announced its intention to create a national park and plans were drawn up in 1940, but they were not finalized until 1950, delayed by the Second World War.

The Serengeti National Park was established in 1950 (fig. 2.6), starting

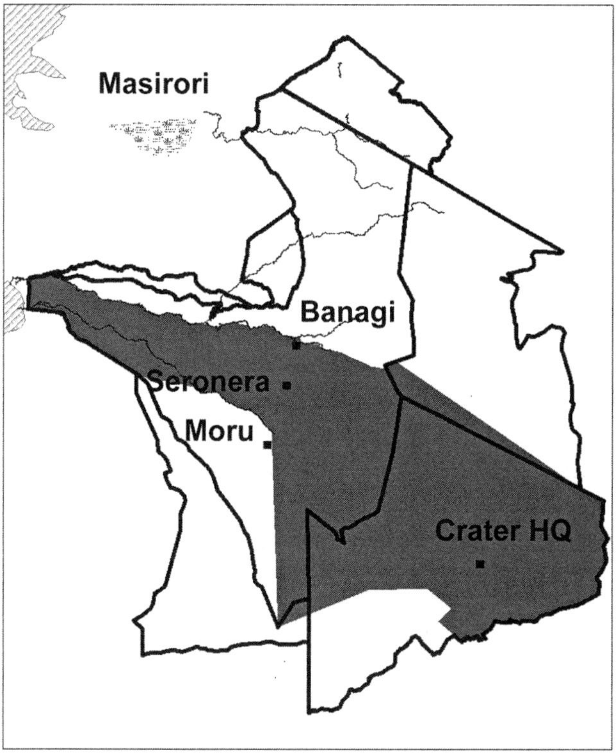

Fig. 2.6 The approximate boundaries (shaded area) of the Serengeti National Park in the period 1950–59 (from the map in Grant 1957) compared to the boundaries after 1959 (solid line) for Serengeti National Park and Ngorongoro Conservation Area (as in fig. 2.1). Seronera became the new headquarters in 1959 while Banagi became the first research center. Masirori swamp was the only place where elephant occurred in the period 1910–30.

from the Western Corridor to the Mwanza-Musoma road, east along the Orangi River to the Ngorongoro highlands. The southern border followed the present NCA roughly to the village of Makao at the western edge of the plains then north to Seronera. The park headquarters was based at Ngorongoro with a western outpost situated at Banagi. In the early 1950s it became clear that the park as it was with people living in it was becoming untenable, and so the Tanganyika government decided to excise those areas where the Maasai had traditionally lived, leaving the remainder as a national park without people—the area that never had residents except for nomadic hunters as far back as records existed into the mid-1800s (fig. 2.3).

To decide where to redraw the new boundaries one needed information on both the location of pastoralists and the movements of the migrating wildlife herds about which almost nothing was known. A commission of inquiry in 1956 employed Professor Pearsall to make the study (Pearsall 1957). He reported that there were two groups of migrating wildebeest, one from the Ngorongoro Crater that moved onto the eastern Serengeti plains in the wet season, and one from the corridor that used the western plains. In between the two there was a gap not used by either, a gap that could conveniently be used for the boundary of the new Serengeti National Park. This advice was accepted and the new boundary running north-south across the middle of the plains was gazetted in 1959 (figs. 2.1, 2.6). The information that Pearsall used was more hearsay than fact and was incorrect—the Ngorongoro wildebeest never went on to the eastern plains but remained on the highlands or in the crater itself. Instead the main migratory wildebeest of Serengeti used the whole of the plains including the Salai plains at the far eastern edge of the Gregory Rift Valley. Ever since the new boundary alignment was created, the migratory wildebeest have moved outside the Serengeti National Park each year during the wet season. These movements have caused a confrontation with the Maasai because wildebeest carry malignant catarrh fever, a disease which kills cattle (Plowright 1968). Maasai herds move away from the plains when the wildebeest are there as the peak periods for transmission occur each spring over the first few months of life of each year's cohort of calves; unfortunately, this coincides with the time when the long rains have made the grasses of the southern short-grassed planes highly nutritious fodder for both wildebeest and cattle.

In 1958 Bernard and Michael Grzimek conducted the first wildlife survey of the Serengeti ecosystem. Part of their work documented the movements of the migrants and they showed up the fallacy in Pearsall's report. However, by this time it was too late and the boundaries had been settled despite strong protests from the Grzimeks (Grzimek and Grzimek 1960a, b; Stewart 1962). To compensate the Serengeti National Park for the loss of the eastern plains the authorities changed the western and northern boundaries. The southwestern boundary, originally running along the edge of the plains, was moved westward to include Moru kopjes and the Nyaraboro and Itonjo hills. Moru was the only area that was then used by some 100 Ndorobo hunters. The Maasai did not live there but used the area for dry season grazing. At the request of the Maasai elders the administrators moved the Ndorobo elsewhere—they were the only people to be moved on the plains.

In 1954 there were 194 Maasai that lived in the Gol Mountains but used the western plains on a seasonal basis (Grant 1957). Since the mid-1930s Maasai from the Gol Mountains had developed a seasonal grazing pattern: as the wildebeest left the western plains in June or July the Maasai followed them to the edge of the plains along the Ngare Nanyuki River in the north and toward Moru in the west to make use of spring water. They used these areas for some three months until the wildebeest returned in November, whereupon they retreated back to the Gol Mountains. This ebb and flow was constrained by ecological factors, namely the tsetse fly which prevented their further movement into the savanna, and malignant catarrh fever carried by wildebeest, which caused the Maasai to move east again at the start of the rains. After 1959 their grazing routine was changed to make use of the Olduvai and the upper Ngare Nanyuki River water sources. The new park boundaries were designed to suit the other sections of the Maasai living in the Gol Mountains and at Olduvai (Grant 1957). A few others lived on the north bank of the Orangi River near Hembe Hill and these were moved to Robanda village.

The northern extension—the whole area north of the Orangi River to the Mara River—was added (fig. 2.6); this area had never had inhabitants due to the presence of the tsetse fly; thus, no people were moved out. It was a remarkable stroke of luck, perhaps the greatest piece of luck in conservation history, because this area turned out to be the essential dry season refuge for the migration, indeed the most important area of the whole ecosystem—without it the migration would have collapsed and the Serengeti reduced to just another sample of savanna with resident animals. Grzimek's (1960) distribution maps showed the animals moving north but far to the west and he opposed this addition because it appeared unused by the migration—his maps were simply too inaccurate and he misplaced the migration routes. The correct routes were only fully appreciated when Lee and Martha Talbot spent several years following the wildebeest herds from the ground and in the air in the early 1960s (Talbot and Talbot 1963), and when Murray Watson conducted his aerial surveys shortly after (Watson 1967).

The Serengeti park headquarters, originally at Ngorongoro, were moved to Seronera where they remained until 1998 when they moved again to Fort Ikoma, outside the park. The Maasai Mara Reserve in Kenya was gazetted in 1961 under Maasai administration from Narok. The Lamai wedge in Tanzania, between the Mara River and Kenya border, was added in 1966, thus creating a continuous protected corridor for the wildebeest migration from the Serengeti plains in the south to the Loita plains in the north. A small area north of the Grumeti River in the corridor was added in 1967. Subsequently the Grumeti and Ikorongo Game Reserves were established in 1993, and the wildlife management areas were decreed in 2003.

LESSONS FROM HISTORY

Events in the nineteenth century had a profound influence on both the ecology of Serengeti and the political decisions affecting the area. Thus, one consequence of the ivory trade of the mid-nineteenth century is that elephant populations were first reduced to almost nothing by the early twentieth century and then rebounded exponentially when hunting was controlled. The increase in numbers had at one time been thought unstable and aberrant (Pienaar, van Wyk, and Fairall 1966; Lamprey et al. 1967; Watson and Bell 1969). Now we see this as a natural rebound after a severe reduction (Sinclair 2012b); we are continuing to monitor the population to look for regulating mechanisms that could stabilize numbers (Sinclair and Metzger 2009). The flexible eight- to twenty-year time delay between birth and age at first reproduction in elephants creates a significant destabilizing influence in their population regulation; this contributes to their potential to overshoot available resources when recovering from a significant population reduction. More importantly the rapid changes in elephant numbers in the mid-twentieth century influenced policies on natural area management throughout Africa; there was a general trend toward the culling of elephants (Pienaar, van Wyk, and Fairall 1966; Laws, Parker, and Johnstone 1975; Pienaar 1983).

Second, the historical record is instructive because it emphasizes that the cause of changes in ungulate populations, their predators, and in the vegetation as we first recorded them in the 1960s were not understood until they were placed in the context of the long-term events of the rinderpest panzootics. Both the historical account of the 1890s and the blood antibody data in the 1960s were integral to explaining these events. History has also allowed us to see how the changes in tree densities affected elephant numbers, which were rebounding as we relate above. Rinderpest resulted in the exodus of the relatively sparse human inhabitants from around Serengeti, the consequent reduction of burning, and a pulse of tree regeneration in 1890. This pulse in young trees produced the mature trees some 70 years later that provided the food for the increasing numbers of elephants in the 1960s (Sinclair et al. 2008); as these trees reach the end of their natural lives, they become weaker and are more readily knocked down by elephants.

Third, from the expeditions of Baumann, Smith, and White we see a far more widespread distribution of the great migration in the 1910s commensurate with a lack of human settlement. The wildebeest were distributed from Waso in the east almost to Musoma in the west, using the Mara River and the Masirori swamp as their dry season water sources. Rarer ungulates such as rhino, roan, and lesser kudu were more widespread and abundant. In contrast, elephants were very rare, consistent with the aftereffects of the ivory trade during the years of 1840–90.

Finally, the events that led to the formation of the modern Serengeti National Park boundaries are instructive. We see that in the absence of good information on the numbers and movements of the wildlife the boundaries were misplaced. The eastern boundary across the plains effectively excises most of the essential wet season grazing for the wildebeest and conservation is now dependent on the policies of the Ngorongoro Conservation Authority (Fosbrooke 1972), and the unnatural shape of the boundary around the village of Robanda could prevent the migration if the probable increase in human settlement prevents the movement of animals. We also see that increasing human settlement on the boundaries is leading to political pressures to change boundaries, allow development, and threaten the integrity of the ecosystem (Dobson et al. 2010). It is argued that humans had once lived in the area but were moved out when the park was declared in 1950. We show that this was not the case—boundaries were in large part set to leave people where they had previously lived and the natural area was demarcated where people had never lived—at least in the past several centuries.

ACKNOWLEDGMENTS

We thank Gerald Rilling for finding the early books, Eric Hinze with translations of Baumann, Stephen Makacha and Elijah Awino for documenting the early movements and history of their tribes, and Kay Turner for her history of Banagi and information on the 1950s. Ann Tiplady and Catherine Sease kindly provided the unpublished diaries and films of the Lieurance hunting expeditions of 1928–29. We thank Rhodes House and the Bodleian Library at Oxford University for the use of their library, and the archivist Lucy McCann for help in finding material.

REFERENCES

Baumann, O. 1894. *Durch Massailand zur nilquelle.* Berlin. Reprinted 1968. New York: Johnson Reprint Corporation.

Dobson, A. 1995. The ecology and epidemiology of rinderpest virus in Serengeti and Ngorongoro Conservation Area. In *Serengeti II: Dynamics, management and conservation of an ecosystem,* eds. A. R. E. Sinclair and P. Arcese, 485–505. Chicago: University of Chicago Press.

Dobson, A. P., M. Borner, A. R. E. Sinclair, P. J. Hudson, T. M. Anderson, G. Bigurube, T. R. B. Davenport, J. D. Deutsch, S. M. Durant, and C. Foley, et al. 2010. Road will ruin Serengeti. *Nature* 247:272–74.

Farler, J. P. 1882. Native routes in East Africa from Pangani to the Masai country and the Victoria Nyanza. *Proceedings of the Royal Geographical Society,* (new series) 4:730–42.

Ford, J. 1971. *The role of trypanosomiases in African ecology.* Oxford: Clarendon Press.

Fosbrooke, H. 1972. *Ngorongoro: The eighth wonder.* London: Deutsch.

Grant, H. St. J. 1957. *A report on human habitation in the Serengeti National Park.* Government Printer, Dar es Salaam.

Grzimek, B., and M. Grzimek. 1960a. *Serengeti shall not die.* London: Hamish Hamilton, Ltd.

Grzimek, M., and B. Grzimek. 1960b. A study of the game of the Serengeti plains. *Zeitschrift fur Saugetierkunde* 25:1–61.

Huxley, J. 1936. *Africa view.* London: Chatto & Windus.

Johnson, M. 1929. *Lion.* New York: G. P. Putnam's Sons.

Lamprey, H. F., P. E. Glover, M. Turner, and R. H. V. Bell. 1967. Invasion of the Serengeti National Park by elephants. *East African Wildlife Journal* 5:151–66.

Laws, R. M., I. S. C. Parker, and R. C. B. Johnstone. 1975. *Elephants and their habitats.* Oxford: Oxford University Press.

Lowe, H. J. 1942. Rinderpest in Tanganyika territory. *Empire Journal of Experimental Agriculture* 10:189–202.

Moore, A. 1937. *Serengeti.* London: Country Life Ltd.

Pankhurst, R. 1966. The Great Ethiopian famine of 1888–1892: a new assessment. Part 2. *Journal of the History of Medicine* (July): 271–94.

Patterson, J. H. 1907. *The Maneaters of Tsavo.* Reprinted 1996. New York: Pocket Books.

Pearsall, W. H. 1957. Report on an ecological survey of the Serengeti National Park, Tanganyika. *Oryx* 4:71–136.

Percival, A. B. 1928. *A game ranger on safari.* London: Nisbet & Co.

Pienaar, U. de V. 1983. Management by intervention. The pragmatic option. In *Management of large mammals in African conservation areas*, ed. N. R. Owen-Smith, 23–26. Pretoria, South Africa: Haum Educational Publishers.

Pienaar, U. de V., P. W. van Wyk, and N. Fairall. 1966. An aerial census of elephant and buffalo in the Kruger National Park and the implications thereof on intended management schemes. *Koedoe* 9:40–108.

Plowright, W. 1968. Malignant catarrhal fever. *Journal of the American Veterinary Medical Association* 152:795–804.

———. 1982. The effects of rinderpest and rinderpest control on wildlife in Africa. *Symposium of the Zoological Society of London* 50:1–28.

Plowright, W., R. D. Ferris, and G. R. Scott. 1960. Blue wildebeest and the aetiological agent of bovine malignant catarrhal fever. *Nature* 188:1167–69.

Rossiter, P. B., W. P. Taylor, B. Bwangamoi, A. R. H. Ngereza, P. D. S. Moorhouse, J. M. Haresnape, J. S. Wafula, J. F. C. Nyange, and I. D. Gumm. 1987. Continuing presence of rinderpest virus as a threat in East Africa 1983–1985. *Veterinary Record* 120:59–62.

Sheriff, A. 1987. *Slaves, spices and ivory in Zanzibar.* Oxford: James Currey.

Simon, N. 1962. *Between the sunlight and the thunder. The wildlife of Kenya.* London: Collins.

Sinclair, A. R. E. 1977. *The African buffalo. A study of resource limitation of populations.* Chicago: University of Chicago Press.

———. 2012a. *Serengeti story.* Oxford: Oxford University Press.

———. 2012b. Ecological history guides the future of conservation: Lessons from Africa. In *Historical environmental variation in conservation and natural resource management*, ed. J. A. Wiens, G. D. Hayward, H. D. Safford, and C. M. Giffen, 265–72. Oxford: Wiley-Blackwell.

Sinclair, A. R. E., J. G. C. Hopcraft, H. Olff, S. A. R. Mduma, K. A. Galvin, and G. J. Sharam. 2008. Historical and future changes to the Serengeti ecosystem. In *Serengeti III: Human impacts on ecosystem dynamics*, ed. A. R. E. Sinclair, C. Packer, S. A. R. Mduma, and J. M. Fryxell, 7–46. Chicago: University of Chicago Press.

Sinclair, A. R. E., and K. Metzger. 2009. Advances in wildlife ecology and the influence of Graeme Caughley. *Wildlife Research* 36:8–15.

Smith, G. E. 1907. From the Victoria Nyanza to Kilimanjaro. *The Geographical Journal* 29: 249–69.

Spinage, C. A. 1973. A review of ivory exploitation and elephant population trends in Africa. *East African Wildlife Journal* 11:281–89.

———. 2003. *Cattle plague, a history*. Dordrecht: Kluwer Academic/Plenum Press.

Stewart, D. R. M. 1962. Census of wildlife on the Serengeti, Mara and Loita plains. *East African Agricultural and Forestry Journal* 28:58–60.

Talbot, L. M., and M. H. Talbot. 1963. The wildebeest in western Masailand. *Wildlife Monographs No. 12*. Washington, DC: The Wildlife Society.

Turner, M. 1987. *My Serengeti years*. London: Elm Tree Books/ Hamish Hamilton Ltd.

Wakefield, T. 1870. Routes of native caravans from the coast to the interior of East Africa. *Journal of the Royal Geographical Society* 11:303–38.

———. 1882. New routes through Masai country. *Proceedings of the Royal Geographical Society* 4:742–47.

Waller, R. D. 1988. Emutai: crisis and response in Maasailand 1883–1902. In *The ecology of survival: Case studies from northeast African history*, ed. D. Johnson and D. Anderson, 73–114. Boulder, CO: Lester Crook Academic Publishing/Westview Press.

Watson, R. M. 1967. The population ecology of the wildebeest (*Connochaetes taurinus albojubatus*) in the Serengeti. PhD. diss., Cambridge University, Cambridge.

Watson, R. M. and R. H. V. Bell. 1969. The distribution, abundance and status of elephant in the Serengeti region of northern Tanzania. *Journal of Applied Ecology* 6:115–32.

Wheeler, S. 2006. *Too close to the sun*. London: Jonathan Cape.

White, S. E. 1915. *The rediscovered country*. New York: Doubleday, Page.

PART I

Natural Sources of Heterogeneity and Disturbance

THREE

Scales of Change in the Greater Serengeti Ecosystem

Kristine L. Metzger, Anthony R. E. Sinclair, Sandy Macfarlane, Michael Coughenour, and Junyan Ding

For many the Serengeti is imagined as a landscape of gently rolling, treeless plains teeming with lions, elephants, and herds of wildebeest and zebra. This preconceived notion is quickly dispelled for first-time visitors who discover that the Serengeti is a mosaic of plains, hills, mountains, dense riverine forests, and human-modified landscapes in a complex of climatic zones.

Ecosystems change over space and time each at a variety of scales. In this chapter we summarize what is known about the abiotic environment of the Serengeti ecosystem as it changes over these different scales. We deal with the broad scale, including the long timescale of millennia and the spatial scale of the whole system, then look at physical change over a scale of kilometers and over time in decades, and then consider such change over meters and over seasons. We then consider how changes at these different scales affect the ecology of the Serengeti.

The distribution of vegetation and biodiversity over the landscape is influenced by biotic and abiotic factors including geology, topography, climate, fire, predation, and herbivory (Merriam 1890; Whittaker and Levin 1977), among many others. The spatial configuration and interactions of these factors give rise to patterns of heterogeneity within and among landscapes. The development of management and land use strategies for this dynamic ecosystem requires a comprehensive understanding of the biotic and abiotic interactions, and while the biotic interactions in the Serengeti have received considerable attention, far less work has focused on the effects of abiotic heterogeneity on ecosystem structure and diversity (Reed et al. 2009).

The geological template of the Serengeti determines the landscape heterogeneity and is the foundation from which parent material, and ultimately soil, is derived. Topography is the geophysical pattern arising from geodynamics, tectonics, and the region's disturbance history, including human-made features. Climate, the composite of the region's long-term or prevailing weather (Bailey 1996), influences biogeographical patterns through the distribution of energy from sunlight and water as rain. In addition it modifies landforms through geomorphic processes that create physical relief and soil. Geological "parent materials" are constantly being broken down by erosion from climatic effects and further are refined and recycled by flora and fauna through physical and chemical mechanisms. Vegetation patterns arise and contribute to the soil patterns of the landscape. In turn, organisms respond to the vegetation template of the landscape and contribute their own disturbance patterns. All of these processes are interrelated to varying degrees over a wide range of spatial and temporal scales.

GEOLOGICAL ORIGINS OF HETEROGENEITY

In order to fully understand Serengeti landscapes, it is necessary to understand changes that occur over geological timescales. Geological processes have affected the current structure and functioning of the ecosystem by creating landforms and topography, which in turn affect hydrology, and through the creation of raw or "parent" materials from which current soils have been derived. The spatially heterogeneous nature of the Serengeti is a direct consequence of the interplay of diverse geological processes and events. The principal geological units in the Serengeti National Park are outlined in figure 3.1, which is a digest of the regional maps produced by the Tanzanian Geological Survey (MADINI) in the 1930s (Stockley 1936) and 1950–60s (Pickering 1958; Pickering 1960; Horne 1962; Pickering 1964; Naylor 1965; Awadallah 1966; Lounsberry and Thomas 1967; Macfarlane 1967; Gray, MacDonald, and Thomas 1969). Reference has also been made to subsequent fieldwork in the park in the late 1960s (Macfarlane 1970), and to a regional study of the Lake Victoria goldfields in 1984–85 (Barth 1990).

Causes of Geological Heterogeneity

The heterogeneity of the Serengeti arises from its location on the northern margin of the Tanzania craton (a stable portion of the Earth's crust) at the intersection of two major orogenic belts (mountain ranges); the E-W Nyanzian-Kavirondian Orogenic Belt (NKOB) of Archean age (2.5–3.8 Ga) in the west, and the later Proterozoic (0.5–2.5 Ga) to Cambrian (0.49–0.54 Ga) N-S Mozambique Orogenic Belt (MOB) in the east. The Nyanzian System is a neo-Archean (2.5–2.8 Ga) geological formation covering much of the area east of Lake Victoria. In the northwest, the NKOB encloses an extensive *inlier* of Pre-Nyanzian rocks of uncertain age and origin. Overlying these older substrates toward the park boundaries in the north, south, and west, respectively, are a much younger Neogene (23 Ma–0.4 Ma) to recent (<0.4 Ma) covers of lava sheets, calcareous tuffs and extensive river alluvium.

Pre-Nyanzian (2.9–3.2 Ga). Crystal-containing gneisses, schists, and amphibolites at the NW boundary of the Serengeti National Park are believed to predate NKOB events. They may be a root facies of the Nyanzian System or, as is suggested here, a remnant of an earlier phase of greenstone belt development synchronous perhaps with the Dodoman Orogenic Belt of central Tanzania. The inlier contains the oldest rocks in the park whose age is in the order of 2.9 Ga or 3.2 Ga.

Nyanzian-Kavirondian Orogenic Belt (NKOB, 2.5–2.7 Ga). The NKOB in NW Tanzania comprises an extensive assemblage of sinuous, mainly E-W trending greenstone belts of Nyanzian volcanic rocks and sediments together with clastic deposits, granitic gneisses, and granites. The greenstone belts were assembled in weak fracture zones or down-warps in pre-Nyanzian crust and subsequently subjected to intense deformation and regional granitization and associated gold mineralization. Colloquially, the NKOB is known as the Lake Victoria goldfields, the eastern limit of which extends into the Serengeti National Park. Dating of the NKOB has yielded ages of 2.5–2.7 Ga. The NKOB is comprised of several different formations that contribute to the geological heterogeneity of the region (see the appendix to this chapter). These include: the Nyanzian System, a metamorphosed lava with metasedimentary components; Synorogenic granites: the Kavirondian System, comprised of basal boulder conglomerates, pebble grits, quartzites, and minor pelitic beds deposited following uplift and erosion; late-orogenic granite, an initial intrusion of both greenstones and older granites by late-orogenic material.

Mozambique Orogenic Belt (530–760 Ma). Development of the N-S trending MOB (760–730, 660–610, 570–530 Ma) was the result of progressive closure of the "Mozambique Ocean," a continental collision with the Tanzania craton and assembly of Gondwanaland (Bingen et al. 2009). Associated tectonic and metamorphic events in the Serengeti National Park are attributed to the final phase, the "Pan-African" orogeny (570–530 Ma), when crustal thickening was at a maximum and led to westward-directed deformation of the Archean foreland margin. These events can be traced across much of the Serengeti National Park in both the Archean floor and a later and once continuous cover of Neoproterozoic shelf sediments, deposited on the western shoreline and shelf of the "Mozambique Ocean." These sediments are now represented by the Ikorongo and Metasedimentary groups. The MOB is comprised of several components, including recrystallized granitic rocks, the Ikorongo sedimentary group in the central part of Serengeti National Park, a Metasedimentary group in the NE, and the results of Pan-African deformation and metamorphism extending westward with decreasing intensity across the width of the park (see the appendix to this chapter).

Neogene to Recent Deposits (23Ma–0.4 Ma). Flat lying phonolite lava sheets extend above and below the Isuria escarpment in the NW corner of the park and are the product of fissure eruptions connected with the rift faulting in that area. In the south, the Serengeti Plains owe their character to an infilling cover of calcareous tuffs blown over the area from carbonatitic volcanoes, including the still-active Oldoinyo Lengai (Dawson 2008), to the east. Isolated coarse grained granite and granitic gneiss "inselbergs" (in the Serengeti traditionally, but incorrectly referred to as "kopjes") rise through the calcareous cover. Their rounded and domed appearance is the result principally of "exfoliation" aided to a lesser extent by "spheroidal weathering" caused by exposure to alternating extremes of climatic conditions. They vary in size from a few meters long and high, to small hills almost a kilometer in length and 50 m high, the latter seen especially in the Moru kopjes on the western edge of the plains. A distinctly different group of small rocky hills that resemble kopjes occur in the far south of the ecosystem, especially in Maswa. Kopjes are found in both the plains and the savanna. Kopjes, particularly in the savanna, are islands of habitat supporting a distinctive vegetation of shrubs, candelabra trees (*Euphorbia*), and fig trees. In the west along the Grumeti corridor extensive alluvial plain deposits accentuate the descent of the main park river drainage lines into Lake Victoria.

GEOMORPHOLOGY AND LANDFORMS

The geologic processes described above, combined with hydrological and erosion-deposition processes, have given rise to a diverse array of landforms across the region. The Serengeti ecosystem has been divided into major landforms by Gerresheim (1971) (fig. 3.1), and Epp (1981). Landform (elevation, slope, aspect) affects climate, water redistribution on the landscape, the frequency and spatial distribution of fire and wind, and physical weathering and redistribution of organic and inorganic substrates. Together, these

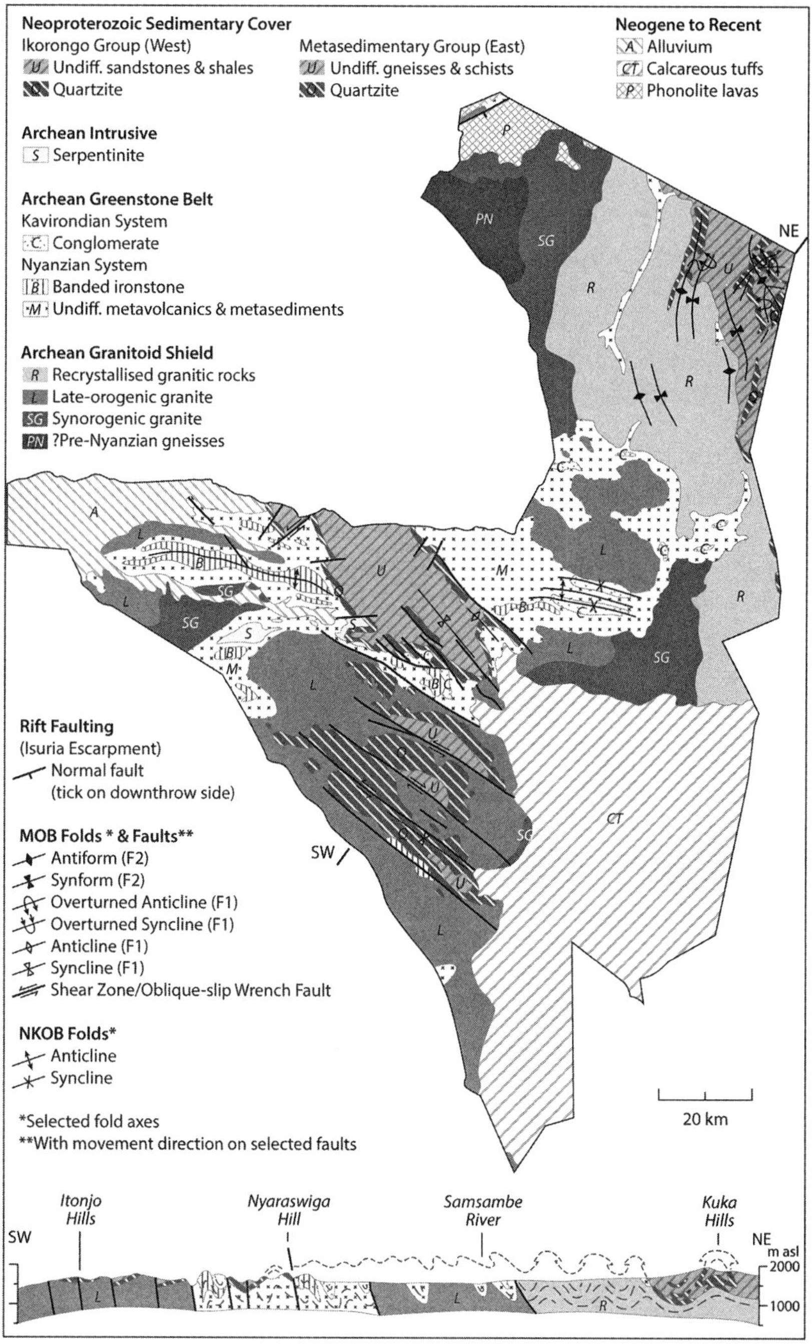

Fig. 3.1 Geological map of the Serengeti National Park including a composite vertical section across the park from the Itonjo hills in the southwest to the Kuka hills in the northeast.

processes are fundamentally responsible for the development and maintenance of spatially heterogeneous abiotic "driving variables" which affect biotic processes.

Originally, the Serengeti region was part of an ancient drainage that flowed from east to west, the outlet being on the west coast of Africa, some 14 million years ago in the Miocene, similar to the Congo drainage of today. Gradual uplifting of the Albert rift began in the Miocene, but it was not until the early Pleistocene, between 1.6 and 0.8 Ma, that river flow was blocked, creating a megalake after 0.8 Ma some 100 m deeper than Lake Victoria today (Johnson et al. 1996). Outflow changed from the Congo drainage to that of the Nile, thus initiating the hydrology of northeastern Africa that we see today.

Large-scale landform characteristics in the Serengeti significantly influence ecosystem functioning. A large-scale altitudinal gradient, sloping from 1800 m in the east (excluding the top of hills) to 1200 m at Lake Victoria in the west, constitutes the remains of the ancient Miocene drainage into the Congo. This sloping surface drives the direction of river flows and has influenced watershed formation. In addition, the northern part of the ecosystem is approximately 200 m higher than the southern part with a rapid change in altitude just north of the Orangi River that bisects the system in the centre. Both the E-W and N-S gradients influence air temperatures. There is a large-scale divergence of landforms between the mainly flat, treeless plains in the south and east, and the heterogeneous landscape mosaics in the west and north which are more wooded.

Landforms range from nearly flat plains in the southeast, to rolling irregular plains, to hills, and low mountains in the north and west (fig. 3.2). The great migration of ungulates alternates seasonally between these broadly divergent landforms, driven mainly by the seasonal rainfall gradient and nutrient availability (for more detail see Hopcraft, chapter 6). The Serengeti plains in the southeast are expansive, flat areas formed by volcanic ash deposition. Northern parts of the Serengeti plains are uplands typified by gently rolling hills intercepted by segments of drainage areas. Large areas of alluvial floodplains occur in the far west near Lake Victoria. The north and particularly the western "corridor" of the park are topographically diverse. The northern landscapes are typified by gently rolling hills dissected by many small streams and rivers. Although much of the north and west is covered with savannas and woodlands, there are several large treeless grasslands having similar landform. There are also several areas with steep-sided rocky hills, as described in the geology above. These occur in the northeast, reaching 2000 m and along the eastern boundary of the park, as far east as the Gol Mountains, and a central group of hills extending in a line west

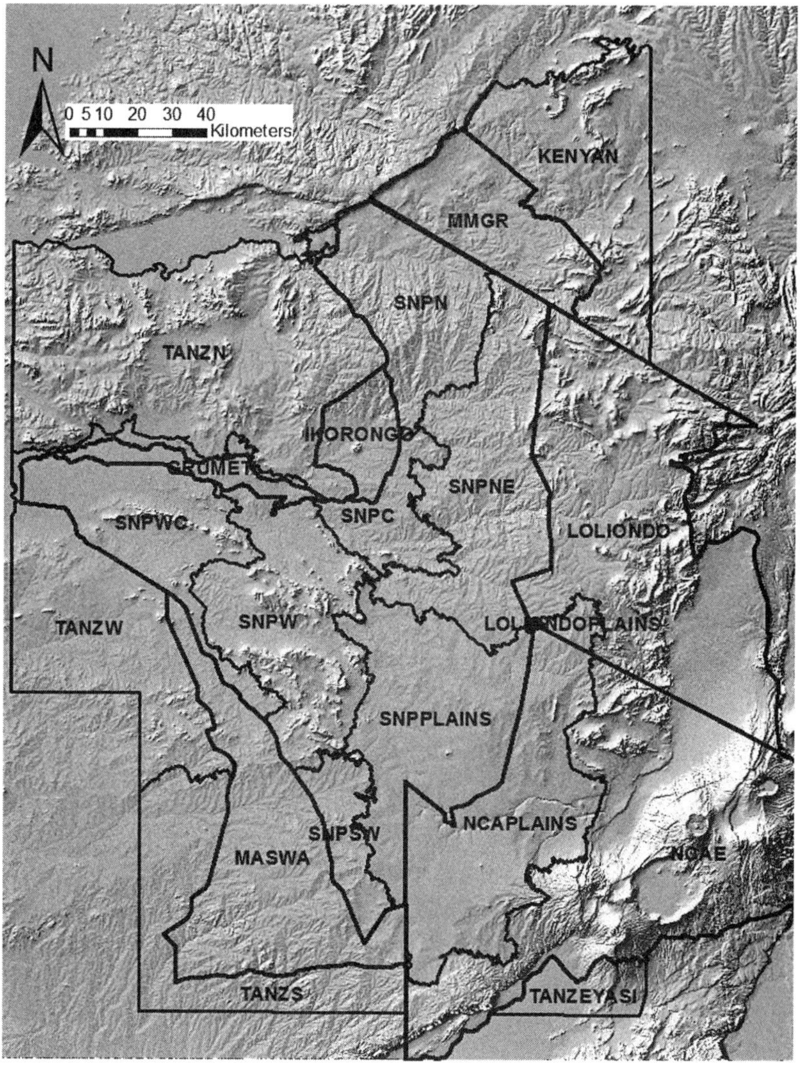

Fig. 3.2 Landform across the Serengeti ecosystem developed from the digital elevation model (USGS SRTM).

toward the lake. At a finer scale, many valleys have narrow (10 m wide) seasonally flowing gullies, usually with steep banks. These flow into the larger rivers. A few water springs flow throughout the year. These derive from the base of the larger hills such as Kuka in the northeast, Lobo on the eastern boundary, Nyaraboro in the south, and Nyamuma in the central hills.

At the mesoscale, topographic diversity influences habitat heteroge-

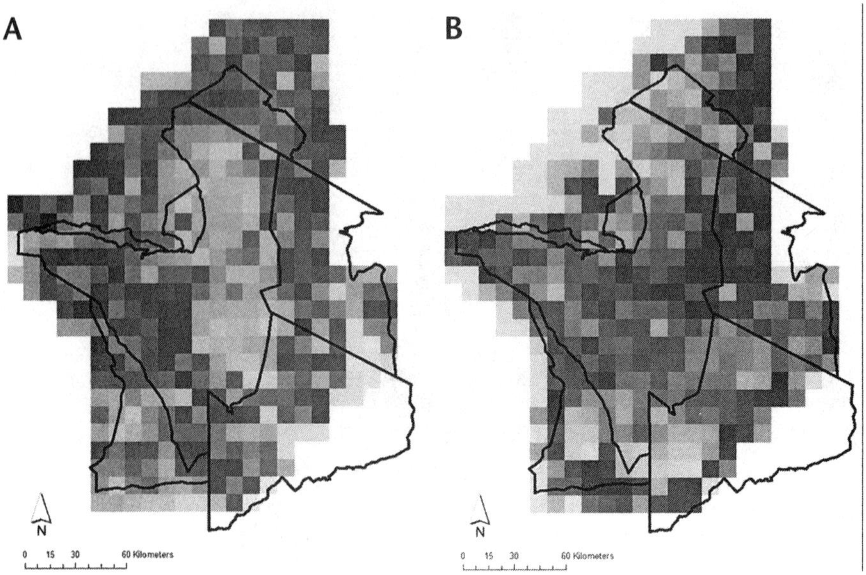

Fig. 3.3 (A) Topographic H′ diversity is an index that ranges from low (light areas) to high (dark areas). (B) Similarly ungulate H′ diversity index was low (light areas) to high (dark areas), developed from systematic aerial surveys (Norton-Griffiths 1978).

neity and the diversity of potential niches, and thus biota, per unit area. Topographic diversity can be quantified as a combination of evenness and richness for each region in the system (table 3.1, fig. 3.3a), as calculated from the digital elevation model provided by the United States Geological Survey Shuttle Radar Topography Mission (USGS SRTM). Loliondo was the highest diversity and Serengeti National Park southwest and northern areas had low diversity. Within the Serengeti National Park the southwest had the lowest topographic diversity. Outside the park, Loliondo and Ngorongoro Conservation Area had the highest diversity which reflects the mountainous landscapes of those areas. Areas with high diversity in topography also have higher diversity of ungulate species (fig. 3.3b). For example, where hills occur in juxtaposition to riverine forest, ungulate diversity is likely to include specialists on hills (Chandler's reedbuck, klipspringer), those in riverine areas (waterbuck, bohor reedbuck, buffalo), as well as the generalist savanna ungulates. The flat open plains will have a reduced number of resident ungulates capable of tolerating those climates, such as Grant's gazelle and oryx. A similar pattern is also likely to be seen with bird species diversity. Along the riverbanks there is a narrow band of thicket which is used by lions to ambush prey (Hopcraft, Sinclair, and Packer 2005; Metzger et al. 2010), thus creating a fine scale gradient of predation risk for ungulates.

Inselbergs, or "kopjes," found throughout the southeastern portion of the ecosystem, are another important example of fine scale landform variation. Kopje-scale heterogeneity carries over into animal communities. Kopjes are used by lions to survey potential prey. Kopjes are used by animals dependent on rocks, such as agama lizards, klipspringer antelope, hyrax species, rodents such as *Acomys* (Byrom, chapter 12), and several birds. Kopjes occur in groups, although individual kopjes often occur within a few hundred meters apart; there may be ten or more kilometers between kopje groups so migration for smaller animal species between these groups is probably limited. Kopjes, particularly in savanna, are islands of habitat supporting a distinctive vegetation of shrubs, candelabra trees (*Euphorbia*), and fig trees. On the plains kopjes have little or no vegetation that is seen in the savanna and consequently animal species diversity is limited.

Soil Formation and the Distribution of Soil Properties

Present day soils are formed through pedogenesis, the combined outcome of geologic, climatic, and biotic factors acting over long time periods. Volcanic activity that began 4 Ma and continues on a small scale today produced the volcanic soils creating the plains and the gradient of soil nutrients. Volcanic soils of the southeast (Serengeti plains and the southwestern Serengeti) (table 3.1) are alkaline and nutrient-rich compared to soil of the northwest Serengeti ecosystem. Volcanic ash deposits were more pronounced in the southeast and they result in the treeless plains, because tree roots cannot penetrate the calcareous hardpan that formed from the ash. Thus, the volcanic ash boundary is marked by the sharp appearance of trees at the northern and western boundary of the plains.

The broad scale distributions of soil nutrient availability and water-holding capacity are important determinants of broadscale variations in plant productivity. The distributions of these properties are mapped in figure 3.4. Total exchangeable bases was used as a proxy for soil nutrient availability, and water-holding capacity was calculated as the difference between volumetric water contents at field capacity and wilting point applied to the depth of the soil. The Serengeti plains and the southwest have the highest nutrient availabilities and relatively high water-holding capacity (table 3.1). In contrast, areas west of and along the Grumeti River and outside the protected area and in the northern portion of the Serengeti have low water-holding capacity. Variations from the northwest to the southeast create a nutrient and water-holding capacity gradient that contributes to the spatial heterogeneity of habitats (Anderson et al. 2008). Importantly, areas where

Table 3.1. Summary of abiotic and land use variables for the various regions within the Serengeti Ecosystem

Game reserves	**Area (km^2)**	**Agriculture (%)**	**Topographic diversity**	**Topographic richness**	**Topographic evenness**	**Total exchangeable bases ($cMolc*kg^{-1}$)**	**Soil water holding capacity**
Grumeti	417	0	97	280	0.34	8.87	212.7
Ikorongo	602	25	101	318	0.32	11.5	218.5
Maswa	2,883	10	99	315	0.31	14.6	239.2
Conservation areas:							
NCA	5,397	14	131	434	0.3	23.9	249.5
NCA plains	2,858	0	137	335	0.4	43.2	218.9
Loliondo	5,201	2	161	408	0.39	23.2	229.1
Loliondo plains	164	0	86	279	0.3	44.5	217.3

Protected areas:							
MMNR	1,525	1	117	365	0.32	*	
Serengeti center	789	1	88	314	0.28	11.6	219.5
Serengeti north	1,816	2	85	339	0.25	11.6	196.6
Serengeti northeast	2,864	0	85	352	0.24	20.7	207.7
Serengeti southwest	649	0	80	285	0.28	43.8	231.8
Serengeti west	2,493	0	139	365	0.38	13.2	224.3
Serengeti West Corridor	1,794	2	126	339	0.37	11.2	221.4
Serengeti plains	2,661	0	97	297	0.33	43.8	219.5
Tanz:							
Tanzania Eyasi	518	25	9	339	0.027	24.7	257
Tanzania north	5,485	69	130	381	0.34	4.06	221
Tanzania south	1,963	12	96	311	0.3	13.8	239.7
Tanzania west	2,855	14	115	320	0.36	4.6	233.3
Kenya:							
Kenya north	2,666	8	129	368	0.35		

Notes: Ngorongoro Conservation Area = NCA. Maasai Mara National Reserve = MMNR

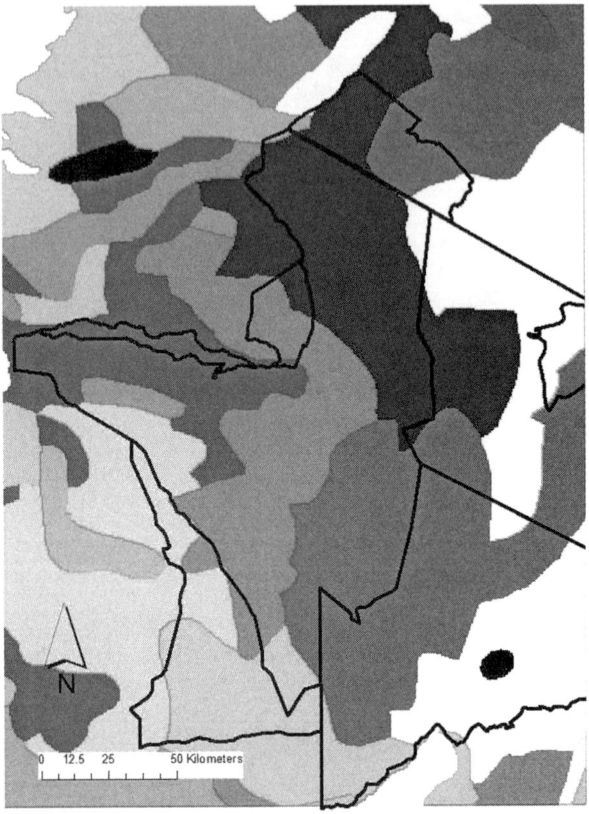

Fig. 3.4 Soil water holding capacity. Total exchangeable bases was used as a proxy for soil nutrient availability, and water holding capacity was calculated as the difference between volumetric water contents at field capacity and wilting point applied to the depth of the soil. Water holding capacity ranged from a high of 2.6 mm × cm^{-1} (dark areas) to a low (light areas) of 1.3 mm × cm^{-1}.

water-holding capacity and nutrient availability are high occur in areas of the lowest rainfall, thus potentially offsetting effects of low rainfall on plant productivity.

Fine scale variation in soil properties has been produced through hill slope processes, giving rise to "catenas." A catena is a soil sequence that varies from ridge top, to hill slope, to foot slope, to valley. On ridge tops, soils are shallow and sandy, often showing bedrock or a calcareous hardpan. Soils become progressively deeper and heavier further downslope, reaching maximum depths and levels of organic matter on valley floors. Classically, the height difference from top to bottom is at maximum 50 m and often only 10 m over a distance of a few hundred meters. However, catenas with much less relief, without ridges and hill slopes per se, are important for the structure and functioning of different vegetation communities on the Serengeti plains (McNaughton 1985).

Catenas are defining features of the landscape in that they determine the locations of near monospecific stands of different *Acacia* species. For example, *A. tortilis* and *A. senegal* are found on the crests, while *A. robusta* and *A. gerardii* are most common on the hill slopes and in valley bottoms where drainage is impeded *A. drepanalobium* and *A. seyal* prevail. *A. polyacantha* thrives where riverine areas are seasonally flooded, and along riverbanks the dominant species are *A. kirkii* and *A. xanthophloea*. *A. mellifera* occurs mainly on foot slopes and at the base of hills where sandy soils have been formed. These differential tree species distributions subsequently determine the distributions of animal species. The catena is also one of the dimensions of niche partitioning among nonmigratory grazing ungulates. Grazers such as Thomson's gazelle and wildebeest prefer the short grasses that grow on ridge tops; zebra feed on taller grasses on hill slopes as do topi and kongoni. Larger-bodied buffalo and waterbuck feed on tall grasses in the valley bottoms. As the dry season progresses and the grasses are eaten, the smaller-bodied species gradually move down the catena (Bell 1971; Murray 1995).

Fine scale spatial variations in soil are also associated with termite mounds, or "termitaria" (for details see deVisser, chapter 10). Termites build mounds by bringing up mineral-rich material from deeper in the soil. Over time these mounds leach out the minerals which flow downhill, thus forming a teardrop-shaped halo around the mound of a few meters in length. The almost even spacing of these mounds produces a characteristic pattern of high mineral patches seen clearly on aerial photographs. The mounds themselves are some two meters across and anywhere from flat to three meters in height, depending on species.

Termitaria are islands on a very small scale, for they support dense vegetation in the *Terminalia* woodlands of the northwest. Elsewhere in the savanna and on the plains they are bare, but the runoff areas surrounding them have different grass species such as *Cynodon dactylon*, which differs from the surrounding grassland complex. Termitaria are home to many animal species that live in them—snakes (cobras), rodents such as the very common *Arvicanthis*, agamas, and some birds such as the anteater chat and sooty chat.

Vegetation also contributes to fine scale soil variability. Large *Acacia* trees, especially *A. tortilis* and *A. robusta*, create small scale soil mosaics. First, through allelopathy they inhibit the growth of younger trees under their canopy so a halo of open ground exists near mature trees. This remains after the mature tree dies. Second, the microclimate under the canopy differs from the outside by being cooler, and the soils moister (Bell 1971; Belsky

1987, 1988; Murray 1995; Ludwig et al. 2003; Treydte et al. 2007). Tree canopies in the savanna create small islands of different herbaceous vegetation. These remain green long after the matrix of grassland outside has dried out, and such areas are then fed upon by ungulates (Treydte et al. 2007). Such canopies also provide shade for ungulates.

CLIMATE

Climatic Variation over a Broad Spatial Scale

The precipitation of the Serengeti ecosystem varies across a SE-NW gradient, ranging from <550 mm mean annual precipitation in the southeast, due to the rain shadow created by the Crater Highlands, to 1,200 mm per year in

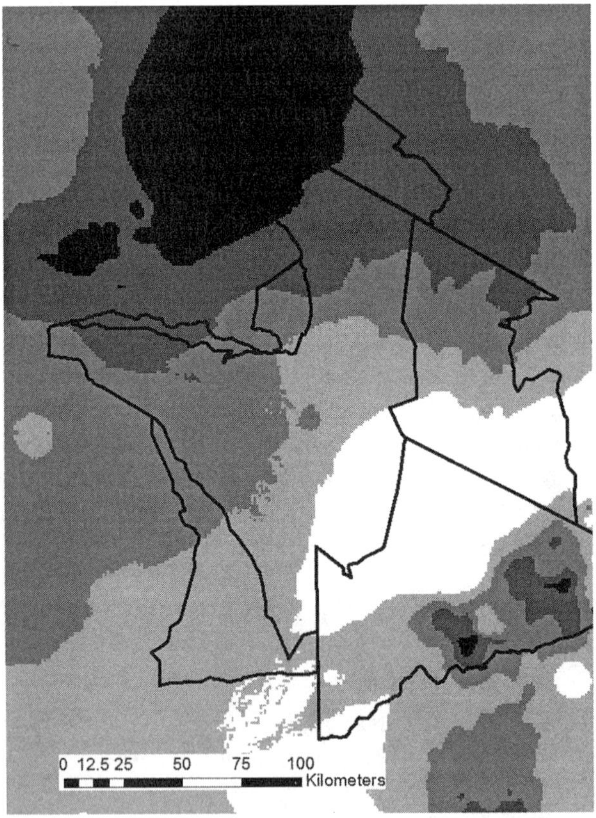

Fig. 3.5 Mean annual precipitation ranges from 500 mm in the southeast (light areas) to 1,200 mm in the north (dark areas).

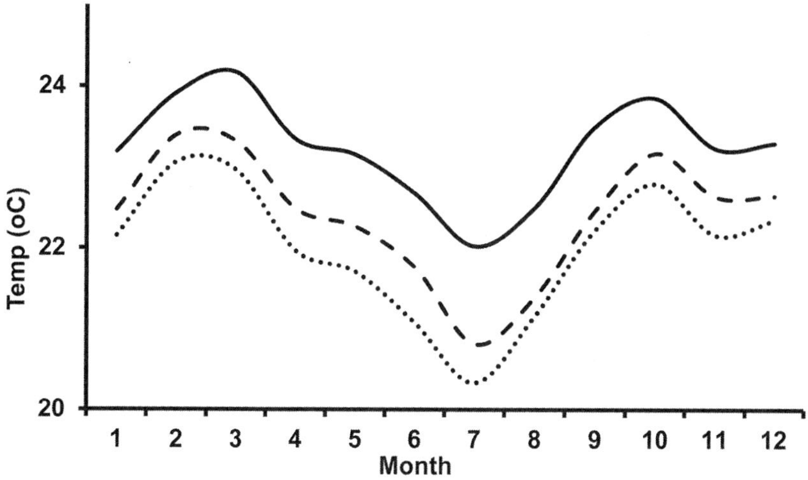

Fig. 3.6 Average daily temperature through the year in three zones in the park; solid line is the west, thick dashed line is the south, and dotted line is the north. Data are from Climate Research Unit (CRU) from 2000–2006.

the northwest, and 1,000 mm per year in the far west, due to the influence of Lake Victoria. Prevailing winds from the east carry moisture originating from the Indian Ocean that is released as air masses cross the Crater Highlands, thus giving rise to a rain shadow north of the highlands. In contrast, winds originating from the northwest and west carry moisture inland from Lake Victoria, counteracting this rain shadow effect, thus giving rise to the broadscale precipitation gradient (fig. 3.5). As a result, the semiarid climate of the southeast plains supports short grasslands, while the northwest is subhumid, with sufficient moisture to support forests if tree cover was not reduced by fire and ungulate herbivory (Turkington, chapter 9).

There is a broad scale temperature gradient associated with altitudinal changes, varying from cooler in the east to warmer in the west at approximately 1°C per 100 km. Temperatures are higher during the wet season (November–June) with February–March and September–October peaking at an average of 23.1°C in the west to 24.2°C in the north. June and July generally are the coolest months with average temperature in the north at 22.1°C and 20.2°C for the west (fig. 3.6).

The strong rainfall gradient that drives broad scale variation in vegetation composition as well as the ungulate migrations originated some 1.5 Ma. In the southeast, the Olduvai Gorge area changed from dense woodland to open treeless plains over the past four million years, a period that began with intense volcanic activity, which has since subsided (Peters et al. 2008). The appearance of the volcanoes forming the Crater Highlands created a rain shadow in this area that caused a shift in vegetation composition to arid grasslands. In the northwest, the development of Lake Victoria created the zone of high rainfall that has prevailed to current times. Indeed, Lake Victoria is so large that it produces its own weather systems.

Climate Variability and Patterns of Drought

Serengeti climate has been highly variable for the past 100,000 years. Periods of drought have occurred that have lasted several thousands of years, and even in the past 1,000 years, droughts have lasted some 100 years. Some of the changes from wet to dry have occurred suddenly, within a few years. Thus, there is no reason to expect that the climate in the future will remain as it is today.

Despite the moderating influence of Lake Victoria, there has been great variability in rainfall over long temporal scales. Sedimentary cores from Lake Malawi, a large, deep African rift valley lake south of Lake Victoria, have provided a record of the climate during the past 135,000 years (Cohen et al. 2007). There were very low lake levels from 135–127 Ka and again from 115–95 Ka, with the lake dropping some 600 m in depth. This indicates that Lake Victoria would have been completely dry in these periods. These levels indicated extreme aridity, more severe than any period in the past 100,000 years. There was insufficient vegetation to support fires which is indicative of a rainfall of <400 mm/yr (Cohen et al. 2007), and a cool semidesert prevailed. The cool, dry conditions during the arid periods were favorable for the spread of montane forest. In highland areas near Lake Malawi (and one can suppose above 2,000 m in the montane forests of northeastern Serengeti), there was an increase in montane trees of *Podocarpus, Olea, Juniperus, and Ilex*, all species found in the Serengeti montane forests today. Aridity diminished after 95 Ka and reached near-modern conditions after 60 Ka, and fire frequency returned to something similar to pre-European conditions of the 19th century.

In general, climatic variability was far greater before 70 Ka than since. However, there has been one period of severe aridity during the last glaciation 35 to 15 Ka. This was identified from lake cores in both Lake Malawi (Cohen et al. 2007) and Lake Victoria (Johnson et al. 1996). Lake Victoria dried out during this period and only started to fill again after 12,000 ya. During the last glaciation there was no lake-derived rainfall, which accounts for some 50% of present-day rain. Peak aridity was at 19 Ka and 18–16 Ka, and there were wet periods before 35 Ka and during 22–20 Ka (Bonnefille and Chalié 2000).

As the environment warmed after 12 Ka there was a period of rapid shifts in climate. Thus, from 12–11 Ka there were warmer and wetter conditions with advancement of lowland forest, a cold dry episode 11–10 Ka, followed by a sustained wet period to about 5 Ka, termed the "African Humid Period" (Thompson et al. 2002). Data from ice cores on Mt. Kilimanjaro show a particularly wet period at 6.5–5.2 Ka. Precipitation was possibly double that of today, so lakes rose some 100 m above present levels. Lake Chad was 25 times greater in area and it was about the same size as the present Caspian Sea. Closer to Serengeti a paleo-lake covered the whole Natron-Magadi basin to a depth of 50 m. During this humid period there were three sharp shifts to drought conditions at 8.3, 5.2 and 4 Ka. The one at 5.2 Ka coincided with the advance of the Sahara Desert and the rise of human societies in the Nile valley (Cullen et al. 2000), but it lasted a mere 100 years. The drought at 4 Ka was very dry and lasted some 300 years (Gillespie, Sweet-Perrott, and Switsur 1983; Thompson et al. 2002).

During the last two millennia there have been several fluctuations in climate. Water levels at Lake Naivasha, Kenya, were high during the three solar minima, named the Wolf, Sporer, and Maunder minima. The earliest (Wolf) period showed high levels from 1290 to 1370 AD, while the last Maunder period showed lake overflow during the 1670–1780 AD period that was interrupted by three prolonged dry episodes (Riehl and Meitin 1979; Verschuren, Laird, and Cumming 2000).

In the last century, the African continent has been warming at a rate of 0.5°C per century (Hulme et al. 2001), and rainfall has gradually increased, learned from the combined records at Shirati and Musoma (1902–2009), on the east side of Lake Victoria. The wettest year in 1961–62 raised the lake level several meters, and this has gradually dropped as rainfall has declined since 1961. However, this recent downward trend in rainfall is misleading. If the record for 1961 is removed, the overall trend has been an increase in the past 100 years, a trend reflected in the rainfall record at Banagi, in the center of the Serengeti.

Even more recently, temperatures have been increasing. Between 1960–2007 a gradual increase in maximum daily temperature has occurred throughout the ecosystem (fig. 3.7a). Dry season and wet season daily maximum temperature has been increasing by 0.017°C/year ($R^2 = 0.27$, p-value <0.0001) while daily minimum temperatures have remained stable (fig. 3.7b).

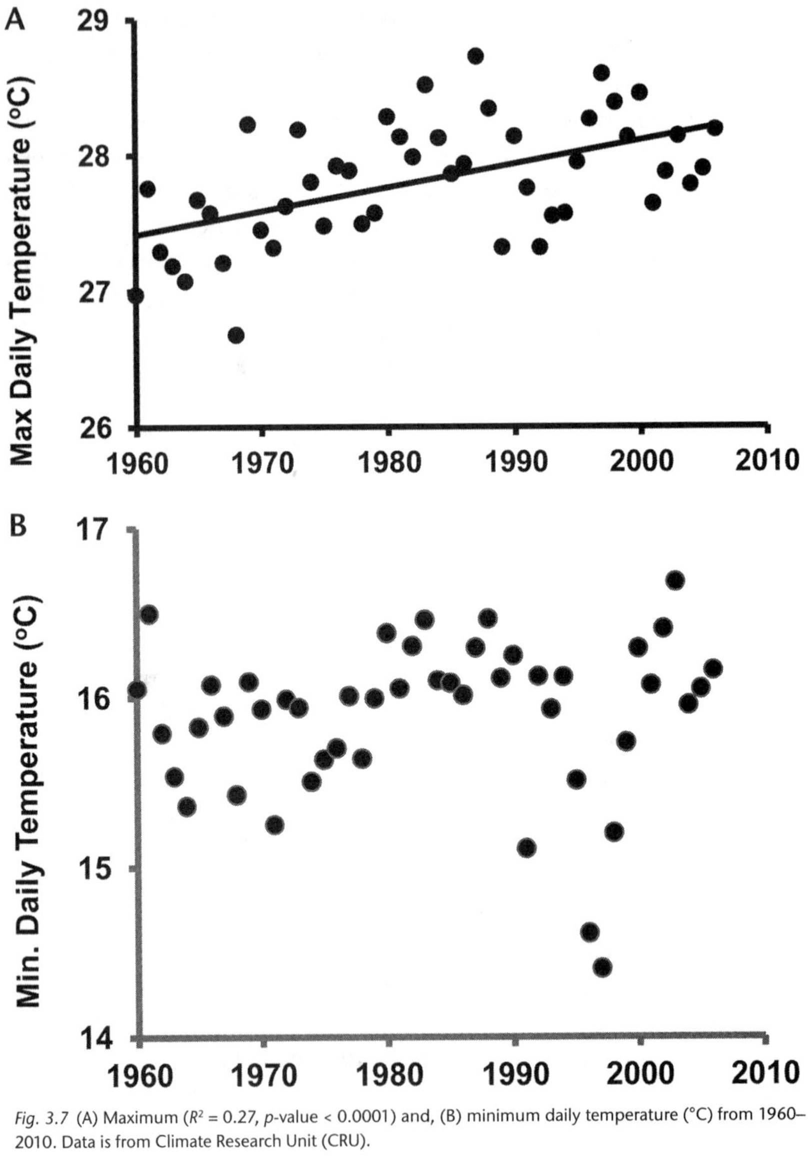

Fig. 3.7 (A) Maximum ($R^2 = 0.27$, p-value < 0.0001) and, (B) minimum daily temperature (°C) from 1960–2010. Data is from Climate Research Unit (CRU).

Climatic Variability Impacts on Flora and Fauna

Climatic changes would have significantly altered vegetation composition. Pollen records at Tsavo National Park, Kenya, show an alternation between savanna and grassland (Gillson 2004). While savanna has prevailed since 1700, vegetation composition shifted from savanna to grassland around 1300 and grassland prevailed 1300–1700, probably as a result of changes in rainfall regimes. In the Serengeti we would expect that during the wet period of 1670–1780 the lowland forest that occurred around Lake Victoria and along the Grumeti and Mbalageti Rivers would have expanded eastward upriver. In contrast, during the cool dry periods, and particularly since 1780, the highland forest would have expanded westward from the Loita forests of Kenya and northern Tanzania, and the Crater Highlands, downstream along the Mara River and across much of northern Serengeti and the Maasai Mara National Reserve (MMNR). The warming and wetter trend in the past century may underlie the decline of these northern forests (Turkington, chapter 9), not seen in the lowland western forests.

Fluctuating hydrology greatly influences the large mammal migrations for which Serengeti is famous. Rises in lake levels during the wet periods would have inundated the Ndabaka floodplains in the west. If the northern Serengeti was forested, then the migration pattern of wildebeest would have been more likely east-west rather than north-south as it is now. The north may well have been cut off by a barrier of thicket and forest and the migration would have moved west toward the lake to access the higher rainfall produced by the lake, and to reach the lakeshore itself. Indeed, Baumann (1894) witnessed the wildebeest migration at the shore of Speke Gulf in 1891 (Sinclair, chapter 1).

Perturbations that occur every few decades can produce rapid changes in wildlife populations, followed by periods of slow recovery. Droughts have occurred every 30 years or so. In 1993 a drought that lasted one year resulted in approximately 25% of the wildebeest and 40% of the buffalo dying from drought-induced starvation. The wildebeest population took some six years to recover their numbers, while buffalo numbers took ten years in areas where illegal hunting was at a minimum (Metzger et al. 2010).

Traditional pastoralism and agriculture was adapted to the seasonal and yearly drought cycles. Pastoralists who dwell to the east of the Serengeti (see Galvin, chapter 17, and Reid, chapter 25) depend upon rainfall for their livelihood, the grazing of livestock. Agriculture is becoming the main income source for pastoralists. However, pastoralists are still highly dependent on the productivity of their herds which are intricately tied to precipitation (Galvin, chapter 17). In the west, drought impacts people through crop failure. With no crops to harvest, the demand for inexpensive food protein increases so that illegal hunting becomes more appealing (E. Knapp, chapter 16). This increase in illegal hunters further burdens an already struggling ungulate population (Arcese, Hando, and Campbell 1995).

Climatic Changes over Short Time Scales

Interannual variability in the form of drought and flood years is a persistent reoccurring feature of the Serengeti ecosystem. The El Niño/Southern Oscillation (ENSO) as measured through the Southern Oscillation Index drives the seasonality of precipitation and is correlated with drought and flood events (Ogutu and Owen-Smith 2003). A drought year is defined as having less than 75% of the 30 year water year mean rainfall (1970–2000). Since 1961 drought events for the water year (October to September) have occurred three times in the north, only once in the western corridor, and six times on the short grass plains. Drought and flood events occur but only occasionally do they occur in the same region in the same year.

The seasonal change in the environment determines the nature of the Serengeti ecosystem. Rainfall is bimodal, with periods of short rains in November–December and long rains in March–June. The short rains are less predictable and often fail on the plains. The seasonal rainfall pattern is the fundamental driver of the wildebeest migration which follows the rainfall gradient as the dry season progresses (Hopcraft, chapter 6). Through their impacts on vegetation and soils and their role as food for predators and carrion feeders, the seasonally driven migration has cascading effects throughout the ecosystem. However, it is the spatial changes in nutrient availability in food which drive the reverse migration back to the plains at the start of the wet season. Thus, migrant ungulates rely on dry season grass production in the north and the west.

The long dry season lasts from July to October. The lowest rainfall month is July, but since there is residual soil moisture, grass can sometimes regrow after fires, which are most extensive in this month. This regrowth, sometimes called "green flush" only a few centimeters in height, is the preferred height for wildebeest. The dry season coincides with the coolest time of year, effectively during austral winter. Strong easterly winds prevail throughout the dry season, and these contribute to the drying out of grass and to driving fires across the landscape. Later in the dry season the soil dries, and green flush occurs only after thunderstorms. Such storms become more frequent in September and October, a precursor to the first rains of November. Strong storm events are most frequent in the northwest and close to Lake Victoria, and are absent on the plains.

The short dry season months of January and February can be dry or wet, varying from year to year. By January and February the wildebeest migration has moved south onto the plains due to the onset of the short rains. If these two months are wet the migration remains on the plains, but if they are dry then the animals move west into the Maswa Game Reserve. Precipitation during the long dry season between July–October is a strong determinant of wildebeest survival rates. Low precipitation during the wet season in the north will have little impact on the migration, while failure of rains in the dry season or in consecutive dry seasons can have devastating effects. Rainfall is spatially variable within seasons and often low rainfall in one area does not mean low rainfall in another. Many species (wildebeest, zebra, Thompson gazelle) migrate and are therefore able to move over smaller spatial scales in response to interseasonal rainfall variability and associated spatial heterogeneity. Additionally, if rainfall fails in one area, animals can adjust through migration. Thus, the plains and Maswa areas are critical for the migratory herds during the short dry season (January–February), while the northwest Serengeti and Mara Reserve is critical for survival during the long dry season (July–October).

THE MULTISCALED HUMAN FOOTPRINT ON THE LANDSCAPE

Land Use in the Serengeti Ecosystem

Patterns of land use can alter both the rate and direction of natural processes, and land use patterns interact with the abiotic template. Land use refers to the way and the purposes for which humans employ the land and its resources (Meyer 1995), while land cover refers to the habitat or vegetation type present. Land use change encompasses all the ways in which humans use the land through time. Land cover patterns are altered principally by direct human use: agriculture, raising of livestock, forest harvesting, and construction (Meyer 1995). Land use and land cover change in the Serengeti ecosystem is most prevalent in areas of unrestricted land use and is important for reasons unique to being adjacent to a protected area, such as water usage (Reid, chapter 25), wildfire (Eby, chapter 4), illegal harvest (Rentsch, chapter 22), human-wildlife conflict (Hampson, chapter 21), and disease transmission (Lembo, chapter 19).

The greater Serengeti ecosystem includes several conservation administrations and additional areas that lack conservation designation (fig. 3.8).

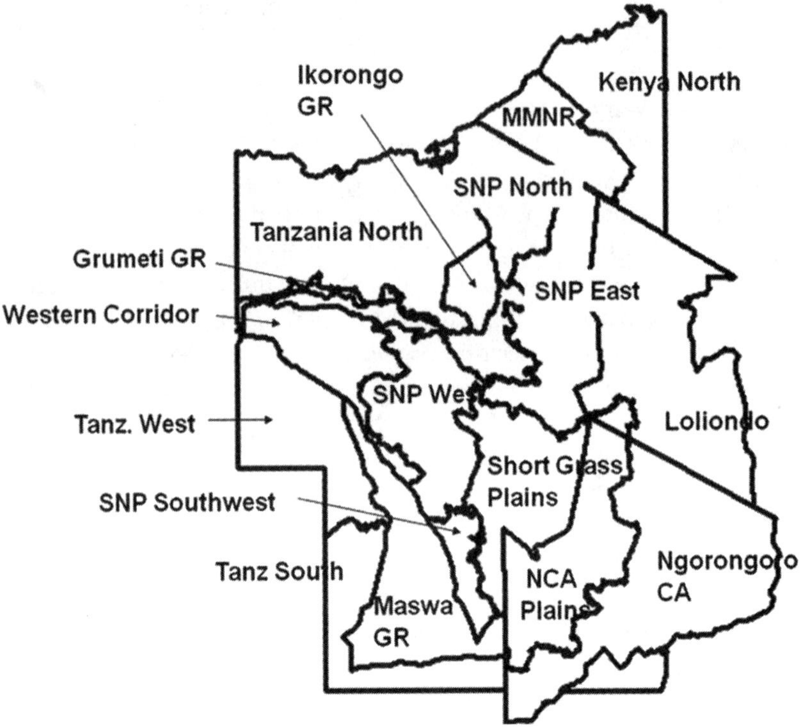

Fig. 3.8 Ecological and political boundaries of the Serengeti ecosystem.

The Maasai Mara National Reserve and the Serengeti National Park are International Union for Conservation of Nature (IUCN) class II designated protected areas. The Ngorongoro Conservation Area, southeast of the park, covers roughly half of the short grass plains, and includes Olduvai Gorge and the Gol Mountains. North of Ngorongoro Conservation Area is Loliondo Game Controlled Area that forms the eastern boundary of the Serengeti ecosystem and is comprised of plains in the south and savanna toward the north. In these areas a mandate for the preservation of both cultural heritage and biodiversity exists. The Maswa Game Reserve (IUCN Cat.VI) lies to the SW of the park. North of the western corridor are the Grumeti and Ikorongo Game Reserves. On the Kenyan side of the ecosystem north of MMNR are the Loita plains and Loita hills, which are the traditional lands of the Kenya Maasai. This area is now devoted to fenced "group ranches" (see Reid, chapter 25). Areas to the north, west, and south of the main protected area can be subdivided based on geographical positions, and are treated separately because they are in different precipitation zones, and have differing parent material. For example, some areas are better for grazing while others are better suited for agriculture, and thus land use and the human population become important designations.

Changes in Human Population Density

The fundamental driver of land use change is human population growth. The growth rate of human settlement adjacent to the protected areas is higher than the average rate of rural population growth (fig. 3.9a). This increase has largely been through immigration (Schmitt 2010). Human population density is important for the Serengeti because the density of humans living adjacent to the park is directly proportional to natural resource extraction within the protected areas (Rentsch, chapter 22). Illegal hunting through its effects on wildlife populations reduces the effective size of the protected area (Dobson and Lynes 2008). Additionally humans dwelling adjacent to the protected area experience costs associated with living close to wildlife (E. Knapp, chapter 16). The human population to the east of the park (population ~136,000) (fig. 3.9b) is primarily comprised of pastoralists who consume little wild game meat for cultural reasons (Homewood, Rodgers, and Arhem 1987; Bourn and Blench 1999).

In contrast, the agricultural population (1.9 million) that resides between Lake Victoria and the western border of the park derives a proportion of their diet from wildlife consumption (Rentsch, chapter 22; Hofer et al. 1996). Therefore, illegal hunting activities within the park originate from the human population dwelling to the west of the game reserves (GR) and Serengeti National Park. The northwest region has an annual population growth rate of 3.1% but areas adjacent to the park (within a 10 km area) are increasing at a faster rate (4% per year), and most of this increase is due to immigration (Schmitt 2010) with some areas reporting rate of increases at close to 10% annually. The area of cultivated land has increased in the northern part of the ecosystem since 1975 (Serneels and Lambin 2001; Sitati et al. 2003; Ogutu, Bhola, and Reid 2005), with advancing cultivation right up to the boundary of the Maasai Mara Nature Reserve (Ogutu et al. 2009). The highest areas of growth are in the east where human populations started out at low levels and are still low when compared to the western part of the ecosystem. The east had the lowest level of immigration while the northwest and the southwest had almost two-thirds to four-fifth of residents being immigrants respectively (Schmitt 2010). The southwest area's population growth was 3.2% while Ngorongoro Conservation Area district had an annual rate of 4.5% (Polansky et al. 2008).

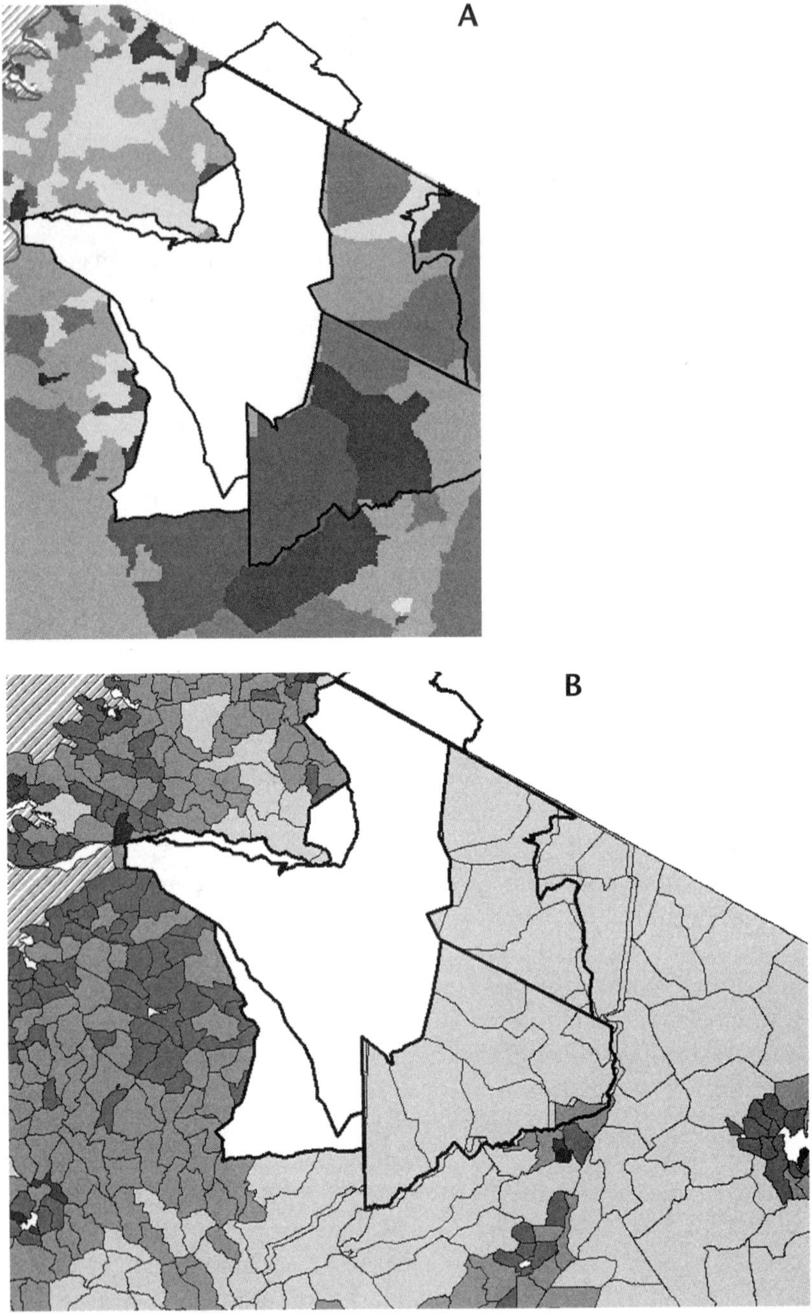

Fig. 3.9 (A) Human population rate of increase, range from –2 (light areas) to +5; (B) human population density ranged from 0 (light areas) to ~500 people per km² (dark areas).

Agricultural Activity

We determined the area of each land use unit that was under cultivation (using FAO Africover data) (table 3.1) with the exclusion of the Serengeti National Park, Maasai Mara National Reserve and GRs. Agriculture is allowed in the conservation area and in all areas adjacent to the protected area and GRs. At the time of this writing, with exclusion of the plain in Ngorongoro Conservation Area, 14% of the conservation area was cultivated. Agriculture in Loliondo was much less at only 2%. For areas that have no political conservation designation, the area to the northwest of Serengeti National Park (TanzN; see fig. 3.1 for land use area naming designations) had 69% of the land cover under cultivation. That was by far the highest of all the areas. TanzS and TanzW had approximately equal cultivated land (12 and 14%) and the TanzEyasi region had 25% agriculture. Outside the protected area boundary, a mix of interacting climatic and edaphic variables define livelihood strategies. West of the protected area, precipitation is high enough to grow crops (Olff and Hopcraft 2008). East of the protected area where precipitation is not plentiful and too variable to support consistent crops, livestock agriculture dominates (Galvin, chapter 17). We used Moderate Resolution Imaging Spectrometer (MODIS) derived Normalized Difference Vegetation Index (NDVI) (Carroll et al. 2004) data as a surrogate for vegetation production to develop a habitat quality map (fig. 3.10). Pixel values greater than 0.5 were considered high productivity, between 0.5 and 2.5 were scored as medium productivity, and monthly NDVI values <0.25 were scored as low productivity. Monthly areas with high, medium, and low productivity were summed to create a composite map displaying yearly habitat productivity. Areas in the NW (TanzN and W) can support plant growth for 12 months a year in contrast to areas in the east which can be between one to three months per year (fig. 3.10). In the TanzS area production is marginal and the impact that humans have on the landscape is evident from the NDVI analysis. In this area the change in habitat quality between inside and outside the park is more pronounced with higher production inside compared to outside (fig. 3.10). This is likely the result of slash and burn agricultural in an area that is not able to sustain it.

Tree Harvesting

Tree harvesting is a major land cover modifier occurring in western Serengeti, particularly in the southwest adjacent to the Maswa game reserve. Using vegetation continuous field (VCF) data (Hansen et al. 2003) collected

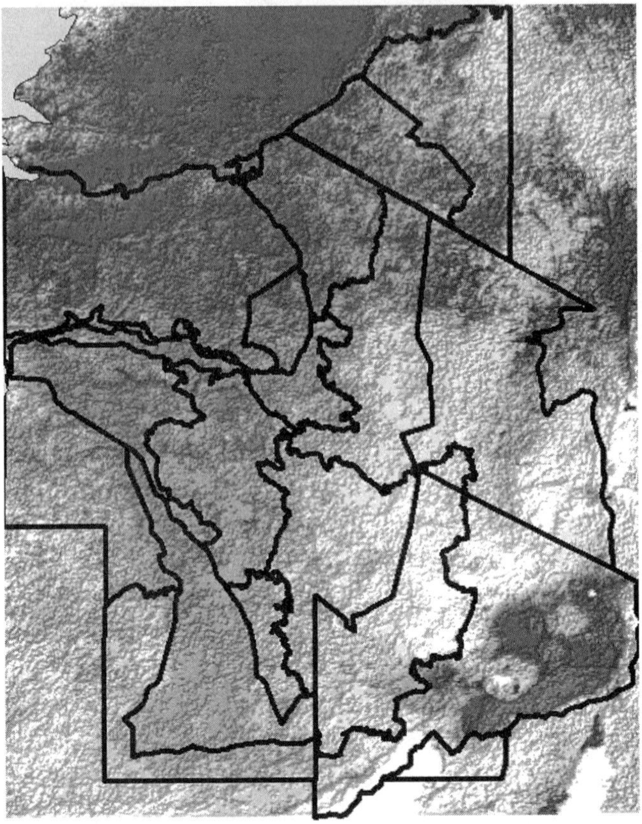

Fig. 3.10 Habitat quality, or the spatial variation in vegetation productivity across the ecosystem, is derived using MODIS Normalized Difference Vegetation Index (NDVI). Darker areas indicate higher productivity (8–12 months*year^{-1}), while light areas are productive 1–3 months*year^{-1}.

in 2001 we calculated differences in tree cover between the differing land use areas (we excluded the plains from this). Vegetation continuous field measures tree canopy cover on a percentage basis. The average tree cover for the Serengeti ecosystem is 11%. Unprotected areas adjacent to the Serengeti in Tanzania were between 2 and 5% below average while immediately adjacent but within the protected area boundary, the tree cover was between 3–7% above the landscape average (fig. 3.11). Examining the VCF one can see the political boundary on the western edge of the park. In addition, areas in the unprotected areas (TanzN, TanzS and TanzW) had a higher percentage of bare ground (1.7, 3.6, and 4.1 respectively). In contrast, the highest amount of bare ground in the adjacent protected areas (0.6 in the Western Corridor) and most having less than 0.05% of area bare.

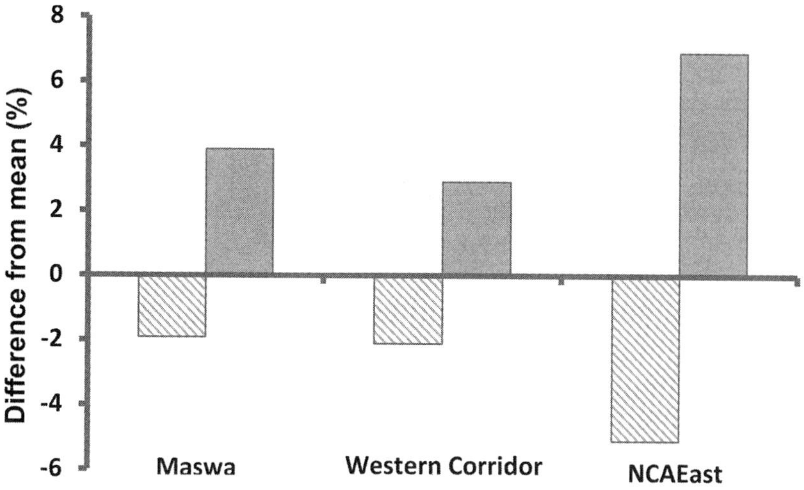

Fig. 3.11 Difference in tree cover (%) between adjacent areas inside (solid bars) and outside protected areas (hashed bars).

Hydrology

Land use change in this area is responsible for changes in the hydrology of the Serengeti ecosystem, particularly within the Mara River drainage (Mati et al. 2008; Gereta, Mwangomo, and Wolanski 2009). The Mara River originates from the Napuiyapi Swamp (2,932 m), with the main perennial tributaries being the Amala and the Nyangores, which drain from the western Mau escarpment. This part of the basin supports forests, agriculture, and medium-sized farms (often tea farms up to 40 acres). The Mara River is the only permanent water source for the protected area and is a dry season and drought refugia for the wildlife of the Serengeti ecosystem (Gereta, Mwangomo, and Wolanski 2009). Over the last 50 years, this catchment has undergone major land use changes (Mati et al. 2008). Forest and grasslands have been cleared for agriculture and charcoal production (IUCN 2000). In 1973 rangelands covered 79% of the total basin area (Mati et al. 2008). This area has now been reduced to 52% and areas that were previously forested have been reduced by 32% of these 1973 cover amounts. Most of the land use change has been attributed to the doubling of land used for agriculture since 1973 (203% change). Deforestation and water diversion for irrigation are implicated in the changes in the hydrology of the Mara River. Impacts on the Mara River include decreases in the dry season flow by 68% since 1972 (Gereta, Mwangomo, and Wolanski 2004, 2009) and the Mara River is now thought to be an undependable source of flowing water during times of drought (Gereta, Mwangomo, and Wolanski 2009). Conversely, wet season flows are more intense, potentially leading to flooding and increased erosion. While trees and dense vegetation in the upper reaches of the watershed, particularly in the Mau forest region, once attenuated peak runoff rates and conserved soil moisture into the dry season to prolong flows, the hydrological regime is now characterized by high amplitude fluctuations in flow rates. The ecological implication of changes to the hydrology in the Mara River are discussed in chapter 26.

Livestock Corrals

Humans have created fine scale spatial heterogeneity by building corrals (bomas) where livestock are kept overnight. Typically, they are 20–30 m in diameter. Large quantities of nutrients are deposited into the corrals in feces and urine. Seeds of shrubs and small trees eaten by the livestock also germinate and establish in these areas (Verchot, chapter 8). Although these corrals were abandoned and disintegrated as much as a century ago, their legacy in the soil has remained. Old bomas with their high-nutrient soils result in differences in microbe activity, and regeneration of dense stands of *Dicrostachys*, a thorny shrub. Such stands are still prevalent today and are used by predators to hide.

Disease

The great perturbation of the rinderpest epizootic in 1890 resulted in decreases in human population density as well as decreases in grazing herbivore densities, which in turn resulted in pulse of *Acacia* recruitment. The resulting even-aged cohort took some 90 years to reach senescence before a new pulse of recruitment occurred in about 1980 (Dublin 1986; Dublin 1991; Sinclair 1995; Sinclair and Arcese 1995; Gereta, Mwangomo, and Wolanski 2009). This periodicity is determined to some extent by the longevity of these tree species. Small *Acacia* species such as *A. drepanalobium* have a much shorter longevity in Serengeti (about 30 years), and pulses have been observed to appear and disappear at least twice in the past 50 years (Shaw et al. 2010). The appearance of these pulses of trees alters the animal communities that depend on them at the same time. Thus, canopy feeding insectivorous birds such as warblers, flycatchers, and shrikes, and hole-nesting birds such as hornbills, woodpeckers, and starlings, would have changed in numbers according to the availability of mature trees.

Fire

Fires in the Serengeti are primarily anthropogenic and frequency is tightly tied to variation in seasonal precipitation patterns. In years with high wet-season rainfall and low dry-season rainfall, high biomass accumulation followed by rapid drying increases the probability of fire occurring in any given location (Holdo, Holt, and Fryxell 2009; Holdo et al. 2009). Fires are far more prevalent in the higher rainfall areas due to the higher grass fuel biomass, and so fires are almost absent on the short grass plains. Fires are more extensive in years that have higher rainfall in the preceding long rains (Eby, chapter 4) and are almost absent outside the western boundaries of the protected area due to active fire suppression and agricultural boundaries. The significance for the migration is that ungulates will move outside the protected area and into cultivation if there has been too much burning inside, especially during July. This exposes the animals to a very high risk of hunting (Rentsch, chapter 22).

CONCLUSION

To understand the functioning of the Serengeti ecosystem and the biodiversity that it harbors, we must consider biotic and abiotic factors acting on multiple spatial and temporal scales and how all these factors intersect and interact. At long temporal and large spatial scales, geological events shape the physical landscape and the parental material forms the chemical properties of the soil. The diversity of topography and habitats from mountains to floodplains and deserts to riverine forests contribute to the high biological diversity. Topography at finer scales develops and maintains the soil catena, providing the opportunities for niche partitioning of grasses, trees, ungulates, and birds.

Climate events are important in shaping the system and also act on many scales. These events, such as droughts, can last decades and in extreme cases, centuries. These events are linked to redistribution of humans and the demise of civilizations and historically have caused Lake Victoria to dry up altogether. Secondary effects of these drought events also cause changes to the weather patterns affecting the Serengeti ecosystem and result in the alteration of the entire ecosystem. On shorter time scales localized drought events have large ecosystem implications, such as animal redistributions and die-off events. During these past drought events it is known that wildebeest have, in the last resort, used Lake Victoria as their water source. In more recent history, because of human land use changes, the park is cut off from the lake as the park protected boundary falls short of the lake by about 5 kilometers near Ndabaka gate in the Western Corridor. One conclusion is a recommendation to allow wildebeest and other migrants access to the lakeshore. A second conclusion, based on the longtime scale of climate trends, is that the present-day conditions will not remain as they are, but will change, possibly quite suddenly, and drought, for example could prevail for a century or more.

The Serengeti ecosystem has been modified by human use. Human population increases add increased demands on the ecosystem, such as water use, increased demand for agricultural land, and resource extraction that are threatening the ecosystem (Reid, chapter 25, and Sinclair, chapter 26). One example is the change in the water flow of the Mara River due to increase agricultural use upstream. If left unmanaged, this will have devastating impacts on the system by converting the Mara River from a drought and dry season dependable source of water to an ephemeral one. Other human land use land change in the Serengeti ecosystem is underway and occurring at multiple scales, and the sustainability of the ecosystem is in question without careful planning and management.

APPENDIX

GEOLOGIC FORMATIONS COMPRISING THE NYANZIAN-KAVIRONDIAN OROGENIC BELT

Nyanzian System

Lightly metamorphosed submarine basaltic-andesite lava extrusions with pillow structures are the oldest rocks and are followed by more eruptive phases of acid-intermediate (rhyolite to dacite) types. Metasedimentary intercalations within the acid-intermediate volcanics include quartzites, siltstones, mudstones, marls, acid tuffs, and banded ironstones. The ironstones are interbanded ferruginised quartzites, cherts, and jaspilites with thin layers of hematite and magnetite. They are highly resistant to erosion and occupy long and discontinuous hill ranges near Banagi and along the Grumeti River corridor. Toward the contact with granite the level of metamorphism increases and, depending on their original composition, the Nyanzian rocks become recrystallized as quartz-sericite, chlorite, mica, actinolite, or hornblende schists. Accessory chloritoid is occasionally present in the first three schist types and garnets become increasingly common eastward in all Nyanzian rocks, particularly in the area around the confluence of the Orangi, Gaboti, and Bololedi Rivers.

The large silicified and carbonated serpentinite bodies between Nyamahanga and Beramanga are altered peridotites or dunites. They are included here within the Nyanzian system (ultrabasic lavas perhaps), as their emplacement/extrusion appears to have been controlled by the general E-W structural trend and they are cut by metadolerite dykes, apparent feeders to the Nyanzian basic volcanics.

Synorogenic Granite

Subsequent deformation and minor uplift of the Nyanzian system during a period of regional stress were accompanied by widespread mobilization and invasion of synorogenic granitic material into both greenstones and the pre-Nyanzian floor. The granites are contaminated, hybrid rocks generally granodioritic with microcline porphyroblasts, but becoming more dioritic and migmatitic toward the contact with Nyanzian rocks. A foliation delineated by mineral alignment and mafic banding is common across the synorogenic mass but is most strongly developed at the greenstone contact and is parallel to the adjacent Nyanzian schistosity.

Kavirondian System

Basal boulder conglomerates, pebble grits, quartzites, and minor pelitic beds were deposited with slight unconformity on a Nyanzian floor following uplift and erosion. The conglomerates are usually monomictic, the composition of the boulders whether from synorogenic granite, Nyanzian metalavas, or metasediments, indicating derivation from a nearby source. Distribution of Kavirondian material is sporadic and outcrops in the Katamono Hill area, between Banagi and Kilimafedha, and in the Orangi, Bololedi, and Samsambe Rivers where the younger clastics are infolded with older Nyanzian rocks and share a parallel metamorphic fabric. Mineral assemblages include accessory chloritoid and kyanite in quartzites south of Kilimafedha.

Late-Orogenic Granite

Initial intrusion of both greenstones and older granites by late-orogenic material was synchronous with deformation of the Kavirondian system, both events controlled by the earlier Nyanzian structural lines. With later stress, release intrusion continued and culminated in late-stage acid minor intrusions and quartz veining associated with epigenetic gold mineralization. The late-orogenic granite is more homogenous than synorogenic types, and is formed mainly of coarse biotite leucogranite, commonly with potash feldspar phenocrysts. Contacts with the Nyanzian-Kavirondian are locally gradational, and the granite contaminated with mafic xenoliths, but more generally are sharp and the country rocks exhibit hornfels textures typical of thermal metamorphism.

Structure and Metamorphism

Major folds in the Nyanzian are isoclinal with E-W trending axes and subvertical axial planes inclined to the north. In the main Nyanzian mass a regional fabric is usually absent in the more massive and rigid metalavas but is expressed as a slatey cleavage parallel to bedding in the less competent metatuffs and metasediments, and is axial planar to the major folds. Toward the contact with synorogenic granite the cleavage becomes a strongly penetrative schistosity and the regional metamorphic grade increases from lower to upper greenschist/lower amphibolite facies. Kavirondian rocks were infolded within the major Nyanzian structures as deformation continued during late orogenic granite intrusion. The final intrusive phase was stress free and imposed contact metamorphism at the granite/country rock interface. Indicator minerals including andalusite, cordierite, and hornblende suggest albite-epidote-hornfels to hornblende-hornfels facies grade.

GEOLOGIC FORMATIONS IN THE MOZAMBIQUE OROGENIC BELT

Recrystallized Granitic Rocks

The strongly foliated N-S trending band of granitic gneisses in the north and east is a consequence of Mozambique tectonism superimposed on the granitic rocks of the Archean floor. The contact between the two is gradational and marks a transitional change from highly sheared, to totally recrystallized granites with a strong gneissosity marked by mineral alignment. Granite gneisses are the most common, but locally become granodioritic and pass into gneissose migmatite distinguished by large streaks and inclusions of Nyanzian material.

Ikorongo Group

The sedimentary sequence, some 4,500 ft thick, comprises rudaceous, arenaceous, and argillaceous sediments that rest with angular unconformity on a foreland floor of older Nyanzian-Kavirondian and granitic rocks.

Discontinuous lenses of a poorly sorted basal conglomerate are succeeded by mature and well sorted purple orthoquartzite frequently striped with magnetite-rich layers. The quartzite is both massive and well-bedded and 50–300 ft thick. It is the dominant lithology within the group and a distinctive marker horizon. Its hard and flinty nature makes it strongly resistant to erosion and gives topographic expression to the sharp ridges of Nyaraswiga and Kamarishe and the high plateaus defining the Mbalageti River valley.

A dominantly argillaceous sequence succeeds the orthoquartzite. It is poorly exposed and confined to isolated outcrops on the Musabi plains and the low-lying country between the Nyaraboro and Itonjo hills. Hard red sandstones and quartzites follow and form a protective cover to the argillaceous beds at the southeast end of the Itonjo hills. They are the remnant of a more extensive sequence that once capped the Ikorongo group.

Metasedimentary Group

The metasediments are confined to the northeast of the Serengeti National Park but belong to a much more extensive sequence, variously named, that outcrops between the park boundary and the younger rift-related volcanic province some 35 miles to the east. They are included here in a "Metasedimentary group" for narrative convenience only. Mozambique tectonism has obliterated the original contact between the Metasedimentary group and the Archean floor. However, an unconformity can be inferred as no NKOB events have been recognized within the metasedimentary sequence.

The sequence is characterized by a distinctive mature white or red-brown quartzite intercalated with quartz mica schists. It is highly resistant to weathering and forms the hill ranges in the NE, the most notable being Kuka and Nyamalumbwa. Fine magnetite banding, cross bedding, and current ripple marks are common and highlight its resemblance to and correlation with the Ikorongo group orthoquartzite. The remainder of the sequence, metasediments of psammitic and pelitic origin, includes biotite and quartzo-feldspathic gneisses locally with hornblende, garnet, and microcline porphyroblasts, and kyanite schists with staurolite and garnet.

STRUCTURE AND METAMORPHISM

Pan-African deformation and metamorphism extends westward with decreasing intensity across the width of the park as far as the Mumughia and Itonjo hills. Three episodes of deformation (D1–D3) are recognized, of which D1 is the most widespread, while D2 and D3 are confined to the eastern sector. Events traced from east to west are as follows.

Cover and Basement to the W Limit of Recrystallized Granitic Rocks

Large-scale recumbent F1 folding of the metasedimentary cover is the principal consequence of the *D1 Deformation*. This is best observed immediately to the east of the park boundary where the fold axes are NW-SE trending and the axial planar S1 schistosity is inclined gently to the NE. In the basement the Archean granites have been reconstituted as granitic gneisses with a strong S1 mineral foliation, and the Nyanzian fabric has been isoclinally folded (F1) and a new axial planar S1 schistosity developed.

Large-scale F2 antiforms and synforms, with generally N-S trending axes and subvertical axial planes inclined to the east, are superimposed by subsequent *D2 Deformation* on the earlier F1 recumbent folds in the cover—the interference patterns so produced being well illustrated by the quartzites of Kuka and Nyamalumbwa hills. The D2 extends into the basement as large-scale F2 folding of the S1 fabric in the granitic gneisses and in the greenstone inliers as tight minor F2 folds of both the Nyanzian and S1 fabrics.

ENE-WSW trending F3 cross folding is associated with the *D3 Deformation* and is highlighted on a regional scale by sinuous warping of earlier D1 and D2 fabrics in both cover and basement, and at outcrop level by small-scale F3 warping and crenulation cleavage.

Mineral assemblages, such as kyanite-staurolite-garnet-muscovite in the metasedimentary cover and hornblende-plagioclase-epidote-garnet in the Nyanzian schists indicate that regional metamorphism associated with the deformation reached amphibolites facies grade.

Basement West to the E Margin of the Ikorongo Group

Only the *D1 Deformation* extends into this sector. The S1 fabric swings from NW-SE, to a more northerly trend between Kilimafedha and the Samsambe River and is steeply inclined to the NE and east, respectively. It has now been "downgraded" to a fracture cleavage in the rigid metalavas and granites, and a slatey cleavage axial planar to small-scale open to tight F1 folds and crenulations of the Nyanzian fabric in the less competent metasediments and schists.

The regional metamorphic grade has also decreased westward, from upper to lower greenschist facies characterized by S1 biotite growth near the contact with the recrystallized granitic rocks and replaced by S1 chlorite growth between Kilimafedha and Banagi.

Cover and Basement in the SW

The western limit of the *D1 Deformation* is marked by NW- SE trending narrow zones of shearing and oblique-slip wrench faulting associated with large-scale F1 flexuring of the Ikorongo group. An S1 slatey cleavage, parallel to the shear zones, is restricted to the argillaceous and gritty beds in the cover and is axial planar to the F1 anticlinal and synyclinal flexures delineated by the arcuate quartzite ridges between Nyaraswiga and Bingwe hills.

Both cover and basement have undergone only incipient regional metamorphism of lower greenschist facies grade marked by cataclasis, recrystallization, and growth of S1 sericite and chlorite.

REFERENCES

Anderson, T. M., J. Dempewolf, K. L. Metzger, D. Reed, and S. Serneels. 2008. Generation and maintenance of heterogeneity in the Serengeti ecosystem. In *Serengeti III: Human impacts on ecosystem dynamics*, ed. A. R. E Sinclair, C. Packer, S. A. R. Mduma, and J. M. Fryxell, 135–82. Chicago: University of Chicago Press.

Arcese, P., J. Hando, and K. Campbell. 1995. Historical and present-day anti-poaching efforts in Serengeti. In *Serengeti II: Dynamics, management and conservation of an ecosystem*, ed. A. R. E. Sinclair and P. Arcese, 506–33. Chicago: University of Chicago Press.

Awadallah, G. W. 1966. Kakesio-Makow geological map. In *QDS 51*, Scale 1:125,000. Tanzania: Mineral Resources Division.

Bailey, R. G. 1996. *Ecosystem geography*. New York: Springer-Verlag.

Barth, H. 1990. Explanatory notes on the 1:500 000 provisional geological map of the Lake Victoria goldfields. *Geologisches Jahrbuch* 72.

Baumann, O. 1894. *Durch Massailand zur nilquelle*. Berlin. Reprinted 1968. New York: Johnson Reprint Corporation.

Bell, R. H. V. 1971. A grazing ecosystem in the Serengeti. *Scientific American* 225:86–93.

Belsky, A. J. 1987. The effects of grazing: Confounding of ecosystem, community, and organism scales. *American Naturalist* 129:777–83.

———. 1988. Regional influences on small-scale vegetational heterogeneity within grasslands in the Serengeti National Park, Tanzania. *Vegetatio* 74:3–10.

Bingen, B., J. Jacobs, G. Viola, I. H. C. Henderson, O. Skar, R. Boyd, R. J. Thomas, A. Solli, R. M. Key, and E. X. F. Daudi. 2009. Geochronology of the Precambrian crust in the Mozambique belt in NE Mozambique and implications for Gondwana assembly. *Precambrian Research* 178:231–55.

Bonnefille, R., and F. Chalié. 2000. Pollen-inferred precipitation time-series from equatorial mountains, Africa, the last 40 kyr BP. *Global and Planetary Change* 26:25–50.

Bourn, D., and R. Blench. 1999. *Can livestock and wildlife co-exist? An interdisciplinary approach: Livestock, wildlife, and people in the semi-arid rangeland of East Africa*. London: Overseas Development Institute.

Carroll, M. L., C. M. DiMiceli, R. A. Sohlberg, and J. R. G. Townshend. 2004. 250m MODIS Normalized Difference Vegetation Index, edited by University of Maryland. College Park, Maryland. http://glcf.umd.edu/data/ndvi/.

Cohen, A. S., J. R. Stone, K. R. M. Beuning, L. E. Park, P. N. Reinthal, D. Dettman, C. A. Scholz, T. C. Johnson, J. W. King, and M. R. Talbot, et al. 2007. Ecological consequences of early late Pleistocene megadroughts in tropical Africa. *Proceedings of the National Academy of Sciences* 104:16422–27.

Cullen, H. M., P. B. deMenocal, S. Hemming, G. Hemming, F. H. Brown, T. Guilderson, and F. Sirocko. 2000. Climate change and the collapse of the Akkadian empire: Evidence from the deep sea. *Geology* 28:379–82.

Dawson, J. B. 2008. *The Gregory rift valley and Neogene-recent volcanoes of northern Tanzania*, vol. 33, Geological Society Memoires. Geological Society of America.

Dobson, A., and L. Lynes. 2008. How does poaching affect the size of national parks? *Trends In Ecology & Evolution* 23:177–80.

Dublin, H. T. 1986. Decline of the Mara woodlands: The role of fire and elephants, PhD diss., University of Brisish Columbia, Vancouver, Canada.

Dublin, H. T. 1991. Dynamics of the Serengeti-Mara woodlands. *Forest and Conservation History* 35:169–78.

Epp, H. 1981. A natural resource survey of Serengeti National Park, Tanzania. *S. R. I. Publication No. 237*.

Gereta, E., E. Mwangomo, and E. Wolanski. 2004. The influence of wetlands in regulating water quality in the Seronera River, Serengeti National Park, Tanzania. *Wetland Ecology and Management* 12:301–07.

———. 2009. Ecohydrology as a tool for the survival of the threatened Serengeti ecosystem. *Ecohydrology and Hydrobiology* 9:115–24.

Gerresheim, K. 1971. Serengeti ecosystem landscape classification units. Arusha, Tanzania Litho Ltd.

Gillespie, R., F. A. Street-Perrott, and R. Switsur. 1983. Postglacial arid episodes in Ethiopia have implications for climate prediction. *Nature* 306:680–83.

Gillson, L. 2004. Testing non-equilibrium theories in savannas: 1400 years of vegetation change in Tsavo National Park, Kenya. *Ecological Complexity* 1:281–98.

Gray, J. M., A. S. MacDonald, and C. M. Thomas. 1969. East Mara geological map. QDS 6 & 14 Scale 1:125 000. Tanzania, Ministry of Resources Division.

Hansen, M. C., R. S. DeFries, J. R. G. Townshend, M. Carroll, C. Dimiceli, and R. A. Sohlberg. 2003. Global percent tree cover at a spatial resolution of 500 meters: First results of the MODIS vegetation continuous fields algorithm. *Earth Interactions* 7: 1–15.

Hofer, H., K. L. Campbell, M. L. East, and S. A. Huish. 1996. The impact of game meat hunting on target and nontarget species in the Serengeti. In *The exploitation of mammal populations*, ed. V. J. Taylor and N. Dunstone, 117–46. London: Chapman and Hall.

Holdo, R. M., R. D. Holt, and J. M. Fryxell. 2009. Opposing rainfall and plant nutritional gradients best explain the wildebeest migration in the Serengeti. *The American Naturalist* 173:431–45.

Holdo, R. M., A. R. E. Sinclair, A. P. Dobson, K. L. Metzger, B. M. Bolker, M. E. Ritchie, and R. D. Holt. 2009. A disease-mediated trophic cascade in the Serengeti and its implications for ecosystem C. *PLoS Biol* 7 (9): e1000210.

Homewood, K., W. A. Rodgers, and K. Arhem. 1987. Ecology of pastoralism in Ngorongoro Conservation Area, Tanzania. *Journal of Agricultural Science, Cambridge* 108: 47–72.

Hopcraft, J. G. C., A. R. E. Sinclair, and C. Packer. 2005. Planning for success: Serengeti lions seek prey accessibility rather than abundance. *Journal of Animal Ecology* 74: 559–66.

Horne, R. G. 1962. Bunda Geological Map. QDS 23 Scale 1:125,000. Tanganyika, Geological Survey.

Hulme, M., R. Doherty, T. Ngara, M. New, and D. Lister. 2001. Africa climate change: 1900–2100. *Climate Research* 17:145–68.

Johnson, T. C., C. A. Scholz, M. R. Talbot, K. Kelts, R. D. Ricketts, G. Ngobi, K. Beuning, I. Ssemmanda, and J. W. McGill. 1996. Late Pleistocene desiccation of Lake Victoria and rapid evolution of cichlid fishes. *Science* 273:1091–93.

Lounsberry, W. A., and C. M. Thomas. 1967. Kirawira. QDS 24 Scale 1:125,000. Tanzania, Ministry of Resources Division.

Ludwig, F., T. E. Dawson, H. de Kroon, F. Berendse, and H. H. T. Prins. 2003. Hydraulic lift in *Acacia tortilis* trees on an East African savanna. *Oecologia* 134:293–300.

Macfarlane, A. 1967. Seronera geology map. QDS 25 Scale 1:125,000. Tanzania, Ministry of Resources Division.

———.1970. Serengeti National Park northern Tanzania. Scale 1:250,000: Unpublished geology map prepared for Serengeti Research Institute.

Mati, B. M., S. Mutie, H. Gadain, P. Home, and F. Mtalo. 2008. Impacts of land-use/cover changes on the hydrology of the transboundary Mara River, Kenya/Tanzania. *Lakes & Reservoirs: Research & Management* 13:169–77.

McNaughton, S. J. 1985. Ecology of a grazing system: the Serengeti. *Ecological Monographs* 55:259–94.

Merriam, C. H. 1890. Results of a biological survey of the San Francisco mountain region and desert of the Little Colorado, Arizona. *North American Fauna* 3:1–136.

Metzger, K. L., A. R. E. Sinclair, R. Hilborn, J. G. C. Hopcraft, and S. A. R. Mduma. 2010. Evaluating the protection of wildlife in parks: the case of African buffalo in Serengeti. *Biodiversity and Conservation* 19:3431–44.

Meyer, W. B. 1995. Past and present land use and land cover in the USA. *Consequences: The Nature and Implications of Environmental Change* 1:25–33.

Murray, M. G. 1995. Specific nutrient requirements and migration of wildebeest. In *Serengeti II: Dynamics, management and conservation of an ecosystem*, ed. A. R. E. Sinclair and P. Arcese, 91–114. Chicago: University of Chicago Press.

Naylor, W. I. 1965. Bumera geology map. QDS 36 Scale 1:125,000. Tanganyika, Geological Survey.

Ogutu, J. O., N. Bhola, and R. Reid. 2005. The effects of pastoralism and protection on the density and distribution of carnivores and their prey in the Mara ecosystem of Kenya. *Journal of Zoology* 265:281–93.

Ogutu, J. O., and N. Owen-Smith. 2003. ENSO, rainfall and temperature influences on extreme population declines among African savanna ungulates. *Ecology Letters* 6: 412–19.

Ogutu, J.O., H. P. Piepho, H. T. Dublin, N. Bhola, and R. S. Reid. 2009. Dynamics of Mara-Serengeti ungulates in relation to land use changes. *Journal of Zoology* 278:1–14.

Olff, H., and G. Hopcraft. 2008. The resource basis of human-wildlife interaction. In *Serengeti III: Human impacts on ecosystem dynamics*, ed. A. R. E Sinclair, C. Packer, S. A. R. Mduma, and J. M. Fryxell, 95–134. Chicago: University of Chicago Press.

Peters, C. R., R. J. Blumenschine, R. L. Hay, D. A. Livingstone, C. W. Marean, T. Harrison, M. Armour-Chelu, P. Andrews, R. L. Bernor, and R. Bonnefille, et al. 2008. Paleoecology of the Serengeti-Mara ecosystem. In *Serengeti III: Human impacts on ecosystem dynamics*, ed. A. R. E. Sinclair, C. Packer, S. A. R. Mduma, and J. M. Fryxell, 47–94. Chicago: University of Chicago Press.

Pickering, R. 1958. Oldoinyo Ogol (Serengeti Plain—East) Geology map. QDS 38 Scale 1: 125,000. Tanganyika: Geological Survey.

———. 1960. Moru (Serengeti Plain—West) Geology map. QDS 37 Scale 1:125,000. Tanganyika: Geological Survey.

———. 1964. Endulen geology map. QDS 52 Scale 1:125,000. Tanganyika: Geological Survey.

Polansky, S., J. Schmitt, C. Costello, and L. Tajibaeva. 2008. Larger-scale influences on the Serengeti ecosystem: National policy, economics, and human demography. In *Serengeti III: Human impacts on ecosystem dynamics*, ed. A. R. E Sinclair, C. Packer, S. A. R. Mduma, and J. M. Fryxell, 347–78. Chicago: University of Chicago Press.

Reed, D. N., T. M. Anderson, J. Dempewolf, K. L. Metzger, and S. Serneels. 2009. The spatial distribution of vegetation types in the Serengeti ecosystem: The influence of rainfall and topographic relief on vegetation patch characteristics. *Journal of Biogeography* 36:770–82.

Riehl, H., and J. Meitin. 1979. Discharge of the Nile River: A barometer of short-period climate variation. *Science* 206:1178–79.

Schmitt, J. 2010. Improving conservation efforts in the Serengeti ecosystem, Tanzania: An examination of knowledge, benefits, costs and attitudes. PhD diss., University of Minnesota.

Serneels, S. and E. F. Lambin. 2001. Impact of land-use changes on the wildebeest migration in the northern part of the Serengeti–Mara ecosystem. *Journal of Biogeography* 28:391–407.

Shaw, P., A. R. E. Sinclair, K. Metzger, A. Nkwabi, S. A. R. Mduma, and N. Baker. 2010. Range expansion of the globally vulnerable Karamoja apalis *Apalis karamojae* in the Serengeti ecosystem. *African Journal of Ecology* 48:751–58.

Sinclair, A. R. E. 1995. Population limitation of resident herbivores. In *Serengeti II: Dynamics, management and conservation of an ecosystem*, ed. A. R. E. Sinclair and P. Arcese, 3–30. Chicago: University of Chicago Press.

Sinclair, A. R. E., and P. Arcese, eds. 1995. *Serengeti II: Dynamics, management and conservation of an ecosystem*. Chicago: University of Chicago Press.

Sitati, N. W., M. J. Walpole, R. J. Smith, and N. Leader-Williams. 2003. Predicting spatial aspects of human-elephant conflict. *Journal of Applied Ecology* 40:667–77.

Stockley, G. M. 1936. Geology of the south and southeastern regions of the Musoma district. In *Short paper*, 48. Geological Survey of Tanganyika.

Thompson, L. G., E. Mosley-Thompson, M. E. Davis, K. A. Henderson, H. H. Brecher, V. S. Zagorodnov, T. A. Mashiotta, P.-N. Lin, V. N. Mikhalenko, and D. R. Hardy, et al. 2002. Kilimanjaro ice core records: Evidence of Holocene climate change in tropical Africa. *Science* 298:589–93.

Treydte, A. C., I. M. A. Heitkönig, H. H. T. Prins, and F. Ludwig. 2007. Trees improve grass quality for herbivores in African savannas. *Perspectives in Plant Ecology, Evolution and Systematics* 8:197–205.

Verschuren, D., K. R. Laird, and B. F. Cumming. 2000. Rainfall and drought in equatorial East Africa during the past 1,100 years. *Nature* 403:410–14.

Whittaker, R. H., and S. A. Levin. 1977. The role of mosaic phenomena in natural communities. *Theoretical Population Biology* 12:117–39.

FOUR

Fire in the Serengeti Ecosystem: History, Drivers, and Consequences

Stephanie Eby, Jan Dempewolf, Ricardo M. Holdo, and Kristine L. Metzger

FIRE IN SAVANNA AND GRASSLAND ECOSYSTEMS

Fire is an important factor in the ecology and evolution of grassland and savanna ecosystems (Fuhlendorf and Engle 2004), modifying the tree-grass balance determined ultimately by precipitation (Sankaran et al. 2005). It has been suggested that many savannas would eventually become forests without its continued presence (Bond, Woodward, and Midgley 2005). Fire affects nutrient cycling (Hobbs and Schimel 1984; Singh 1993; Van de Vijver, Poot, and Prins 1999), soil organic N and soil C storage (Anderson et al. 2007; Holdo et al. 2007), modifies plant species composition (Gibson and Hulbert 1987; Fuhlendorf and Engle 2004), and diversity (Gibson 1988; Belsky 1992), and drives changes in tree cover (Pellew 1983; Dublin, Sinclair, and McGlade 1990; Sankaran, Ratnam, and Hanan 2008; Holdo et al. 2009). Fire affects the distributions of large mammalian herbivores (Wilsey 1996; Gureja and Owen-Smith 2002; Tomor and Owen-Smith 2002), impacts the diversity (O'Reilly et al. 2006), and density (Woinarski 1990) of bird species, and affects invertebrate abundance (Ferrar 1982; Spickett et al. 1992; Davidson, Siefken, and Creekmore 1994; Chambers and Samways 1998; Stafford, Ward, and Magnarelli 1998; Cully 1999).

Some researchers consider tropical savannas to be fire climax communities (Gillon 1983). Fire has been occurring for millennia in African grasslands, and because of seasonal dry periods, Africa is prone to natural fires caused by lightning storms (Trollope, Trollope, and Austin 2005). Additionally, humans have influenced fire in the sub-Sahara landscape at least since the Holocene (Bird and Cali 1998) and possibly as early as 1.4 Myr BP (Gowlett et al. 1981; Brain and Sillen 1988), and there has been an intensification of fire use during the past 2,000 years (Olindo 1971). The constant presence of fire in these ecosystems has resulted in the evolution of fire-resistant communities of plants and animals that are dependent on periodic burning for their existence (Olindo 1971). Today fire is accepted as a valuable tool for the management of grassland vegetation (Edwards 1984; van Wilgen, Biggs, and Potgieter 1998).

THE HISTORY OF FIRE IN THE SERENGETI

The Serengeti ecosystem has been described in great detail elsewhere (Sinclair 1979a, 1979b; chapters 2, 3). The salient feature we note here is the pronounced rainfall gradient that exists across the park, with the northwest being the wettest and the southeast the driest parts of the ecosystem, with mean annual rainfall ranging from <500 to >1,000 mm (Sinclair 1979a). This gradient has a strong impact on fuel buildup and thus the spatial heterogeneity of fire frequency.

There is a long history of fire management in the ecosystem and how fire should be used in the ecosystem has been a topic of discussion throughout this history (Vesey-Fitzgerald 1972). In the 1930s there was low fire activity in the park, most likely due to low human population numbers (Sinclair 2004). By the 1960s, up to 90% of the park was burned annually, most likely due to an increase in human population and anthropogenic ignition events, and a relatively low wildebeest population density, which resulted in high fuel loads (Mduma, Sinclair, and Hilborn 1999; Sinclair 2004). It wasn't until the 1970s that fire declined in areas removed from the park boundary (figs. 4.1 and 4.2). The decline in fire extent followed the removal of rinderpest from the system, the resulting increase in the wildebeest population, and the consequent reduction in grass biomass (Norton-Griffiths 1979; McNaughton 1992; Mduma, Sinclair, and Hilborn 1999; Holdo et al. 2009). In recent years, as part of a policy of early burning by the park management, over 30% of the park burns annually (Dempewolf et al. 2007). As with other ecosystems, there appears to have been a shift from fire being ignition-limited in the 1930s to being fuel-limited at present (Guyette, Muzika, and Dey 2002).

The current fire management plan for the park, the Stronach fire management plan, was written in 1988 (Stronach 1988). One third of the park is to be burned annually at the very end of the wet season/beginning of the

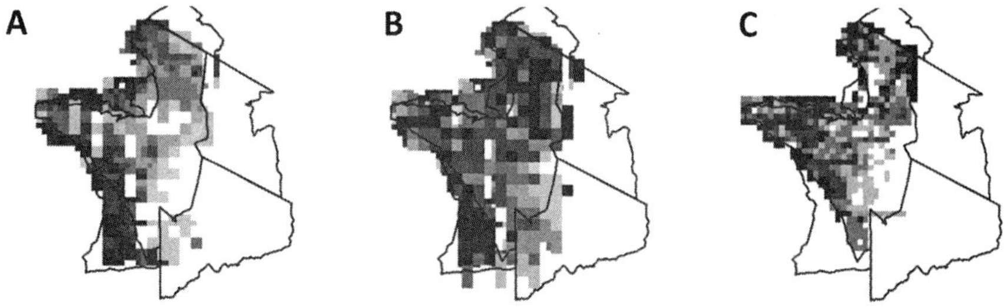

Fig. 4.1 Fire frequency was determined by scoring the years an area was burned and dividing that score by the total number of years that area had been sampled. Sample grain size for the 1960s and 1970s was a 10 km × 10 km area and in the 1980s it was a 5 km × 5 km area. Sampled data were divided up by decades. Dark areas indicate areas of high fire frequency. The 1960s and 1970s data were collected by M. Norton-Griffiths (see Norton-Griffiths 1979 for more details). The 1980s data were collected by N. Stronach (Stronach 1989). (A) Fire frequency in the 1960s. Years sampled: 1963, 1964, and 1966–1969. (B) Fire frequency in the 1970s. Years sampled: 1971–1973, 1975, and 1976. (C) Fire frequency in the 1980s. Years sampled: 1985–1987 and 1988.

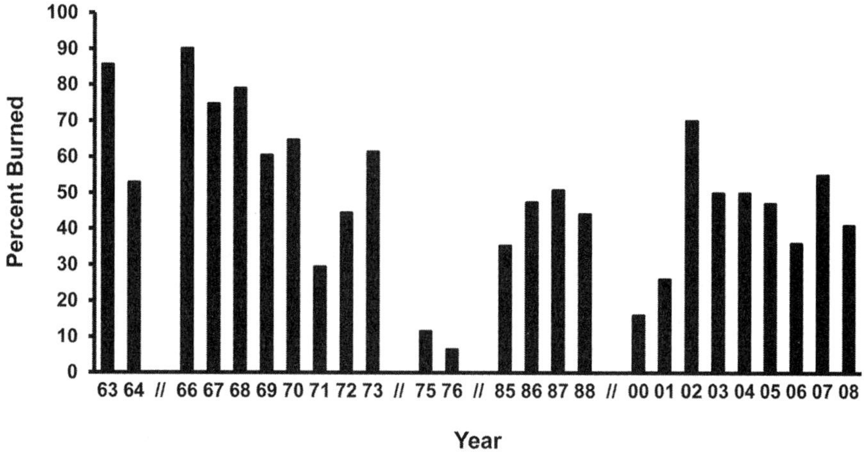

Fig. 4.2 The percentage of the Serengeti National Park that burned, for the years of available data, from 1963 to 2008 (Norton-Griffiths 1979; Stronach 1989; MODIS burned area product MCD45A1).

dry season, in order to create cool, controlled burns (E. Mwangomo, head park ecologist, pers. comm.). The main objectives are to reduce the impact of fire on regenerating woodlands, to exclude fires from rocky hills, forests, and thickets in order to prevent their decline, to prevent large fires, to keep vegetation aesthetically pleasing in areas of high tourism (Stronach 1989), and to control wildfires caused by surrounding communities (Trollope, Trollope, and Austin 2005). Fire maps for the years 2000–09 show yearly variation in the percentage of the ecosystem that is burned and site differences in fire frequency within the ecosystem (table 4.1 and figs. 4.2 and 4.3) (Roy

Table 4.1 Yearly (May 1 to April 30) percent area burned for different zones of the Serengeti-Mara ecosystem

Year	MGR	NCA	MMNR	IGR	GGR	LGCA
2000/1	57	0	4	21	51	1
2001/2	43	10	26	14	21	17
2002/3	85	6	36	83	83	20
2003/4	72	4	18	69	43	20
2004/5	72	2	13	71	73	3
2005/6	58	3	12	82	44	4
2006/7	63	2	4	30	73	4
2007/8	79	14	20	82	40	14
2008/9	61	1	6	45	37	3

Notes: Zones are Maswa Game Reserve (MGR), Ngorongoro Conservation Area (NCA), Maasai Mara National Reserve (MMNR), Ikorongo Game Reserve (IGR), Grumeti Game Reserve (GGR), and Loliondo Game Controlled Area (LGCA). Data were derived from the MODIS burned area product MCD45A1.

et al. 2005; Dempewolf et al. 2007). Most of the burning in the Serengeti occurs between May and August, but there is yearly variation in the percentage of burning that occurs in any given month (fig. 4.4).

SCALE-DEPENDENT DRIVERS OF FIRE FREQUENCY IN THE SERENGETI

Savanna fires are fueled by grass biomass produced in large amounts during the wet season, followed by favorable burning conditions during the dry season (Norton-Griffiths 1979; Gillon 1983; Holdo, Holt, and Fryxell 2009). The amount of grass biomass available for burning in savannas is controlled by the total amount and seasonal distribution of rainfall, and as the dry season progresses, by herbivory and decomposition (Stronach and McNaughton 1989; Higgins, Bond, and Trollope 2000; Sinclair et al. 2007; Holdo, Holt, and Fryxell 2009), as well as by other factors influencing productivity such as soil nutrients (McNaughton 1985). In addition, tree cover may play an important role in regulating grass biomass and fire frequency in areas with significant tree cover (Archibald et al. 2009), given the fact that grass cover declines as a function of woody biomass in savannas (Scholes and Archer 1997) (fig. 4.5a).

Extensive historical records, derived from aerial reconnaissance data (Norton-Griffiths 1979), and by traversing the park by vehicle or on foot (Stronach 1989) (fig. 4.1), document the extent of burning starting in the 1960s. These records suggest a decline in area burned associated with the

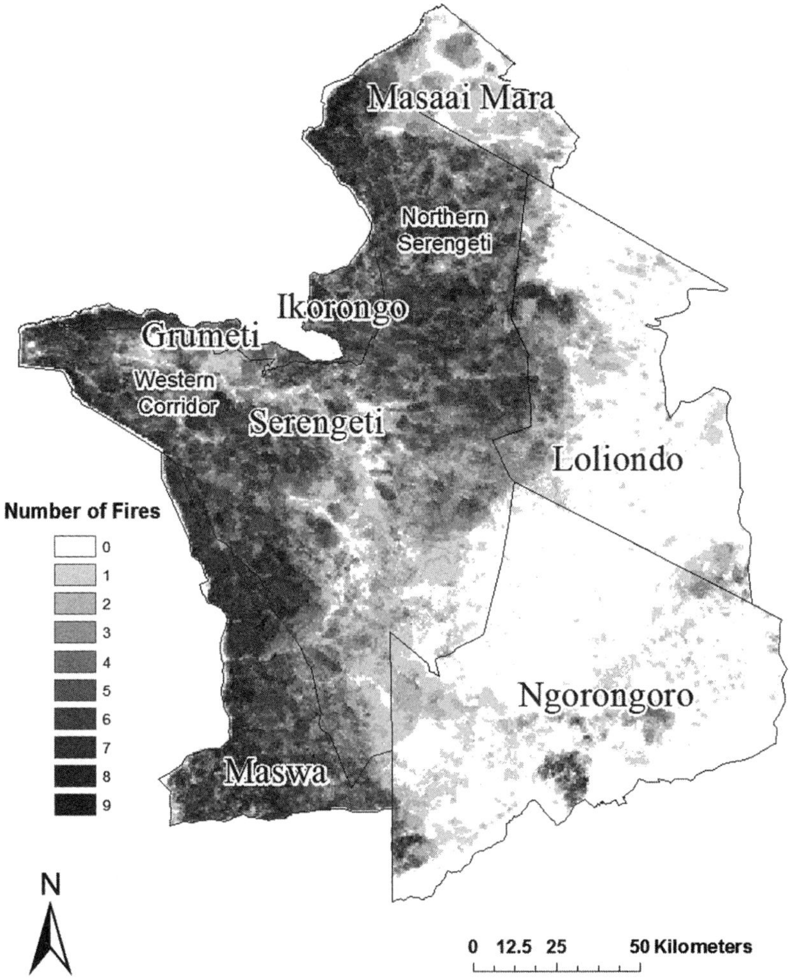

Fig. 4.3 Number of fires in the Serengeti-Mara ecosystem from 2000 to 2009 with the protected areas and the locations of the northern Serengeti and the Western Corridor labeled. Data were derived from the MODIS burned area product MCD45A1.

rapid increase in wildebeest numbers following rinderpest eradication in the 1960s. At the whole-ecosystem scale, the grazing pressure imposed by the wildebeest population has thus been the key determinant of changes in fire frequency over the past few decades (Sinclair et al. 2007; Holdo et al. 2009) (fig. 4.5b). As the wildebeest ceased to be regulated by disease, the ensuing population explosion caused a reduction in grass biomass, and a decline in the amount of area burned annually in the Serengeti from 70% in the 1960s to less than 50% in the 1980s (fig. 4.2). Additional variation in

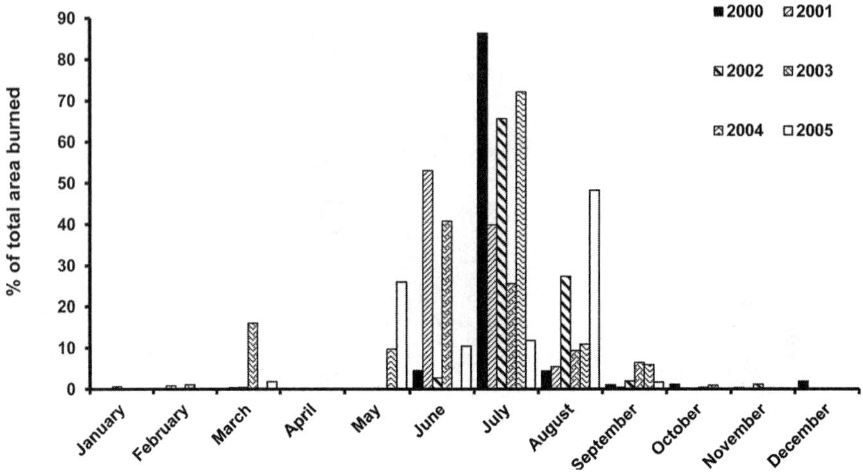

Fig. 4.4 Percentage of yearly total area burned per month within a given year (May 1 to April 30), in the Serengeti National Park (data derived from Dempewolf et al. 2007).

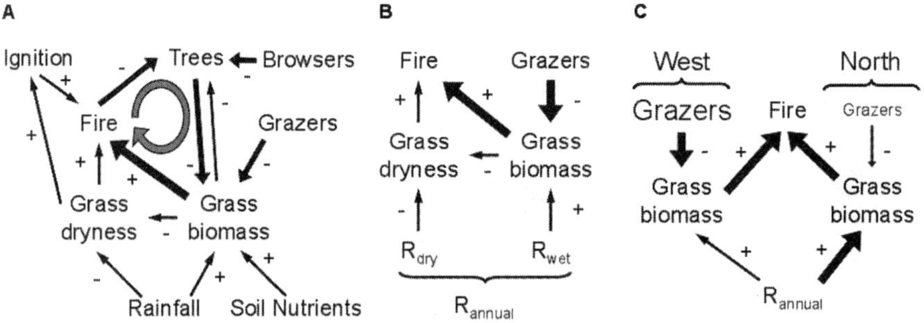

Fig. 4.5 Drivers of fire in savanna ecosystems. A general model is depicted where (A) fire occurrence depends on fuel loads as determined by grass biomass, grass dryness, and ignition events. Grass biomass can in turn vary as a function of rainfall (R), grazing intensity, tree cover, and soil nutrients (soil N). The drivers of fire occurrence in the Serengeti based on spatiotemporal trends in fire extent at (B) the whole-ecosystem level, where grazing effects dominate, and (C) at smaller spatial scales, where grazing intensity is more variable, leading to grazing (top-down) control of fire at some sites, west, and rainfall (bottom-up) control at others, north. The thickness of the arrows signifies the relative strength of influences between different factors.

interannual patterns of fire extent is explained by the seasonal distribution of rainfall (Norton-Griffiths 1979; Stronach 1989; Holdo et al. 2009) (fig. 4.5b). This can be explained as follows: in years with abundant wet-season rainfall and low dry-season rainfall, high biomass accumulation followed by rapid drying increases the probability of fire occurring in any given location after controlling for the effects of wildebeest grazing (Holdo et al. 2009). Total precipitation, however, was not found to contribute to global patterns of fire extent in the Serengeti. It is notable that, as human population density around the boundaries of protected areas increased dramatically over this time period, the amount of area burned declined, suggesting that fires for the whole Serengeti are fuel-limited rather than ignition-limited (Holdo et al. 2009). However, ignition events (fig. 4.5a) are bound to be far more important at local spatial scales.

Analyses conducted at smaller spatial scales suggest a far more variable impact of grazers, however, with the importance of rainfall varying in accordance with the timing and intensity of grazing. We compared the northern Serengeti (NS) with the Western Corridor (WC) to investigate the relationship between fire return intervals and annual precipitation (AP) at two sites that differed in the timing of grazing by large migratory wildebeest herds. The NS and WC regions were delineated based on ecologically meaningful land regions identified by Gerresheim (1974) and aggregated by Pennycuick (1975). NS comprises Gerresheim regions 3, 4, 7, and 8, and WC comprises regions 9, 12, and 13 (fig. 4.3). The average annual rainfall for the NS from 2000 to 2008 derived from the Famine Early Warning System (FEWS) Dekadal Rainfall Estimates (RFE) version 2.0 dataset was 897 mm ± 151 mm and for the WC 833 mm ± 140 mm. FEWS RFE data are produced by an automated algorithm using observations from the Meteosat satellite in the infrared spectral band, rain gauge reports, and microwave satellite observations (Xie and Arkin 1997).

Satellite-derived burned area maps produced from data collected by the Moderate Resolution Imaging Spectroradiometer (MODIS) satellite sensor (MODIS product MCD45A1) were acquired for 9 years from 2000 to 2008. The burned area maps were combined with the spatially explicit layers of rainfall from the FEWS RFE. The FEWS RFE data were used to calculate yearly precipitation from May 20 to May 19 the following year, the approximate start of the dry season.

Results in the NS showed a significant relationship between area burned and AP ($r^2 = 0.91, p < 0.001$). The relationship was weaker in the WC ($r^2 = 0.44, p = 0.055$) (fig. 4.6).

The migratory wildebeest herds occupy the Serengeti short grass plains in the southeast from about November to April, the WC in May and June, move northward in July, and occupy the NS and the Maasai Mara from August to October before returning to the short grass plains (Pennycuick 1975; Campbell 1989; Mduma 1996; Thirgood et al. 2004). Figure 4.7 shows the average cumulative monthly area burned for the NS and WC as a percentage of the average annual area burned in each one year period from May to April. The averages were calculated for the years 2000 to 2008 and differentiated as areas burned during wildebeest occupancy and nonoccupancy. The results show that 77% of burned areas in the WC burn during or directly

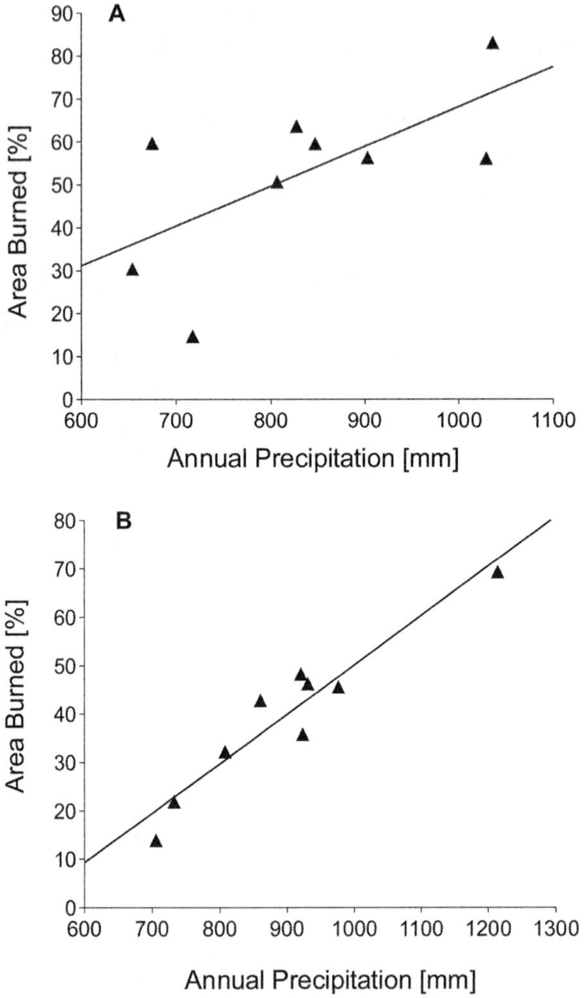

Fig. 4.6 Annual precipitation from May 20 to May 19 for the years 2000 to 2008 in the Western Corridor (A) and northern Serengeti (B) and area burned in the subsequent year.

after the time of wildebeest occupation, while in the NS only 27% of the burned areas burn during this time frame.

This analysis highlights the importance of the timing of wildebeest grazing on fire dynamics in the Serengeti. The amount of area burned in the NS is primarily driven by AP because fires tend to occur before the passage of the wildebeest. In the WC the effect of AP is masked by the effect of wildebeest grazing. The grazing pressure exerted by wildebeest in the WC during the early dry season significantly reduces grass biomass (McNaughton 1985)

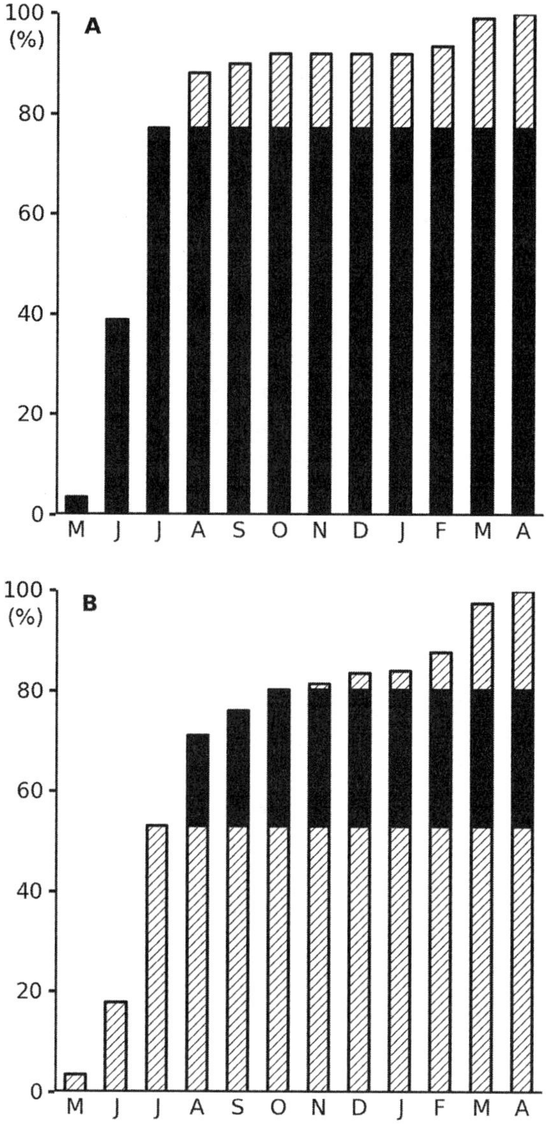

Fig. 4.7 Average cumulative area burned per month for the years 2000 to 2008 in the Western Corridor (A) and northern Serengeti (B) as a percentage of average annual area burned from May to April. The mean and standard deviation of annual rainfall are similar in both areas with 833 mm ± 140 mm in the Western Corridor and 897 mm ± 151 mm in the northern Serengeti. The percent area burned is higher in the Western Corridor with 53% ± 20% than in the northern Serengeti with 40% ± 16%. Areas burned during months with wildebeest occupancy are shown in black, areas burned during nonoccupancy in gray.

and less fire fuel is available for burning. This effect might explain the lower correlation between area burned and AP in the WC compared to the NS (fig. 4.5c). This suggests that, at intermediate spatial scales within the Serengeti ecosystem, the timing of grazing in relation to the timing of the wet-dry season transition (which can affect fire occurrence and spread) can be of fundamental importance (fig. 4.5c).

Thus, because wildebeest abundance patterns vary in both time and space as a result of the migration, the role of wildebeest grazing as a driver of fire frequency likewise varies within the Serengeti. In intensively grazed areas, wildebeest effectively compete with fire for biomass. Where wildebeest grazing intensity is low, grass biomass, and therefore fuel load, is regulated by soil moisture availability, which is primarily driven by precipitation.

The differences seen in the yearly average percent area burned, which is lower in the NS (40% ± 16%) than in the WC (53% ± 20%) might be explained by the higher ruggedness of the terrain in the north. Increased topographic variability is likely to cause different drying patterns across the catena, effectively creating fire barriers of green vegetation with high moisture content. A difference in management practices might also play a role with less aggressive burning in the NS due to the risk of fires escaping across the international border into the Kenyan part of the ecosystem. The larger amount of anthropogenic ignition events in the WC may also explain the lower correlation between area burned and AP in the WC compared to the NS.

A less-explored factor driving fire frequency in savannas is tree cover (fig. 4.5a). At the continental scale, annual precipitation, rainfall seasonality, and tree cover have been found to be the primary factors controlling the extent of fire in sub-Saharan Africa, with grazing being important on a site-specific basis and ignition events being far less important than fuel loads (Archibald et al. 2009). In the Serengeti, relatively little empirical work has been conducted to explore the relationship between tree cover and grass biomass, let alone between tree cover and fire frequency, yet model simulations suggest that trees can potentially play an important role in regulating fire through their moderating influence on grass biomass (Holdo, Holt, and Fryxell 2009) (fig. 4.5a).

Any other reduction in grass biomass (e.g., caused by grazing as discussed above) could result in less frequent fire and an increase in tree cover. If a certain threshold is reached, the dominant effect of trees over grasses could exacerbate this effect and trap the system in an alternative stable state where thickets or woodland patches are formed that are difficult to revert to more open savanna or grassland. The occurrence of such alternative stable states have been proposed for the Serengeti based on theoretical models

(Dublin, Sinclair, and McGlade 1990; Holdo, Holt, and Fryxell 2009), but as yet have not been shown empirically to occur. There is strong evidence from field studies, however, that increases in tree cover have the potential to lead to runaway local woody encroachment as a result of reductions in grass cover and facilitation of tree recruitment (Sharam, Sinclair, and Turkington 2006), as in areas where *Acaci polyacantha* has been observed to increase (Sharam et al. 2009).

In addition to the general factors driving fire occurrence (e.g., grazing and rainfall), there are site-specific differences that may be related to the community composition of vegetation communities across the Serengeti (e.g., tree cover as discussed above). The fire return interval (FRI) per vegetation type was determined for the entire Serengeti-Mara ecosystem. FRI was calculated from the MODIS burned area time series as the average time period between fire events. Vegetation types were derived from a vegetation map based on Landsat satellite data from the year 2000, and field surveys (Reed et al. 2009) that used a classification system developed by Grunblatt, Ottichilo, and Sinange (1989). The vegetation map and FRI layers were converted to the same projection and spatial resolution of 30 m and the FRI per vegetation type calculated, weighted by area.

Figure 4.8 shows FRI per vegetation type at two hierarchical levels. The results show a shortening fire return interval from grasslands (2.8 years) to shrublands (2.2 years) to woodlands/forests (2.0 years). The longest fire return intervals were recorded for grasslands with sparse grass canopy cover (2–19% canopy cover, 28.7 years, 2% of the study area) and grasslands with closed grass canopy cover (80–100% canopy cover, 4.1 years, 14% of the study area). The shortest fire return interval was observed for dense savanna woodlands (50–79% canopy cover, 1.7 years, 5% of the study area).

The shorter FRI in shrublands, woodlands, and forests compared to grasslands is counterintuitive given what we know about how tree cover affects grass biomass, and may be due to wildebeest grazing. Wildebeest prefer grasslands and open savanna habitat and avoid woodlands with greater predation risk (Darling 1960). It is therefore reasonable to assume that wildebeest reduce fire fuel load through grazing to a larger degree in grasslands than in shrublands and woodlands/forests, resulting in longer FRI in the grasslands. A similar effect of grazing by wildebeest on FRI is suggested by the fact that after sparse grasslands (which are fuel-limited due to low grass cover), closed grasslands show the longest fire return interval. Closed grasslands provide larger amounts of forage for wildebeest and may experience heavier grazing levels, reducing fuel load to a larger degree than in dense (50–79% canopy cover) or open (20–49% canopy cover) grasslands.

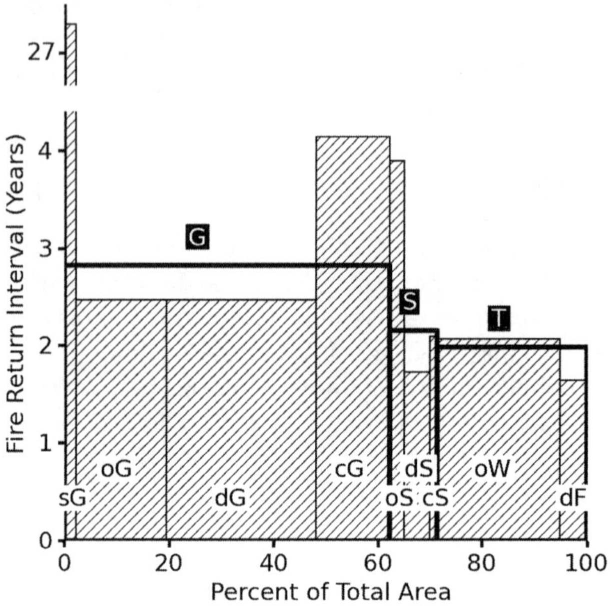

Fig. 4.8 Fire return interval (FRI) per vegetation type in the Serengeti-Mara ecosystem at two hierarchical levels. The first level vegetation types are grasslands (G), shrublands (S), and woodlands or forests (T). The second level vegetation types use the three first level classes along with modifiers describing canopy coverage of the dominant life form. The modifiers are sparse (s, 2–19%), open (o, 20–49%), dense (d, 50–79%), and closed (c, 80–100%) canopy coverage. The width of the bars corresponds to the percent area of the study region.

FIRE INTENSITY

Fire intensity is a key factor of savanna fire regimes. It controls the impact of fire on ecosystems, often referred to as fire severity, and is typically measured in terms of tree mortality (Frost and Robertson 1987; Govender, Trollope, and van Wilgen 2006). Fire intensity is a driving factor for generating heterogeneity in woody plant distributions in savannas (Higgins, Bond, and Trollope 2000). It is a function of fuel load, fuel moisture, wind speed, and slope and therefore highly dependent on local conditions at the time of burning and very challenging to measure or predict, especially over large areas.

Stronach (1989) and Stronach and McNaughton (1989) present measurements of fire temperature and combustion completeness at several locations in Serengeti in June 1986. Both parameters can be used as indicators of fire intensity. The measurements were not repeated though, and did not extend over areas comparable in size to typical dry season fires in the Serengeti. An empirical relationship for estimating fire intensity from grass biomass, fuel moisture, relative air humidity, and wind speed was developed by Trollope (1998) from 200 monitored fires in South African savannas. However, using the same approach for Serengeti would require the model to be parameterized for local conditions. The importance of fire for vegetation structure and heterogeneity in the Serengeti ecosystem calls for further research on estimating and monitoring fire intensity and its controlling factors.

THE INFLUENCE OF FIRE ON ECOSYSTEM STRUCTURE AND FUNCTION

Fire as a Driver of Woody Cover Change in the Serengeti

The relative impact of climate, fire, and herbivores (elephants in particular) on tree cover in African savannas, including the Serengeti, has long been the subject of debate (Laws 1970; Caughley 1976; Dublin, Sinclair, and McGlade 1990; Sankaran, Ratnam, and Hanan 2004; Bond, Woodward, and Midgley 2005; Sankaran et al. 2005). Elephants, fire, and wildebeest have variously been proposed as factors driving changes in tree density in the Serengeti-Mara ecosystem (Croze 1974; Norton-Griffiths 1979; Pellew 1983; Dublin, Sinclair, and McGlade 1990; Ruess and Halter 1990; Sinclair et al. 2007; Estes, Raghunathan, and Van Vleck 2008). In the Maasai Mara Reserve, adjacent to the Serengeti, Dublin, Sinclair, and McGlade (1990) used a mathematical model to suggest that fire (not elephants) drove woodland decline in the 1960s, and browsing prevented tree regeneration in the 1980s. In the Serengeti woodlands, conditions are somewhat different, as the large-scale disappearance of woodlands occurring in the Kenyan Mara have not yet materialized. Here, targeted short-term studies in areas with high elephant population density between the 1960s and 1980s have implicated elephants in sharp declines in mature tree densities (Croze 1974; Pellew 1983; Ruess and Halter 1990), but longer-term, more recent, and more widespread studies are lacking.

Spatio-temporal patterns of woody cover change in Serengeti in relation to fire. An analysis of woodland change over a 10-year period in the 1960s by Norton-Griffiths (1979) suggested that elephants were the primary drivers of tree cover change in central Serengeti, but fire was dominant in NS. There was a great deal of concern over the role of elephants as agents of tree cover loss in the 1960s and 1970s, particularly in the central woodlands around Seronera (Croze 1974; Lamprey et al. 1967), but the role of elephants diminished as their numbers were decimated by poaching beginning in the late 1970s (Dublin and Douglas-Hamilton 1987).

Apart from this 10-year span, until quite recently (Holdo et al. 2009) no long-term statistical analysis had been conducted to rigorously examine the factors driving changes in tree dynamics in Serengeti. Longitudinal photopanorama data show that tree density in the Serengeti has followed a pattern of a decline in cover during much of the twentieth century up until around 1980, whereupon it began to recover (Sinclair et al. 2007). This long-term pattern was mainly driven by the aforementioned changes in fire frequency that are primarily mediated by wildebeest grazing at the whole-ecosystem scale (Sinclair et al. 2007; Holdo et al. 2009), rather than by changes in elephant population density and climate (fig. 4.9).

Simulated effects on woody cover and tree size distributions. To investigate how changing FRIs may impact future patterns of woody cover in the Serengeti, we conducted a series of simulations with the Savanna Dynamics (SD) model (Holdo, Holt, and Fryxell 2009). The tree dynamics component of the SD model is largely based on an earlier model by Pellew (1983), with modifications to account for density-dependence, recruitment, and effects of rainfall on variation in growth rates (Holdo, Holt, and Fryxell 2009). The model simulates the integrated dynamics of grazers, fire, grass, and tree cover across spatially-realistic templates of rainfall and soil nutrients across the Serengeti ecosystem. Both trees and grasses are simulated as generic species, with data from the widespread *Acacia tortilis* used in the case of trees. Under realistic conditions, fire interacts with a number of other factors in affecting tree population dynamics, including browsing by giraffe and elephants, and rainfall (Pellew 1983; Holdo, Holt, and Fryxell 2009). The frequency and distribution of fire is also strongly dependent on fuel loads as determined by grazing patterns (Sinclair et al. 2007; Holdo, Holt, and Fryxell 2009; Holdo et al. 2009). To isolate the independent effects of fire in the system, in this exercise we treated FRI as a fixed factor (with equal probability of occurrence across space and over time, independent of variation in fuel loads), and assumed a constant level of giraffe browsing (but no elephant browsing), which in this model translates into reduced growth rates (Pellew 1983). We conducted 100-year simulations, with 10 runs per simulation.

Our results suggested strong nonlinear effects of FRI on the steady-state amount of tree cover (fig. 4.10a) and the tree size-class distribution (fig. 4.10b) in the Serengeti. FRIs below about five years are predicted to reduce canopy cover below the initial (roughly present-day) value of 30%. Fire results in a strong shift in tree height classes from being well distributed across all height classes when fire is infrequent, to a relative shift toward the smallest and largest height classes when fire is frequent (fig. 4.10b). This occurs because large trees are able to escape the effects of fire (and are therefore largely unaffected, except by reduced recruitment), and trees of intermediate height are frequently top-killed and revert to the smallest size class.

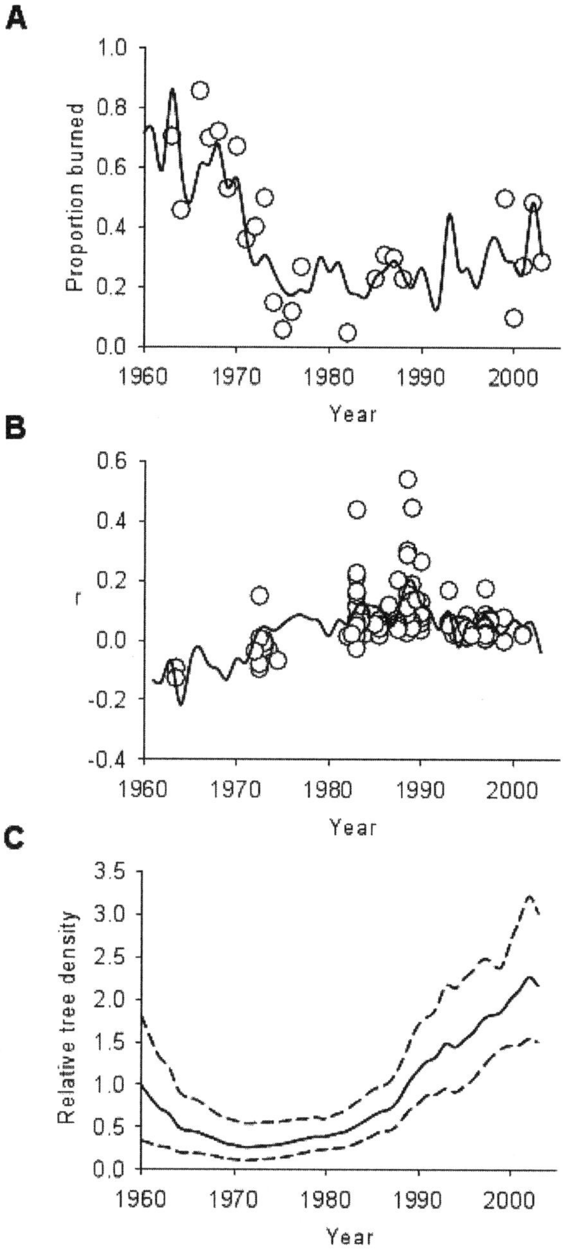

Fig. 4.9 Observed (○) and predicted (–) (A) fire occurrence (as a proportion of the ecosystem burning each year) and (B) rate of change (expressed as $r = \ln[N_t + \Delta t/N_t]/\Delta t$) in tree density (N) in the Serengeti. The predicted values are based on a best-fit Bayesian state-space model of rainfall, herbivore, fire, and tree dynamics (Holdo et al. 2009); (C) modeled (mean ± 95% credible intervals) changes in tree density (with 1960 density fixed at an arbitrary value of 1) based on the state-space model.

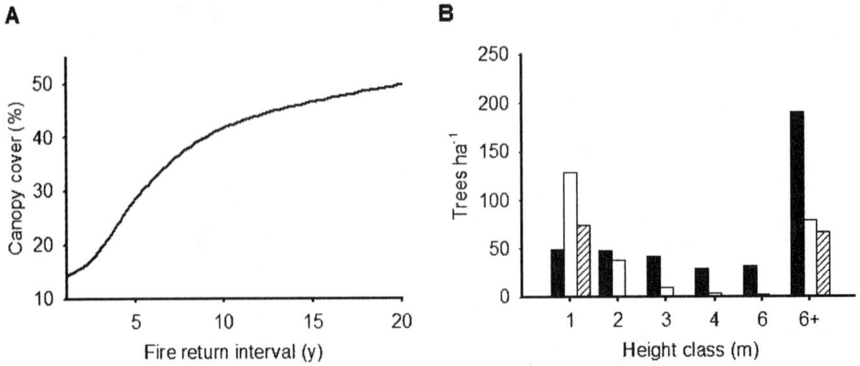

Fig. 4.10 (A) Simulated canopy cover and (B) tree size class distributions in the Serengeti woodlands as a function of fire return interval (20 y [solid], 3 y [open] and 1 y [cross-hatched]) in the absence of elephant herbivory. Values represent steady state (after 100-year simulations), conditions averaged across space (means for 10 runs), with an initial canopy cover of 30%.

The Effects of Fire on Herbaceous Communities and Leaf Nutrients

Much has been studied on the impacts of fire on grassland communities. Plant regrowth following fire is higher in Ca, P, N, Mg and K (Mes 1958; Komarek 1969; Christensen 1977; Batmanian and Haridasan 1985; Singh 1993; Van de Vijver, Poot, and Prins 1999). In addition, the plant regrowth is also higher in water (Mes 1958) and crude protein content (Lemon 1968; Komarek 1969; Oliver 1978). Van de Vijver, Poot, and Prins (1999) suggested that these increases in nutrient concentrations were due to rejuvenation of plant material, the distribution of nutrients over less biomass, and an increase in leaf-to-stem ratios. Furthermore, fire causes a decrease in litter and standing dead material (Dwyer and Pieper 1967; McNaughton 1985; Singh 1993), an increase in plant species diversity (Gibson and Hulbert 1987; Gibson 1988; Belsky 1992), and changes in plant species composition (Box, Powell, and Drawe 1967; Gibson and Hulbert 1987; Fuhlendorf and Engle 2004).

In the Serengeti, fire has been shown to increase seedling density (McNaughton 1983) and cause a slight increase in production in the first growing season postfire (McNaughton 1985). Fire also increased grass species diversity in areas that had burned infrequently, twice in seven years, as compared to areas that had not burned or had burned frequently, six times in seven years (McNaughton 1983). Fire did not have as strong an effect on grass community heterogeneity as other factors such as subsurface soil sodicity (Belsky 1988), nor did fire have a strong effect on the relative abundance of individual grass species (Belsky 1992).

Anderson et al. (2007) found that increased fire frequency (from zero times to five times burned within a 5-year period) has a negative effect on leaf N levels through the changes it causes in plant community composition. They also found that fire has a direct positive influence on leaf Na, but has an indirect negative effect by decreasing the abundance of Na-rich grasses in a community. This indirect negative effect was stronger than the direct positive effect so that the overall impact of fire on leaf Na was negative. Additionally, fire caused an indirect negative impact on leaf P through its influence on plant community composition. However, its direct impact on leaf P was unimodal with leaf P increasing with fire frequencies up to three years after which it declined. Lastly, they found a strong positive correlation between fire frequency and *Themeda triandra* abundance. They argue that increases in *T. triandra* could lead to herbivore nutrient deficiencies as average P and Na for *T. triandra* do not meet the dietary requirements of lactating female wildebeest. This may be especially true in the *T. triandra* dominated grasslands of the Western Corridor where there is a resident herd of wildebeest (Murray 1995) and which serves as a dry-season transitional zone for migratory wildebeest and zebra.

The Effects of Fire on Soil Nutrient Dynamics

In addition to its effects on ecosystem structure, composition, and biodiversity, the changing nature of the fire regime in the Serengeti has important implications for ecosystem function, including N cycling and soil C storage (Anderson et al. 2007; Holdo et al. 2007). Using a model of coupled differential equations describing the dynamics of soil N pools, plant biomass, and herbivores, Holdo et al. (2007) predicted that fire (by increasing volatile N losses) can lead to substantial declines in soil organic N (and ultimately net primary productivity) in the Serengeti. These negative effects of fire were predicted to be partially offset by positive effects of grazing, partly due to increased N turnover, but mainly because grazers reduce the probability of fire occurrence by reducing fuel loads. A separate empirical study suggested that soil organic matter declines with increasing fire frequency (Anderson et al. 2007). These results have potentially long-term implications for soil fertility and C storage in parts of the Serengeti ecosystem (both within and outside protected areas) that are presently subjected to frequent burning (Holdo et al. 2009).

The Effects of Fire on Invertebrates

Little work has been done testing the impacts of fire on invertebrates (Parr and Chown 2003). What has been done has shown mixed results (Ferrar 1982; Gandar 1982; Chambers and Samways 1998; Parr, Bond, and Robertson 2002; Parr et al. 2004; O'Reilly et al. 2006). Nkwabi et al. (2010) trapped for ants, spiders, beetles, flies, wasps, bees, hemiptera (aphids among others), and grasshoppers in recently burned (1 to 20 weeks before sampling) and unburned habitats in the Serengeti. In general, they found a greater abundance of insect groups in the unburned sites as compared with the burned sites. The only exception was wasps which were marginally more abundant in burned sites than in unburned sites. Conversely, Eby (2010) trapped for both tsetse flies (*Glossina pallidipes* and *Glossina swynnertoni*) and non-tsetse flies (Diptera) and did not find a significant difference in abundance between burned and unburned areas. Additionally, Eby (2010) found no difference in tick densities between burned and unburned areas.

The Effects of Fire on Bird Distributions

The impact of fire on bird distributions is highly variable. In some ecosystems, fire destroys bird habitats (Gottschalk 2002) while in others it creates them (Wooller and Brooker 1980). Fire can decrease species richness (Jansen et al. 1999) or have no effect (O'Reilly et al. 2006), and can increase bird diversity (Woinarski 1990; O'Reilly et al. 2006), density (Woinarski 1990), and abundance (O'Reilly et al. 2006; Valentine et al. 2007). Season of burn and time since burning can also influence bird assemblages (Valentine et al. 2007).

Only one study has looked at the impacts of fire on birds in the Serengeti (Nkwabi et al. 2010). It found increased species richness and abundance in recently burned sites (1 to 20 weeks before sampling) as compared to unburned sites. It also found differences in the most frequent species present in each habitat type. For example, of the top ten most frequent species in unburned sites only two appeared in the top ten for burned sites (Nkwabi et al. 2010).

Fire as a Factor Influencing Large Herbivore Distributions

Numerous studies on both domestic and wild animals have found that herbivores are attracted to recently burned areas (Oliver 1978; Rowe-Rowe 1982;

Moe, Wegge, and Kapela 1990; Vinton et. al. 1993; Mduma and Sinclair 1994; Wilsey 1996; Tomor and Owen-Smith 2002; Gureja and Owen-Smith 2002; Vermeire et al. 2004; Eby et al. 2014). Herbivores will even move away from other resources in order to utilize these burned areas. For example, cattle will travel over 1,600 m from a water source in order to utilize burned patches (Vermeire et al. 2004). This preference for burned areas has mainly been attributed to the attraction of ungulates to the new postfire plant growth (Frost 1984; Wilsey 1996; Eby et al. 2014).

Within the Serengeti Sinclair (1977) found avoidance of burned areas by buffalo (*Syncerus caffer*) while Mduma and Sinclair (1994) found that oribi (*Ourebia ourebi*) were attracted to burned areas a few weeks old. Additionally, Thomson's gazelles (*Gazella thomsoni*), Grant's gazelles (*Gazella granti*), impala (*Aepyceros melampus*), and warthog (*Phacochoerus aethiopicus*), prefer burned areas, while zebra (*Equus burchelli*), show no preference (Wilsey 1996; Eby et al. 2014). Meanwhile for both wildebeest (*Connochaetes taurinus*) and topi (*Damaliscus lunatus*), the results are conflicting showing that they sometimes prefer burned areas and sometimes show no preference (Wilsey 1996; Eby et al. 2014).

In general smaller herbivores show a stronger preference for burned areas than larger ones. The reason for this attraction appears to be due to the more nutritious forage that sprouts after a fire (Wilsey 1996; Eby et al. 2014).

Fire Effects on Carnivore Distributions

What little work has been done on the impacts of fire on carnivore distributions in a landscape has shown mixed results (Ballard et al. 2000; Blanchard and Knight 1990; Dees, Clark, and Manen 2001; Cunningham, Kirkendall, and Ballard 2006). In a savanna ecosystem, Ogen-Odoi and Dilworth (1984) found an increase in medium-sized carnivores (<9 kg) in areas after they had burned, which they attributed to increased hare abundance in these areas. They saw no difference in the number of large carnivores (≥9 kg) observed before and after burning, although they did see a decrease in the number of species observed from four (lion, *Panthera leo*, leopard, *Panthera pardus*, spotted hyena, *Crocutta crocutta*, and African civet, *Civettictis civetta*) to two (lion and spotted hyena). Eby et al. (2013) looked at the impact of fire on lion distribution. They found that lions avoid burned areas, possibly because the decrease in cover leads to decreases in hunting success.

IMPLICATIONS FOR THE FIRE MANAGEMENT PLAN

Fire is the main management tool utilized for the management of vegetation in the Serengeti and in other parks throughout the savannas of Africa. As such it is important for managers to know the impacts of controlled burning on the ecosystems, which they are charged with protecting. Recently, the park has formulated a new fire management plan, which has not yet been implemented, that is designed to accomplish eight major goals, each with specific objectives (Trollope, Trollope, and Austin 2005).

Goal 1: To create suitable conditions for tourism and enhance visitor experience.

1. Improve visibility for tourists' game-viewing circuits
2. Attract animals to green grass flushes within the park
3. Minimize large unsightly burnt areas

The research presented in this chapter only allows us to address objective two. Research conducted in the Serengeti has shown that many species of herbivores are attracted to newly-burned areas (Mduma and Sinclair 1994; Wilsey 1996; Eby et al. 2014). However, implementation of burning may not completely accomplish this goal because not all herbivores are attracted to burned areas, and some of the large herbivores avoid these areas (Sinclair 1977; Wilsey 1996; Eby et al. 2014). Therefore, care should be taken to leave some areas along tourist tracks unburned so as not to decrease the ability of tourists to see certain, generally large herbivores.

While there is currently no data available to evaluate the efficacy of fire management in regards to objectives one and three, it is reasonable to assume that their implementation is possible with the lighting of strategically placed fires. In order to improve visibility, fires need to be lit in areas along the game circuits. However, as vegetation height will return to preburn levels about six months after burning, this modification will be temporary. In order to minimize large burnt areas, cool fires, as is already the practice, need to be lit around the areas where large burns are not desired so as to keep hot fires from burning these areas later in the dry season.

Goal 2: To create suitable conditions for patrols.

1. Improve visibility and accessibility in high poaching areas
2. Improve visibility in areas with security concerns

This will be straightforward to implement if the goal is only to reduce grass height and not shrub or tree density. Lighting cool fires, as is already the practice, in high poaching and security concern areas will reduce grass cover; however, hotter fires lit later in the dry season may be necessary to reduce shrubs and trees.

Goal 3: To preserve sensitive vegetation communities and associated habitats.

1. Preserve riverine forests
2. Preserve hilltop thickets
3. Preserve hill vegetation and reduce erosion of exposed soils
4. Preserve kopje habitat
5. Preserve wetland vegetation

All of these objectives can be met by burning fires around these habitats in order to prevent unwanted burning within the habitats. However, the current FRI of two years for woodlands/forests and shrublands could be causing a decrease in those habitats and would therefore not be meeting objectives one and two.

Goal 4: To protect rare and sensitive species.

1. Protect herps including tortoises, pancake tortoises, snakes, lizards, Moru toads
2. Protect rhino habitat
3. Protect red patas and black and white colobus habitats
4. Protect predator rearing areas
5. Protect ground nesting bird habitats (although some birds, such as night jars, pipits, and larks require burnt areas for nesting)
6. Preserve cover for shy animals, specifically bushbuck, reedbuck, and leopard

Many of these objectives would be met by meeting the objectives of goal three.

Goal 5: To maintain adequate foraging resources for animals.

1. Preserve adequate dry season forage for migrant species
2. Preserve adequate early dry season forage for migrant species
3. Preserve adequate forage in areas with large populations of resident ungulates
4. Stimulate new grass growth to attract animals to stay within the park

A mosaic of burning should be conducted to allow for continuous availability of forage for consumption. Specifically to objective four, postfire regrowth has been shown to be higher in nutrients (Van de Vijver, Poot, and Prins 1999; Eby et al. 2014) and many species of herbivores are attracted to these areas (Mduma and Sinclair 1994; Wilsey 1996; Eby et al. 2014). However buffalo and lions avoid these burned areas (Sinclair 1977; Eby et al. 2013). Since the desired goal is to keep animals within the park, patches of burned and unburned areas are recommended as burning too large an area near the boundary of the park might drive buffalo and lions out of the park, potentially leading to increases in human-wildlife conflict.

It is also important to remember that fire can have unintended consequences on the flora of the ecosystem which can in turn impact the fauna. For instance, decreased FRIs could cause decreases in tree cover and subsequent loss of habitat for browsers and other tree-dependent species (Turkington et al., chapter 9; Anderson et al., chapter 5). Additionally, increased fire frequency has an overall negative effect on grass leaf N and Na which could hinder the ability of herbivores to get sufficient nutrients (Anderson et al. 2007). Thus managers should increase FRIs in areas with high amounts of browsers and other tree-dependent species to allow for the continued existence of the trees on which these species depend. Furthermore, managers should try to decrease fire frequency in high fire frequency areas (fig. 4.3) to try and minimize decreases in grass leaf nutrients.

Goal 6: To control disease vectors.

1. Control tsetse flies in high-use areas
2. Control tick load

The data on the use of fire to control tsetse flies in high human use areas, in order to decrease exposure to African trypanosomiasis (sleeping sickness), and to control tick load, to protect wildlife from tick related costs such as blood loss and disease, is inconclusive. Eby (2010) found no significant difference in tsetse fly numbers between burned and unburned habitats. This leads us to believe that the cool controlled burning practiced by management will not be an effective tool against tsetse fly numbers. However, tsetse flies are attracted to areas covered with brush, woody vegetation, or shrubs (Kitron et al. 1996) and it is possible that hot fires, set late in the dry season, would reduce shrub, thereby possibly reducing tsetse fly numbers (Lamprey and Reid 2004). However, fires set late in the dry season would have to be carefully controlled to prevent them from unintentionally spreading.

Several studies have found decreases in tick loads with burning (Spickett et al. 1992; Davidson, Siefken, and Creekmore 1994; Stafford, Ward, and Magnarelli 1998; Cully 1999; Fyumagwa et al. 2007); however, the only study conducted in the Serengeti found no difference (Eby 2010). Additionally, other studies have found no effect of burning on tick loads (Zieger, Horak, and Cauldwell 1998), and the use of fire as a management tool to control ticks has been questioned (Minshull and Norval 1982; Spickett et al. 1992). This is an area that needs further work. However, based on what we currently know, it does not appear that burning will be an effective management tool against tick loads in the Serengeti.

Goal 7: To minimize the impact of uncontrolled and unprescribed fires.

1. Prevent the spread of unprescribed fires from adjacent communities surrounding the park
2. Reduce risk of wildfires on park infrastructure
3. Reduce the risk of poachers' fires
4. Reduce risk of wildfires on long grass plains
5. Reduce risk of wildfire in tall grass woodlands

All of the objectives under this goal can probably be reached by creating a patch mosaic of burns throughout the ecosystem which will create a scattered distribution of fuel loads, thus minimizing the spread of wildfires. Additionally, specifically to meet objective one, fire breaks can be burnt at the edges of the ecosystem.

Goal 8: To protect facilities and park infrastructure.

1. Prevent park facilities from being burnt

In order to effectively achieve this goal, vegetation should be kept short around infrastructure so as to prevent fires from getting close enough to burn the infrastructure.

Additional Data Needed to Evaluate the Fire Management Plan

Fire used as a management tool within the Serengeti would benefit from the implementation of adaptive ecosystem management either passively or actively. Passively, where existing management decisions are evaluated through research and management, and plans changed if the goals are not being reached. Actively, where different management practices are applied and tested to see which one gives the desired result (Walters and Hilborn 1978; Wilhere 2002; van Wilgen and Biggs 2010). From a passive direction several of the above objectives could be evaluated with the collection of more data. For example, when management fires are set there should be a record of where and when the fire was set, as well as its intensity. This would allow for management fires to be distinguished from wildfires and would allow for the evaluation of outcomes. Studies should be conducted in sensitive habitats to monitor if fires are burning within these habitats, and if so, how they are affecting the flora and fauna within the habitats. Aerial surveys could be conducted during the burning season to determine if large burns are occurring. Actively, there is a need for the establishment of long-term fire experiments within different habitats in order to study the impacts of fire on the flora and fauna of the ecosystem.

SUMMARY

Here we highlight what is currently known about the history of fire and its impacts on the Serengeti ecosystem. While much is known about the role that fire plays, many questions are still unanswered. For instance, there have been no studies on the effects of fire on small mammals, although impacts have been seen elsewhere (Gandar 1982; Bigalke and Willan 1984). In addition, fire frequency has been shown to impact herbaceous and woody species in the Serengeti (Anderson et al. 2007; Holdo, Holt, and Fryxell 2009; Strauch and Eby 2012), yet we know little about how it impacts the fauna. Some studies have untangled the complex interaction between fire and habitat modification and the subsequent impact on species (Sharam, Sinclair, and Turkington 2009). However, it is unlikely that all habitats react in a linear way to being burnt and frequency and/or intensity is likely important. Furthermore, it is important to understand the way fire interacts with other factors (e.g., grazing, rainfall). This has been touched on above, but most of this chapter focuses only on the impact of fire by itself. This is partly due to the fact that this chapter is specifically about the role that fire plays in the ecosystem. However, in some cases it is also because no research has been conducted on fire's interaction with other ecosystem processes. All of these are potential areas for future research.

Fire is one of the main management tools used in the Serengeti ecosystem. Despite what we know about the role of fire in this ecosystem, it is still difficult to assess the efficacy and outcome of the management plan outlined above and many questions remain. Questions for further research should include whether and how much prescribed fire is needed to achieve the management goals above, and do management actions conflict with goals? Managers would benefit from tracking their actions to determine if outcomes and goals were achieved. This calls for an adaptive management approach to assess if management actions are effective.

ACKNOWLEDGMENTS

The authors would like to thank the Tanzania Wildlife Research Institute, Tanzania National Parks and Tanzania Commission for Science and Technology for permission to conduct research in the Serengeti. This manuscript was greatly improved by insightful comments from N. Stronach and B. van Wilgen. J. Dempewolf was supported by the National Aeronautics and Space Administration (NASA) headquarters under the Earth System Science Fellowship Grant NGT530459.

REFERENCES

Anderson, T. M., M. E. Ritchie, E. Mayemba, S. Eby, J. B. Grace, and S. J. McNaughton. 2007. Forage nutritive quality in the Serengeti ecosystem: The roles of fire and herbivory. *American Naturalist* 170:343–57.

Archibald, S., D. P. Roy, B. W. van Wilgen, and R. J. Scholes. 2009. What limits fire? An examination of drivers of burnt area in southern Africa. *Global Change Biology* 15: 613–30.

Ballard, W., P. Krausman, S. Boe, S. Cunningham, and H. Whitlaw. 2000. Short-term response of gray wolves, *Canis lupis*, to wildfire in northwestern Alaska. *Canadian Field-Naturalist* 114:241–47.

Batmanian, G. J., and M. Haridasan. 1985. Primary production and accumulation of nutrients by the ground layer community of cerrado vegetation of central Brazil. *Plant and Soil* 88:437–40.

Belsky, A. 1988. Regional influences on small-scale vegetational heterogeneity within grasslands in the Serengeti national park, Tanzania. *Plant Ecology* 74:3–10.

———. 1992. Effects of grazing, competition, disturbance and fire on species composition and diversity in grassland communities. *Vegetation Science* 3:187–200.

Bigalke, R. C., and K. Willan. 1984. Effects of fire regime on faunal composition and dynamics. In *Ecological effects of fire in South African ecosystems*, ed. P. de V. Booysen and N. M. Tainton, 255–71. New York: Springer-Verlag.

Bird, M. I., and J. A. Cali. 1998. A million-year record of fire in sub-Saharan Africa. *Nature* 394:767–69.

Blanchard, B., and R. Knight. 1990. Reactions of grizzly bears, *Ursus arctos horribilis*, to wildfire in Yellowstone National Park, Wyoming. *Canadian field-naturalist* 104: 592–94.

Bond, W. J., F. I. Woodward, and G. F. Midgley. 2005. The global distribution of ecosystems in a world without fire. *New Phytologist* 165:525–37.

Box, T. W., J. Powell, and D. L. Drawe. 1967. Influence of fire on south Texas chaparral communities. *Ecology* 48:955–61.

Brain, C. K., and A. Sillen. 1988. Evidence from the Swartkrans cave for the earliest use of fire. *Nature* 336:464–66.

Campbell, K. L. I. 1989. Programme report. Serengeti ecological monitoring programme, September. Frankfurt Zoological Society.

Caughley, G. 1976. The elephant problem—an alternative hypothesis. *East African Wildlife Journal* 14:265–83.

Chambers, B. Q., and M. J. Samways. 1998. Grasshopper response to a 40-year experimental burning and mowing regime, with recommendations for invertebrate conservation management. *Biodiversity and Conservation* 7:985–1012.

Christensen, N. L. 1977. Fire and soil-plant nutrient relations in a pine-wiregrass savanna on the coastal plain of North Carolina. *Oecologia* 31:27–44.

Croze, H. 1974. The Seronera bull problem II: the trees. *East African Wildlife Journal* 12: 29–47.

Cully, J. F. 1999. Lone star tick abundance, fire, and bison grazing in tall-grass prairie. *Journal of Range Management* 52:139–44.

Cunningham, S. C., L. Kirkendall, and W. Ballard. 2006. Gray fox and coyote abundance and diet responses after a wildfire in central Arizona. *Western North American Naturalist* 66:169–80.

Darling, F. F. 1960. *An ecological reconnaissance of the Mara plains in Kenya Colony,* vol. 5, Wildlife Monographs. Washington, D C: The Wildlife Society.

Davidson, W. R., D. A. Siefken, and L. H. Creekmore. 1994. Influence of annual and biennial prescribed burning during March on the abundance of *Amblyomma america. Journal of Medical Entomology* 31:72–81.

Dees, C. S., J. D. Clark, and F. T. V. Manen. 2001. Florida panther habitat use in response to prescribed fire. *Journal of Wildlife Management* 65:141–47.

Dempewolf, J., S. Trigg, R. S. DeFries, and S. Eby. 2007. Burned area mapping of the Serengeti-Mara region using MODIS reflectance data. *Geoscience and Remote Sensing Letters, IEEE* 4 (2): 312–16.

Dublin, H. T., and I. Douglas-Hamilton. 1987. Status and trends of elephants in the Serengeti-Mara ecosystem. *African Journal of Ecology* 25:19–33.

Dublin, H. T., A. R. E. Sinclair, and J. McGlade. 1990. Elephants and fire as causes of multiple stable states in the Serengeti-Mara woodlands. *Journal of Animal Ecology* 59: 1147–64.

Dwyer, D. D., and R. D. Pieper. 1967. Fire effects on blue grama-pinyon-juniper rangeland in New Mexico. *Journal of Range Management* 20:359–62.

Eby, S. L. 2010. Fire and the reasons for its influence on mammalian herbivore distributions in an African savanna ecosystem. PhD diss., Syracuse University.

Eby, S., T. M. Anderson, E. P. Mayemba, and M. E. Ritchie. 2014. The effect of fire on habitat selection of mammalian herbivores: the role of body size and vegetation characteristics. *Journal of Animal Ecology*. DOI: 10.1111/1365-2656.12221.

Eby, S., A. Mosser, A. Swanson, C. Packer, and M. Ritchie. 2013. The impact of burning on lion *Panthera leo* habitat choice in an African savanna. *Current Zoology* 59 (3):335–39.

Edwards, P. J. 1984. The use of fire as a management tool. In *Ecological effects of fire in South African ecosystems*, ed. P. de V. Booysen and N. M. Tainton, 349–62. New York: Springer-Verlag.

Estes, R. D., T. E. Raghunathan, and D. Van Vleck. 2008. The impact of horning by wildebeest on woody vegetation of the Serengeti ecosystem. *Journal of Wildlife Management* 72:1572–78.

Ferrar, P. 1982. Termites of a South African savanna. III. Comparative attack on toilet roll baits in subhabitats. *Oecologia* 52:139–46.

Frost, P. G. H. 1984. The responses and survival of organisms in fire-prone environments. In *Ecological effects of fire in South African ecosystems*, ed. P. de V. Booysen and N. M. Tainton, 273–309. New York: Springer-Verlag.

Frost, P. G. H., and F. Robertson. 1987. The ecological effects of fire in savannas. In *Determinants of tropical savannas*, ed. B. H. Walker, 93–140. Harare, Zimbabwe: IRL Press.

Fuhlendorf, S. D., and D. M. Engle. 2004. Application of the fire-grazing interaction to restore a shifting mosaic on tallgrass prairie. *Journal of Applied Ecology* 41:604–14.

Fyumagwa, R. D., V. Runyoro, I. G. Horak, and R. Hoare. 2007. Ecology and control of ticks as disease vectors in wildlife of the Ngorongoro Crater, Tanzania. *South African Journal of Wildlife Research* 37:79–90.

Gandar, M. V. 1982. *Description of a fire and its effects in the Nylsvley nature reserve: A synthesis report.* South African National Scientific Programmes Report no. 63. Pretoria: Cooperative Scientific Programmes.

Gerresheim, K. 1974. *The Serengeti landscape classification-map and manuscript.* Nairobi: African Wildlife Leadership Foundation.

Gibson, D. J. 1988. Regeneration and fluctuation of tall-grass prairie vegetation in response to burning frequency. *Bulletin of the Torrey Botanical Club* 115:1–12.

Gibson, D. J., and L. C. Hulbert. 1987. Effects of fire, topography and year-to-year climatic variation on species composition in tall-grass prairie. *Vegetatio* 72:175–85.

Gillon, D. 1983. The fire problem in tropical savannas. In *Ecosystems of the world 13: Tropical savannas*, ed. F. Bourliere, 617–41. New York: Elsevier Scientific Publishing Company.

Gottschalk, T. 2002. Birds of Grumeti river forest in Serengeti National Park, Tanzania. *African Bird Club Bulletin* 9:101–04.

Govender, N., W. S. W. Trollope, and B. W. van Wilgen. 2006. The effect of fire season, fire frequency, rainfall and management on fire intensity in savanna vegetation in South Africa. *Journal of Applied Ecology* 43:748–58.

Gowlett, J., J. Harris, D. Walton, and B. Wood. 1981. Early archaeological sites, hominid remains and traces of fire from Chesowanja, Kenya. *Nature* 294:125–29.

Grunblatt, J., W. K. Ottichilo, and R. K. Sinange. 1989. A hierarchical approach to vegetation classification in Kenya. *African Journal of Ecology* 27:45–51.

Gureja, N., N. and Owen-Smith. 2002. Comparative use of burnt grassland by rare antelope species in a lowveld game ranch, South Africa. *South African Journal of Wildlife Research* 32:31–38.

Guyette, R. P., R. M. Muzika, and D. C. Dey. 2002. Dynamics of an anthropogenic fire regime. *Ecosystems* 5:472–86.

Higgins, S. I., W. J. Bond, and W. S. W. Trollope. 2000. Fire, resprouting and variability: a recipe for grass-tree coexistence in savanna. *Journal of Ecology* 13:295–99.

Hobbs, N. T., and D. S. Schimel. 1984. Fire effects on nitrogen mineralization and fixation in mountain shrub and grassland communities. *Journal of Range Management* 37: 402–5.

Holdo, R. M., R. D. Holt, M. B. Coughenour and M. E. Ritchie. 2007. Plant productivity and soil nitrogen as a function of grazing, migration and fire in an African savanna. *Journal of Ecology* 95:115–28.

Holdo, R. M., R. D. Holt, and J. M. Fryxell. 2009. Grazers, browsers, and fire influence the extent and spatial pattern of tree cover in the Serengeti. *Ecological Applications* 19: 95–109.

Holdo, R. M., A. R. E. Sinclair, K. L. Metzger, B. M. Bolker, A. P. Dobson, M. E. Ritchie, and R. D. Holt. 2009. A disease-mediated trophic cascade in the Serengeti and its implications for ecosystem C. *PloS Biology* 7:e1000210.

Jansen, R., R. M. Little, T. M. Crowe, and D. Ikanda. 1999. Implications of grazing and burning of grasslands on the sustainable use of francolins (*Francolinus* spp.) and on overall bird conservation in the highlands of Mpumalanga province, South Africa. *Biodiversity and Conservation* 8:587–602.

Kitron, U., L. H. Otieno, L. L. Hungerford, A. Odulaja, W. U. Brigham, O. O. Okello, M. Joselyn, M. M. Mohamed-Ahmed, and E. Cook. 1996. Spatial analysis of the distribution of tsetse flies in the Lambwe Valley, Kenya, using Landsat TM satellite imagery and GIS. *Journal of Animal Ecology* 65:371–80.

Komarek, E. V. 1969. Fire and animal behavior. *Tall Timbers Fire Ecology Conference Proceedings* 9:161–207.

Lamprey, H. F., P. E. Glover, M. Turner, and R. H. V.Bell. 1967. Invasion of the Serengeti National Park by elephants. *East African Wildlife Journal* 5:151–66.

Lamprey, R. H., and R. S. Reid. 2004. Expansion of human settlement in Kenya's Maasai Mara: What future for pastoralism and wildlife? *Journal of Biogeography* 31: 997–1032.

Laws, R. M. 1970. Elephants as agents of habitat and landscape change in East Africa. *Oikos* 21:1–15.

Lemon, P. C. 1968. Fire and wildlife grazing on an African plateau. *Tall Timbers Fire Ecology Conference Proceedings* 8:71–88.

McNaughton, S. J. 1983. Serengeti grassland ecology: The role of composite environmental factors and contingency in community organization. *Ecological Monographs* 53: 291–320.

———. 1985. Ecology of a grazing ecosystem: The Serengeti. *Ecological Monographs* 55: 259–94.

———. 1992. The propagation of disturbance in savannas through food webs. *Journal of Vegetation Science* 3:301–14.

Mduma, S. A. R. 1996. Serengeti wildebeest population dynamics: Regulation, limitation and implications for harvesting. PhD diss., University of British Columbia.

Mduma, S. A. R., and A. R. E. Sinclair. 1994. The function of habitat selection by oribi in Serengeti, Tanzania. *African Journal of Ecology* 32:16–29.

Mduma, S. A. R., A. R. E. Sinclair, and R. Hilborn. 1999. Food regulates the Serengeti wildebeest: A 40-year record. *Journal of Animal Ecology* 68:1101–22.

Mes, M. G. 1958. The influence of veld burning or mowing on the water nitrogen and ash content of grasses. *South African Journal of Science* 54:83–86.

Minshull, J. I., and R. A. I. Norval. 1982. Factors influencing the spatial distribution of *Rhipicephalus appendiculatus* in Kyle recreational park, Zimbabwe. *South African Journal of Wildlife Research* 12:118–23.

Moe, S. R., P. Wegge, and E. B. Kapela. 1990. The influence of man-made fires on large wild herbivores in Lake Burungi area in northern Tanzania. *African Journal of Ecology* 28:35–43.

Murray, M. G. 1995. Specific nutrient requirements and migration of wildebeest. In *Serengeti II: dynamics, management, and conservation of an ecosystem*, ed. A. R. E. Sinclair and P. Arcese, 231–56. Chicago: University of Chicago Press.

Nkwabi, A. K., A. R. E. Sinclair, K. L. Metzger, and S. A. R. Mduma. 2010. Disturbance, ecosystem function and compensation: The influence of wildfire and grazing on the avian community in the Serengeti ecosystem, Tanzania. *Austral Ecology* 36:1–9.

Norton-Griffiths, M. 1979. The influence of grazing, browsing, and fire on the vegetation dynamics of the Serengeti. In *Serengeti: dynamics of an ecosystem*, ed. A. R. E. Sinclair and M. Norton-Griffiths, 310–52. Chicago: University of Chicago Press.

Ogen-Odoi, A. A., and T. G. Dilworth. 1984. Effects of grassland burning on the savanna hare-predator relationships in Uganda. *African Journal of Ecology* 22:101–6.

Olindo, P. M. 1971. Fire and conservation of the habitat in Kenya. *Tall Timbers Fire Ecology Conference Proceedings* 11:243–56.

Oliver, M. D. N. 1978. Population ecology of oribi, grey rhebuck and mountain reedbuck in Highmoor State Forest Land, Natal. *South African Journal of Wildlife Research*. 8: 95–105.

O'Reilly, L., D. Ogada, T. Palmer, and F. Keesing. 2006. Effects of fire on bird diversity and abundance in an East African savanna. *African Journal of Ecology* 44:165–70.

Parr, C., W. Bond and H. Robertson. 2002. A preliminary study of the effect of fire on ants (Formicidae) in a South African savanna. *African Entomology* 10:101–11.

Parr, C. L., and S. L. Chown. 2003. Burning issues for conservation: A critique of faunal fire research in southern Africa. *Austral Ecology* 28:384–95.

Parr, C. L., H. G. Robertson, H. C. Biggs, and S. L. Chown. 2004. Response of African savanna ants to long-term fire regimes. *Journal of Applied Ecology* 41:630–42.

Pellew, R. A. P. 1983. The impacts of elephant, giraffe and fire upon the Acacia-Tortilis woodlands of the Serengeti. *African Journal of Ecology* 21:41–74.

Pennycuick, L. 1975. Movements of migratory wildebeest population in the Serengeti area between 1960 and 1973. *East African Wildlife Journal* 13:65–87.

Reed, D., T. M. Anderson, J. Dempewolf, K. Metzger, and S. Serneels. 2009. The spatial distribution of vegetation types in the Serengeti ecosystem: the influence of rainfall and topographic relief on vegetation patch characteristics. *Journal of Biogeography* 36: 770–82.

Rowe-Rowe, D. T. 1982. Influence of fire on antelope distribution and abundance in the Natal Drakensberg. *South African Journal of Wildlife Research* 12:124–29.

Roy, D. P., Y. Jin, P. E. Lewis, and C. O. Justice. 2005. Prototyping a global algorithm for systematic fire-affected area mapping using MODIS time series data. *Remote Sensing of Environment* 97:137–62.

Ruess, R. W., and F. L. Halter. 1990. The impact of large herbivores on the Seronera woodlands, Serengeti National Park, Tanzania. *African Journal of Ecology* 28:259–75.

Sankaran, M., N. P. Hanan, R. J. Scholes, J. Ratnam, D. J. Augustine, B. S. Cade, J. Gignoux, S. I. Higgins, X. Le Roux, and F. Ludwig, et al. 2005. Determinants of woody cover in African savannas. *Nature* 438:846–49.

Sankaran, M., J. Ratnam, and N. P. Hanan. 2004. Tree-grass coexistence in savannas revisited—insights from an examination of assumptions and mechanisms invoked in existing models. *Ecology Letters* 7:480–90.

———. 2008. Woody cover in African savannas: the role of resources, fire and herbivory. *Global Ecology and Biogeography* 17:236–45.

Scholes, R. J., and S. R. Archer. 1997. Tree-grass interactions in savannas. *Annual Review of Ecology and Systematics* 28:517–44.

Sharam, G., A. R. E. Sinclair, and R. Turkington. 2006. Establishment of broad-leaved thickets in Serengeti, Tanzania: The influence of fire, browsers, grass competition, and elephants. *Biotropica* 38:599–605.

———. 2009. Serengeti birds maintain forests by inhibiting seed predators. *Science* 325:51.

Sharam, G. J., A. R. E. Sinclair, R. Turkington, and A. L. Jacob. 2009. The savanna tree *Acacia polyacantha* facilitates the establishment of riparian forests in Serengeti National Park, Tanzania. *Journal of Tropical Ecology* 25:31–40.

Sinclair, A. R. E. 1977. *The African Buffalo*. Chicago: University of Chicago Press.

———. 1979a. Dynamics of the Serengeti ecosystem. In *Serengeti: dynamics of an ecosystem*, ed. A. R. E. Sinclair and M. Norton-Griffiths, 1–30. Chicago: University of Chicago Press.

———. 1979b. The Serengeti Environment. In *Serengeti: dynamics of an ecosystem*, ed. A. R. E. Sinclair and M. Norton-Griffiths, 31–45. Chicago: University of Chicago Press.

———. 2004. History of fire management in the Serengeti National Park. Proceedings workshop I: Fire management plan for the Serengeti National Park. Unpublished manuscript, Seronera, Tanzania.

Sinclair, A. R. E., S. A. R. Mduma, J. G. C. Hopcraft, J. M. Fryxell, R. Hilborn, and S. Thirgood. 2007. Long-term ecosystem dynamics in the Serengeti: Lessons for conservation. *Conservation Biology* 21:580–90.

Singh, R. S. 1993. Effect of winter fire on primary productivity and nutrient concentration of a dry tropical savanna. *Vegetatio* 106:63–71.

Spickett, A. M., I. G. Horak, A. Van Niekerk, and L. E. O. Braack. 1992. The effect of veld-burning on the seasonal abundance of free-living Ixodid ticks as determined by drag-sampling. *Onderstepoort Journal of Veterinary Research* 59:285–92.

Stafford, K. C., J. S. Ward, and L. A. Magnarelli. 1998. Impact of controlled burns on the abundance of *Ixodes scapularis* (Acari; Ixodidae). *Journal of Medical Entomology* 35: 510–13.

Stronach, N. 1988. The management of fire in Serengeti National Park: Objectives and prescriptions. Unpublished manuscript. Arusha: Tanzania National Parks.

———. 1989. Grass fires in Serengeti National Park, Tanzania: Characteristics, behaviour and some effects on young trees. PhD diss., University of Cambridge.

Stronach, N. R. H., and S. J. McNaughton. 1989. Grassland fire dynamics in the Serengeti ecosystem, and a potential method of retrospectively estimating fire energy. *Journal of Applied Ecology* 26:1025–33.

Strauch, A. M., and S. Eby. 2012. The influence of fire frequency on the abundance of *Maerua subcordata* in the Serengeti National Park, Tanzania. *Journal of Plant Ecology* 5: 400-406.

Thirgood, S. J., A. Mosser, S. Tham, J. G. C. Hopcraft, E. Mwangomo, T. Mlengeya, M.

Kilewo, J. Fryxell, A. R. E. Sinclair, and M. Borner. 2004. Can parks protect migratory ungulates? The case of the Serengeti-Mara wildebeest. *Animal Conservation* 7:113–20.

Tomor, B. M., and N. Owen-Smith. 2002. Comparative use of grass regrowth following burns by four ungulate species in the Nylsvley Nature Reserve, South Africa. *African Journal of Ecology* 40:201–4.

Trollope, W. S. W. 1998. *Effect and use of fire in the savanna areas of Southern Africa*. Unpublished report. Department of Livestock and Pasture Science, Faculty of Agriculture, University of Fort Hare, Alice, South Africa.

Trollope, W. S. W., L. A. Trollope, and C. de B. Austin. 2005. Unpublished report. Recommendations on Fire Management in the Serengeti National Park, Tanzania. Arusha: Tanzania National Parks.

Valentine, L. E., L. Schwarzkopf, C. N. Johnson, and A. C. Grice. 2007. Burning season influences the response of bird assemblages to fire in tropical savannas. *Biological Conservation* 137:90–101.

Van de Vijver, C. A. D. M., P. Poot, and H. H. T. Prins. 1999. Causes of increased nutrient concentrations in post-fire regrowth in an East African savanna. *Plant and Soil* 214: 173–85.

van Wilgen, B. W., and H. C. Biggs. 2010. A critical assessment of adaptive ecosystem management in a large savanna protected area in South Africa. *Biological Conservation* 144:1179–87.

van Wilgen, B., H. Biggs, and A. Potgieter. 1998. Fire management and research in the Kruger National Park, with suggestions on the detection of thresholds of potential concern. *Koedoe* 41:69–87.

Vermeire, L. T., R. B. Mitchell, S. D. Fuhlendorf, and R. L. Gillen. 2004. Patch burning effects on grazing distribution. *Journal of Range Management* 57:248–52.

Vesey-Fitzgerald, D. 1972. Fire and animal impact on vegetation in Tanzania National Parks. *Tall Timbers Fire Ecology Conference Proceedings* 11:297–317.

Vinton, M. A., D. C. Hartnett, E. J. Finck, and J. M. Briggs. 1993. Interactive effects of fire, bison (*Bison bison*) grazing and plant community composition in tall-grass prairie. *American Midland Naturalist* 129:10–18.

Walters, C. J., and R. Hilborn. 1978. Ecological optimization and adaptive management. *Annual Review of Ecology and Systematics* 9:157–88.

Wilhere, G. F. 2002. Adaptive management in habitat conservation plans. *Conservation Biology* 16:20–29.

Wilsey, B. J. 1996. Variation in use of green flushes following burns among African ungulate species: The importance of body size. *African Journal of Ecology* 34:32–38.

Woinarski, J. C. Z. 1990. Effects of fire on the bird communities of tropical woodlands and open forests in northern Australia. *Austral Ecology* 15:1–22.

Wooller, R. D., and K. S. Brooker. 1980. The effects of controlled burning on some birds of the understory in Karri forests. *Emu* 80:165–66.

Xie, P. and A. Arkin. 1997. A 17-year monthly analysis based on gauge observations, satellite estimates, and numerical model outputs. *Bulletin of the American Meteorological Society* 78:2539–58.

Zieger, U., I. G. Horak, and A. E. Cauldwell. 1998. Dynamics of free-living ixodid ticks on a game ranch in the central province, Zambia. *Onderstepoort Journal of Veterinary Research* 65:49–59.

FIVE

Spatial and Temporal Drivers of Plant Structure and Diversity in Serengeti Savannas

T. Michael Anderson, John Bukombe, and Kristine L. Metzger

The Serengeti is one of the earth's great storehouses of terrestrial mammalian biodiversity and home to one of the most impressive mammalian migration events. Because of these unique features, it is easy to see why plant diversity within the Serengeti ecosystem is often overlooked. However, much more is known about the structure and diversity of vegetation in the Serengeti than is often realized. Here, we aim to answer some basic questions with regard to the vegetation in the Serengeti and bring attention to the critical role of vegetation in the ecosystem. For example, what are the predominant patterns of plant diversity within the Serengeti and what are the main temporal and spatial drivers of that diversity? Moreover, how does the diversity of plants within Serengeti savannas compare with that of other grassland or savanna systems? Is it highly diverse, like that of the mammals and other tropical forested biomes, or is it comparatively species poor in terms of plant diversity?

The main objective of this chapter is to address these questions by summarizing results from the literature and presenting some new results of our own. First, we describe the patterns of herbaceous and woody plant species diversity across the savanna grassland habitat that typifies the ecosystem. Next, we look at spatial and temporal factors that regulate and act as the key drivers of diversity and variation across the system. We also discuss structural diversity across the system, and by this we mean the relative contribution of the two dominant plant functional types: herbaceous and woody plants. We conclude by discussing the future of plant diversity research in the ecosystem and highlighting key missing elements from the current vegetation monitoring program.

As suggested by our title, we focus on typical savanna/grassland because it is in this habitat where the vast majority of the data have been collected and accounts for the majority of the ecosystem spatially. However, with that said, we recognize that the bulk of the plant diversity within the system likely resides in the rare habitats, such as riverine gallery forests, natural springs, kopjes, and escarpments, which we ignore here. Clearly these rarer habitats are important to our understanding of plant diversity in the Serengeti and will hopefully be the focus of future study. Another important issue is that of scale, as patterns of plant diversity depend on aspects of scale, such as spatial grain, extent, and focus (Anderson, Metzger, and McNaughton 2007). Much of the data discussed within this chapter were collected within tenth hectare plots (20 × 50 m). Clearly, other plot sizes may be better suited for understanding particular processes or studying diversity in a particular habitat. However, this spatial grain has been shown to be both efficient and effective for the study of grassland and savanna plant diversity (Stohlgren, Falkner, and Schell 1995).

PATTERNS OF PLANT DIVERSITY IN SPACE AND TIME

We start with a simple question: Spatially, where does the greatest plant diversity reside in Serengeti savannas? We then ask if patterns of plant species richness are similar for grasses, the most abundant plant life-form in the Serengeti, and forbs, which contain more species and are composed of monocots (e.g., lilies), dicots, herbaceous, and woody plants. A logical follow-up question follows: Can we provide some simple predictors of plant diversity across the Serengeti ecosystem? Here we focus on broad factors such as climate, topographic heterogeneity, and fire frequency. Subsequently, we turn to the issue of compositional stability through time and space and what factors may influence the turnover of species through time. We conclude the chapter by comparing plant diversity in the savanna/grassland habitats of the Serengeti to other, similar grassland and savanna systems around the world.

Diversity Patterns in Space

To address our first question we reanalyzed data from two previously published studies (Anderson, Metzger, and McNaughton 2007; Anderson et al.

2007) that enumerated plant species richness in 134 tenth hectare plots across the Serengeti. A map of the results created with inverse distance weighted interpolation shows that the greatest plant species richness is located in the northwest hills between the Grumeti and Mara Rivers (fig. 5.1). The number of forb species is strongly correlated with total species richness ($r = 0.95$; $p < 0.001$), so when mapped, both forb and graminoid (grasses + sedges) species richness is essentially identical to that of total plant species richness (data not shown).

The drivers of plant species richness across the 134 plots were investigated with structural equation modeling (SEM), a technique that builds on path analysis but enables a more sophisticated evaluation of model fit (Grace 2006). Our a priori SE model included: (1) average annual rainfall, (2) wet season Normalized Difference Vegetation Index (NDVI), (3) fire frequency, (4) topographic heterogeneity, and (5) potential evapotranspiration (PET) as predictor variables and tested for both direct and indirect effects of predictors on plant species richness. The NDVI data is measured remotely by the Moderate Resolution Imaging Spectroradiometer (MODIS) and quantifies the concentration of green leaf vegetation on the earth's surface; it is an index of general site conditions that integrates water availability and soil fertility. Fire frequency is also derived from MODIS data (Dempewolf et al. 2007) and represents the main disturbance factor in the model, as we have no similar measure of grazing pressure. Rainfall represents the main resource input in the model and can influence plant richness directly or indirectly through its effects on fire frequency and NDVI. Topographic heterogeneity represents habitat variation in the form of soil and catena variation. Finally, PET represents the potential water evaporated from the system and is a surrogate measure for energy availability. More complete descriptions of the predictors can be found elsewhere (Anderson, Metzger, and McNaughton 2007; Anderson et al. 2007; Anderson 2008). The analysis was conducted in AMOS version 18.0 (Arbuckle 2008); model significance was evaluated with a chi-square test and the final model was selected by minimizing the Akaike's information criteria (AIC).

The final accepted model ($\chi^2 = 2.27$, df = 4, $P = 0.69$, AIC = 36.3; fig. 5.2) explained 29% of the variation in plant species richness across the plots (note that P-values > 0.05 suggest an acceptable model at $\alpha = 0.05$). The model suggests that PET had the largest direct standardized effect on plant species richness (–0.34) but that it also had an indirect positive effect, which was mediated by fire frequency. The next largest direct effect was topographic variation, which was positively associated with plant species richness (0.27), followed by NDVI, which has both a positive direct effect (0.19) and an indirect positive effect that had an influence via fire frequency. Fire frequency

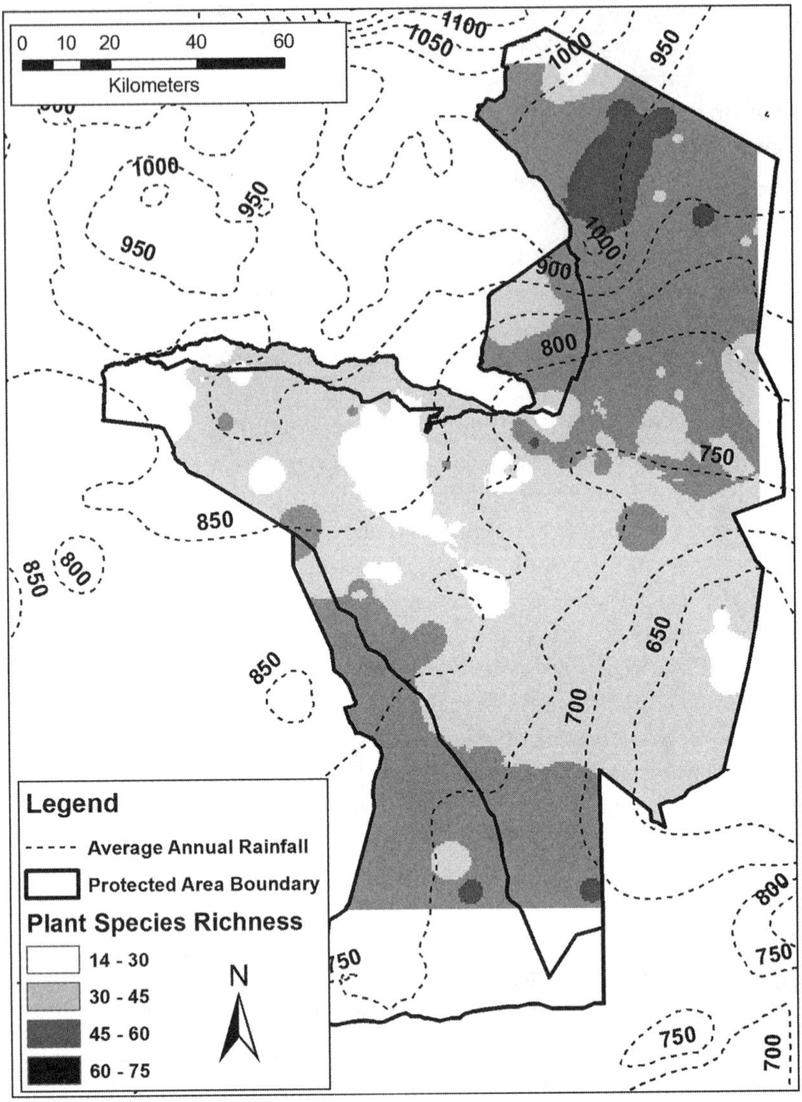

Fig. 5.1 Map of predicted plant species richness per 1,000 m² derived from the SEM results of 134 plots and interpolated across the Serengeti ecosystem based on landscape predictors (see fig. 5.2).

had a slightly weaker positive association with species richness (0.16). Rainfall had only indirect effects, which were relatively weak (0.10) in comparison to the other drivers in the model. In general, the model results are consistent with the hypothesis that energy (i.e., PET) and climate (i.e., rainfall and NDVI) drive patterns of plant species richness at regional scales (Field,

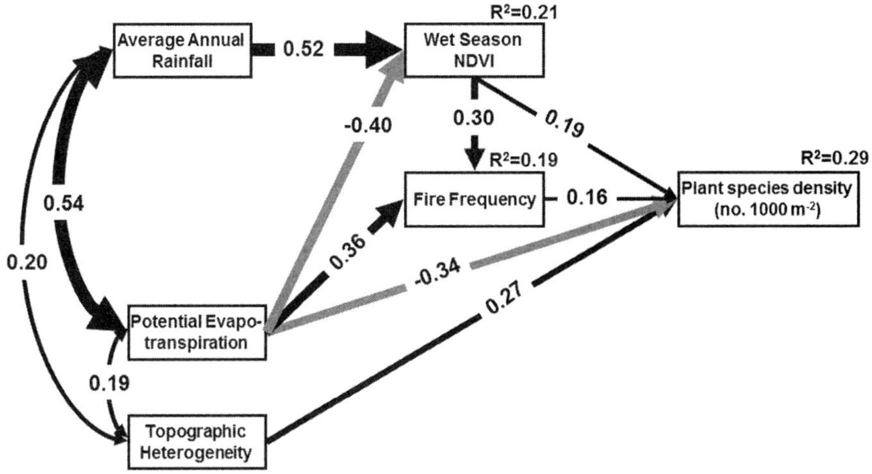

Fig. 5.2 Results of the SEM predicting plant species richness per 1,000 m^2 across Serengeti National Park from 134 plots (data from Anderson, Metzger, and McNaughton 2007; Anderson et al. 2007).

O'Brien, and Whittaker 2005) and that topographic relief provides heterogeneity that increases the availability of different habitat types for different species (Anderson, Dong, and McNaughton 2006; Anderson, Metzger, and McNaughton 2007). Moreover, the results provide direct evidence that fires are associated with increased plant species richness across the Serengeti landscape, even after controlling for correlations between fire and climate, most likely because disturbance reduces the dominance of superior competitors.

Notably, a significant proportion of the variation in plant species richness across plots (71%) was left unexplained. As a result, we wished to investigate the reliability of our analysis by comparing our plant species richness map, which was created using the results of the SEM (fig. 5.2), to a completely independent data set. We compared the predicted plant species richness from our map to the plant species richness observed in tenth hectare (20 × 50 m) plots sampled by Metzger (2002). There was a strong positive correlation between predicted and observed plant species richness values ranging from 23–66 (fig. 5.3; $r^2 = 0.83$, L95%CI = 0.79, U95%CI = 1.60, $p < 0.0001$). The plot of observed versus predicted suggests that the reliability of the map may decline as species richness increases (i.e., the points above the 1:1 line at higher species richness); this effect may be explained by greater temporal plant community dynamics in the species-rich mesic grasslands of the Serengeti (see below), which tend to have greater compositional turnover through time than drier grasslands in the Serengeti plains (Anderson 2008).

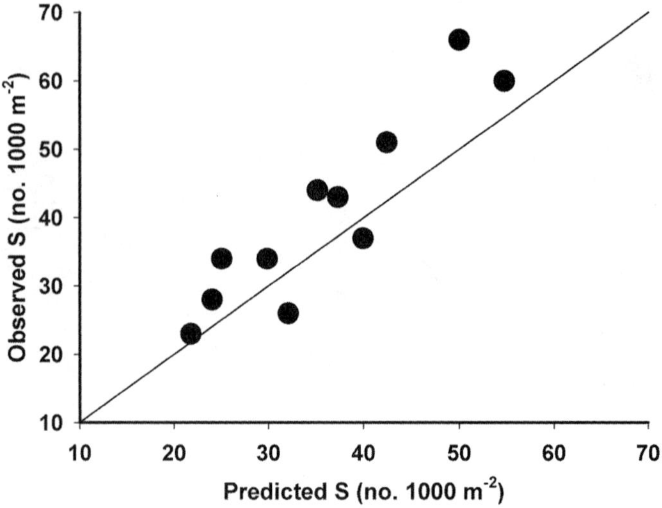

Fig. 5.3 Relationship between species richness (S) per 1,000 m^2 predicted from the SEM analysis and species richness observed in an independent data set collected by Metzger (2002).

Caveats to the Model Results

Data recently collected by our group (Anderson, Schütz, and Risch 2014) suggest that seed germination in the dung of large mammalian herbivores in the Serengeti is more prevalent than previously documented and likely affects plant species richness through both local and potentially long-distance dispersal events. Moreover, herbivores have direct and indirect affects on plant species richness through a variety of mechanisms linked to resources availability (water, nutrients, and light) and plant tolerance of defoliation (Anderson et al. 2007). Therefore, grazing intensity by mammalian herbivores would likely be an important explanatory predictor in the model. Unfortunately, estimating grazing intensity across a system as large and dynamic as the Serengeti is untenable at all but for very small (i.e., relatively few exclosure plots) and very large (i.e., regional movements of migrant herbivores) spatial scales.

Exclosure experiments suggest that local grazing (16 m^2 plots) increases plant species richness across the Serengeti rainfall gradient, but effects are more pronounced at rainfall between 600 and 800 mm yr^{-1} (Anderson, Ritchie, and McNaughton 2007). Additionally, compositional turnover between grazed and ungrazed plots is strongly related to rainfall because forbs persist inside and sedges persist outside exclosures at high rainfall (fig.

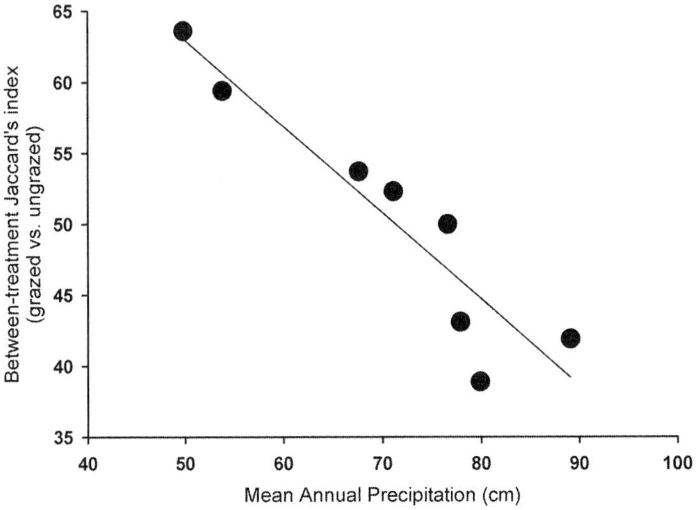

Fig. 5.4 Compositional difference measured by Jaccard's index between grazed and ungrazed plots across eight experimental sites as a function of rainfall. Low rainfall sites showed greater compositional similarity, and similarity declined as a linear function of rainfall (Anderson, Ritchie, and McNaughton 2007).

5.4; Anderson, Ritchie, and McNaughton 2007). The effects of grazing on plant diversity and composition is also seen by comparing regions of the Serengeti ecosystem where hunting occurs. These areas have reduced herbivores densities, compared to adjacent areas within the core protected area (Anderson et al. 2007). While correlative, results suggest that differences in disturbance at large scales can shift the processes that regulate plant composition and diversity from top-down driven by grazers to top-down driven by fire (Anderson et al. 2007; Anderson 2010).

McNaughton (1994) found that the beta diversity, or patch differences in species composition, is a prominent feature of the hierarchical plant diversity within Serengeti grasslands. He attributed this to soil, geology, and catena variation, thus supporting topographic heterogeneity as a determinant of plant diversity in our SE model and the idea that landscape variation is a key source of plant diversity in the Serengeti (Anderson, Metzger, and McNaughton 2007; Reed et al. 2009). In addition to landscapes, there are several other prominent processes in the Serengeti that likely generate and maintain heterogeneity among vegetation patches including termites, grazing, fire, trees, and so forth, although these influences have been reviewed elsewhere (Anderson et al. 2008).

Compositional Turnover through Time

Studies by Belsky (1985) conducted in the Serengeti plains suggested that, despite interannual differences in climate, plant species composition did not change over a ten-year period. These results imply that Serengeti grasslands are relatively resistant to interannual fluctuations in climate, and thus future climate change. Yet the Serengeti plains are only one component of this system and it is important to understand how grasslands across the entire system may respond to climatic fluctuations. Repeated sampling at eight grassland sites across the Serengeti rainfall gradient between 2000 and 2007 (Anderson 2008) demonstrated that the plains are indeed relatively stable over this time period. However, northern and western sites at higher rainfall (>700 mean annual precipitation) experienced greater plant compositional turnover through time. Plant compositional turnover was related to resource inputs in the form of dry season rainfall rather than disturbance due to fire. Moreover, the results suggested that plots in dry regions of the Serengeti may experience compositional turnover at small scales (e.g., within tenth hectare plots), but that the species pool in the plains is restricted to a relatively small number of species that can tolerate seasonal drought and extreme herbivory. As a result, within-plot turnover produces no significant change in richness or composition at larger spatial scales through time. This result is significant because it implies that the stability of the plains vegetation observed by Belsky (1986) may be threatened by increases in rainfall or the introduction of invasive species that are drought and grazing tolerant (see Summary and Future Directions).

Comparing the Diversity of Serengeti Savannas to Other Systems

How does plant diversity in Serengeti savannas compare to that in other African savannas and other herbivore-dominated savannas? To answer this question we first created a species accumulation curve for all 134 modified Whittaker plots sampled in the Serengeti (fig. 5.5) using the package "vegan" in R. The lack of a clear asymptote in the curve demonstrates that our sampling is approaching, but has not yet reached, a stable estimate of the total plant species number in the savanna habitat; as a result, our current predicted value may be an underestimate. Using an index derived by Chao et al. (2006) we estimated the total species pool, or the number of observed species plus the estimated number of unobserved species. The Chao estimate for the total species pool in savanna habitats is 533.0 ± 27.1,

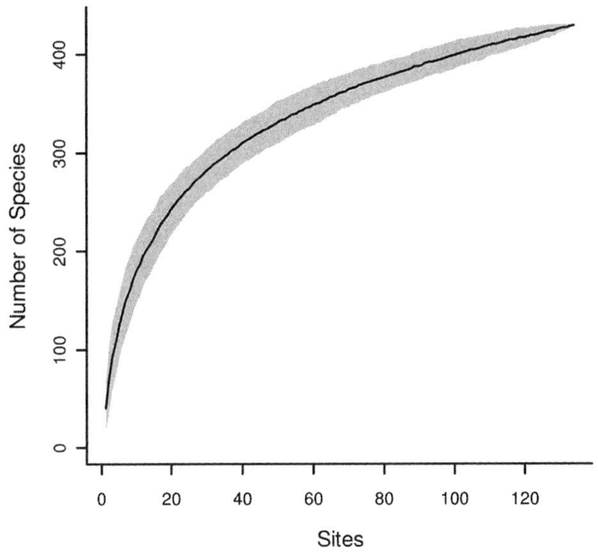

Fig. 5.5 Rarefaction analysis from 134 tenth hectare plots predicting total savanna plant species richness in the Serengeti ecosystem (see Anderson, Metzger, and McNaughton [2007] for more information). The lack of a significant asymptote shows that the sampling has approached but not yet arrived at an estimate of total plant species richness. Shaded area shows ± 1 standard error.

with the caveat that this estimate does not include unique features or rare habitats, such as riverine forests, where much of the plant diversity is likely to reside.

Comparisons of plant diversity among different ecosystems are replete with technical issues stemming from differences in collection techniques, sampling effort, and scale (Palmer, McGlinn, and Fridley 2008), so no rigorous comparison will be attempted here. However, it is interesting to note differences among the floral checklist from other savanna, grassland, or grazing ecosystems. Palmer (2007) found 671 native plant species in the prairie and savanna-like woodland habitats of the 154.1 km^2 Tallgrass Prairie Preserve in Oklahoma (a total of 763 plants species were identified, 12% of which were nonnative). In the Laikipia district of Kenya, Muasya, Young, and Okebiro (1994) documented 708 plant species on the 360 km^2 Ol Ari Nyiro ranch. Plant diversity in some South African savannas is impressively high: Kruger National Park in South Africa reports 1,990 plant taxa in the protected area (http://www.sanparks.org/parks/kruger/conservation/scientific/ff/biodiversity_statistics.php), almost 300 of which are graminoids alone (224 grasses and 74 sedges). Thus, it may come as a surprise that the plant diversity of savannas of the Serengeti, especially when compared to large mammalian herbivores and carnivores, is ordinary at best and perhaps even sparse relative to other comparable grassland ecosystems.

WOODY SPECIES DENSITY AND DIVERSITY

The density and diversity of woody plants is important for savanna ecosystems for two primary reasons. First, the proportion of woody versus herbaceous plant species is a fundamental property of savannas that influences ecosystem function, energy flow, hydrology, and nutrients cycling (Scholes and Archer 1997). Second, the tree/grass balance has important direct and indirect consequences for animal communities by determining relative forage availability, providing cover and shade, and relative predation risk for ungulate herbivores (Riginos and Grace 2008).

Landscape Patterns of Tree Composition and Diversity

Trees belonging to the family Fabaeceae comprise more than 84% of all trees under 600 mm yr^{-1} precipitation and 73% of all trees under 800 mm yr^{-1} precipitation (fig. 5.6; Metzger 2002). In above 800 mm yr^{-1} precipitation, Fabaeceae comprises less than half of the trees found in the plots (47%), with the remaining tree species more evenly distributed among four families: Combretaceae, Ebenaceae, Euphorbiaceae, and Burseraceae (fig. 5.6). Thus, there is a gradual change in the taxonomic dominance of trees across the rainfall gradient, with the greatest change observed at ~800 mm yr^{-1} annual rainfall (Metzger 2002). A logical hypothesis is that the availability of water is controlling, either directly or indirectly, the observed changes in tree diversity and composition across the Serengeti landscape. We will not explore further whether this is a direct or indirect effect, except to note that the SEM analysis revealed that rainfall was an indirect driver of total plant diversity (fig. 5.2) and that more research is required to know whether the same is true for tree species composition. Like total plant diversity, the tree species diversity in a typical patch of savanna in the Serengeti appears low in comparison to other systems. For example, the vast majority of individual trees encountered in our surveys (Metzger 2002; Anderson, Metzger, and McNaughton 2007) were members of the Fabaeceae family (fig. 5.6), belong to the genus *Acacia*, of which only approximately 20 species occur in the Serengeti.

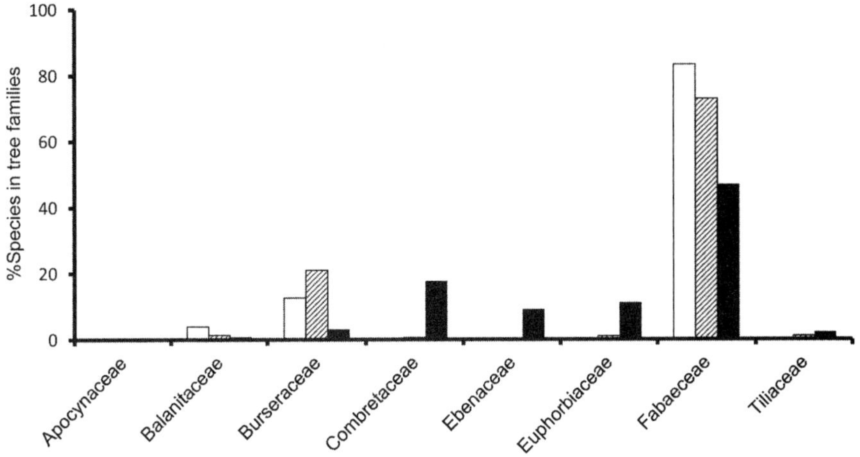

Fig. 5.6 The percentage of trees belonging to the eight major plant families by precipitation zone; data sums to 100% within each precipitation zone (data from Metzger [2002]). Empty bars <600 mm, dashed bars = 600–800 mm, and solid bars = >800 mm mean annual precipitation.

Landscape Heterogeneity in Tree Cover

The determinants of tree density and woody cover in savannas are complex and, in general, can be grouped into bottom-up factors, such as competition for water and nutrients, and top-down factors, such as fire and herbivory (Bond 2008; Anderson 2010). While no general model or consensus exists, continent-wide analysis of tree cover suggests that the upper boundary of maximum woody vegetation cover in savannas is set by rainfall, and is thus under climatic control (Sankaran et al. 2005). The data suggested that < 650 mm yr^{-1} rainfall maximum tree cover in savannas is constrained by moisture and cannot reach full closure, while > 650 mm yr^{-1} closed canopy savanna is possible. Underneath the maximum bound set by rainfall, variation in fire return interval, soil nutrients (especially P and N) and soil texture create variation in tree cover (Sankaran et al. 2005). In particular low fire return intervals, high soil N mineralization, and clay-rich soils are negatively associated with tree cover across African savannas, while soil P has a complex, nonlinear association (Sankaran, Ratnam, and Hanan 2008).

A ground-truthed, satellite-based vegetation map of the Serengeti was used to analyze the diversity of vegetation types (grassland or woodland patches) and the spatial distribution of vegetation patches across the landscape in relation to climate and topography (Reed et al. 2009). Climate was represented by average annual rainfall and the coefficient of variation

(CV) in annual rainfall using 40 years of rain-gauge data collected by the Serengeti Monitoring Programme. Topography was represented by the topographic moisture index (TMI), which is a common index used in watershed analysis integrating slope and upstream watershed area that drains to a particular point (Beven and Kirkby 1979). As an example, a point with a high TMI has a large upstream area (i.e., abundant water drainage), is relatively flat, or both. Diverse landscape patches, composed of similar parts grassland, woodland, and shrubland as measured by Simpson's diversity index, occurred at intermediate rainfall and diversity decreased as TMI increased. Woodland patches were larger at intermediate rainfall and decreased in size with TMI, while the distance among woodland patches was lowest at intermediate rainfall and increased in distance as TMI increased.

Consistent with Sankaran et al. (2005), the limit to woody cover in the Serengeti shows the same upper boundary set by rainfall, although the breakpoint is slightly higher at 675 mm yr^{-1} (Reed et al. 2009) compared to 650 mm yr^{-1} in the continent-wide data set. This is not surprising, since there is overlap between the two data sets. However, there is another important transition that should be noted that occurs between 600 and 700 mm yr^{-1} rainfall in the Serengeti: below this precipitation range, soil physical properties are associated with the shallow petrocalcic layer that inhibits tree establishment, limits maximum soil depth, and is largely impermeable to tree roots (de Wit 1978; Jager 1982). Anyone who has driven into the Serengeti through Naabi Hill Gate knows that tree cover can be quite high on the Serengeti plains at a rainfall below 650 mm yr^{-1} (fig. 5.7). It is unknown whether this effect is because of openings in the petrocalcic layer (e.g., soil depth), variation in soil physical properties associated with the geology of the exposed bedrock, or suppressed fire and herbivory. The lack of tree establishment in long-term exclosures (pers. obs.; Anderson, Ritchie, and McNaughton 2007) seems to suggest that tree recruitment is associated with soil properties or water limitation rather than herbivory. In addition, if soil depth was primarily responsible for the absence of adult trees that are abundant on Naabi Hill but lacking from the Serengeti plains, one would expect to see many stunted juvenile saplings and older "gullivers" (sensu Bond and van Wilgen 1996) in the plains; this, however, is not the case as *Acacia* seedlings in the plains are rare (pers. obs.; Anderson, Metzger, and McNaughton 2007).

Spatial Patterns of Tree Density and Basal Area

Woody basal area and tree density varies widely in response to dynamic drivers such as fire and herbivory across African savannas (Bond 2008). To

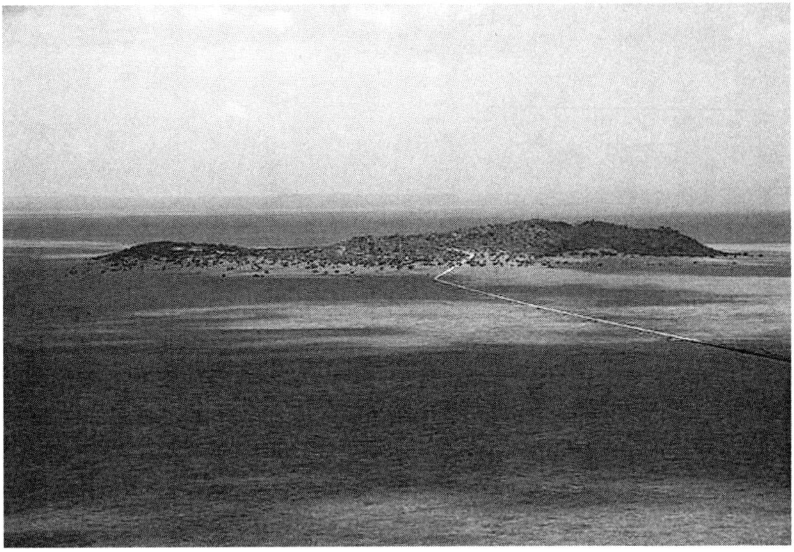

Fig. 5.7 Aerial photograph of Naabi Hill Gate in the southern Serengeti plains, clearly demonstrating that tree cover can be substantial in the plains despite highly seasonal rainfall. Photo credit: James Wolstencroft.

understand the causes of this variation and monitor future changes in the Serengeti, we established a series of 32 tenth hectare plots in 2009 in which the density, basal area, and height of all woody stems, from seedlings to adult trees, are measured annually. Across plots, we found no relationship between mean annual precipitation and either total basal area (fig. 5.8, *top left*) or the density of trees over >1 m height (fig. 5.8, *top right*). In contrast, the density of trees <1 m height increases with rainfall (fig. 5.8, *bottom left*), suggesting that water is limiting small trees at the dry end of the gradient and that constraint is somewhat released as rainfall increases. As a result of these patterns, the bivariate plot between the mean density of trees >1 m and those <1 m is also relatively weak (fig. 5.8, *bottom right*). This implies that there is relatively weak coupling between the processes that encourage germination, establishment, and the growth of small trees and the processes that maintain the density of adult trees across the Serengeti. More research is required to understand these patterns, but potential explanations lie in the size-specific effects of competition with grasses, fire, herbivory, and elephant damage (Bond 2008; Holdo, Holt, and Fryxell 2009).

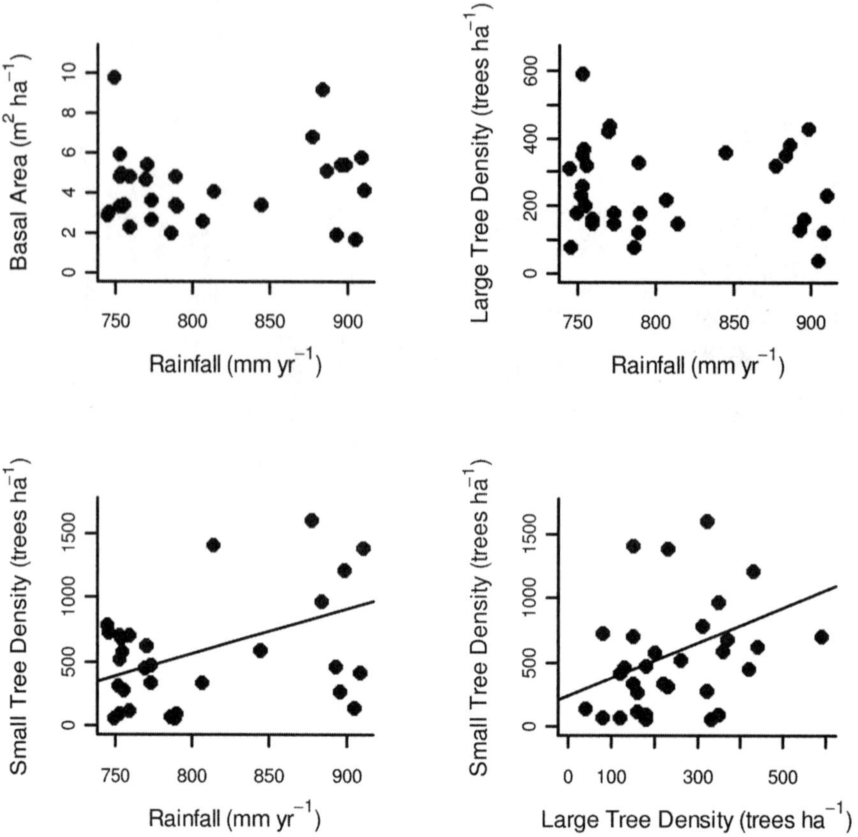

Fig. 5.8 Graphical data from 32 tenth hectare plots sampled across the Serengeti environmental gradient: (*top left*) total basal area of woody trees versus mean annual rainfall; (*top right*) density of large trees (>1 m height) versus mean annual rainfall; (*bottom left*) density of small trees (<1 m height) versus mean annual rainfall (small tree density = –2250.1 + 3.52*rainfall, $F = 6.44$, $P = 0.02$, $R^2 = 0.18$); *bottom right*, density of small trees versus large trees (Pearson correlation coefficient = 0.35, P = 0.05). Solid lines in the *bottom panels* show statistically significant least squares linear fit between predictor and response.

Temporal Dynamics in Tree Cover

Tree communities in the Serengeti are highly dynamic in space and time (Sinclair et al. 2007). Fire is identified as the main factor determining *Acacia* tree seedling pulses and cohort recruitment events (Sinclair et al. 2008; Holdo, Holt, and Fryxell 2009). However, the dominant grazers in the system, wildebeest, are responsible for determining the extent and frequency of fire and thus may be the ultimate drivers of tree recruitment events. For *Acacias*, the dominant trees of the savanna woodlands, there have been two seedling recruitment pulse events since records were first collected. The first cohort of dominant Acacia trees occurred around 1890 and the second cohort more recently in the mid-1970s.

Interestingly, both recruitment events have been linked to rinderpest, the viral epizootic disease that first appeared in east Africa in the late 1880s and decimated the cattle and wildebeest population in the early 1900s. This first tree cohort recruitment event is thought to coincide with the rinderpest's arrival, which drove people living within the ecosystem to abandon their agricultural lands and move elsewhere in search of food. The disappearance of people is believed to be associated with the loss of fire from many parts of the system, which then triggered the first tree regeneration event in the late 1880s. Drivers of the second tree regeneration event can be traced back to the eradication of rinderpest. This disease greatly reduced the abundance of grazers from the ecosystem. During the period of reduced wildebeest population abundance, plant biomass increased and resulted in larger fuel loads and thus large and frequent fires. The high frequency and large spatial extent of fires suppressed tree regeneration in areas that would have otherwise experienced natural tree regeneration. As the wildebeest population recovered, the standing dead biomass that served as fuel was reduced by grazing and so was fire. Tree regeneration resulted, and the *Acacia* tree cohort we see today is the result of the impact of disease on grazers and the trophic cascade that followed (Holdo et al. 2009).

In an attempt to map temporal changes in tree cover across the ecosystem, we used a photo series taken at fixed locations and known time points to reconstruct changes in relative cover across the Serengeti between 1920 and 2000. Relative changes in tree density were determined by returning to photo-point locations, rephotographing the site, and estimating tree densities by species (Sinclair, pers. comm.). Recent tree densities (e.g., in the year 2000) were measured via the point-center-quarter method (Metzger 2002). This data was then used to create a calibration from which tree densities could be estimated from the photo points. Finally, relative changes in tree density from three time points (1920, 1960, and 2000) were then extrapolated across the landscape by applying estimates to a habitat distribution map developed by Herlocker (1974). While not all the savanna tree species occurring in the Serengeti were included in the analysis, all the dominant species were measured, which included: *Acacia tortilis, A. robusta, A. senegal, A. drepanalobium, A. gerarrdhii, A. mellifora, A. polycantha, Combretum molle, and Terminalia mollis.*

Results of the tree reconstruction exercise largely confirmed the patterns suggested above. Between the 1920s and 1960s tree density decreased (fig. 5.9), most likely because of increased fuel load and fire intensities associated with low wildebeest numbers. From 1960 to the present trees have been in-

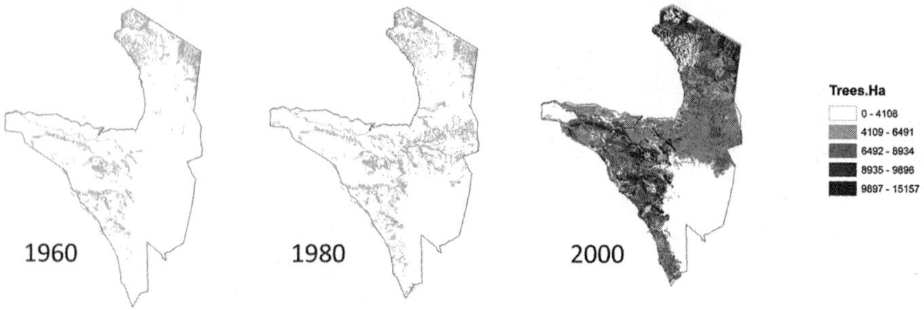

Fig. 5.9 Tree density change through time: (a) circa 1960 density ranged from 0 to 5,700 tree ha^{-1}, (b) circa 1980 density ranged from 0 to 5,700 tree ha^{-1}, and (c) recent tree densites range from 0 to 15,000 tree ha^{-1}.

creasing, but still may not have regenerated to the previous level at the turn of the twentieth century.

Riverine Forests

Fire, via a different trophic cascade than the one that involved rinderpest and wildebeest, has been identified as impacting tree regeneration in riverine forest trees. Riverine forests are a specialized forest type made of mixed broadleaf trees occurring along the Mara and the Grumeti Rivers. Riverine forests have been rapidly declining in density and extent (70–80%) since 1950 (Turkington et al., chapter 9). The cause of this decline is a complicated interaction between fire, birds, insects, and seedling regeneration (Sharam, Sinclair, and Turkington 2009). Forest-adapted frugivorous birds consume tree seeds and then regurgitate them; the act of seed regurgitation protects the seeds from beetle attack. Beetle-attacked seeds do not germinate and therefore birds, through a protective mechanism, play an integral role in tree seedling establishment. Fire creates canopy openings in these dense riverine forests, which reduces their suitability for forest-adapted birds. This chain of events has resulted in low seedling regeneration and, ultimately, riverine forest conversion to grassland (Sharam, Sinclair, and Turkington 2009).

SUMMARY AND FUTURE DIRECTIONS

Woody Cover Change in African Savannas

Given the prominence of savannas worldwide, which account for ~20% of the terrestrial biosphere, and their threatened status in many human-dominated ecosystems, those attempting to protect this endangered biome would benefit tremendously from a broad understanding of the controls on savanna vegetation dynamics. It is rather striking that, despite four decades of research on the topic, the direct and indirect effects of fire, soil moisture, grass competition, and herbivory on the germination, recruitment, growth, and survival of trees in savannas is still largely unknown. In the Serengeti, in which the study of vegetation has lagged sorely behind animal research, the best work on tree dynamics in the Serengeti comes from modeling (Holdo, Holt, and Fryxell 2009; Holdo et al. 2009) rather than empirical studies. This is astounding for an ecosystem as widely popular, well studied, and critical as the Serengeti. Other sites throughout Africa, especially protected areas in South Africa, have far better data and understanding on the dynamic controls over tree recruitment (Bond 2008). In the Serengeti, like in other savanna ecosystems, there is an increasing need to address this data gap, as impending climate change threatens to make a complicated and dynamic system even more complex. For example, changes in temperature, rainfall, and rainfall seasonality, all likely in future climate change scenarios, may disrupt the fire—grass fuel—herbivory interaction web, which currently drives dynamics in the system (Holdo, Holt, and Fryxell 2009; Holdo et al. 2009).

Recent research has begun to explore the possibility that increased atmospheric CO_2 is having a "fertilization effect" in savannas that is tipping the competitive balance between grasses and trees, ultimately favoring woody over herbaceous vegetation (reviewed in Bond 2008). The current thinking is that if tree seeds can germinate successfully, elevated CO_2 levels will increase growth rates so that seedlings have a higher probability of escaping fire and herbivory during their first few seasons, which are crucial for establishment. In addition, elevated CO_2 will enhance growth rates for trees resprouting after fire and herbivore damage, which will also favor woody vegetation. An alternative mechanism is that higher CO_2 concentrations will lead to reduced rates of plant transpiration, allowing water to percolate to greater depths in the soil profile. Deeper water percolation will favor trees over grasses as woody species tend to be more deeply rooted than herbaceous species. Either way, the predictions are that savannas will undergo significant changes in the near future, with the outcome being a much higher woody rather than herbaceous component.

Invasive Species

When one ponders the future of Serengeti National Park and its biota, one worrisome aspect is the lack of an invasive plant monitoring program within the park and surrounding game reserves. Although invasive species were largely absent from the 134 tenth hectare plots described previously, personal observation and discussions with Tanzania National Parks (TANAPA) ecologists suggest that invasive species are largely confined to areas disturbed by road maintenance and water diversion. Unfortunately, the outlook for the future of invasive plant species in the Serengeti is concerning: several species, including *Parthenium hysterophorus* and *Chromolaena odorata*, are encroaching southward from Kenya, which could prove devastating for savannas in the Serengeti should they become established (Witt 2011). Heavily disturbed areas in the nearby Ngorongoro Conservation Area are already invaded by noxious weeds, which spread along drainage lines and roads (Estes et al. 2006; Foxcroft et al. 2006). Future road development in the northern part of the Serengeti (e.g., Dobson et al. 2010) could serve as a major conduit for future plant invasion from Kenya. We finish our chapter with a strong recommendation to heed the advice of Foxcroft et al. (2006) and Witt (2011), who suggest that we pay close attention to the threat that invasive plant species pose for the Serengeti while there is still time to deal with the problem, lest we witness the decline in one of the great natural protected areas on earth because of human neglect and managerial oversight.

REFERENCES

Anderson, T. M. 2008. Plant compositional change over time increases with rainfall in Serengeti grasslands. *Oikos* 117:675–82.

———. 2010. Community ecology: Top-down turned upside down—How ants stabilize tree cover in African savannas. *Current Biology* 20:R854–55.

Anderson, T. M., J. Dempewolf, K. L. Metzger, D. N. Reed, and S. Serneels. 2008. Generation and maintenance of heterogeneity in the Serengeti ecosystem. In *Serengeti III: Human impacts on ecosystem dynamics*, ed. A. R. E. Sinclair, C. Packer, S. A. R. Mduma, and J. M. Fryxell, 135–82. Chicago: University of Chicago Press.

Anderson, T. M., Y. Dong, and S. J. McNaughton. 2006. Nutrient acquisition and physiological responses of dominant Serengeti grasses to variation in soil texture and grazing. *Journal of Ecology* 94:1164–75.

Anderson, T. M., K. L. Metzger, and S. J. McNaughton. 2007. Multi-scale analysis of plant species richness in Serengeti grasslands. *Journal of Biogeography* 34:313–23.

Anderson, T. M., M. E. Ritchie, E. Mayemba, S. Eby, J. B. Grace, and S. J. McNaughton. 2007. Forage nutritive quality in the Serengeti ecosystem: The roles of fire and herbivory. *American Naturalist* 170:343–57.

Anderson, T. M., M. E. Ritchie, and S. J. McNaughton. 2007. Rainfall and soils modify plant community response to grazing in Serengeti National Park. *Ecology* 88:1191–201.

Anderson, T. M., M. Schütz, and A. Risch. 2014. Endozoochorous seed dispersal and the evolution of germination strategies in Serengeti plants. *Journal of Vegetation Science* 25:636–47.

Arbuckle, J. L. 2008. Amos 18.0 update to the Amos user's guide. Chicago: Smallwaters Corporation.

Belsky, A. J. 1985. Long-term vegetation monitoring in the Serengeti National Park, Tanzania. *Journal of Applied Ecology* 22:449–60.

———. 1986. Population and community processes in a mosaic grassland in the Serengeti, Tanzania. *Journal of Ecology* 74:841–56.

Beven, K., and M. Kirkby. 1979. A physically based, variable contributing area model of basin hydrology. *Hydrological Sciences Bulletin* 24:43–69.

Bond, W. J. 2008. What limits trees in C4 grasslands and savannas? *Annual Review of Ecology, Evolution and Systematics* 39:641–59.

Bond, W. J., and B. van Wilgen. 1996. *Fire and plants*. London: Chapman & Hall.

Chao, A., R. L. Chazdon, R. K. Colwell, and T. J. Shen. 2006. Abundance-based similarity indices and their estimation when there are unseen species in samples. *Biometrics* 62: 361–71.

Dempewolf, J., S. Trigg, R. S. DeFries, and S. Eby. 2007. Burned-area mapping of the Serengeti–Mara region using MODIS reflectance data. *IEEE Geoscience and Remote Sensing Letters* 4:312–16.

de Wit, H. A. 1978. Soils and grassland types of the Serengeti plain (Tanzania): Their distribution and interrelations. PhD diss., Agricultural University, Wageningen, The Netherlands.

Dobson, A. P., M. Borner, A. R. E. Sinclair, P. J. Hudson, T. M. Anderson, G. Bigurube, T. B. B. Davenport, J. Deutsch, S. M. Durant, R. D. Estes, et al. 2010. Road will ruin Serengeti. *Nature* 467:272–73.

Estes, R. D., J. L. Atwood, and A. B. Estes. 2006. Downward trends in Ngorongoro Crater ungulate populations 1986–2005: conservation concerns and the need for ecological research. Biological Conservation 131:106—120.

Field, R., E. M. O'Brien, and R. J. Whittaker. 2005. Global models for predicting woody plant richness from climate: Development and evaluation. *Ecology* 86:2263–77.

Foxcroft, L. C., W. D. Lotter, V. A. Runyoro, and P. M. C. Mattay. 2006. A review of the importance of invasive alien plants in the Ngorongoro Conservation Area and Serengeti National Park. *African Journal of Ecology* 44:404–6.

Grace, J. B. 2006. *Structural equation modeling and natural systems*. Cambridge: Cambridge University Press.

Herlocker, D. 1974. *Map of the woody vegetation of the Serengeti National Park*. College Station: Texas A&M University.

Holdo, R. M., R. D. Holt, and J. M. Fryxell. 2009. Grazers, browsers, and fire influence the extent and spatial pattern of tree cover in the Serengeti. *Ecological Applications* 19: 95–109.

Holdo, R. M., A. R. E. Sinclair, K. L. Metzger, B. M. Bolker, A. P. Dobson, M. E. Ritchie, and R. D. Holt. 2009. A disease-mediated trophic cascade in the Serengeti: Implications for ecosystem C. *PLOS Biology* 7:e1000210. Doi: 10.1371/journal.pbio.1000210.

Jager, T. J. 1982. *Soils of the Serengeti woodlands, Tanzania*. Wageningen, Holland: Agricultural University.

McNaughton, S. J. 1994. Conservation goals and the configuration of biodiversity. In *Systematics and conservation evaluation*, ed. P. L. Forey, C. J. Humphries, and R. I. Vane-Wright, 41–62. Oxford: Clarendon Press.

Metzger, K. L. 2002. The Serengeti ecosystem: Species richness patterns, grazing, and land-use. PhD diss., Natural Resource Ecology Laboratory, Colorado State University.

Muasya, J. M., T. P. Young, and D. N. Okebiro. 1994. Vegetation map and plant checklist of Ol Ari Nyiro Ranch and the Mukutan Gorge, Laikipia, Kenya. *Journal of East African Natural History* 83:143–97.

Palmer, M. W. 2007. The vascular flora of the Tallgrass Prairie Preserve, Osage County, Oklahoma. *Castanea* 72:235–46.

Palmer, M. W., D. J. McGlinn, and J. F. Fridley. 2008. Artifacts and artifictions in biodiversity research. *Folia Geobotanica* 43:245–57.

Reed, D., T. M. Anderson, J. Dempewolf, K. Metzger, and S. Serneels. 2009. The spatial distribution of vegetation types in the Serengeti ecosystem: The influence of rainfall and topographic relief on vegetation patch characteristics. *Journal of Biogeography* 36: 770–82.

Riginos, C. and J. B. Grace. 2008. Savanna tree density, herbivores, and the herbaceous community: Bottom-up vs. top-down effects. *Ecology* 89:2228–38.

Sankaran, M., M. Sankara, N. P. Hanan, R. J. Scholes, J. Ratnam, D. J. Augustine, B. S. Cade, J. Gignoux, S. I. Higgins, X. Le Roux, et al. 2005. Determinants of woody cover in African savannas. *Nature* 438:846–69.

Sankaran, M., J. Ratnam, and N. Hanan. 2008. Woody cover in African savannas: The role of resources, fire and herbivory. *Global Ecology and Biogeography* 17:236–45.

Scholes, R. D., and S. R. Archer. 1997. Tree–grass interactions in savannas. *Annual Review of Ecology and Systematics* 28:517–44.

Sharam, G. J., A. R. E. Sinclair, and R. Turkington. 2009. Serengeti birds maintain forests by inhibiting seed predators. *Science* 325:51.

Sinclair, A. R. E., J. C. G. Hopcraft, H. Olff, S. A. R. Mduma, K. A. Galvin, and G. J. Sharam. 2008. Historical and future changes to the Serengeti ecosystem. In *Serengeti III: Human impacts on ecosystem dynamics*, ed. A. R. E. Sinclair, C. Packer, S. A. R. Mduma, and J. M. Fryxell, 7–46. Chicago: University of Chicago Press.

Sinclair, A. R. E., S. A. R. Mduma, J. G. C. Hopcraft, J. M. Fryxell, R. Hilborn, and S. Thirgood, S. 2007. Long-term ecosystem dynamics in the Serengeti: Lessons for conservation. *Conservation Biology* 21:580–90.

Stohlgren, T. J., M. B. Falkner, and L. D. Schell. 1995. A modified-Whittaker nested vegetation sampling method. *Vegetatio* 117:113–21.

Witt, A. 2011. Silent invader may threaten biggest wildlife migration on planet. *Swara* April-June:18–21.

SIX

Why Are Wildebeest the Most Abundant Herbivore in the Serengeti Ecosystem?

J. Grant C. Hopcraft, Ricardo M. Holdo, Ephraim Mwangomo, Simon A. R. Mduma, Simon J. Thirgood, Markus Borner, John M. Fryxell, Han Olff, and Anthony R. E. Sinclair

Wildebeest (*Connochaetes taurinus*) are the cornerstone of the Serengeti ecosystem and influence virtually every dynamic we observe. In this chapter we ask: Why is it that wildebeest (and not one of the other 29 species of large herbivore) dominate the ecosystem? The wildebeest population alone outnumbers all the other mammal populations in the Serengeti combined; it is six times larger than the next most abundant herbivore (zebra [*Equus burcheli*]) and 100 times more abundant than their closest taxonomic relative (hartebeest [*Alcelaphus bucelaphus*]). To answer the question of why wildebeest are so abundant in Serengeti we divide the chapter into three sections, each of which compares specific attributes of wildebeest with other sympatric herbivores. Each section builds on the results of the previous sections to form a complete story that combines biology, behavior, and the geomorphology of the Serengeti ecosystem; it explains how a single species can outnumber all other species, which is a question commonly asked in ecology.

We begin this chapter by comparing wildebeest anatomy, digestive physiology, reproduction, and demography with that of their closest taxonomic relatives, topi (*Damaliscus korrigum*) and hartebeest. In particular, we investigate what specific aspects of wildebeest biology might give them a competitive advantage over their sister clades. The second section of the chapter compares the migratory population of wildebeest with four resident subpopulations in the Serengeti ecosystem. This section specifically asks: If fundamental biological traits (other than migratory behavior, which appears to be plastic) confer on wildebeest a competitive advantage over other herbivores, then why are the resident subpopulations of wildebeest not as abundant as the migrant population? In the third section of the chapter we investigate the advantages conferred specifically by migration. Here we ask, If migration enables a population to escape local regulation and become superabundant, then why are migratory zebra not as abundant as wildebeest? This logical progression of questions pieces together many critical parts of the wildebeest story that has been uncovered by years of research and leads us to an answer that combines several components; it is precisely the combination of wildebeest biology, their adaptive migratory behavior, and the unique geomorphological features of the landscape that makes the Serengeti particularly well suited for wildebeest. In other words, the superabundance of wildebeest in the Serengeti is most likely the result of a tight feedback between the organism and its environment, where we find a highly adapted species living in a system that closely matches its requirements.

The observation that ecosystems are composed of a few common species and many rare ones has driven several aspects of ecological research (Hubbell 2001), and is not a unique characteristic of the Serengeti. Although we do not attempt to resolve the causes of this pattern, understanding exactly why certain species dominate an ecosystem enables us to pinpoint the reasons that could cause this pattern to change and this knowledge enables us to better manage wildlife populations. This chapter aims to unify 40 years of wildebeest research in the Serengeti and, by logical deduction, provide an answer as to why wildebeest in particular are so abundant in the Serengeti. We hope the answers may be applicable to similar questions in other ecosystems and will assist the managers with their task of protecting the Serengeti.

DOES WILDEBEEST BIOLOGY EXPLAIN THEIR ABUNDANCE OVER OTHER ALCELAPHINES?

Topi and Coke's hartebeest are the closest taxonomic relatives to wildebeest in the Serengeti. All three of these species belong to the tribe Alcelaphini, which is at least 10 million years old (Vrba 1979; Georgiadis 1995); wildebeest branched off from the topi and Coke's hartebeest lineage about four million years ago. All three species can exhibit migratory behavior, but wildebeest are the only alcelaphine to migrate in the Serengeti. In southern Sudan, topi (where they are called tiang) migrate in large numbers in the Boma-Jonglei system (Fryxell and Sinclair 1988b), and Coke's hartebeest previously migrated on the Athi-Kapiti plains of Kenya.

In this section we compare wildebeest diet and reproduction with that of topi and Coke's hartebeest to understand how wildebeest biology might provide them with a competitive advantage. Specifically, we explore how wildebeest, topi, and Coke's hartebeest differ in their ability to acquire resources (i.e., bottom-up regulation) and how they differ in predator avoidance (i.e., top-down regulation).

Digestion and Intake Rates

In theory, a species that maximizes the rate at which it consumes food and converts this energy into growth and reproduction should outperform less efficient competitors and become dominant. Ruminants use a complex fermentation process that harnesses microbial decomposition to break down cellulose into volatile fatty acids that are readily absorbed through the stomach wall. The digestive efficiency of the rumen varies between species and this imposes restrictions on the quality and abundance of food the animal can ingest. Wildebeest and topi ingest approximately equal quantities of grass per unit of metabolic body weight (65.7 g/kg $W^{0.75}$/ day and 67.0 g/kg $W^{0.75}$/ day, respectively), however the slightly larger Coke's hartebeest consumes less (56.5 g/kg $W^{0.75}$/ day) but has greater digestive efficiency (Murray 1993). Because of their similar dietary requirements we might expect wildebeest, topi, and Coke's hartebeest to compete for the same food resources.

The maximum rate at which each species can ingest forage depends on the architecture of the mouth and the structure of the grass sward. The wide dental pad and incisor row of wildebeest enables them to rapidly consume prostrate grasses that are approximately 3 cm high (Wilmshurst et al. 1999). These matt-forming grasses also tend to have the highest protein content (Fryxell 1995; Gordon, Illius, and Milne 1996; Wilmshurst et al. 1999). This differs from both topi and Coke's hartebeest, whose narrow muzzle and dental arcade is best for plucking individual leaves from stems of tall grass swards, which tend to be low quality (Murray and Brown 1993; Murray 1993; Murray and Illius 2000). Therefore, it is likely that competition for forage has lead wildebeest, topi, and Coke's hartebeest to divide the resource based on the architecture of the grass.

Diet and Water Requirements

Wildebeest seek nutrient-rich grass patches (Ben-Shahar and Coe 1992; Wilmshurst et al. 1999; Hopcraft 2010); however, there is no evidence to sug-

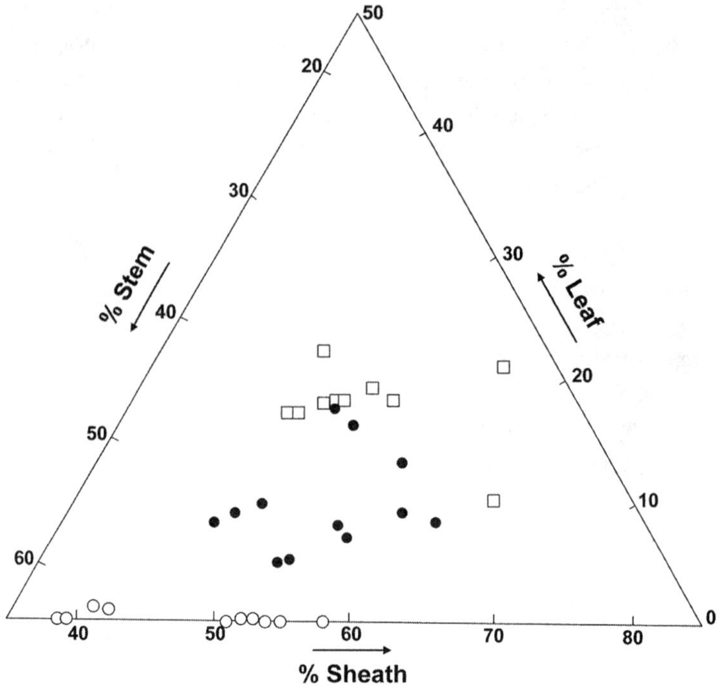

Fig. 6.1 Analysis of the stomach contents from wildebeest (open squares), topi (closed circles), and zebra (open circles) illustrates the selection for different components of the grass structure (data from Bell [1970]; Gwynne and Bell [1968]).

gest that they select specific species of grass that are more nutritious than those sought by topi or hartebeest. Approximately 90–100% of the diet of wildebeest, topi, and Coke's hartebeest is composed of C_4 grasses (the remainder consisting of C_3 forbs and legumes). However, the analysis of stomach contents indicates wildebeest eat primarily leaf material while topi or Coke's hartebeest consume a mixture of leaves and stems (fig. 6.1), and this might provide wildebeest with a competitive advantage over their closest taxonomic relatives (Bell 1970; Gwynne and Bell 1968; Casebeer and Koss 1970; Cerling et al. 1997; Codron et al. 2007; Tieszen and Imbamba 1980).

There is no evidence to suggest that wildebeest are more resilient to drought or water stress than topi or Coke's hartebeest. The population dynamics of all three species are closely linked with rainfall and the availability of drinking water (Ogutu et al. 2008; Mduma, Sinclair, and Hilborn 1999; Gereta, Mwangomo, and Wolanski 2009). A resource selection analysis shows that both topi and Coke's hartebeest select areas in proximity to water; however, topi select areas with more grass biomass than those

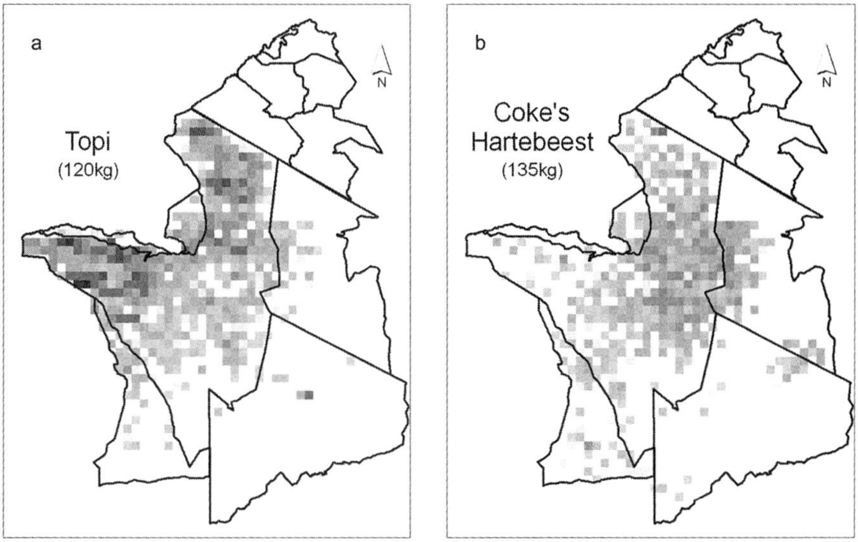

Fig. 6.2 The distribution of (a) topi and (b) Coke's hartebeest is segregated roughly by rainfall. Topi tend to select areas with higher grass biomass in the wetter areas of the Western Corridor and northern woodlands, while hartebeest are more common in the drier central woodlands (Hopcraft et al. 2012). Resident herds of wildebeest occur sympatrically with topi and Coke's hartebeest (compare to fig. 6.8). (Note that the Mara and Kenyan group ranches were not included in these surveys.)

selected by hartebeest (Hopcraft et al. 2012). Topi are more abundant in the high-rainfall areas of the Western Corridor and the northern woodlands, while the hartebeest are distributed further to the east in the dry central and eastern woodlands (fig. 6.2). Wildebeest essentially migrate between all these habitats. Therefore, we cannot conclude that access to water enables wildebeest to become superabundant while limiting the topi and Coke's hartebeest populations.

Reproduction

Calving synchrony. Perhaps the most striking feature about wildebeest reproduction in the Serengeti is the calving synchrony of the migrant population. Over 250,000 calves are born within a three-week period starting in mid-February, which accounts for approximately 80% of the year's calves. This equates to 500 calves being born every hour, or 12,000 calves born a day. The Serengeti has approximately 2,500 lions, of which only about 300 might have access to the calving grounds, and about 7,000 hyena that commute across the ecosystem (Hanby, Bygott, and Packer 1995; Hofer and East

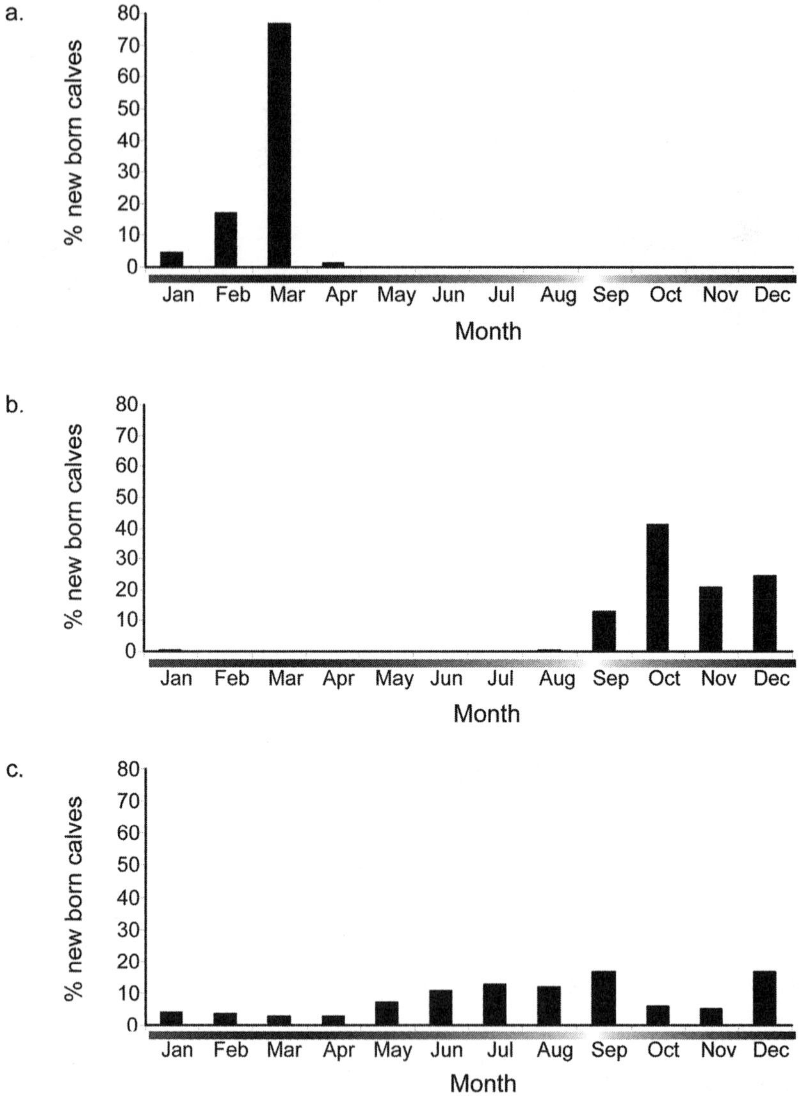

Fig. 6.3 The number of calves per month (displayed as a percentage of the entire year) illustrates the strong calving synchrony in (a) wildebeest. Topi (b) are less synchronous and there is no discernable calving peak for (c) Coke's hartebeest, which tend to be asynchronous (data from Sinclair, Mduma, and Arcese [2000]). Dark shading along the x-axis represents wet season months and light shading represents dry season months.

1995). The huge synchronous pulse of wildebeest calves in the wet season completely outstrips the predators' ability to limit wildebeest recruitment.

The calving period for topi and hartebeest is much less synchronized than that of the wildebeest (fig. 6.3). Approximately 70% of topi calves are born over a three-month period at the very end of the dry season (Sinclair, Mduma, and Arcese 2000). Peak calving for topi generally occurs in October and November, but this varies between years. For instance, topi calving can be delayed or advanced by up to three months as a result of rainfall conditions during the preceding season (Ogutu et al. 2010). Hartebeest are even less synchronous and calve throughout the year, although there is a vague peak in August and September during the height of the dry season (Sinclair, Mduma, and Arcese 2000; Ogutu et al. 2008). Both topi and Coke's hartebeest cannot swamp predators with a huge synchronous pulse of calves; greater rates of calf predation might limit their recruitment and might partially explain why these populations are not as abundant as wildebeest.

If synchronous calving enables wildebeest to swamp predators and boost their recruitment rate beyond what topi and Coke's hartebeest are capable of, how do female wildebeest synchronize conception so precisely? Although the exact physiological mechanisms are not entirely known, we summarize the most likely explanation together with what is known about wildebeest reproductive biology, as this paves the way for future research ideas.

Female wildebeest undergo multiple estrus cycles, each lasting approximately 23 days and repeated for up to 203 days, unless mating occurs. Following this extended period of estrus cycling (termed diestrus), non-pregnant females shed the lining of the uterus during a period of ovarian quiescence before starting a new diestrus cycle (Clay et al. 2010). The first estrus starts about 102 days postpartum and the first ovulation is probably silent (i.e., cows ovulate normally but do not display overt behavioral signs) (Watson 1967; Clay et al. 2010). The beginning of the diestrus cycle is probably triggered by declining rainfall and reduced concentration of crude protein in the diet (Sinclair 1977b; Clay et al. 2010). It is hypothesized that the combination of this population-wide cue of declining crude protein (i.e., the ultimate cause) and the diestrus cycle (i.e., the proximate cause) best explain synchronous calving in wildebeest (illustrated in fig. 6.4). Specifically, if the first ovulation in the diestrus cycle is initiated by a drop in the crude protein content of forage and is silent, this could cue the bulls without engaging them in copulation, while simultaneously inducing other females to begin ovulating (conspecific cueing is common in several synchronously breeding species [Ims 1990]). Males cued by early estrus females become more territorial and start to separate small harems from the main herd. This

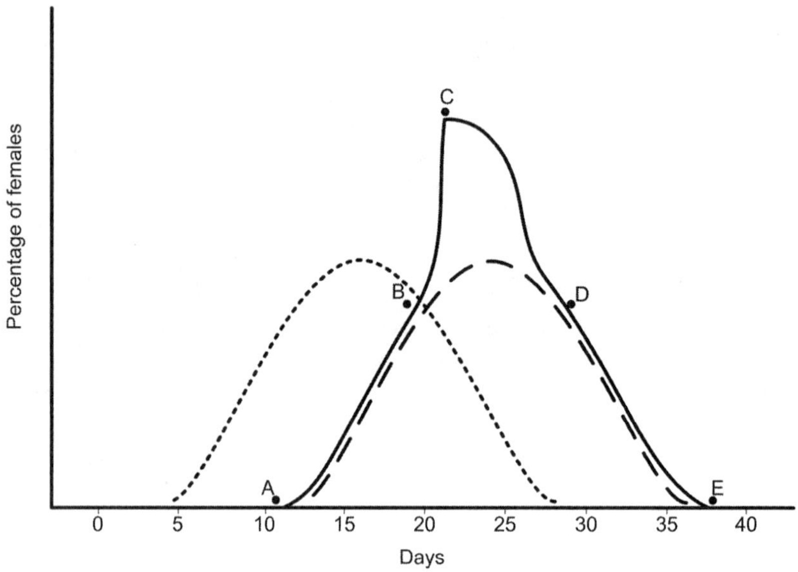

Fig. 6.4 The diestrus cycle of female wildebeest could lead to intense male competition, which might account for their high degree of breeding synchrony. The first ovulation in the diestrus cycle is silent (dotted line); females do not display overt behavioral signs and do not permit the males to mate. The first ovulation cues the bulls and increases male-male competition, as well as inducing other females to start ovulating. During the second ovulation (dashed line) females display overt mating behaviors and permit bulls to mate. The fertilization rate (solid line) increases synchronously with the second ovulation (between points AB), during which time the competition between bulls increases in intensity. Between points BC the intermale competition becomes so intense that bulls no longer wait for overt behavioral signals from females. During this time, females in both their first and second ovulations are harassed until they are mated, resulting in a peak fertilization rate at point C. From points CD the fertilization rate declines as the number of unmated females decreases. By the end of the rut (DE) only the remaining second ovulating females are fertilized (as suggested by Watson [1967]).

social disruption could also lead to higher endocrine function in both males and females (Mysterud, Coulson, and Stenseth 2002). As the crude protein declines further, the number of females in the first silent ovulation increases and the intensity of the rut builds in a positive feedback loop, with males becoming more and more excited. Only during the second ovulation do females become behaviorally receptive to copulation and allow the males to mount (i.e., a nonsilent ovulation). Bulls mate rapidly with any and all receptive females and the rut quickly approaches its climax. During the peak of the rut, the competition between bulls becomes so intense that they no longer wait for overt behavioral cues from females and will also continuously attempt to copulate with females in their first silent ovulation.

The rut demands a large amount of energy from bulls and selects for the fittest males. Males commonly lose up to 80% of their kidney fat and

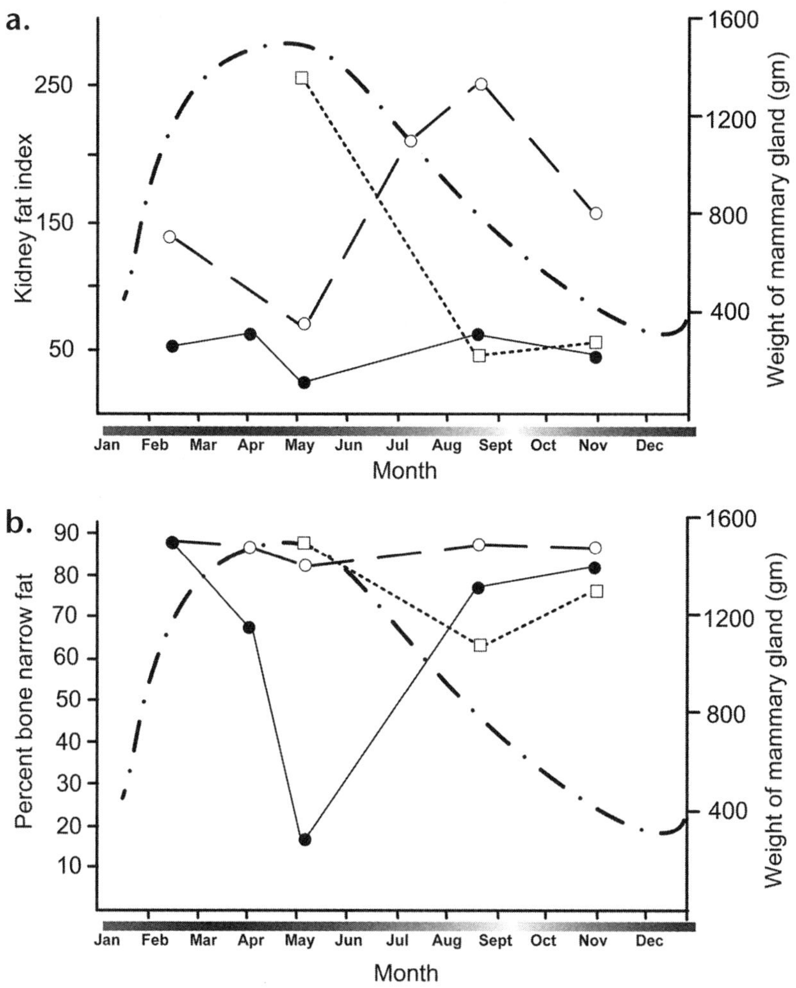

Fig. 6.5 The seasonal fluctuations of (a) kidney fat and (b) bone marrow fat for lactating females (closed circles with solid line), nonlactating females (open circles with dashed line), and male wildebeest (open squares with dotted line). Lactating females consistently have lower amounts of kidney fat than nonlactating females and male wildebeest. Declines in male kidney fat are associated with the large energy demands of the rut in June. The loss of bone marrow fat, which is the body's final reserve, occurs in females during peak lactation (note the weight of the mammary glands [dot-dash line]) and illustrates the extreme nutritional demands of reproduction, especially in April and May (data from Sinclair [1977a]; Watson [1967]). Dark shading along the x-axis represents wet season months and light shading represents dry season months.

as much as 26% of their bone marrow fat during this period (fig. 6.5; Sinclair 1977a). Furthermore, bulls can only sustain this level of activity for short periods of time. Exhausted bulls fall out of the competition and are quickly replaced by fresh bulls, so that females are constantly harassed until they breed, even if they are in their first silent ovulation. The result is that during the peak of the rut 80% of reproductively active females (approximately 200,000 females) are mated within a two- to three-week period (postmortem data from [Watson 1967]). This mass breeding event leads to synchronous calving eight months later (fig. 6.6), which then swamps the predator base (Estes, Raghunathan, and Van Vleck 2008), and may partially explain the superabundance of wildebeest in the Serengeti. Note, this strategy only works when there are sufficient numbers to swamp the predators (as in wildebeest or large herds of topi); if numbers are low then synchrony of births would work in reverse and attract predators and thereby increase the predation rates on newborns. Hence, synchronous calving is probably not an evolutionarily advantageous strategy for resident ungulates living in low-density conditions.

An alternative hypothesis is that the tight timing of the rut and calving periods may simply be an adaptive consequence of the migration itself, rather than being an adaptive response to reduce the predation on calves. A defining feature of wildebeest behavioral ecology in the Serengeti is the capacity for redistribution across the landscape. Unlike resident ungulate species, wildebeest track areas of high resource availability throughout the annual cycle, and have enormous "freedom of movement" (Boone, Thirgood, and Hopcraft 2006). It is possible, given the extent of their movement, that reproductive synchronicity may simply be a by-product of continually moving large distances; February might be the only period during the year when individual females have sufficient time and resources to give birth and, therefore, selection has favored all females to deliver at the same time.

Breeding and the rut. The wildebeest rut occurs as the herds are moving off the plains and into the western corridor (fig. 6.6). Males defend a stationary territory, sometimes very briefly and often many times per day, and this is repeated hundreds of times as the migration moves (Estes 1966). Males continually attempt to recruit females into their territory. Presumably most mating occurs at night around the first full moon in late May or early June (Sinclair 1977b) because it is not commonly observed.

In comparison, the topi rut typically occurs between March and May and lasts about 1.5 months. Males form leks from which they display and compete for female copulation, although some satellite males opt to defend territories and coerce females into copulation through harassment (Bro-Jørgensen 2003; Bro-Jørgensen and Durant 2003). By contrast, female Coke's hartebeest are only receptive for a single day and will mate several times within a few minutes during this period (Estes 1992).

The main difference in the breeding behavior between these three species is that wildebeest and topi reproduction is associated with strong male competition over several weeks, while Coke's hartebeest is not. The com-

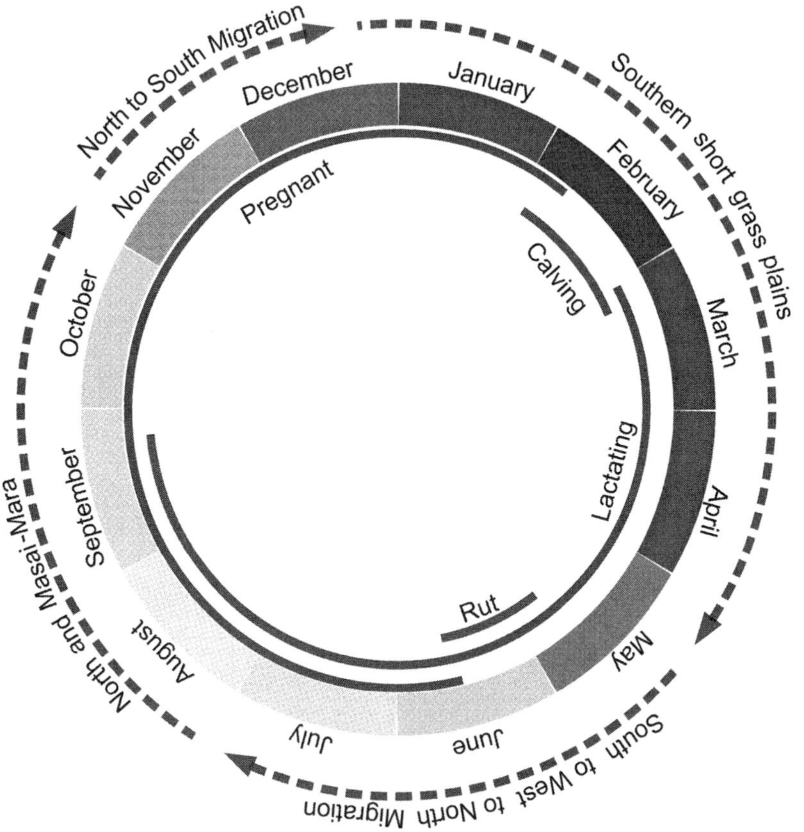

Fig. 6.6 The reproductive cycle of Serengeti wildebeest in relation to the seasonal annual migration (adapted from Watson [1967]; Clay et al. [2010]). Female wildebeest are either pregnant or lactating (or both) all year round. Dark shading represents wet season months and light shading represents dry season months.

petitive mating displays by sexually mature bulls associated with territorial and lekking behavior leads to breeding synchrony and to potential predator swamping when the calves are born. Although the behaviors associated with reproduction might facilitate calf survival, it is unlikely this accounts for the difference in the size of the population of these three species.

Gestation and calving. Wildebeest gestation lasts eight months (240 days), which is approximately the same in all alcelaphines (Estes 1992; Clay et al. 2010). Wildebeest, topi, and Coke's hartebeest only produce a single calf per season (Ogutu et al. 2010; Watson 1967; Clay et al. 2010). There are no reliable reports of twinning, although this is sometimes confused with small crèches or the strong follower instincts of lost orphans. There is no

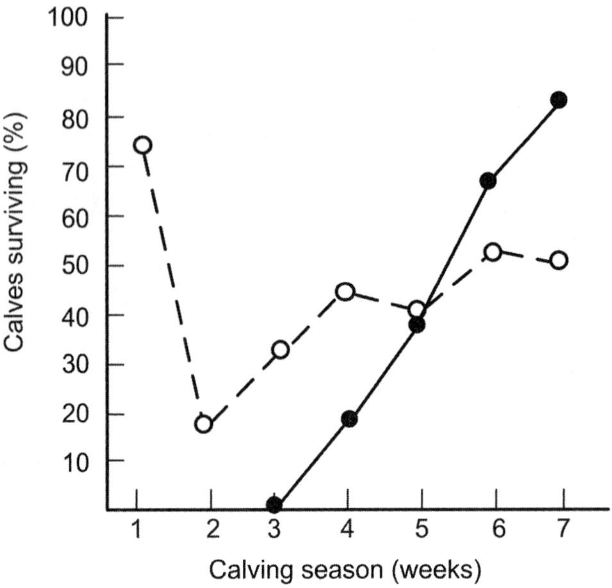

Fig. 6.7 Wildebeest calves have the highest survival when born into large herds (solid points) later in the calving season. Calves born into small herds of less than 10 females (open points) have a lower survival because predators can easily single out and attack solitary calves. This suggests that predator swamping is an effective strategy that decreases a wildebeest calf's risk of mortality (data from Estes [1976]).

evidence that orphaned calves are adopted by other cows, even if a female has lost her own calf (Estes and Estes 1979; Estes 1966). Therefore, neither gestation nor the number of offspring can explain why wildebeest are more abundant than topi or Coke's hartebeest (i.e., the intrinsic reproductive capacity is the same in all species).

Female wildebeest rarely calve in isolation but instead prefer to form crèches, which are joined by other expecting females. Calves born in crèches are defended against small predators by cooperative females (Estes 1992). Studies from the Ngorongoro Crater suggest that calf survival declines drastically from 84% in crèche herds with more than 10 females to 50% in herds with fewer than 10 females, suggesting communal calving is an effective antipredation strategy for wildebeest (fig. 6.7; Estes 1976; Sinclair, Mduma, and Arcese 2000). Unlike wildebeest, topi and hartebeest do not calve communally. Instead, they give birth in isolation and discretely visit their hidden calves periodically for many days before rejoining the herd.

Wildebeest calves express two important behaviors almost immediately postpartum that differ from topi and Coke's hartebeest, both of which are critically important for survival in large herds: imprinting and a "follower" instinct (Estes and Estes 1979; Estes 1992). Mothers and calves imprint on each other within minutes and are capable of individually recognizing each other's calls (Estes and Estes 1979; Estes 1966). Their mutual recognition is confirmed by smell. The "follower" instinct in young wildebeest is so great that calves will often be no more than a meter from their mothers in the herd. As a result, predators have difficulty isolating and killing individual wildebeest calves. Calves that have been separated from their mothers will sometimes follow vehicles or even predators without discerning danger because of their overpowering follower instinct.

Topi and hartebeest calves are less precocial than wildebeest calves and take up to 30 minutes or more before they stand, and as long as 45 minutes before they can follow their mothers for short distances. By comparison, wildebeest calves are able to stand within an average of 6 minutes and walk within 30 minutes, which is an adaptive trait for such a mobile species (Estes 1992; Sinclair, Mduma, and Arcese 2000). Wildebeest calves are capable of outrunning hyena within a day, whereas hartebeest calves are unable to keep up with their mothers until they are more than a week old (Kruuk 1972; Estes and Estes 1979). The marked difference in the precocial nature and follower behavior of wildebeest calves enables them to evade predators much more effectively than either topi or Coke's hartebeest. This release from top-down regulation provides further clues as to why wildebeest are the most abundant of the three species.

Lactation. The energetic demands associated with lactation could potentially compromise female survival, and if large differences existed in the lactation periods between wildebeest, topi, and Coke's hartebeest, this could explain the differences in the size of the populations. The peak lactation for wildebeest is between February and May, which coincides with the greatest amount of rainfall and fresh grazing on the short-grass plains (fig. 6.6; Kreulen 1975). Although females continue to lactate until September (the core of the dry season; Watson [1967]), they lose up to 60% of their bone marrow fat in the first three months of lactation and they consistently have lower amounts of kidney fat than nonlactating females or males (fig. 6.5; Sinclair 1977a). For four months (June to October), female wildebeest are both pregnant and lactating, as illustrated in figure 6.6. This extended demand means all the females must constantly have access to water and adequate grazing. Lactating wildebeest require 30% more energy per day, five times more calcium, three times more phosphorous, and twice as much sodium than females in early pregnancy (table 6.1; Murray 1995). Females are especially vulnerable to predation and starvation during the last stages of pregnancy and into peak lactation (Sinclair and Arcese 1995). Because topi and Coke's hartebeest reproduce less synchronously than wildebeest,

Table 6.1 The dietary requirements of adult female wildebeest during early pregnancy as opposed to peak lactation

	MJ/day (kg equivalent of *T. triandra*)	Ca g/day	P g/day	Na g/day
Early pregnancy	22.3 (~3.10 kg)	3.59	5.76	1.07
Peak lactation	32.7 (~4.54 kg)	15.51	17.61	2.35

Source: Murray 1995.

their nutritional demands associated with lactation are spread throughout the year. Instead, individual topi and hartebeest can find sufficient resources locally rather than being forced to migrate long distances. For wildebeest, the cost of calving synchronously is the sudden and urgent demand for the same food and water resources by all the lactating females (approximately 20–30% of the population). Wildebeest evade this nutritional bottleneck by migrating and exploiting high-quality ephemeral resources in the Serengeti plains during the wet season. In summary, the timing of lactation between these three species cannot explain the difference in their abundances because females can avert being regulated by nutrition by either (a) calving asynchronously and reducing intraspecific competition, or by (b) calving synchronously and reducing intraspecific competition by migrating between high-quality patches.

Sexual maturity. The age of sexual maturity does not differ significantly between wildebeest, topi, and Coke's hartebeest (Estes 1992). Males and females reach sexual maturity after two and three years, respectively. However, most males do not become sexually active until their fifth year because they are excluded from breeding by intense male competition (Estes 1966, 1992). Therefore, difference in the age of sexual maturity cannot account for the differences in abundance between wildebeest, topi, and Coke's hartebeest.

Summary: Wildebeest Reproductive and Digestive Physiology Cannot Entirely Account for Their Dominance in the Serengeti

There are several aspects of wildebeest biology that might explain why they are more abundant than topi or Coke's hartebeest in the Serengeti. First, their flat and wide dental arcade enables wildebeest to rapidly select leaves from mat-forming grasses while avoiding the lower-quality stems and sheaths, whereas topi and hartebeest are better suited to pick individual leaves from vertically complex grasses. The prostrate grasses selected by wildebeest are typically in a young and nutritious growth stage, which means they are easier to digest and provide more energy than tall grasses (Wilmshurst et al. 1999). Therefore, wildebeest are more efficient at processing high-quality grass and may be less regulated by the availability of food than topi or Coke's hartebeest, and this might partially explain why wildebeest are more abundant (i.e., bottom-up regulation).

Second, the breeding and calving synchrony of wildebeest coincides with the wet season when grass supply on the short-grass plains is the most nutritious (Kreulen 1975; McNaughton 1985; fig. 6.6). In addition, the strong synchrony of wildebeest calving swamps the predator guild and reduces the overall predation rates on neonates (Sinclair, Mduma, and Arcese 2000; Ims 1990). The peak calving in topi varies interannually, but is somewhat synchronized with the early green flush in long-grass areas. Historically, very large concentrations of topi (up to 14,000 animals in a single herd; A. Sinclair, pers. obs.) breeding somewhat synchronously would have had a predator swamping effect; however, there is no evidence for this in smaller herds. Coke's hartebeest also occur in small herds and are completely asynchronous. Hartebeest gain no advantage from the seasonal abundance of high-quality food (Sinclair, Mduma, and Arcese 2000; Ogutu et al. 2008), nor do they gain any of the predator swamping advantages of synchronous breeders (Ogutu et al. 2010). Therefore, the advantage of predator swamping is only realized by synchronous breeders living in large herds, such as wildebeest, which reduces the overall effects of top-down predator regulation.

Wildebeest calves are more precocial than either topi or hartebeest calves and have a strong follower instinct, which provides them with the additional protection of large herds. Wildebeest calves are exposed to predators for a brief period of time postpartum, after which they accompany their mother and can escape predators by losing themselves in large herds. By comparison, the "hider" calves of topi and Coke's hartebeest do not gain the protection afforded by a large herd (Ims 1990), so their populations are more regulated by predation (i.e., top-down affects) than wildebeest.

It is possible that the combined interaction of top-down and bottom-up factors regulating populations of wildebeest, topi, and Coke's hartebeest have very different effects on each of these populations. A "trait-mediated" effect, whereby predators do not directly affect the abundance of prey population numerically, but rather by modulating their use of resources, could have disproportionately large effects on small populations. For instance, topi may form large local herds to evade predators even if the density of predators is low, but as a result, individual topi may be strongly regulated by the availability of local resources. Individual wildebeest might not be subject to this constraint because the very large herds (i.e., often in excess of 50,000 animals) dilute an individual's risk of being predated and allows them to exploit high-quality grazing across the landscape without the fear of predation.

Our analysis so far focuses on the differences in digestive and reproductive physiology. However, if this alone enabled wildebeest to out-compete all other sympatric species, then we should expect the resident subpopulations of wildebeest to be as abundant as the migrants, which is not the case. Therefore, while wildebeest biology might partially account for their dominance in the ecosystem, it does not explain it entirely.

WHY ARE THERE SO FEW RESIDENT WILDEBEEST IN SERENGETI?

In this section we compare the attributes of the main migratory wildebeest population with four other resident subpopulations of wildebeest in the Serengeti ecosystem. Specifically, we ask what accounts for the differences in abundance between the resident and migratory herds, given their comparable biology.

Four Populations of Resident Wildebeest in Serengeti

There are four resident subpopulations of wildebeest, which are distinct yet seasonally sympatric with the main migratory population. These resident subpopulations are the Kirawira, Maasai Mara, Loliondo, and Ngorongoro Crater wildebeest herds (fig. 6.8a). Genetic evidence suggests these smaller subpopulations were once part of a larger interbreeding pool, which included the main Serengeti migratory population (Georgiadis 1995).

The Kirawira residents in the Western Corridor move seasonally between the open grasslands of the Ndabaka, Ndoha, Musabi, and Sibora plains, but do not move beyond the western arm of the ecosystem (fig. 6.8). The separate population of semi-migratory wildebeest in the Maasai Mara moves a short distance to the Loita plains during the wet season. The Mara wildebeest could be the remnants of a larger pre-rinderpest population that perhaps migrated to the northeast. The Loliondo wildebeest are residents of the Ngata Kheri plains near Waso (Pennycuick 1975; Watson 1967).

The wildebeest in the Ngorongoro Crater are primarily resident, al-

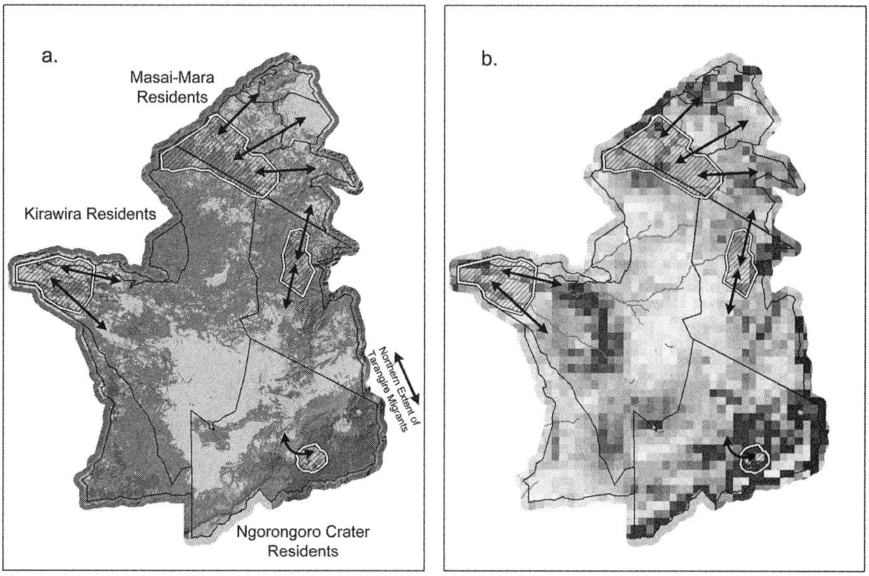

Fig. 6.8 There are (a) four resident subpopulations of wildebeest in the Serengeti (white hatched areas with arrows indicating approximate movement) that are seasonally sympatric with the main migratory population (grasslands, woodlands, and forests are shown in light, medium, and dark shading, accordingly). In (b), the heterogeneity of grass nitrogen (estimated by the standard deviation) suggests that resident wildebeest tend to occur in areas where resources vary at local scales. Darker cells indicate areas that have a mixture of nitrogen-rich and nitrogen-poor grass. Grass nitrogen in the south tends to be homogenously rich for a short period during the wet season, while it is homogeneously poor in the north. Grass quality varies similarly at local and regional scales, which supports two alternative foraging strategies in the Serengeti: grazers can either be resident and move locally at fine scales in specific areas, or migrate over large distances at regional scales.

though some members of the population emigrate out of the crater during the wet season (Talbot [1964]; Watson [1967]; Estes and Small [1981]; observations by Tony Mence as referenced by Grzimek and Grzimek [1960b]; fig. 6.8a). Between 10–20% of the Ngorongoro population leave the crater and potentially mix with migrants in the Olbalbal Depression and at the base of the Crater Highlands between February and April, although this would be rare (from unpublished tagging studies by Orr and Watson; Watson [1967]; Estes and Estes [1979]; Estes [1966]).

The distance between the Selai plains (where the Serengeti migrants occasionally graze during the wet season) and the Natron-Lengai area (the northern extent of the Tarangire migrants [Bolger et al. 2008; Morrison 2011; Morrison and Bolger 2012]) is only 30 km (fig. 6.8a). However, the steep escarpment of the Rift Valley and the thick forests of the Ngorongoro highlands act as a barrier between these populations, which have not interbred for thousands of years. Instead, the Tarangire wildebeest, which currently do not exceed 6,000 animals (Tanzania Widlife Research Institute

2001), are more related to Nairobi National Park wildebeest 300 km north (Georgiadis 1995). This suggests that all the wildebeest east of the Rift Valley were once a single breeding population but are now effectively separated from the Serengeti populations.

Constraints on the Extent of Resident Wildebeest Distributions

A comparison of the habitats occupied by the resident subpopulations of wildebeest versus the average habitat of the Serengeti ecosystem (i.e., the habitat available to the migrant population) suggests that resident wildebeest only occur in areas that receive 900 mm or more of rainfall per year with permanent access to water (table 6.2). The concentration of grass nitrogen in the ranges of resident wildebeest is no greater than what is available to migrants (table 6.2). However, grass nitrogen occurs much more heterogeneously within the ranges of resident wildebeest than beyond their ranges, as can be seen by the juxtaposition of high- and low-quality patches in figure 6.8b. When resources are distributed between small localized patches, herbivore populations tend to stabilize (Owen-Smith 2004), which might explain why resident herbivores only persist in areas where water and good grazing occur adjacently at fine scales. Local access to resources negates the need to migrate between distant patches. Therefore, it is possible that resident wildebeest are not as abundant as migratory wildebeest simply because there are relatively few areas where they can persist.

Differences between Resident and Migrant Wildebeest

There are four key differences between the resident and migratory wildebeest populations in the Serengeti that might provide clues as to why there are fewer resident wildebeest than migrant wildebeest. First, migrant wildebeest have a significantly smaller stature than resident wildebeest. The average male and female body sizes for Serengeti migrants are 160 ± 13.4 kg and 120 ± 10.5 kg, followed by the semi-migratory Maasai Mara wildebeest, which average 171 ± 11.4 kg and 145 ± 17.8 kg, the Ngorongoro wildebeest are approximately 184 ± 11.8 kg and 128 ± 6.8 kg, and the Kirawira wildebeest are the largest at 192 ± 16.0 kg and 145 ± 9.5 kg (Talbot and Talbot 1963; Watson 1967). This suggests that small gracile morphs are better suited for migration while the large robust morphs are resident, perhaps as a means to reduce their exposure to predation.

Second, the calving period for resident wildebeest tends to be earlier

Table 6.2 The average grass nitrogen content, annual rainfall, and the average distance to water in the ranges of the four resident wildebeest populations in comparison to the average availability in the Serengeti ecosystem

	Grass nitrogen (%)	Rainfall (mm)	Distance to water (km)
Ngorongoro range (40 wildebeest/km^2)	1.04	995	2.5
	(0.12)	(42)	(1.4)
Kirawira range (6 wildebeest/km^2)	0.766	940	2.6
	(0.07)	(39)	(1.4)
Maasai-Mara range (21 wildebeest/km^2)	1.17	1042	4.6
	(0.12)	(109)	(3.4)
Loliondo (8 wildebeest/km^2)	1.08	873	15.1
	(0.08)	(27)	(5.6)
Average across the Serengeti ecosystem	1.04	805	20.1
(52 wildebeest/km^2)	(0.21)	(147)	(16.4)

Notes: The data suggest that resident wildebeest only occur at lower densities and in areas that receive approximately 900 mm or more of rainfall, and near permanent water. Values in brackets are standard deviations. Range distributions are estimated based on personal observations; data for grass nitrogen, rainfall, and distance to water as per (Hopcraft 2010).

and less synchronous than migrants (fig. 6.9). Calving typically starts in November and can last until March depending on the subpopulation (Watson 1967; Ndibalema 2009). The extended calving period means that resident wildebeest might be exposed to higher calf mortality due to predation, which could limit the populations.

Third, the sex ratio between resident and migratory populations differs. The male-to-female ratio in the Kirawira population is 0.26 as opposed to almost 1 in the migratory population (Ndibalema 2009; Mduma, Sinclair, and Hilborn 1999), suggesting there could be a substantial male-biased mortality in resident herds. Although carcass evidence does not suggest a male-biased mortality, this does not rule out the effects of poaching (Ndibalema 2009). Furthermore, there is no evidence that resident bulls join the migration even though their larger body size would give them a competitive advantage during the rut. The rut is usually over by the time the migration reaches the Kirawira area, so it is unlikely that resident Kirawira males could be lured into the frenzy of the migrant rut and away from the resident population. Therefore, it is improbable that switching from "resident" to "migrant" explains why the resident herds of wildebeest are small.

The fourth major difference that distinguishes resident wildebeest from

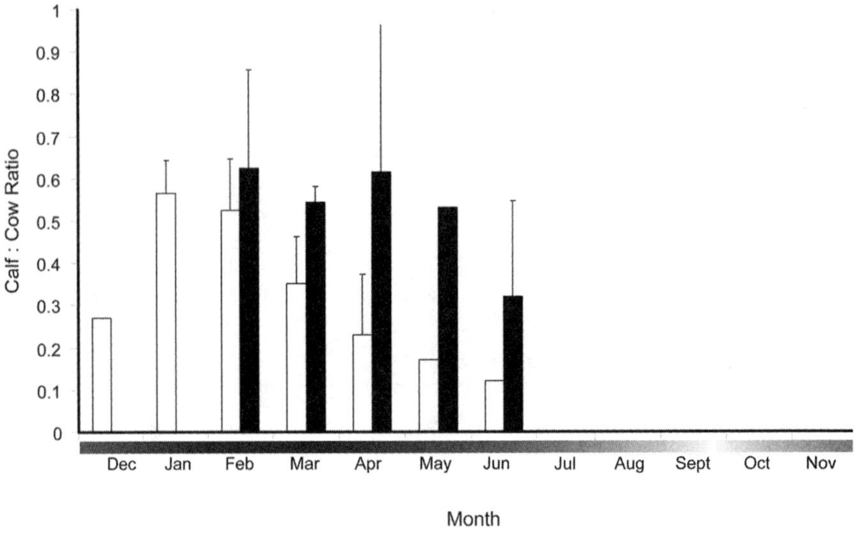

Fig. 6.9 Resident wildebeest (open bars) tend to calve earlier and less synchronously than migrant wildebeest (closed bars), which results in higher calf mortality. Migrant calves have better survival, but yearling mortality tends to be greater (data from Ndibalema [2009]). Dark shading along the x-axis represents wet season months and light shading represents dry season months.

migrants is the age of first reproduction. Watson (1967) found that resident wildebeest tend to reproduce earlier; 50–75% of two-year-old resident females were pregnant, while only 37% of two-year-old migrant females were pregnant. Early maturation in female wildebeest is a response to lower grazing competition, and therefore more resources per individual (Mduma, Sinclair, and Hilborn 1999), which suggests resident wildebeest might not be regulated by food supply. Additionally, the resident bulls tend to participate in the rut earlier than their migrating counterparts (Watson 1967). If resident herds have fewer bulls, it is possible that the male competition during the rut may be less intense in resident herds than in migrant herds, which might allow resident bulls to mate while they are young. Less male competition might also explain why there is less breeding synchrony in resident herds, assuming the intense male competition during the rut explains the sharp leptokurtic peak in fertilization (fig. 6.4).

Population Dynamics of Resident versus Migrant Wildebeest

Population size, density, and instantaneous growth rates. Table 6.3 summarizes the abundance of the main migratory population starting with the earliest

Table 6.3 The population estimates for migratory wildebeest

Year	Original estimate	Accepted estimate	Standard error	Comment
1956	101,000[a]			Rough estimate based on ground transects
1957	190,000[b]			
1958	99,481[c,d]			Census during exodus off plains; many animals missed in woodlands
1961	221,699[e,f,g]	263,362[h]		
1963	322,000[i]	356,124[h]		
1965	381,875[i]	439,124[h]		
1966	334,425[i]			In woodlands
1967	390,000[h]	483,292[h]		
1971	583,188[h]	692,777[h]	28,825	
1972		773,014[j]	76,694	
1977		1,400,000[j]	200,000	
1978		1,248,934[j]	354,668	
1980		1,337,979[j]	80,000	
1982		1,208,711[k]	271,935	
1984		1,337,879[l]	138,135	
1986		1,146,340[l]	133,862	
1991		1,221,783[l]	177,240	
1994		917,204[l]	173,632	
1999		1,296,944[l]	300,072	
2000		1,245,222[l]	144,934	
2003		1,183,966[l]	128,371	
2006		1,239,164[l]	263,536	
2009		1,272,233[l]	66,261	
2012		1,361,548[m]		

Sources: [a] Pearsall (1959); [b] Swynnerton (1958); [c] Grzimek and Grzimek (1960a); [d] Grzimek and Grzimek (1960b); [e] Talbot (1964); [f] Talbot and Talbot (1963); [g] Stewart (1962); [h] estimate corrected by Sinclair (1973); [i] Watson (1967); [j] Sinclair and Norton-Griffiths (1982); [k] Sinclair, Dublin, and Borner (1985); [l] Tanzania Wildlife Research Institute (2010); [m] preliminary estimates based on analysis of raw data by R. Hilborn and A. R. E. Sinclair (pers. comm.).

crude guesstimates in 1956 up to the most recent estimates in 2012. The most striking pattern in the abundance of the wildebeest in the Serengeti ecosystem is the dramatic increase in the migratory population following the large-scale livestock rinderpest vaccination campaign surrounding the park in the late 1950s (Dobson 1995; Taylor and Watson 1967; Plowright and McCulloch 1967; fig. 6.10). The abundance of the migratory wildebeest oscillates around 1.2 million animals, which equates to approximately

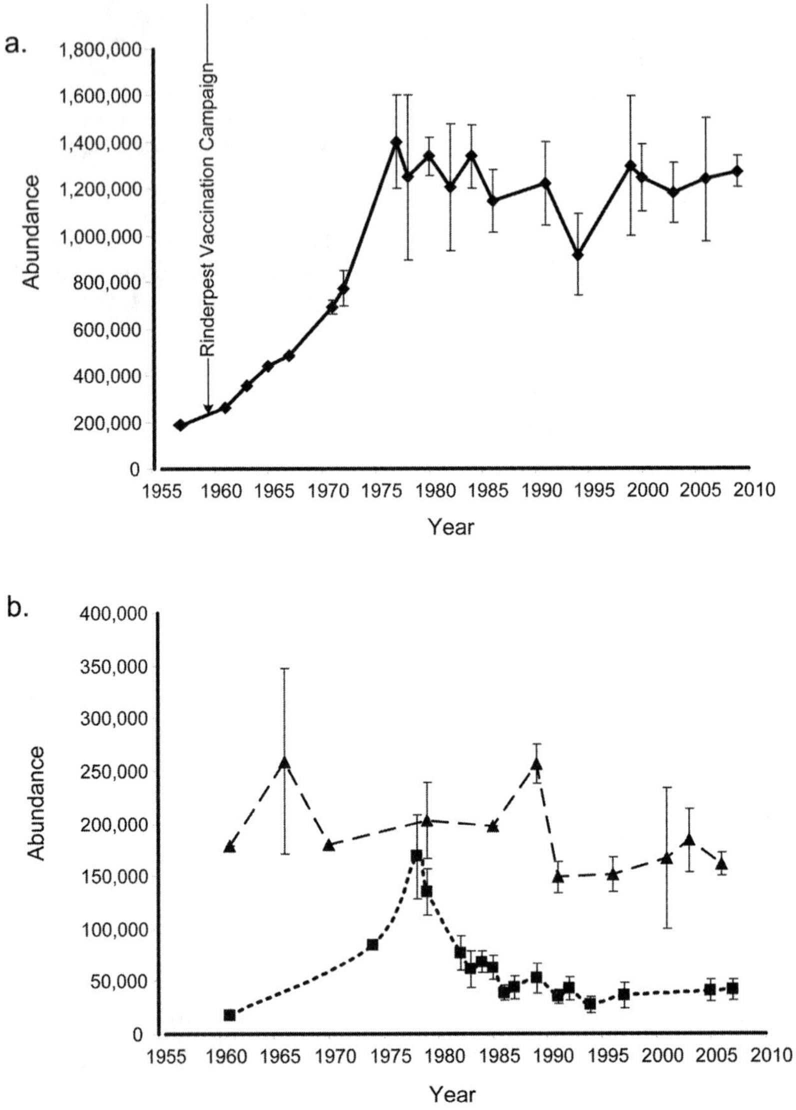

Fig. 6.10 (a) The migratory wildebeest population increased dramatically after the rinderpest vaccination campaign. (b) The Maasai Mara population (solid squares with dotted line) increased slightly after rinderpest was eliminated from the ecosystem, but subsequently declined due to mechanized agriculture blocking access to their wet season calving grounds (data from Serneels and Lambin [2001]; Kenyan DRSRS; J. Ogutu, pers. comm.). The Serengeti migrant zebra population (closed triangles with dashed line) has essentially remained unchanged during this period.

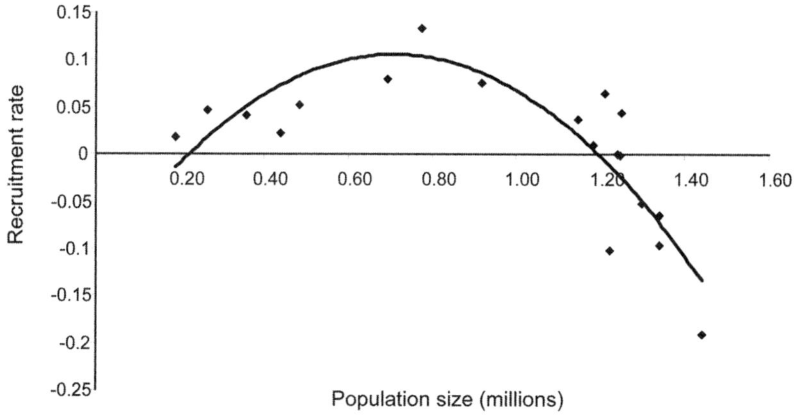

Fig. 6.11 The recruitment rate of migratory wildebeest as a function of population size suggests that the population is stable at approximately 1.2 million animals. Furthermore, the maximum rate of growth of the population is about 10% per annum (Mduma, Sinclair, and Hilborn [1999]; data from table 6.3).

52 wildebeest/km^2. When the abundance is less than 1.2 million the recruitment rate is positive and the population grows; however, when the population exceeds 1.2 million it tends to decline (fig. 6.11). The maximum rate of recruitment for the wildebeest population is approximately 10% per annum (Mduma, Sinclair, and Hilborn 1999) and occurred when the population was recovering from rinderpest (fig. 6.11).

In comparison, the largest resident population of wildebeest in the Serengeti is the Maasai Mara population that varies around 40,000 animals, which equates to roughly 22 wildebeest/km^2 (fig. 6.10; Serneels and Lambin 2001). The Western Corridor herd probably does not exceed 7,000 animals (about 6 wildebeest/km^2), and the Ngorongoro resident wildebeest number about 10,000 (about 40 wildebeest/km^2). The Loliondo herds are no more than 5,000 animals (about 8 wildebeest/km^2). The resident populations of wildebeest have essentially remained unchanged during the last 50 years, with the exception of the Maasai Mara population, which has been reduced by 70%. The decline of the Maasai Mara wildebeest population is primarily due to the expansion of mechanized agriculture on the Loita plains in the late 1970s, which displaced wildebeest from the high-quality grazing of their wet season calving grounds (Serneels and Lambin 2001; Ottichilo, de Leeuw, and Prins 2001).

Population regulation. What determines the abundance of resident versus migrant wildebeest in the Serengeti (fig. 6.10)? The migration essentially enables wildebeest to maximize their intake of seasonally available high-quality grass, which means they can avoid being regulated by food quality

(Hopcraft, Olff, and Sinclair 2010; Fryxell and Sinclair 1988a). Furthermore, the superabundance of animals in these herds reduces any single individual's exposure to predation and eliminates the role of top-down regulation via predation (Fryxell, Greever, and Sinclair 1988). Mduma, Sinclair, and Hilborn (1999) comprehensively showed that the abundance of dry season forage regulates the migratory population of wildebeest through adult and calf survival, as was suggested in earlier papers (Sinclair and Norton Griffiths 1982; Sinclair, Dublin, and Borner 1985). Although pregnancy rates show density dependence (i.e., a decline as the population increases) and neonatal mortality accounts for the greatest numerical loss, neither of these factors are key in population regulation (Mduma, Sinclair, and Hilborn 1999).

Resident populations of wildebeest have not changed dramatically since the elimination of rinderpest in 1963 (with the exception of the Maasai Mara), which suggests that something other than disease regulates their abundance at low numbers. The recent decline of the Maasai Mara population provides a convenient (but unfortunate) natural experiment that gives some valuable insights as to why resident wildebeest are less numerous than migrants. The sudden expansion of Kenya's industrialized agriculture starting in the 1980s excluded the wildebeest from the prime wet season grazing on their calving grounds and resulted in an overall reduction of the food per capita. This semi-migrant population of wildebeest was effectively forced to become resident; in essence, they became tied to the local rainfall dynamics and could no longer track the seasonal availability of food across the region (Serneels and Lambin 2001). The Maasai Mara supports exceptionally high densities of resident predators and prey (Ogutu and Dublin 2002). It is likely that this undernourished population of wildebeest experienced greater predation than they had previously, which further compromised their survival. Therefore, the Mara example suggests that resident wildebeest are probably regulated at lower densities by a combination of predation as well as seasonal shortages of the local food supply (Fryxell, Greever, and Sinclair 1988).

Recruitment into the adult stage. The recruitment into the adult cohorts (i.e., beyond the age of two) is largely dependent on the survivorship of the calves rather than conception rates or yearling survival. Yearlings die less frequently than calves (Mduma, Sinclair, and Hilborn 1999), however, there is a sex bias; yearling males die more frequently than yearling females (Watson 1967; table 6.4 and fig. 6.12). Extreme food limitation during droughts, as occurred in 1993, can lead to spontaneous abortions and fewer births than in normal years (Mduma, Sinclair, and Hilborn 1999). However, under typical conditions the majority of conceptions are carried to term.

The first six weeks of life are the most critical for wildebeest calves (fig.

Table 6.4 Life tables for male and female wildebeest

	Male			Female			
Age	n_x	d_x	q_x	n_x	d_x	q_x	m_x
0	1000	0		1000	0		0
1	746	254	0.25	746	254	0.25	0
2	540	460	0.28	660	340	0.12	0.19
3	508	492	0.06	571	429	0.13	0.42
4	470	530	0.07	523	477	0.08	0.48
5	432	568	0.08	450	550	0.14	0.48
6	373	627	0.14	379	621	0.16	0.48
7	314	686	0.16	306	694	0.19	0.48
8	252	748	0.20	217	783	0.29	0.48
9	232	768	0.08	160	840	0.26	0.48
10	190	810	0.18	87	913	0.46	0.48
11	145	855	0.24	79	921	0.09	0.48
12	124	876	0.14	71	929	0.10	0.48
13	107	893	0.14	63	937	0.11	0.48
14	82	918	0.23	47	953	0.25	0.48
15	76	924	0.07	39	961	0.17	0.48
16	61	939	0.20	39	961	0.00	0.48
17	53	947	0.13	39	961	0.00	0.48
18	36	964	0.32	31	969	0.21	0.48
19	24	976	0.33	23	977	0.26	0.48
20	13	987	0.46	8	992	0.65	0.48
21	8	992	0.38	0	1,000	1.00	0.48
22	3	997	0.63				
23	3	997	0.00				
24	0	1,000	1.00				

Source: Data from 1962 to 1965, Watson (1967).

Notes: n_x = number of wildebeest alive at age x; d_x = number of deaths at age x; q_x = per capita death rate; m_x = age specific fecundity (number of female offspring produced at age x).

6.13). As much as 20% of the calves die if the grazing conditions are poor or widely spaced during the wet season (Watson 1967; Mduma, Sinclair, and Hilborn 1999). If the rain fails on the short-grass plains, migrant cows will give birth in the woodlands. However, these females cannot sustain lactation in the woodlands because of the poor nutritional quality of the grass (Murray 1995; Kreulen 1975). Furthermore, the calves are more susceptible to predators in the woodlands because there is more cover for ambush pred-

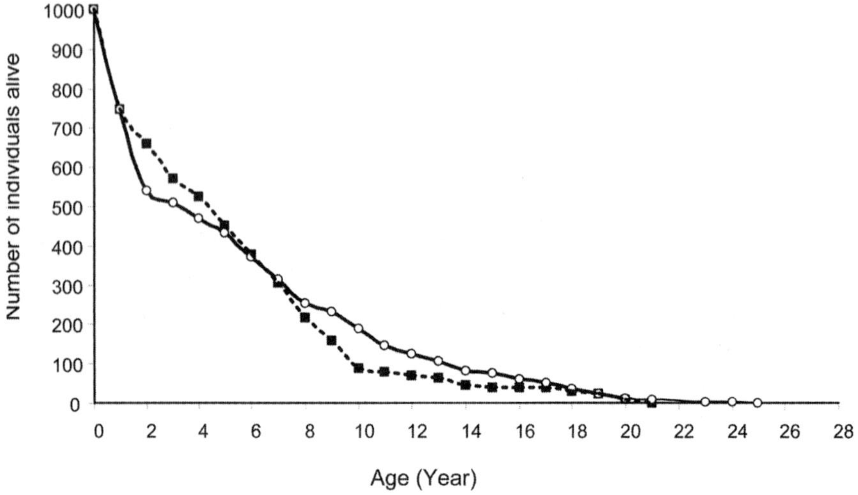

Fig. 6.12 Wildebeest age-based survival illustrates that the highest mortality occurs within the first two years, during which males (open circles with solid line) die more than females (closed squares with dotted line). Females beyond the age of two have a lower survival than males, probably due to the high energetic costs associated with reproduction (data from 1962–1965; Watson [1967]).

ators and this leads to even lower recruitment. The short-grass plains provide very little cover for ambush predators such as lions (Hopcraft, Sinclair, and Packer 2005), which reduces the susceptibility of migrant wildebeest calves to predation. Resident calves do not have access to the short-grass plains and remain susceptible to ambush predators. Furthermore, the synchronous parturition of migratory wildebeest swamps the predators with an abundance of prey resulting in relatively fewer attacks on calves than in resident herds.

As the grazing conditions decline during the transition from the wet season to dry season, migrant cows are forced to travel long distances in order to find adequate food and water, which can result in many calves being severely weakened or abandoned. The severity of the transition and the spatial distribution of resources varies between years, which influences the rates of calf mortality around July (Mduma, Sinclair, and Hilborn 1999; fig. 6.13). By comparison, resident wildebeest are always in close proximity to food and water (table 6.2 and fig. 6.8) and the calves do not have to travel large distances. Resident calves are probably not as susceptible as migrant calves to sudden environmental changes. Furthermore, calf survival does not appear to be impacted by weaning or the rut, even though the rut is very disruptive and many calves become temporarily separated from their mothers (Watson 1967; fig. 6.13).

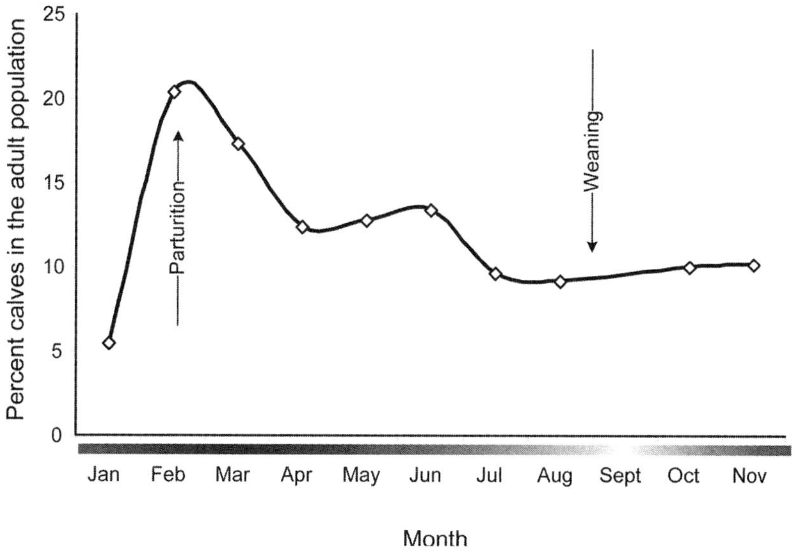

Fig. 6.13 The percentage of calves in the adult population of migrant wildebeest over the year illustrates high calf mortality in the first two months postpartum, after which the survivorship is relatively constant (data from 1965 aerial surveys using vertical photos from Watson [1967] and corroborated by Mduma, Sinclair, and Hilborn [1999]). Dark shading along the x-axis represents wet season months and light shading represents dry season months.

Resident calves have equivalent rates of survival to migrant calves (fig. 6.14), despite being less synchronous (Ndibalema 2009) and potentially more susceptible to predators. Instead, the largest difference in survival between these populations occurs after their first dry season (fig. 6.14). Resident yearlings have better access to food but die more frequently than migrant yearlings; this suggests that recruitment in the resident populations is probably determined by chronic predation (A. Sinclair, unpublished data). Conversely, the survival of yearling migrants is most likely determined by the availability of dry season forage between August and October (Mduma, Sinclair, and Hilborn 1999).

Causes of mortality. One of the primary causes of mortality in wildebeest is starvation. In 1964 there were about 350,000 migratory wildebeest, and at this point Watson (1967) estimated that 16% of the mortality could be attributed to density-dependent causes such as starvation. However, by 1994, when the population exceeded a million animals, Mduma estimated 75% of the mortality was caused by density-dependent starvation (at the peak of the 1993 drought approximately 3,000 wildebeest were starving to death per day; Mduma, Sinclair, and Hilborn [1999]).

A severe loss of body condition because of starvation can lead to in-a.

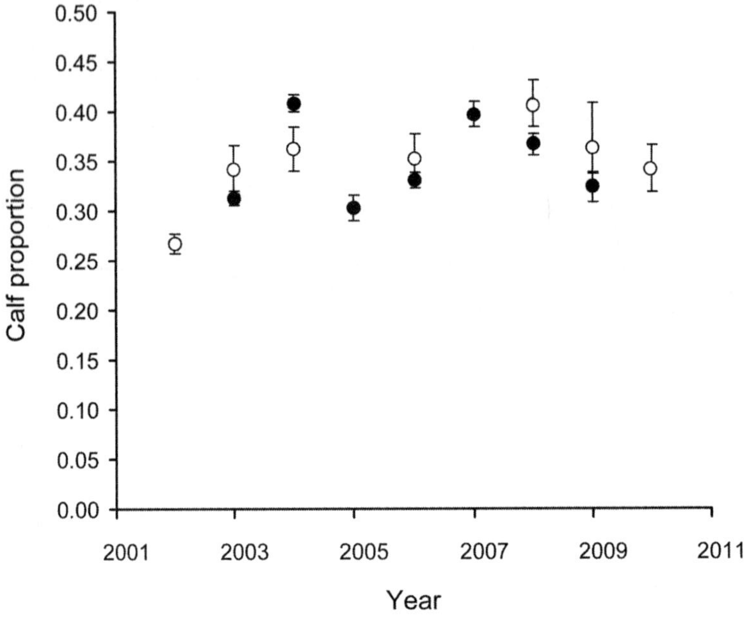

b.

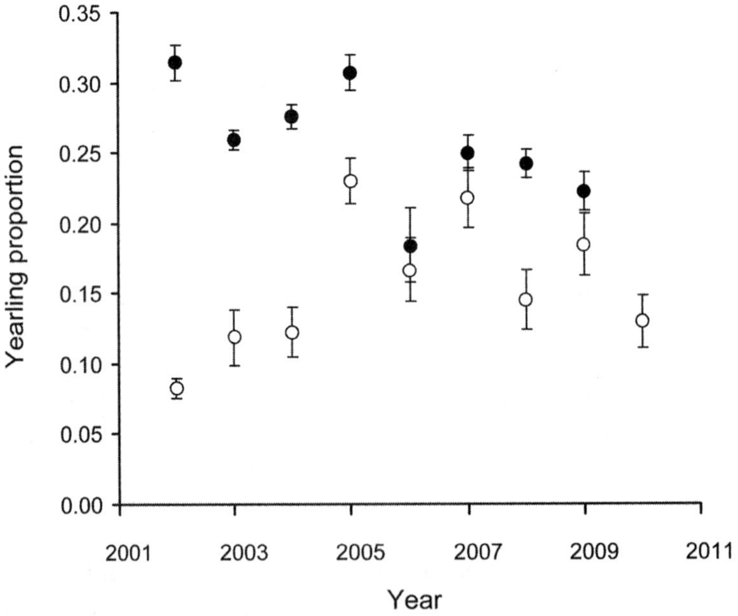

Fig. 6.14 Ground transects surveying wildebeest herds suggest there are (a) no differences in the proportion of calves between resident (open circles) and migrant populations (closed circles), indicating an equal recruitment rate. However, (b) there are relatively fewer yearlings in the resident herds, which suggests they are exposed to greater mortality than migrants, probably because of predation (unpublished data from A. R. E. Sinclair).

creased predation or infection, especially during drought years. Females are more vulnerable to predation and starvation than males and are most vulnerable at the end of the dry season, through the calving period, and into peak lactation (October to June; Sinclair and Arcese 1995). The nutritional content of the forage during the wet season is especially critical for females during which time they are in peak lactation and have the greatest energy demands. The annual patterns of starvation in resident wildebeest are probably similar to migrants.

A comparison of predation rates between resident and migrant wildebeest suggests predation might partially account for the differences in abundances. Predation accounts for little of the overall mortality in the migratory population (2.5–5% per year). The most vulnerable cohort are calves; however, adults in moderate or poor condition are more likely to be killed than healthy individuals (i.e., those that are nutritionally compromised; Mduma, Sinclair, and Hilborn [1999]). Predation has a greater effect on resident populations because they do not move beyond the home ranges of resident predators and do not have the security afforded by large herds (Sinclair 1995).

Disease. The most important disease to influence the abundance of both resident and migrant wildebeest in the Serengeti was rinderpest. Rinderpest reduced the Serengeti wildebeest population by sixfold (assuming the pre-rinderpest population was similar to the current population), which equates to approximately 85% mortality. There is no evidence to suggest that resident and migrant wildebeest are exposed to different diseases, or that infection rates differ. Therefore, disease cannot account for the differences in their abundance. (Diseases are covered in more detail in chapter 19.)

Summary: Suitable Habitat and Predation Most Likely Limit the Abundance of Resident Wildebeest

In summary, we return to the question, If wildebeest biology gives them a competitive advantage over their closest taxonomic relative, then why are resident wildebeest less numerous than migratory wildebeest? A comparison of the resident and migrant wildebeest ranges in the Serengeti indicates that resident wildebeest only occupy areas where the local heterogeneity of grass quality and quantity occurs at fine scales, and where the annual rainfall exceeds 900 mm with access to permanent water. There are only a few locations in the Serengeti were this is possible, which implies that the abundance of resident wildebeest is constrained by the availability of suitable habitat. Migrants escape these local constraints by moving long distances between resource-rich patches, thereby maximizing their access to high-quality grazing. Additionally, migratory wildebeest move beyond the home ranges of resident predators, which means the numerical response of predators is limited by the less numerous resident prey and not by the abundance of migrants. Resident wildebeest likely suffer greater predation because predators switch between migrant and resident prey (a type II functional response). Therefore, migrants can exhaust the local food supply without being regulated by it, as well as uncouple themselves from any regulation by resident predators, which probably accounts for their greater abundance.

However, zebra, eland, Thomson's gazelle, and Grant's gazelle also migrate in the Serengeti and yet their populations remain far less numerous than wildebeest (fig. 6.10). Therefore, migration alone cannot explain why wildebeest are the most abundant herbivore in the Serengeti. In the next section we investigate why zebra (the next most numerous herbivore in the Serengeti) are not as abundant as wildebeest.

WHY ARE MIGRATORY ZEBRA NOT AS ABUNDANT AS MIGRATORY WILDEBEEST?

There are several aspects of wildebeest biology, such as their grazing selection and birthing synchrony, which might give them a competitive advantage over other closely related taxa like topi or Coke's hartebeest. Additionally, by migrating, wildebeest can access more high-quality food while avoiding the impacts of local predators, which might otherwise cap their overall abundance in a similar fashion to the resident wildebeest populations. If the advantages of migration are so profound, then why are the other migratory species in the Serengeti not as abundant as wildebeest?

In this section we compare the migration of zebra in the Serengeti to that of wildebeest. A brief overview of zebra biology highlights the key digestive features that enable two species with similar body sizes to specialize on different attributes of the grass. There have been several explanations proposed for drivers of the migration, which we assess from the perspective of a generalist grazer like the zebra, as opposed to a more specialized grazer like the wildebeest. We conclude by summarizing the features of the Serengeti ecosystem that favors wildebeest migrations over zebra migrations.

Zebra: Hindgut Fermenters versus Ruminants

Like wildebeest, zebra are obligate grazers consuming at least 90% C4 grasses (Tieszen and Imbamba 1980; Casebeer and Koss 1970; Codron et al.

2007); however, their digestive systems are very different. Zebra are hindgut fermenters, which means the majority of nutrient extraction occurs in a specialized section of the cecum and colon rather than in the foregut (i.e., reticulorumen) like wildebeest. Ruminants tend to be more efficient than hindgut fermenters at extracting energy across a broader spectrum of moderate plant quality (Foose 1982; Duncan et al. 1990; Maloiy and Clemens 1991). Zebra and other hindgut fermenters compensate for their less efficient digestive systems by consuming more grass biomass and processing it faster. The increased through-put of large amounts of grass means that hindgut fermenters, like zebra, are better than wildebeest at extracting nutrients from poor-quality plant material that tends to be fibrous and lignified. A comparison of diets illustrates that zebra consume proportionately more grass stems than leaves, which wildebeest select (fig. 6.1; Bell 1970; Gwynne and Bell 1968).

The differences in digestive physiology between hindgut fermenters and ruminants make it unlikely that zebra and wildebeest compete for the same attributes of the grass under normal grazing conditions; zebra can tolerate lower-quality food so long as there is sufficient biomass (Illius and Gordon 1992; Duncan 1992), while wildebeest are confined to better-quality grasslands. These differences in digestive physiology give equids an "ecological refuge," which explains how the two species coexist (Duncan et al. 1990; Bell 1971). If grass biomass becomes depleted but the quality remains high, such as on the short-grass plains of the Serengeti, hindgut fermenters have difficulty acquiring sufficient quantities of food and are at competitive disadvantage to ruminants (Duncan et al. 1990). Therefore, although zebra can consume a wider range of plant material than wildebeest, they are probably constrained by depletions in food supply rather than by quality of the food. Strong seasonal fluctuations of food supply during the Miocene likely favored the expansion of ruminants, which might explain why many intermediate-sized equids disappeared during this time, especially in North America (Janis, Gordon, and Illius 1994).

What Is the Extent of the Wildebeest and Zebra Migration?

The digestive differences between ruminants and hindgut fermenters mean that zebra are less restricted by the availability of forage than wildebeest and, therefore, the seasonal movements of zebra tend to be more diffuse (fig. 6.15; Hopcraft, Morales et al. 2014).

Up until the mid-1950s the extent of wildebeest and zebra movement remained unknown. The first ecological surveys of the Serengeti ecosystem

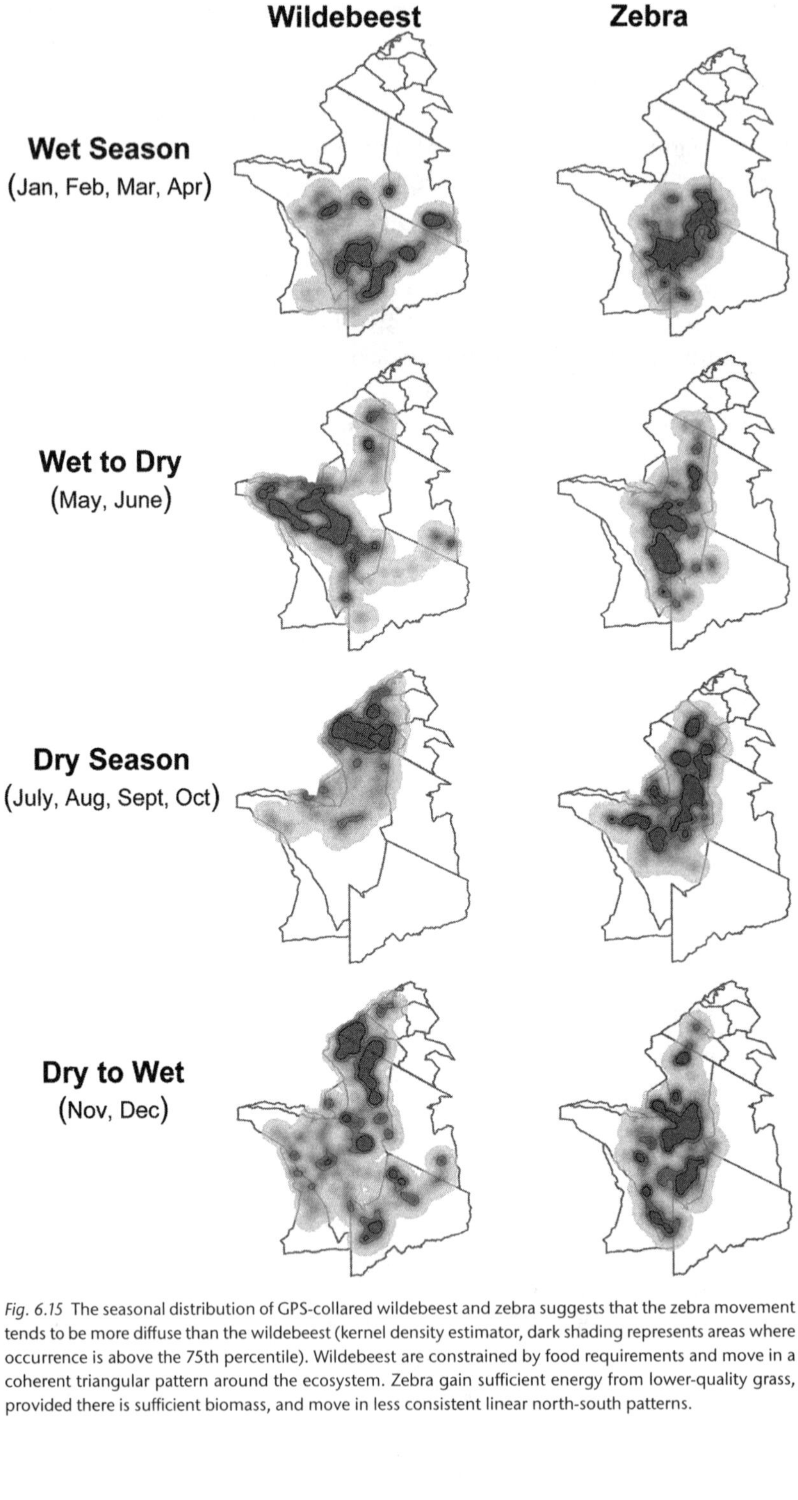

Fig. 6.15 The seasonal distribution of GPS-collared wildebeest and zebra suggests that the zebra movement tends to be more diffuse than the wildebeest (kernel density estimator, dark shading represents areas where occurrence is above the 75th percentile). Wildebeest are constrained by food requirements and move in a coherent triangular pattern around the ecosystem. Zebra gain sufficient energy from lower-quality grass, provided there is sufficient biomass, and move in less consistent linear north-south patterns.

noted the different movement patterns between zebra and wildebeest and attributed these to differences in their grazing selection (Pearsall 1959; Darling 1960; Grzimek and Grzimek 1960b), but it was not until 1957 that the true extent of the migration was studied. The earliest attempts to study the movement patterns of wildebeest used colored plastic collars, ear tags, paint, bleach, and even hair clips to identify individuals. However, as one might imagine, relocating marked animals in dense herds proved to be very difficult, especially over large areas (Talbot and Talbot 1963; Grzimek and Grzimek 1960a, 1960b). Despite this, the early efforts by the Talbots in the Maasai Mara and the Grzimeks in the Serengeti were instrumental in establishing the general movement of wildebeest. They used this information to lobby for the realignment of the national park boundaries to protect the entire migratory route. The complete protection of the migration is without a doubt why the Serengeti is still one of the only large-scale herbivore migratory systems still persisting in the world (Harris et al. 2009) and these early researchers should be credited for their foresight.

The first aerial surveys starting in 1962 produced rough seasonal density maps that improved our understanding of wildebeest movement but left many questions unanswered (Watson 1967). This was followed by intensive monthly aerial surveys of multiple species for 39 months between September 1969 and August 1972 by M. Norton-Griffiths and A. R. E. Sinclair (Pennycuick 1975; Maddock 1979). A more detailed analysis of wildebeest migration began with a radio-collaring program between 1971 and 1973, which tracked 24 animals weekly from an aircraft for 21 consecutive months (Inglis 1976; Wilmshurst et al. 1999). More recently, the Frankfurt Zoological Society funded a GPS radio-collaring program of 35 wildebeest and 17 zebra collared intermittently between 1999 and 2014 (Thirgood et al. 2004; Hopcraft, Morales et al. 2014), which we present in the following sections.

Distance of Wildebeest and Zebra Migrations

The round-trip distance of the wildebeest migration from the southern plains to the Maasai Mara via the Western Corridor and back again is about 650 km, however the total distance a single animal walks is on average 1,550 km per year (Hopcraft 2010). The average daily movement for both wildebeest and zebra is remarkably similar; 4.25 km per day for both species (maximum recorded distance is 58 km/day and 33 km/day, respectively Hopcraft [2010]). This concurs well with models of wildebeest movement based on historic data in which wildebeest are estimated to have a 5 km radius zone of perception (roughly 80 km^2; Holdo, Holt, and Fryxell [2009]).

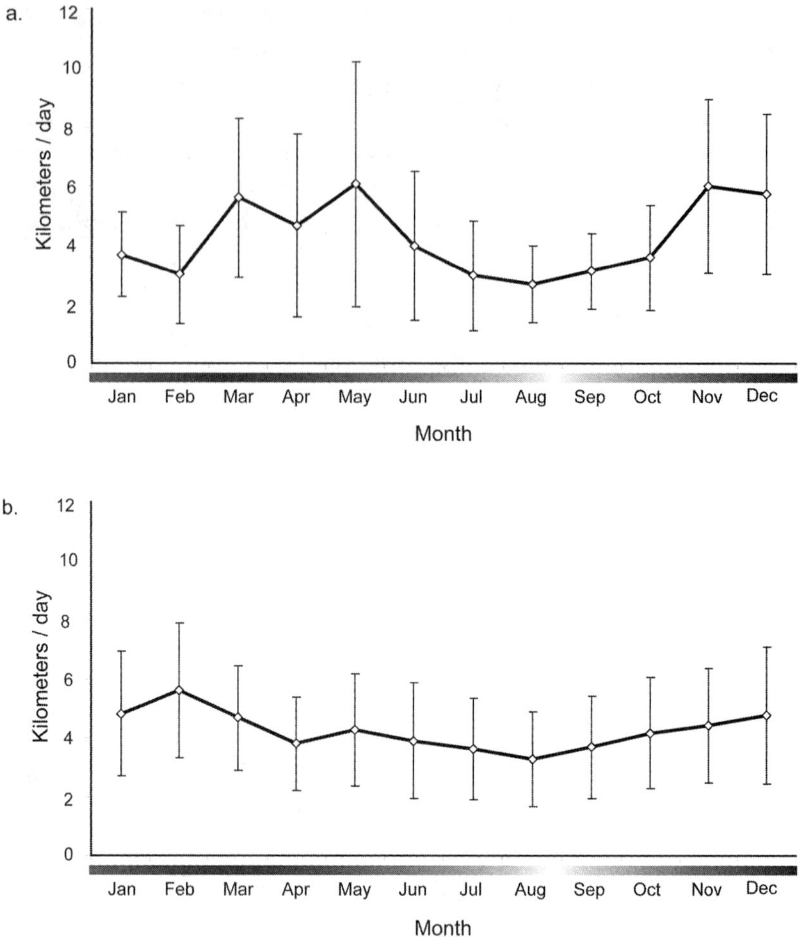

Fig. 6.16 Average daily step lengths of (a) wildebeest and (b) zebra by month. Both species tend to move less during the dry season when resources are scarce. During the wet season when both zebra and wildebeest are on the short-grass plains they tend to move further each day, probably because the grazing conditions change very quickly due to density-dependent competition and the fast greening and drying process associated with the shallow soils. Dark shading along the x-axis represents wet season months and light shading represents dry season months.

The distance animals move each day varies by season, especially for wildebeest (fig. 6.16). Wildebeest tend to move furthest during the wet season directly after calving in February (Inglis 1976; Hopcraft 2010; Hopcraft, Morales et al. 2014). These long movements are somewhat counterintuitive, as one might expect less movement during the wet season when resources are plentiful (i.e., wildebeest should linger in high quality patches). However, the localized rain showers and shallow soils on the short-grass plains result in fast greening and drying processes that attract hundreds of thousands of competing grazers and, combined with the high energy requirements of lactating females, most likely accounts for large daily movements of wildebeest during the wet season (Hopcraft, Morales et al. 2014). During the dry season the movement is less than 4 km per day (Hopcraft, Morales et al. 2014) because food and water are accessible only in local areas around the Mara River. The daily movement of zebra has a similar pattern to wildebeest, however, with much less variance (fig. 6.16; Hopcraft, Morales et al. 2014). Because zebra and wildebeest have very similar movement patterns, it is unlikely that mobility accounts for the difference in abundance between these two migrant species.

IS THE EXTENT OF THE MIGRATION DENSITY DEPENDENT?

If the availability of food determines the extent of the migration, then we should expect the migrants to travel further afield as the population grows, and this might provide clues as to why zebra are less plentiful than wildebeest. The wildebeest population underwent a tenfold increase in numbers between 1959 and 1977 because of the rinderpest vaccination campaign. A comparison of the wet and dry season distributions of wildebeest during this recovery phase illustrates that the extent of their dry season range has indeed shifted considerably (fig. 6.17). In 1958, when Fraser Darling surveyed the Maasai Mara, there was little evidence that the migration crossed the Mara River. Darling noted the movement of resident wildebeest from Lemai to the Loita plains, but he has no accounts of large concentrations moving southward (Darling 1960).

Surveys and distribution maps created by Murray Watson, Myles Turner, and the Grzimeks between 1959 and 1965 (as reported by Pennycuick [1975]; Watson [1967]; Grzimek and Grzimek [1960b]) show the highest dry season densities of wildebeest occurred in the Western Corridor (specifically on the Ndabaka, Dutwa, Ndoha, Musabi, Kirawira, and Nyasirori plains in what is now the Serengeti National Park, and the Kawanga and Sibora plains in what is now the Grumeti Reserve; fig. 6.17a). The wildebeest population at this time was growing from 260,000 to 440,000 animals. The later of these accounts suggest there was some movement north to what is currently the Ikorongo Game Reserve and as far as the Wogakuria and Kogatende areas. Earlier accounts note the majority of the population moved westward from the plains with little northern movement (Darling 1960; Pearsall 1959).

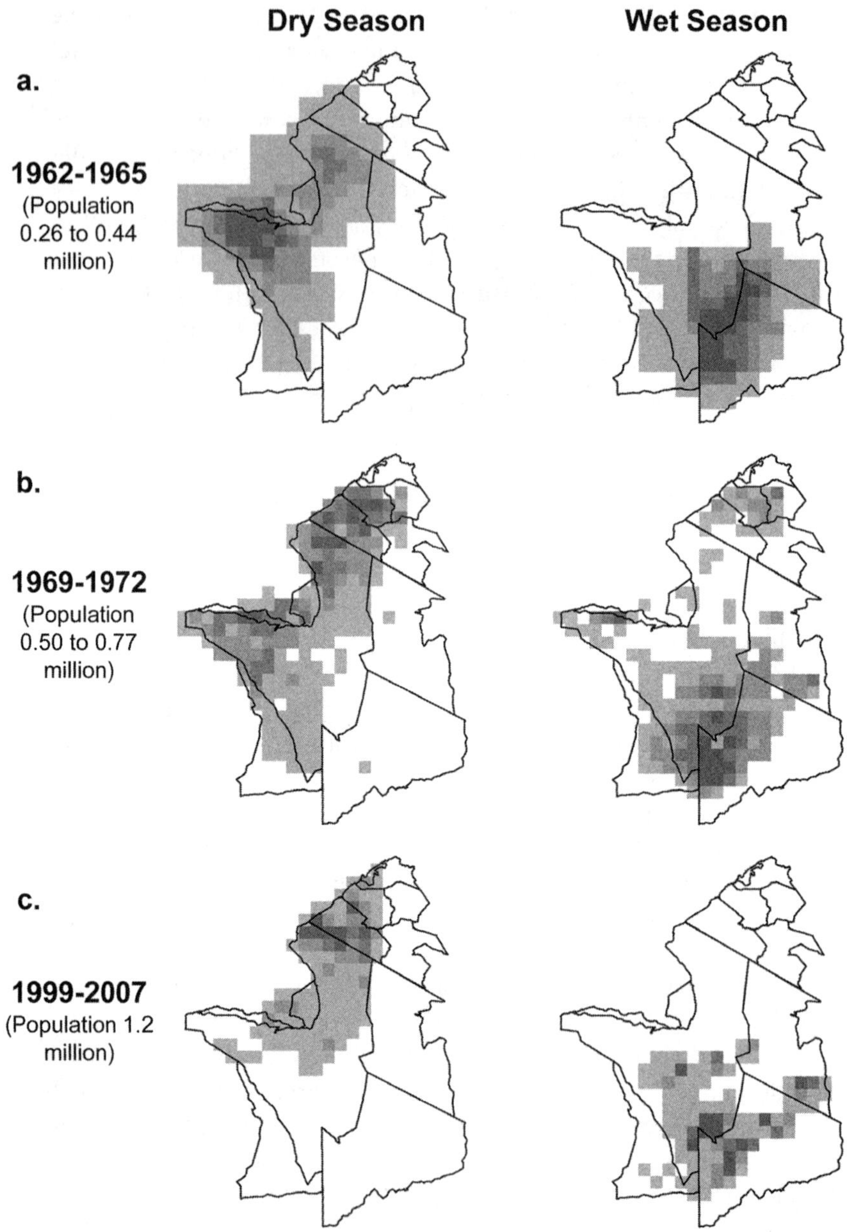

Fig. 6.17 The extent of wildebeest distributions during the dry and wet season from (a) 1962 to 1965, (b) 1969 to 1972, and (c) 1999 to 2007. Dark cells are the locations with the greatest proportion of the population (established by aerial transects previous to 1972, or GPS collars after 1999). The dry season range has extended further north as the population increased from 260,000 animals to over 1.2 million, suggesting that the extent of the migration is density dependent. (Data from Pennycuick [1975]; Thirgood et al. [2004]; Watson [1967]; and FZS, unpublished.)

Between 1969 and 1972 the wildebeest population was growing from approximately 500,000 to 770,000 animals. During this time the migration routinely occupied the Maasai Mara, Lemai Wedge, and Wogakuria areas in the far north of the ecosystem during the dry season (fig. 6.17b; M. Norton-Griffiths, pers. comm.; Pennycuick [1975]). There were occasional pockets of wildebeest remaining on the larger plains in the Western Corridor, but not in densities that are comparable to pre-1965. The small groups of wildebeest that persisted along the southeast edge of Maswa and the southwestern woodlands during the dry season in 1962–65 as recorded by Watson (1967) no longer persisted by 1972. This large shift in dry season distribution of wildebeest was most likely a result of less forage being available, especially since 1969 to 1970 were drier years compared to the previous period.

The most recent data from 25 GPS radio-collared wildebeest between 1999 and 2007 suggest that the most commonly used locations in the dry season are almost exclusively around the Mara River and into the Maasai Mara Reserve in Kenya (fig. 6.17c). These collared individuals were often in groups of 10,000 animals or more (as estimated during radio-tracking flights) so we can reasonably assume that the movement of these 25 animals is representative of the entire population and comparable with previous periods. The abundance of wildebeest between 1999 and 2007 has remained stable at approximately 1.2 million, which is substantially larger than the previous two periods. The area along the northwestern boundary from Ikorongo to Lemai and including Wogakuria poses an interesting and more recent trend in the distribution of the migration. These areas were once heavily used, especially during the dry season between 1969 and 1972; however, the wildebeest currently use this area less frequently than expected. This could be due to intensive poaching in the area (Hilborn et al. 2006).

Unlike the dry season, the wet season distributions have remained relatively unchanged from 1962 to 2009. It was not until the population exceeded about 750,000 animals that the wildebeest migrated in a more cohesive fashion using the full extent of the Serengeti ecosystem. This supports the hypothesis that wildebeest are limited by the availability of food, especially during the dry season (Mduma, Sinclair, and Hilborn 1999).

The abundance of zebra in the Serengeti has remained stable since the early 1960s (fig. 6.10) and there is no evidence that the extent of the zebra migration changed during this period (Maddock 1979; Watson 1967). The stability of the zebra population during this time suggests that zebra are regulated by something other than food, because if competition for food were important we would expect the zebra population to decline as the wildebeest population expanded. The zebra population is most likely regulated by predation, especially on the juvenile age classes (Grange et al.

2004), and by migrating with the wildebeest zebra could be diluting their exposure to predators (Hopcraft 2010; Hopcraft, Morales et al. 2014). Therefore migrating zebra probably gain the advantage of predator evasion, while migrating wildebeest gain both food and predator advantages, which might partially explain why zebra are not as abundant as wildebeest.

What Drives the Wildebeest and Zebra Migration?

Before we can address the question of what drives the herbivore migrations in the Serengeti, it might be prudent to first ask: Why migrate? Grazing experiments have shown that solitary grazers gain energy faster than those living in groups; therefore, a migratory system of superabundant herds should be evolutionarily unstable because of intraspecific competition (Fryxell 1995). However, there are multiple advantages to migrating: (a) individuals dilute their chances of being killed by a predator when they are in a herd, (b) migration allows prey to seasonally escape predation because the predator population cannot respond numerically to an ephemeral prey base, and (c) rotational grazing by herbivores can increase the quality of the grass (Fryxell 1995; Fryxell 1991; Fryxell, Greever, and Sinclair 1988; Braun 1973; McNaughton 1990; McNaughton et al. 1989). This general framework has lead to multiple explanations for drivers of migrations in the Serengeti.

There is little doubt that the dynamics of the Serengeti ecosystem are rainfall dependent. Rainfall simultaneously determines both top-down and bottom-up processes by influencing predation rates as well as food quality and quantity (Hopcraft, Olff, and Sinclair 2010). Therefore, rainfall is the premise that underlies the following explanations of drivers of the migration since rainfall is intrinsically tied to everything else.

The strong rainfall gradient in the Serengeti is matched by a countergradient of soil nutrients, which results in a clear regional sequence of grass quality and abundance (see chapter 3, this volume). The seasonal pulsing of high-quality grass on the southern plains during the wet season attracts many grazer species (Holdo, Holt, and Fryxell 2009; Boone, Thirgood, and Hopcraft 2006; Kreulen 1975; McNaughton 1985; Wilmshurst et al. 1999; Hopcraft, Morales et al. 2014). However, the shallow soils on the plains dry quickly, which in the absence of continued rain forces wildebeest and zebra back into the moister long-grass areas of the woodlands. During the core of the dry season, the Mara River is the only permanent source of water and huge numbers of wildebeest concentrate in this area for several months waiting for the rains.

If wildebeest and zebra were only responding to rainfall and the subsequent green flush of food, we would expect their movement to be linear (i.e., directly up and down the rainfall gradient). However, the true movement is more triangular (fig. 6.15), which suggests there are additional factors that influence the movement of migrants.

There are several alternate explanations for the drivers of migrations in the Serengeti based on seasonal gradients of food resources. Most of these drivers have equal benefits for both wildebeest and zebra, and do not provide explanations as to why migratory wildebeest are more abundant than migratory zebra. For instance, there is a distinct gradient in the concentration of grass protein from the rich southeast plains to the poor northern woodlands (Braun 1973; Murray 1995; Kreulen 1975), so by migrating to the southern plains during the wet season, wildebeest and zebra can maximize their total protein consumption. Furthermore, by repeatedly grazing the short-grass plains, wildebeest and zebra induce the grasses to tiller, which further increases the abundance of high-quality leaf material (compensatory vegetation production; Augustine and McNaughton [2006]; Braun [1973]; Murray [1995]; McNaughton [1976]). Additionally, the migrants frequently rotate between patches allowing the grass to recover, and therefore the same area is capable of supporting more migratory herbivores than resident herbivores (Fryxell 1995). The short-grass plains also have high concentrations of calcium and phosphorous, which are required by lactating females, but are only seasonally available to both zebra and wildebeest (McNaughton 1990; Murray 1995; Kreulen 1975). Physiological restrictions make it impossible for large herds of wildebeest and zebra to remain in the north where calcium and phosphorous are low, and therefore lactating females must find suitable forage further afield. Grass sodium concentrations are highest in the short-grass plains and the Western Corridor (Hopcraft 2010). Sodium is especially important for lactating females and this might partially explain why the migration tends to swing west, making a triangular shape, before going north. The timing of departure from the short-grass plains coincides with the increasing salinity in the ephemeral water holes (i.e., evaporation concentrates the calcium carbonate) (Wolanski et al. 1999; Gereta and Wolanski 1998). However, because the water salinity is dependent on rainfall, and rainfall also determines the availability of forage, it is not possible to separate the effects of water from food. It is likely that some combination of these factors accounts for the timing of departure and the variation in the annual migratory route.

The advantages gained by moving onto the short-grass plains in terms of reduced predation are probably greater for zebra than for wildebeest and provide some insights as to why zebra are not as abundant as wildebeest in the Serengeti (Darling 1960; Fryxell and Sinclair 1988a). There is little vegetative cover that could conceal predators on the short-grass plains, which significantly reduces a predator's efficiency at catching prey (Hopcraft, Sinclair, and Packer 2005). As much as 59–74% of the mortality in the zebra population is due to predation (Sinclair and Norton Griffiths 1982), as opposed to less than 25% for wildebeest (Mduma, Sinclair, and Hilborn 1999). This is supported by studies of radio-collared lions (Packer, pers. comm.), which show that lions kill zebra more often than expected; zebra compose about 18% of kills, whereas we expect them to be about 8% based on their abundance in the ecosystem (by comparison, wildebeest compose 25% of lion kills but based on their abundance we would expect this to be about 60%; Hopcraft [2002]). Furthermore, equids spend up to 15 hours per day grazing because of the lower digestive efficiency associated with hindgut fermentation (ruminants spend about 8 hours per day grazing). Therefore, zebra often continue grazing into the night and might be exposed to greater risks of predation (Duncan et al. 1990). As a result, the relative safety of the short-grass plains could be very beneficial for zebra. Evidence suggests that zebra movement on the plains is determined by predation risk more than wildebeest, which supports this hypothesis (Hopcraft 2010; Hopcraft, Morales et al. 2014).

Summary: Migration Alone Cannot Explain the Superabundance of Wildebeest

There are several reasons why we might expect zebra to be as plentiful as wildebeest in the Serengeti even though they are not: (1) by migrating, zebra can maximize their access to high-quality forage without exhausting the local supply, and therby become more abundant than would be possible otherwise (i.e., migration should favor zebra and wildebeest equally); (2) zebra are generalist grazers and better at digesting low-quality grass than wildebeest, and as a result zebra should not be limited by food (unlike wildebeest); and (3) the seasonal movement and distribution of zebra is less constrained than wildebeest, probably because of their generalist diet (Hopcraft, Morales et al. 2014), therefore they should be freer to relocate to better patches. However, contrary to our expectations, zebra are at least six times less abundant than wildebeest in the Serengeti (i.e., opposite to what we expect). The most plausible explanation for this is that the zebra population is regulated by predation (instead of food), especially on the juvenile age classes (Grange et al. 2004). In addition, wildebeest can outcompete zebra on moderate- to high-quality grass because of their superior digestive efficiency. Therefore, the combination of predation and digestive physiology probably explains why zebra are not as abundant as wildebeest. We conclude that migration alone cannot explain superabundance; however, the zebra comparison does provide some valuable clues as to why wildebeest in particular dominate the Serengeti ecosystem.

CONCLUSION: WHY ARE WILDEBEEST NUMERICALLY DOMINANT IN THE SERENGETI ECOSYSTEM?

To conclude, we return to our original question: Why are wildebeest so abundant in the Serengeti? There are specific aspects of wildebeest biology that give them a competitive advantage over other species, and partially explains their abundance. These biological attributes include synchronous birth of precocial calves, which coincides with seasonal abundance of food and swamps the predator community, and a strong selection for high-quality, matt-forming grasses on which they have the highest intake rates. However, these factors alone do not explain wildebeest abundance because we should expect to see similar densities of wildebeest in other ecosystems, which is not the case. The Serengeti ecosystem has a unique countergradient of soil fertility and rainfall that creates exceptionally large areas of high-quality grasses (more than 3,500 km^2). Furthermore, this nutritional gradient in the grass occurs at a scale that is accessible to migratory wildebeest. Therefore, specific aspects of wildebeest diet and reproduction, combined with their capacity to move long distances in an ecosystem with a predictable nutrient gradient, enables migrant wildebeest to escape regulation by predation or food quality; it is this combination that enables wildebeest to dominate the ecosystem beyond the capacity of any other competitor species. In addition, the large densities of wildebeest in the Serengeti might facilitate themselves in a positive feedback loop by heavily grazing large areas, which further stimulates production and increases the protein concentration in the grass.

Wildebeest are superabundant in the Serengeti because the ecosystem closely matches their requirements. This general conclusion can be applied to other migratory systems, which share key commonalities: (1) refuge areas for migrating animals during times of stress; (2) ephemeral and unusually high quality of resources, which means animals are better off moving than remaining resident; and (3) the areas are connected, allowing animals to freely move between patches. In the Serengeti, this occurs for wildebeest but not for buffalo, impala, or topi because their resources are not spread out in such a way that matches their requirements. However, in the Boma-Jonglei system of southern Sudan, kob (about the size of impala) and tiang (a race of topi) dominate the system and also migrate, but wildebeest do not (Fryxell

Box 6.1

Key Questions Remaining to Be Answered

Long-term research in the Serengeti has answered many questions regarding the wildebeest migration, but several questions still remain unanswered. Here we suggest some future directions for research in the Serengeti:

1. What regulates the resident wildebeest subpopulations?
2. How much interbreeding occurs between the resident and the migrant subpopulations?
3. Could improvements to the quantitative analysis of the census data provide greater insights into causes of mortality and recruitment rates (especially the reanalysis of historic data)?
4. What are the current off-take rates of wildebeest due to illegal hunting?
5. To what extent does predation as opposed to food regulate the zebra population in the Serengeti?
6. How are the physiological requirements of wildebeest linked to their migratory behavior and do physiologically informed models improve our understanding of the drivers of migration?
7. Does the diestrus cycle and intense male competition during the wildebeest rut account for breeding synchrony?
8. Do the rates of sex-based and age-based mortality vary during different phases of the migration?
9. What accounts for the variation in the migratory routes of wildebeest?
10. Do other migratory species have similar movement patterns to wildebeest (i.e., zebra, Thomson's gazelle, Grant's gazelle, and eland)?
11. What are the expected consequences of global climate change for wildebeest migrations (in particular, less predictable weather events) and how can we best contend with these scenarios?

Box 6.2

Threats to the Serengeti Wildebeest

There are several issues threatening the wildebeest population that could lead to its collapse and cause permanent changes in the dynamics of the Serengeti ecosystem. These include:

1. Less predictable rainfall patterns as a result of climate change could result in drying and greening processes that occur out of phase across the ecosystem. If wildebeest cannot reliably track the changes in grass quality (Holdo, Holt, and Fryxell 2009), this could result in greater rates of starvation and could reduce the population below its current carrying capacity.
2. Uncontrolled deforestation in the Mau Forests and large-scale irrigation schemes are causing the Mara River to have less water volume and become more ephemeral (Gereta, Mwangomo, and Wolanski 2009). The drying of the Mara River, which is the only permanent source of water in the ecosystem, would lead to a collapse of the wildebeest population.
3. Plans for developing infrastructure in the northern Serengeti, such as a national highway (Dobson et al. 2010; Hopcraft, Mduma et al., forthcoming), could potentially block animals from reaching their dry season refuge in the Maasai Mara. If wildebeest are blocked from accessing water and grazing during the dry season, the population would collapse (Holdo et al. 2011).
4. Increasing human density on the perimeter of the ecosystem are turning soft boundaries and buffer areas into hard boundaries. Fencing the national park would restrict the migration and could lead to a collapse of the population. Furthermore, fences give the false impression of agricultural security on the edge of the national park and could lead to increased human-wildlife conflicts (Creel et al. 2013; Packer et al. 2013).
5. Greater exposure to shared diseases with livestock could potentially increase infection rates and reduce the wildebeest population, similar to the rinderpest epidemic in the early 1900s.
6. Any increase in the amount of illegal hunting would lead to further declines in the population. Current estimates indicate that 80,000–115,000 wildebeest are poached every year, which equates to 3–8%

of the population and accounts for about one-third to three quarters of the total mortality (see Rentsch et al., chapter 22).

7. The provision of artificial watering points (i.e., boreholes and dams) could result in overgrazing, which could lead to erosion and an overall decline in productivity. Furthermore, permanent water induces migratory species to move less, which increases the prey base for predators and results in unsustainable predation rates on resident drought-tolerant species (Owen-Smith and Mills 2008).
8. There are several alien plant species that are invading the Serengeti ecosystem, such as thorn apple (*Datura stramonium*), Mexican poppy (*Argemone mexicana*), feverfew (*Tanacetum parthenium*), Mexican marigold (*Tagetes minuta*), and Parthenium (*Parthenium hysterophorus*) (Hoeck 2010; Witt 2011). Although the exact extent of these invasions is not known, these species have taken over grasslands in other ecosystems and could exclude wildebeest from accessing critical areas in the future.

and Sinclair 1988b). The resources in the Boma-Jonglei system are probably distributed in such a way that matches the requirements of kob and tiang in a similar fashion that makes the Serengeti ecosystem very well suited for wildebeest (Fryxell and Sinclair 1988a). Therefore, if any of these attributes were slightly different in the Serengeti, the system might favor zebra, topi, or Coke's hartebeest, which are the dominant species in other ecosystems.

Many questions still remain regarding wildebeest in the Serengeti (box 6.1) and their future remains uncertain (box 6.2). However, what is clear is that the Serengeti would have completely different dynamics if not for wildebeest, and reciprocally, wildebeest would not be so abundant if not for the unique features of the Serengeti. Therefore the two are inextricably intertwined, making the Serengeti synonymous with the wildebeest migration.

REFERENCES

Augustine, D. J., and S. J. McNaughton. 2006. Interactive effects of ungulate herbivores, soil fertility, and variable rainfall on ecosystem processes in a semi-arid savanna. *Ecosystems* 9:1242–56.

Bell, R. H. V. 1970. The use of herb layer by grazing ungulates in the Serengeti. In *Animal populations in relation to their food resources*, ed. A. Watson, 111–24. Oxford: Blackwell.

———. 1971. A grazing ecosystem in the Serengeti. *Scientific American* 225:86–94.

Ben-Shahar, R., and M. J. Coe. 1992. The relationships between soil factors, grass nutrients and the foraging behavior of wildebeest and zebra. *Oecologia* 90:422–28.

Bolger, D. T., W. D. Newmark, T. A. Morrison, and D. F. Doak. 2008. The need for integrative approaches to understand and conserve migratory ungulates. *Ecology Letters* 11: 63–77.

Boone, R. B., S. J. Thirgood, and J. G. C. Hopcraft. 2006. Serengeti wildebeest migratory patterns modeled from rainfall and new vegetation growth. *Ecology* 87:1987–94.

Braun, H. M. H. 1973. Primary production in Serengeti. *Annales de l'Université d'Abidjan. Série E: Écologie* 6:171–88.

Bro-Jørgensen, J. 2003. No peace for estrous topi cows on leks. *Behavioral Ecology* 14: 524–25.

Bro-Jørgensen, J., and S. M. Durant. 2003. Mating strategies of topi bulls: Getting in the centre of attention. *Animal Behaviour* 65:585–94.

Casebeer, R. L., and G. G. Koss. 1970. Food habits of wildebeest, zebra, hartebeest and cattle in Kenya Masailand. *African Journal of Ecology* 8:25–36.

Cerling, T. E., J. M. Harris, S. H. Ambrose, M. G. Leakey, and N. Solounias. 1997. Dietary and environmental reconstruction with stable isotope analyses of herbivore tooth enamel from the Miocene locality of Fort Ternan, Kenya. *Journal of Human Evolution* 33:635–50.

Clay, A. M., R. D. Estes, K. V. Thompson, D. E. Wildt, and S. L. Monfort. 2010. Endocrine patterns of the estrous cycle and pregnancy of wildebeest in the Serengeti ecosystem. *General and Comparative Endocrinology* 166:365–71.

Codron, D., J. Codron, J. A. Lee-Thorp, M. Sponheimer, D. de Ruiter, J. Sealy, R. Grant, and N. Fourie. 2007. Diets of savanna ungulates from stable carbon isotope composition of faeces. *Journal of Zoology* 273:21–29.

Creel, S., M. S. Becker, S. M. Durant, J. M'Soka, W. Matandiko, A. J. Dickman, D. Christianson, E. Dröge, T. Mweetwa, N. Pettorelli, et al. 2013. Conserving large populations of lions—The argument for fences has holes. *Ecology Letters* 16:1413-16.

Darling, F. 1960. An ecological reconnaissance of the Mara plains in Kenya colony. *Wildlife Monographs* 5:5–41.

Dobson, A. 1995. The ecology and epidemiology of rinderpest virus in Serengeti and Ngorongoro Conservation Area. In *Serengeti II: Dynamics, management, and conservation of an ecosystem*, ed. A. R. E. Sinclair and P. Arcese, 485–505. Chicago: University of Chicago Press.

Dobson, A. P., M. Borner, A. R. E. Sinclair, P. J. Hudson, T. M. Anderson, G. Bigurube, T. B. B Davenport, J. Deutsch, S. M. Durant, R. D. Estes, et al. 2010. Road will ruin Serengeti. *Nature* 467:272–73.

Duncan, P. 1992. *Horses and grasses: The nutritional ecology of equids and their impact on the Camargue*. London: Springer-Verlag.

Duncan, P., T. J. Foose, I. J. Gordon, C. G. Gakahu, and M. Lloyd. 1990. Comparative nutrient extraction from forages by grazing bovids and equids—A test of the nutritional model of equid bovid competition and coexistence. *Oecologia* 84:411–18.

Estes, R. D. 1966. Behaviour and life history of the wildebeest (*Connochaetes taurinus burchell*). *Nature* 212:999–1000.

———. 1976. The significance of breeding synchrony in the wildebeest. *African Journal of Ecology* 14:135–52.

———. 1992. *The behavior guide to African mammals: Including hoofed mammals, carnivores, primates*. Los Angeles: University of California Press.

Estes, R. D., and R. K. Estes. 1979. The birth and survival of wildebeest calves. *Zeitschrift für Tierpsychologie* 50:45–95.

Estes, R. D., T. E. Raghunathan, and D. Van Vleck. 2008. The impact of horning by wildebeest on woody vegetation of the Serengeti ecosystem. *Journal of Wildlife Management* 72:1572–78.

Estes, R. D., and R. Small. 1981. The large herbivore populations of Ngorongoro Crater. *African Journal of Ecology* 19:175–85.

Foose, T. J. 1982. Trophic strategies of ruminant versus nonruminant ungulates. PhD diss., University of Chicago.

Fryxell, J. M. 1991. Forage quality and aggregation by large herbivores. *American Naturalist* 138:478–98.

———. 1995. Aggregation and migration by grazing ungulates in relation to resources and predators. In *Serengeti II: Dynamics, management, and conservation of an ecosystem*, ed. A. R. E. Sinclair and P. Arcese, 257–73. Chicago: University of Chicago Press.

Fryxell, J. M., J. Greever, and A. R. E. Sinclair. 1988. Why are migratory ungulates so abundant? *American Naturalist* 131:781–98.

Fryxell, J. M., and A. R. E. Sinclair. 1988a. Causes and consequences of migration by large herbivores. *Trends in Ecology & Evolution* 3:237–41.

———. 1988b. Seasonal migration by white-eared kob in relation to resources. *African Journal of Ecology* 26:17–31.

Georgiadis, N. 1995. Population structure of wildebeest: Implications for conservation. In *Serengeti II: Dynamics, management, and conservation of an ecosystem*, ed. A. R. E. Sinclair and P. Arcese, 473–84. Chicago: University of Chicago Press.

Gereta, E., E. Mwangomo, and E. Wolanski. 2009. Ecohydrology as a tool for the survival of the threatened Serengeti ecosystem. *Ecohydrology & Hydrobiology* 9:115–24.

Gereta, E., and E. Wolanski. 1998. Wildlife-water quality interactions in the Serengeti National Park, Tanzania. *African Journal of Ecology* 36:1–14.

Gordon, I. J., A. W. Illius, and J. D. Milne. 1996. Sources of variation in the foraging efficiency of grazing ruminants. *Functional Ecology* 10:219–26.

Grange, S., P. Duncan, J.-M. Gaillard, A. R. E. Sinclair, P. J. P. Gogan, C. Packer, H. Hofer, and M. East. 2004. What limits the Serengeti zebra population? *Oecologia* 140: 523–32.

Grzimek, M., and B. Grzimek. 1960a. Census of plains animals in the Serengeti National Park, Tanganyika. *Journal of Wildlife Management* 24:27–37.

———. 1960b. A study of the game of the Serengeti plains. *Zeitschrift fur Saugetierkunde* 25: 1–61.

Gwynne, M. D., and R. H. V. Bell. 1968. Selection of vegetation components by grazing ungulates in the Serengeti National Park. *Nature* 220:390–93.

Hanby, J. P., J. D. Bygott, and C. Packer. 1995. Ecology, demography, and behavior of lions in two contrasting habitats: Ngorongoro Crater and the Serengeti plains. In *Serengeti II: Dynamics, management, and conservation of an ecosystem*, ed. A. R. E. Sinclair and P. Arcese, 315–31. Chicago: University of Chicago Press.

Harris, G., S. Thirgood, J. G. C. Hopcraft, J. P. G. M. Cromsigt, and J. Berger. 2009. Global decline in aggregated migrations of large terrestrial mammals. *Endangered Species Research* 7:55–76.

Hilborn, R., P. Arcese, M. Borner, J. Hando, J. G. C. Hopcraft, M. Loibooki, S. Mduma, and A. R. E. Sinclair. 2006. Effective enforcement in a conservation area. *Science* 314:1266.

Hoeck, H. 2010. Invasive plant species in the Serengeti ecosystem and Ngorongoro crater. *Tanzania Wildlife Magazine (Kakakuona)* 58:21–22.

Hofer, H., and M. East. 1995. Population dynamics, population size, and the commuting system of Serengeti spotted hyenas. In *Serengeti II: Dynamics, management, and conservation of an ecosystem*, ed. A. R. E. Sinclair and P. Arcese, 332–63. Chicago: University of Chicago Press.

Holdo, R. M., J. M. Fryxell, A. R. E. Sinclair, A. Dobson, and R. D. Holt. 2011. Predicted impact of barriers to migration on the Serengeti wildebeest population. *PLoS ONE* 6: 1–7.

Holdo, R. M., R. D. Holt, and J. M. Fryxell. 2009. Opposing rainfall and plant nutritional gradients best explain the wildebeest migration in the Serengeti. *American Naturalist* 173:431–45.

Hopcraft, J. G. C. 2002. The role of habitat and prey density on foraging by Serengeti lions. MS thesis, University of British Columbia, Vancouver.

———. 2010. Balancing food and predation risk: Ecological implications for large herbivores in the Serengeti. PhD diss., University of Groningen.

Hopcraft, J. G. C., T. M. Anderson, S. Perez Vila, E. Mayemba, and H. Olff. 2012. Body size and the division of niche space: Food and predation differentially shape the distribution of Serengeti grazers. *Journal of Animal Ecology* 81:201–13.

Hopcraft, J. G. C., S. A. R. Mduma, M. Borner, G. Bigurube, A. Kijazi, D. T. Haydon, W. Wakilema, D. Rentsch, A. R. E. Sinclair, A. P. Dobson, and J. D. Lembeli. Forthcoming. Conservation and economic benefits of a road around the Serengeti. *Conservation Biology*.

Hopcraft, J. G. C., J. M. Morales, H. L. Beyer, M. Borner, E. Mwangomo, A. R. E. Sinclair, H. Olff, and D. T. Haydon. 2014. Competition, predation and migrations: Individual choice patterns of Serengeti migrants captured by heirarchical models. *Ecological Monographs* 84:355–72.

Hopcraft, J. G. C., H. Olff, and A. R. E. Sinclair. 2010. Herbivores, resources and risks: Alternating regulation along primary environmental gradients in savannas. *Trends in Ecology & Evolution* 25:119–28.

Hopcraft, J. G. C., A. R. E. Sinclair, and C. Packer. 2005. Planning for success: Serengeti lions seek prey accessibility rather than abundance. *Journal of Animal Ecology* 74: 559–66.

Hubbell, S. P. 2001. *The unified neutral theory of biodiversity and biogeography*. Princeton, NJ: Princeton University Press.

Illius, A. W., and I. J. Gordon. 1992. Modeling the nutritional ecology of ungulate herbivores: Evolution of body size and competitive interactions. *Oecologia* 89:428–34.

Ims, R. A. 1990. The ecology and evolution of reproductive synchrony. *Trends in Ecology & Evolution* 5:135–40.

Inglis, J. M. 1976. Wet season movements of individual wildebeests of the Serengeti migratory herd. *East African Wildlife Journal* 14:17–34.

Janis, C. M., I. J. Gordon, and A. W. Illius. 1994. Modeling equid-ruminant competition in the fossil record. *Historical Biology: An International Journal of Paleobiology* 8:15–29.

Kreulen, D. 1975. Wildebeest habitat selection on the Serengeti plains Tanzania in relation to calcium and lactation: A preliminary report. *East African Wildlife Journal* 13:297–304.

Kruuk, H. 1972. *The spotted hyena: A study of predation and social behavior*. Chicago: University of Chicago Press.

Maddock, L. 1979. The migration and grazing succession. In *Serengeti I: Dynamics of an ecosystem*, ed. A. R. E. Sinclair and M. Norton-Griffiths, 104–29. Chicago: University of Chicago Press.

Maloiy, G. M. O., and E. T. Clemens. 1991. Aspects of digestion and invitro fermentation in the cecum of some East-African herbivores. *Journal of Zoology* 224:293–300.

McNaughton, S. J. 1976. Serengeti migratory wildebeest facilitation of energy flow by grazing. *Science* 191:92–94.

———. 1985. Ecology of a grazing ecosystem: The Serengeti. *Ecological Monographs* 55: 259–94.

———. 1990. Mineral nutrition and seasonal movements of African migratory ungulates. *Nature* 345:613–15.

McNaughton, S. J., M. Oesterheld, D. A. Frank, and K. J. Williams. 1989. Ecosystem-level patterns of primary productivity and herbivory in terrestrial habitats. *Nature* 341: 142–44.

Mduma, S. A. R., A. R. E. Sinclair, and R. Hilborn. 1999. Food regulates the Serengeti wildebeest: A 40-year record. *Journal of Animal Ecology* 68:1101–22.

Morrison, T. A. 2011. Demography and movement of migratory wildebeest (*Connochaetes taurinus*) in northern Tanzania. PhD diss., Dartmouth College.

Morrison, T. A., and D. T. Bolger. 2012. Wet season range fidelity in a tropical migratory ungulate. *Journal of Animal Ecology* 81:543–52.

Murray, M. G. 1993. Comparative nutrition of wildebeest, hartebeest and topi in the Serengeti. *African Journal of Ecology* 31:172–77.

———. 1995. Specific nutrient requirements and migration of wildebeest. In *Serengeti II: Dynamics, management, and conservation of an ecosystem*, ed. A. R. E. Sinclair and P. Arcese, 231–56. Chicago: University of Chicago Press.

Murray, M. G., and D. Brown. 1993. Niche separation of grazing ungulates in the Serengeti: An experimental test. *Journal of Animal Ecology* 62:380–89.

Murray, M. G., and A. W. Illius. 2000. Vegetation modification and resource competition in grazing ungulates. *Oikos* 89:501–08.

Mysterud, A., T. Coulson, and N. C. Stenseth. 2002. The role of males in the dynamics of ungulate populations. *Journal of Animal Ecology* 71:907–15.

Ndibalema, V. G. 2009. A comparison of sex ratio, birth periods and calf survival among Serengeti wildebeest sub-populations, Tanzania. *African Journal of Ecology* 47:574–82.

Ogutu, J. O., and H. T. Dublin. 2002. Demography of lions in relation to prey and habitat in the Maasai Mara National Reserve, Kenya. *African Journal of Ecology* 40:120–29.

Ogutu, J. O., H. P. Piepho, H. T. Dublin, N. Bhola, and R. S. Reid. 2008. Rainfall influences on ungulate population abundance in the Mara-Serengeti ecosystem. *Journal of Animal Ecology* 77:814–29.

———. 2010. Rainfall extremes explain interannual shifts in timing and synchrony of calving in topi and warthog. *Population Ecology* 52:89–102.

Ottichilo, W. K., J. de Leeuw, and H. H. T. Prins. 2001. Population trends of resident wildebeest and factors influencing them in the Masai Mara ecosystem, Kenya. *Biological Conservation* 97:271–82.

Owen-Smith, N. 2004. Functional heterogeneity in resources within landscapes and herbivore population dynamics. *Landscape Ecology* 19:761–71.

Owen-Smith, N., and M. G. L. Mills. 2008. Shifting prey selection generates contrasting herbivore dynamics within a large-mammal predator-prey web. *Ecology* 89:1120–33.

Packer, C., A. Loveridge, S. Canney, T. Caro, S. T. Garnett, M. Pfeifer, K. K. Zander, A. Swanson, D. MacNulty, G. Balme, et al. 2013. Conserving large carnivores: Dollars and fence. *Ecology Letters* 16:635–41.

Pearsall, W. H. 1959. Report on an ecological survey of the Serengeti. London, The Fauna Preservation Society.

Pennycuick, L. 1975. Movements of the migratory wildebeest population in the Serengeti area between 1960 and 1973. *East African Wildlife Journal* 13:65–88.

Plowright, W., and B. McCulloch, B. 1967. Investigations on incidence of rinderpest virus infection in game animals of northern Tanganyika and southern Kenya 1960–63. *Journal of Hygiene* 65:343–58.

Serneels, S., and E. F. Lambin. 2001. Impact of land-use changes on the wildebeest migration in the northern part of the Serengeti-Mara ecosystem. *Journal of Biogeography* 28: 391–407.

Sinclair, A. R. E. 1973. Population increases of buffalo and wildebeest in the Serengeti. *East African Wildlife Journal* 11:93–107.

———. 1977a. Lunar cycle and timing of mating season in Serengeti wildebeest. *Nature* 267:832–33.

———. 1977b. *The African buffalo: A study of resource limitation of populations*. Chicago: University of Chicago Press.

———. 1995. Population limitation of resident herbivores. In *Serengeti II: Dynamics, management, and conservation of an ecosystem*, ed. A. R. E. Sinclair and P. Arcese, 194–219. Chicago: University of Chicago Press.

Sinclair, A. R. E., and P. Arcese. 1995. Population consequences of predation-sensitive foraging: The Serengeti wildebeest. *Ecology* 76:882–91.

Sinclair, A. R. E., H. Dublin, and M. Borner. 1985. Population regulation of Serengeti wildebeest: A test of the food hypothesis. *Oecologia* 65:266–68.

Sinclair, A. R. E., S. A. R. Mduma, and P. Arcese. 2000. What determines phenology and synchrony of ungulate breeding in Serengeti? *Ecology* 81:2100–11.

Sinclair, A. R. E., and M. Norton-Griffiths. 1982. Does competition or facilitation regulate migrant ungulate populations in the Serengeti, Africa. A test of hypotheses. *Oecologia* 53:364–69.

Stewart, D. R. M. 1962. Census of wildlife on the Serengeti, Mara and Loita plains. *East African Agricultural and Forestry Journal* 28:58–62.

Swynnerton, G. H. 1958. Fauna of the Serengeti National Park. *Mammalia* 22:435–50.

Talbot, L. M. 1964. First wildlife census of the entire Serengeti—Mara region, East Africa. *Journal of Wildlife Management* 28:815–27.

Talbot, L. M., and M. H. Talbot. 1963. The wildebeest in western Masailand, East Africa. *Wildlife Monographs* 12:3–88.

Tanzania Widlife Research Institute. 2001. Aerial census in the Tarangire ecosystem. Arusha, Tanzania, Conservation Information Monitoring Unit.

———. 2010. Survey information system (SISTA) database. Arusha, Tanzania, Conservation Information Monitoring Unit.

Taylor, W. P., and R. M. Watson. 1967. Studies on the epizootiology of rinderpest in blue wildebeest and other game species of northern Tanzania and southern Kenya, 1965–7. *Journal of Hygiene* 65: 537–45.

Thirgood, S., A. Mosser, S. Tham, J. G. C. Hopcraft, E. Mwangomo, T. Mlengeya, M. Kilewo, J. Fryxell, A. R. E. Sinclair, and M. Borner. 2004. Can parks protect migratory ungulates? The case of the Serengeti wildebeest. *Animal Conservation* 7:113–20.

Tieszen, L. L., and S. K. Imbamba. 1980. Photosynthetic systems, carbon isotope discrimination and herbivore selectivity in Kenya. *African Journal of Ecology* 18:237–42.

Vrba, E. S. 1979. Phylogenetic analysis and classification of fossil and recent Alcelaphini Mammalia: Bovidae. *Biological Journal of the Linnean Society* 11:207–28.

Watson, R. M. 1967. The population ecology of the wildebeest in the Serengeti. PhD diss., Cambridge University.

Wilmshurst, J. F., J. M. Fryxell, B. P. Farm, A. R. E. Sinclair, and Henschel. 1999. Spatial distribution of Serengeti wildebeest in relation to resources. *Canadian Journal of Zoology-Revue Canadienne De Zoologie* 77:1223–32.

Witt, A. 2011. Silent invader may threaten biggest wildlife migration on planet. *Swara* 34: 18–21.

Wolanski, E., E. Gereta, M. Borner, and S. Mduma. 1999. Water, migration and the Serengeti ecosystem. *American Scientist* 87:526–33.

SEVEN

Climate-Induced Effects on the Serengeti Mammalian Food Web

John M. Fryxell, Kristine L. Metzger, Craig Packer, Anthony R. E. Sinclair, and Simon A. R. Mduma

Climatic variation can have substantial effects on the demography of large terrestrial herbivores (Post and Forchammer 2002; Coulson et al. 2001; Tyler, Forchhammer, and Oritsland 2008) and carnivores (Post et al. 1999). Such climatic effects can be realized through direct effects on survival (via exposure, prolonged heat stress, accidents, or lack of water) or indirect effects mediated through changes in metabolic costs, vegetation availability, vegetation quality, predation rates, or disease risk. In the latter case, demographic effects can trace from bottom-up or top-down food-web interactions. In this chapter, we consider long-term variability in the demographic rates of several species of large herbivores and lions in Serengeti National Park. We primarily concentrate on juvenile recruitment, since this is recognized as the primary driver of population change in large mammals in general (Saether 1997; Gaillard, Festa-Bianchet, and Yoccoz 1998).

Opportunities for direct studies of marked individuals are understandably limited in a national park setting; hence, we use the ratio of juveniles to adult females as an index of juvenile recruitment. This demographic interpretation obviously depends on the critical assumption that adult survival rates are far less variable than offspring production or juvenile survival, which is often the case in large mammals (Saether 1997; Gaillard, Festa-Bianchet, and Yoccoz 1998). Age-structure data were obtained over a 40-year span, in most years by visual counts by a single individual (A. R. E. Sinclair, unpublished data). Age determination was based largely on horn length and shape for antelope species (wildebeest, kongoni, topi, waterbuck, impala) and body height in other large herbivores (warthogs, zebra, and giraffe). These data have proven useful in comparing patterns of breeding seasonality among Serengeti ungulates (Sinclair, Mduma, and Arcese 2000). For the purposes of this chapter, however, we restrict our attention to yearling fractions of each population, operationally defined as individuals 9–14 months of age.

We elected to use the Southern Oscillation Index (SOI), downloaded from the Australian Government Bureau of Meteorology website, as a general predictor of year-to-year climatic conditions (Stenseth et al. 2003; Holmgren et al. 2006; Sinclair et al. 2013). Variation in SOI has been linked to patterns of primary productivity (Oba, Post, and Stenseth 2001; Ogutu et al. 2007) and wildlife population dynamics (Ogutu and Owen-Smith 2003) in southern Africa. Extreme SOI values relate strongly to El Niño and La Niña events (Anyamba, Tucker, and Mahoney 2002) through global circulation patterns. One would hope, however, to translate any general index, such as the SOI, into direct environmental effects. One of the best ways to identify which variables are most influential within a complex set of time-series data is structural equation modeling, an augmented form of path analysis (Grace 2006; Anderson et al. 2010). This procedure allows one to identify the most likely network of factors that could contribute to variation in a state variable of interest, while parsing out covariation with other variables in the model. A particularly clear way to visualize these interactions is by graphical depiction of partial correlation coefficients for pairs of standardized variables linked in a larger web (fig. 7.1). Some climatic variables have been recorded for a long time, such as SOI or rainfall. Others are of more recent origin, owing to innovations in remote sensing technology, such as NDVI (Near-Infrared Difference Vegetation Index), which is a standardized measure of green vegetation abundance assessed from satellites circling the globe on a regular basis. Data were available from 1991 to the present for the following variables most likely of importance to Serengeti large herbivores and carnivores: SOI, maximum temperature, minimum temperature, annual rainfall, dry season rainfall, wet season rainfall, dry season NDVI, and wet season NDVI.

Variation in SOI is associated with a number of different climatic and environmental variables that co-vary to a considerable degree (Sinclair et al. 2013). Years with high SOI values tend to be hotter, with more rainfall during the dry season, but less rainfall during the wet season, than years with low SOI scores (fig. 7.1). Grass growth in the Serengeti, like other savanna grasslands, is usually limited by soil moisture (McNaughton 1985). Higher temperatures and higher rainfall in high SOI years contribute to increased dry season NDVI, resulting in augmented supplies of food during

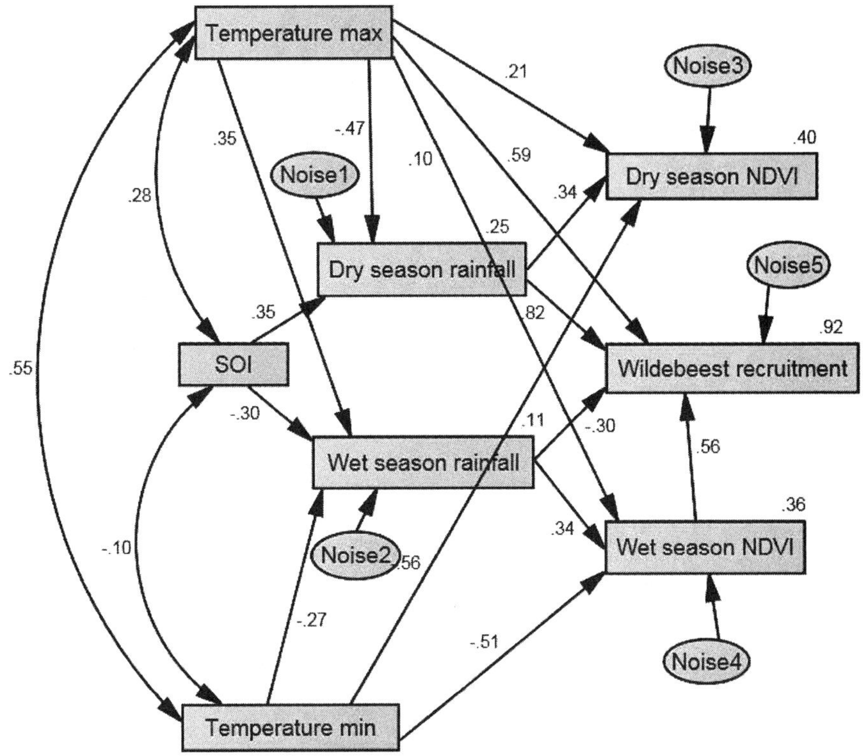

Fig. 7.1 Best-supported structural equation model for wildebeest juvenile recruitment in relation to climatic and environmental variables. The numbers along arrows indicate standardized partial correlation coefficients and the numbers above each variable indicate the proportion of variance explained by the combined set of factors.

the dry season (fig. 7.1). In contrast, years with lower than average SOI values experience more rainfall and increased grass growth during the wet season (fig. 7.2).

HERBIVORE JUVENILE RECRUITMENT IN RELATION TO CLIMATE

Considerable variation was observed for all species with respect to juvenile recruitment (fig. 7.2). Only wildebeest and topi, however, showed statistically significant correlations between demographic parameters and SOI scores (fig. 7.2). This is perhaps surprising, given that all herbivores depend on suitable vegetation conditions to meet their energetic needs. Wildebeest and topi differ from the other large herbivore species, however, in having short, well-defined breeding seasons in the Serengeti, whereas other large

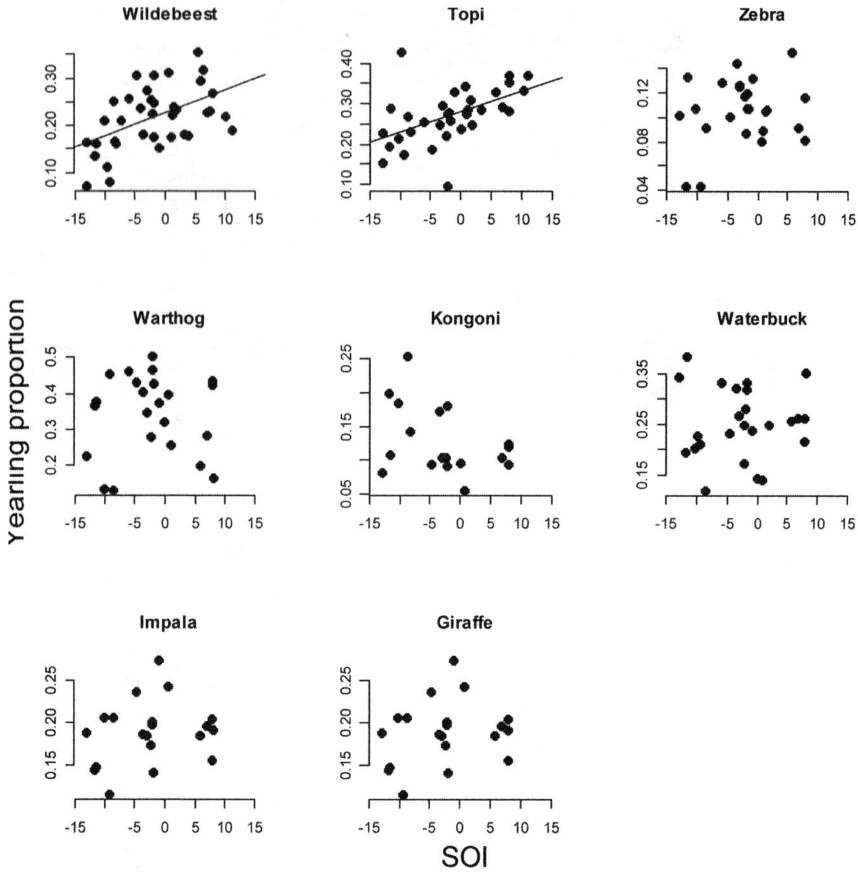

Fig. 7.2 Scatterplots of juvenile recruitment and SOI level in the year of birth for eight large herbivore species in the Serengeti from 1960 to 2008. Lines indicate statistically significant regression relationships for wildebeest and topi ($P < 0.05$).

herbivores breed throughout the year. One possible reason most species are apparently insensitive to climatic variation stems from the contrasting effects of variation in SOI on environmental conditions during the dry season versus wet season. Although years of above average SOI should improve feeding conditions during the dry season, which should lead to increased recruitment, it would also have the opposite impact on juveniles during the wet season. Trade-offs between wet and dry season reproductive success in herbivore species that breed year round may cancel out to a considerable degree, buffering most Serengeti herbivores against year-to-year variation in SOI.

Another possible explanation for the weak influence of SOI on juvenile recruitment in most herbivore species is that recruitment is routinely limited by predation, disease, or other nonnutritional factors. It has been suggested that zebra are limited largely by predators rather than food resources (Grange et al. 2004). Population dynamics of the other small to mid-sized grazing ungulates in the Serengeti (kongoni, warthogs, waterbuck, and impala) have not yet been thoroughly analyzed, but recent habitat selection work suggests that zebra and kongoni tend to avoid locations (dense woodlands, riverine savannas) with elevated predation risk, which would be consistent with a predator limitation hypothesis (Anderson et al. 2010; Hopcraft et al.2012). On the other hand, there is good evidence that megaherbivores such as giraffe and elephant are rarely threatened by predators (Sinclair, Mduma, and Brashares 2003). It is certainly conceivable, however, that browse availability may be less sensitive to SOI variation than is that of graminoids and small ground-level forbs fed upon by smaller herbivores.

The positive correlation between SOI score and recruitment patterns of topi and wildebeest suggests that dry season conditions are particularly critical. To explore possible causal factors more thoroughly, we enlarged the structural equation model to include wildebeest recruitment and then compared alternate structural equation models using Akaike information criterion (AIC) scores as a measure to model plausibility (Burnham and Anderson 1998). The resulting model suggests that wildebeest recruitment is positively related to dry season rainfall (fig. 7.3), with wet season rainfall, wet season NDVI, and maximum temperature playing more minor roles (table 7.1). The SOI dropped out as a direct impact, but still exerted enormous influence on wildebeest recruitment patterns through its effect on annual variation in dry season rainfall.

The situation for topi is somewhat more complex. The most plausible structural equation model for topi recruitment included SOI, temperature maxima, temperature minima, rainfall in both the dry and wet seasons, as well as NDVI during the dry season (table 7.1). This suggests that while dry season conditions are probably of particular importance to juvenile recruitment, the causal factors are less clearly articulated.

A positive effect of rainfall on juvenile recruitment in wildebeest is consistent with other known aspects of their ecology (Hopcraft et al., chapter 6). Wildebeest calving is tightly synchronized to occur while animals are on the short-grass plains, usually in late January to early March (Sinclair 1977). The young-growth stage of short grasses on the plains, coupled with high soil concentrations of minerals produces excellent nutritional conditions for small to mid-sized species of Serengeti grazers like gazelles (Wilmshurst, Fryxell, and Colucci 1999) and wildebeest (Wilmshurst et al. 1999; Holdo, Holt, and Fryxell 2009; Hopcraft et al. 2012). Several months later, however,

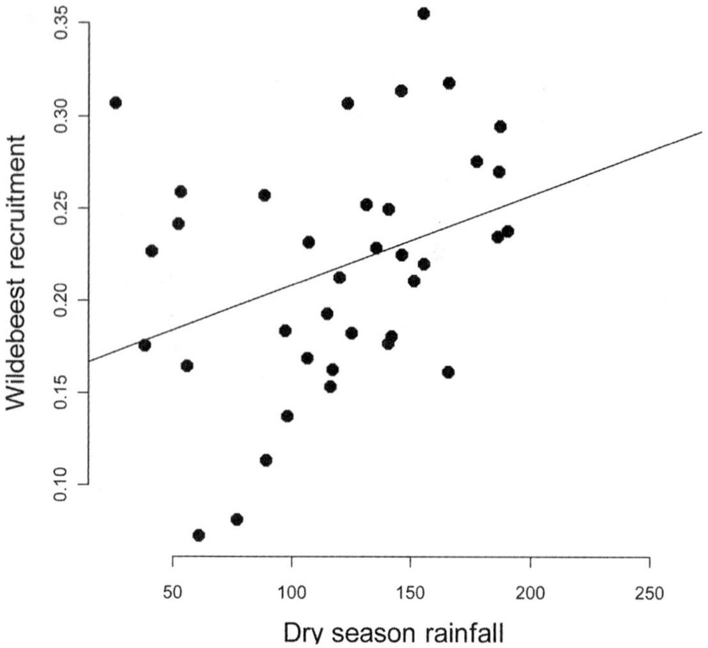

Fig. 7.3 Juvenile recruitment of wildebeest in relation to dry season rainfall during the preceding year.

feeding conditions often decline substantially. By this stage of the migration, wildebeest have left the short-grass plains and are found largely in savanna woodlands in the north or west (Maddock 1979; Wilmshurst et al. 1999; Hopcraft et al., chapter 6), dominated by different communities of grasses that are considerably taller, with typically lower concentrations of nutrients and digestible energy (Anderson, Metzger, and McNaughton 2007; Anderson et al. 2010; Hopcraft et al. 2012). The only remedy for nutritional stress that typifies the dry season is fresh regrowth stimulated by sporadic rainfall events. Annual monitoring of dry season mortality has demonstrated a strong inverse relationship between wildebeest mortality and dry season rainfall, largely mediated via the supply of green grass (Mduma, Sinclair, and Hilborn 1999).

In the best of all possible worlds, one would compare recruitment rates with patterns of population change, to verify that temporal variability in juvenile recruitment has a long-lasting effect at the population level. Aerial censuses of wildebeest are conducted every few years, not annually, so it is not possible to directly assess annual variation in SOI and wildebeest population abundance. Instead, we calculated the average exponential rate of increase ($\ln[N_{t+\tau}/N_t]/\tau$) and regressed the rate of wildebeest increase

Table 7.1 Structural equation model competition for recruitment of wildebeest (top section), lions (middle section), or topi (bottom section)

SOI	Temp max.	Temp min.	Dry rain	Wet rain	Dry NDVI	Wet NDVI	AIC
x	x	x	x	x	x	x	92.37
	x		x	x	x	x	88.41
	x		x	x		x	86.76
	x		x			x	89.88
	x		x	x			92.58
			x	x			97.26
x							92.46
x	x	x	x	x	x	x	93.01
	x		x	x	x		87.04
	x		x	x			85.79
	x			x			86.85
	x		x				89.33
			x	x			88.23
x							87.88
x	x	x	x	x	x	x	92.73
x	x	x	x	x	x		90.79
x	x	x			x		92.66
x		x			x		95.22
x					x		97.4
x							95.64

Note: The most plausible model has the lowest AIC score.

against average SOI scores over each time interval. After removing density-dependent effects, the residual rate of increase by wildebeest was positively related to average SOI score (fig. 7.4), suggesting that climate-induced effects on juvenile wildebeest recruitment are also detectable in the longer term at the population level.

LION JUVENILE RECRUITMENT IN RELATION TO CLIMATE

Given that there is firm evidence that topi and wildebeest respond demographically to annual variation in climate, we go on to consider population recruitment patterns in lions, based on the long-term demographic studies that have been conducted in a roughly 2,000 km^2 study spanning the short-grass plains to the woodland area around Seronera (Packer et al. 2005). Lion

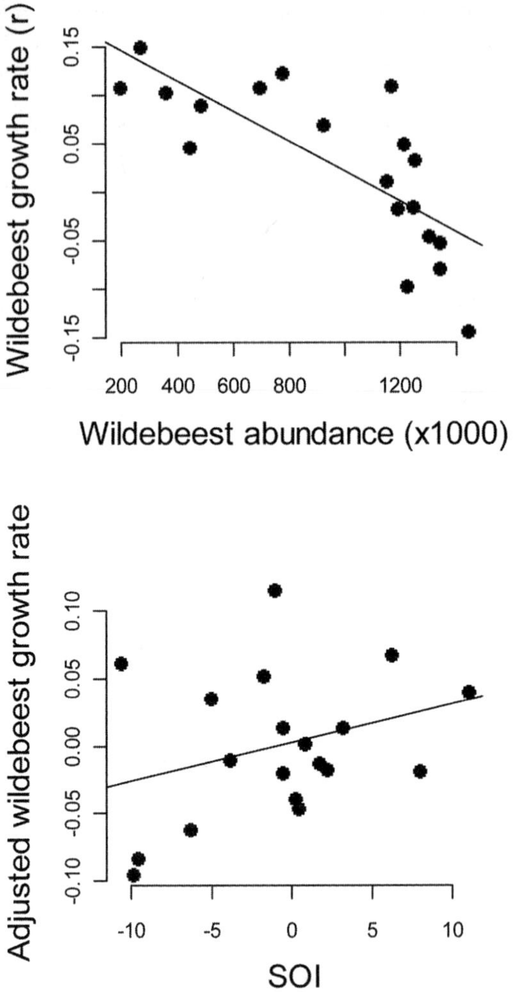

Fig. 7.4 Exponential rate of wildebeest population growth in relation to wildebeest abundance at the start of each time interval (top panel) and the residual rate of population growth in relation to SOI averaged over each time interval between censuses (bottom panel).

prides are monitored more or less continuously throughout the year. Here we restrict ourselves to the recruitment of cubs born throughout the year.

The social system of Serengeti lions is organized around permanent territories occupied by closely related groups of females and their cubs, with short periods of tenure by coalitions of males. Takeover events by new male coalitions are typically followed by episodes of infanticide on young cubs, which bring females rapidly into estrous again. As a consequence, a

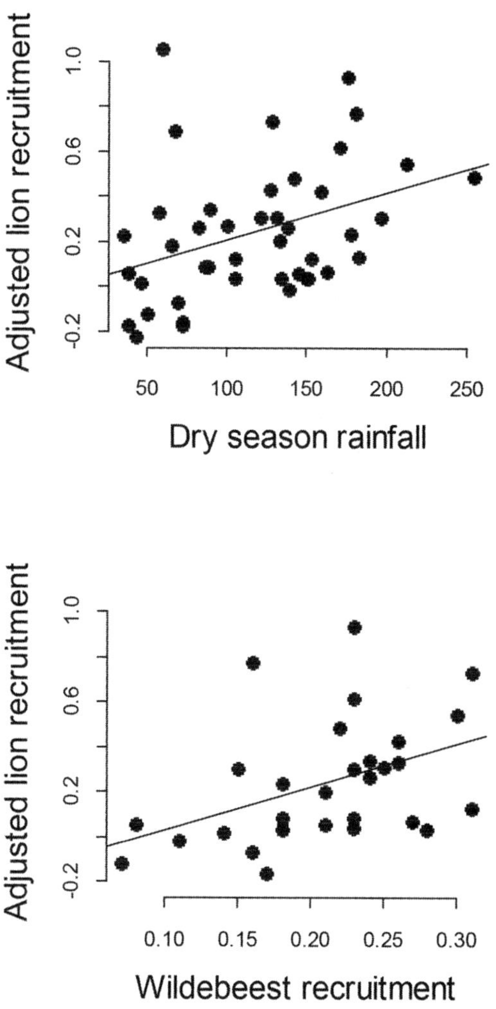

Fig. 7.5 Residual lion recruitment (measured as cubs per female) in relation to dry season rainfall (top panel) and the level of wildebeest juvenile recruitment (bottom panel).

key variable influencing cub recruitment is the probability of takeover by a new set of males. After statistically removing the effect of male takeovers, the residual rate of cub recruitment is positively related to dry season rainfall (fig. 7.5, *top panel*). The residual rate of cub recruitment is also positively related to juvenile recruitment rates in wildebeest (fig. 7.5, *bottom panel*).

The proposed impact of climate on lion demography can be explained in at least two different ways. Many mammalian carnivores preferentially kill prey that are very young or very old or individuals in poorer condition rather than prime-aged adults (Peterson 1977; Temple 1987; Wright et. al. 2006). If this is true of Serengeti lions, then the climatic link may be circumstantial, arising simply because lions have higher feeding success in years with a larger fraction of vulnerable juvenile wildebeest. On the other hand, state-of-art movement models suggest that the wildebeest migration is cued to seasonal changes in weather (Maddock 1979; Boone, Thirgood, and Hopcraft 2006; Holdo, Holt, and Fryxell 2009), with evidence that animals respond on a spatial scale of tens of km to episodic rainfall events (Holdo, Holt, and Fryxell 2009). Monthly systematic censuses conducted during the 1970s clearly demonstrated that the dry season movement away from the short-grass plains can be slowed or even reversed when rainstorms occur in the south. This has the effect of keeping the bulk of the wildebeest population in savanna woodlands adjoining the Serengeti plains, where they are in close proximity to the Seronera lion population. Previous work has shown that improved access to prey during at least part of the year has a demonstrable effect on cub survival and, therefore, population recruitment (Mosser et al. 2009). Teasing apart these competing hypotheses will probably require more detailed data on variation in seasonal lion diets in relation to wildebeest age composition, spatial distribution during the dry season, and climatic variation.

Given that there is evidence that recruitment of both wildebeest and lions is enhanced in years with substantial dry season rainfall, we compared patterns of population growth of both species. Time-series data suggest rather similar general trends, with both species increasing following the elimination of rinderpest in the wildebeest population in the early 1960s, followed by a leveling off in later decades (fig. 7.6). Linear regression suggests a marginally significant ($0.05 < P < 0.10$) positive relationship between the exponential rate of population growth by wildebeest and that of lions (fig. 7.7). This suggests that a proper accounting of trophic dynamics in the Serengeti requires recognition that both predators and prey have patterns with some degree of positive covariation.

CONSEQUENCES OF SOI RESPONSES FOR PREDATOR-PREY INTERACTIONS

Given the climate-induced demographic effects we have identified here, we developed a simple mathematical model to explore the implications with respect to long-term dynamics. For simplicity, the model assumes species of prey, wildebeest (x) and other herbivores (y), fed upon by lions (z). Both herbivore populations are partitioned into juvenile and adult segments. Juvenile recruitment is assumed to follow a Ricker logistic formula aug-

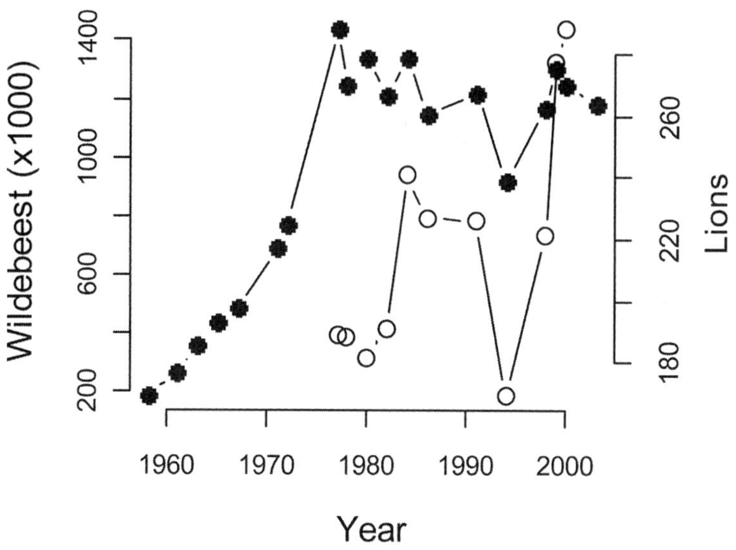

Fig. 7.6 Population dynamics of lions and wildebeest over identical time intervals.

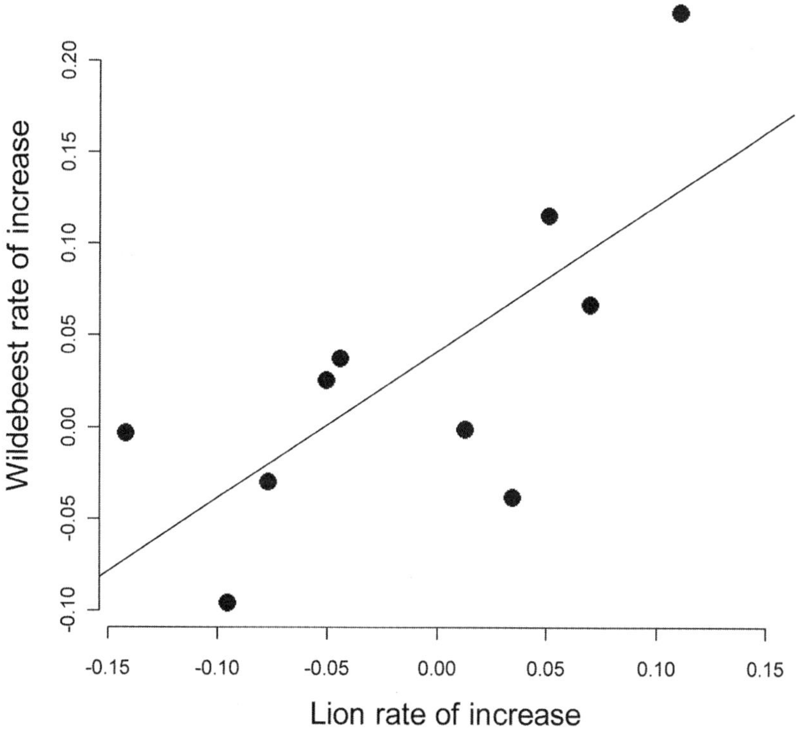

Fig. 7.7 Exponential rate of increase by lions plotted against the rate of wildebeest increase over the same time interval.

mented by a Gaussian environmental variability term ε to mimic episodic variation in climate, with a = the maximum exponential recruitment rate for wildebeest, f = the maximum exponential recruitment rate for other herbivores, b = the density-dependent effect of adults on the exponential rate of wildebeest recruitment, g = the density-dependent effect of adults on the exponential rate of recruitment by other herbivores, c = the constant probability of attack by an individual predator, d = the density-independent mortality rate of adult herbivores, h = a coefficient converting prey killed into new predator offspring, and j = the density-dependent mortality rate of predators. Note that adult herbivores are assumed to have a constant risk of mortality, whereas juvenile herbivores have mortality rates dependent on their own density and that of predators. Lions have density-dependent mortality, which is plausibly assumed as a consequence of territorial strife. To keep things as simple as possible, we assume that lions feed only on juveniles of either herbivore species, but not adults, with feeding rates proportionate to prey density (i.e., a type I functional response). Because juvenile recruitment of herbivores is stochastic, so is the prey population potentially available to lions. Equations are as follows:

$$x_{t+1} = x_t(1 + e^{a-bx_t+\varepsilon_t}(1 - cz_t))e^{-d}$$

$$y_{t+1} = y_t(1 + e^{f-gy_t+\varepsilon_t}(1 - cz_t))e^{-d}$$

$$z_{t+1} = z_t(1 + h(x_t e^{a-bx_t+\varepsilon_t} + y_t e^{f-gy_t+\varepsilon_t})e^{-jz_t}$$

Results suggest some interesting outcomes worthy of further consideration. First, in the absence of environmental stochasticity, such trophic models based on predators specializing on juvenile segments of the population are typically quite stable (Smith and Mead 1974; Hastings 1983). Given the effects of stochastic SOI events on wildebeest recruitment, however, there are years when both lions and wildebeest prosper, generating swings in abundance over time (fig. 7.8a). Since wildebeest were reduced in the first half of the twentieth century because of rinderpest, the model predicts that the subsequent increase in wildebeest should have been matched by a similar increase in lions, mirroring the historical record (fig. 7.6).

In years of high wildebeest recruitment due to extreme positive SOI events, new recruits swamp predators, augmenting the adult population. In years with low SOI, predators may remove a much larger fraction of the potential pool of recruits, but otherwise have little impact on adults. Hence, the model suggests that wildebeest can better weather changes in predation numbers over time than would a population without age-specific variation

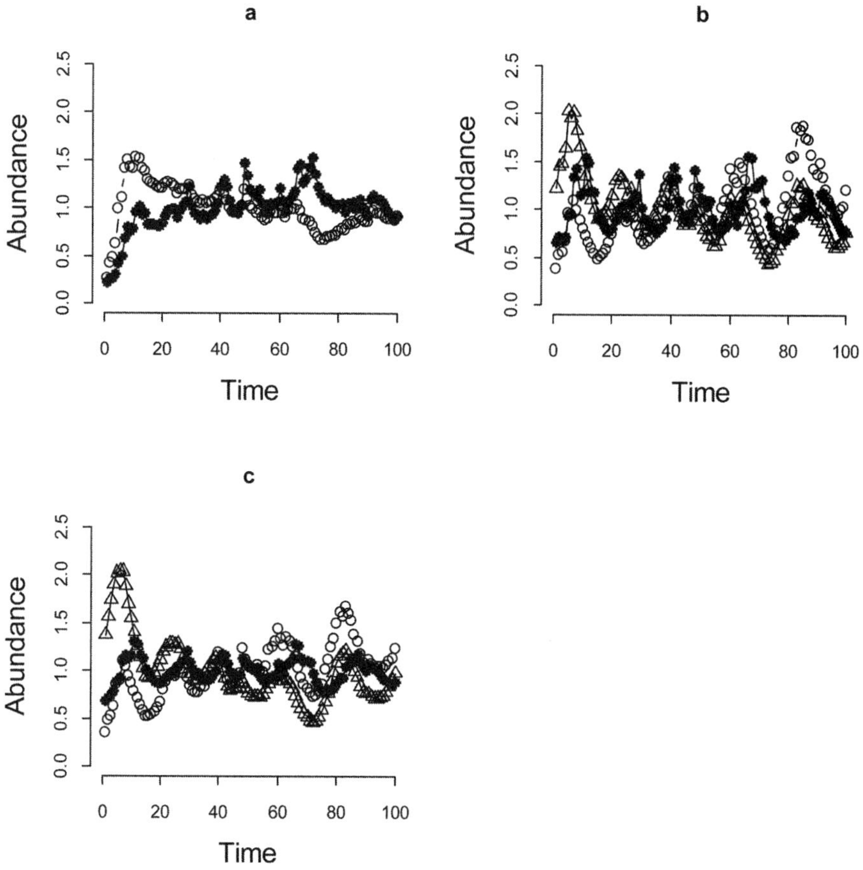

Fig. 7.8 Simulated time dynamics of a model with (a) wildebeest (open circles) and lions (filled circles); (b) wildebeest (open circles); topi (open triangles), and lions (filled circles); or (c) wildebeest (open circles), other prey (open triangles), and lions (filled circles). All sets of simulations had the same parameters for the equations expressed in the text: $a = f = 0.2$, $b = g = 0.3$, $c = 0.3$, $d = 0.1$, $h = 0.18$, $j = 0.1$, ε is normally distributed with a standard deviation = 0.4 and mean = 0.0 except in panel (c), where other prey experience no environmental stochasticity.

in vulnerability to predators, simply because adults are basically in a life stage "refuge" inaccessible to predators. Nonetheless, the model predicts that we should still often see quasi-cyclic variation (extended cycles with slow rates of damping) due to stochastic events and the coupled interactions between predators and prey.

Adding other herbivores of similar size to such a food web would inevitably increase variation in the abundance of both predators and prey, simply because more individuals are available to feed the predator population. Note that the model assumes no direct competition between herbivores, although apparent competition is inherent. Under such circumstances, the fact that all herbivore species do not respond similarly to environmental stochasticity has noteworthy implications. If all herbivore species prosper during years with large SOI values, this would set off predator-prey quasi cycles of large amplitude (fig. 7.8b). When only one or two species (wildebeest and topi) respond simultaneously, but recruitment variation by all other species is asynchronous, then alternate prey buffer to some degree against boom and bust cycles for predators (fig. 7.8c) and the disastrous risk of extinction that might otherwise occur at low points of the cycle due to chance demographic events.

CONCLUSION

Long-term data support the hypothesis that global variation in climate, as measured by SOI, has substantial effects on juvenile recruitment of wildebeest, topi, and lions in the Serengeti ecosystem, whereas the majority of herbivore species are apparently insensitive to SOI variation. This environmental variability perturbs these populations around their long-term means, particularly in El Niño or La Niña years. The effect of SOI is complex, with influences on a wide range of environmental variables, including temperature, rainfall, fire frequency, and grass growth. Of these factors, however, dry season rainfall seems to have the strongest impact on wildebeest, probably because survival during the dry season is highly dependent on adequate food supplies largely generated by dry season rain showers. Similar effects on lions could stem from either increased access to juvenile wildebeest mediated by improvements in juvenile survival or longer periods of access to wildebeest herds in high SOI years because herds linger longer on the plains and wooded savanna ecotone before initiating migratory movements to more northerly woodlands. The implications of these stochastic links are explored in simple trophic models, suggesting they have important implications for long-term variability and probability of population persistence. There is some evidence that dry season rainfall may be increasing in the Serengeti (Ritchie 2008). If this trend were to continue, our data imply that wildebeest and topi recruitment should benefit, as would that of lions. This suggests that deeper appreciation of global controls on climate change may have important long-term implications for population abundance and persistence in the Serengeti ecosystem.

ACKNOWLEDGMENTS

The Natural Science and Engineering Research Council of Canada and the Frankfurt Zoological Society funded the long-term monitoring. We thank the Tanzania National Parks and Tanzania Wildlife Research Institute for permissions and support.

REFERENCES

Anderson, T. M., J. G. C. Hopcraft, S. Eby, M. Ritchie, J. B. Grace, and H. Olff. 2010. Landscape-scale analyses suggest both nutrient and antipredator advantages to Serengeti herbivore hotspots. *Ecology* 99:1519–29.

Anderson, T. M., K. L. Metzger, and S. J. McNaughton. 2007. Multi-scale analysis of plant species richness in Serengeti grasslands. *Journal of Biogeography* 34:313–23.

Anyamba, A., C. J. Tucker, and R. Mahoney. 2002. From El Niño to La Niña: Vegetation response patterns over east and southern Africa during the 1997–2000 period. *Journal of Climate* 15:3096–103.

Australian Government Bureau of Meteorology. http://www.bom.gov.au/climate/current //soihtml1.shtml.

Boone, R. B., S. J. Thirgood, and J. G. C. Hopcraft. 2006. Serengeti wildebeest migratory patterns modeled from rainfall and new vegetation growth. *Ecology* 87:1987–94.

Burnham, K. P., and D. R. Anderson. 1998. *Model selection and multimodel inference*. New York: Springer.

Coulson, T., E. Catchpole, S. D. Albon, B. J. T. Morgan, J. M. Pemberton, T. H. Clutton-Brock, M. J. Crawley, and B. T. Grenfell. 2001. Age, sex, density, winter weather, and population crashes in Soay sheep. *Science* 292:1528–31.

Gaillard, J. M., M. Festa-Bianchet, and N. G. Yoccoz. 1998. Population dynamics of large herbivores: Variable recruitment with constant adult survival. *Trends in Ecology and Evolution* 13:58–63.

Grace, J. B. 2006. *Structural equation modeling and natural systems*. Cambridge: Cambridge University Press.

Grange, S., P. Duncan, J. M. Gaillard, A. R. E. Sinclair, P. J. P. Gogan, C. Packer, H. Hofer, and M. East. 2004. What limits the Serengeti zebra population? *Oecologia* 140:523–32.

Hastings, A. 1983. Age-dependent predation is not a simple process. *Theoretical Population Biology* 23:347–62.

Holdo, R. M., R. D. Holt, and J. M. Fryxell. 2009. Opposing rainfall and fertility gradients explain the wildebeest migration in the Serengeti. *American Naturalist* 173:431–45.

Holmgren, M., P. Stapp, C. R. Dickman, C. Gracia, S. Graham, J. R. Gutiérrez, C. Hice, F. Jaksic, D. A. Kelt, M. Letnic, et al. 2006. Extreme climatic events shape arid and semiarid ecosystems. *Frontiers in Ecology and the Environment* 4:87–95.

Hopcraft, J. G. C., T. M. Anderson, S. Perez Vila, E. Mayemba, and H. Olff. 2012. Body size and the division of niche space: Food and predation differentially shape the distribution of Serengeti grazers. *Journal of Animal Ecology* 81:201–13.

Maddock, L. 1979. The "migration" and grazing succession. In *Serengeti: Dynamics of an ecosystem*, ed. A. R. E. Sinclair and M. Norton-Griffiths, 104–29. Chicago: University of Chicago Press.

McNaughton, S. J. 1985. Ecology of a grazing ecosystem: The Serengeti. *Ecological Monographs* 55:259–94.

Mduma, S. A. R., A. R. E. Sinclair, and R. Hilborn. 1999. Food regulates the Serengeti wildebeest population: A 40-year record. *Journal of Animal Ecology* 68:1101–22.

Mosser, A. J. M. Fryxell, L. Eberly, and C. Packer. 2009. Serengeti real estate: Density vs. fitness-based indicators of lion habitat quality. *Ecology Letters* 12:1050–60.

Oba, G., E. Post, and N. C. Stenseth. 2001. Sub-saharan desertification and productivity are linked to hemispheric climate variability. *Global Change Biology* 7:241–46.

Ogutu J. O., and N. Owen-Smith. 2003. ENSO, rainfall and temperature influences on extreme population declines among African savanna ungulates. *Ecology Letters* 6: 412–19.

Ogutu, J. O., H. P. Piepho, H. T. Dublin, N. Bhola, and R. S. Reid. 2007. El Niño-Southern Oscillation, rainfall, temperature and Normalized Difference Vegetation Index fluctuations in the Mara-Serengeti ecosystem. *African Journal of Ecology* 46:132–43.

Packer, C., R. Hilborn, A. Mosser, B. Kissui, M. Borner, G. Hopcraft, J. Wilmshurst, S. Mduma, and A. R. E. Sinclair. 2005. Ecological change, group territoriality and population dynamics in Serengeti lions. *Science* 307:390–93.

Peterson, R. O. 1977. *Wolf ecology and prey relationships on Isle Royale*. US National Park Service Scientific Monograph Series no. 11, US Government Printing Office, Washington, DC.

Post, E., and M. Forchhammer. 2002. Synchronization of animal population dynamics by large-scale climate. *Nature* 420:168–71.

Post, E., R. O. Peterson, N. C. Stenseth, and B. McLaren. 1999. Ecosystem consequences of wolf behavioural response to climate. *Nature* 401:905–07.

Ritchie, M. 2008. Global environmental changes and their impact on the Serengeti. In *Serengeti III: Human impacts on ecosystem dynamics*, ed. A. R. E. Sinclair, C. Packer, S. A. R. Mduma, and J. M. Fryxell, 183–208. Chicago: University of Chicago Press.

Saether, B. E. 1997. Environmental stochasticity and population dynamics of large herbivores: A search for mechanisms. *Trends in Ecology and Evolution* 12:143–49.

Sinclair, A. R. E. 1977. Lunar cycle and timing of mating season in Serengeti wildebeest. *Nature* 267:832–33.

Sinclair, A. R. E., S. A. R. Mduma, and P. Arcese. 2000. What determines phenology and synchrony of ungulate breeding in Serengeti? *Ecology* 81:2100–11.

Sinclair, A. R. E., S. A. R. Mduma, and J. S. Brashares. 2003. Patterns of predation in a diverse predator-prey system. *Nature* 425:288–90.

Sinclair, A. R. E., K. L. Metzger, J. M. Fryxell, C. Packer, A. E. Byrom, M. E. Craft, K. Hampson, T. Lembo, S. M. Durant, G. J. Forrester, et al. 2013. Asynchronous food web pathways could buffer the response of Serengeti predators to El Niño Southern Oscillation. *Ecology* 94:1123–30.

Smith, R. H., and R. Mead. 1974. Age structure and stability in models of predator-prey systems. *Theoretical Population Biology* 6:308–22.

Stenseth, N. C., G. Otterson, J. W. Hurrell, A. Mysterud, M. Lima, K. S. Chan, N. G. Yoccoz, and B. Adlansvik. 2003. Studying climate effects on ecology through the use of climate indices: The North Atlantic Oscillation, El Niño Southern Oscillation and beyond. *Proceedings of the Royal Society of London (B)* 270:2087–96.

Temple, S. A. 1987. Do predators always capture substandard individuals from prey populations? *Ecology* 68:669–74.

Tyler, N. J., M. C. Forchhammer, and N. A. Oritsland. 2008. Nonlinear effects of climate and density in the dynamics of a fluctuating population of reindeer. *Ecology* 89: 1675–86.

Wilmshurst, J. F., J. M. Fryxell, and P. E. Colucci, P. E. 1999. What constrains intake in Thomson's gazelles? *Ecology* 80:2338–47.

Wilmshurst, J. F., J. M. Fryxell, B. P. Farm, A. R. E. Sinclair, and C. P. Henschel. 1999. Spatial distribution of Serengeti wildebeest in relation to resources. *Canadian Journal of Zoology* 77:1223–32.

Wright, C. J., R. O. Peterson, D. W. Smith, and T. O. Lemke. 2006. Selection of northern Yellowtone elk by gray wolves and hunters. *Journal of Wildlife Management* 70: 1070–78.

PART II

Response of Biodiversity to Disturbance

EIGHT

From Bacteria to Elephants: Effects of Land-Use Legacies on Biodiversity and Ecosystem Processes in the Serengeti-Mara Ecosystem

Louis V. Verchot, Naomi L. Ward, Jayne Belnap, Deborah Bossio, Michael Coughenour, John Gibson, Olivier Hanotte, Andrew N. Muchiru, Susan L. Phillips, Blaire Steven, Diana H. Wall, and Robin S. Reid

One of the most powerful ideas of Western conservation is the separation of people from wildlife, with wildlife-only parks and human-dominated landscapes (Peterson 2001). This approach to landscape management was brought to East Africa and the Mara-Serengeti ecosystem by colonists from southern Africa and Europe about a century ago (MacKenzie 1987). With growing human populations, many conservationists point to an even greater need to save the "last wild places on earth" (Terborgh 1999). Recently, some native East Africans and social scientists have articulated an opposing view—that people and wildlife can coexist on the same landscapes in a sustainable and beneficial way (Brockington and Homewood 2001; Neumann 2002). According to this view, the removal of people to establish wildlife-only parks is a violation of ancient land and human rights.

There is support today for both points of view in East Africa. Supporting the humancentric view is that, as far as we know, humans and their ancient relatives have mingled with diverse assemblages of large mammals longer in the Mara-Serengeti ecosystem than anywhere else on earth—at least 3.7 million years (Leakey and Hay 1979). This ancient way of using the land, with a mixture of hominins and large mammals, weathered the hominid development of progressively more effective tools (Schick and Toth 1993), increasingly sophisticated hominin social organization and hunting ability (Rose and Marshall 1996) and, eventually, the evolution of *Homo sapiens*. Much more recently, about 3,000 years ago, herders migrated down from the drying Sahara with taurine cattle and settled in and around the ecosystem, launching a new land use—mixed wildlife and livestock grazing (Marshall 2000). About 2,500 years ago Bantu farmers migrated in from the west, bringing farming methods from the rainforest and creating another land use in the region—small-holder farming (Newman 1995)—which mixed with the older hunting and herding cultures in the Mara-Serengeti (Shetler 2007). At this time, the farmers generally cultivated the looser soils on the hills, the hunters used the woodlands, and the herders grazed their livestock on the plains. About 300–400 years ago (Homewood and Rodgers 1991), Maasai pastoralists moved down from the north and displaced the ancient herders. Today diverse groups of farmers, many of whom also hunt, are concentrated west of the ecosystem while the Maasai dominate the north, east, and south of the ecosystem in both Kenya and Tanzania. In the early 1900s the most recent form of land use arrived in this ecosystem: wildlife-only parks in the form of the Maasai Mara National Reserve (MMNR) and Serengeti National Park (other reserves were established more recently). Establishing these parks required the removal of people.

Maasai leaders claim that it is no coincidence that the most diverse assemblages of large mammals are in the lands that they inhabit in northern Tanzania and southern Kenya (Parkipuny 1991). However, even inside Maasailand, the growth of human populations, changes in land tenure, and the expansion of small towns are reducing wildlife in some places (Reid et al. 2008). For example, in the MMNR, poaching, growing human populations, pastoral homesteads, and small towns around the edge of the reserve caused about a 25% decline in seven ungulate wildlife species from 1989 to 2003 (Ogutu et al. 2009). These land-use changes outside reserves affect wildlife inside reserves; across Kenya wildlife populations are in decline both inside and adjacent to reserves (Ogutu et al. 2009; Western, Russell, and Cuthill 2009). These trends support the view that wildlife-only parks are essential for conservation and that use of the land surrounding the parks has profound effects.

Given this long history of land use, we sought to understand how different types (and intensities) of wildlife and livestock use, both ancient (mixed wildlife and livestock systems) and more modern (intensifying pastoralism and wildlife-only parks), influence biodiversity and soils in the Mara part of the ecosystem. Our comparisons do not distinguish the type and intensity of grazing by wildlife and livestock, but do represent contrasting and realistic options for grazing land use. To deepen our understanding, we chose to take a very broad view of biodiversity, to include soil organisms (bacteria, fungi, and nematodes), plants, and wildlife. We compared not only biodiversity and soils across types of grazing use, but also interactions within each grazing-use type. Although this study cannot definitively resolve the long-standing conflict of pastoralism (people) versus parks (wildlife), it can contribute to our understanding of how these land uses affect key elements of ecosystem structure in East African grasslands. Our main goal in writing this chapter is to encourage future research that integrates aboveground and belowground effects to answer questions surrounding sustainable management of these grasslands and ensure that humans and wildlife can continue to coexist and thrive for millennia to come.

STUDY SITE DESCRIPTION

The Mara part of the Mara-Serengeti ecosystem lies entirely in southwestern Kenya and consists of the MMNR (1°13′–1°45′ S, 34°45′–35°25′ E) and the surrounding private ranches, group ranches, and trust land. The MMNR represents about 25% of the 25,000 km^2 ecosystem and is a critical dry season grazing reserve for migrating wildlife. The central part of the landscape lies on the Mara plains, at an elevation of 1,550–1,650 m. To the west, the Siria escarpment forms part of the rift wall and rises 300 m above the Mara plains. The Talek, Sand, and Mara Rivers are the only perennial rivers that drain this landscape.

Annual rainfall in this part of the ecosystem is high, ranging from about 600 mm in the southeast to about 1,000 mm in the northwest, and is strongly influenced by the Lake Victoria weather system (Norton-Griffiths, Herlocker, and Pennycuick 1975). Rainfall is weakly bimodal with short rains in October and November and "long" rains from March to May (Lamprey and Reid 2004). However, rain often falls on and off throughout the year with a drier period in January and February. The long dry season lasts from June to October. Rainfall varies with a five-year quasi periodicity, with five wet years generally followed by five dry years. Dry to very dry years occur about once every four years, and wet to very wet years occur one year in ten. Long-term rainfall records in the town of Narok show rainfall increased from 1914 to 1966 and declined from 1966 to 2003. From 1989 to 2003, the wet seasons were the hottest in the 43-year period from 1960 to 2003, suggesting a warming trend. Similarly, the Normalized Difference Vegetation Index fell during the 1989–2003 period, particularly during the wet season, suggesting a drying trend (Ogutu et al. 2007).

Lower-lying areas are dominated by "black cotton" soils, which are dark brown to black, deep, sandy clay loam Vertisols with impeded drainage, while lighter and shallower sandy loams dominate hilltop areas (Lamprey and Waller 1990). Since the 1960s, when Maasai herders repeatedly burned the Mara landscape to drive away the tsetse fly, the landscape has remained open with grasslands on the plains, woodlands on hilltops, and ribbons of forest along the rivers (Lamprey and Waller 1990). Grasslands are dominated by *Themeda triandra* (Forssk) and *Eragrostis tenuifolia* (Steud.), and hilltops are often covered by *Croton dichogamus* (Pax) shrubs (Lamprey and Waller 1990).

The MMNR, along with the Serengeti, supports the most diverse herds of migrating ungulates in the world (Sinclair 1995). Two wildebeest migrations (including wildebeest, *Connochaetes taurinus*; zebra, *Equus burchelli*; and Thomson's gazelle, *Gazella thomsonii*) mix from July to October, one moving up from the Serengeti in the south and one moving down from the Loita plains in the north (Sinclair and Norton-Griffiths 1979). In the MMNR, these two herds mix with resident herds of wildlife, but only the resident herds are present in the ecosystem from about November until June.

In November 2002, 10 months after we collected the data described in this chapter, a team of 84 researchers conducted a fine-resolution ground count of people, settlements, and animals. They counted over 400,000 animals (or 30 million kg of animal biomass) in 1,457 km^2, including 43 species of wildlife and 4 species of livestock (and dogs) within the Mara ecosystem (Reid et al. 2003). There were significant numbers of cattle in the reserve at the time of this count, as well as a few settlements. As might be expected, the MMNR had a higher concentration and biomass of wildlife species, and the ranches had a higher concentration and biomass of livestock (table 8.1). The MMNR had a higher concentration of biomass overall. This wildlife difference is due in large part to the seasonal wildebeest migration; if we remove this species (ignoring the associated zebra and Thomson's gazelle) from the calculation, the wildlife density is similar for the group ranches and the MMNR. Despite this similarity, however, the composition of the wildlife community differed between the ranches and the MMNR; the ranch community had much higher densities of giraffe, impala, and Thomson's gazelle and lower densities of all other taxa. Human settlement density was more than one order of magnitude greater outside the reserve, and the human population density was two orders of magnitude greater.

METHODS

Natural Treatments

In January and February 2002, we selected sites featuring four different grazing-use types on one soil type—heavy, black, sandy clay loam Vertisols (fig. 8.1). Our plots were located as far as possible from obvious ecotones,

Table 8.1 Density of the 12 most common wildlife species, livestock types, humans and settlements (bomas), and biomass for all wildlife and livestock in the Maasai Mara National Reserve (reserve) and the pastoral group ranches (ranches) in mid-November 2002

Species of wildlife or livestock, humans, settlements	Reserve	Ranches
Density	No. km^{-2}	
Baboon	0.86 (0.16)	0.98 (0.28)
Buffalo	**1.30 (0.49)**	**0.08 (0.05)**
Eland	**1.37 (0.26)**	**0.15 (0.07)**
Elephant	**0.55 (0.11)**	**0.09 (0.04)**
Giraffe	**0.25 (0.05)**	**0.65 (0.09)**
Grant's gazelle	**2.72 (0.20)**	**1.96 (0.17)**
Impala	**6.08 (0.46)**	**12.22 (0.90)**
Thomson's gazelle	**21.30 (0.93)**	**28.12 (1.51)**
Topi	4.21 (0.29)	3.79 (0.24)
Warthog	**1.38 (0.90)**	**0.74 (0.07)**
Wildebeest	108.35 (8.83)	25.58 (1.80)
Zebra	21.01 (1.65)	15.80 (1.09)
Cattle	**15.89 (2.86)**	**34.30 (3.25)**
Sheep and goats	**9.19 (2.14)**	**61.96 (4.83)**
Donkeys	**0.01 (0.01)**	**0.26 (0.06)**
Domestic dogs	**0.02 (0.01)**	**0.33 (0.04)**
Humans (estimate)	0.72 (no SE)	13.71 (no SE)
Bomas	**0.017 (0.005)**	**0.360 (0.023)**
Biomass	kg km^{-2}	
Wildlife	**21,106 (11,075)**	**8,892 (3,516)**
Livestock	**3,043 (516)**	**7,321 (599)**
Total	**24,149**	**16,213**

Sources: Reid et al. (2003). Biomass for each species was based on Coe, Cumming, and Phillipson (1976).

Notes: Numbers in bold are significantly different at $P < 0.05$ (reserve vs. ranches) with the same results using Kruskal-Wallis, Wilcoxon, and repeated measures ANOVA

within similar vegetation physiognomy (apart from differences due to grazing), and at least 300 m from recently abandoned settlement sites and corrals, which are visible in the landscape for about 15 years after abandonment. We established and sampled four grazing-use types that constitute "natural treatments":

1. Livestock-dominated use (livestock): Plots were located about 300 m from the nearest pastoral settlements at the northern edge of the MMNR. Cattle dominated grazing, but other small ruminants were present.
2. Mixed livestock and wildlife grazing (mixed): Plots were located just inside and outside the northern reserve boundary.
3. Wildlife-only grazing (wildlife): Plots were located deep inside the MMNR boundary.
4. Ungrazed (seasonal): Plots had little to no visible use by either livestock or wildlife at the time of sampling and were grazed only during the seasonal wildebeest migration.

We established four replicate blocks for each grazing type spread from northwest to southeast along the northern border of the MMNR with four sets (or blocks, labeled A, D, E, and H) of the four grazing-use types aligned as plots (as much as was practical) perpendicular to a slight NW–SE-oriented rainfall gradient of about 750–900 mm mean annual rainfall. This plot placement allowed us to control for rainfall. Careful field sampling of soils allowed us to control for soil texture and type, as well as landscape position, with plots generally located mid-slope. One weakness in our design is that some of the

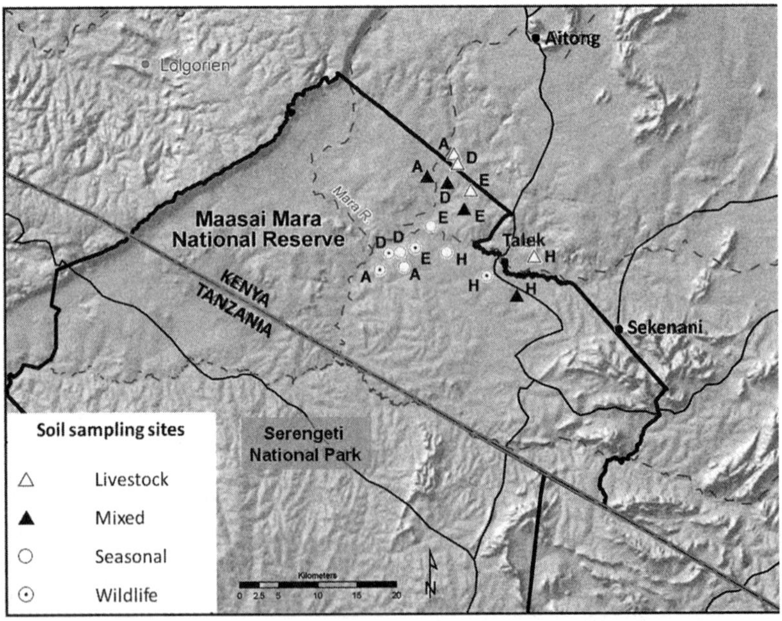

Fig. 8.1 The Maasai Mara National Reserve with study plots mapped along the northeastern border. Lettering as described in the text.

plots for each grazing-use type are closer to each other (but usually several kilometers apart) than they are to the plots of a different grazing-use type within the same block. The sampling plots ranged from 1,550 to 1,640 m elevation and were spread across 17.7 km from east to west and 19.6 km from north to south; the average distance between plots was 9.9 km with a minimum distance of 1.3 km between plots. Distances between plots within one grazing-use type (livestock, wildlife, mixed, and seasonal) ranged from 5 km to 10 km, and distances between plots within each block (A, D, E, H) ranged from 6 km to 11 km.

At the time of our study, the MMNR was mostly comprised of wildlife, seasonal, and mixed landscapes, whereas the adjacent ranches included only mixed and livestock landscapes. Subsequent research showed that wildlife use the livestock sites extensively at night when livestock are inside the settlement corrals (Reid et al. 2008) Three of the livestock plots were located in ranch land outside the northern reserve boundary, but within 2.5 km of the MMNR boundary; the other plot was located 1 km inside the reserve boundary. The remaining plots were all inside the reserve, with the mixed plots located 2–3 km from the nearest reserve boundary, the wildlife plots 3–13 km, and the seasonal plots 3–9 km from the boundary. The wildlife plots were close enough to the northern reserve border to be used by cattle (but not sheep or goats) in the dry season and droughts (Butt et al. 2009; Reid et al. 2003).

Dung and Vegetation

We quantified aboveground plant species richness, lichen cover, standing crop, and vegetation cover in each of the four land-use types within 20 × 20 m plots. At each plot, five linear transects, 20 m long and 5 m apart, were placed at a 45° angle to the direction of the slope. Within each plot, we randomly sampled 25 quadrats of 15 × 30 cm in size, in which we clipped all aboveground plant biomass to ground level and separated it into green grass, dry grass, shrubs, herbaceous forbs (mostly green), and litter. Within each quadrat, we estimated the percent cover of vegetation, litter, and bare ground such that the sum of the three categories totaled 100%. The percent cover of any lichens, mosses, liverworts, and/or visible cyanobacteria observed was also estimated. We collected dung samples at the same points as plant biomass. All the dung on a plot was aggregated and the total number of frames sampled was recorded. All plant and dung samples were dried at 100°C until they reached a constant weight (about 24 h) and then weighed.

In April 2002, we sampled plant species diversity in each of the 20 ×

20 m plots using a pin frame (McNaughton 1979) randomly placed within the plots. Eight pins, each 2.1 mm in diameter, were passed through the vegetation at a 65° angle. Plants intercepting the pins were identified and the interceptions were counted. On the completion of this data collection, each plot was carefully searched for additional rare species missed during the pin-frame sampling. The percent cover of these additional species was recorded by visual estimation.

We tested differences between means using ANOVA and a post hoc Student-Newman-Keuls test when the data were normal; otherwise, we used a Kruskal-Wallis test with Dunn's method post hoc. To analyze community composition using principal components analysis (PCA), we filtered the vegetation data and included only species that were present on 50% of the plots to ensure rare species did not unduly influence the results. This resulted in low plant cover (<80%) in a few cases. To remedy this, we retained data for any species that accounted for at least 5% of the cover on one or more plots. This resulted in the inclusion of at least 91% of the plant cover on any plot for the ordination exercise.

Soil Sampling

We took the soils for DNA sampling immediately after establishing the plot boundaries and before beginning any other activity inside the plot. A sterilized (washed with deionized water and rinsed with rubbing alcohol) African hoe (*jembe*) was used to dig two holes along each of the five 20-m-long transects. One hole was dug between the 2 and 3 m marks, and the other between the 16 and 20 m marks, with the exception of the third transect, along which the holes were dug at the center (10 m mark) because the same hole was used for bulk-density sampling. The holes were at least 15 cm deep and 20 cm wide. The researchers, wearing a hairnet and latex gloves, used a sterilized flour scoop (washed with soap and deionized water, autoclaved twice, and sealed in sterile plastic bags) to scrape a small, evenly distributed (down the soil profile) amount of soil from the top 10 cm of the soil into a fresh Whirl-Pak® bag. After all 10 holes per plot were sampled, the composite sample was immediately chilled in a cooler with ice. On returning to the laboratory in the evening, the samples were frozen on dry ice.

Soil samples for chemical and phospholipid fatty acid (PLFA) analysis were collected using tube samplers (2 cm diameter). Soil cores were collected to a depth of 10 cm, approximately 1 m to the right of each transect at 1-m intervals. The 100 soil samples were put into one container and mixed well after crushing the clods, and subsamples for the different analyses were prepared and transferred to plastic bags for storage. All but the chemistry sample were cooled immediately and kept chilled; the chemistry sample was left open to air dry. The samples to be used for PLFA were frozen upon return to the laboratory in the evening.

Soil Chemistry

Total C and N were analyzed by dry combustion in a Europa ANCA-GSL CNS analyzer. The $\delta^{13}C$ and $\delta^{15}N$ isotope ratios were measured by mass spectrometry using a Europa Hydra 20/20 isotope ratio mass spectrometer at the University of California, Davis. Extractable P, available K, and micronutrients were measured using a modified Olsen procedure for extraction (bicarbonate extraction) followed by colorimetric analysis. Exchangeable bases were extracted with 1*N* KCl (10:1 solution to soil). Exchangeable Ca and Mg were determined on an atomic absorption spectrophotometer; exchangeable Na was analysed on a flame photometer. Exchangeable K was obtained with a neutral NH_4OAc extraction solution and analyzed with a flame photometer. Soil texture was determined by sedimentation using the hydrometer method after dispersion with Na_2PO_7 (Day 1965).

Phospholipid Fatty Acid (PLFA)–Based Analysis of Microbial Community Composition

We studied the active microbial communities at these sites by characterizing the relative presence of short-chain phospholipids (C10–C20), which derive uniquely from microbes, to provide a fingerprint of the community. We used well-established "signature" lipid biomarkers from the cell membrane and wall of microorganisms to identify important functional groups within the community (White and Ringelberg 1998; Waldrop, Blaser, and Firestone 2000; Bünemann et al. 2004; McKinley, Peacock, and White 2005; Bossio et al. 2006). See Bossio and Scow (1998) for details of the methods for PLFA extraction. Samples were analyzed using a Hewlett Packard 6890 Gas Chromatograph and peaks were identified using bacterial fatty acid standards and MIDI peak identification software. Peak identification was verified by mass spectrometry. Fatty acid nomenclature is as follows: total number of C atoms refers to the number of double bonds, followed by the position of the double bond from the methyl end of the molecule; cis and trans geometry are indicated by the suffixes c and t; the prefixes a and i refer to anteiso- and iso-branching; 10me indicates a methyl group on the tenth

C atom from the carboxyl end of the molecule; positions of the hydroxygroups are noted; and cy indicates cyclopropane fatty acids.

To analyze community composition we used PCA. The original observations from the chromatograms were in units of nmole g-soil^{-1}. To evaluate community composition, we converted these observations to mol% to account for differences caused by extraction efficiencies. We filtered the data and included only PLFAs that were present on 50% of the plots to ensure that rare PLFAs did not unduly influence the results.

Ribosomal RNA Gene Sequencing–Based Analysis of Microbial Community Composition

Bulk genomic DNA was extracted and purified from soil samples using the UltraClean® Soil DNA Isolation Kit according to the manufacturer's instructions. The 16S ribosomal RNA genes were PCR-amplified using 27F (5′-GAGTTTGATCCTGGCTCAG–3′) and 1525R (5′-AGAAAGGAGGTGATCCAGCC–3′) primers, specific to the domain Bacteria, and Platinum® Taq PCR supermix. Amplifications were performed using a PTC–225 DNA Engine Tetrad Thermal Cycler with an initial denaturation of 2 min. at 94°C, followed by 29 cycles of 30 s at 94°C, 30 s at 55°C and 2 min. at 72°C, with a final extension of 5 min. at 72°C. The resulting PCR product was cloned using the TOPO® TA Cloning Kit, according to the manufacturer's instructions, and the cloning vector primers M13F and M13R were used to generate sequencing data. Overlapping 16S rRNA gene sequences of at least 500 base pairs were compiled and aligned using the Ribosomal Database Project (RDP) aligner (Cole et al. 2009). Potential sequence chimeras were identified using the Mallard software package (Ashelford et al. 2006), and all putative chimeras were removed from further analysis. Quality-checked sequences were realigned in RDP and assigned to operational taxonomic units (OTUs) using the Mothur software package (Schloss et al. 2009). Representatives of each OTU were assigned to taxonomic groups using the RDP Classifier software (Cole et al. 2009). Sequences assigned to phyla at greater than 80% confidence are reported; other sequences are designated as unclassified. Principal component clustering of bacterial 16S rRNA clone libraries was performed using the web-based UniFrac program (Lozupone, Hamady, and Knight 2006). Phylogenetic trees for input into UniFrac were generated using the neighbor program in the PHYLIP–3.68 software package (Felsenstein 1993). Community clustering dendrograms were generated using the abundance-based Jaccard similarity index (Bonin, Ehrich, and Manuel 2007) as implemented in the Mothur software package (Schloss et al. 2009). Differences in the count data between grazing use-type were analyzed using a generalized linear model (Proc GEN-MOD), assuming a Poisson distribution in SAS (Littell, Stroup, and Freund 2002).

Nematodes

Soils from each site and treatment were gently mixed in a plastic bag prior to shipping to Colorado State University, where they were stored at 4°C. Before enumerating the nematodes, we used subsamples to compare nematode abundance from two extraction methods: the Baermann funnel (Coleman et al. 1999), and sugar flotation and centrifugation (Jenkins 1964; Coleman et al. 1999). As the sugar flotation and centrifugation method resulted in greater nematode abundance, we used it for all sample analyses. Subsamples of 100 g were used to extract nematodes, followed by killing and preservation using the 5% hot-cold formalin procedure. The extracted nematodes were enumerated and identified to a trophic-group level under an inverted microscope. Trophic groups were assigned according to Yeates et al. (1993) and included bacterial and fungal feeders, plant parasites (nematodes that feed on plant roots), omnivores (nematodes that feed at several trophic levels), predators, algal feeders, and root associates (feeding habits are unconfirmed and can include root grazers or fungal feeders). A subset of each treatment was randomly selected and plant parasites were identified to genus and species where possible. To determine soil moisture, 50 g from each soil sample was oven-dried at 110°C for 24 h (Jackson, Anderson, and Pockman 2000). Nematode abundance is reported as the total number of live organisms kg^{-1} of soil oven-dry mass equivalent. Estimates of nematode abundance were log $(x + 1)$ transformed to satisfy assumptions of normality and uniformity of variance. Data were analyzed by analysis of variance (ANOVA). The Shannon (H′) and Simpson (1-D) trophic diversity indices were analyzed using untransformed data. The ratio of secondary consumers—fungal-feeding (FF) to bacterial-feeding nematodes (BF)—was used as an index of the relative importance of the fungal and bacterial energy channels.

ABOVEGROUND COMMUNITY STRUCTURE (PLANTS, WILDLIFE, LIVESTOCK)

The amounts of dung in the four grazing-use types did not differ ($F = 2.03$; $P = 0.144$), but variability within each grazing-use type was high (table 8.2).

Table 8.2 Mean ± SE of the biomass of green and dry grass, forbs, shrubs, litter and dung in the four grazing-use plots

Grazing regime	**Green grass**	**Dry grass**	**Forbs**	**Litter**	**Shrubs**	**Dung**
	t ha^{-1}					kg ha^{-1}
Livestock	9.3 (0.3)^b	6.9 (0.3)^b	5.7 (0.7)	7.1 (0.3)^b	4.4 (2.6)	168 (111)
Mixed	8.9 (0.5)^b	7.1 (0.2)^b	3.6 (0.5)	7.7 (0.3)^b	6.5 (0.8)	387 (112)
Wildlife	10.4 (0.4)^b	7.4 (0.3)^b	4.2 (0.5)	8.5 (0.4)^b	4.0 (1.1)	280 (127)
Seasonal	18.2 (1.4)^a	16.0 (1.5)^a	4.2 (0.5)	19.1 (1.7)^a	10.0 (7.5)	49 (28)
P-value	<0.0001	<0.0001	0.0712	<0.0001	0.3441	0.1389

Notes: Values in a column followed by the same letter are not significantly different from each other. *P*-values associated with the land-use term in the ANOVA are given in the final row of the table.

The dung data might suggest higher recent-grazing levels in the mixed plots and the lowest levels in the seasonal plots, but this measure may underestimate the true recent-grazing intensity because livestock are removed from the measurement areas and corralled at night. Seasonal plots supported about twice as much plant biomass as the three other grazing-use types (Kruskal-Wallis: $H = 59.0$, $P < 0.001$; table 8.2). The biomass of grass (both green and dry), litter, and shrubs was greatest on the seasonal plots. The livestock plots supported a greater biomass of forbs than the other plots. Differences in shrub biomass were difficult to discern because of the high variability in shrub presence from plot to plot.

The pin-frame measurements showed that livestock plots supported about 10% less total plant cover than the other three treatments ($H = 56.9$; $P < 0.001$; table 8.3). Lichen cover was much greater in the livestock and mixed plots than in the wildlife or seasonal plots ($\text{H} = 183.1$; $P < 0.001$; table 8.3). This is likely because these plots have more bare ground to colonize and more sunlight reaches the ground where the cover and height of plants are lower; lichens are able to withstand hoof pressure in this ecosystem, as they mostly colonize dead but flexible grass bases and thus bend with hoof pressure. Litter cover followed the same pattern as the total plant biomass, with about twice as much litter in the seasonal plots as in the other plots ($\text{H} = 66.9$; $P < 0.001$; table 8.3). Bare soil differed across all the plots in the order livestock > mixed > wildlife > seasonal ($\text{H} = 224.4$; $P < 0.001$; table 8.3). Finally, vegetation height was lowest in the livestock and mixed plots, with slightly taller vegetation in the wildlife plots and much taller vegetation in the seasonal plots ($\text{H} = 285.7$; $P < 0.001$; table 8.3).

Table 8.3 Ground cover and plant height for the four grazing-use types

Grazing regime	Vegetation (%)	Litter (%)	Lichens (%)	Bare ground (%)	Vegetation height (m)
Livestock	53.7	15.9	18.2	31.8	5.4
	(2.1)[b]	(1.7)[b]	(1.7)[a]	(1.9)[a]	(0.2)[d]
Mixed	69.1	14.1	16.3	17.0	7.3
	(1.5)[a]	(0.7)[b]	(2.0)[a]	(1.3)[c]	(0.4)[c]
Wildlife	73.9	14.8	1.5	11.4	13.9
	(1.3)[a]	(0.7)[b]	(0.4)[b]	(1.3)[b]	(0.5)[b]
Seasonal	68.7	26.6	0.0	1.1	27.2
	(2.1)[a]	(1.5)[a]	(0.0)[b]	(0.3)[d]	(0.8)[a]
P-value =	<0.0001	<0.0001	<0.0001	<0.0001	<0.0001

Notes: Values are means (SE). Values followed by the same letter are not significantly different from each other. *P*-values associated with the land-use term in the ANOVA are given in the final row of the table.

The PCA of the dominant plant species revealed clear differences between the plant communities, which clustered closely by grazing regime, with the exception of the wildlife plots (fig. 8.2). The first axis was weighted on the negative side of the axis by five grasses, *Themeda triandra, Sporobolus pyramidalis, Setaria sphacelata, Aristida adoensis,* and *Chloris gayana,* and on the positive side by four forbs, *Sida massaica, Dyschoriste radicans, Mariscus circumclusus,* and *Fimbristylis ovate,* and by the crabgrass *Digitaria macroblephara.* The second axis was weighted on the positive side by four grasses, *Bothriochloa insculpta, Setaria sphacelata, Aristida adoensis,* and *Chloris gayana,* and by one forb, *Fimbristylis ovata.* The negative side of the axis was weighted by the forb *Cynoctonum tuberculatum,* and four grasses *Sporobolus stapfianus, Panicum coloratum, Eragrostis tenuifolia,* and *Cynodon dactylon; C. dactylon* tolerates heavy grazing (Mwendera, Saleem, and Woldu 1997) and is often found on old pastoral settlement sites (Muchiru, Western, and Reid 2009).

SOIL CHEMISTRY

We are only just beginning to appreciate the importance of human activities for nutrient redistribution in East African grasslands. This redistribution creates edaphic heterogeneity that alters the vegetation composition locally and leads to the development of persistent shrub-free glades of nutrient-rich grasses, which increase the carrying capacity for large herbivores (Young,

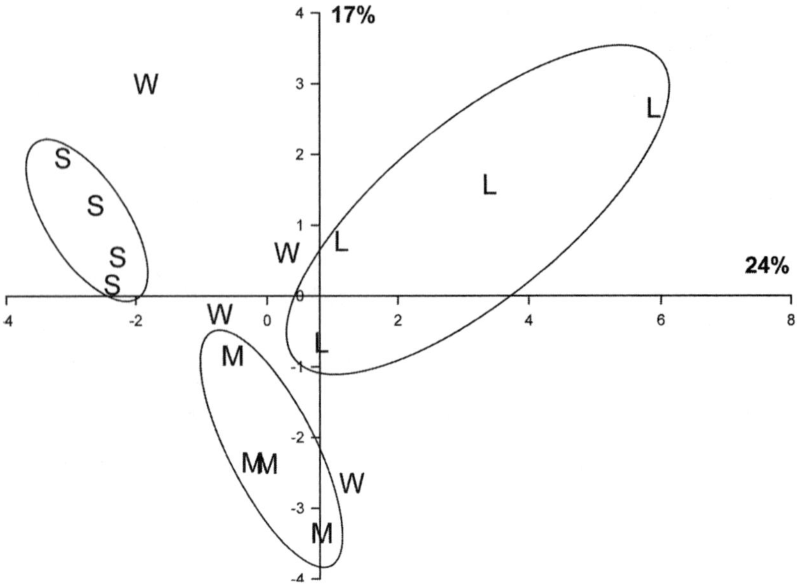

Fig. 8.2 Results of PCA of vegetation data showing first two component axes. Three of the grazing regimes cluster closely, whereas the wildlife grazing sites are spread out. S = seasonal, W = wildlife, M = mixed, L = livestock.

Partridge, and Macrae 1995; Augustine 2003). Intensive grazing can create positive feedback that maintains high levels of key nutrients; these glades can persist for up to 40 years in this ecosystem and longer in drier climates (McNaughton 1983; Young, Partridge, and Macrae 1995; McNaughton, Banyikwa, and McNaugton 1997; Augustine 2003; Muchiru, Western, and Reid al. 2009).

We found the livestock and mixed plots had elevated levels of most of the extractable elements measured, compared with seasonal plots (table 8.4), although we were unable to find statistically significant differences for any of the soil variables due to low replication and intersite variability. Even with our sparse sampling, *P*-values were particularly low for several parameters. As expected, this included available P, which is known to be redistributed by grazing animals via their feces (Belnap 2011); values tended to be lower in the mixed and livestock plots. The exchangeable bases Ca and Mg were slightly lower on the livestock sites than on the seasonal sites, and the *P*-values were relatively low. Thus, future work might look at the effects of grazing animals on these elements. We found no statistically significant difference in soil texture across the sites (table 8.4), nor any significant difference in soil water content or effect of the rainfall gradient.

Table 8.4 Soil chemical and physical properties by grazing-use type

Grazing type	**Extractable P**	**Available K**	**Extractable Zn**	**Extractable Fe**	**Extractable Mn**
	$\mu g\ g^{-1}$				
Livestock	2.03 (0.31)	346 (16)	1.45 (0.23)	122 (19)	46 (8)
Mixed	3.30 (0.97)	422 (76)	1.64 (0.31)	130 (26)	48 (5)
Wildlife	1.26 (0.25)	286 (28)	1.07 (0.15)	103 (11)	51 (5)
Seasonal	1.56 (0.17)	279 (37)	1.03 (0.04)	94 (19)	47 (6)
P-value	0.1065	0.2133	0.1216	0.3602	0.9150
	Extractable Cu	**Exchangeable Ca**	**Exchangeable Mg**	**Exchangeable K**	**Exchangeable Na**
	$\mu g\ g^{-1}$				
Livestock	3.53 (1.04)	2481 (265)	371 (27)	651 (40)	96 (10)
Mixed	2.60 (0.76)	3095 (275)	480 (30)	887 (159)	109 (27)
Wildlife	2.06 (0.28)	3721 (399)	461 (35)	614 (24)	125 (13)
Seasonal	1.76 (0.42)	3334 (433)	429 (38)	566 (82)	118 (14)
P-value	0.3481	0.1460	0.1207	0.1954	0.5046
	Sand	**Silt**	**Clay**	**Gravimetric water content**	
	%			$g\ kg^{-1}$	
Livestock	57.4 (3.0)	15.0 (1.7)	27.6 (1.3)	218	
Mixed	52.5 (3.1)	15.4 (1.2)	32.1 (2.2)	239	
Wildlife	53.5 (2.2)	12.7 (1.8)	33.8 (2.1)	205	
Seasonal	56 (3.7)	12.9 (1.5)	31.1 (4.3)	247	
P-value	0.7304	0.3604	0.4743	0.8006	

Notes: Values are means (SE). The *P*-values associated with the land-use term in the ANOVA are given in the final row of each section.

Table 8.5 Total organic carbon and nitrogen concentrations and isotope values

Grazing regime	Total C	Total N	C:N ratio	$\delta^{15}N$	$\delta^{13}C$
	g kg^{-1}			‰	
Livestock	23.2 (3.3)	1.83 (0.21)	12.6 (0.3)	7.20 (0.18)[a]	-12.51 (0.33)
Mixed	25.7 (2.0)	2.05 (0.19)	12.6 (0.4)	7.11 (0.53)[a]	-12.21 (0.17)
Wildlife	21.7 (1.8)	1.80 (0.16)	12.3 (0.3)	6.30 (0.13)[ab]	-12.43 (0.23)
Seasonal	20.2 (3.2)	1.58 (0.27)	12.7 (0.2)	5.56 (0.19)[b]	-12.30 (0.12)
P-value	0.5924	0.5551	0.7618	0.0060	0.8174

Notes: Values are means (SE). Values followed by the same letter are not significantly different from each other. The *P*-values associated with the land-use term in the ANOVA are given in the final row of the table.

We found no statistically significant differences in soil organic carbon (SOC) content between grazing-use types (table 8.5). Furthermore, the SOC pools had very similar $\delta^{13}C$ signatures across sites ($P = 0.61$), and there was no effect of grazing intensity. The $\delta^{13}C$ values indicate that the soil organic matter pool is predominantly derived from C_4 grasses. We also found no statistically significant effect of grazing intensity on the total N stocks in the soils. The absolute values of the stocks followed those of total organic carbon, with the highest values on the mixed plots and the lowest values on the seasonal sites. The ratios of C:N were fairly constant and unaffected by grazing.

The clearest finding from the soil measurements was that more intensive livestock grazing in the region is altering the N cycle, as we found $\delta^{15}N$ was significantly higher on the livestock and mixed sites compared with the seasonal sites. Sites intensively grazed by wildlife had slightly elevated $\delta^{15}N$ content compared with seasonal sites. This was likely due to two factors. First, urine has a high $\delta^{15}N$ signal due to partitioning within the animal. Ungulates in tropical and temperate grassland ecosystems consume around 50% of the aboveground plant production; in rangelands with livestock, consumption can be as high as 75% of the production (Frank, Evans, and Tracy 2004). Grazers return a large proportion of the ingested N (>65%) to the ecosystem through their urine, which represents a major N flux in this ecosystem and other grasslands (Mould and Robbins 1981; Ruess 1987; McNaughton, Ruess, and Seagle 1988; Hobbs 1996; Frank, Evans, and Tracy 2004). Second, enhanced urease activity on these sites (L. Verchot, unpublished) suggests that volatilization is an important loss pathway in this ecosystem, which would contribute to an increase in $\delta^{15}N$ values on more intensively grazed sites. The heavier isotope tends to remain in the soil during volatilization.

BELOWGROUND COMMUNITY STRUCTURE

PLFA-Based Analysis of Microbial Community Structure

The various components of the microbial community respond differently to changes in the quality of organic matter inputs associated with management and land-use change. Plant species that are adapted to fertile conditions return high-quality litter (low C:N ratios, low phenolics, low lignin, and structural carbohydrates) and support soil food webs in which energy transfers are accomplished through bacterial channels. By contrast, low-quality organic matter inputs from the plant community condition food webs on infertile soils, in which fungal energy channels predominate (de Ruiter, Neutel, and Moore 1995; Wardle et al. 2004). Thus, there is tight linkage between the trajectories of the microbial community and the aboveground vegetation, and microbial communities respond rapidly to stress and disturbance (Sowerby et al. 2005).

The PCA revealed good separation of the microbial community composition between the group of livestock, mixed, and wildlife plots and the less heavily grazed seasonal plots (fig. 8.3). The first component had positive loading with monounsaturated PLFAs indicative of Gram-negative bacteria, which also indicates well-aerated conditions, and high negative loading of branched and normal saturated PLFAs indicative of Gram-positive bacteria. The second component is more difficult to interpret. It has high positive weighting for one of the fungal markers (18:1ω9c) and two actinomycete markers (10me 18:0 and 10me 17:0), which were correlated ($R = 0.71$, $P = 0.002$). The component had high negative weighting for the third uncorrelated actinomycete marker (10me 16:0) and for one of the protozoa markers (20:2 ω6,9c). The component also had negative weighting for a number of biomarkers for Gram-positive bacteria and positive weighting for a number of Gram-negative bacteria biomarkers. Thus, this axis sorts sites by the nonbacterial components of the community and provides some detailed sorting between Gram-positive and Gram-negative bacteria, perhaps indicating a combination of soil moisture and carbon source conditions. The greatest separation along this second component was between the livestock and seasonal plots. The mixed-grazing plot in block A stands out from the rest and had a very different community from all other plots examined here, with extreme values for a number of the PLFAs.

We explored the functional groups in greater detail (table 8.6). Seasonal sites had higher values for Gram-positive biomarkers and lower values for Gram-negative, fungi, sulfate reducers, and protozoa markers. These sites also had higher bacteria-to-fungi ratios. The livestock sites showed almost the opposite picture with lowered Gram-positive markers, elevated markers

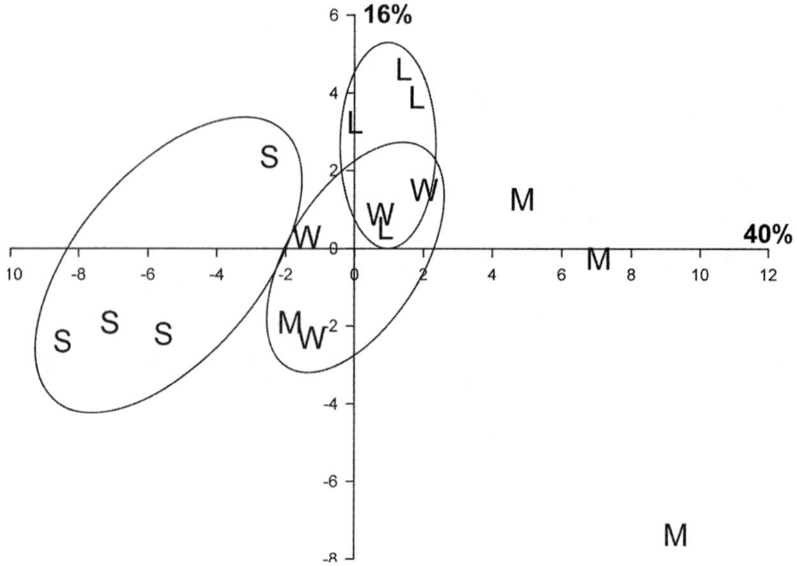

Fig. 8.3 Results of PCA of PLFA data showing the first two component axes. Three of the grazing regimes cluster closely, whereas the mixed-grazing system is more spread out. S = seasonal, W = wildlife, M = mixed, L = livestock.

for fungi, and somewhat elevated markers for Gram-negative bacteria, sulfate reducers, and protozoa. The mixed and wildlife sites were intermediate in many of the functional groups, but the mixed sites stood out as having the highest levels of protozoa and sulfate reducer markers. The livestock sites had the highest values for Gram-negative and fungal biomarkers and the lowest values for Gram-positive. These sites also had lower bacteria-to-fungi ratios. The three ratios that indicate stress in the bacterial community (Bossio et al. 2006; McKinley, Peacock, and White 2005; Ponder and Tadros 2002) gave inconsistent results. The sum of the ratios of the cyclopropanes to their precursors was not significantly different between grazing-use types. The ratio of normal saturated to monounsaturated PLFAs was lower on the livestock plots, while the ratio of iso to anteiso 15:0 and 17:0 PLFAs was lower on the seasonal plots.

Thus, the clearest result emerging from the PLFA analysis is that the seasonal sites had a significantly different microbial community composition from the others, with bacteria tending to dominate the microbial community of seasonal sites more. In particular, there was a higher bacterial-to-fungal fatty acid ratio, indicating more fertile soil conditions (Bardgett et al. 2006). A greater abundance of bacteria relative to fungi is typically

Table 8.6 Total PLFA (nmoles g-soil^{-1}), microbial community composition (mole% for each biomarker), and index ratios of the microbial community biomarkers

	Livestock		Mixed		Wildlife		Seasonal		$P > \|F\|$
Total PLFA	103	(12.73)	135.3	(9.89)	110.13	(8.14)	125.65	(6.50)	0.1215
Biomarkers									
Gram-positive	27.00	(0.45)^{c}	27.22	(0.84)^{c}	28.72	(0.30)^{b}	31.03	(0.78)^{a}	0.0002
Gram-negative	23.32	(0.30)^{a}	22.50	(0.33)^{a}	21.86	(0.15)^{a}	19.63	(0.22)^{b}	0.0008
Actinomycete	7.67	(0.19)	8.22	(0.32)	7.42	(10.00)	7.45	(0.26)	0.1474
Fungi	14.77	(0.72)^{a}	12.02	(1.08)^{ab}	12.16	(0.70)^{ab}	11.11	(0.33)^{b}	0.0452
AMF	2.33	(0.03)	2.51	(0.19)	2.53	(0.03)	2.26	(0.09)	0.2456
Total fungi	17.1	(0.74)^{a}	14.53	(0.93)^{ab}	14.70	(0.70)^{ab}	13.37	(0.32)^{b}	0.0347
Protozoa	0.44	(0.01)^{ab}	0.55	(0.08)^{a}	0.42	(0.02)^{ab}	0.33	(0.02)^{b}	0.0448
Sulfate reducers	0.59	(0.03)^{ab}	0.75	(0.09)^{a}	0.62	(0.02)^{ab}	0.47	(0.03)^{b}	0.0318
Ratio indices									
Bacteria:fungi*	3.43	(0.08)^{b}	4.20	(0.19)^{ab}	4.19	(0.10)^{ab}	4.57	(0.06)^{a}	0.0366
Sum of cyclopropyl to precursor ratios†	0.57	(0.01)	0.53	(0.02)	0.56	(0.02)	0.61	(0.01)	0.0806
Normal saturated:monounsaturated‡	1.02	(0.02)^{b}	1.15	(0.04)^{a}	1.12	(0.02)^{a}	1.16	(0.03)^{a}	0.0201
(i15:0 + i17:0)/(a15:0 + a17:0)§	1.41	(0.03)^{a}	1.55	(0.19)^{a}	1.30	(0.11)^{a}	0.96	(0.11)^{b}	0.0142

Notes: ANOVA results are reported in the last column. In rows where a significant effect of grazing intensity was found, results of the means separation using the Student-Newman-Keuls test are indicated by letters. Values followed by the same letter in each row are not significantly different from each other.

* This ratio was calculated using Gram-positive and Gram-negative biomarkers and the two traditional fungal markers,18:2ω6c and 18:1ω9c. We did not include the biomarkers for actinomycetes in the total bacteria or the AMF marker in the estimate of total fungi.

† This ratio assesses the growth rate of the Gram-negative community. Lower values indicate rapid growth; higher values indicate a stress situation and slow growth.

‡ This ratio compares general bacteria to Gram-negative biomarkers and has been proposed as an indicator of nutrient stress (McKinley, Peacock, and White 2005). Higher values indicate nutrient stress.

§ This ratio is also an indicator of the growth stage of the Gram-negative microbial community (Bossio et al. 2006; McKinley, Peacock, and White 2005). Lower values indicate rapid growth; higher values indicate a stress situation and slow growth.

associated with increased rates of nutrient cycling (Bardgett et al. 2006). In contrast to the seasonal sites, the livestock plots had higher fungi values and lower bacterial-to-fungal fatty acid ratios. These results may indicate that these plots are being overgrazed and that fungal energy pathways are more important in these soils. This is consistent with observations in the nematode community (see below).

Ribosomal RNA Gene Sequencing–Based Analysis of Microbial Community Composition

Analysis of 16S ribosomal (r) RNA genes is currently the standard molecular genetics approach for determining the composition of soil microbial communities. These genes are present in all Bacteria and Archaea and can be retrieved easily from environmental samples. The 16S rRNA genes can be analysed by sequencing, or by profiling methods such as denaturing gradient gel electrophoresis (Muyzer and Smalla 1998) or terminal restriction fragment length polymorphism (Liu et al. 1997).

We determined the number of OTUs (a species equivalent for microbes) by clustering sequences at 99% identity. We then used the frequency of these OTUs to describe the diversity of the microbial communities (table 8.7). The Shannon diversity (H′) index, a metric that assesses both the community richness and evenness, showed moderate variability between soils. However, soils with lower numbers of sequences generally showed lower diversity, suggesting a sampling effect.

The PCA of the bacterial populations (fig. 8.4A) suggested that there was an effect of grazing-use type and rainfall gradient. For example, the first component showed positive values for three of the wildlife sites and two seasonal sites, and negative values for all livestock and mixed sites. The second component showed positive values for all sites in blocks A and D, and negative values for all sites in blocks E and H. It is important to note that the first and second components accounted for only 13% and 8% of the variation in the data, respectively, and that no particular microbial group was found to significantly contribute loading to either component. To further compare the effect of grazing treatment on bacterial communities, we compiled 16S rRNA gene sequences from sites under the same grazing use, and then compared them using the abundance-based Jaccard similarity coefficient (Bonin, Ehrich, and Manuel 2007; fig. 8.4B). The seasonal grazing-use type appeared to be the most distant from the other treatments.

At the phylum level, all of the soil microbial communities are dominated by the same five phyla: Acidobacteria, Actinobacteria, Proteobacteria,

Table 8.7 The 16S rRNA gene sequencing and diversity statistics for the total number of clones identified in each sample

Plot/block	No. of clones	No. of OTUs*	H′*	ChaoI*
Seasonal				
A	191	180	5.1	3190
D	n.d.	n.d.	n.d.	n.d.
E	147	135	4.9	996
H	201	165	5.0	832
Wildlife				
A	166	159	5.1	1593
D	159	156	5.0	3063
E	160	131	4.8	553
H	163	125	4.7	365
Mixed				
A	170	162	5.1	1867
D	192	168	5.1	977
E	187	159	5.0	929
H	130	104	4.5	460
Livestock				
A	130	117	4.7	912
D	166	158	5.0	1400
E	170	147	4.9	755
H	136	123	4.8	632

Notes: Operational taxonomic units (OTUs) are the species equivalents for bacteria. The OTUs were used to compute the Shannon diversity index (H′). The ChaoI estimator was used to predict the total number of OTUs present in each grazing-use type. Note that some data are missing for block D because of a problem with the seasonal sample.

* Calculated at 0.01 sequence difference.

Verrucomicrobia, and Firmicutes (fig. 8.5A). These are five of the nine dominant soil phyla identified in a 2006 meta-analysis of soil microbial community studies (Janssen 2006), and partially overlap (Acidobacteria, Actinobacteria, and Proteobacteria) with the four recurring dominant phyla listed in a more recent meta-analysis (Fierer et al. 2009). The other four dominant phyla described by Janssen (Bacteroidetes, Chloroflexi, Planctomycetes, and Gemmatimonadetes) are numerically the next most abundant groups in our data. Thus, it appears that soils of the Mara-Serengeti ecosystem adhere to the remarkably consistent phylum-level compositions observed in other soils worldwide. The similar overall composition of the bacterial communities under different treatments suggests that the primary factor leading

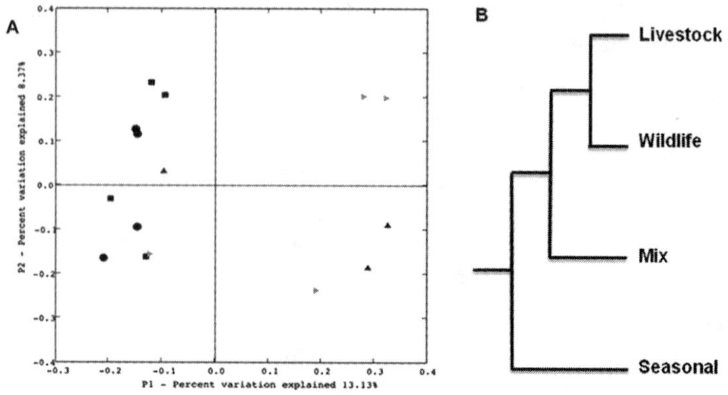

Fig. 8.4 Community clustering of plots. (A): UniFrac-generated PCA of 16S rRNA clone library sequences from each plot. Letters represent the blocks. The percent diversity described by each component is indicated on the axis. Mixed plots are indicated by filled circles (●); livestock plots, squares (■); seasonal plots, dark triangles (▲); and wildlife plots, light triangles (▲). (B): clustering of 16S rRNA clone libraries based on grazing type. Branch lengths were generated using the abundance-based Jaccard similarity coefficient as implemented in the Mothur software package (Schloss et al. 2009). The community dendrogram was generated using the Treeview program in the PHYLIP package (Felsenstein 1993).

Note: Data for the seasonal plot in block D are missing because of a problem with the sample.

to treatment effects is the relative abundance of individual phyla, rather than the presence of unique phyla. This is supported by comparison of the relative abundance of the seven most abundant phyla across the grazing regimes (fig. 8.5B). There was a significant difference in the relative abundance of phyla, particularly with lower abundance of Firmicutes ($P < 0.0001$) and higher abundance of Verrucomicrobia ($P < 0.0001$) and Acidobacteria ($P = 0.0432$) in seasonal plots compared with the other treatments. This result for Verrucomicrobia is particularly interesting, as Philippot et al. (2009) showed Verrucomicrobia were significantly less abundant in areas of high livestock impact. In that study, soil nitrate and ammonium were the variables most strongly correlated with the relative abundance of the Verrucomicrobia. In further contrast to our results, Philippot et al. (2009) showed no grazing effect for Firmicutes, which they attributed to the inclusion of spore-formers—able to survive environmental fluctuations—in this group.

We examined the subphylum composition for Verrucomicrobia and Firmicutes to discover whether particular lower taxa were responsible for the differences. In the case of Verrucomicrobia, all four treatments were dominated by members of the family Xiphinematobacteriaceae, which are reported to engage in a specific symbiotic relationship with nematodes of the genus *Xiphinema* (Vandekerckhove et al. 2000). Surprisingly, no members of this nematode group were recovered (see next section), suggesting either

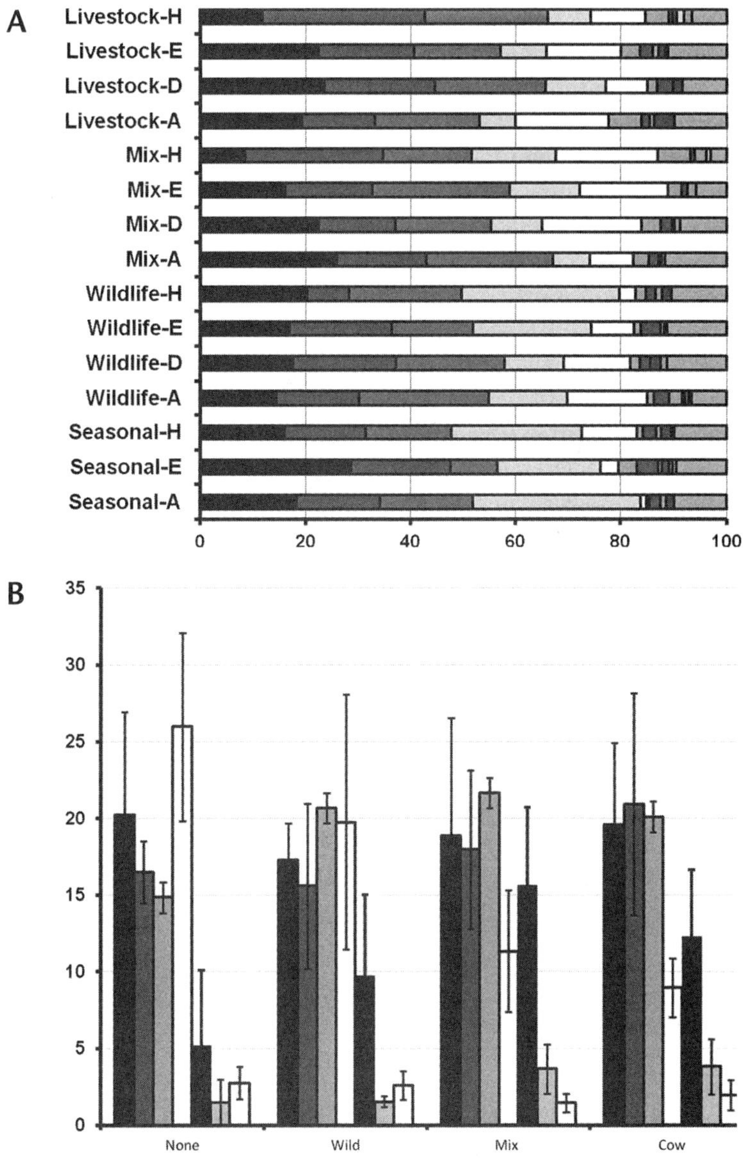

Fig. 8.5 Relative abundance of bacterial phyla identified in the 16S rRNA gene clone libraries.

Notes: (A) Abundance of 16S rRNA gene sequences in each plot. Shaded regions from left to right represent: Acidobacteria, Actinobacteria, Proteobacteria, Verrucomicrobia, Firmicutes, Bacteroidetes, Gemmatimonadetes, Planctomycetes, BRC1, Chloroflexi, Cyanobacteria, Nitrospira, OP10, Thermomicrobia, unclassified. (B) Relative abundance of the seven dominant bacterial phyla identified in the plots, grouped by grazing type. The phyla are from left to right: Acidobacteria, Actinobacteria, Proteobacteria, Verrucomicrobia, Firmicutes, Gemmatimonadetes, Planctomycetes. Error bars represent the standard error of the mean relative abundance for each phylum. Asterisk indicates significant difference ($P < 0.05$) between treatments. Statistical differences were determined using pair-wise ANOVA of the variation of the mean relative abundance. Data are missing from the Seasonal plot in block D.

that another nematode genus in the soils harbors verrucomicrobial symbionts, or that the Xiphinematobacteriaceae sequences recovered represent a nonsymbiotic verrucomicrobial taxon. The relative abundance of these sequences was two to three times greater in the wildlife and seasonal plots than in the livestock and mixed plots. Within the Firmicutes, all grazing treatments were dominated by members of the family Bacillaceae, containing sequences mostly affiliated with the genus *Bacillus* that occurred two to three times more frequently in the livestock and mixed plots than in the wildlife and seasonal plots.

We also examined the subphylum composition of the phylum Acidobacteria, as it is the second most abundant soil group in Janssen's (2006) analysis and is poorly characterized compared with other abundant groups such as Actinobacteria (actinomycetes), Firmicutes (Bacillus and Clostridium), and Proteobacteria. Acidobacteria were affiliated with 13 of the 26 groups described by Barns et al. (2007); groups 1, 3, 4, and 6 were the most abundant in our study, as found in previous studies (Barns, Takala, and Kuske 1999; Jones et al. 2009). There were no significant differences in acidobacterial phylum or subphylum composition across the different grazing-use types. This suggests that Acidobacteria may be relatively resistant to the disturbances introduced by grazing, perhaps because of their reported metabolic flexibility and ability to grow in diverse environments (Quaiser et al. 2003; Bryant et al. 2007; Pankratov et al. 2008; Ward et al. 2009).

Nematodes

Nematodes are a major microinvertebrate group in grassland soils (Ingham and Detling 1990; Wall-Freckman and Huang 1998) and the nature of this community affects decomposition, belowground herbivory, and nutrient cycling (Seastedt, Todd, and James 1987; Porazinska et al. 2003). In our study, the average nematode abundance ranged from 3,788 kg dry soil^{-1} in mixed plots to 7,194 kg dry soil^{-1} in wildlife plots. Total nematode abundance was similar across the four treatments. Predators and algal feeders had significantly lower densities than other trophic groups in all sites (fig. 8.6). Plant parasites were the dominant trophic group and had significantly higher densities than other trophic groups in seasonal ($P < 0.0001$) and mixed plots ($P < 0.0001$). In seasonal plots, plant parasites were significantly more abundant than bacterial feeders, followed by root associates (not significantly different from bacterial feeders). Sites grazed only by wildlife followed a similar pattern, but there were no significant differences between the three trophic groups. However, in the wildlife sites, plant parasites were more

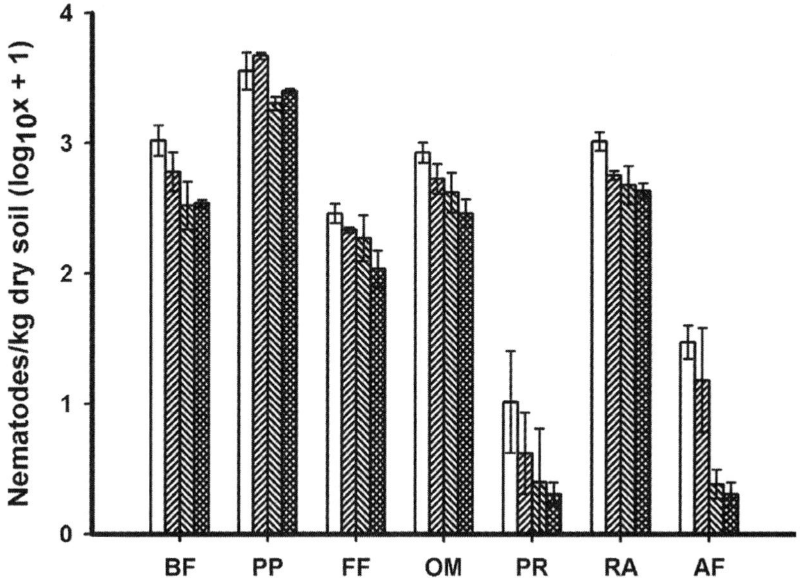

Fig. 8.6 Nematode trophic composition in four grazing-use types (wildlife = white bar, seasonal = downward left diagonal, livestock = downward right diagonal, and mixed = diamond hatch). BF = bacterial feeders, PP = plant parasites, FF = fungal feeders, OM = omnivores, PR = predators, RA = root associates, AF = algal feeders. Vertical bars are one SE.

abundant than fungal feeders, algal feeders, and predators ($P < 0.0001$). In livestock and mixed plots, root associates, rather than bacterial feeders, were the second most abundant group, but the difference from plant parasites was significant only in the mixed sites.

Plant-parasitic nematodes were represented by members of the *Helicotylenchus, Hoplolaimus, Criconemoides*, and *Tylenchus* genera, which vary in their potential impact on root biomass and plant productivity. This representation is comparable to that found in other ecosystems, but we cannot confirm the presence of any endemic species because we did not identify to the species level within trophic groups. The plant-parasitic genus *Xiphinema* was not found, although members of the bacteria family *Xiphinematobacteriaceae*, previously known only as symbiotic associates of *Xiphinema* (see 16S rRNA section above), occurred widely. The Shannon index suggested that trophic group diversity is more evenly distributed in wildlife and livestock sites than in seasonal and mixed sites. The Simpson index also indicated greater trophic group diversity in wildlife and livestock plots.

Belowground herbivory by nematodes—not only aboveground herbivory—is important in ecosystems (Stanton 1983). Ingham and Detling

(1990) showed that, in moderately grazed pastures, cattle, arthropods, and nematodes each accounted for approximately 33% of herbivory, with lagomorphs and rodents contributing less than 1%. As noted above, plant parasites, root associates (both having a role in the herbivore functional group), and bacterial feeders were the three dominant trophic groups in wildlife and seasonal plots. This distribution could indicate either high herbivory or a resilience of the plant species to belowground grazing. Ingham and Detling (1990) suggested that heavy aboveground grazing promoted belowground grazing by root herbivores and decreased net primary production. However, Bardgett and Wardle (2003) found that results of studies on belowground herbivory were idiosyncratic and suggested additional mechanisms by which aboveground grazing influences soil food webs and nutrient cycling (see also Porazinska et al. 2003). Our results reflect those of other systems, such as grazed areas in a semiarid steppe (Wall-Freckman and Huang 1998) or the US semiarid grasslands, where low annual precipitation and historical grazing by bison resulted in a large proportion of belowground plant biomass and herbivory (Milchunas and Lauenroth 1993).

Nematodes have been used as a biological index of soil disturbance for many food web pathways (Bongers and Ferris 1999). Nematode trophic composition reflects the multiple living food sources in soils and provides information on the functioning of the soil food web (Moore and de Ruiter 1991). Hunt et al. (1987) estimated that 83% of the total organic matter mineralization in a semiarid steppe attributable to soil invertebrates was due to bacterial-feeding nematodes and protozoa. The decomposition pathways and nutrient retention in soils are indicated by the composition of the microbial community (see PLFA section) and its invertebrate consumers, shown by the FF:BF ratio, which can also indicate soil disturbance (Beare 1997). Bacterial-feeding nematodes initially stimulate mineralization by bacteria through faster recycling of nutrients and decomposition than via fungal pathways (Hunt et al. 1987). For example, Bongers and Ferris (1999) found that bacterial feeders and other opportunistic nematodes increase in numbers four days after manure was added to soil but that fungal decomposition followed some days later. In contrast, we did not find nematode patterns consistent with our dung data, which suggests that other sources (e.g., the rhizosphere) were supplying the labile carbon.

The FF:BF nematode ratio results (table 8.8) suggest more energy flows through the bacterial decomposition pathway for all sites except those grazed by livestock exclusively, where fungi appear to play a slightly greater role; this is consistent with the PLFA results showing higher relative abundance of fungal:bacterial PLFA on livestock plots. These findings suggest that these plots may be overgrazed. The difference between our results and

Table 8.8 Nematode trophic diversity measured using the Shannon (H′) and Simpson (1-D) indices and ratio of fungal-to bacterial-feeding nematodes (FF:BF)

Site	Shannon (H′)	Simpson (1-D)	FF:BF
Livestock	1.32	0.654	0.54
Mixed	1.02	0.490	0.34
Wildlife	1.34	0.659	0.32
Seasonal	1.05	0.498	0.29

those of Bongers and Ferris (1999) could be that their study included high-productivity cattle in the Netherlands, presumably with a different food source, and hence there were more labile feces patches in their grazing experiment. However, in this study, the ratio of FF:BF nematodes (table 8.8) appears to indicate more disturbance to the livestock site, perhaps due to the greater area of bare ground (table 8.3), lower vegetation height (table 8.4), and highest percentage of sand (table 8.4), all of which create microclimatic effects in soil habitat that may alter decomposition pathways.

Overall, the results of our study indicate differences between grazing-use types in soil nematode diversity, evenness between species within the community, and distribution of trophic groups. Particularly striking in terms of trophic group composition was the finding that bacterial feeders were the second most abundant group in the wildlife and seasonal plots, which may indicate that plant herbivory and decomposition were reduced in the livestock and mixed plots.

CORRELATIONS BETWEEN MEASURES

Our study includes a wide range of measures of the structures of different components of the ecosystem. Within the boundaries of our natural experiment, there is sufficient variation between factors to investigate how several factors might co-vary. In the following analysis, we explore significant correlations and draw upon current ecological knowledge of tropical grasslands to propose hypotheses for more systematic testing in the future.

We begin this analysis by examining the effect of soil fertility on the structure of aboveground and belowground communities. Ruess and Seagle (1994) showed that soils in the northern part of the Mara-Serengeti ecosystem have some of the lowest total N and extractable P contents along the eutrophic-dystrophic gradient of soils that runs from south to north in the ecosystem. In our study, the relationships between the PCA scores for

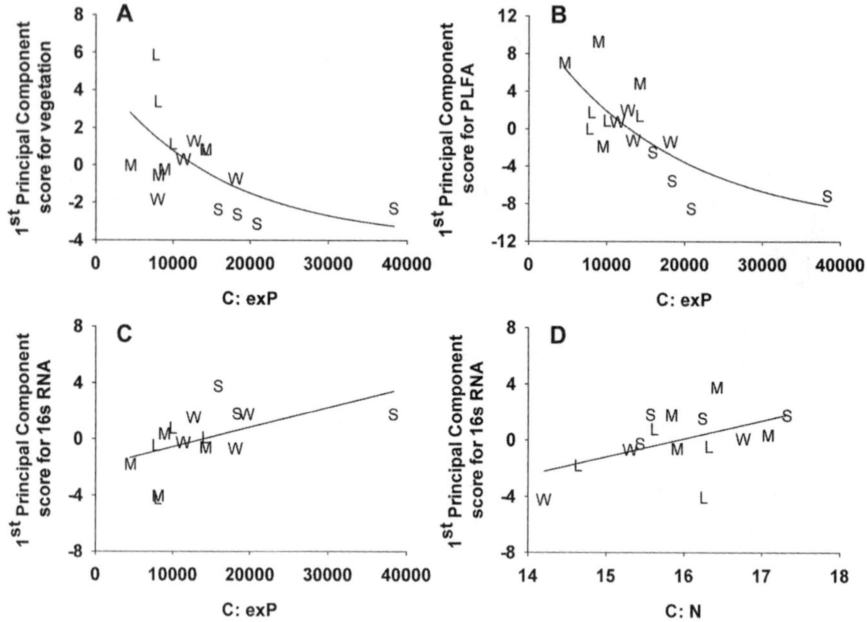

Fig. 8.7 Relationships between the first principal component scores of individual sites for vegetation, PLFA, and 16S rRNA and soil nutrient ratios. Nonlinear negative relationships indicate nutrient limitations, as the second element in the ratio becomes limiting. Regression statistics are: Panel A: $R^2 = 0.38$ ($P = 0.0411$); panel B: $R^2 = 0.55$ ($P = 0.0052$); panel C: $R^2 = 0.28$ ($P = 0.0442$); panel D: $R^2 = 0.25$ ($P = 0.0562$). S = seasonal, W = wildlife, M = mixed, and L = livestock.

both vegetation data and PLFA data and the nutrient ratios are nonlinear (fig. 8.7). This suggests that, in this ecosystem, P is the limiting element that affects the structure of both plant and microbial communities. Even though one of the seasonal sites, in which P is particularly limiting, influences the relationships to some extent, the other data show a clear negative nonlinear trend. We found a marginally significant correlation between the vegetation principal component scores of the sites and total C content ($R = -0.44$; $P = 0.083$) and N content ($R = -0.42$; $P = 0.100$), but no relationship with the C:N ratio ($R = -0.19$; $P = 0.461$). There was no significant correlation between the PLFA component scores of the sites and total C content ($R = -0.16$; $P = 0.544$), N content ($R = -0.12$; $P = 0.643$), or C:N ratio ($R = -0.06$; $P = 0.813$). However, there was a significant correlation ($R = 0.60$; $P = 0.015$) between extractable P and the PLFA component scores.

Exchangeable bases, particularly Na, are important limiting factors for wildlife, particularly for nutrition during gestation and lactation (McNaughton, Banyikwa, and McNaughton 1997). We found that the first

PLFA axis correlated significantly and negatively with both exchangeable Na (R =–0.55; P = 0.028) and exchangeable Ca (R =–0.61; P = 0.012). Ca was the only exchangeable base with a significant and negative correlation (R =–0.54; P = 0.030) with the first component of vegetation. It is not clear whether nutrients alone are responsible for the variation shown by the second PLFA axis. However, we found significant correlation between the second axis scores and both exchangeable Ca (R =–0.47; P = 0.066) and Mg (R =–0.61; P = 0.011). There were no significant correlations for the second principal component of vegetation. The 16S rRNA bacterial data suggest that the community is responding to both N and P availability (fig. 8.7).

A second set of relationships worth exploring is between-group covariance. As both the microbial and the herbaceous plant communities were responding to the same resource-availability gradients, we might expect relationships between the two groups. Indeed, correlation analysis of the results obtained from 16S rRNA gene sequencing and vegetation analysis showed two bacterial groups were correlated with plant biomass. First, the relative abundance of Planctomycetes was positively correlated with all biomass parameters except shrub biomass, the strongest correlation being with green grass biomass (R = 0.65; P = 0.009). Second, the relative abundance of Proteobacteria was negatively correlated with all biomass parameters except shrub biomass, the strongest correlation being with total biomass (R =–0.72; P = 0.003). We need to examine these correlations more closely to determine whether they are meaningful. For example, although Planctomycetes remain enigmatic, recent work has shown that they respond to changes in Ca and pH (Buckley et al. 2006). Some of these microbes possess the genetic capability to undertake C_1 metabolism (metabolism of methyl compounds), which are at the heart of methyltrophy and methanogenesis (Chistoserdova et al. 2004). Additionally, members of one Planctomycete group perform the recently discovered anammox reaction (Jetten et al. 2001). Thus, these organisms may be indicators of anaerobic conditions. Furthermore, Proteobacteria are a very diverse group that includes important plant and animal pathogens as well as legume-nodulating bacteria. The meaning of the negative correlations observed with Proteobacteria is difficult to determine because of the diversity within the phylum, but α-Proteobacteria dominated the soils in this study, as is generally the case. Even this class is very diverse, with species that fix N, are photosynthetic, oxidize ammonia, and are methyltrophic (Williams, Sobral, and Dickerman 2007).

Differences in plant community composition were also associated with the relative abundances of several bacterial groups. We found that two bacterial phyla, Firmicutes and Gemmatimonadetes, were positively correlated with the first vegetation principal component (R = 0.60; P = 0.019 and

$R = 0.52$; $P = 0.046$, respectively), and that Verrucomicrobia and Thermomicrobia were negatively correlated with the first vegetation axis ($R = -0.69$; $P = 0.004$ and $R = -0.57$; $P = 0.026$, respectively). Chloroflexi were positively correlated with the second vegetation axis ($R = 0.76$; $P = 0.001$). These are all recently described bacterial phyla and the functional significance of these relationships requires further examination.

Comparisons between the bacterial community composition as revealed by 16S rRNA gene sequencing and that obtained by PLFA analysis are also interesting. In this data set, there was little congruence between the two methods with respect to grazing treatments. Whereas 16S rRNA-based analysis indicated fewer Gram-positive bacteria (Actinobacteria and Firmicutes) in wildlife and seasonal plots (23–26% relative abundance) than in the livestock and mixed plots (35% abundance), the PLFA findings suggested the opposite. It is not necessarily surprising to uncover discrepancies between 16S rRNA- and PLFA-based analyses, given the bias introduced by selective cell lysis and PCR amplification inherent in the 16S rRNA approach, and the bias toward active microbes associated with PLFA. Furthermore, there is the difficulty of reconciling the broad chemotaxonomic categories used in PLFA (Gram-negative, Gram-positive, etc.) with the phylogeny-derived categories used in 16S rRNA-based analysis. Nevertheless, we did find some interesting correlations between the relative PLFA data and the relative abundance of certain bacterial groups (table 8.9).

The assessments of actinomycetes by the two procedures were positively correlated. For Verrucomicrobia (one of the two bacterial phyla showing the greatest effects of grazing treatment on their relative abundance), several associations with P-values <0.05 were observed. Relative abundance of Verrucomicrobia was positively correlated with PLFA markers for Gram-positive bacteria and negatively correlated with those signifying Gram-negative bacteria; relative abundance of Verrucomicrobia was also positively correlated with one of the cy/ω7c PLFA indicators for nutritional stress ($P = 0.004$).

Nematode populations have top-down effects on the microbial community through predation; in turn, the bottom-up effects of the microbes on nematodes can also be significant. However, correlation analysis of absolute numbers of sequences belonging to 16S rRNA-determined bacterial groups and the various functional groups showed only a few significant correlations. The most notable result was the negative correlation between Gemmatimonadetes and bacterial feeder, plant parasites, and algal feeder nematode groups ($P < 0.074$). The significance of this is unclear; Ladygina et al. (2009) found no association between Gemmatimonadetes and nematodes in a grassland in eastern Netherlands. They found an association between

Table 8.9 Extract of the correlation matrix showing significant correlations (in bold) between the 16S rRNA and PLFA data

	Gram-positive	Gram-negative	Actinomycetes	Fungi	Protozoa	Sulfate reducers	17cy: 16:1ω7c	19cy: 18:1ω7c
Actinobacteria	–0.082	–0.0510	**0.508**	0.039	0.031	0.133	–0.279	0.054
	0.0719	(0.8568)	**(0.0534)**	(0.8899)	(0.9129)	(0.6368)	(0.3131)	(0.8491)
Proteobacteria	–0.335	0.287	–0.012	0.009	0.425	0.411	–0.294	**–0.684**
	(0.2226)	(0.2999)	(0.9667)	(0.9744)	(0.1146)	(0.1281)	(0.2868)	**(0.0049)**
Verrucomicrobia	**0.769**	**–0.675**	–0.304	–0.300	–0.462	–0.462	**0.703**	0.205
	(0.0008)	**(0.0058)**	(0.2706)	(0.2779)	(0.0827)	(0.0833)	**(0.0035)**	(0.4636)
Gemmatimonadetes	–0.376	0.358	0.450	**0.514**	0.210	0.186	–0.213	0.136
	(0.1675)	(0.1889)	(0.0926)	**(0.0498)**	(0.4530)	(0.5061)	(0.4460)	(0.6289)
Chloroflexi	0.183	0.097	**–0.590**	0.377	–0.478	–0.582	0.481	**0.574**
	(0.5128)	(0.7299)	**(0.0205)**	(0.1662)	(0.0715)	(0.0227)	(0.0696)	**(0.0252)**

Note: Values are *R* (*P*).

bacterial feeders and both Firmicutes and Proteobacteria, but we did not observe significant correlations here.

Finally, with respect to higher trophic levels in the soil microbial community, we found a significant correlation between bacterial-feeding nematodes and Gram-positive bacteria and a marginally significant negative relationship with actinomycetes ($R = 0.52$; $P = 0.040$ and $R = -0.45$; $P = 0.077$, respectively). We found no relationship with Gram-negative bacteria ($R = -0.405$; $P = 0.120$) or between PLFA fungal markers and the fungal-feeding nematodes ($R = -0.05$; $P = 0.828$). Protozoa had one plot with very high values that we removed from the analysis to look at the relationships. These organisms were negatively correlated with Gram positive bacteria (R =–0.81; P < 0.001) and positively correlated with both Gram-negative bacteria and actinomycetes ($R = 0.70$; $P = 0.004$ and $R = 0.49$; $P = 0.063$, respectively). Protozoa are also very strongly correlated with sulfate reducers ($R = 0.88$; $P < 0.001$, respectively), the significance of which is unclear.

CONCLUSION

Generally, ecological research has considered the aboveground and belowground components of ecosystems separately. Consequently, frameworks for integrating the two components are not well developed. Integrating the microbial components into ecosystem ecology requires different approaches from those offered by plant ecology, partly because of the scales at which microbial processes operate and partly because of measurement constraints (Verchot 2010). However, during the past decade or so, more studies have attempted to integrate the two components to understand how changes in one affect the other. These studies often treat the belowground components of the ecosystem as a black box and do not relate structural changes to ecological function. More recently, however, studies have begun to relate microbial community structure to ecosystem function. The work to date has largely taken advantage of "natural experiments"—as we have done here—but we are beginning to see deliberate manipulations of the microbial community to examine the relationship between structure and function (e.g., Lucas et al. 2007).

Having compared the results of these disparate measurements, we have presented a number of correlations that we find interesting. In some cases, we have reasonable hypotheses about the nature of the relationships our analysis revealed, but in many others, we have no idea whether these are simply correlations, with two segments of an ecosystem responding to another underlying variable, or whether they represent a cause-and-effect relationship. One of the constraints of natural experiments is that several factors often vary simultaneously. We hope that the presentation of these correlations will stimulate further research with deliberate manipulations to elucidate the ecological significance (or insignificance) of these observations.

The integration that we have achieved here reveals several similarities between the responses by microbial and plant communities to the environmental gradients that have been created in this East African grassland landscape by both natural and anthropogenic forces. The creation of the gradients we observed is partly due to the separation of livestock and wildlife achieved through the establishment of the MMNR. If we exclude data from the phase where wildlife biomass is very high within the MMNR—that is, during the seasonal wildebeest migration—the animal biomass is about the same in the reserve and the ranches. The main difference is the nature of the grazing community, with greater numbers of close (Thomson's gazelles, sheep, and goats) and ripping (cattle) grazers in the ranches. The other key element in the creation of these gradients is the seasonal wildebeest migration. Wildebeest grazing is heavy in this ecosystem from July to October and thus its effect, although transient, is important. Thus, we conclude that our comparison reflects differences in both grazing habits and intensity; intensity is largely a function of the seasonal migrations, whereas grazing habits are a function of the nature of the animal community.

The different grazing habits had significant and different effects on the structure of the plant and soil microbial communities. With intensive livestock grazing, plant communities shifted from the lush *Themeda* and *Sporobolus* community to a community dominated by forbs and poorer-quality grasses. The PLFA and nematode data suggest that, with heavy grazing, the microbial community also shifted to a higher proportion of fungi. Belowground food webs dominated by fungal energy flows indicate degraded ecosystems (de Ruiter, Neutel, and Moore 1995; Wardle et al. 2004). This degradation is perhaps reflected in the N cycle, as the intensive grazing may have led to significant N losses from the system. Thus, there is reason for concern over the long-term sustainability of the more intensive grazing systems where wildlife and livestock are segregated; this needs to be monitored.

Looking at the soil microbial community as a whole, we see much smaller effects of grazing intensity on community structure than in the plant community. It is striking that there is a large overlap between the wildlife and livestock treatments that we observed among plant species composition; this is reflected in both 16S rRNA and PLFA data. This observation leads us to conclude that intensive wildlife and livestock grazing have remarkably similar impacts on plants, microbes, and soils in this ecosystem. The story for nematodes is rather different—here, it appears that the presence of cattle in the grazing community affects populations. Sites with intensive livestock only or mixed wildlife and livestock grazing had lower populations of almost all categories of nematodes observed. This impact could have long-term effects on carbon turnover and other biogeochemical processes in the soil.

The similar overall composition of the bacterial communities under different treatments suggests that the primary factor leading to treatment effects is the relative abundance of individual phyla, rather than the presence of unique phyla. The big difference was in the seasonal plots, which had more Verrucomicrobia and fewer Firmicutes than the other plots. Interestingly, there was no statistical difference among the different bacterial groups between the sites with heavy grazing (wildlife, mixed, and livestock). It is therefore surprising, particularly in light of the differences in the PLFA data, that the 16S rRNA data do not show a stronger grazing effect. A previous study (Bossio et al. 2005), in which both PLFA and 16S rRNA profiling were conducted on soils in western Kenya, found that soil type was the primary determinant of 16S rRNA profiles, and therefore of the total bacterial community. The PLFA data in that study, which measured the active microbial community, were found to be more sensitive to management activities. In other work soil pH, which is related to soil parent material and thus soil type, has been shown to explain a large portion of the variation in bacterial community profiles based on 16S rRNA gene sequencing data (Fierer et al. 2009). Our results, based on the molecular genetics approach of 16S rRNA sequencing, demonstrated differences caused by management even within the same soil type. That PLFA and 16S rRNA sequence data were not in agreement as to the specific impact of grazing treatments demonstrates that the different approaches captured information on different portions of the community.

Our observations, particularly of sizable effects in places with no current (i.e., seasonal) grazing, indicate that the presence or absence of grazing has large impacts on plants, nematodes, microbes, and soils in this ecosystem. These results suggest that excluding people and livestock from the MMNR, or preventing heavier livestock from grazing around settlements, may not change the general structure of the ecosystem (soils, plant structure), but can change the numbers and diversity of wildlife, nematodes, and microbes in this ecosystem in subtle ways.

ACKNOWLEDGMENTS

Field guidance and research support was kindly provided by Mike Rainy, Cathy Wilson, Teresa Borelli, and the late John Rakwa. We thank Russ Kruska for creating fig. 8.1 and Judd Hill for analyzing the plant and dung data. Imogen Badgery-Parker edited the manuscript. Soil analyses were carried out by the staff of the plant and soils lab of the World Agroforestry Centre. Microbial function analyses were conducted by Teresa Borelli, Edith Anyango, Margaret Thiongo, and Mercy Nyambura. Joel Mwakaya and Kevin Penn generated and sequenced 16S rRNA gene clone libraries. PLFA analysis was done in the laboratory of Dr. Kate Scow at UC-Davis. This research was partly supported by core funds from the CGIAR to ILRI, IWMI, ICRAF, and CIFOR and by a grant to ICRAF from the Rockefeller Foundation (grant number 2001-FS-089). Support for Naomi Ward and Blaire Steven was provided by NSF award EPS-0447681. Any use of trade, product, or firm names is for descriptive purposes only and does not imply endorsement of the products or firms by the US government.

REFERENCES

Ashelford, K. E., N. A. Chuzhanova, J. C. Fry, A. J. Jones, and A. Weightman. 2006. New screening software shows that most recent large 16S rRNA gene clone libraries contain chimeras. *Applied and Environmental Microbiology* 72:5734–41.

Augustine, D. J. 2003. Long-term, livestock-mediated redistribution of nitrogen and phosphorus in an East African savanna. *Journal of Applied Ecology* 40:137–49.

Bardgett, R. D., R. S. Smith, R. S. Shiel, S. Peacock, J. M. Simkin, H. M. Quirk, and P. J. Hobbs. 2006. Parasitic plants indirectly regulate below-ground properties in grassland ecosystems. *Nature* 439:969–72.

Bardgett, R. D., and D. A. Wardle. 2003. Herbivore-mediated linkages between aboveground and belowground communities. *Ecology* 84:2258–68.

Barns, S. M., E. C. Cain, L. Sommerville, and C. R. Kuske. 2007. Acidobacteria phylum sequences in uranium-contaminated subsurface sediments greatly expand the known diversity within the phylum. *Applied and Environmental Microbiology* 73:3113–16.

Barns, S. M., S. L. Takala, and C. R. Kuske. 1999. Wide distribution and diversity of members of the bacterial kingdom *Acidobacterium* in the environment. *Applied and Environmental Microbiology* 65:1731–37.

Beare, M. 1997. Fungal and bacterial pathways of organic matter decomposition and nitrogen mineralization in arable soils. In *Soil ecology in sustainable agricultural systems,* ed. L. Brussard and R. Ferrera-Cerrato, 37–70. Boca Raton, Florida: CRC Lewis Publishers.

Belnap, J. 2011. Biological phosphorus cycling in dryland regions. In *Phosphorus in action—Biological processes in soil phosphorus cycling,* ed. E. K. Bünemann, A. Oberson, and E. Frossard, 371–406. Berlin: Springer.

Bongers, T., and H. Ferris. 1999. Nematode community structure as a bioindicator in environmental monitoring. *Trends in Ecology and Evolution* 14:224–28.

Bonin A., D. Ehrich, and S. Manuel. 2007. Statistical analysis of amplified fragment length polymorphism data: A toolbox for molecular ecologists and evolutionists. *Molecular Ecology* 16:3737–58.

Bossio, D. A., J. A. Fleck, K. M. Scow, and R. Fujii. 2006. Alteration of soil microbial communities and water quality in restored wetlands. *Biology and Biochemistry* 36: 889–901.

Bossio, D. A., M. S. Girvan, L. Verchot, J. Bullimore, T. Borelli, A. Albrecht, K. M. Scow, A. S. Ball, J. N. Pretty, and A. M. Osborn. 2005. Soil microbial community response to land use change in an agricultural landscape of western Kenya. *Microbial Ecology* 49:50–62.

Bossio, D. A., and K. M. Scow. 1998. Impacts of carbon and flooding on soil microbial communities: Phospholipid fatty acid profiles and substrate utilization patterns. *Microbial Ecology* 35:265–78.

Brockington, D., and K. Homewood. 2001. Degradation debates and data deficiencies: The Mkomazi Game Reserve, Tanzania. *Africa* 71:449–80.

Bryant, D. A., A. M. Costas, J. A. Maresca, A. G. Chew, C. G. Klatt, M. M. Bateson, L. J. Tallon, J. Hostetler, W. C. Nelson, J. Heidelberg, et al. 2007. Candidatus *Chloracidobacterium thermophilum*: An aerobic phototrophic Acidobacterium. *Science* 317: 523–26.

Buckley, D. H., V. Huangyutitham, T. A. Nelson, A. Rumberger, and J. E. Thies. 2006. Diversity of *Planctomycetes* in soil in relation to soil history and environmental heterogeneity. *Applied and Environmental Microbiology* 72:4522–31.

Bünemann, E. K., D. A. Bossio, P. C. Smithson, E. Frossard, and A. Oberson. 2004. Microbial community composition and substrate use in a highly weathered soil as affected by crop rotation and P fertilization. *Soil Biology & Biochemistry* 36:889–901.

Butt, B., A. Shortridge, and A. WinklerPrins. 2009. Pastoral herd management, drought coping strategies, and cattle mobility in southern Kenya. *Annals of the Association of American Geographers* 99:309–34.

Chistoserdova, L., C. Jenkins, M. G. Kalyuzhnaya, C. J. Marx, A. Lapidus, J. A. Vorholt, J. T. Staley, and M. E. Lidstrom. 2004. The enigmatic planctomycetes may hold a key to the origins of methanogenesis and methylotrophy. *Molecular Biology and Evolution* 21:1234–41.

Coe, M. J., D. H. Cumming, and J. Phillipson. 1976. Biomass and production of large African herbivores in relation to rainfall and primary production. *Oecologia* 22:341–54. DOI: 10.1007/BF00345312.

Cole, J. R., Q. Wang, E. Cardenas, J. Fish, B. Chai, R. J. Farris, A. S. Kulam-Syed-Mohideen, D. M. McGarrell, T. Marsh, G. M. Garrity, et al. 2009. The ribosomal database project: Improved alignments and new tools for rRNA analysis. *Nucleic Acids Research* 37: D141–45.

Coleman, D. C., E. T. Elliott, J. M. Blair, and D. W. Freckman. 1999. Soil invertebrates. In *Standard soil methods for long-term ecological research*, ed. G. P. Robertson, D. C. Coleman, C. S. Bledsoe, and S. Phillips, 349–77. New York: Oxford University Press.

Day, P. R. 1965. Particle fractionation and particle size analysis. In *Methods of soil analysis, part 1, agronomy monograph no. 9.*, ed. C. A. Black, 545–67. Madison, WI: ASA and SSSA.

de Ruiter, P. C., A. M. Neutel, and J. C. Moore. 1995. Energetics, patterns of interaction strengths, and stability in real ecosystems. *Science* 269:1257–60.

Felsenstein, J. 1993. PHYLIP (PHYLogenetic inference package) version 3.5.1. Department of Genetics, University of Washington, Seattle.

Fierer N, M. S. Strickland, D. Liptzin, M. A. Bradford, and C. C. Cleveland. 2009. Global patterns in belowground communities. *Ecology Letters* 12:1238–49.

Frank, D. A., R. D. Evans, and B. F. Tracy. 2004. The role of ammonia volatilization in controlling the natural ^{15}N abundance of a grazed grassland. *Biogeochemistry* 68:169–78.

Hobbs, N. T. 1996. Modification of ecosystems by ungulates. *Journal of Wildlife Management* 60:695–713.

Homewood, K. M., and Rodgers, W. A. 1991. *Maasailand ecology: Pastoralist development and wildlife conservation in Ngorongoro, Tanzania.* Cambridge: Cambridge University Press.

Hunt, H. W., D. C. Coleman, E. R. Ingham, R. E. Ingham, E. T. Elliott, J. C. Moore, S. L. Rose, C. P. P. Reid, and C. R. Morley. 1987. The detrital food web in a short-grass prairie. *Biology and Fertility of Soils* 3:57–68.

Ingham, R. E., and J. K. Detling. 1990. Effects of root-feeding nematodes on aboveground net primary production in a North American grassland. *Plant and Soil* 121:279–81.

Jackson, R. B., L. J. Anderson, and W. T. Pockman. 2000. Measuring water availability and uptake in ecosystem studies. In *Methods in ecosystem science*, ed. O. E. Sala, R. B. Jackson, H. A. Mooney, and R. W. Howarth, 199–214. New York: Springer-Verlag.

Janssen, P. H. 2006. Identifying the dominant soil bacterial taxa in libraries of 16S rRNA and 16S rRNA genes. *Applied and Environmental Microbiology* 72:1719–28.

Jenkins, W. R. 1964. A rapid centrifugal-floatation technique for separating nematodes from soil. *Plant Disease Reporter* 48:692.

Jetten M. S. M., M. Wagner, J. Fuerst, M. van Loosdrecht, G. Kuenen, and M. Strous. 2001. Microbiology and application of the anaerobic ammonium oxidation ('anammox') process. *Current Opinion in Biotechnology* 12:283–88.

Jones, R. T., M. S. Robeson, C. L. Lauber, M. Hamady, R. Knight, and N. Fierer. 2009. A comprehensive survey of soil acidobacterial diversity using pyrosequencing and clone library analyses. *ISME Journal* 3:442–53.

Ladygina, N., T. Johansson, B. Canbäck, A. Tunlid, and K. Hedlund. 2009. Diversity of bacteria associated with grassland soil nematodes of different feeding groups. *FEMS Microbiology Ecology* 69:53–61.

Lamprey, R., and R. S. Reid. 2004. Expansion of human settlement in Kenya's Maasai Mara: What future for pastoralism and wildlife? *Journal of Biogeography* 31:997–1032.

Lamprey, R., and R. Waller. 1990. The Loita-Mara region in historical times: Patterns of subsistence, settlement and ecological change. In *Early pastoralists of south-western Kenya*, ed. P. Robertshaw, 16–35. Nairobi, Kenya: British Institute in Eastern Africa

Leakey, M. D., and R. L. Hay. 1979. Pliocene footprints in the Laetolil beds at Laetoli, northern Tanzania. *Nature* 278:317–23.

Littell, R. C., W. W. Stroup, and R. J. Freund. 2002. *SAS for linear models*. SAS Institute, Cary, NC.

Liu, W. T., T. L. Marsh, H. Cheng, and L. J. Forney. 1997. Characterization of microbial diversity by determining terminal restriction fragment length polymorphisms of genes encoding 16S rRNA. *Applied and Environmental Microbiology* 63:4516–22.

Lozupone, C., M. Hamady, and R. Knight. 2006. UniFrac--an online tool for comparing microbial community diversity in a phylogenetic context. *BMC Bioinformatics* 7:371.

Lucas, R. W., B. B. Casper, J. K. Jackson, and T. C. Balser. 2007. Soil microbial communities and extracellular enzyme activity in the New Jersey Pinelands. *Soil Biology and Biochemistry* 39:2508–19.

MacKenzie, J. M. 1987. Chivalry, social Darwinism and ritualised killing: The hunting ethos in Central Africa up to 1914. In *Conservation in Africa: People, policies and practice*, ed. D. M. Anderson and R. Grove, 41–61. Cambridge: Cambridge University Press.

Marshall, F. 2000. The origins and spread of domesticated animals in East Africa. In *The origins and development of African livestock: Archaeology, genetics, linguistics and ethnography*, ed. R. M. Blench and K. C. McDonald, 191–221. London: University College Press.

McKinley, V. L., A. D. Peacock, and D. C. White. 2005. Microbial community PLFA and PHB responses to ecosystem restoration in tallgrass prairie soils. *Soil Biology and Biochemistry* 37:1946–58.

McNaughton, S. J. 1979. Grassland-herbivore dynamics. In *Serengeti: Dynamics of an ecosystem*, ed. A. R. E. Sinclair and M. Norton-Griffiths, 46–81. Chicago: University of Chicago Press.

———. 1983. Serengeti grassland ecology: The role of composite environmental factors and contingency in community organization. *Ecological Monographs* 53:291–320.

McNaughton, S. J., F. F. Banyikwa, and M. M. McNaughton. 1997. Promotion of the cycling of diet-enhancing nutrients by African grazers. *Science* 278:1798–800.

McNaughton, S. J., R. W. Ruess, and S. W. Seagle. 1988. Large mammals and process dynamics in African ecosystems. *BioScience* 38:794–800.

Milchunas, D. G., and W. K. Lauenroth. 1993. Quantitative effects of grazing on vegetation and soils over a global range of environments. *Ecological Monographs* 63:327–66.

Moore, J. C., and P. C. de Ruiter. 1991. Temporal and spatial heterogeneity of trophic interactions within belowground food webs. *Agriculture, Ecosystems and Environment* 34:371–97.

Mould, E. D., and C. T. Robbins. 1981. Nitrogen metabolism in elk. *Journal of Wildlife Management* 45:323–34.

Muchiru, A. N., D. J. Western, and R. S. Reid. 2009. The impact of abandoned pastoral settlements on plant and nutrient succession in an African savanna ecosystem. *Journal of Arid Environments* 73:322–31.

Muyzer G., and K. Smalla. 1998. Application of denaturing gradient gel electrophoresis (DGGE) and temperature gradient gel electrophoresis (TGGE) in microbial ecology. *Antonie Van Leeuwenhoek* 73:127–41.

Mwendera, E. J., M. A. M. Saleem, and Z. Woldu. 1997. Vegetation response to cattle grazing in the Ethiopian highlands. *Agriculture, Ecosystems and Environment* 64:43–51.

Neumann, R. 2002. *Imposing wilderness: Struggles over livelihood and nature preservation in Africa*. Berkeley: University of California Press.

Newman, J. L. 1995. *The peopling of Africa*. New Haven, CT: Yale University Press.

Norton-Griffiths, M., D. Herlocker, and L. Pennycuick. 1975. The patterns of rainfall in the Serengeti ecosystem, Tanzania. *East African Wildlife Journal* 13:347–74.

Ogutu, J. O., H. P. Piepho, H. T. Dublin, N. Bhola, and R. S. Reid. 2007. El Niño-Southern Oscillation, rainfall, temperature and Normalized Difference Vegetation Index fluctuations in the Mara-Serengeti ecosystem. *African Journal of Ecology* 46:132–43.

———. 2009. Dynamics of Mara-Serengeti ungulates in relation to land use changes. *Journal of Zoology* 278:1–14.

Pankratov, T. A., Y. M. Serkebaeva, I. S. Kulichevskaya, W. Liesack, and S. N. Dedysh.

2008. Substrate-induced growth and isolation of Acidobacteria from acidic Sphagnum peat. *International Society for Microbial Ecology Journal* 2:551–60.

Parkipuny, M. L. 1991. Pastoralism, conservation and development in the greater Serengeti region. Paper no. 26, Drylands Network Programme. London: International Institute for Environment and Development.

Peterson, A. L. 2001. *Being human: Ethics, environment and our place in the world*. Berkeley: University of California Press.

Philippot, L., D. Bru, N. P. A. Saby, J. Cuhel, D. Arrouays, M. Simek, and S. Hallin. 2009. Spatial patterns of bacterial taxa in nature reflect ecological traits of deep branches of the 16S rRNA bacterial tree. *Environmental Microbiology* 11:3096–104.

Ponder, F., and M. Tadros. 2002. Phospholipid fatty acids in forest soil four years after organic matter removal and soil compaction. *Applied Soil Ecology* 19:173–82.

Porazinska, D. L., R. D. Bardgett, M. B. Blaauw, H. W. Hunt, A. N. Parsons, T. R. Seastedt, and D. H. Wall. 2003. Relationships at the aboveground-belowground interface: Plants, soil biota, and soil processes. *Ecological Monographs* 73:377–95.

Quaiser, A., T. Ochsenreiter, C. Lanz, S. C. Schuster, A. H. Treusch, J. Eck, and C. Schleper. 2003. Acidobacteria form a coherent but highly diverse group within the bacterial domain: Evidence from environmental genomics. *Molecular Microbiology* 50:563–75.

Reid, R. S., H. Gichohi, M. Y. Said, D. Nkedianye, J. O. Ogutu, M. Kshatriya, P. Kristjanson, S. C. Kifugo, J. L. Agatsiva, S. A. Adanje, et al. 2008. Fragmentation of a peri-urban savanna, Athi-Kaputiei Plains, Kenya. In *Fragmentation in semi-arid and arid landscapes: Consequences for human and natural systems*, ed. K. A. Galvin, R. S. Reid, R. H. Behnke, and N. T. Hobbs, 195–224. Dordrecht, Germany: Springer.

Reid, R. S., M. Rainy, J. Ogutu, R. L. Kruska, M. Nyabenge, M. McCartney, K. Kimani, M. Kshatriya, J. Worden, L. N'gan'ga, et al. 2003. *Wildlife, people, and livestock in the Mara ecosystem, Kenya: The Mara count 2002*. Nairobi, Kenya: International Livestock Research Institute.

Rose, L., and F. Marshall. 1996. Meat eating, hominid sociality, and home bases revisited. *Current Anthropology* 37:307–38.

Ruess, R. W. 1987. The role of large herbivores in nutrient cycling of tropical savannas. In *Determinants of tropical savannas*, IUBS Monograph Series no. 3, ed. B. H. Walker, 67–91. Oxford: Oxford University Press.

Ruess, R. W., and S. W. Seagle. 1994. Landscape patterns in soil microbial processes in the Serengeti National Park, Tanzania. *Ecology* 75:892–904.

Schick, K. D., and N. Toth. 1993. *Making silent stones speak: Human evolution and the dawn of technology*. New York: Simon and Schuster.

Schloss P. D., S. L. Westcott, T. Ryabin, J. R. Hall, M. Hartmann, E. B. Hollister, R. A. Lesniewski, B. B. Oakley, D. H. Parks, C. J. Robinson, et al. 2009. Introducing Mothur: Open-source, platform-independent, community-supported software for describing and comparing microbial communities. *Applied and Environmental Microbiology* 75: 7537–41.

Seastedt, T. R., T. C. Todd, and S. W. James. 1987. Experimental manipulations of the arthropod, nematode and earthworm communities in a North American tallgrass prairie. *Pedobiologia* 30:9–17.

Shetler, J. B. 2007. *Imaging Serengeti: A history of landscape memory in Tanzania from earliest times to the present*. Athens: Ohio University Press.

Sinclair, A. R. E. 1995. Serengeti past and present. In *Serengeti II: Dynamics, management and conservation of an ecosystem*, ed. A. R. E. Sinclair and P. Arcese, 3–30. Chicago: University of Chicago Press.

Sinclair, A. R. E., and M. Norton-Griffiths, eds. 1979. *Serengeti: Dynamics of an ecosystem*. Chicago: University of Chicago Press.

Sowerby, A., B. Emmett, C. Beier, A. Tietema, J. Peñuelas, M. Estiarte, M. J. M. Van Meeteren, S. Hughes, and C. Freeman. 2005. Microbial community changes in heathland soil communities along a geographical gradient: Interaction with climate change manipulations. *Soil Biology and Biochemistry* 37:1805–13.

Stanton, N. L. 1983. The effect of clipping and phytophagous nematodes on net primary production of blue grama, *Bouteloua gracilis*. *Oikos* 40:249–57.

Terborgh, J. 1999. *Requiem for nature*. Washington, DC: Island Press.

Vandekerckhove T., A. Willems, M. Gillis, and A. Coomans. 2000. Occurrence of novel verrucomicrobial species, endosymbiotic and associated with parthenogenesis in *Xiphinema americanum*-group species (Nematoda, Longidoridae). *International Journal of Systematic and Evolutionary Microbiology* 50:2197–205.

Verchot, L. V. 2010. Impacts of forest conversion to agriculture on microbial communities and microbial function. In *Soil biology and agriculture in the tropics*, ed. P. Dion, 45–63. Heidelberg, Germany: Springer.

Waldrop M. P., T. C. Blaser, and M. K. Firestone. 2000. Linking microbial community composition to function in a tropical soil. *Soil Biology and Biochemistry* 32:1837–46.

Wall-Freckman, D., and S. P. Huang. 1998. Response of the soil nematode community in a shortgrass steppe to long-term and short-term grazing. *Applied Soil Ecology* 9:39–44.

Ward, N. L., J. F. Challacombe, P. H. Janssen, B. Henrissat, P. M. Coutinho, M. Wu, G. Xie, D. H. Haft, M. Sait, J. Badger, et al. 2009. Three genomes from the phylum Acidobacteria provide insight into the lifestyles of these microorganisms in soils. *Applied and Environmental Microbiology* 75:2046–56.

Wardle, D. A., R. D. Bardgett, J. N. Klironomos, H. Setälä, W. H. van der Putten, and D. H. Wall. 2004. Ecological linkages between aboveground and belowground biota. *Science* 304:1629–33.

Western, D., S. Russell, and I. Cuthill. 2009. The status of wildlife in protected areas compared to nonprotected areas of Kenya. *PLoS ONE* 4:1–6.

White, D. C., and D. B. Ringelberg. 1998. Signature lipid biomarker analysis. In *Techniques in microbial ecology*, ed. R. S. Burlage, R. Atlas, D. Stahl, G. Geesey, and G. Sayler, 255–72. New York: Oxford University Press.

Williams, K. P., B. W. Sobral, and A. W. Dickerman. 2007. A robust species tree for the alphaproteobacteria. *Journal of Bacteriology* 189:4578–86. Doi 10.1128/JB.00269-07.

Yeates, G. W., T. Bongers, R. G. M. de Goede, D. W. Freckman, and S. S. Georgieva. 1993. Feeding habits in soil nematode families and genera—An outline for soil ecologists. *Journal of Nematology* 25:315–31.

Young, T. P., N. Partridge, and A. Macrae. 1995. Long-term glades in acacia bushland and their edge effects in Laikipia, Kenya. *Ecological Applications* 5:97–108.

NINE

Biodiversity and the Dynamics of Riverine Forests in Serengeti

Roy Turkington, Gregory Sharam, and Anthony R. E. Sinclair

THE PROBLEM

Forests fringe the three major rivers of the Serengeti. These riverine forests are an exceptionally vulnerable habitat in Africa because of their location. People come to rivers to obtain water, build their houses, plant vegetable gardens, and graze domestic animals. In the process the undergrowth, which includes regenerating seedlings and saplings, is removed. Eventually the forest decays if it has not already been cut down for timber and firewood. In many areas, particularly arid areas, riparian forests are the only closed-canopy forests, and they are often heavily degraded.

In the Serengeti, forests form a narrow strip along rivers, often on one bank only, about 50 m wide. This strip can be as narrow as one tree in width (10 m), and rarely as extensive as 200 m on both sides. The forest is maintained by groundwater that seeps in from the river. The water table is highly seasonal, reaching the surface during heavy rain seasons. The canopy in intact forest is closed and overhangs the river. Hence, the name "gallery forest" is sometimes used.

The problem with these forests within the Serengeti protected area is that they have been declining for at least the past 50 years despite the absence of direct human exploitation. Comparisons of 1960s aerial photos with ground surveys in 2002 showed that 50–70% of forests on the Mara River have been converted to grasslands. The forests support a large number of animal species—mammals, birds, and probably arthropods and molluscs—that are confined to them. Some species, such as the Schalow's turaco (*Turaco schalowi [persa]*), have very restricted ranges within Africa and even have local populations restricted to the area around Serengeti. Loss of these forests would endanger such populations. With this in mind a research program began in 1998 on the regeneration and survival dynamics of the forests.

This chapter describes the disturbances affecting the forests. Grassland fires are thought to be the primary cause of forest decline, but the removal of fire alone does not allow forests to reestablish. Hence, this chapter examines the large-scale question of the processes of forest regeneration by mass-establishment (state and transition dynamics) and/or succession. The mechanisms that control tree establishment and govern forest regeneration are also examined, including the effects of fire, soil moisture, shade, competition with grasses, and browsing by elephants and impala. The two forest types on the Grumeti and Mara Rivers differ in almost every aspect of their dynamics and serve as a laboratory to examine the theories of forest regeneration.

THE FORESTS

At a broad scale Serengeti represents the intersection between the two primary types of broad-leaved forests in East Africa. The first is lowland forest, epitomized by the Congo and Niger River drainage basins forming the heart of equatorial center and west Africa. Its eastern extent reaches the western shores of Lake Victoria in Tanzania and Lake Tanganyika in Rwanda and Burundi. The island of Rubondo at the south end of Lake Victoria provides an example of lowland rainforest although with less tree diversity than the main Congo forest. In Pliocene and Pleistocene times there were several periods when lowland forest spread, leading to long-term alterations between forested and grassland-dominated ecosystems in the Lake Victoria region, including Serengeti (Jolly et al. 1997). This forest extension would have occurred during wet periods when the lake itself expanded eastward some 30 km across what is now the Ndabaka floodplains. In dry periods the lake contracted westward some 100 km so that what is now Speke Bay, which is very shallow, would have become an extension of the Ndabaka floodplains. (See Metzger et al., chapter 3 for a more detailed account of the paleoclimate of Serengeti).

Lowland forest is found in Serengeti along the lower reaches of the Grumeti-Orangi, Mbalageti, and Duma Rivers in western Serengeti (fig. 9.1). The most extensive forest is on the Grumeti from Kirawira in the west

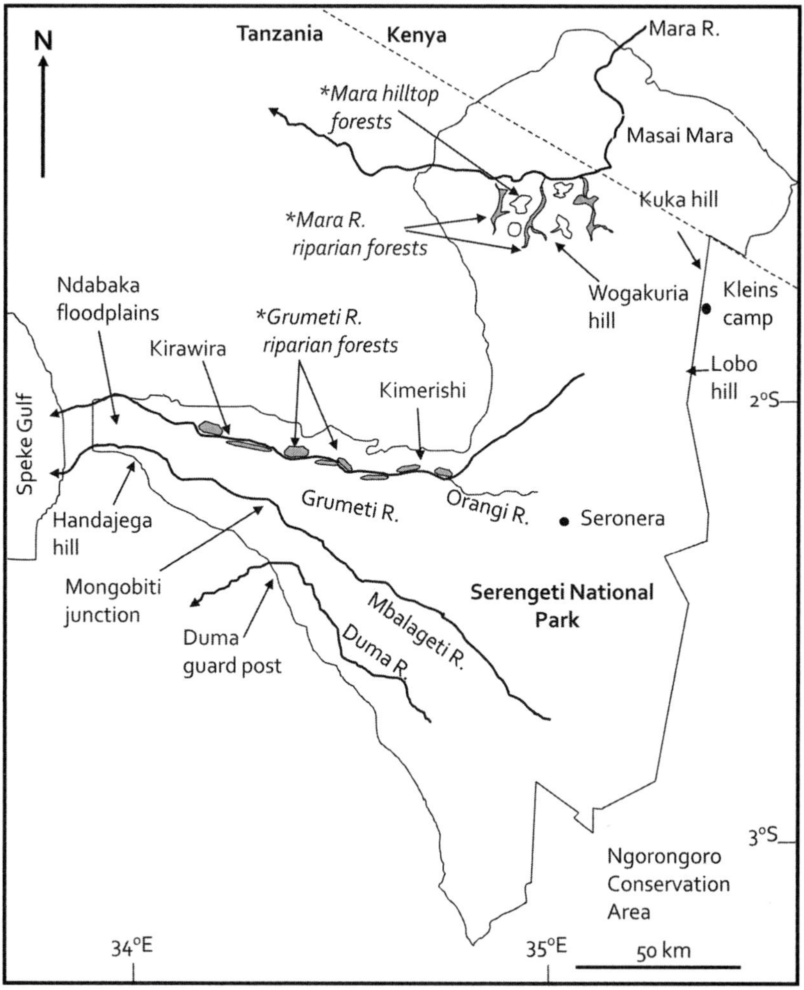

Fig. 9.1 Location of riparian forests along the Grumeti River system of western Serengeti National Park, and of riparian and hilltop forests near the Mara River in northern Serengeti. Asterisks indicate the location of the three study areas (modified from Sharam, Sinclair, and Turkington 2006; Sharam et al. 2009a).

to Kimerishi on the Orangi tributary in the east. In parts it can be 200 m wide. Very small stands are now found on the Mbalageti between the Mongobiti junction and Handajega hill, and even smaller stands are found on the Duma near Duma guard post. Riparian forests along the Grumeti have a dense canopy of *Lecaniodiscus fraxinifolius* Baker, *Elaeodendron buchananii* Loes., and *Ziziphus pubescens* Oliv., with lesser amounts of *Grewia bicolour* Juss. (table 9.1).

Table 9.1 Tree species observed in riparian forest patches along the Grumeti and Mara Rivers (G. Sharam, pers. obs. 1998–2001)

Grumeti riparian forests	Mara riparian and hilltop forests
Lecaniodiscus fraxinifolius Baker	*Croton dichogamus* Pax.
Elaeodendron buchananii Loes.	*Teclea trichocarpa* Engl.
Ziziphus pubescens Oliv.	*Diospyros abyssinica* (Hiern) F. White
Grewia bicolor Juss.	*Drypetes gerrardii* Hutch.
	Ekebergia capensis Sparrm.
	Euclea divinorum (Hiern) subsp. *keniensis* (R. E. Fries) de Wit
	Olea africana Mill.
	Strychnos spp.

The second type of forest is found on mountains at a higher altitude. An isolated stand is found on Mount Cameroon in west Africa but the majority occurs on the eastern high plateau running from Ethiopia to southern Africa. In Tanzania it occurs on the Eastern Arc Mountains, Kilimanjaro, Meru, the Crater Highlands, and the Loliondo Mountains extending north to the Loita hills east of the Mara Park of Kenya. In the Serengeti ecosystem montane forest is found as riverine forest along the Mara River in Kenya, and the several tributaries flowing north into the Mara in northwest Serengeti (fig. 9.1; table 9.1). Small remnants are found on top of Kuka hill at 2000 m and the hills east of Lobo hill; here are the remains of a much more extensive forest. Stands of very degraded forest occur on ridges between the tributaries indicating that at one time the whole *Terminalia* woodland of northwest Serengeti could have been montane forest. It is uncertain when this extensive forest occurred. S. E. White (1915), while on a hunting trip in 1913, records that they had to cut their way from Wogakuria hill 10 km north to the Mara River. At least since the 1960s this area has been open grassland with small stands of relict forest easily circumvented. White's account suggests, therefore, that more extensive forest occurred in the late nineteenth century before the great rinderpest epidemic but there were still open areas according to the boundary surveyor G. E. Smith (1907; see chapter 2). Forest has been declining in extent since the nineteenth century.

In summary, the Serengeti ecosystem supports both types of riverine forest running along parallel rivers a mere 50 km apart. The tree species in these two forest types are very different, with only one tree species being commonly found in the canopy of both forests. Seedlings of some canopy trees occur in the each forest, but are extremely rare outside of their "home" forest (<1% of seedlings). Few areas in Africa have such close proximity of these forest types, and as such, they contribute to the diversity of biota in the Serengeti ecosystem. They each show similarity with Mara to montane forest and Grumeti to Congo forests via Rubondo Island (G. Sharam, unpublished data).

THE FAUNA

Bird species found in the two forest types are given in the appendix at the end of this chapter. A number of species are found in both forest types, such as Schalow's turaco, Narina trogon, wattle-eye, and black-headed oriole. However, there are also species that are confined to one or the other forest. In lowland forest, there is another turaco, the eastern grey plantain-eater. Although in its total range it inhabits a broad range of woodlands, in Serengeti it is confined to the narrow lowland forest. Other species found only in lowland forest are black-headed gonolek, red-capped robin chat, and rufous chatterer.

The montane forest supports both widespread forest birds (black-and white casqued hornbill, crowned hornbill, ashy flycatcher, grey cuckoo-shrike, black-and-white manikin) and restricted birds (Ross's turaco, blue flycatcher, snowy-capped robin chat, moustached warbler), none of which occur in lowland forest.

Mammals also differ between the forests. The lowland forest has the lowland form of black-and-white colobus (*Colobus guereza*) on both Grumeti and Mbalageti, but neither lowland nor highland form of colobus are found on the Mara. However, lowland colobus have also been observed in the Serengeti highlands, those areas above 2,000 m, east of Lobo (A. Kilpin, pers. comm.). This area is on the Grumeti watershed and indicates that in times past, lowland forest extended at least 50 km further upriver than it presently does and connected with the mountains on the eastern side of the ecosystem. The lowland colobus have subsequently been isolated on mountaintops, which are not their normal habitat. Isolated specimens of the lowland forest tree, *Cordia goetzii*, are presently found on termite mounds near the Grumeti River at Kleins camp, a clear indication of lowland forest in eastern Serengeti in previous times. In contrast, the Mara montane forest supports tree hyrax (*Dendrohyrax arboreus*) not found in lowland forest.

The montane forest on the Serengeti highlands dominated by *Podocarpus* species, although related to the Mara forests, supports a highland fauna not seen on the Mara River, such as the purple-throated cuckoo shrike (A. Kilpin pers. comm.). We do not consider this habitat here. A full list of the birds in the two habitats is given in the appendix at the end of this chapter.

DYNAMICS OF THE LOWLAND RIVERINE FOREST OF THE GRUMETI

Lowland forests in Serengeti are being converted to grasslands at an increasing rate, and subsequent reestablishment of trees in grasslands is poor. Photographs of riparian forests on the Orangi River (a tributary of the Grumeti) were taken in 1980 by A. R. E. Sinclair and photographs were taken from the same positions in 2001. In 1978, we observed patches of *Acacia polyacantha* Willd. thickets establishing in grasslands adjacent to the Grumeti and Orangi Rivers, and today it occurs as dense stands of varying size, from single canopy trees up to 2–5 ha. *Acacia polyacantha* is a tall semideciduous tree, from 20 to 25 m tall, with a long, straight bole having 4–5 cm thorns. Some regeneration of forest trees is associated with these stands, which often contain abundant seedlings and support a variety of understory vegetation, from dense grasses to herbs and spiny shrubs. We decided to investigate whether stands of this savanna tree may facilitate the establishment of riparian forests, and examine the conditions that allow forest trees to establish under *A. polyacantha*. Specifically, we tested the influence of stand size, the presence of canopy trees, fires, grass, and shrubs on tree seedling survival and abundance, and the potential of *A. polyacantha* as a tool for forest restoration.

In Serengeti, establishment of forest trees is limited in riparian grassland by fires, browsing, and competition with grasses (Sharam, Sinclair, and Turkington 2006). In contrast, *Acacia* trees establish easily in grasslands when fires (Sinclair et al. 2007) and/or browsers (Prins and van der Jeugd 1993) are removed or dry-season rainfall is more abundant (Higgins, Bond, and Trollope 2000), although the relative importance of these variables is debated. In recent history, *Acacia* trees have undergone two pulses of natural establishment in Serengeti. The first occurred from 1890 to the 1930s when, starting in the 1890s, the epizootic rinderpest devastated wildlife in Serengeti and domestic cattle of the surrounding pastoralist people (Mallet 1923; Sandford 1919; Sinclair 1979). The subsequent famine reduced the number of pastoralists as people died and emigrated from the area (Ford 1971), leading to a reduction in fires. The second occurred during the late 1970s when a wide range of *Acacia* species, including *A. polyacantha,* established over much of Serengeti (Sinclair et al. 2008). This pulse of establishment occurred when a series of wetter than average dry seasons coincided with increasing wildebeest (*Connochaetes taurinus*) numbers that removed sufficient grass to reduce fire frequency (Packer et al. 2005). This combination of reduced fire and wet dry seasons led to establishment of a variety of *Acacia* species over broad areas of Serengeti (Sinclair et al. 2007).

Savanna trees may be able to facilitate forest establishment into adjacent grasslands (Kellman and Miyanishi 1982). Savanna trees can provide shade and perches for birds, which increases soil moisture and seed rain respectively. Trees also reduce grass density (Holl 2002) that may ultimately exclude fire (Biddulph and Kellman 1998). Despite these altered conditions, recruitment of forest tree seedlings is low under single savanna trees and within small stands of trees (Holl et al. 2000). However, forest trees have regenerated under large continuous stands of savanna trees in Uganda (Lejju, Oryem-Origa, and Kasenene 2001) and South Africa (Ben-Shahar 1991), suggesting that the conditions occurring inside larger stands may be important.

Methods

We surveyed 75 *A. polyacantha* stands ranging from individual trees to 5 ha stands for stand size, canopy cover, interbole distance, shrub content, herbaceous cover, grass cover, and grass height. The understory of each stand was surveyed for seedlings of riparian forest trees. We tagged 362 seedlings and monitored them for survival every six months for two years; we recorded seedlings of *Lecaniodiscus fraxinifolius* (0.49 ± 0.17 m^{-2}), *Elaeodendron buchananii* (0.36 ± 0.13 m^{-2}), and *Ziziphus pubescens* (0.17 ± 0.09 m^{-2}). In addition, we conducted experimental burns during the mid-dry season (July) in the adjacent grassland 25 m from the edge of stands and monitored survival of tagged seedlings. To test the effects of grass, we created artificial canopy gaps by removing *A. polyacantha* on the periphery of each stand and recorded the time taken for grass to invade the newly created openings from monthly visits. We conducted experimental burns in riparian grassland opposite these gaps and measured how far fires progressed into the gaps during the following dry season and the survival of seedlings there. The influence of shrubs on seedling survival was tested by tagging tree seedlings below shrubs in large *A. polyacantha* stands. Half of the shrubs were removed by cutting at ground level and subsequent seedling survival was monitored. Finally, bulk soil moisture was measured in both the wet and dry seasons 10 m from the *A. polyacantha* stands' boundary inside each stand and at 10 m into the adjacent grassland.

Results

Our experimental results show that *A. polyacantha* can establish in grassland and facilitate the establishment of forest trees. During four years of fire suppression, no seedlings of forest species established in the grasslands. In contrast, seedlings of *A. polyacantha* were established during 1999 when there was an elevated dry season rainfall, similar to the establishment event during the late 1970s. These results suggest that the existing riparian forests in Serengeti are the result of the 1890–1930s *Acacia* establishment event when rinderpest removed much of the wildlife and cattle, and therefore people and fires from the area. In addition, while some seedlings of *A. polyacantha* were observed in the grassland, none were found inside *A. polyacantha* stands or inside riparian forests. This suggests that forest establishment is a unidirectional process by which *A. polyacantha* establishes into grasslands when conditions allow, followed by establishment of forest tree species, producing a riparian forest.

Once *A. polyacantha* trees have established in grasslands, stands of these trees can facilitate the establishment of forest trees by excluding grasses and fires. This effect is dependent on the stand size (fig. 9.2). Forest tree seedlings were not found under single *A. polyacantha* trees (fig. 9.2d), smaller stands that contained more grasses, or in the grassland proper, and occurred only in the larger stands with less grass (fig. 9.2a). The 1 m diameter area around seedlings also contained less grass than the average for each stand. When canopy trees were removed, grasses rapidly invaded canopy gaps, leading to reduced seedling survival. Fires did not burn under large stands of *A. polyacantha*, but did burn under single trees and in experimental canopy gaps with grass. Larger stands of *A. polyacantha* also supported a dense understory of herbs (fig. 9.2b) and thorny shrubs (fig. 9.2c). These shrubs protected seedlings from browsing by antelope, especially impala. When shrubs were removed, the rate of browsing increased on exposed seedlings and their survival declined. Browsers are particularly important in savannas, where impala and other antelope can reach high numbers and prevent regeneration of *Acacia* spp. over large areas, including Serengeti National Park (NP) (Belsky 1984) and Manyara NP, Tanzania (Prins and van der Jeugd 1993). In Serengeti, browsing by antelope severely reduced growth and survival of forest trees inside riparian forests (Sharam 2005) and limited survival of seedlings in the grasslands (Sharam, Sinclair, and Turkington 2006).

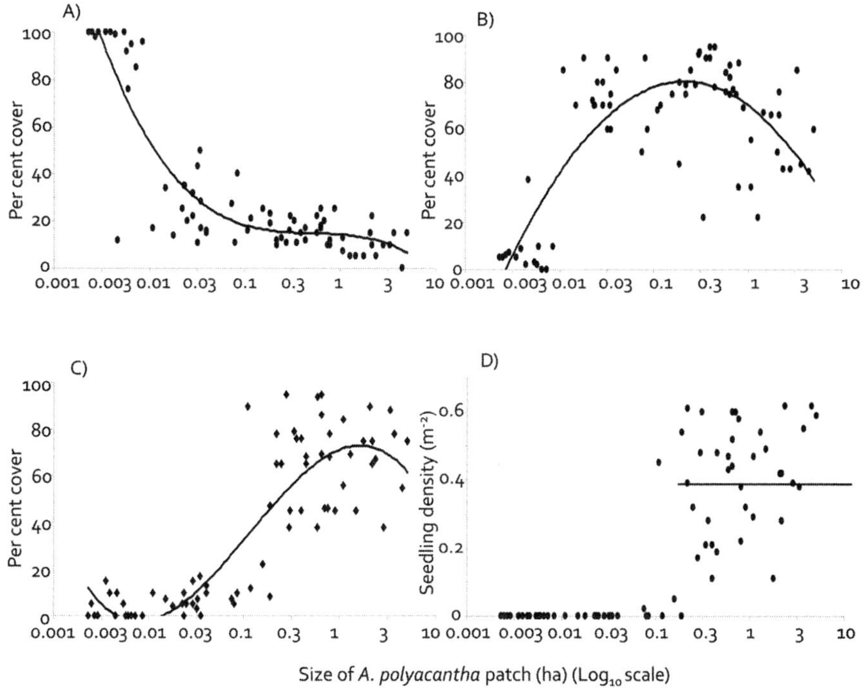

Fig. 9.2 Percent cover of (A) grass, (B) herbs, (C) shrubs, and (D) density of seedlings of forest trees relative to the size of *Acacia polyacantha* stands. Curves are best-fit lines (modified from Sharam et al. 2009a). (A) $y = -7.31x^3 - 6.28x^2 - 2.96x + 14.4$; ($R^2 = 0.79$). (B) $y = -22.2x^2 - 29.9x + 70.3$; ($R^2 = 0.61$). (C) $y = -12.0x^3 - 33.9x^2 + 16.7x + 71.1$; ($R^2 = 0.74$). (D) $y = 0.4$; ($R^2 = 0.02$).

Regeneration of Lowland Forests

Soil moisture was higher under *A. polyacantha* stands than in the adjacent grassland and this likely increased the probability of seedling survival. In Serengeti, seedlings of forest trees established inside riparian forests only during years with floods (Sharam, Sinclair, and Turkington 2006). This indicates that the abiotic conditions under *A. polyacantha* stands are also important to the survival of forest tree seedlings. Although elephants have been implicated in preventing *Acacia* woodland regeneration (Dublin 1995), we have no evidence of elephants feeding or damaging forest tree seedlings in *A. polyacantha* stands. Elephants appear to avoid contact with *A. polyacantha* trees, although they rub against other tree species nearby, damaging bark and covering tree trunks with mud. *Acacia polyacantha* trees are probably protected from elephants by the large, broad-based spines on their trunks. Consequently, the impact of elephant trampling was much reduced within larger stands of *A. polyacantha*.

We have shown that seedlings of forest trees can establish in stands of *A. polyacantha* because these stands have four interrelated factors: (a) reduced grass density, (b) reduced fire frequency, (c) dense thorny shrubs that exclude impala, and (d) increased soil moisture during the dry season. However, there is a minimum stand size of *A. polyacantha* where these conditions occur and seedlings of forest trees are found. In addition, *A. polyacantha* can establish directly into grasslands when both fires are suppressed and dry season rainfall is higher than normal. In contrast, seedlings of forest trees cannot by themselves establish in riparian grassland. Therefore, we hypothesize that riparian forests of the lowland forest type in Serengeti establish in a two-step process: *Acacia polyacantha* first establishes in the grassland, and forest tree seedlings then establish in the stands of *A. polyacantha*. Our results also suggest that the existing riparian forests in Serengeti are the product of an earlier *A. polyacantha* establishment event during the period of 1890–1930s.

DISTURBANCES AND REGENERATION ON THE MARA MONTANE FORESTS

The second major type of forest in Serengeti occurs in the north of the ecosystem along the Mara River, its tributaries, and on some hilltops in the region (fig. 9.1). Riparian forests there are typically composed of two layers. The core area has a dense, closed canopy with similar structure, but different species, to those along the Grumeti River (table 9.1), while the outer layer intersects the grassland and is composed of common, partially fire-tolerant thicket species, with a dense shrub layer and a canopy of *Euclea divinorum* Hiern. Hilltop stands are thickets of this outer, more fire-tolerant band. Some larger hilltop thickets also contain emergent canopy trees from the core area of riparian forests.

Forests and thickets in this area have been declining for the last century. S. E. White (1915) records that his expedition had to cut its way from Wogakuria hill 10 km north to the Mara River during a hunting trip in 1913. As recently as the early 1950s many of the riparian forests and hilltop thickets were continuous with each other, but they were becoming fragmented by frequent grassland fires by the late 1960s (Norton-Griffiths 1979). Since the mid-1960s approximately 70% of the forests have been removed presumably by fire (Sharam 2005). Despite the role of fire in converting forests to grasslands, forests and thickets have not reestablished when fire was reduced in the 1970s. The dynamics of thickets are not entirely understood, particularly the processes that lead to their establishment. Nevertheless they are important refuges for many species of animals including mammals, birds, and butterflies.

Typically, the establishment of woodlands and *Acacia* thickets are controlled by fire, including thickets composed of common species such as *Dichrostachys cinerea* (L.) Wight & Arnott (Tobler, Cochard, and Edwards 2003), *Acacia tortilis* (Forsk.) Hayne, *Grewia flava* DC. and *Terminalia sericea* Burch. ex. DC., which establish when fire is removed (Moleele et al. 2002). This process is often facilitated by grazing, which removes grass biomass, reducing fire intensity and increasing shrub establishment. In Serengeti *Acacia* woodlands experienced major episodes of regeneration during 1890–1930 and the late 1970s, as described earlier.

Acacia establishment in grassland is also reduced by browsers (Prins & van der Jeugd 1993) and elephant (*Loxodonta africana*) disturbance (Dublin 1995). Antelope reduce *Acacia* growth and survival in Serengeti NP (Belsky 1984). Forest tree establishment is also limited by fire and by competition with grass (Chapman et al. 1999). Grass competes with tree seedlings for water, light, and nutrients (Hoffmann, Orthen and Franco 2004). High buffalo numbers in Serengeti may have removed grass competitors and facilitated thicket establishment historically (Sinclair, 1979). Understanding the role of these factors in the establishment of these hilltop thickets and riparian forests in grasslands is essential for their future conservation.

Hilltop thickets cover areas from 1 to 5 ha on hilltops south of the Mara River in northern Serengeti. They consist of a dense matrix of 2–3 m *Croton dichogamus* Pax. and *Teclea trichocarpa* (Engl.) Engl. shrubs with interspersed emergent (6–15 m) *Euclea divinorum* trees and smaller numbers of *Diospyros abyssinica* (Hiern) F. White and *Olea africana* Mill. Riparian forests have a similar structure and extend 50–100 m from and 100–500 m along Mara River tributaries. The core of riparian thickets sits adjacent to the river and contains a dense area of closed-canopy forest with increased species richness, including the canopy trees *Diospyros abyssinica* (Hiern) F. White, *Drypetes gerrardii* Hutch., *Ekebergia capensis* Sparrm., *Olea africana* and *Ficus* spp., with an understory of *C. dichogamus, T. trichocarpa*, and *Strychnos* spp. All species studied are bird-dispersed (Beentje 1994).

Euclea divinorum is common in thorn scrub in riparian areas and in rocky areas and hilltops in Tanzania (Dale and Greenway 1961). *Euclea divinorum* occurs in both hilltop thickets and riparian forests and as seedlings and stunted root stocks in the adjacent grassland. This species has two features characteristic of pioneer species: it is fire resistant by way of its thick bark, and reproduces via root-coppicing following fires (Beentje 1994). This species is defended against herbivory with a variety of emetic compounds

(Mebe, Cordell, and Pezzuto 1998). We examined the pattern and process of forest and thicket establishment in grassland via *E. divinorum*.

Methods

We surveyed grassland sites for seedlings of thicket trees by searching plots adjacent to both hilltop thickets and riparian forests (table 9.2). Seedlings were also surveyed inside thickets and forests along belt transects extending from the forest edge to center (table 9.2) (Sharam, Sinclair, and Turkington 2006). Of the seedlings found inside thickets and forests, five species occurred under *E. divinorum* canopy trees. These species included a high density of two shrub species (*C. dichogamus* and *T. trichocarpa*) and a lower density of three canopy tree species (*D. gerrardii, E. buchananii, and D. abyssinica*) (table 9.3). Seedlings were recorded along with the distance to and identity of the nearest canopy tree. Forty *Euclea divinorum* seedlings in grassland sites were tagged and their height measured. We also removed the grass from around half of these seedlings using herbicide. Seedling survival, height, and condition were monitored every two months for two years. An additional 150 *E. divinorum* seedlings were tagged in the grassland and half were protected by small thorn fences that were effective at excluding all small browsing antelope (i.e., impala, bushbuck, reedbuck, and dik-dik). Half of the fenced plots, and adjacent unfenced plots were burned during the mid-dry season (June 2000 and 2001). Seedlings were monitored before and after fires for survival, height, and browsing every two months for two years.

Table 9.2 Seedling density ha^{-1} (± SD) of canopy tree species in grassland, hilltop thickets, and riparian forests in northern Serengeti (Sharam 2005)

	Grassland sites	Hilltop thickets	Riparian forests
Euclea divinorum	14.1 ± 4.3	56 ± 23	126 ± 75
Croton dichogamus	0	29 ± 11	31 ± 20
Teclea trichocarpa	0	44 ± 21	68 ± 31
Diospyros abyssinica	0	16 ± 8	112 ± 59
Drypetes gerrardii	0	18 ± 12	49 ± 38
Ekebergia capensis	0	48 ± 37	45 ± 41
Ficus spp.	0	3 ± 2.5	12 ± 7
Olea africana	0	0	2
Strychnos spp	0	40 ± 23	24 ± 11

Table 9.3 Seedling density ha^{-1} (± SD) under the canopy of *Euclea divinorum* in hilltop thickets and riparian forests in northern Serengeti

	Hilltop thickets	**Riparian forests**
Euclea divinorum	154 ± 42	186 ± 55
Croton dichogamus	129 ± 49	164 ± 23
Teclea trichocarpa	156 ± 63	138 ± 59
Diospyros abyssinica	17 ± 9	58 ± 17
Drypetes gerrardii	9 ± 4	32 ± 11
Ekebergia capensis	18 ± 11	36 ± 28

Results

Euclea divinorum has attributes of a pioneer species that can facilitate the establishment of hilltop thickets and montane riparian forests in Serengeti: it establishes into grassland areas and when mature, other tree species can establish beneath it. However, under the present conditions of frequent fires and abundant browsers and grass, seedling survival is low and yearly growth rates are negative. Only when both browsers are excluded and burning stopped are seedling growth rates positive. Removal of grass also allows for positive growth, even in the presence of browsers. Thus, establishment of new thickets and forests in East Africa via *E. divinorum* establishment may depend on reducing browsing and fire, or removing grass.

Our studies showed that fire reduces both seedling survival and growth rate (fig. 9.3). Seedlings are scorched and burned when the surrounding grassland burns. There is no compensatory growth or survival of seedlings following fires, and thus no positive effect of fire in removing competitors or increasing growth rates. Removal of fire may be insufficient, however, to allow *E. divinorum* to establish in grasslands, as seedling growth rates were negative when fires were removed, but browsers remained (fig. 9.3). Moreover, no establishment occurred when grass was removed from riparian areas of grassland for four years in other experiments. Thus, reduction in fire frequency alone does not permit *E. divinorum* to establish in grasslands of Serengeti, and successful establishment likely requires removal of browsing or grass competition.

The main browsers are impala and they decrease seedling survival by 70%, but their principal effect is to reduce seedling growth to negative rates. Browsers clip the apical bud and youngest shoots from seedlings. Browsers also limited the height of *Acacia* seedlings to below 32 cm in Serengeti, but do not affect survival (Belsky 1984). Moreover, no thicket or forest establish-

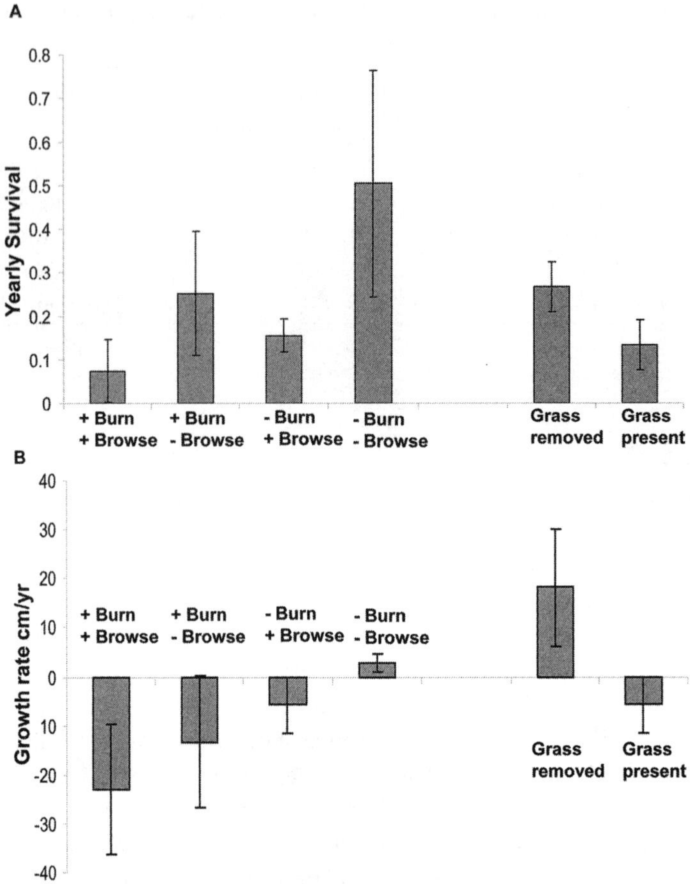

Fig. 9.3 (A) Survival (± 1SD) and (B) growth rates (± 1SD) of *Euclea divinorum* seedlings in riparian areas along the Mara River with and without fire and browsing, and with and without grass present (modified from Sharam, Sinclair, and Turkington 2006)

ment was observed when fires were reduced in Serengeti in the late 1970s and early 1980s due to wildebeest grazing and higher dry season rainfall (Sinclair 1995), but browser numbers remained high (Campbell and Borner 1995).

For browsers to successfully limit thicket regeneration, they must maintain both high and constant browsing pressure on seedlings. Impala numbers do decline periodically due to diseases, such as anthrax in Manyara NP, Tanzania (Boshe and Malima 1986). Thus, conditions necessary for *E. divinorum* establishment could occur naturally if increased dry season rainfall reduced fire frequency and disease reduced browser abundance.

It was surprising that elephants were not observed to consume *E. divinorum* seedlings. We expected them to do so because we observed them consuming *Acacia* seedlings in nearby savanna areas, and elephants appear to prefer seedlings to larger trees. Dublin (1986) reported that elephants prefer *Acacia* seedlings that are 0.5–2 m tall. Elephants may avoid feeding on *E. divinorum* due to chemical defenses (Mebe, Cordell, and Pezzuto 1998) and prefer seedlings that are not chemically defended. *Acacia* seedlings are physically defended, but not heavily chemically defended (Gowda 1997).

Frequent fires can remove seeds from the seed bank (Williams 2000), such that insufficient seedlings are present to take full advantage of favorable growing conditions. Dry season rainfall of at least once every 30 days is considered essential for *Acacia* seedling establishment in South Africa (Higgins, Bond, and Trollope 2000). Thus, while *E. divinorum* must have both fire and browsing removed for growth to occur on the time scale of this study, establishment may be possible if fire alone is reduced for longer periods of time (i.e., more than four years), or by increasing dry season rainfall.

Competition from grasses decreases seedling survival by 50% and produces negative growth rates. Seedlings follow a pattern of rapid growth following germination (Sharam 2005), which may be followed by repeated browsing and a slow decrease in height. We did not observe rapid growth of young *E. divinorum* seedlings in grassland, but it was observed inside forest areas where seedling abundance was higher (Sharam 2005). Regardless of the cause of negative growth, the highest growth rates found in the study occurred when grass was removed.

Thicket and forest establishment in grasslands via a facilitating species has been reported in west Africa (Favier, Namur, and Dubois 2004), South Africa (Dean, Milton, and Jeltsch 1999), South America (Kellman and Miyanishi 1982) and India (Puyravaud, Dufour, and Aravajy 2003). Typically, the trees that establish first are fire resistant savanna trees around which rain forest species established via increased seed rain (Dean, Milton, and Jeltsch 1999), higher nutrients, and reduced competition with grasses (Holl 2002), which consequently reduced fires (Sharam 2005). Our results suggest that *E. divinorum* functions as a pioneer species when conditions are suitable, establishing in grassland and facilitating the establishment of other forest trees. Such conditions existed during the 1890–1930s, as described earlier. With reduced browsing and fires, *E. divinorum* likely established on hilltops and in riparian areas and facilitated thicket and forest establishment. These results also suggest that grass removal by buffalo is not the mechanism by which forests establish in grasslands. First, buffalo were present in Serengeti in high numbers in the 1950–1970s, a period during which no thicket or forest establishment was observed (Norton-Griffiths 1979). Second, long periods with high buffalo numbers and low grass would lead to a long period of establishment rather than a pulse of establishment.

Demographic evidence from riparian forests supports the hypothesis that forests established via a pulse of recruitment. Diameter measurements of canopy tree trunks indicate that trees are of similar age. *Euclea* trees in riparian areas of Zambia attained 30 cm diameter at breast height (DBH) after 100 years (Chitondo 1996), a size similar to those found in Serengeti. These data suggest a pulse of establishment in Serengeti about 1900. Most DBH growth models in Africa are based on arid or semiarid sites with slow growth rates (Geldenhuys 1998), which may not be applicable to riparian areas, as trees in riparian and hilltop areas grow faster and are taller than trees elsewhere (Beentje 1994). In this study, DBH measures were not different within and between species suggesting that all trees are of a common age class. There was also a trend for *E. divinorum* DBH to be higher, though not significantly so, suggesting that it established before other species.

Dynamics of Montane Forests

In addition to DBH data, community composition also suggests a pulsed recruitment event. First, individual forest stands have different dominant canopy trees in addition to *E. divinorum* (Sharam 2005), suggesting a founder effect. Second, islands within streams contain a higher number of tree species and those trees are larger and are likely older than forests in comparable riparian habitats (Sharam 2005). We propose that during the period 1890–1920s, reduced browsers and fire allowed *E. divinorum* to establish in grassland areas. Existing trees on islands then provided seeds of the other thicket and forest species, which established under the *E. divinorum* canopy and formed the current thicket and forest communities.

Of fundamental concern is whether an initial loss of these forests, and the species they contain, leads to further positive feedback decline in communities. Despite fragmentation, if the forest canopy is undisturbed, then stands can support high wildlife diversity, including birds. However, disturbances such as fire or browsing by animals open the forest canopy, and in these circumstances seedling germination and abundance declines (Terborgh et al. 2001; Silman, Terborgh, and Kiltie 2003; Green, O'Dowd, and Lake 2008) and is insufficient to maintain the forest even when disturbances are removed (Sharam 2005). In the following section we describe a mechanism that contributes to the regeneration of riverine forests in the Serengeti and the destabilizing consequences when disturbance alters that mechanism by opening the forest canopy.

MECHANISMS FOR REGENERATION OF THE MONTANE FOREST

When disturbances such as fire or browsing by animals open the forest canopy, they produce a trophic cascade through the loss of frugivorous bird species that eventually compromise the regeneration capacity of the forests. Here, we combine evidence from long term changes in forest structure and bird communities with experiments on seed viability to explain their consequences on the abundance of forest tree seedlings and subsequent survival of the forest.

Methods

Our first step was to monitor changes in forest stands and bird communities over time in riverine forests on the Mara River; stands range from 0.1–400 ha. In these stands we examined the relationship between forest stem density and bird species diversity in two ways. First, a single 1.1 ha stand was monitored for birds for 40 years from 1966 to 2006 as structure changed from a closed canopy to an open thicket of shrubs without trees. Second, we observed 18 stands at the same time, for 10 years from 1997 to 2006, 1.1—200 ha in size and differing in structure, scoring tree density and bird diversity. The stands were selected to cover the range of canopy integrity found in our 1966–2006 stand, with samples of six stands in each of closed canopy, open canopy, and thicket.

Results and Discussion

In our long-term study plot, tree stem density, estimated from a combination of photographs and belt transects, declined from a dense forest of large trees and understory shrubs of 1,094 stems/ha to a thicket of shrubs with only 2 stems/ha (fig. 9.4). Bird species were recorded by point counts from several days to eight weeks at a time. Birds were divided into frugivores and insectivores, and classified by their preferred habitat of closed forest, thicket, or savanna. The total number of bird species remained roughly constant. As tree density and canopy cover declined, the number of forest bird species declined from 26 species in 1966 to 3 in 2006 (fig. 9.4). The decline in species number occurred when gaps in the canopy allowed grasses and herbs to encroach upon the forest floor. Under the open canopy (292 stems/ha) the frugivore diversity declined from 13 to 4 species, while insectivores showed little change (13 to 10 species). The number of thicket bird species

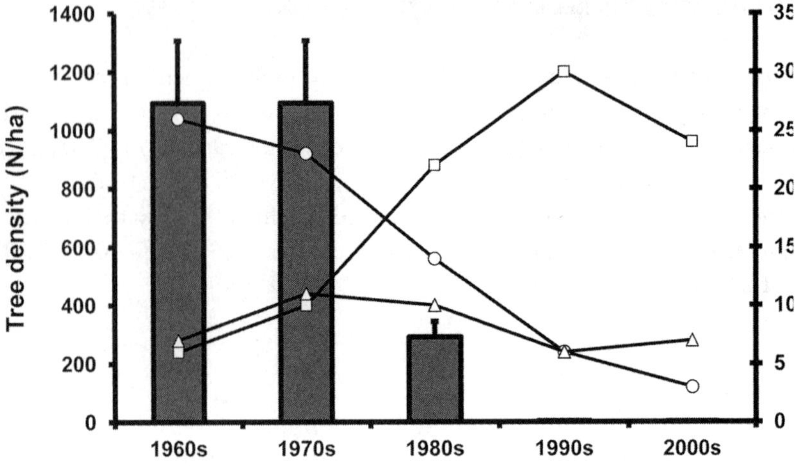

Fig. 9.4 Decline in forest tree density (±1 SE) in a 1 ha forest stand along the Mara River (histogram), from 1966 to 2006. The number of forest bird species (circles) declined, thicket species (triangles) remained relatively constant, and savanna species (squares) increased during the same period (modified from Sharam, Sinclair, and Turkington 2009b).

remained constant at 7–10 species as tree density declined but savanna species, largely insectivores, increased from 6–10 in closed canopy and 22–30 in open canopy and thicket stages.

The distribution of forest, thicket, and savanna bird species in the 18 forest stands during 1997–2006 paralleled the change in species from 1966 to 2006. The total number of forest bird species in all large, dense stands was 33. This number almost halved in open forest stands, a significant decline in diversity. This decline in species number was largely due to the loss of frugivores (primarily hornbills, turacos, and greenbuls) from 16 to 6 species, whereas insectivores declined from 17 to 12 species. Thicket bird species varied between 14 and 18 and did not show significant changes as tree density declined. In contrast, savanna species increased from 27 in closed to 32 in open forest and 41 in thicket. The total number of insect and fruit eaters therefore remained similar at 74, 68, and 67 in the three forest types.

Between 1997 and 2006 we investigated how the decline in number of forest birds and insects affected seed viability, germination, and seedling recruitment. Frugivores ingested fruits, removed the pericarp, and either regurgitated or excreted the seed. The proportion of all seeds on the forest floor that had been previously fed upon and dropped by birds declined from 80% in stands with closed canopy and high stem density to 3% in open forest and thickets with low stem density (fig. 9.5B).

We collected at random 250 individuals each of (a) uneaten fruits from

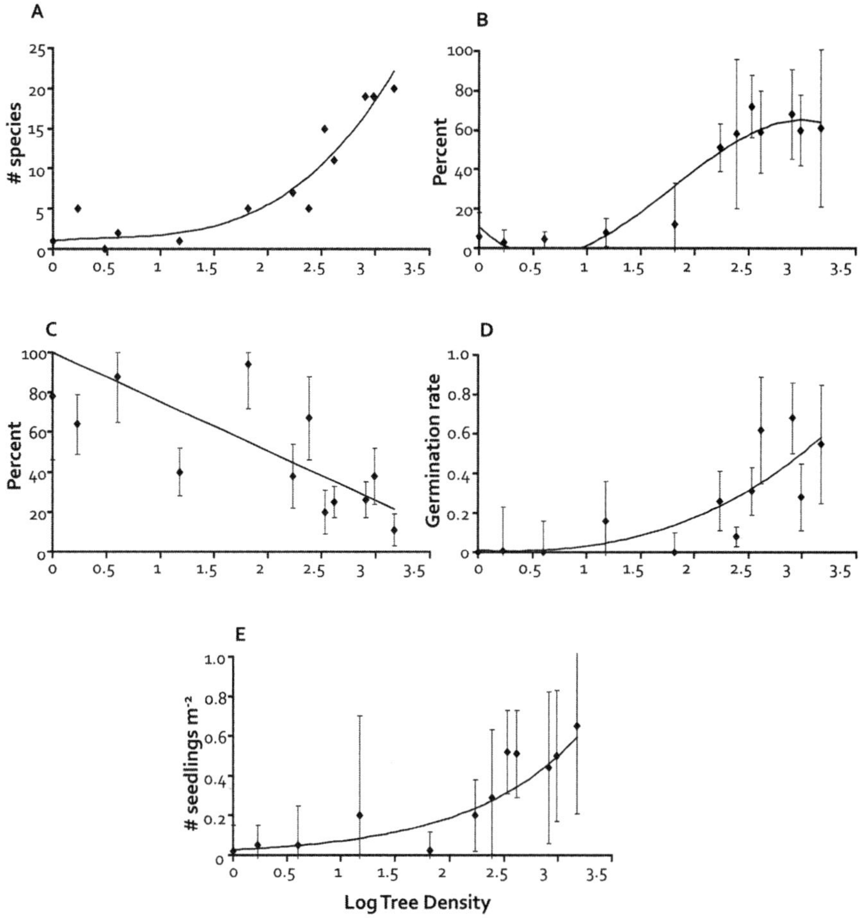

Fig. 9.5 Forest or thicket stands of differing tree density relative to (A) the number of forest bird species, (B) percent of fruits eaten by birds, (C) percent of seeds attacked by beetles, (D) germination rate of tree seeds/m^2/year, and (E) tree seedling density/m^2, during 2000–2001 (modified from Sharam, Sinclair, and Turkington 2009b). (A) $y = 1.03x^3 - 1.57x^2 + 1.27x + 1$; ($R^2 = 0.90$). (B) $y = -9.84x^3 + 53.0x^2 - 52.3x + 11.7$; ($R^2 = 0.92$). (C) $y = -17.6x - 82.4$; ($R^2 = 0.49$). (D) $y = 0.76x^2 - 0.66x + 0.08$; ($R^2 = 0.71$). (E) $y = 0.027\ e^{0.976x}$; ($R^2 = 0.73$).

the trees, (b) seeds without the pericarp deposited on the ground by birds, and (c) uneaten fruits that had fallen to the ground. Bruchid beetles are the main insect seed predators (Pellew and Southgate 1984). Seeds collected directly from trees were not attacked by beetles. The attack rate by beetles on seeds previously fed upon by birds was low and constant (~5%) across stand sizes. The attack rate on whole seeds that had fallen to the ground, however, was high and varied from 68% at low stem density to 92% at high stem density. The overall attack rate by beetles on all seeds on the ground varied from 30% to 75% as stem density decreased because the feeding by frugivores declined as tree density declined (fig. 9.5C). Seeds from the tree canopy and free of the effects of birds and beetles germinated at 56%. Seeds collected from the forest floor without beetle holes had a similar germination rate with (i.e., without bird feeding, 42%) or without the pericarp and so had been fed upon by birds (48%). Seeds that had been attacked by beetles did not germinate whether the fruit had been processed by birds or not. Therefore, while beetles had a major effect on seed viability, birds only did so indirectly through reducing beetle attack (fig. 9.5C).

Finally, we related seed viability to seed germination rate (fig. 9.5D) and seedling density (fig. 9.5E) from 1999 to 2001. The density of new seedling recruits declined from a high of 3–7 new seedlings/m^2/yr in forest stands with high stem density and a closed canopy to zero in forest stands with less than 20 stems/ha (fig. 9.5D). This trend paralleled that for forest bird diversity (fig. 9.5A) and was in contrast to the proportion of seeds attacked by beetles (fig. 9.5C). The number of new recruits varied between plots within forest stands, but did not differ significantly between tree species. The density of seedlings, those plants that had grown beyond cotyledon emergence but <50 cm, followed the same pattern as that for new tree recruits, namely high density in dense stands (0.65 seedlings/m^2) and low density (0.02 seedlings/m^2) in stands with <20 stems/ha.

Regeneration of Montane Forests

We have shown that the loss of particular bird species, the frugivores, resulted in a loss of the processing of seeds by birds. This led to a higher attack rate by beetles, reduced seed germination rates, and reduced seedling density and recruitment. Although consumption by birds and beetle attack contributed to the decline in recruitment and hence forest replacement, other factors such as fire and browsing by ungulates may also have reduced recruitment and seedling survival (Sharam 2005; Mduma, Sinclair, and Turkington 2007) and most likely all of these worked in concert. Absolute seed density was high in forest stands with high canopy cover but fell rapidly at low canopy cover. Also recruitment rates increased when more seeds were available (Sharam 2005) although the rate depended on tree species (Zanne and Chapman 2001). Thus, conservation could reverse this unraveling of the forest by reconstituting the closed canopythrough both planting of seedlings in canopy gaps, where seedlings have shown high growth and survival (Paul et al. 2004), and improving viable seed rain by promoting

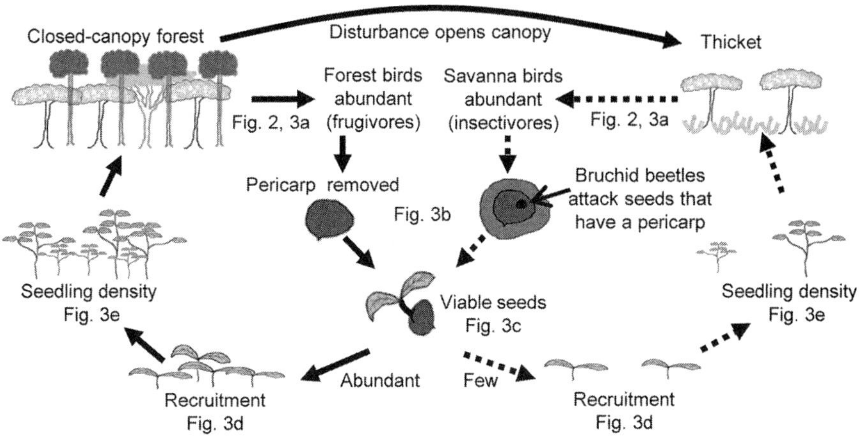

Fig. 9.6 Schematic demonstrating how disturbance that opens forest canopy leads to an unraveling of the processes that maintain the forest. The mechanism involves loss of frugivorous bird species so that unconsumed fruits are attacked by beetles. A high attack rate leads to low germination rate and eventually a progressive loss of tree regeneration and decay of forest. Once initiated this process continues without further disturbance (modified from Sharam, Sinclair, and Turkington 2009b)

the return of frugivores (Holl 2002). The link between bird frugivores and germination in Serengeti forests is paralleled by mammal granivores and *Acacia* germination in the adjacent savanna. Bruchid beetles kill over 95% of *Acacia* seeds (Okello and Young 2000), but if ungulates consume the seeds before beetle attack, seed viability increases (Lamprey, Halevy, and Makacha 1974; Or and Ward 2003).

We conclude that beetles are the main factor determining the germination rate of seeds. One ecosystem function of birds is to counter beetle predation and contribute to stability in the system. Birds protect seeds through their feeding from beetle attack, and thereby indirectly improve the germination of seeds and ultimately regeneration of forest tree species. Therefore, disturbances that lead to the opening of the canopy and reduction in stem density initiate a series of interactions that contribute to reduced recruitment and the disappearance of forest stands in a positive feedback loop (fig. 9.6). First, once the forest structure is opened, key frugivorous bird species drop out, beetle attack increases, germination drops, and seedling recruitment declines. Over a timescale of decades this leads to further opening of the canopy. Secondly, the initial external disturbance is no longer necessary to enforce this positive feedback; once begun, the process inexorably leads to forest fragmentation and progressive unraveling in the community until it disappears, as we have recorded from photographs over the past 50 years (Sinclair et al. 2008).

There are several implications for this research. First, the results imply that there is a minimum patch size and/or total forest cover that is required for a forest patch to remain viable. Once fire has opened the canopy and caused the loss of these key bird species, the forest will not regenerate even if fire is removed. Hence, the conservation and restoration of forest patches must be conducted at a scale that addresses the habitat requirements of the key bird species, not at the scale of individual patches.

Second, in the debate over the ecosystem services provided by species diversity, this is an example where the loss of a particular suite of bird species had an unexpected and catastrophic effect on the viability of the ecosystem as a whole. Finally, the change between forested and grassland states can best be described as a state and transition model; once the forested ecosystem has been pushed to a particular point, a positive feedback process accelerates the transition to the alternative state, in this case grassland.

CONCLUSIONS

We have shown that both *Acacia polyacantha* and *Euclea divinorum* can act as nurse species to facilitate the establishment of other canopy tree species. That forest stands can establish when facilitated by *A. polyacantha*, as we have shown occurred during the 1970s, has three wide-reaching implications. First, riparian forests in East Africa may not be remnants of central African forests that have survived 8,000 years since forest cover was reduced by changing climate and the immigration of agriculturalists (Jolly et al. 1997; Livingstone 1982). Instead, individual forest stands may be ephemeral, which suggests that complex forest communities can establish quickly. Second, efforts to restore degraded land or fragmented forests must consider the effects of fire and browsers and may be accomplished using *A. polyacantha* as a management tool. Third, this research also suggests that the establishment of new forest areas must occur close enough to and on a large enough scale that habitat values are provided for the seed-transporting and processing bird species to occur. These findings represent a major shift in our understanding of riparian forest dynamics in East Africa and indicate that the ecology of forest establishment is dynamic. This dynamic ecology may have important implications for protected areas. While riparian forests occur inside Serengeti, they are rare outside the park where the trees are harvested for timber. Managers may need to consider that protected areas are a seed source for the surrounding areas, and once forests are lost inside protected areas, there may be limited potential for reestablishment.

We have also shown that *E. divinorum* can establish in grasslands when either fire and antelope or grass are removed, and provide conditions for the establishment of hilltop thickets and riparian forest trees. However, under the current conditions of frequent fires, abundant browsers, and grass in Serengeti, *E. divinorum* growth and survival are very poor and thicket and forest regeneration is not occurring. Current thickets likely established during the 1890–1920s period following the rinderpest epizootic, but did not establish in the 1970s when only fire was reduced. Thus, *E. divinorum* is a less aggressive pioneer species than *Acacia* species such as *A. tortilis* and *A. robusta* that were able to establish in the 1970s. The lack of an aggressive pioneer species that can facilitate establishment of thickets may help to explain the rapid decline of thickets on and near the Mara River during the last 35 years; that is, despite yearly variability in climate and animal abundance, this species has been unable to establish in grasslands and reclaim areas lost to seasonal grass fires. Future conservation efforts need to consider the interacting roles of fire and grazing pressure against a backdrop of likely climatic change and characteristics of the pioneer trees species found in a particular region.

The findings reported in this chapter represent a major shift in our understanding of riparian forest dynamics in East Africa. Lowland forests along the Grumeti are composed of tree species from the large, continuous forests in central Africa and west Africa, and montane species along the Mara are common in the highlands surrounding Lake Victoria (Medley and Hughes 1996). It has been thought that forest cover expanded from central Africa during a warm, wet period 10,000–8,000 BP (Jolly et al. 1997) and today's individual riverine forest fragments are remnants of this expansion (Medley and Hughes 1996). Our studies challenge this and indicate that forests can establish rapidly under suitable conditions via the facilitative effects of *A. polyacantha* and *E. Divinorum*. Moreover, the conditions suitable for the establishment, reduced fire, browsers, and grass can be altered through management practices on a local scale. Hence, managers are not dependent on long-term shifts in climate or groundwater. These results suggest that the ecology of forest establishment is much more dynamic than previously thought.

ACKNOWLEDGMENTS

This work was supported by the Canadian Natural Sciences and Engineering Research Council and the Frankfurt Zoological Society. It was conducted with permission from the Tanzania Wildlife Research Institute and Tanzania National Parks. We thank Simon Mduma for logistical support and S. Makacha for assistance in Serengeti.

APPENDIX

Table 9A.1 Frugivorous and insectivorous bird species observed in forests, thickets, and savanna grasslands in Serengeti National Park from 1966 to 2006

Forest birds		Thicket birds		Savanna birds	
Frugivores		**Frugivores**		**Frugivores**	
Common name	***Latin name***	**Common name**	***Latin name***	**Common name**	***Latin name***
Black-headed oriole	*Oriolus larvatus*	African golden oriole	*Oriolus auratus*	Afri'can citril	*Serinus citrunelloides*
Blue spotted wood dove	*Turtur afer*	Bronze mannikin	*Lonchura cucullata*	Black-cheeked (faced) waxbill	*Estrilda erythronotos*
Black & white casqued hornbill	*Ceratogymna subcylindricus*	Brown parrot	*Poicephalus meyeri*	Bare-faced go-away bird	*Corythaixoides personata*
(Cabanis's greenbul)	*Pyllastrephus cabanisi*	Emerald-spotted wood dove	*Turtur chalcospilos*	Black-and-white mannikin	*Lonchura bicolor*
Crested guineafowl	*Guttera pucherani*	Red-eyed dove	*Streptopelia semitorquata*	Cardinal Quelea	*Quelea cardinalis*
Crowned hornbill	*Tochus alboterminatus*	Red-fronted tinker bird	*Pogoniulus psisillus*	Golden-breasted bunting	*Emberiza flaviventris*
Grey-olive greenbul	*Phyllastrephus cerviniventris*	Spotted-flanked barbet	*Tricholaema lacrymosa*	Grey-headed sparrow	*Passer griseus*
Green pigeon	*Treron calva*	Speckled mousebird	*Colius striatus*	Green-wing pytilia	*Pytilia melba*
Narina's trogon	*Apaloderma narina*	Thick-billed seedeater	*Serinus burtoni*	Helmeted guineafowl	*Numida meleagris*
Olive pigeon	*Columba arquatrix*	Violet backed starling	*Cinnyricinclus leucogaster*	Hildebrandt's starling	*Spreo hildebrandti*
Ross's turaco	*Musophaga rossae*	Waxbill	*Estrilda astrild*	Laughing dove	*Streptopelia senegalensis*
Schalow's turaco	*Turaco schalowi (persa)*	White-headed barbet	*Lybius leucocephalus*	Namaqua dove	*Oena capensis*
Shelley's francolin	*Francolinus shelleyi*	Yellow-vented bulbul	*Pycnonotus barbatus*	Purple grenadier	*Uraeginthus ianthinogaster*

Tambourine dove	*Turtur tympanistria*			Red-cheeked cordon-bleu	*Uraeginthus bengalus*
Yellow white-eye	*Zosterops senegalensis*			Ring-necked dove	*Streptopelia capicola*
Yellow-throated greenbul	*Chlorocichla flavicollis*			Red-throated spurfowl	*Francolinus afer*
				Speckled pigeon	*Columba guinea*
				Superb starling	*Spreo superbus*
				Wattle starling	*Creatophora cinerea*
				Yellow-fronted canary	*Serinus mozambicus*
				Yellow-mantled widowbird	*Euplectes macrourus*
				Yellow-spotted petronia	*Petronia pyrgita*
Insectivores		**Insectivores**		**Insectivores**	
Ashy flycatcher	*Muscicapa caerulescens*	Arrowmarked babbler	*Turdoides jardinei*	Abyssinian nightjar	*Caprimulgus poliocephalus*
Black-backed puffback shrike	*Dryoscopus cubla*	African scops owl	*Otus senegalensis*	Banded tit-flycatcher	*Parisoma boehmi*
Cinnamon-chested bee-eater	*Merops oreobates*	Grey-backed camaroptera	*Camaroptera brachyura*	Black-breasted apalis	*Apalis flavida*
Collared sunbird	*Anthreptes collaris*	Moustached warbler	*Melocichla mentalis*	Buff-bellied warbler	*Phyllolais pulchella*
Grey-headed bushshrike	*Malaconotus blanchoti*	Paradise flycatcher	*Terpsiphone viridis*	Bearded woodpecker	*Dendropicos namaquis*
Grey apalis	*Apalis cinerea*	Sulpher-breasted bushshrike	*Malaconotus sulfureopectus*	Brown-headed bush-shrike	*Tchagra australis*
Grey cuckooshrike	*Coracina caesia*	Slate-coloured boubou	*Laniarius funebris*	Black-headed bush-shrike	*Tchagra senegala*
Holub's golden weaver	*Ploceus xanthops*	Tawny flanked prinia	*Prinia subflava*	Black coucal	*Centropus grillii*
Little spotted woodpecker	*Compethera cailliautii*	Woodland kingfisher	*Halcyon senegalensis*	Black cuckooshrike	*Campephaga flava*
Red-faced cisticola	*Cisticola erythrops*	White-faced scops owl	*Otus leucotis*	Black-throated honeyguide	*Indicator indicator*
Snowy-headed robin-chat	*Cossypha niveicapilla*			Cardinal woodpecker	*Dendropicos fuscescens*
Tropical boubou	*Laniarius ferrugineus*			Croaking cisticola	*Cisticola natalensis*

continued

Table 9A.1 continued

Forest birds		Thicket birds		Savanna birds	
Frugivores		**Frugivores**		**Frugivores**	
Common name	***Latin name***	**Common name**	***Latin name***	**Common name**	***Latin name***
Trilling cisticola	*Cisticola woosnami*			Chin-spot puffback flycatcher	*Batis molitor*
Wahlberg's honey-guide	*Prodotiscus regulus*			Drongo	*Dicrurus adsimilis*
Wattleye	*Platysteira cyanea*			Dusky nightjar	*Caprimulgus fraenatus*
White-browed robin-chat	*Cossypha heuglini*			Flappet lark	*Mirafra rufocinnamomea*
White-bellied tit	*Parus albiventris*			Green-capped eremomela	*Eremomela scotops*
				Grey flycatcher	*Bradornis microrhynchus*
				Grey woodpecker	*Dendropicos goertae*
				Green wood-hoopoe	*Phoeniculus purpureus*
				Hoopoe	*Upupa epops*
				Lilac-breasted roller	*Coracias caudata*
				Lesser honeyguide	*Indicator minor*
				Little bee-eater	*Merops pusillus*
				Mariqua sunbird	*Nectarinia mariquensis*
				Northern black flycatcher	*Melaenornis edolioides*
				Nubian woodpecker	*Compethera nubica*
				Olivaceous warbler	*Hippolais pallida*
				Plain-backed pipit	*Anthus leucophrys*

Pale flycatcher	*Bradornis pallidus*
Rattling cisticola	*Cisticola chiniana*
Red-backed shrike	*Lanius collurio*
Red-backed scrub robin	*Erythropygia leucophrys*
Red-faced crombec	*Sylvietta whytii*
Rock-thrush	*Monticola saxatilis*
Ruppell's starling	*Lamprotornis purpuropterus*
Scimitarbill	*Phoeniculus cyanomelas*
Sooty chat	*Myrmecocichla nigra*
Spotted flycatcher	*Muscicapa striata*
Striped kingfisher	*Halcyon chelicuti*
White-browed coucal	*Centropus superciliosus*
Winding cisticola	*Cisticola galactotes*
Yellow-throated longclaw	*Macronyx croceus*

REFERENCES

Beentje, H. 1994. *Kenya trees, shrubs and lianas*. Nairobi, Kenya: National Museums of Kenya.

Belsky, A. J. 1984. Role of small browsing mammals in preventing woodland regeneration in the Serengeti National Park Tanzania. *African Journal of Ecology* 22:271–80.

Ben-shahar, R. 1991. Successional patterns of woody plants in catchment areas in a semiarid region. *Vegetatio* 93:19–28.

Biddulph, J., and M. Kellman. 1998. Fuels and fire at savanna-gallery forest boundaries in southeastern Venezuela. *Journal of Tropical Ecology* 14:445–61.

Boshe, J. I., and C. Malima. 1986. Impact of anthrax outbreak on the impala population of Lake Manyara National Park Tanzania. *African Journal of Ecology* 24:137–40.

Campbell, K., and M. Borner. 1995. Population trends and distribution of Serengeti herbivores: implications for management. In *Serengeti II: Dynamics, management, and conservation of an ecosystem*, ed. A. R. E. Sinclair and P. Arcese, 117–45. Chicago: University of Chicago Press.

Chapman, C. A., L. J. Chapman, L. Kaufman, and A. E. Zanne. 1999. Potential causes of arrested succession in Kibale National Park, Uganda: Growth and mortality of seedlings. *African Journal of Ecology* 37:81–92.

Chitondo, P. L. W. 1996. *New silvicultural guidelines for joint forest management planning*. Provincial forestry action programme, Ndola, Zambia.

Dale, I. R., and P. J. Greenway. 1961. *Kenya trees and shrubs*. London: Buchanan's Kenya Estates.

Dean, W. R. J., S. J. Milton, and F. Jeltsch. 1999. Large trees, fertile islands, and birds in arid savanna. *Journal of Arid Environments* 41:61–78.

Dublin, H. T. 1986. *Decline of the Mara woodlands: the role of fire and elephants*. PhD diss., University of British Columbia, Vancouver.

———. 1995. Vegetation dynamics in the Serengeti-Mara Ecosystem: The role of elephants, fire, and other factors. In *Serengeti II: Dynamics, management, and conservation of an ecosystem*, ed. A. R. E. Sinclair and P. Arcese, 71–90. Chicago: University of Chicago Press.

Favier, C., C. D. Namur, and M.-A. Dubois. 2004. Forest progression modes in littoral Congo, central Atlantic Africa. *Journal of Biogeography* 31:1445–61.

Ford, J. 1971. *The role of the trypanosomiases in African ecology*. Oxford: Clarendon Press.

Geldenhuys, C. J. 1998. Growth, ingrowth and mortality patterns over stands and species in the Groenkop forest study site, George. Division of Water, Environment and Forest Technology, CSIR, Pretoria, South Africa.

Gowda, J. H. 1997. Physical and chemical response of juvenile *Acacia tortilis* trees to browsing. Experimental evidence. *Functional Ecology* 11:106–11.

Green P. T., J. T. O'Dowd, and P. S. Lake. 2008. Recruitment dynamics in a rainforest seedling community: context-independent impact of a keystone consumer. *Oecologia* 156:373–85.

Higgins, S. I., W. J. Bond, and W. S. W. Trollope. 2000. Fire, resprouting and variability: A recipe for grass-tree coexistence in savanna. *Journal of Ecology* 88:213–29.

Hoffmann, W. A., B. Orthen, and A. C. Franco. 2004. Constraints to seedling success of savanna and forest trees across the savanna-forest boundary. *Oecologia* 140:252–60.

Holl, K. D. 2002. Effect of shrubs on tree seedling establishment in an abandoned tropical pasture. *Journal of Ecology* 90:179–87.

Holl, K. D., M. E. Loik, E. H. V. Lin, and I. A. Samuels. 2000. Tropical montane forest restoration in Costa Rica: Overcoming barriers to dispersal and establishment. *Restoration Ecology* 8:339–49.

Jolly, D., D. Taylor, R. Marchant, A. Hamilton, R. Bonnefeille, G. Bouchet, and G. Riollet. 1997. Vegetation dynamics in central Africa since 18,000 yr BP: Pollen records from the interlacustrine highlands of Burundi, Rwanda, and western Uganda. *Journal of Biogeography* 24:495–512.

Kellman, M. and K. Miyanishi. 1982. Forest seedling establishment in neotropical savannas: observations and experiments in the mountain pine ridge savanna, Belize. *Journal of Biogeography* 9:193–206.

Lamprey, H. F., G. Halevy, and S. Makacha. 1974. Interactions between *Acacia*, bruchid seed beetles and large herbivores. *East African Wildlife Journal* 12:81–85.

Lejju, J. B., H. Oryem-Origa, and J. M. Kasenene. 2001. Regeneration of indigenous trees in Mgahinga Gorilla National Park, Uganda. *African Journal of Ecology* 39:65–73.

Livingstone, D. A. 1982. Quaternary geography of Africa and the refuge theory. In *Biological Diversification in the Tropics*, ed. G. T. Prance, 523–36. New York: Columbia University Press.

Mallet, M. 1923. *A white woman among the Masai*. New York: E. P. Dutton & Co.

Mduma, S. A. R., A. R. E. Sinclair, and R. Turkington. 2007. What is the role of seasonality and synchrony in reproduction of savanna trees in Serengeti? *Journal of Ecology* 95: 184–96.

Mebe, P. P., G. A. Cordell, and J. M. Pezzuto. 1998. Pentacyclic triterpenes and naphthoquinones from *Euclea divinorum*. *Phytochemistry* 47:311–13.

Medley, K. E., and F. M. R. Hughes. 1996. Riverine forests. In *East African ecosystems and their conservation*, eds. T. R. McClanahan and T. P. Young, 361–84. New York: Oxford University Press.

Moleele, N. M., S. Ringrose, W. Matheson, and C. Vanderpost. 2002. More woody plants? The status of bush encroachment in Botswana's grazing areas. *Journal of Wildlife Management* 64:3–11.

Norton-Griffiths, M. 1979. The influence of grazing, browsing, and fire on the vegetation dynamics of the Serengeti. In *Serengeti: Dynamics of an ecosystem*, ed. A. R. E. Sinclair and M. Norton-Griffiths, 310–52. Chicago: University of Chicago Press.

Okello, B. D. and T. P. Young. 2000. Effects of fire, bruchid beetles and soil type on the germination and seedling establishment of *Acacia drepanolobium*. *African Journal of Range and Forage Science* 17:46–51.

Or, K., and D. Ward. 2003. Three way interaction between Acacia, large mammalian herbivore and bruchid beetles. *African Journal of Ecology* 41:257–65.

Packer, C., R. Hilborn, A. Mosser, B. Kissui, M. Borner, G. Hopcraft, J. Wilmshurst, S. Mduma, and A. R. E. Sinclair. 2005. Ecological change, group territoriality, and population dynamics in Serengeti lions. *Science* 307:390–93.

Paul, J. R., A. M. Randle, C. A. Chapman, and L. J. Chapman. 2004. Arrested succession in logging gaps: Is tree seedling growth and survival limiting? *African Journal of Ecology* 42:245–51.

Pellew, R. A., and B. J. Southgate. 1984. The parasitism of *Acacia tortilis* seeds in the Serengeti. *African Journal of Ecology* 22:73–75.

Prins, H. H. T., and H. P. van der Jeugd. 1993. Herbivore population crashes and woodland structure in East Africa. *Journal of Ecology* 81:305–14.

Puyravaud, J. P., C. Dufour, and S. Aravajy. 2003. Rain forest expansion mediated by successional processes in vegetation thickets in the Western Ghats of India. *Journal of Biogeography* 30:1067–80.

Sandford, G. R. 1919. *An administrative and political history of the Masai Reserve*. London: Waterlow and Sons.

Sharam, G. J. 2005. *The decline and restoration of riparian and hilltop forests in Serengeti National Park, Tanzania*. PhD diss., Zoology Department, University of British Columbia, Vancouver.

Sharam, G., A. R. E. Sinclair, and R. Turkington. 2006. Establishment of broad-leaved thickets in Serengeti, Tanzania: The influence of fire, browsers, grass competition, and elephants. *Biotropica* 38:599–605.

Sharam, G., A. R. E. Sinclair, R. Turkington, and A. L. Jacob. 2009a. The savanna tree *Acacia polyacantha* facilitates the establishment of riparian forests in Serengeti National Park, Tanzania. *Journal of Tropical Ecology* 25:31–40.

Sharam, G., A. R. E. Sinclair, and R. Turkington. 2009b. Serengeti birds maintain forests by inhibiting seed predators. *Science* 325:51.

Silman M. R., J. Terborgh, and R. A. Kiltie. 2003. Population regulation of a dominant rain forest tree by a major seed predator. *Ecology* 84:431–38.

Sinclair, A. R. E. 1979. The eruption of the ruminants. In *Serengeti: Dynamics of an ecosystem,* ed. A. R. E. Sinclair and M. Norton-Griffiths, 82–103. Chicago: University of Chicago Press.

———. 1995. Equilibria in plant-herbivore interactions. In *Serengeti II: dynamics, management, and conservation of an ecosystem,* ed. A. R. E. Sinclair and P. Arcese, 91–113. Chicago: University of Chicago Press.

Sinclair, A. R. E., J. G. C. Hopcraft, H. Olff, S. A. R. Mduma, K. A. Galvin, and G. J. Sharam. 2008. Historical and future changes to the Serengeti ecosystem. In *Serengeti III: Human impacts on ecosystem dynamics,* ed. A. R. E. Sinclair, C. Packer, S. A. R. Mduma, and J. M. Fryxell, 7–46. Chicago: University of Chicago Press.

Sinclair, A. R. E., S. Mduma, G. Hopcraft, J. M. Fryxell, R. Hilborn, and S. Thirgood. 2007. Long term ecosystem dynamics in the Serengeti: Lessons for conservation. *Conservation Biology* 21:580–90.

Smith, G. E. 1907. From the Victoria Nyanza to Kilimanjaro. *The Geographical Journal* 24: 249–69.

Terborgh, J., L. Lopez, P. Nuñez, M. Rao, G. Shahabuddin, G. Orihuela, M. Riveros, R. Ascanio, G. H. Adler, and T. D. Lambert, et al. 2001. Ecological meltdown in predator-free forest fragments. *Science* 294:1923–26.

Tobler, M. W., R. Cochard, and P. J. Edwards. 2003. The impact of cattle ranching on large-scale vegetation patterns in a coastal savanna in Tanzania. *Journal of Applied Ecology* 40:430–44.

White, S. E. 1915. *The Rediscovered Country*. London: Hodder & Stouten.

Williams, P. R. 2000. Fire-stimulated rainforest seedling recruitment and vegetative regeneration in a densely grassed wet sclerophyll forest of north-eastern Australia. *Australian Journal of Botany* 48:651–58.

Zanne, A. E., and C. A. Chapman. 2001. Expediting reforestation in tropical grasslands: Distance and isolation from seed sources in plantations. *Ecological Applications* 11: 1610–21.

TEN

Invertebrates of the Serengeti: Disturbance Effects on Arthropod Diversity and Abundance

Sara N. de Visser, Bernd P. Freymann, Robert F. Foster, Ally K. Nkwabi, Kristine L. Metzger, Andrew W. Harvey, and Anthony R. E. Sinclair

The graceful pace of the long-legged giraffes and the impressive moving mass of the elephants attract the attention of the visitor's eye as it searches the horizon. The thousands of tiny marching steps of Serengeti's arthropods and the dazzling flight of Serengeti's butterflies only become visible to the attentive observer who also searches right next to his feet. In an effort to draw attention to Serengeti's smaller animals we document previous work on invertebrates conducted around Serengeti since 1884. We then analyze information on three major groups: termites, dung beetles, and grasshoppers, major players in ecosystem processes important to savannas.

Second, as repercussions to perturbations on the ecosystem become a focus in mammalian studies, we present the invertebrate side of the story in this mammal-dominated ecosystem. We analyze the effect of three different types of disturbances on arthropod diversity and abundance: fire, grazing pressure, and agricultural land conversion. Third, we illustrate spatial patterns of three main arthropod groups and their interseasonal variation; we contrast this to the relatively constant activity we find in termites. In conclusion, we generate hypotheses and outline future research directions, focusing on invertebrates and their interactions with the system and its larger-sized mammals. Ecological interactions between and responses of insect guilds may turn out to be just as diverse to the observer as those of the larger vertebrates.

GLOBAL INVERTEBRATE DIVERSITY

To a first approximation, most multicellular species on earth are insects (May 1986). In fact, only about 5% (51,000 described species) are vertebrates (Ruppert and Barnes 1994), and over half of all described species are insects (Mayhew 2007). However, most of the insects, some 80–95% of all insect species, are estimated to still be unknown to science; not collected, named, or described (Stork 2007). Within beetles alone this number is said to be some 70–95% (Grove and Stork 2000). The tropics contain overall a high diversity of animal and plant life, and it is therefore not surprising that in earlier days it was estimated that the tropical arthropod groups alone consisted of some 30 million species (Erwin 1982). However, this value has been revised to a four- to fivefold lower estimate (Novotny et al. 2002) after it was discovered that Erwin used unrealistic proportions of herbivorous versus predaceous arthropods, thereby overestimating total diversity. Besides the clear bias in taxonomy toward vertebrates a further bias exists within the described arthropods. This bias lies heavily on the herbivorous insect guilds, mainly due to ecological (and economic) interest in plant-herbivore interactions (Stork 2007). The proportion, though, of other guilds such as decomposing species, predators, and parasites may be as high as 50–70% of all insects, compared to the 16% suggested earlier by Erwin (1982). Nevertheless, there is still little data available on arthropod diversity for a single tropical region, and this data is only for a single taxonomic group (Noyes 1989; Hammond 1990; Hodkinson and Casson 1991). Despite considerable research on savannas there are virtually no estimates of insect diversity available for these regions (Lewinsohn and Price 1996). Without much more comprehensive sampling to monitor insect density and phenology the functional roles of insects in ecosystem processes in savannas will remain unclear.

SERENGETI INVERTEBRATES: HISTORICAL BACKGROUND

Going back in time to the colonial period, several expeditions were sent to explore the Maasai land, near and around the area that is nowadays named Serengeti National Park. This region was primarily interesting to geologists, who studied the Ol Doinyo Lengai volcano and surroundings of Olduvai Gorge (Fischer 1885a), and to cartographers who mapped this uncharted region from Lake Victoria up to Mount Kilimanjaro (Smith 1907, 250).The latter describes literally "It must be recollected that the journey we were about to undertake from the Victoria Nyanza to Laitokitok was across unknown country, a distance of 273 miles as the crow flies." The region was also of interest to bird collectors (Fischer 1885b) and insect collectors (Gerstaecker 1884). Gerstaecker (1884) recorded a total of 167 species of beetles, many for which he gives precise morphological descriptions. Regions around Lake Victoria, Lake Natron, and the Ol Doinyo Lengai volcano were extensively explored by insect collectors during these times, but the center of Serengeti remained mostly untouched. Of the many research groups active in Serengeti today few focus on invertebrates. What do we actually know about the invertebrate communities in Serengeti National Park?

Historically, research in Serengeti National Park, like in most other terrestrial ecosystems, focused on the larger vertebrates. There are several reasons why this is the case. As described in the first Serengeti volume (Sinclair and Norton-Griffiths 1979), the original objectives of the park were conservation of the large-mammal fauna and its habitats. This is reflected by the boundaries of Serengeti National Park, which were determined by the yearly migration of the wildebeest (*Connochaetes taurinus*), the extent of which was first recognized by Bernhard Grzimek and Michael Grzimek (Grzimek and Grzimek 1959). Wildebeest specifically are thought to play an important role in the system. Disease studies reported severe impacts on the entire ecosystem caused by the declines of the wildebeest and buffalo population under the rinderpest outbreak (Sinclair 1979; Dobson 1995). Other studies reported large landscape effects caused by the exclusion and concurrent return of the elephant (Norton-Griffiths 1979). However, two of the most important processes that occur in savannas concern the availability of nutrients and the process of decomposition (Baruch et al. 1996), processes that are mostly governed by invertebrates. Additionally, invertebrates in their high abundance are important resources to many larger savanna animals. Therefore, in order to understand the entire ecosystem with all its inhabitants, we evidently need to include the invertebrates; a future research priority was set by Sinclair in both the first (1979) and the second Serengeti volume (1995). These animals have been severely neglected and until now little or nothing is known about them from Serengeti.

SERENGETI INVERTEBRATES: RESOURCES

The few studies on Serengeti's invertebrates describe insects as avian or mammalian food resources. Sinclair (1978) describes the presence of localized high abundances of insects caused by rainfall patterns and their attractiveness to migrating palearctic birds. He hypothesized that the onset of high insect abundances after the rains determines the timing of the breeding seasons of birds in this area. Whether insect abundance determines bird abundance directly was investigated a few years later by Folse (1982) in the woodlands around Seronera and in the southern plains. He recorded abundances of herbivorous and insectivorous birds and a wide variety of invertebrates present in Serengeti's plains; the invertebrates included were beetles (Coleoptera), flies (Diptera), ants (Hymenoptera, Formicidae), bees and wasps (Hymenoptera), butterflies (Lepidoptera), spiders (Arachnida), termites (Isoptera), grasshoppers (Orthoptera, Acrididae), bugs (Hemiptera, Heteroptera), and leaf- and planthoppers (Hemiptera, Homoptera). His findings indicated that in the Serengeti plains, insect abundance was a poor descriptor in explaining short-term variations in bird population densities. Instead, low vegetation biomass and well-developed vertical vegetation structure provided cover and greater ground-level mobility for most species of birds. Within vegetation types (woodland, medium-height grass, and short-grass plains), the green vegetation biomass was the main factor associated with increased bird densities (Folse 1982). In 2002, Sinclair and coworkers conducted a survey of insect and bird abundances, comparing agricultural areas outside the park with protected areas inside the park. They focused on three insect groups: leaf- and planthoppers (Hemiptera, Homoptera), flies (Diptera), and wasps (Hymenoptera). They showed that the abundance of these insects declined drastically in agricultural areas; in fact it was only 20–50% of the abundances found in the protected savanna areas. With the decline of their food resources, densities of insectivorous birds were reduced outside the protected areas. Nkwabi et al. (2011) investigated in more detail the effects of burning and grazing on insectivorous bird densities via a change in insect abundances. This study included ants, spiders, and beetles all caught by pitfalls, Orthoptera and Hemiptera caught by sweep net, and flies, bees, and wasps caught by tray traps. Interestingly, they found that not the absolute number of insects predicted bird diversity and abundance, but the greater availability of insects as the grass structure changed from long to short grass due to burning or grazing. Although bird species richness increased in burnt and grazed sites, replacement of functional equivalent species took place. Obviously, birds are limited in the time they can spend searching for food and show preferences for certain habitat types or food items. If abundances of their food items become too low they may search for food elsewhere. An observation of how plastic insectivorous birds can be in finding their food in Serengeti National Park was reported by Freymann and Olff (2007); here a cardinal woodpecker (*Dendropicos fuscescens*) was observed repeatedly searching for moth grubs of the genus Ceratophaga (Tineidae, Lepidoptera) not in its usual foraging substrate of dead trees, but inside the outer keratine layers of wildebeest carcass horns.

Many savanna mammals include invertebrates in their diet (Estes 1991; Kingdon 1974, 1977). Kruuk and Sands (1972) studied food selection of aardwolves (*Proteles cristatus*) in Serengeti by investigating the number of insect samples in their feces. They found mostly prey items of various termite species (*Trinervitermes* sp., *Odontotermes* sp., *Macrotermes* sp., and *Hodotermes* sp.) as well as some ants (*Camponotus* sp., *Dorylus* sp., *Pheidole* sp., and *Odontomachus* sp.) and unidentified beetles. Aardwolves are specialist insectivores, but many more mammal species incorporate invertebrates into their diet. A recent study on a newly compiled Serengeti food web included the many known invertebrate groups present in this ecosystem and their predators, including many mammals that are omnivores or insectivores (de Visser, Freymann, and Olff 2011). In total, nearly half of all feeding links in this food web structure of the Serengeti involved invertebrates.

SERENGETI INVERTEBRATES: RECORDS AND OBSERVATIONS

Otte and Cade (1984) describe two new species of the genus *Platygryllus* (Orthoptera: Gryllidae) that so far are known only from the Serengeti plains, these being *Platygryllus atritus* and *P. serengeticus* found in 1980. These were identified among nine other African crickets collected elsewhere from South Africa, Kenya and Tanzania. The species *P. serengeticus* and *P. atritus* are found on the surface, in soil cracks, and are in fact especially numerous inside mammal skulls of wildebeest and buffalo scattered everywhere over the plains.

Freymann and Krell (2011) report mass dying of dung beetles on the surface of the volcanic originated black sand of one of the few remaining "shifting sand" dunes of the barchan type near Olduvai Gorge. They describe the find of high amounts of intact dung beetle carcasses. It is hypothesized that these beetles either actively searched for digging material or were blown by the prevailing west wind into the boiling hot surface of this black sand dune that turned out to be a death trap.

Another interesting find is described by Pedgley et al. (1989) and concerns the African armyworm (*Spodoptera exempta*: Lepidoptera, Noctuidae). This species migrates with the Intertropical Convergence Zone (ITCZ), exploiting the first flush of grass growth at the beginning of the rainy season. This species whose life cycle is about six weeks with no diapause, survives the dry season in the perennially moist coastal region. With the onset of the rains, the moths are carried inland by the prevailing winds to central Tanzania around Morogoro and are concentrated by the wind systems of rainstorms so that the larvae are found in dense patches. They are exclusively graminivorous. Successive generations are carried north by the ITCZ, often as far as Eritrea and Yemen. Outbreaks of armyworm frequently occur in the park and conservation area (Pedgley et al. 1989). Murray (pers. comm.) observed "lawns" in the long-grass plains of Serengeti that had been created by armyworm and subsequently kept grazed down by topi. Armyworm outbreaks provide an abundant and accessible supply of food to insectivorous birds (especially storks) and mammals. Hoffer (pers. comm.) observed spotted hyenas feeding on them.

These finds illustrate just how much more there is to be discovered in specialized niches common in Serengeti but perhaps rare elsewhere.

ECOSYSTEM IMPORTANCE OF INVERTEBRATES IN SERENGETI

Although large herbivores are considered major organizers of nutrient cycling on the Serengeti plains and generally in tropical savannas (Ruess 1987) it must be recognized that dung beetles and termites also have significant effects on ecosystem processes. Savanna soils are generally deficient in nitrogen, phosphorus, and other nutrients, which are abundant in herbivore dung (Foster 1993). The return of these nutrients to the system is performed by the main dung decomposers, the dung beetles, and surprisingly also by termites. Here we describe the activity and spatial distribution of these small insect taxa to strengthen our understanding of the system processes.

Dung Beetles

Dung beetles (Coleoptera; Scarabaeidae) are an important food source for a wide variety of Serengeti predators, in particular bat-eared foxes (*Otocyon megalotis*), Egyptian mongooses (*Herpestes ichneumon*), jackals (*Canis* spp.), and ratels (*Mellivora capensis*); many birds such as kori bustards (*Ardeotis kori*), sacred ibis (*Threskiornis aethiopicus*), white storks (*Ciconia ciconia*), blackheaded herons (*Ardea melanocephala*), and helmeted guineafowls (*Numidia meleagris*) were also seen to feed on dung beetles or their larvae (Foster 1993). However, dung beetles are also known to play a significant functional role in important ecological processes in savanna ecosystems. Dung beetles are responsible for the removal and burial of almost all dung during the wet season (e.g., Weir 1971). Therefore, dung beetles are said to be generally responsible for nutrient cycling from dung and are also important in bioturbation, plant growth enhancement, secondary seed dispersal, and parasite control (Nichols et al. 2008).

Dung beetles are coprophages, meaning they feed on dung of vertebrates, and do so both as larvae and adults (Halffter and Matthews 1966). Adult dung beetles feed exclusively on the liquid component of the dung by means of specialized filtering mouthparts (Holter 2000), thereby relying on the water content of the dung. Larvae have stronger chewing mandibles and can consume coarser, drier dung (Hata and Edmonds 1983). Different dung beetle species use the dung source in different ways (e.g., burying it right under the dung pat or first rolling it away before burying: paracoprids and telocoprids respectively), and process the dung at different rates (fast-burying versus slow-burying). Approximately 2,000 species of Scarabaeinae are found in the Ethiopian region (Bornemissza 1979), with some 320 species of Scarabaeidae known from Kenya and northern Tanzania alone (Davis and Dewhurst 1993). Foster and Bresele (1992) studied dung beetles near Lake Lagarja (also called Lake Ndutu) and recorded 105 species of Scarabaeinae. Serengeti's dung beetle species richness is thereby comparable to other African savannas (Foster 1993).

Serengeti provides unique opportunities for dung beetle studies because of the annual migration of ungulates, resulting in the spatial and temporal heterogeneous distribution of huge amounts of dung. Wildebeest densities may exceed 2,500/km^2 on the plains during the wet season (Ruess and McNaughton 1984). McNaughton (1983) found that dung deposition is clustered; while wildebeest often foraged in the woodland, dung deposition was over eight times higher on adjacent open grassland patches where they rested and ruminated. Although dung beetles (Scarabaeinae) did not seem to show a preference for particular vegetation types, their species composition and correlated dung removal rates were affected by spatial heterogeneity of soil type and texture. Sandier soils (in the grasslands) housed mostly large telocoprids and fast-burying paracoprids, whereas the dung beetle community on finer soils (in the woodlands) was dominated by slow-burying paracoprids (Foster 1993). Although the larger herbivores deposit the dung, it is the dung beetle community that plays the major role in dung removal and burial and they, therefore, determine the rate and spatial distribution at which the nutrients are returned to the system.

Another ecological study on a different beetle group was conducted in Serengeti by Pellew and Southgate (1984). They report on a beetle species of the family Bruchidae (*Bruchidius spadiceus* [Fahr.]) that parasitizes seeds of *Acacia tortilis* subsp. *spirocarpa* in the central woodlands of the Serengeti National Park. The authors studied the interplay of the beetles, *Acacia* seeds, and the main seed-eating herbivores. Seed production in Serengeti is confined to mature canopy trees. Germination of *Acacia* seeds is facilitated through ingestion by herbivores, mainly giraffe and elephant. However, parasitized seeds drop off the trees when the young adult beetle emerges and thus will not be eaten unless the process is prematurely interrupted by ingestion of herbivores. Pellew and Southgate studied the Serengeti woodlands after the increased impact of elephants during the 1970s when mature canopy trees significantly declined to about 9 trees ha^{-1}. The decline in mature canopy caused isolation of trees and decline in seed infestation rates by *Bruchidius* beetles. While *Bruchidius spadiceus* still had very high parasitism rate of 95.6–99.6% in 1972 (Lamprey, Halevy, and Makacha 1974), this rate declined noticeably to 73.6% in 1973 (Southgate 1981) and more drastically to 5.1–8.4% in 1979–1980 (Pellew and Southgate 1984). This study indicates an intricate balance between *Acacia* trees, beetle, and herbivores that had been disturbed by elephants.

Termites

Termites are known as ecosystem engineers by building aboveground mounds and digging extensive subterranean tunnels (Coventry, Holt, and Sinclair 1988; Lobry de Bruyn and Conacher 1990; Scholes and Walker 1993; Holt, Bristow, and McIvor 1996; Dangerfield, McCarthy, and Ellery 1998). They act as herbivores as well as decomposers, feeding on a wide range of living, dead, and decaying plant material (Adamson 1943; Noirot and Noirot-Thimothee 1969; Lee and Wood 1971; Wood 1976, 1978; Bignell and Eggleton 2000; Traniello and Leuthold 2000), including the consumption and turnover of large volumes of soil rich in organic matter and fungi. What is less known is that termites play a significant role in the recycling of mammalian herbivore dung complementary to that of the better known dung-removing dung beetles (Freymann et al. 2008). Termites can quickly remove large amounts of mammalian dung, especially in the dry season (when dung beetles are scarce), when on average about one-third of the dung deposited in a given habitat is removed by termites within one month (Freymann et al. 2008). Even more surprising is the fact that termites not only decompose plant material but also utilize mammalian remains. In Serengeti, termites of the genus *Odontotermes* (most likely *Odontotermes badius*) are a frequent decomposer of mammalian hooves from dead animals (Freymann et al. 2007). None of the scavenging vertebrates in Serengeti uses these remains, not even vultures (e.g., *Gyps africanus, G. rueppellii)* or hyenas (*Crocuta crocuta*).

Since the early observations in Serengeti of "mosaic grass patches" attributed to termite activity (H. de Wit 1978 pers. comm., cited in Sinclair 1979) no detailed studies were performed on the further-reaching ecosystem role of termites in Serengeti until recently. Because of their wide distribution throughout the tropical and subtropical regions of the world (Eggleton 2000), and on a landscape scale throughout the Serengeti National Park (Freymann et al., n.d.), termites play a central role in savanna nutrient cycling through their decomposition activity. Recent work in the Serengeti has shed some light on the ecological role of termites concerning decomposition. Freymann, de Visser, and Olff (2010) identified spatial and temporal hotspots of termite-driven decomposition in this region using litter and dung bait experiments in the long-grass plains. The region is characterized by an undulating topography with small ridges and catenae showing an elevational difference of 15–25 m between hilltops and bases (see also chapter 3 for the catena description). At eight locations litter and dung baits, packed inside mesh bags to keep dung beetles out, were placed on the top and base of the catena. The bags had close contact to the soil where termites could directly access them. Baits were collected at day 1, 2, 3, 4, 5, 6, 10, 15, and 20 of the experiment and the material consumed by termites was measured. This sampling method proved the assumed constant rate of bait mass loss by termites over a time period of 20 days. The experiment was conducted in the dry season and repeated in the following wet season. The number of termite mounds (mapped in 100 × 100 m blocks) around the experimental plots remained constant over these several months and showed a higher density at the top of the catena compared to its base. In the dry season termites removed more bait (litter as well as dung) at the top catena positions than from the bottom positions, but there was no effect of catena position in the wet season. In their paper, Freymann, de Visser, and Olff (2010) propose a delicate balance of factors affecting termite building and feeding activity in this region ruled mainly by soil depth and available moisture (via rainfall and runoff). Interestingly, the spatial hotspots of termite activity coincided with those of higher densities of both mammalian herbivores and predators.

Termite mounds, or termitaria, are the known ecosystem engineering products of termites. These mounds are thought to play a keystone role in providing habitat to several small mammals and reptiles as well as other invertebrates (Redford 1984). This keystone role for invertebrates has been quantified for termite mounds present in the southern plains of Serengeti by de Visser, Freymann, and Schnyder (2008). They show with the use of $\delta^{15}N$ and $\delta^{13}C$ stable isotopes that the diverse invertebrates within termite mounds in the Serengeti plains are trophically connected. The top predators present in these termite mounds were spiders, such as ground spiders (Prodidomidae), crab spiders (Thomisidae), and web-building cellar spiders (Pholcidae). However, the spiders do not feed directly on termites but predominantly on termitophagous invertebrates; this raises the spiders 2–3 trophic levels higher than the termites (one trophic level shift of ca. + 3‰). Therefore, termitaria provide spatial foci of invertebrate trophic interactions, especially in the relatively homogeneous grasslands such as Serengeti's long-grass plains (de Visser, Freymann, and Schnyder 2008).

Grasshoppers

Sinclair (1975) showed that green grass consumption of grasshoppers could be as large as or larger than that of large ungulates and other small mammalian herbivores in grasslands. An estimated 27.7% of total grass consumption per year was computed for grasshoppers versus 68.2% for the large ungulates and just 4.2% for small mammals (rats and mice) in the long-grass plains. However, the relative consumption differences increase in favor of the invertebrates during the wet season months. In the kopjes, invertebrate grass consumption is higher than ungulate grass consumption throughout all months. This illustrates the ecosystem-wide influences of these small but highly abundant insects. Grasshoppers have been the focus of two main samplings within the Serengeti ecosystem, one by A. Harvey in 1985–1989, and the other by the Serengeti Biodiversity Program, 1997–2010. The former sample was collected across habitats within the Serengeti ecosystem and specimens were identified to 71 species by A. Harvey, N. D. Jago, and J. R. M. Ritchie. Some species, especially Gomphocerinae, were not fully determined and were assigned to unnamed species. The latter sample was collected both within Serengeti and outside in the villages; these species were identified by M. Mungai at the National Museum, Nairobi, in 2010.

Both sets of samples were used to compare diversity across habitats within native savanna. The more recent sample was used to compare savanna with agriculture. Comparisons used the Bray-Curtis similarity measure (1-*B*) (Krebs 1989). There are two ways that species diversity in two habitats can differ; one is by habitats differing in their species, the other by one habitat having a subset of the species in another more diverse habitat. In the latter condition if sample size is very different the measure can be distorted a small amount toward dissimilarity (the greatest distortion occurs when B is close to 0.5—one sample being 10% the size of the more diverse will give a value of $B = 0.4$ instead of 0.5. If $B = 0.1$ then the distortion results in $B = 0.09$). The B statistic was corrected by adjusting sample size to that of the smaller sample and recalculating species abundance.

A total of 102 species of grasshopper have been recorded from the two surveys (see appendix 2 at www.press.uchicago.edu/sites/serengeti/). The

Table 10.1 Bray-Curtis values of similarity (1-*B*) for comparing species diversity in the dominant Acacia habitat, *A. robusta*, with other habitats in upper catena, lower catena, grassland types and agriculture

Habitat	*N* alone	Combined with	*N* combined	Bray-Curtis similarity
A. robusta	41	*A. tortilis*	41	0.78
(upper catena)	41	Combretum hills	49	0.55
	41	*A. gerrardii*	45	0.41
	41	*A. senegal*	42	0.47
A. robusta	41	*A. drepanalobium*	45	0.25
(lower catena)	37	*A.seyal*	37	0.31
	41	Balanites	41	0.18
	41	Upper grassland	49	0.53
	37	Floodplain (west)	37	0.34
	37	Plains long	37	0.50
	37	Plains intermediate	37	0.49
	41	Plains short	47	0.31
	37	Terminalia	45	0.32
	41	Forest	43	0.20
	37	Agriculture	41	0.47

Note: The difference between the number of species in *A. robusta* alone and when combined with the other habitat shows the number of new species added by the second habitat.

grasshopper fauna in the dominant Acacia savanna, *A. robusta*, was compared with those in other types of Acacia divided into upper and lower catena habitats (table 10.1). Greatest similarity was with the other major tree habitat, *A. tortilis*, (78%); and in general *A. robusta*, which is an upper catena type, was more similar to other upper catena Acacia habitats (hill habitat with *Combretum*, *A. senegal*, *A. gerrardii*) than with the wetter lower catena (*A. seyal*, *A.drepanolobium*, *Balanites*). *A. robusta* also showed a gradient of similarity with different types of grassland (habitats without trees). Thus, grasslands within the Acacia zone were most similar in grasshopper fauna, followed by long-grass plains, intermediate plains, and finally the short-grass plains. The floodplains had a fauna more similar to that of the adjacent lower catena *A. seyal*. Overall these results corroborate the analysis by Harvey & Ritchie (1990) using Detrended Correspondence Analysis (DECORANA) that there were distinct clusters of species associated with a wet-dry cline and these corresponded to the north-south distributions found in the Sahel of northern Africa. This is particularly clear in the samples from the short-grass plains. In the long-grass plains, woodlands, and riverine habitats further west the situation is more complex and not fully resolved by the samples obtained. It is tentatively suggested that the second axis corresponds to the cline in soil types down the catena, with *Jasomenia sansibara* having the most extreme score and being associated with wallows in black cotton soil. Soil is an important determinant of grasshopper distribution as egg pods require distinct adaptation to different soil types (Phipps 1959). However, the grasshopper fauna in Acacia savanna differed markedly from the broad-leaved Terminalia (32% similarity) and from riverine forest (20%), while both forest and Terminalia supported different species (14% similarity). Agriculture had a grasshopper fauna similar to *A. robusta*, the native habitat that preceded the modification 50–60 years ago; only four extra species were recorded in agriculture not yet found in native habitat.

Synopsis

This synopsis summarizes our knowledge about invertebrates in Serengeti up to now. It does not cover all existing groups. It is also clear that invertebrates are normally studied in the context of resources to larger animals, mainly to birds. The ecological role of invertebrates, the processes they influence, and the ecological interactions they affect are understudied. Clearly, the taxa we have discussed here in more detail, dung beetles, termites and grasshoppers, demonstrate the potential of small insects to affect the entire ecosystem. The studies mentioned here report responses of invertebrates to their mammalian counterparts (reacting to elephant population numbers, decomposing dung, decomposing mammalian hooves, occurring in or feeding on mammalian skulls, grazing effects of insects) and to habitat changes (agricultural land use, declining tree density). In the following two sections we investigate the responses of invertebrates to three different types of disturbances (section 2) and to spatial and temporal variation (section 3).

DISTURBANCE EFFECTS ON ARTHROPOD DIVERSITY AND ABUNDANCE

Natural disturbances such as fire and herbivory play an important role in sustaining the open woodland structure that dominates the Serengeti. However, human-caused disturbances are able to threaten this very balance by permanently converting natural savanna into agricultural areas. How these factors affect the larger vertebrates in the Serengeti-Mara ecosystem has been studied for some groups, such as the large herbivores (Hassan et al.

2007; Serneels and Lambin 2001), but the effects of such disturbances on the invertebrate community are largely unknown. Invertebrates are important resources for larger vertebrates, are present in high numbers, and may play an essential role in ecosystem processes. Therefore, the lack of knowledge on the responses of invertebrates to disturbances may have consequences for our understanding of the responses of the ecosystem to future perturbations. Two studies known from the Serengeti (Sinclair, Mduma, and Arcese 2002; Nkwabi et al. 2011), near the villages of Seronera and Robanda and the southern plains respectively, have shown insect responses to disturbances in the context of avifaunal resources.

Over a period of eight years (1999–2006) arthropods were systematically sampled with different methods (by sweep net, pitfall, window trap, and tray trap), effective for specific arthropods, along transects throughout the park and across the borders in the agricultural areas by the Serengeti Biodiversity Program. Woody versus grassland sites, burnt versus unburnt sites, grazed versus ungrazed sites, and agricultural sites versus natural savanna sites were all sampled at different locations across the Serengeti. Within the park several sampling sites in the north (Kogatenda, Terminalia woodlands), west (Ndabaka plains, Nyasirori), south (long-grass plains, short-grass plains, Naabi), and mid-region (Serengeti Wildlife Research Centre, Seronera woodlands, Banagi, Kubukubu) were visited multiple times a year. Sampling areas outside of the park were located near the villages of Robanda, Nata, and Bwitenge (fig. 10.1). In the Serengeti Biodiversity Program different sampling methods were used in order to collect as many arthropod taxa as possible. We briefly outline each of the methods used: (A) the sweep net (30 sweeps per sampling effort) catches the grass-layer arthropods: mainly individuals of the order Orthoptera (grasshoppers, crickets, katydids), Heteroptera (true bugs), Homoptera (leaf- and planthoppers), Diptera (flies), Lepidoptera (moths and butterflies) and some Arachnida (spiders); (B) pitfall trapping (8.7 cm diameter) is used for the ground wandering arthropods, mainly Arachnida (spiders), Hymenoptera (Formicidae; ants), and Coleoptera (beetles); (C) using a yellow tray trap (28 cm diameter); or (D) a window-trap with formalin at the bottom catches both flying and jumping insects, mainly Hymenoptera (wasps and bees) and some Orthoptera (grasshoppers).

The database consists of 51,000 data lines in which arthropods have been classified at a taxonomic order and family level and then typed as morphospecies. This means that morphologically similar individuals were grouped under one type and dissimilar individuals were assumed to be of a different species and thus of separate morphospecies. The number of individuals per morphospecies has been counted and documented.

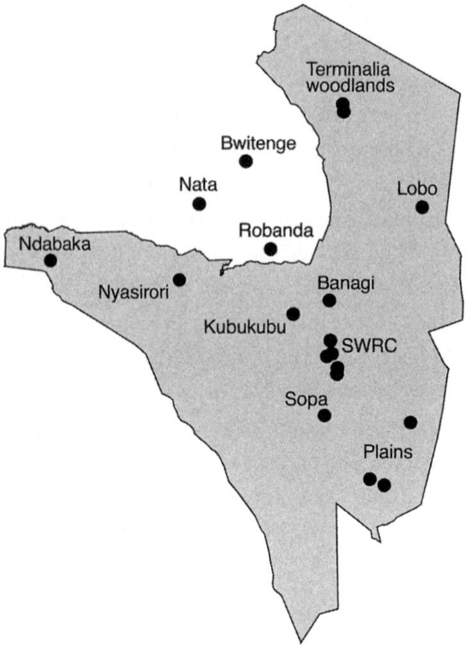

Fig. 10.1 Serengeti National Park, showing the 24 locations of the invertebrate field sampling in and around the park visited over a period of eight years (1999–2006).

The aim of the following section was to analyze the extensively collected data on the effect of different types of disturbances on the arthropod community in Serengeti. The study done by Nkwabi et al. (2011) in the southern Serengeti Plains region clearly shows that the abundance of certain insects is affected by fire and grazing pressure. Sinclair, Mduma, and Arcese (2002) showed the effect of agricultural land conversion on some insect groups. For purposes of this chapter, we investigated whether across the entire Serengeti these three disturbances affected total arthropod diversity (distinguishing between the grass-layer and ground-layer) as well as the abundance of the three grass-layer insect groups, Orthoptera, Homoptera, and Heteroptera. We concentrate on the overall effects of these disturbances on the arthropod community and therefore perform the analyses of the full database across locations as well as years.

Disturbance Effects: Hypotheses

A perturbation can be defined as a sudden and relatively short change in the prevalent environmental conditions (Silva 1996). Some common perturbations in savannas are changes in the prevalent climate or fire regime and in the patterns of grazing. Human land-use change is a maintained perturbation with a longer lasting and permanent change. Disturbances can have two kinds of effects: first, those that modify community structure directly (by differential removal of species as in grazing, or total replacement as in agriculture); second, those that have indirect effects, modifying some of the prevailing physical conditions (such as fire regime and plant available moisture and nutrients) (Silva 1996). These indirect effects may in turn change community structure.

Fire, grazing pressure, and land-use change (agriculture) are the three main disturbances present in and around Serengeti. We briefly elaborate on each of these perturbations. Results on how these perturbations can influence arthropod diversity and abundance are summarized from literature that is not exclusively restricted to savannas in the following paragraphs.

Fire. Arthropods are very much dependent on their habitat, especially the vegetation structure. In the short term fire can have devastating effects on the vegetation, causing direct mortality to plants. In the long term fire regenerates aboveground shoots, and through ash deposition, fertilizes the soil. The intensity and duration of the fire are therefore predicted to play an important role in affecting arthropod communities. Literature studies show a variety of often contrasting results. Arthropod abundance and richness have been reported to decrease with frequency and intensity of fires (Anderson, Leahy, and Dhillion 1989; York 1999; Sackmann and Farji-Brener 2006). However, other work suggests highest diversities on sites subjected to an intermediate frequency of fire (Evans 1984; Moretti et al. 2002; Moretti, Obrist, and Duelli 2004) or were reported even to increase with burning (Orgeas and Andersen 2001). For several invertebrate groups no effect at all has been found (Dunwiddie 1991; Andersen and Müller 2001). Some studies suggest that the latter may be a mere short-term result and that in the longer term a change in activity would be found (Collett 2003). A recent study by Nkwabi et al. (2011) on avifaunal distributions showed for the plains region of the Serengeti that disturbances like fire reduced the abundance of several grass-layer as well as ground-layer arthropod groups, although the differences in abundance depended on the insect group and the time sampled after the fire (one, four, and twenty weeks).

Grazing. In general, grazing reduces the vertical structure of the vegetation. On a small spatial scale, grazing-induced changes in the plant community can affect diversity and composition of the associated arthropod communities, though its effects can be diverse for different arthropod groups (Gibson et al. 1992). Following grazing disturbance, arthropod abundance has been reported to show no difference at all (Mysterud et al.

2005), to decrease (Rambo and Faeth 1999; Nkwabi et al. 2011), or even to increase (Suominen et al. 2003). This response seemed to be dependent on the feeding type of the invertebrate groups and their sensitivity to microclimate conditions (Suominen et al. 2003). Grazed sites experience warmer and dryer conditions because of the more open, sun-exposed vegetation layers. Open habitats are, therefore, known to be favored by some ground-dwelling arthropods (Lindroth 1985, 1986). Studies of grazing effects on insect diversity show a similar variety of results. Arthropod diversity either did not change (Rambo and Faeth 1999; Mysterud et al. 2005), decreased (Kruess and Tscharntke 2002), or for specific groups such as Carabid and Curculionid beetles, increased (Suominen et al. 2003) in grazed as opposed to ungrazed sites. However, relating the diversity of macroarthropods to a continuous gradient of grazing intensity resulted in a bell-shaped relationship (Milchunas, Lauenroth, and Burke 1998; Suominen et al. 2003).

Agriculture. Monocultures and constant human disturbance via agricultural fields may have similar multidirectional effects. Reduction of vegetational heterogeneity may cause decrease in the diversity of insects. The type of crop is said to be of importance to insect community composition (Booij and Noorlander 1992) and monocultures may, therefore, have low insect diversity potential. It may also be assumed that the use of chemical fertilizers and insecticides will affect insects in a detrimental way. However, factors such as reduction of predators by human influence may counteract the negative effects mentioned above (Wilson et al. 1999). The avifaunal study conducted by Sinclair, Mduma, and Arcese (2002) inside Serengeti National Park and in the agricultural areas outside of the park showed that the abundance of avian food sources (leaf- and planthoppers, flies and wasps), declined in agricultural areas. No predictions exist for the effects on overall arthropod diversity, though it may be hypothesized to follow the predictions for grazing disturbances.

Arthropod Groups and Collection Methods

For this analysis we report the effect of the previous described disturbances on total grass-layer and ground-layer arthropod diversity (including spiders, ticks, butterflies, moths, flies, damselflies, ants, wasps, bees, beetles, weevils, grasshoppers, crickets, katydids, stick insects, ant lions, thrips, bugs, mantises, cockroaches). Grass-layer insects are caught by sweep nets, ground-layer by using pitfalls. We selected three insect groups to investigate the effect of disturbances on the abundance: Orthoptera, Homoptera, and Heteroptera. These three orders are the dominant insect groups and generally occur in relatively large numbers compared to other arthropod groups. By choosing orders that are caught best and numerously with the same method (in this case by sweep net), the potential of inter-group comparisons is ensured. For example, the absolute number of arthropods from a small pitfall that is revisited a second day cannot be compared to a 30-paces-long sweeping of the vegetation. The role of the chosen three insect groups in the system is that of abundant prey, but also their own feeding activities may have effects on the system. The order of Orthoptera includes crickets and katydids, but mainly grasshoppers. Grasshoppers are pure herbivores and in great numbers can reduce the vegetation biomass extensively (Hewitt 1977). Homoptera and Heteroptera are suborders and belong to the order of true bugs, Hemiptera. Homoptera comprises small-to-large, terrestrial, plant-feeding bugs. Some examples are snout bugs and leafhoppers. Heteroptera are the carnivorous and herbivorous bugs, which in our case are all terrestrial. Examples of this order are assassin bugs, lady-bird bugs, and stinkbugs (Picker, Griffiths, and Weaving 2004). The latter groups are important in pest control (Chang & Snyder 2004).

Diversity analysis. We chose to investigate the grass-layer arthropods sampled by the sweep net and the ground-layer arthropods sampled by pitfalls. These methods were the most consistently performed across locations and years. The total number of samples across eight years is 935 for the sweep net method and 658 for the pitfall traps (these include all treatments). Diversity indices were calculated for all arthropods per sample (i.e., 30 sweeps with the net or one pitfall trap), and assuming morphospecies to be equivalent to species. The number of individuals per morphospecies per sample enabled us to compute the proportion of species *i* (p_i) within the total number of species (S). The Shannon-Wiener index of diversity (H′) and Pielou's index of species evenness (J) incorporate the abundance data and show potential dominance of particular species. These diversity indices were then compared between disturbed and undisturbed areas.

$$H' = -\sum_{i=1}^{S} p_i \ln p_i \tag{1}$$

$$J = \frac{H'}{\ln(S)} \tag{2}$$

Here, a higher value H′ (index of diversity) indicates higher species diversity. Pielou's index of species evenness indicates the degree of structuring of a community. The species evenness index is constrained between zero and one. A value close to one indicates low variation of species abundances within communities; that is, all species occur in relatively similar proportions. On the other hand, a value near zero means that a single species may be dominant with the others very rare.

Similarity of family abundance between two treatments *j* and *k* was measured using the Bray-Curtis value (1-*B*) so that 1 = similar, 0 = dissimilar (Krebs 1989).

Abundance. Abundance data in disturbed and undisturbed areas was computed per sample for the three main insect groups at a taxonomic level of order: Orthoptera, Homoptera, and Heteroptera. These groups are mainly living in the grass layer and all best caught by sweep net ($N = 935$). Therefore, as methodologies are similar, this allows comparisons between and within these groups.

Statistical analyses. Statistical effects were tested using general linear models (single-factor ANOVA) and two-way *T*-tests, STATISTICA version 6.1 (2003).

Results: Disturbance Effects on Arthropod Species Diversity

Fire. We found across years and across the entire Serengeti that fire had no effect on the grass-layer invertebrate diversity (Shannon-Wiener index H′, $P = 0.36$) nor on Pielou's index of evenness J ($P = 0.11$), although on average, burnt areas showed a lower diversity index. For example, burnt: H′ = 1.66 ± 1.65 (SD), J = 0.95 ± 0.41 (SD), $N = 51$; unburnt: H′ = 1.87 ± 0.54 (SD), J = 0.86 ± 0.11 (SD), $N = 60$ (fig. 10.2A). The diversity of ground-dwelling invertebrates was equally unaffected by fire (H′ ($P = 0.85$); J ($P = 0.22$)): burnt: H′ = 1.61 ± 0.62 (SD), J = 0.82 ± 0.16 (SD), $N = 100$; unburnt: H′ = 1.63 ± 0.18 (SD), J = 0.84 ± 0.13 (SD), $N = 99$ (fig. 10.2B).

The Bray-Curtis similarity values were high for all three types of trap (table 10.2), but total abundance of arthropods on burn sites was twice that of unburnt sites in pitfalls, but the reverse in sweep nets. In trays abundance was even across treatment. Burn had little effect on ground-living arthropods, with ground spiders, dung beetles, and other ground beetles showing no effect in either pitfall or tray. Ants were marginally more abundant in burn sites. In the herb layer sampled by sweeps, orthoptera and homoptera were more abundant in unburned sites as might be expected by the greater vegetation biomass.

Grazing pressure. Grass-layer invertebrate species diversity was significantly higher in ungrazed sites (H′ = 1.78 ± 0.62 (SD), $N = 158$) than in grazed sites (H′ = 1.56 ± 0.13 (SD), $N = 130$) ($P = 0.004$), with no difference in the index of evenness, J ($P = 0.45$) (fig. 10.3A). No significant differences in species diversity ($P = 0.43$) and evenness ($P = 0.15$) were found for the

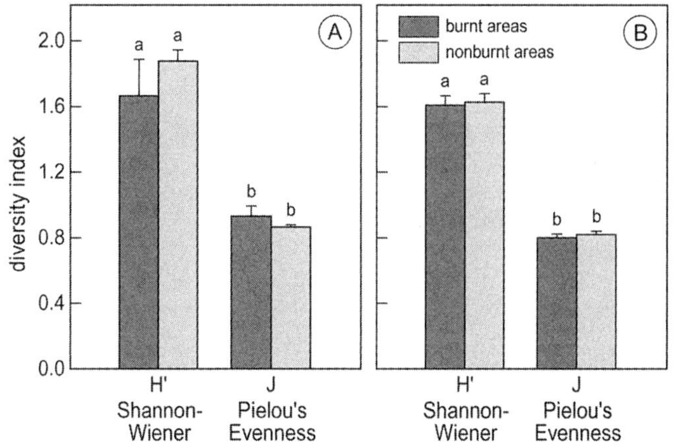

Fig. 10.2 The effect of fire on the invertebrate species diversity (Shannon-Wiener index, H') and species evenness (Pielou's index of evenness, J). Dark gray bars, burnt areas; light gray bars, nonburnt areas. Bars indicate means, whiskers indicate ± SE. Identical letters indicate nonsignificant ($P > 0.05$) differences, (A) grass-layer invertebrates, and (B) ground-dwelling layer invertebrates.

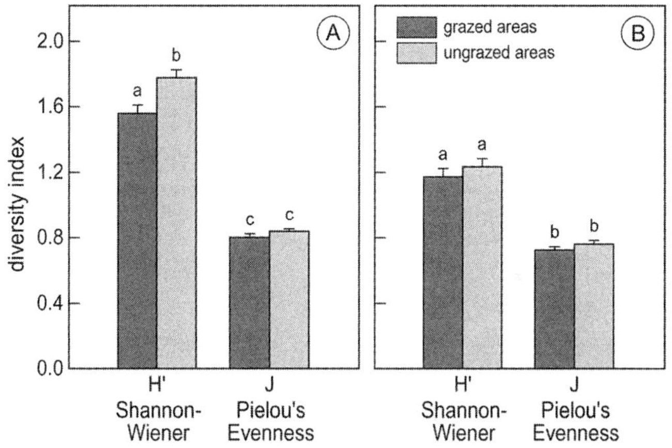

Fig. 10.3 The effect of grazing on the invertebrate species diversity (Shannon-Wiener index, H') and species evenness (Pielou's index of evenness, J). Dark gray bars, grazed areas; light gray bars, ungrazed areas. Bars indicate means, whiskers indicate ± SE. Identical letters indicate nonsignificant ($P > 0.05$) differences, (A) grass-layer invertebrates, and (B) ground-dwelling layer invertebrates.

ground-dwelling invertebrates between grazed ($N = 180$) and ungrazed sites ($N = 179$) (fig. 10.3B).

The Bray-Curtis similarity values for pitfall, tray, and sweep nets (table 10.2) showed high values of similarity because all the main families appeared in both grazed and ungrazed areas. Despite this, there were large

Table 10.2 Bray-Curtis values of similarity (1-*B*) for families of arthropods

Method	Grazed-ungrazed	*N*	Burnt-unburnt	*N*
Pitfall	0.84	(795, 1,082)	0.71	(1,038, 577)
Tray	0.79	(1,730, 1,185)	0.85	(1,064, 1,006)
Sweep	0.73	(696, 1,147)	0.65	(700, 1,351)

Note: Sample sizes of each treatment in parentheses.

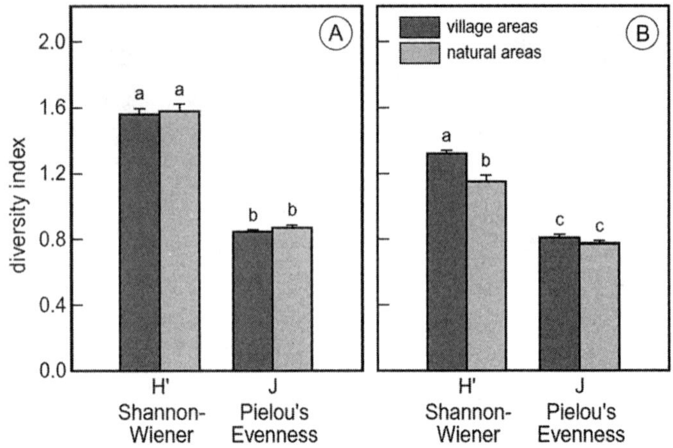

Fig. 10.4 The effect of agricultural land use on the invertebrate species diversity (Shannon-Wiener index, H') and species evenness (Pielou's index of evenness, J). Dark gray bars, village areas; light gray bars, natural areas. Bars indicate means, whiskers indicate ± SE. Identical letters indicate nonsignificant ($P > 0.05$) differences, (A) grass-layer invertebrates, and (B) ground-dwelling layer invertebrates.

differences in abundance in different families. Thus spiders, ground beetles, and ants (found in pitfall and trays) were more abundant in ungrazed sites, though dung beetles showed no difference with grazing. Flies which are attracted to trays were almost twice as abundant in grazed areas, whereas orthoptera (sampled by sweep nets) were twice as abundant in ungrazed areas.

Agriculture. Agricultural sites ($N = 335$) did not differ in grass-layer invertebrate diversity compared to natural savanna sites ($N = 229$), as indicated by the Shannon-Wiener index ($P = 0.75$). Also, the index of species evenness, J, did not differ ($P = 0.88$) (fig. 10.4A). However, the diversity of ground-dwelling invertebrates was significantly higher in agricultural sites (1.32 ± 0.62 (SD), $N = 435$) than in natural savanna sites (1.15 ± 0.65 (SD), $N = 224$) ($P = 0.002$). Again, the index of evenness showed no difference ($P = 0.16$) (fig. 10.4B). No Bray-Curtis similarity values were computed.

Disturbance Effects on the Abundance of Orthoptera, Homoptera, and Heteroptera

Fire. Including all investigated sites across Serengeti and all eight years of data, there was no effect of fire on the number of orthopterans ($P = 0.25$), the number of homopterans ($P = 0.58$), or the number of heteropterans ($P = 0.69$) (fig. 10.5).

Grazing pressure. Also, on average, grazing had no overall effect on the abundances of orthopterans ($P = 0.84$), homopterans ($P = 0.34$), or heteropterans ($P = 0.57$) across Serengeti over eight years (fig. 10.6). However, all grazed sites had a slightly higher number of insects than the ungrazed sites.

Agriculture. Orthoptera and Heteroptera showed statistically significant differences between areas inside and outside the park, though exactly opposite. Orthopterans occurred in higher numbers in natural savanna areas in the center of Serengeti (8.67 ± 8.69 (SD)) compared to agricultural areas (4.59 ± 5.11 (SD)) ($P < 0.0001$). Heteropterans, though, were more abundant in agricultural areas (4.56 ± 8.12 (SD)) than in areas in the center of Serengeti (2.14 ± 1.92 (SD)) ($P = 0.004$). Homopterans tended to follow the pattern of the group of heteropterans, though not statistically significant ($P = 0.12$) (fig. 10.7).

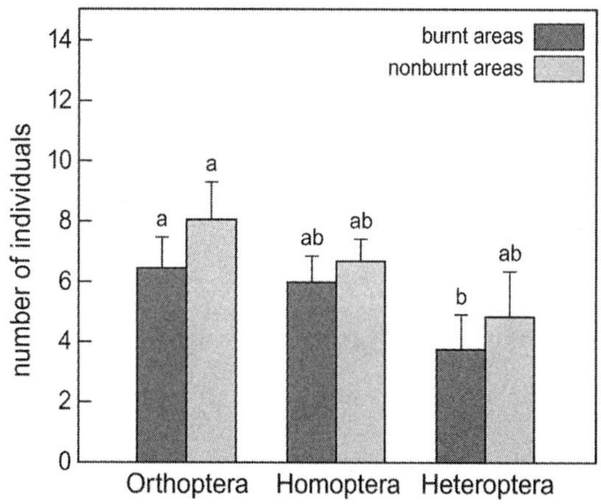

Fig. 10.5 The effect of fire on the abundance (average number of individuals per sample) of three main insect groups within Serengeti National Park: Orthoptera, Homoptera, and Heteroptera. Dark gray bars, burnt areas; light gray bars, nonburnt areas. Bars indicate means, whiskers indicate ± SE. Identical letters indicate nonsignificant ($P > 0.05$) differences.

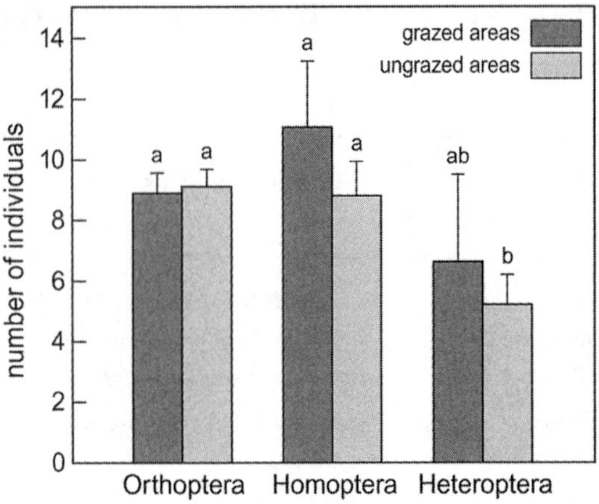

Fig. 10.6 The effect of grazing on the abundance (average number of individuals per sample) of three main insect groups within Serengeti National Park: Orthoptera, Homoptera, and Heteroptera. Dark gray bars, grazed areas; light gray bars, ungrazed areas. Bars indicate means, whiskers indicate ± SE. Identical letters indicate nonsignificant ($P > 0.05$) differences.

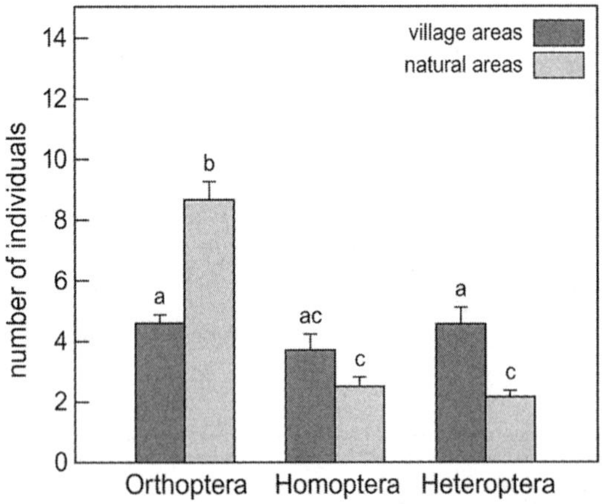

Fig. 10.7 The effect of agricultural land use on the abundance (average number of individuals per sample) of three main insect groups within Serengeti National Park: Orthoptera, Homoptera, and Heteroptera. Dark gray bars, village areas; light gray bars, natural areas. Bars indicate means, whiskers indicate ± SE. Identical letters indicate nonsignificant ($P > 0.05$) differences.

Synopsis

We analyzed the effects of three different types of disturbances on the grass-layer invertebrate diversity, the ground-dwelling invertebrate diversity, and the abundance of the three most common insect taxonomic orders. The most severe disturbances in the Serengeti region, also known to affect the larger vertebrates (McNaughton 1979; Anderson et al. 2007) are fire, grazing, and agricultural land use. Of these three disturbances Serengeti's invertebrates respond the strongest to agricultural land use, though contradictory results were found for different strata as well as for different taxonomic groups.

In our analysis across sites all over the Serengeti over all eight years (dry and wet seasons) the perturbation caused by fire resulted in no significant differences in the diversity of the grass-layer and the ground-dwelling invertebrates, or in the abundances of orthopterans, homopterans, and heteropterans. However, burning seems to negatively affect insects caught by sweep net, but positively affects insects caught in pitfall traps, and overall diversity is slightly (not significantly) lower in burn sites. In the southern plains Nkwabi et al. (2011) found for the dry seasons of the years 2001–04 that fire caused a decline in the abundance of the insect fauna, but with variable intensities dependent on the time since the fire occurred. Also, the literature provides studies with various results of fire effects on insect species diversity (also shown by the Bray-Curtis similarity analyses) and abundance, which when averaged may result in no clear effects. Therefore, whether the direction of the effect on invertebrates is positive or negative depends on the timing of the sampling after the fire (Nkwabi et al. 2011) and the intensity of the fire, and may depend on the species' feeding type and sensitivity to the microclimate conditions. In addition, carnivorous invertebrates may be attracted to burn sites by the high number of burnt, dead prey. For example, some mammalian vertebrates such as baboons (*Papio cynocephalus*) and banded mongooses (*Mungos mungo*) forage easily on insects killed in the fire while they pass through a burned site (S. de Visser, pers. obs.). Sampling arthropods soon after a fire or after a longer recovery time may also result in different abundances and diversities dependent on the ability of regrowth of the vegetation. Herbivorous invertebrates may be attracted to the young shoots after a fire. However, the intensity as well as the timing after a fire was not recorded in our surveys.

Nkwabi et al. (2011) showed that grazing in the dry season caused a decline in the abundance of the insect fauna. Insect biodiversity changes due to grazing were not found (see also the Bray-Curtis analysis). Our analysis shows lower arthropod diversity in grazed sites, though only the species diversity of the grass living arthropods showed a statistically significant difference. The latter can be easily explained by the reduction of habitat structure; grazed sites contain fewer strata for different types of arthropods to live in. However, we did not find a negative effect of grazing on the individual abundances of the three insect groups mainly living in the grass-layer stratum when we analyzed the data over all eight years (all dry and wet seasons), indicating once more specific seasonal and perhaps year effects too. Generally, abundances from sweep net and pitfall sampling were higher in ungrazed sites. But, homopterans and heteropterans showed a slight but not significant preference for grazed sites, as did flies caught by tray. Heteroptera and Homoptera may perhaps favor the dryer and warmer conditions in the open habitats.

The disturbance that is most affecting the invertebrate community is agricultural land conversion. Once more, this type of disturbance resulted in differential effects for the three insect groups: it decreased abundance of orthopterans, increased abundance of heteropterans, and showed no effect on the abundance of homopterans. The grass-layer arthropod diversity showed a slight, though nonsignificant decrease over all years in agricultural areas. However, the ground-layer arthropod diversity significantly increased statistically in these human-dominated areas when compared to savanna areas in the center of Serengeti. This increase in the ground-layer arthropod diversity may be predominantly caused by an increase in beetle species and the occurrence of relatively more flies caught by the pitfalls (i.e., high fly abundance as in grazed sites), as indicated by a preliminary analysis of our data. The ground-layer arthropods could also be less vulnerable to human activities or to harvest practices as they live nearer the soil. Another explanation may be that they actually benefit from the predicted lower number of avian predators outside the park. Sinclair (1978) and Sinclair, Mduma, and Arcese (2002) showed that in the wet season insectivorous bird densities were lower in agricultural fields. How and why this predation aspect alters the abundances differently for orthopterans (negatively) versus heteropterans (positively) remains to be investigated.

Although the Shannon-Wiener diversity index sometimes differed between treatments, the index of species evenness remained alike in all cases. A Bray-Curtis analysis was not performed. For example, the number of species of grass-layer invertebrates decreased in grazed sites, and the ground-living arthropod diversity increased in sites outside the park, but in both cases the evenness (proportion) of the species did not change. This may indicate that under these disturbances some species are affected (addition or removal of a species), but the disturbances did not favor a particular species.

To outline future directions based on our findings, we suggest studies should include different time scales and intensities of the disturbances and measure their effects on arthropods. Nkwabi et al. (2011) mention a dissipation time after a disturbance of around four to five months in the dry season. We do not know if in the long term burnt areas become more attractive to a particular group of arthropods, or if very intense fires are deleterious to some species and not to others. The same investigation may be proposed for the effect of grazing. It is interesting that although the grass-layer diversity is higher in ungrazed sites, we found no difference in abundances of the three main groups living in this stratum. This may be caused by our analysis that grouped all locations and all sampled years, thereby obscuring potential site-specific differences, or it may show a real quantitative limitation of food and competition among invertebrates. Last, the effect of sampling outside the park versus inside the park showed the strongest results in abundances and diversity. However, why do we find differences in diversity of ground-layer arthropods, but not in grass-layer arthropods? Why do we find higher diversity of ground-layer arthropods in agricultural sites? Finally, why are there more heteropterans in agricultural sites? These questions require detailed and taxa-specific experiments in the field. In addition, these studies suggest seasonal differences in insect abundances and potentially their responses to disturbances. In the following section, we will show how insect abundance changes spatially as well as temporally.

SEASONAL EFFECTS ON SPATIAL PATTERNS OF INVERTEBRATE ACTIVITY

Being located close to the equator, Serengeti experiences no strong temperature difference, but more apparent differences in rainfall throughout its seasons occurs. This rainfall pattern is one of the main factors influencing the annual wildebeest migration through its effect on the vegetation. Rainfall and vegetation structure (through food availability) are also main factors influencing insect abundance (Wallner 1987). With the large amount of invertebrate data collected throughout the park over eight years by the Serengeti Biodiversity Program, we are able to show, for the first time, the spatial distribution and seasonal patterns of activity of some of the (mobile) invertebrates present in Serengeti. To complement this, we show data collected on termite activity (sessile, colonial insects) in the dry and wet season. These patterns should be seen as preliminary results. They should stimulate research on these particular study objects by pointing out interesting insect groups and locations in Serengeti to investigate in more detail. We generate several potential hypotheses for the interseasonal patterns found that should advance future research on this small-bodied though important part of the animal fauna in Serengeti.

Methods

Abundance data was extracted from the extensive Serengeti Biodiversity Program database (for details, see previous section) for each location taken over all collected years and grouped for dry and wet season (as defined by Sinclair 1975, 1978). Dry season incorporates the months July–October. Only untreated sites were chosen for this analysis, thereby excluding disturbance effects of fire and (heavy) grazing. The sites outside of the park were considered untreated. We selected three dominant orders, namely Orthoptera, Homoptera, and Heteroptera, caught by the same sweep net technique (details in previous section). This allows comparisons of numbers among the three taxonomic groups as well as comparisons within each group for both seasons. Abundance of insects for a certain location was calculated as the total number of individuals per order per season over all years and divided by the sampling effort per location per season (number of samples taken over all years). This last step is necessary as the sampling effort for dry and wet season and between sites is not equal. Each sweep net sampling effort roughly corresponds to 30 m^2. Thereafter, the average of all sampling locations was taken per region, with Serengeti being divided into a northern, western, central, and south region. Locations with less than two sampling efforts were excluded from analysis in order to avoid singular extreme values.

To estimate termite activity by counting individuals is both labor-intensive and unreliable. Moreover, termites are hard to catch by the usual pitfall traps. Therefore, Freymann, de Visser, and Olff (2010) estimated their activity in the Serengeti plains by measuring feeding through decomposition rates. More specifically, they calculated the percent mass loss of two bait types, litter, and wildebeest dung (ca. 20 g per bait, 9 baits per location and 16 locations), that were put out for 20 days in the long-grass plains. This experiment was conducted in the dry season of 2005 and the wet season of 2006 in the southern long-grass plains (Freymann, de Visser, and Olff 2010). These values (means and standard error percent mass loss) give an indication of the activity of the termites for each season.

Seasonal Abundance of Insect Groups

We examined the relative abundances of orthoptera, homoptera, and heteroptera across seasons (table 10.3). In addition, we present the results of a decomposition rate experiment conducted in the southern plain region of Serengeti. The most striking patterns for each taxonomic order are described below, as well as a discussion of spatiotemporal scales and potential interactions among the invertebrate groups.

Orthoptera. By comparing the average abundance across all locations, among the three insect groups investigated, orthopterans are clearly the

Table 10.3 Seasonal abundance of the three main insect taxa (Orthoptera, Homoptera, and Heteroptera) across the Serengeti and outside the park (number of locations given in parentheses) given by the average number of individuals per 30 m^2 caught by sweep net

Serengeti region (locations)	**Taxa**	**Wet season abundance (± SE)**	**Dry season abundance (± SE)**
Northern region (*n* = 7)	Orthoptera	5.5 ± 1.8	5.0 ± 1.4
	Homoptera	1.5 ± 0.3	5.9 ± 1.8
	Heteroptera	2.2 ± 0.1	1.5 ± 0.2
Western region (*n* = 2)	Orthoptera	14.3	5.2 ± 1.2
	Homoptera	2.5	3.8 ± 2.8
	Heteroptera	0	3
Central region (*n* = 5)	Orthoptera	13.9 ± 2.7	10.3 ± 2.3
	Homoptera	3.5 ± 1.1	2.6 ± 0.3
	Heteroptera	2.9 ± 0.6	2.3 ± 0.3
Southern region (*n* = 7)	Orthoptera	9.5 ± 3.4	3.6 ± 0.6
	Homoptera	4.2 ± 1.2	4.6 ± 1.4
	Heteroptera	4.9 ± 2.4	3.0 ± 0.3
Outside SNP (*n* = 3)	Orthoptera	3.7 ± 0.7	5.8 ± 1.1
	Homoptera	2.4 ± 0.2	4.4 ± 1.2
	Heteroptera	3.8 ± 0.7	4.3 ± 1.5

most abundant in Serengeti. On average seven Orthoptera individuals are caught per sample, as compared to three homopterans and three heteropterans per sample. Orthopterans occured everywhere where sampling took place. An interseasonal comparison, though, shows on average a doubling in Orthoptera abundance in the wet season (nine individuals/sample) as compared to the dry season (four individuals/sample) ($P < 0.0001$, $T = 2.05$, df = 27), with a contrasting decrease in wet season abundance outside the park and no change in abundance in the northern (wetter) regions.

Homoptera. The abundances of homopterans range on average from four individuals per sample in the dry season to three individuals per sample in the wet season, a nonsignificant difference ($P = 0.49$, $T = 2.09$, df = 19). The interseasonal comparison of sampling locations outside the park shows an increase in abundance in the wet season for Bwitenge village, but slight decreases for the villages of Nata and Robanda. Most regions inside the park show a decrease of numbers in the wet season.

Heteroptera. Heteropteran abundances range on average from four individuals per sample in the dry season to three individuals per sample in the wet season, a nonsignificant difference ($P = 0.85$, $T = 2.09$, df = 19). An interseasonal comparison across all regions shows for the wet season a slight decrease in abundance in areas outside the park and in the west, in contrast to small increases in the north, central, and southern region.

Termitoidae. Direct measures of termite abundance have not yet been possible. We show here a proxy of their activity via the decomposition rate of two bait types, litter and wildebeest dung, indicated by the mean daily mass loss (%). Although a slightly higher activity may be seen in the dry season versus the wet season for both bait types (0.16% (± 0.07 SE) and 0.12% (± 0.04 SE) for litter, 0.17% (± 0.08 SE) and 0.15% (± 0.09 SE) for dung respectively for the dry and wet season), no statistically significant difference is apparent between the seasons (Freymann, de Visser, and Olff 2010).

Synopsis

Overall, Orthoptera, Homoptera, and Heteroptera abundances show no apparent spatial unidirectional gradient within Serengeti, although the averaged abundances in the northern and outside regions of the park are a little lower as compared to the southern and central parts of the park. We find indications of seasonal differences interacting with spatial locations. Orthoptera abundance increased in the wet season as compared to the dry season, with a contrasting decrease outside of the park. Homoptera abundance seemed in general to decrease in the wet season inside and outside the park, with an exception for the village of Bwitenge. Heteroptera abundance decreased slightly in the wet season, though not uniformly in all locations, and mostly so at the locations near the villages. A speculative explanation for the higher abundances outside the park in the dry season may be the fact that the vegetation outside the park, including gardens and decorative trees, remains suitable in the dry season (perhaps through irrigation). Similarly, leftover items of harvested crop fields after the wet season may be beneficial to certain insect groups. Unfortunately, most of the sample locations in the southern plains region consisted only of a single trapping exercise and were therefore excluded from the maps. However, the numbers in this region may still be of interest, although the following numbers should be taken with caution as they are based on only one trapping exercise. In the wet season the abundance of Orthoptera in the plains was found to be higher than in the dry season. In contrast, both Homoptera and Heteroptera decreased in abundance in these regions.

Sinclair (1978) described how Lepidoptera, Coleoptera, Orthoptera, and swarming Isoptera are sparse in the dry season but become locally abundant after the first rainstorms and remain abundant in the ensuing rainy season.

Although not all sites were visited in both seasons, a similar result was found for Orthoptera: on average Orthoptera abundance increased in the wet season. An increase of productivity (plant-based as well as animal-based food sources) and suitable microhabitat (temperature and moisture) in the wet season may explain higher abundances. The more stressful time of year, the less productive dry season, is presumably passed by most insects as a stage of quiescence (e.g., as larval, pupal, or adult diapause).

A point of interest in relation to vertebrate studies may be the order of Orthoptera. This order includes some carnivores (most crickets and katydids), but mostly herbivores (grasshoppers and a few crickets and katydids). We may speculate on a potential competition between mammalian herbivores and grasshoppers during the wet season, especially in the nutrient-rich southern plains. The data shown from the 1970s by Sinclair (1975) indicate though that grass consumption in the short-grass plains is considerable higher for the ungulates and might leave the effect of the higher abundance of grasshoppers negligible. However, although we can provide no data to support this assumption, the argument of competition may be strengthened by a decrease in orthopterans in the wet season outside the park, around the villages. High grazing pressure by cattle, goats, and sheep may decrease the food availability there with resulting deceased abundance of orthopterans. However, our experimental analysis of direct grazing pressure (section 2), where we compared abundances of Orthoptera in grazed and ungrazed areas showed only a nonsignificant lower abundance for the grazed sites.

In contrast to the varying seasonal activities of Orthoptera, Homoptera, and Heteroptera, the overall decomposition activity by termites (Isoptera) remained relatively constant between the wet and dry season, at least for the long-grass plain region in the south of Serengeti (Freymann, de Visser, and Olff 2010). This result is comparable to a study performed in Botswana (Schuurman 2006). Here, larger termites such as *Macrotermes* and *Odontotermes* were seen to forage constantly throughout the year, whereas the foraging activity of smaller termites (*Microtermes*, *Allodontermes*) responded more to seasonal conditions. On the Serengeti plains the most dominant termite taxa are the fungus-growing genera *Odontotermes*, a group known to be mostly feeding on woody vegetation, but on the treeless plains preferring litter, and *Macrotermes subhyalinus*, a mixed-feeding species with a preference for grass litter (de Visser, Freymann, and Schnyder 2008). Though direct competition remains speculative, there seems to be a positive interaction between termites and mammalian herbivores, at least in the case of termites utilizing and recycling the dung of larger mammalian herbivores (Freymann et al. 2008).

Further research should focus clearly on the combination of the spatial patterns with the interseasonal variation in insect activities, and maybe more so, the potential relation to their mammalian herbivorous counterparts. Study areas we indicate here to be of interest and in need of more sampling effort are the southern plains (higher abundance of Orthoptera, but lower abundances for Homoptera and Heteroptera in the wet season versus the dry season). Also of particular interest are areas outside the park (especially the decrease in Orthoptera in the wet season) and perhaps, as a control area, the center of the park (high but more or less stable abundances). The presented patterns from the invertebrate surveys may give rise to several questions with respect to invertebrate-vertebrate interactions. First, when herbivorous insects are present in high abundance, could they compete with the large mammalian herbivores? Investigating in more detail spatiotemporal patterns between these different herbivore groups across Serengeti may prove interesting. Secondly, is the presence of arthropods influenced by resources (bottom-up) or by predation risk (top-down)? In order to investigate these questions, more detailed information is needed on the spatial patterns and activities of specific insect groups and their vertebrate and invertebrate predators.

CONCLUSION

Our understanding of Serengeti's invertebrates greatly lags behind our knowledge of Serengeti's vertebrates. However, here we show some examples of interesting ecological interactions between these two groups. For this alone, arthropods deserve a lot more attention. In this chapter we show that arthropods may play important roles in offering resources to vertebrates in the process of decomposition, and potentially other processes important to savannas. We show that arthropod diversity and abundance are sensitive to certain perturbations that are present in and around Serengeti. Of interest here is the apparent lack of competition among arthropods (correcting for the different feeding guilds), as indicated by the discrepancy of change in diversity without change in species evenness, as in grazing treatments. We propose several hypotheses of the documented patterns, some in relation to the larger mammalian herbivores, that could be investigated in more detail. We hope these preliminary results will stimulate further research of invertebrates in Serengeti.

ACKNOWLEDGMENTS

We thank the Tanzanian Wildlife Research Institute (TAWIRI), Tanzania National Parks (TANAPA) and the Tanzanian Commission of Science and Technology (COSTECH) for their permission to work in Serengeti. We are indebted to J. Bukombe for his work in the field, and P. Eggleton for an introduction to termite identification. Margery Hingley assisted with grasshopper collections in 1972. Grasshopper specimens from 1972 were identified by N. Jago of the Antilocust Research Centre, London, while those from 1997 to 2008 were identified by M. Mungai of the National Museum, Nairobi Kenya. We are grateful to D. Visser for the figures and to B. Jaeger for sending us some of the old references. SNdV and BPF were funded by the Netherlands Organization for Scientific Research (NWO) and BPF by the Robert Bosch Foundation (Germany). ARES was funded by the Natural Sciences and Engineering Research Council of Canada (NSERC) program. Andrew Harvey was working for the Overseas Development Administration (British Aid) on the armyworm project in Tanzania when collecting specimens in 1986–89.

REFERENCES

Adamson, A. M. 1943. Termites and the fertility of soils. *Tropical Agriculture* 20:107–12.

Andersen, A. N., and W. J. Müller. 2001. Arthropod responses to experimental fire regimes in an Australian tropical savannah: Ordinal-level analysis. *Austral Ecology* 25:199–209.

Anderson, R. C., T. Leahy, and S. S. Dhillion. 1989. Numbers and biomass of selected insect groups on burned and unburned sand prairie. *American Midland Naturalist* 122:151–62.

Anderson, T. M., M. E. Ritchie, E. Mayemba, S. Eby, J. B. Grace, and S. J. McNaughton. 2007. Forage nutritive quality in the Serengeti ecosystem: The roles of fire and herbivory. *The American Naturalist* 170:343–57.

Baruch, Z., A. J. Belsky, L. Bulla, C. A. Franco, I. Garay, M. Haridasan, P. Lavelle, E. Medina, and G. Sarmiento. 1996. Biodiversity as regulator of energy flow, water use and nutrient cycling in savannas. In *Biodiversity and savanna ecosystem processes: A global perspective*, ed. O. T. Solbrig, E. Medina, and J. F. Silva, 175–96. Berlin: Springer-Verlag.

Bignell, D. E., and P. Eggleton. 2000. Termites in ecosystems. In *Termites: Evolution, sociality, symbioses, ecology*, ed. T. Abe, D. E. Bignell, and M. Higashi, 363–87. Dordrecht: Kluwer Academic Publications.

Booij, C. J. H., and J. Noorlander. 1992. Farming systems and insect predators. *Agriculture, Ecosystems & Environment* 40:125–35.

Bornemissza, G. F. 1979. The Australian dung beetle research unit in Pretoria. *South African Journal of Science* 75:257–60.

Chang, G. C., and W. E. Snyder. 2004. The relationship between predator density, community composition, and field predation of Colorado potato beetle eggs. *Biological Control* 31:453–61.

Collett, N. 2003. Short and long-term effects of prescribed fires in autumn and spring on surface-active arthropods in dry sclerophyll eucalypt forests of Victoria. *Forest Ecology and Management* 182:117–38.

Coventry, R. J., J. A. Holt, and D. F. Sinclair. 1988. Nutrient cycling by mound-building termites in low-fertility soils of semi-arid tropical Australia. *Australian Journal of Soil Research* 26:375–90.

Dangerfield, J. M., T. S. McCarthy, and W. N. Ellery. 1998. The mound-building termite *Macrotermes michaelseni* as an ecosystem engineer. *Journal of Tropical Ecology* 14:507–20.

Davis, A. L. V., and C. F. Dewhurst. 1993. Climatic and biogeographical associations of Kenyan and northern Tanzanian dung beetles (Coleoptera: Scarabaeidae). *African Journal of Ecology* 31:290–306.

de Visser, S. N., B. P. Freymann, and H. Olff. 2011. The Serengeti food web: Empirical quantification and analysis of topological changes under increasing human impact. *Journal of Animal Ecology* 80:484–94.

de Visser, S. N., B. P. Freymann, and H. Schnyder. 2008. Trophic interactions among invertebrates in termitaria in the African savanna: a stable isotope approach. *Ecological Entomology* 33:758–64.

de Wit, H. A. 1978. *Soils and grassland types of Serengeti plains (Tanzania)*. PhD diss., Agricultural University, Wageningen.

Dobson, A. 1995. The ecology and epidemiology of rinderpest virus in Serengeti and Ngorongoro Conservation Area. In *Serengeti: Dynamics, management, and conservation of an ecosystem*, ed. A. R. E. Sinclair and P. Arcese, 485–505. Chicago: University of Chicago Press.

Dunwiddie, P. W. 1991. Comparisons of aboveground arthropods in burned, mowed and untreated sites in sandplain grasslands on Nantucket Island. *American Midland Naturalist* 125:206–12.

Eggleton, P. 2000. Global diversity patterns. In *Termites: Evolution, sociality, symbiosis, ecology*, ed. T. Abe, D. E. Bignell, and M. Higashi, 25–51. Dordrecht: Kluwer Academic Publishers.

Erwin, T. L. 1982. Tropical forests: Their richness in Coleoptera and other arthropod species. *The Coleopterists Bulletin* 36:74–75.

Estes, R. D. 1991. *The behavior guide to African mammals: Including hoofed mammals, carnivores, primates*. Berkeley: University of California Press.

Evans, E. W. 1984. Fire as a natural disturbance to grasshopper assemblages of tallgrass praire. *Oikos* 43:9–16.

Fischer, G. A. 1885a. Bericht über die im auftrage der geographischen gesellschaft in Hamburg unternommene reise in das Massai-Land I: Allgemeiner bericht. *Mitteilungen der Geographischen Gesellschaft in Hamburg* 5 (1882–1883): 36–99.

———. 1885b. Übersicht der in Ostafrika gesammelten vogelarten, mit angabe der verschiedenen fundorte. *Journal für Ornithologie* 33:113–42.

Folse, L. J. 1982. An analysis of avifauna-resources relationships on the Serengeti plains. *Ecological Monographs* 52:111–27.

Foster, R. 1993. Dung beetle community ecology and dung removal in the Serengeti. PhD diss., University of Oxford.

Foster, R., and L. Bresele. 1992. *Dung beetles: Important invertebrates worth considering*. Conservation Monitoring News, no. 5: 6–7. Frankfurt Zoological Society.

Freymann, B. P., T. M. Anderson, E. P. Mayemba, and H. Olff. n.d. "Determinants of spatial variation in termite mound densities in the Serengeti: Importance of resources vs. conditions." Manuscript.

Freymann, B. P., R. Buitenwerf, O. DeSouza, and H. Olff. 2008. The importance of termites (Isoptera) for the recycling of herbivore dung in tropical ecosystems: A review. *European Journal of Entomology* 105:165–73.

Freymann, B. P., S. N. de Visser, E. P. Mayemba, and H. Olff. 2007. Termites of the genus *Odontotermes* are optionally keratophagous. *Ecotropica* 13:143–47.

Freymann, B. P., S. N. de Visser, and H. Olff. 2010. Spatial and temporal hotspots of termite-driven decomposition in the Serengeti. *Ecography* 33:443–50.

Freymann, B. P., and F.-T. Krell. 2011. Dung beetles (Coleoptera: Scarabaeidae) trapped by a moving sand dune near Olduvai Gorge, Tanzania. *The Coleopterists Bulletin* 65:422–24.

Freymann, B. P., and H. Olff. 2007. Cardinal woodpecker (*Dendropicos fuscescens*) utilizes wildebeest carcass horns as a foraging substrate. *Ostrich* 78:653–54.

Gerstaecker, A. 1884. *Bestimmung der von Herrn Dr. G. A. Fischer während seiner reise nach dem Massai-Land gesammelten Coleopteren*. Jahresbericht, vol. 1, 41–63, Naturhistorisch Museum, Hamburg.

Gibson, C. W. D, V. K. Brown, L. Losito, and G. C. McGavin. 1992. The response of invertebrate assemblies to grazing. *Ecography* 15:166–76.

Grove, S. J., and N. E. Stork. 2000. An inordinate fondness for beetles. *Invertebrate Taxonomy* 14:733–39.

Grzimek, B., and M. Grzimek. 1959. *Serengeti darf nicht sterben: 367.000 Tiere suchen einen Staat*. Frankfurt-Main: Verlag Ullstein GmbH.

Halffter, G., and E. G. Matthews. 1966. The natural history of dung beetles of the subfamily Scarabaeinae (Coleoptera: Scarabaeidae). *Folia Entomologica Mexicana* 12 (14): 1–312.

Hammond, P. M. 1990. Insect abundance and diversity in the Dumoga-Bone National Park, N. Sulawesi, with special reference to the beetle fauna of lowland rain forest in the Toraut region. In *Insects and the rain forests of South East Asia (Wallacea)*, ed. W. J. Knight and J. D. Holloway, 197–254. London: Royal Entomological Society of London.

Harvey, A. W., and J. M. Ritchie. 1990. Community ecology of Serengeti grasshoppers. *Boletín Sanidad Vegetal Plagas* (Fuera de Serie) 20:409.

Hassan, S. N., G. M. Rusch, H. Hytteborn, C. Skarpe, and I. Kikula. 2007. Effects of fire on sward structure and grazing in western Serengeti, Tanzania. *African Journal of Ecology* 46:174–85.

Hata, K., and W. D. Edmonds. 1983. Structure and function of the mandibles of adult dung beetles (Coleoptera, Scarabaeidae). *International Journal of Insect Morphology & Embryology* 12:1–12.

Hewitt, G. B. 1977. Review of forage losses caused by rangeland grasshoppers. *USDA-ARS Miscellaneous Publications* no. 1348:22.

Hodkinson, I. D., and D. Casson. 1991. A lesser predilection for bugs: Hemiptera (Insecta) diversity in tropical rain forests. *Biological Journal of the Linnean Society* 43:101–09.

Holt, J. A., K. L. Bristow, and J. G. McIvor. 1996. The effects of grazing pressure on soil animals and hydraulic properties of two soils in semi-arid tropical Queensland. *Australian Journal of Soil Research* 34:69–79.

Holter, P. 2000. Particle feeding in Aphodius dung beetles (Scarabaeidae): Old hypotheses and new experimental evidence. Functional Ecology 14:631–37.

Kingdon, J. 1974. *East African mammals: An atlas of evolution in Africa. vol. 2, part A, Insectivores and bats*. London: Academic Press.

———. 1977. *East African mammals: An atlas of evolution in Africa. vol. 3, part A, Carnivores.* London: Academic Press.

Krebs, C. J. 1989. *Ecological methodology*. New York: Harper & Row.

Kruess, A., and T. Tscharntke. 2002. Contrasting responses of plant and insect diversity to variation in grazing intensity. *Biological Conservation* 106:293–302.

Kruuk, H., and W. A. Sands. 1972. The aardwolf (*Proteles cristatus*, Sparrman) as predator of termites. *East African Wildlife Journal* 10:211–27.

Lamprey, H. F., G. Halevy, and S. Makacha. 1974. Interactions between Acacia, bruchid seed beetles and large herbivores. *East African Wildlife Journal* 12:81–85.

Lee, K. E., and T. G. Wood. 1971. *Termites and soils*. London: Academic Press.

Lewinsohn, T. M., and P. W. Price. 1996. Diversity of herbivorous insects and ecosystem processes. In *Biodiversity and Savanna Ecosystem Processes, a global perspective*, ed. O. T. Solbrig, E. Medina, and J. F. Silva, 143–57. Berlin: Springer-Verlag.

Lindroth, C. H. 1985. *The Carabidae (Coleoptera) of Fennoscandia and Denmark*. Fauna Entomologica Scandinavica 15, part 1. Copenhagen: Scandinavian Science Press Ltd.

———. 1986. *The Carabidae (Coleoptera) of Fennoscandia and Denmark*. Fauna Entomologica Scandinavica 15, part 2. Copenhagen: Scandinavian Science Press Ltd.

Lobry de Bruyn, L. A., and A. J. Conacher. 1990. The role of termites and ants in soil modification: A review. *Australian Journal of Soil Research* 28:55–93.

May, R. M. 1986. Biological diversity: How many species are there? *Nature* 324:514–15.

Mayhew, P. J. 2007. Why are there so many insects? Perspective from fossils and phylogenies. *Biological Reviews* 82:425–54.

McNaughton, S. J. 1979. Grazing as an optimization process: Grass-ungulate relationships in the Serengeti. *The American Naturalist* 113:691–703.

McNaughton, S. J. 1983. Serengeti grassland ecology: The role of composite environmental factors and contingency in community organization. *Ecological Monographs* 53: 291–320.

Milchunas, D. G., W. K. Lauenroth, and I. C. Burke. 1998. Livestock grazing: Animal and plant biodiversity of shortgrass steppe and the relationship to ecosystem function. *Oikos* 83:65–74.

Moretti, M., M. Conedera, P. Duelli, and P. J. Edwards. 2002. The effects of wildfire on ground-active spiders in deciduous forests on the Swiss southern slope of the Alps. *Journal of Applied Ecology* 39:321–36.

Moretti, M., M. K. Obrist, and P. Duelli. 2004. Arthropod biodiversity after forest fires: Winners and losers in the winter fire regime of the southern Alps. *Ecography* 27:173–86.

Mysterud, A., L. O. Hansen, C. Peters, and G. Austrheim. 2005. The short-term effect of sheep grazing on selected invertebrates (Diptera and Hemiptera) relative to other environmental factors in an alpine ecosystem. *Journal of Zoology* 266:411–18.

Nichols, E., S. Spector, J. Louzada, T. Larsen, S. Amezquita, M. E. Favila, and the Scarabaeinae Research Network. 2008. Ecological functions and ecosystem services provided by Scarabaeinae dung beetles. *Biological Conservation* 141:1461–74.

Nkwabi, A. K., A. R. E. Sinclair, K. L. Metzger, and S. A. R. Mduma. 2011. Disturbance, ecosystem function and compensation: The influence of wildfire and grazing on the avian community in the Serengeti Ecosystem, Tanzania. *Austral Ecology* 36:403–12.

Noirot, C., and C. Noirot-Thimothee. 1969. The digestive system. In *Biology of termites, vol. 1*, ed. K. Krishna and F. M. Weesner, 49–88. London: Academic Press.

Norton-Griffiths, M. 1979. The influence of grazing, browsing, and fire on the vegetation dynamics of the Serengeti. In *Serengeti: Dynamics of an ecosystem*, ed. A. R. E. Sinclair and M. Norton-Griffiths, 310–52. Chicago: University of Chicago Press.

Novotny, V., Y. Basset, S. E. Miller, G. D. Weiblen, B. Bremer, L. Cizek, and P. Drozd. 2002. Low host specificity of herbivorous insects in a tropical forest. *Nature* 416:841–44.

Noyes, J. S. 1989. The diversity of Hymenoptera in the tropics with special reference to Parasitica in Sulawesi. *Ecological Entomology* 14:197–207.

Orgeas, J., and A. N. Andersen. 2001. Fire and biodiversity: Responses of grass-layer beetles to experimental fire regimes in an Australian tropical savanna. *Journal of Applied Ecology* 38:49–62.

Otte, D., and W. Cade. 1984. African crickets (Gryllidae). 4. The genus *Platygryllus* from eastern and southern Africa (Gryllinae, Gryllini). *Proceedings of the Academy of Natural Sciences of Philadelphia* 136:45–66.

Pedgley, D. E., W. W. Page, A. Mushi, P. Odiyo, J. Amisi, C. F. Dewhurst, W. R. Dunstan, L. D. C. Fishpool, A. W. Harvey, and T. Megenasa, et al. 1989. Onset and spread of an African army worm upsurge. *Ecological Entomology* 14:311–33.

Pellew, R. A., and B. J. Southgate. 1984. The parasitism of *Acacia tortilis* seeds in Serengeti. *African Journal of Ecology* 22:73–75.

Phipps, J. 1959. Studies on East African Acridoidea (Orthoptera), with special reference to egg-production, habitats and seasonal cycles. *Transactions of the Royal Entomological Society of London* 111:27–56.

Picker, M., C. Griffiths, and A. Weaving. 2004. *Field guide to insects of South Africa*. Cape Town: Struik Publishers.

Rambo, J. L., and S. H. Faeth. 1999. Effect of vertebrate grazing on plant and insect community structure. *Conservation Biology* 13:1047–54.

Redford, K. H. 1984. The termitaria of *Cornitermes cumulans* (Isoptera, Termitidae) and their role in determining a potential keystone species. *Biotropica* 16:112–19.

Ruess, R. W. 1987. The role of large herbivores in nutrient cycling of tropical savannas. In *Determinants of tropical savannas*, IUBS Monograph Series no. 3, ed. B. H. Walker, 6–91. Oxford: IRL Press Ltd.

Ruess, R. W., and S. J. McNaughton. 1984. Urea as a promotive coupler of plant-herbivore interactions. *Oecologia* 63:331–37.

Ruppert, E. E., and R. D. Barnes. 1994. *Invertebrate Zoology*. Fort Worth, TX: Saunders College Publishing.

Sackmann, P., and A. Farji-Brener. 2006. Effect of fire on ground beetles and ant assemblages along an environmental gradient in NW Patagonia: Does habitat type matter? *Ecoscience* 13:360–71.

Scholes, R. J., and B. H. Walker. 1993. *An African savanna: synthesis of the Nylsvley study (South Africa)*. Cambridge: Cambridge University Press.

Schuurman, G. 2006. Foraging and distribution patterns in a termite assemblage dominated by fungus-growing species in semi-arid northern Botswana. *Journal of Tropical Ecology* 22:277–87.

Serneels, S. and E. F. Lambin. 2001. Impact of land-use changes on the wildebeest migration in the northern part of the Serengeti-Mara ecosystem. *Journal of Biogeography* 28: 391–407.

Silva, J. F. 1996. Biodiversity and stability in tropical savannas. In *Biodiversity and savanna ecosystem processes, a global perspective*, ed. O. T. Solbrig, E. Medina, and J. F. Silva, 161–74. Berlin: Springer-Verlag.

Sinclair, A. R. E. 1975. The resource limitation of trophic levels in tropical grassland ecosystems. *Journal of Animal Ecology* 44:497–520.

———. 1978. Factors affecting the food supply and breeding season of resident birds and movements of palaearctic migrants in a tropical African savannah. *Ibis* 120:480–97.

———. 1979. Dynamics of the Serengeti ecosystem: Process and pattern. In *Serengeti: Dynamics of an ecosystem*, ed. A. R. E. Sinclair and M. Norton-Griffiths, 1–30. Chicago: University of Chicago Press.

———. 1995. Serengeti past and present. In *Serengeti: Dynamics, management, and conservation of an ecosystem*, ed. A. R. E. Sinclair and P. Arcese, 3–30. Chicago: University of Chicago Press.

Sinclair, A. R. E., S. A. R. Mduma, and P. Arcese. 2002. Protected areas as biodiversity benchmarks for human impact: Agriculture and the Serengeti avifauna. *Proceedings of the Royal Society of London Series B* 269:2401–05.

Sinclair, A. R. E., and M. Norton-Griffiths. 1979. *Serengeti: Dynamics of an ecosystem*. Chicago: University of Chicago Press.

Smith, G. E. 1907. From the Victoria Nyanza to Kilimanjaro. *The Geographical Journal* 29: 249–69.

Southgate, B. J. 1981. Univoltine and multivoltine cycles: Their significance. In *The Ecology of Bruchids attacking Legumes (Pulses)*, ed. V. Labeyrie, 17–22. The Hague: Junk.

StatSoft Inc. 2003. STATISTICA. Version 6.1. StatSoft Inc., Tulsa, OK.

Stork, N. E. 2007. World of insects. *Nature* 448:657–58.

Suominen, O., J. Niemelä, P. Martikainen, P. Niemelä, and I. Kojola. 2003. Impact of reindeer grazing on ground-dwelling Carabidae and Curculionidae assemblages in Lapland. *Ecography* 26:503–13.

Traniello, J. F. A., and R. H. Leuthold. 2000. The behavior and ecology of foraging in termites. In *Termites: evolution, sociality, symbiosis, ecology*, ed. T. Abe, D. E. Bignell, and M. Higashi, 141–68. Dordrecht: Kluwer Academic Publishers.

Wallner, W. E. 1987. Factors affecting insect population dynamics: Differences between outbreak and non-outbreak species. *Annual Review of Entomology* 32:317–40.

Weir, J. S. 1971. The effect of creating additional water supplies in a Central African national park. In *The scientific management of animal and plant communities for conservation*, ed. E. Duffey and A. S. Watt, 367–85. Oxford: Blackwell Scientific Publications.

Wilson, J. D., A. J. Morris, B. E. Arroyo, S. C. Clark, and R. B. Bradbury. 1999. A review of the abundance and diversity of invertebrate and plant foods of granivorous birds in northern Europe in relation to agricultural change. *Agriculture, Ecosystems and Environment* 75:13–30.

Wood, T. G. 1976. The role of termites (Isoptera) in decomposition processes. In *The role of terrestrial and aquatic organisms in decomposition processes*, ed. J. M. Anderson and A. Macfadyen, 145–68. Oxford: Blackwell Scientific Publications.

———. 1978. Food and feeding habits of termites. In *Production ecology of ants and termites*, ed. M. V. Brian, 55–80. Cambridge: Cambridge University Press.

York, A. 1999. Long term effects of frequent low-intensity burning on the abundance of litter-dwelling invertebrates in coastal blackbutt forests of southeastern Australia. *Journal of Insect Conservation* 3:191–99.

ELEVEN

The Butterflies of Serengeti: Impact of Environmental Disturbance on Biodiversity

Anthony R. E. Sinclair, Ally K. Nkwabi, and Kristine L. Metzger

How environmental heterogeneity impacts biodiversity is relevant to both the functioning and stability of ecosystems and also to their conservation; spatial variation has been examined in African systems (du Toit et al. 2003) while hurricanes, volcanic eruptions, and other disasters are used to study temporal disturbance (Johnson and Miyanishi 2008; Prach and Walker 2011; Walker 2012). In particular, human-induced disturbances through modified landscapes are thought to reduce biodiversity in a range of biota. Butterflies have been used as indicator species for such studies. However, the great majority of studies in tropical regions have focused on disturbances to forests by making comparisons with cleared areas or with agroforestry (Hamer et al. 1997; Lawton et al. 1998; Hamer and Hill 2000; Stork et al. 2003;Benedick et al. 2006; Schulze, Schneeweihs, and Fiedler 2010; Hill et al. 2011). Few studies of tropical savanna have been conducted, though Fitzherbert et al. (2006) have examined diversity in ephemeral habitats at Lake Katavi of Tanzania, and Akite (2008) has compared changes in butterfly diversity with increasing human disturbance in northern Uganda. A general conclusion is that primary forest is important for endemic butterfly species, but secondary (regrowth) forest is also a vital resource for a large proportion of the biota. In contrast, agroforestry has low biodiversity, with the species found there largely those that are widespread and abundant, with the rare endemics usually missing (Schulze, Schneeweihs, and Fiedler 2010). Hence, in this study we examine whether the same conclusions apply to a tropical savanna, the Serengeti ecosystem. Do agricultural crops (a form of modified grassland) show similar reductions in butterfly diversity to those of agroforestry, compared with adjacent native savanna grasslands?

Spatial heterogeneity is expressed as the mosaic of vegetation types within the ecosystem, while temporal heterogeneity is seen as the disturbances in climate over time. Both are constrained by scales relevant to the species living in the ecosystem: too fine a scale results in animals seeing the habitat as uniform—they cannot economically choose between types—and too large a scale results in animals living within a single habitat patch. Scale effects have already been noted as influencing the conclusions concerning disturbance and biodiversity: large-scale agricultural impacts produce decreases of butterfly diversity, while small-scale disturbances result in increases in biodiversity (Hamer and Hill 2000; Hill and Hamer 2004).

The same constraints apply at the temporal level: whether disturbances act to enhance biodiversity through, say, the intermediate disturbance hypothesis (Connell 1978), or reduce it by too much or too little disturbance can be tested by comparing habitats with different disturbance regimes. Similarly whether spatial heterogeneity enhances biodiversity or not can be tested by comparing the richness of species in different habitats: if species see habitats as uniform (too fine or too coarse a scale) then species richness should have very high similarity with few new species between habitats.

We tested these predictions on the role of disturbance and patchiness using the butterfly fauna of the Serengeti ecosystem. The study of environmental change on insect communities is hampered by two problems. First insects are generally difficult to monitor reliably. Techniques for censusing them are crude and their numbers fluctuate markedly. Second their high biodiversity means that many cannot be identified to the species level. Thus conclusions on how species respond to disturbances are confined to generalities about insect groups. Butterflies, however, are now well-known and can with some certainty be identified to species level. They can also be counted, at least as adults, with a degree of repeatability. We have monitored adult butterfly populations in Serengeti since 1998, with some gaps in the record, until 2012. We examine here how numbers and diversity responded to spatial heterogeneity by comparing different habitats and disturbances to habitat by natural modifications through burning and grazing of the grasslands, by human modifications of agriculture, temporal disturbance via seasonal changes, and multiannual changes in weather through rainfall. Fitzherbert et al. (2006) recorded the butterfly fauna in a lake floodplain ecosystem (Lake Katavi) in southern Tanzania and found that ephemeral grasslands supported many species. Such temporal changes in habitat could therefore be important here.

METHODS

Numbers of butterflies were monitored in two ways. One involved counting the number of flying butterflies that crossed a strip of known length (somewhere between 40 and 80 m) and 20 m wide for 10 minutes in the middle of the day when the insects were flying. This was the "visual" method. The observer used a convenient tree as the end marker and recorded individuals that crossed this strip, noting the types as whites, yellows, reds, or individual species if they were easily identifiable. Distance to the end marker was measured later. At any census period and site three counts were conducted.

The second method involved the observer walking 100 meters and counting all butterflies that crossed in front in a strip 2 m each side of the observer (a 4 m strip). This was the "walk" method. Data were recorded in the same way as for the "visual" method. Four such strips were recorded for each time and site.

Sites for both methods focused on the main upper catena Acacia woodland dominated by *A. robusta* (plus samples from *A. tortilis* and *A. gerrardii*) with a total of six sites in the central, western, and eastern savanna. Other habitats were the broad-leaved, *Terminalia-Combretum* woodland (our version of Miombo); open grassland divided into grasslands within savanna (savanna grassland), long-grass plains, and short-grass plains; and riverine forest. There are two forest types, lowland related to the Congo forest in western Serengeti, and montane along the Mara River related to the Loita and Crater Highlands (see Turkington et al. Chapter 9; Sharam 2005; Sharam, Sinclair, and Turkington 2006; Sharam et al. 2009). They are distinct plant communities with little overlap in tree species. In agricultural areas west of the Serengeti ecosystem a total of 12 sites were monitored.

In 2003–04 grazed and burnt sites were counted during the study of the avifauna (see methods in Nkwabi et al. 2011, and chapter 14). Both grazing from the migrating wildebeest and burning perturbations take place in July, at the beginning of the dry season. This is when fires are set by people and when the wildebeest move through on their migration to dry season grazing areas. Wildebeest graze down the grass layer from about 70 cm in height to about 20 cm; they eat only grass and leave the dicots, but many are trampled in the process. Burning removes the dry biomass almost entirely and a short green regrowth of both grass and small dicots takes place a few weeks later. Rains begin around November and full regrowth is complete by the next March. We counted butterflies immediately after the perturbations in July and after the short rains in March.

Samples of butterflies were collected with nets throughout the 14-year study for individual identification. Species were identified using the handbook by Larsen (1996), which includes plates of nearly all Kenya species, most of which are relevant to the Serengeti-Mara ecosystem. Of the 1,400 individuals that we collected, less than 10 could not be clearly assigned to a species. A list of species found in Serengeti is given in appendix 3 at www.press.uchicago.edu/sites/serengeti. Apart from the census sites in native Acacia woodland and agriculture, other habitats were also sampled for species identification; these included riverine forest, open grassland, other Acacia types, and Terminalia woodland.

Within a habitat the replicate counts for each site and time were averaged and the standard error computed. The averages for the several sites were then merged to give a weighted mean using the variances for each site as the weighting.

DISTRIBUTION AND BIODIVERSITY

The 186 species identified so far in Serengeti (and this is certainly incomplete for habitats such as riverine forest and especially hills) are divided into five families sorted by their main habitat preferences as given by Larsen (1996) (table 11.1). These habitats are consistent with those we recorded here. The swallowtails have only five species, all associated with forest although one occurs commonly throughout the savanna. The skippers have 22 species and the others over 50 each. By far the most speciose habitat is the savanna complex, perhaps reflecting the greater time collecting there, with forest being the next most speciose. Other habitats are represented by only a few species.

Plant families and other groups that are the main food base for the butterfly families show a great diversity of flowering plants for larval food (table 11.2). There are 35 plant families represented. Notably, however, only 22 butterfly species feed on grasses, the other 88% feeding on dicots mostly in the herb layer; this is all the more significant given the great preponderance of grass in the herb layer biomass. Other notable butterfly groups are the five species of Lipteninae that specialize on lichens of forests or rocks, and two Miletinae that are insectivorous, feeding on planthoppers.

The geographic distribution of species found in Serengeti as described by Larsen (1996) shows some 45% are widespread being found either across the tropics or across Africa (table 11.3). On the other hand another 45% are confined to eastern or central Africa, and significantly, 19 species are very local. The whites (Pieridae) are generally more widespread than the other families which have some members endemic to the Serengeti area.

Table 11.1 General habitat preferences of species collected in Serengeti by family

Habitat	Papilionidae	Pieridae	Lycaenidae	Nymphalidae	Hesperiidae	Total
Forest	5	10	2	16	5	38
Grassland			1	1	3	5
Hills					1	1
Marsh		1	1	3	2	7
Montane		2	1	2	1	6
Riverine			1	3		4
Savanna		23	39	11	6	79
Savanna dry		9	2	8	3	22
Savanna wet		4	2	9	1	16
Broad-leaved woodland		1		1		2
Rocky hills			3	1		4
Savanna, shamba				2		2
Total	5	50	52	57	22	186

Source: Larsen (1996).

Table 11.2 Food plant family (or other food type) for butterfly species collected in Serengeti sorted by insect family

Food plant family	**Papilionidae**	**Pieridae**	**Lycaenidae**	**Nymphalidae**	**Hesperiidae**	**Total**
Acanthaceae			1	8	3	12
Amaranthaceae			1			1
Arecaceae				1		1
Asclepiadaceae				2		2
Boraginaceae			1			1
Brassicaceae		1				1
Capparidaceae		33				33
Celastraceae		1				1
Combretaceae			1	2		3
Commelinaceae				1		1
Convolvulaceae				1		1
Crassulaceae			2			2
Euphorbiaceae				4		4
homoptera			2			2
Hypericaceae		2				2
Lamiaceae			8	4		12
Leguminosae		7	19	3		29
lichens			5			5
Loranthaceae		3	1			4
Malvaceae				1	2	3

Meliaceae				1		1
Myrtaceae			1			1
Oxalidaceae			1			1
Passifloraceae				1		1
Poaceae				13	8	21
Polyphagous			2	3	1	6
Portulacaceae				1		1
Rhamnaceae			1	1		2
Rutaceae	5					5
Salvadoraceae		1				1
Santalaceae			1			1
Sapindaceae			1	3		4
Sterculiaceae					2	2
Tiliaceae					2	2
Urticaceae				1		1
Verbenaceae			1			1
Violaceae				1		1
Unknown		2	3	5	4	14
Total	**5**	**50**	**52**	**57**	**22**	**186**

Source: Larsen (1996).

Table 11.3 Range of butterfly species collected in Serengeti scored by family and subfamily

Family	Global	African	Regional	Local	Endemic	Total
Papilionidae		**2**	**3**			**5**
Papilioninae		2	3			5
Pieridae	**7**	**18**	**24**	**1**		**50**
Coliadinae	3	3	3			9
Pierinae	4	15	21	1		41
Lycaenidae	**5**	**18**	**21**	**4**	**4**	**52**
Lipteninae		1	3		1	5
Miletinae		1	1			2
Polyommatinae	5	14	12	3	2	36
Theclinae		2	5	1	1	9
Nymphalidae	**7**	**19**	**24**	**2**	**5**	**57**
Acraeinae		6	3	1		10
Charaxinae		1	4		1	6
Danainae	1	1				2
Nymphalinae	4	9	11	1	1	26
Satyrinae	2	2	6		3	13
Hesperiidae	**2**	**7**	**10**	**2**	**1**	**22**
Coeliadinae		1				1
Hesperiinae	1	4	2	1		8
Pyrginae	1	2	8	1	1	13
Grand Total	**21**	**64**	**82**	**9**	**10**	**186**

Source: Larsen (1996).

ABUNDANCE AND BIODIVERSITY

Rare and Endemic Species

General abundance (table 11.4) follows the distribution with widespread species being more abundant. However, some 34% of the species are scarce and 16 species are rare. There are some 14 species that could be considered either rare locals or globally endemic; they are confined to the Serengeti-Mara or a part of southwestern Kenya or Tanzania (table 11.5), with the exception of the global migrant painted lady (*Vanessa cardui*), of which only one specimen has been recorded. Seven are associated with riverine forest patches or swamps, and three with rocky hills or kopjes, both habitats being very restricted in Serengeti. Two endemic species known to occur in Serengeti (Larsen 1996) were not collected in this study. One, *Neocoenyra (Neita) victorii*, is associated with wet areas by the lake, and the other, *Charaxes*

Table 11.4 General abundance of butterfly species collected in Serengeti scored by family and subfamily

Family	Abundant	Common	Scarce	Rare	Total
Papilionidae	**1**	**4**			**5**
Papilioninae	1	4			5
Pieridae	**7**	**29**	**14**		**50**
Coliadinae	1	5	3		9
Pierinae	6	24	11		41
Lycaenidae	**4**	**23**	**18**	**7**	**52**
Lipteninae			2	3	5
Miletinae		1	1		2
Polyommatinae	4	16	13	3	36
Theclinae		6	2	1	9
Nymphalidae	**3**	**24**	**24**	**6**	**57**
Acraeinae	1	4	5		10
Charaxinae		3	2	1	6
Danainae	1	1			2
Nymphalinae	1	12	11	2	26
Satyrinae		4	6	3	13
Hesperiidae		**11**	**8**	**3**	**22**
Coeliadinae		1			1
Hesperiinae		4	3	1	8
Pyrginae		6	5	2	13
Grand Total	**15**	**91**	**64**	**16**	**186**

Source: Larsen (1996).

chepalungu, is found only in the Mara River forests. The other species are found in savanna. All but one of this group are scarce; the other, *Neocoenyra masaica*, a brown, is common in the savanna grass layer. The forest and rock endemics are specialist feeders on lichens. Of the 51 specimens collected in this group only one (a skipper *Sarangesa princei*) has been found in agriculture.

Migrants

The term "migrant" when applied to butterflies generally means species that occasionally erupt and then irrupt; both the whites *Catopsilia florella* and *Belenois aurota* have been recorded as appearing in large numbers and then moving in a particular direction (Williams 1958; Larsen 1988, 1992). Strictly speaking this is emigration. True migration, as applied to birds with

Table 11.5 Frequency of rare butterfly species collected in different habitats (cols. 4–12) in the Serengeti ecosystem

Subfamily: Species	Distribution	Habitat	Acacia	Broad-leaved	Plains long	Plains short	Lake	Forest lowland	Forest montane	Hills	Agric	Total
Lipteninae												
Alaena species	Endemic	Rocky hills								1		1
Liptena xanthostola	Regional	Forest							1			1
Liptena sp.	Regional	Forest						2				2
Theclinae												
Iolaus arborifera	Endemic	Montane						2				2
Polyommatinae												
Neurellipes gemmifera	Regional	Savanna	3									3
Lepidochrysops pterou	Endemic	Rocky hills		1								1
Lepidochrysops elgonae	Endemic	Rocky hills				3						3
Satyrinae												
Neita victoriae	Endemic	Marsh					1*					
Neocoenyra masaica	Endemic	Grassland	28	1								29
Charaxinae												
Charaxes chepalungu	Endemic	Forest							1*			
Nymphalinae												
Euphaedra paradoxa	Endemic	Forest							3			3
Vanessa cardui	Global	Savanna	1									1
Pyrginae												
Sarangesa princei	Endemic	Savanna			1						1	2
Spialia wrefordi	Local	Savanna dry		4				1			1	6
Hesperiinae												
Borbo sirena	Local	Forest							1			1

Source: General distribution and habitat from Larsen (1996).

Note: Asterisk indicates not collected in this study but reported in Larsen.

regular annual round trips as seen in the monarchs of North America, is rare in our area although the painted lady is one of these. Although the great majority of Serengeti species are residents we recorded 17 species of migrants (table 11.6). Most of these have a wide distribution being global or found across Africa; they are either abundant or common and about half of them appear frequently in our samples. The majority (12 species) occur in savanna and seven of these are also found in agriculture. There are some notable exceptions. The meadow white (*Pontia helice*), prefers highland grassland and this subspecies is found only regionally in eastern Africa. In the Serengeti ecosystem it was confined to the short grasslands of the plains where it was abundant. Three migrant species live in forests of which only one, the citrus butterfly (*Papilio demodocus*), is found in agriculture. As mentioned elsewhere the painted lady was collected only once.

DISTURBANCE AND ABUNDANCE

Fluctuations in Rainfall

Butterfly populations were estimated as density per ha from the visual method with a weighted mean merged from the six to ten sites in the Acacia and broad-leaved habitat. Monthly counts were conducted for three years from June 1998 to June 2001 (fig. 11.1). These counts show a clear seasonal difference in numbers with high numbers occurring sometime from January to April following the short wet season rains in November–December. Numbers appear to be responding to the growth of plants following the rain.

After June 2001 we reduced the counts to those at the peak period of the year. Interannual peaks showed considerable fluctuation, with coincidentally the highest being the year we started (fig. 11.2). This happened following the wettest season (1997–98) since the great floods of 1961. When peaks of butterfly abundance (*PA*) are plotted against annual rainfall (*R*) the previous calendar year there is a strong positive relationship ($PA = .006R^{3.114}$, $R^2 = .482$, $n = 12$); higher rainfall leads to greater insect abundance a few months later probably because of greater plant food abundance. An index of butterfly food abundance is the cumulated biomass of grass which is a direct function of cumulated rainfall (Sinclair 1977) (fig. 11.3). Using this relationship the annual peak butterfly numbers is also related to cumulated grass biomass (fig. 11.4). Although the vegetation measures are for grass (which for the most part butterflies do not eat) we see this as an index of total herb layer growth, including the dicots.

Table 11.6 Frequency of migrant butterfly species collected in different habitats (cols. 3–13) in the Serengeti ecosystem

Species	Habitat	Sav Acacia	Sav broad-leaved	Grassland	Plains long	Plains short	Lake	Forest lowland	Forest montane	Hills	Agric	Total
Papilioninae												
Papilio demodocus	Forest	2	1	1				2	2		4	12
Coliadinae												
Catopsilia florella	Savanna	16	2	1					1		9	29
Pierinae												
Belenois aurota	Savanna	84	14	22	11	7		12	1	2	13	166
Belenois creona	Savanna	78	36	23	2	6		14	15	1	24	199
Pontia helice	Montane				2	36						38
Polyommatinae												
Lampides boeticus	Savanna	23	9	1	1		1	2		10		47
Leptotes pirithous	Savanna	40	16	10	5	5	19			20		115
Azanus ubaldus	Savanna	1										1
Danainae												
Danaus chrysippus	Savanna	13	6	3		4		23	16		5	70
Tirumala petiverana	Savanna	1	1					1	1			3
Charaxinae												
Charaxes varanes	Forest		1						2			3
Nymphalinae												
Sallya amulia	Forest								1			1
Hypolimnas misippus	Savanna	22						5		5	2	34
Junonia orithya	Savanna dry	13	5	8	2	4			1		3	36
Vanessa cardui	Savanna	1										1
Phalanta phalantha	Savanna wet				1							1
Hesperiinae												
Pelopidas mathias	Savanna	5	1	1								7

Source: General habitat from Larsen (1996).

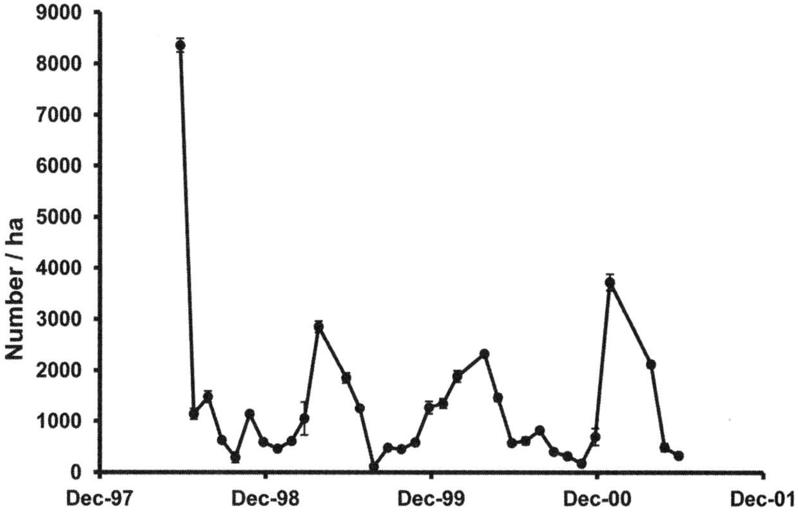

Fig. 11.1 The mean numbers per ha of butterflies merged over six sites in the Acacia woodlands from June 1998 to June 2001. Vertical bars are 1 SE.

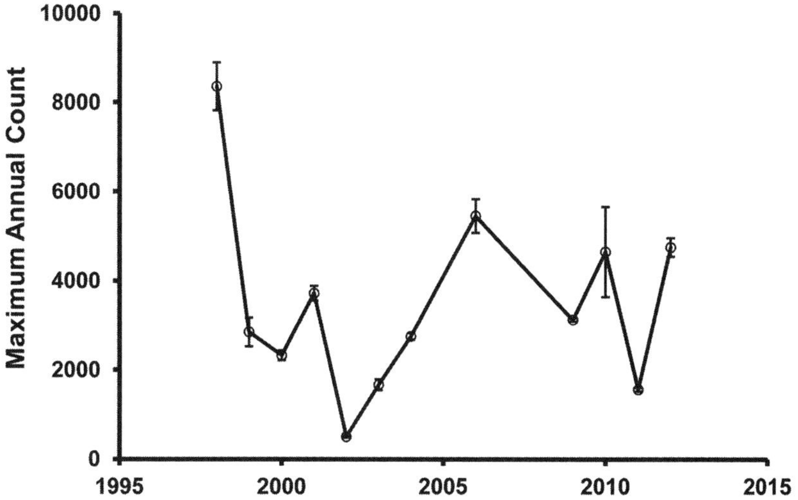

Fig. 11.2 The mean peak counts of butterflies (numbers per ha) each year merged over sites in the Acacia woodlands from 1998 to 2012. Vertical bars are 1 SE.

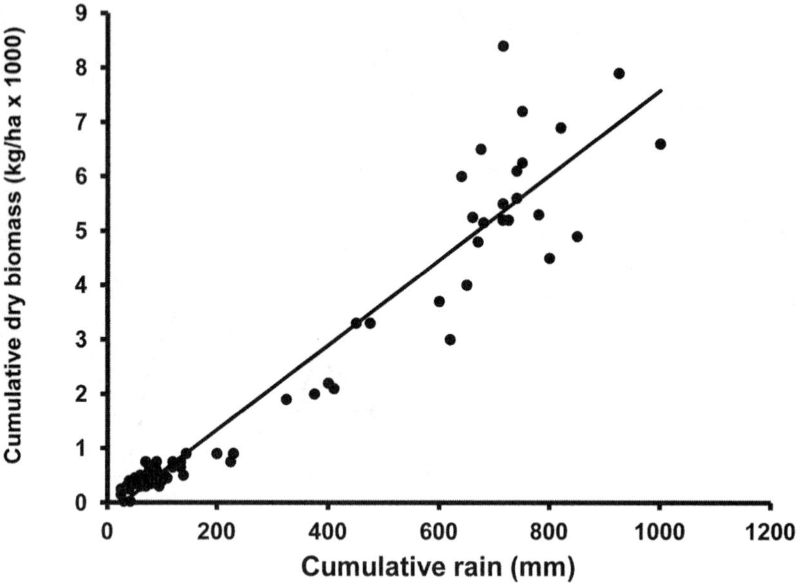

Fig. 11.3 The cumulated dry biomass of grass (*B*) versus the cumulated rainfall (*R*) measured on harvested plots from the start of the rains in November (Sinclair 1977). $B = (0.0078 * R) - 0.2251$. ($R^2 = 0.9225$).

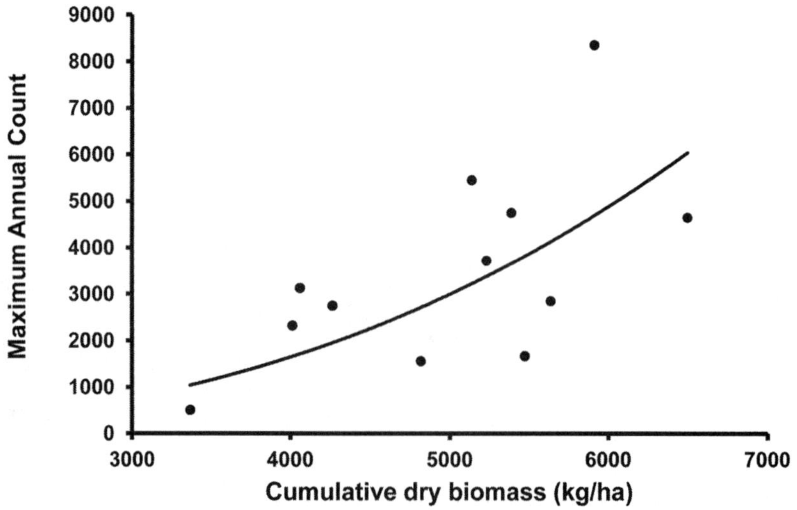

Fig. 11.4 The peak abundance of butterflies is related to the cumulated grass biomass (estimated from fig. 11.3) in the previous calendar year. ($R^2 = 0.4867$, $P < .02$).

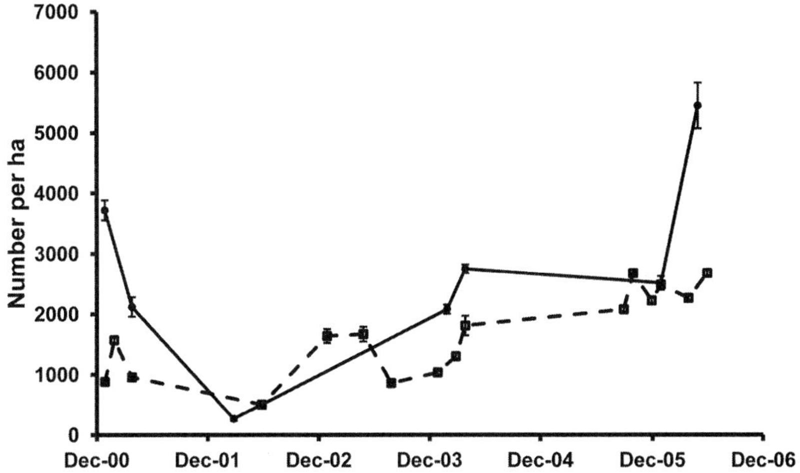

Fig. 11.5 The mean peak densities (numbers per ha) of butterflies in native habitats (solid line) of Serengeti from January 2001 to May 2006 compared with those in agricultural sites adjacent to the protected area (broken line). Vertical bars are 1 SE.

Habitat Modification from Agriculture

Counts of butterflies in agriculture covered some five years (2001–06). Compared with savanna, densities were generally lower. Although they followed the same seasonal and interannual fluctuations as those in savanna they did not show the high peaks of abundance (fig. 11.5).

Grazing and Burning

Counts of butterflies on grazed and burnt plots in July 2003 showed lower density than those in adjacent control plots (fig. 11.6). In February 2004, when both the herb layer had regenerated and butterflies were at peak numbers, densities on the disturbed plots were still significantly lower than those on undisturbed plots. Although the study sites (for counting birds) were only 100 m diameter, areas that were burnt or grazed were much greater, often more than 1 km^2. Nevertheless butterflies are highly mobile so the disturbances were unlikely to have affected the insects directly; more likely the butterflies were responding to changes in their food plants, and these effects carried over into the next growing season—there are delayed effects of disturbance on the insect fauna.

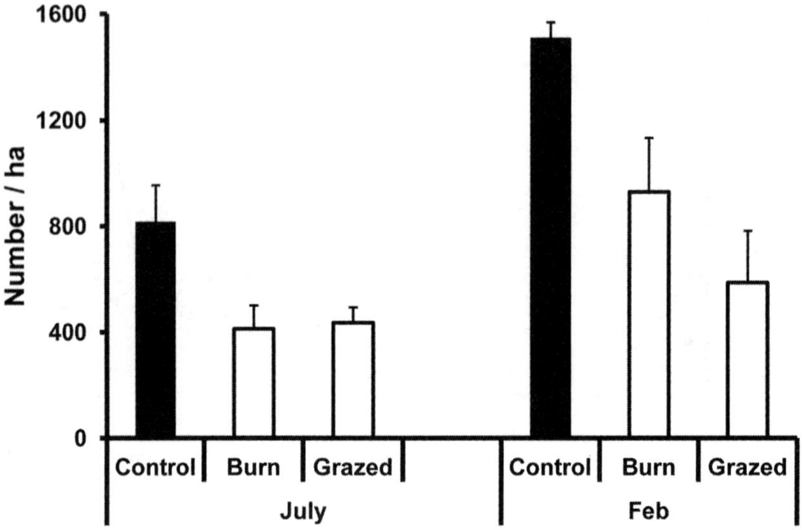

Fig. 11.6 The mean densities of butterflies in the first month of the dry season (July 2003 and 2004) following disturbance by wildebeest grazing and by grass fires compared to adjacent undisturbed sites. The same comparisons are shown for the following February after the short rains of November–December. Vertical bars are 1 SE.

DISTURBANCE AND BIODIVERSITY

Habitat Heterogeneity and Biodiversity

We compared butterfly species richness and similarity across savanna, grassland, and forest habitat types to test whether different habitats result in increased biodiversity. First, each habitat that we identified supported unique species—that is, species found only in that habitat (table 11.7). Acacia savanna had the greatest number, namely 30, which is to be expected because that is where the greatest sampling effort occurred. More importantly, despite low sampling intensity other habitats also supported unique species; broad-leaved savanna (*Terminalia-Combretum*) had nine species and the two forest types each had 14 species that were confined to them.

We made pair-wise comparisons between different habitats using the Bray-Curtis measure of similarity (1-*B*) (1 = similar, 0 = dissimilar) (table 11.8) (Krebs 1989). Thus, Acacia savanna had 105 species but combined with broad-leaved woodland another 16 species were included. However, savanna grassland contributed only another five species to the Acacia butterfly richness, suggesting the butterflies did not see these grasslands, embedded within savanna, as very different from Acacia. In contrast, when Acacia is combined with the total forest fauna, another 51 species are added,

Table 11.7 Number of species confined to the one specified habitat

Habitat	Unique Species
Acacia savanna	33
Broad-leaved savanna	9
Grassland	3
Plains long	2
Plains short	1
Lake	3
Forest lowland	18
Forest montane	14
Hills	2
Agriculture	3

Table 11.8 The increase in species richness when one habitat (*N* species) is combined with a second (*N* combined)

Habitat	*N* alone	Combined with	*N* combined	Bray-Curtis Similarity
Acacia	105	Broad-leaved	121	0.30
		Grassland	110	0.27
		All Forest	151	0.27
		Agriculture	115	0.17
Broad-leaved	56	Agriculture	72	0.32
Plains long	19	Grassland	48	0.33
		Plains short	26	0.22
Forest lowland	57	Forest Montane	82	0.42

Note: Bray-Curtis statistic of similarity (1-*B*) compares the numbers of individuals of each species in the two habitats.

half as much again—forest and savanna are clearly very distinct. Even more surprising is that butterflies see the two forest types as very different: lowland forest has 57 species, and montane 54 species but they share only half of these, with 28 and 25 species respectively not occurring in the other type.

The Effects of Habitat Disturbance: Agriculture

The Bray-Curtis similarity (table 11.8) showed that agriculture had a low value because the species were a small subset of the native Acacia savanna.

Table 11.9 The number and percentage of butterfly species of different distributional range found in agriculture around Serengeti and in native habitats

	Global	Africa	Regional	Local	Endemic	Total
Agriculture	10	18	11	1	1	41
% Agriculture	24.4	43.9	26.8	2.4	2.4	100
Native	11	46	71	8	7	143
% Native	7.7	32.2	49.7	5.6	4.9	100

Agriculture supported 41 species compared with the 105 species in Acacia savanna indicating that this human-modified habitat was far less diverse than the native habitat from which it originated. In addition, agriculture contributed only another 10 species to the Acacia savanna fauna suggesting that biodiversity is not much enhanced by the native-agriculture mosaic.

A far greater proportion of the species found in agriculture are widespread being either global or African in distribution (68%) (table 11.9), compared with 40% of those in native habitats. In contrast, only 2 species (5%) of the local and endemic forms occur in agriculture compared with 15 species in native savanna.

SYNTHESIS

In savanna habitats the abundances of butterflies change with season, related to their food supply which is determined by rainfall. Consequently interannual abundances change markedly depending on wet or dry years. These data reflect general insect abundance for which data are much more difficult to obtain. For example, they act as an index of food abundance for studies of avian breeding seasons. In agriculture, abundances follow similar trends to those in savanna but are generally lower, perhaps reflecting the poorer plant food availability and the fewer butterfly species there.

Habitat heterogeneity within the protected area clearly contributes to biodiversity. Not only do all the identified habitats support unique species (those so far collected only in that habitat) but the total number of species in pairs of habitats often increases over each habitat alone; in particular, forests add considerably to the savanna richness, and moreover, the two forest types are themselves different and support complimentary species. The plains grasslands—long versus short swards—also differ from one another. Thus, the overall diversity of butterflies is enhanced by the heterogeneity of the Serengeti ecosystem. In other words beta-diversity is high.

In contrast, agriculturally modified savanna shows that over half of the species disappear, and the agriculture-savanna mosaic does not increase species richness very much over savanna alone. This result supports earlier work where other insect groups in agriculture were markedly less diverse and abundant than those in savanna (Sinclair, Mduma, and Arcese 2002), a result also found in forest-agriculture comparisons (Lawton et al. 1998; Hill and Hamer 2004). However, this result contrasts with those from other studies, such those in Costa Rica, where there is an enhanced richness of species in a forest-agriculture mix relative to native forest alone (Daily et al. 2003; Mayfield and Daily 2005; Harvey et al. 2008).

Serengeti supports several rare endemic butterflies mostly associated with the restricted habitats of riverine forest and rocky hills. These forests are almost absent outside the protected area, long ago cut down during the expansion of agriculture in the early twentieth century (see chapter 2). The fact that these forests originate from different directions, the lowland from the western Congo forests and the montane from the eastern highlands, makes them doubly valuable as remnants of old, more extensive ecosystems in millennia past, now only occurring inside the park (chapter 9). Rocky hills are clearly a special habitat that needs further exploration as do the ephemeral swamps along some of the rivers in the park and along the edge of Lake Victoria. Both habitats have species restricted to them and some of these species are endemic.

In general we find that environmental variability has marked influences on the biodiversity of butterflies in the Serengeti ecosystem. Temporal fluctuations, both seasonally and interannually, affect their abundance. Habitat heterogeneity enhances species richness considerably, but disturbances both natural and human-induced reduce this richness; grazing by wildebeest, burning, and agriculture all reduce richness. In particular, habitat modification through agriculture results in the loss of most of the restricted range species. Conservationists, therefore, should be aware that the diversity of Serengeti is determined by the many different habitats that occur within it, and that too much disturbance could reduce such diversity.

ACKNOWLEDGMENTS

Daniel Rosengren introduced us to and translated from Swedish the information from Göran Sjöberg of Sweden who went to great lengths to identify the difficult specimens for us. We thank Diane Srivastava and Greg Crutsinger for help with the background to this chapter. John Bukombe, John Mchetto, Anne Sinclair, Evelyn Turkington, Stephen Makacha, and Simon Mduma helped with collecting and counting.

REFERENCES

Akite, P. 2008. Effects of anthropogenic disturbances on the diversity and composition of the butterfly fauna of sites in the Sango Bay and Iriiri areas, Uganda: Implications for conservation. *African Journal of Ecology* 46 (Suppl. 1): 3–13.

Benedick, S., J. K. Hill, N. Mustaffa, V. K. Chey, M. Maryati, J. B. Searle, M. Schilthuizen, and K. C. Hamer. 2006. Impacts of rain forest fragmentation on butterflies in northern Borneo: Species richness, turnover and the value of small fragments. *Journal of Applied Ecology* 43:967–77.

Connell, J. H. 1978. Diversity in tropical rainforests and coral reefs. *Science* 199:1302–10.

Daily, G. C., G. Ceballos, J. Pacheco, G. Suzan, and A. Sanchez-Azofeifa. 2003. Countryside biogeography of Neotropical mammals: Conservation opportunities in agricultural landscapes of Costa Rica. *Conservation Biology* 17:1814–26.

Du Toit, J. T., K. H. Rogers, and H. C. Biggs, eds. 2003. *The Kruger experience*. Washington, DC: Island Press.

Fitzherbert, E., T. Gardner, T. R. B. Davenport, and T. Caro. 2006. Butterfly species richness and abundance in the Katavi ecosystem of western Tanzania. *African Journal of Ecology* 44:353–62.

Hamer, K. C., and J. K. Hill. 2000. Scale-dependent effects of habitat disturbance on species richness in tropical forests. *Conservation Biology* 14:1435–40.

Hamer, K. C., J. K. Hill, L. A. Lace, and A. M. Langan. 1997. Ecological and biogeographical effects of forest disturbance on tropical butterflies of Sumba, Indonesia. *Journal of Biogeography* 24:67–75.

Harvey, C. A., O. Komar, R. Chazdon, B. G. Ferguson, B. Finegan, D. M. Griffith, M. Martinez-Ramos, H. Morales, R. Nigh, and L. Soto-Pinto, et al. 2008. Integrating agricultural landscapes with biodiversity conservation in the Mesoamerican hotspot. *Conservation Biology* 22:8–15.

Hill, J. K., and K. C. Hamer. 2004. Determining impacts of habitat modification on diversity of tropical forest fauna: The importance of spatial scale. *Journal of Applied Ecology* 41:744–54.

Hill, J. K., M. A. Gray, V. K. Chey, S. Benedick, N. Tawatao, and K. C. Hamer. 2011. Ecological impacts of tropical forest fragmentation: How consistent are patterns in species richness and nestedness? *Philosophical Transactions of the Royal Society B* 366:3265–76.

Johnson, E. A., and K. Miyanishi. 2008. Testing the assumptions of chronosequences in succession. *Ecology Letters* 11:419–31.

Krebs, C. J. 1989. *Ecological Methodology*. New York: Harper & Row.

Larsen, T. B. 1988. A butterfly migration in Kenya, May 1987. *Atalanta* 18:291–92.

Larsen, T. B. 1992. Migration of *Catopsilia florella* in Botswana (Lepidoptera: Pieridae). *Tropical Lepidoptera* 3:2–11.

Larsen, T. B. 1996. *The butterflies of Kenya and their natural history*, 2nd ed. Oxford: Oxford University Press.

Lawton, J. H., D. E. Bignell, B. Bolton, G. F. Bloemers, P. Eggleton, P. M. Hammond, M. Hodda, R. D. Holt, T. B. Larsenk, and N. A. Mawdsley, et al. 1998. Biodiversity inventories, indicator taxa and effects of habitat modification in tropical forest. *Nature* 391:72–75.

Mayfield, M. M., and G. C. Daily. 2005. Countryside biogeography of Neotropical herbaceous and shrubby plants. *Ecological Applications* 15:423–39.

Nkwabi, A. K., A. R. E. Sinclair, K. L. Metzger, and S. A. R. Mduma. 2011. Disturbance, ecosystem function and compensation: The influence of wildfire and grazing on the avian community in the Serengeti ecosystem, Tanzania. *Austral Ecology* 36:403–12.

Prach, K., and L. R. Walker. 2011. Four opportunities for studies of ecological succession. *Trends in Ecology and Evolution* 26:119–23.

Schulze, C. H., S. Schneeweihs, and K. Fiedler. 2010. The potential of land-use systems for maintaining tropical forest butterfly diversity. In *Tropical rainforests and agroforests under global change*, ed. T. Tscharntke, C. Leuschner, E. Veldkamp, H. Faust, E. Guhardja, and A. Bidin, 7–96. Berlin: Springer-Verlag.

Sharam, G. J. 2005. *The decline and restoration of riparian and hilltop forests in Serengeti National Park, Tanzania.* PhD diss., University of British Columbia, Vancouver.

Sharam, G., A. R. E. Sinclair, and R. Turkington. 2006. Establishment of broad-leaved thickets in Serengeti, Tanzania: The influence of fire, browsers, grass competition, and elephants. *Biotropica* 38:599–605.

Sharam, G., A. R. E. Sinclair, R. Turkington, and A. L. Jacob. 2009. The savanna tree *Acacia polyacantha* facilitates the establishment of riparian forests in Serengeti National Park, Tanzania. *Journal of Tropical Ecology* 25:31–40.

Sinclair, A. R. E. 1977. *The African buffalo. A study of resource limitation of populations.* Chicago: University of Chicago Press.

Sinclair, A. R. E., S. A. R. Mduma, and P. Arcese. 2002. Protected areas as biodiversity benchmarks for human impacts: Agriculture and the Serengeti avifauna. *Proceedings of the Royal Society, London B* 269:2401–5.

Stork, N. E., D. S. Srivastava, A. D. Watt, and T. B. Larsen. 2003. Butterfly diversity and silvicultural practice in lowland rainforest of Cameroon. *Biodiversity and Conservation* 12:387–410.

Walker, L. R. 2012. *The biology of disturbed habitats.* Oxford: Oxford University Press.

Williams, C. B. 1958. *Insect migration.* London: Collins.

TWELVE

Small Mammal Diversity and Population Dynamics in the Greater Serengeti Ecosystem

Andrea E. Byrom, Wendy A. Ruscoe, Ally K. Nkwabi, Kristine L. Metzger, Guy J. Forrester, Meggan E. Craft, Sarah M. Durant, Stephen Makacha, John Bukombe, John Mchetto, Simon A. R. Mduma, Denne N. Reed, Katie Hampson, and Anthony R. E. Sinclair

Relatively little attention has been focused on the natural history, diversity, and dynamics of small mammals in Serengeti National Park, compared to the larger herbivores and predators. Since the 1960s, small mammal abundance has been recorded occasionally by various researchers in field notes, as has that of birds of prey and the smaller carnivore species. Two early trapping surveys of small mammals were done in the park (Swynnerton 1958; Misonne and Verschuren 1966), and three recent surveys: one comparing agriculture with native habitats (Magige 2013), one at the park boundary (Magige and Senzota 2006), and one in kopjes (Timbuka and Kabigumila 2006). In addition, small mammal remains have been recorded in barn owl (*Tyto alba*) and spotted eagle owl (*Bubo africanus*) regurgitated pellets (Laurie 1971; Reed 2007). Anecdotal evidence suggested that the population dynamics of some birds of prey, such as black-shouldered kites (*Elanus caeruleus*; a resident species of the Serengeti) were linked to small mammal abundance. However, the link between small mammal dynamics and those of their predators had not been explored for the greater Serengeti ecosystem.

In agricultural areas throughout Africa, small mammals, particularly the multimammate rat *Mastomys*, show multiannual fluctuations (large peaks in abundance occurring at intervals of 3–5 years; Crespin et al. [2008]; Leirs et al. [1996], [1997]; Monadjem and Perrin [2003]). Early unpublished field notes and the few published studies mentioned above suggested that outbreaks of small mammals were not confined to agricultural areas of East Africa, but also occurred in unmodified ecosystems inside the park. In 1999, as part of the Serengeti Biodiversity Program, a regime of trapping was instigated with the aim of understanding the spatial and temporal diversity and dynamics of small mammal communities in Serengeti National Park. In 2004, the program was extended to include cultivated agricultural areas and villages northwest of the park.

In this chapter, we combine data on small mammals and their predators occasionally collected from the 1960s to the 1990s with more intensive and systematic trapping data from 2000 to 2010. The chapter is split into two parts: in part 1 we collate informal records of mammals by habitat type, and determine species diversity in a range of rare and common habitats throughout the park. In part 2 we quantify fluctuations in rodent abundance from 1968 to 2010, assess the role of rainfall as a driver of rodent dynamics, and investigate ecological relationships between rodents and their predators. We discuss the implications of our findings for trophic dynamics in the Serengeti ecosystem, as well as the practical implications for human populations living in agricultural areas near the park boundary. Our definition of "small mammals" includes rodents, shrews, lagomorphs, hyraxes, and lemurs in the descriptive sections of part 1. For the diversity analyses in part 1, and for analyses of abundance and dynamics in part 2, we restrict our focus to rodents and shrews.

PART 1: SPATIAL DISTRIBUTION AND SPECIES DIVERSITY OF SMALL MAMMALS BY HABITAT

Records of Small Mammals by Habitat

There are many ways one could partition Serengeti vegetation into different habitats, but from the point of view of small mammals we identified 10 (fig. 12.1). In this section we describe these 10 habitat types and the small mammal species that were recorded there. Data were collated from published records, the unpublished live-trap records of A. R. E. Sinclair (1972–1973) and D. Reed (1999), and monitoring from the Serengeti Biodiversity Program, 2000–2010. Published records were obtained from Misonne and Verschuren (1966), Senzota (1982, 1984), Packer (1983), Magige and Senzota (2006), and Reed (2007). We summed the number of observations recorded in each habitat type as the best measure of the frequency of occurrence of each species, since methods of capture or observation varied slightly between studies. It was not possible to account for variation in sampling effort among habitat types. The species are listed by habitat in table 12.1, with the number of records and frequency in the sample of all species collected.

The most extensive survey of the ecosystem was conducted some 50

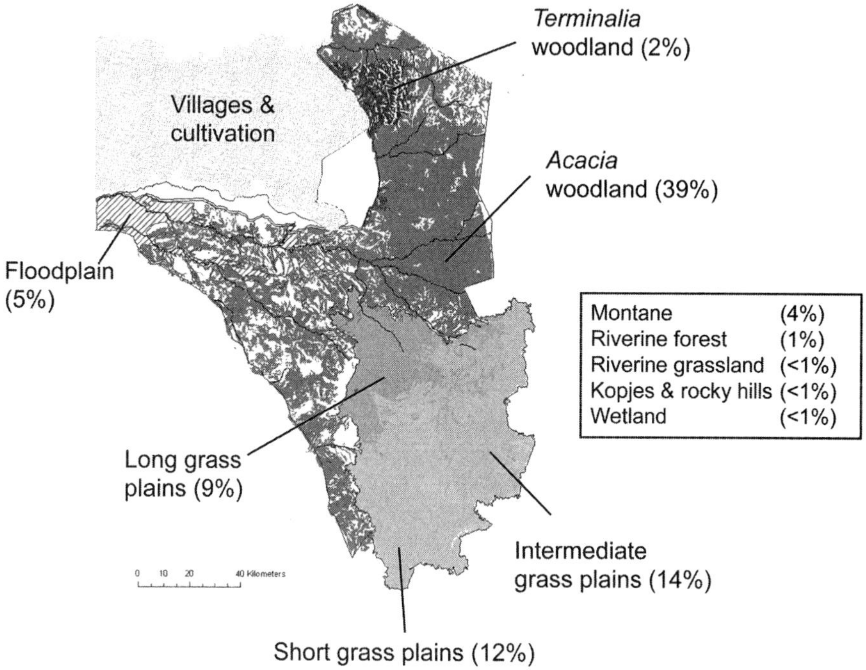

Fig. 12.1 Map of the greater Serengeti ecosystem showing 12 habitat types, 10 of which were surveyed for rodents between 1962 and 2010 (see text for description of habitats). Percentage figures indicate percentage of total area covered by each habitat type; habitats listed in the text box are found throughout the park.

years ago by J. Verschuren, who with X. Misonne, an internationally experienced taxonomist, sampled in 1962–1963 from the Loita plains in Kenya to the south of Maswa, and the eastern short-grass plains to the western Ndabaka floodplains (Misonne and Verschuren 1966). Their total catch was not recorded, and only specimens kept for taxonomic examination were published (consequently, only the taxonomic specimens they recorded are included in table 12.1). Subsequent studies focused on populations at a few sites rather than extensive surveys, the most frequently monitored site being the *Acacia* woodlands at the Serengeti Wildlife Research Centre near Seronera.

Just two species, *Arvicanthis niloticus* and *Mastomys natalensis*, accounted for 60% of the 6,000 or so records (table 12.1). However, species richness was very high, with another 35 species of small rodents or shrews recorded, and of these, 13 species account for another 37% of the remaining records. Thus, some 20 species are infrequent or rare in the greater Serengeti ecosystem. Misonne and Verschuren (1966) recorded some rare and unexpected

Table 12.1 Frequency of occurrence (number of observations) of 37 small mammal species in 10 habitat types in the greater Serengeti ecosystem, from most to least common (1962–2010)

Latin name	Common name	*Acacia* woodland	*Terminalia* woodland	Flood-plain	Riverine forest	Riverine grassland	Long-grass plains	Short-grass plains	Kopje and rocky hills	Villages and cultivation	Montane	Total	%
Arvicanthis niloticus	Grass rat	1,394	30	55	1	2	88	1	74	368		2013	31.43
Mastomys natalensis	Multimammate rat	465	361	419	5	18	35	4	3	556		1866	29.14
Crocidura	Shrew	328	28	20		10	76	73		8		543	8.48
Steatomys parvus	Fat mouse	174	77				63	93		1		408	6.37
Gerbilliscus robusta	Large naked-soled gerbil	115	32	6			11	76	1	152		393	6.14
Mus (Nannomys) musculoides	Mouse	155	20	4	4	3	50	6		3	1	246	3.84
Lemniscomys barbarus	Striped grass mouse	33	70		7	1	9	4		27		151	2.36
Dendromus melanotis	Tree mouse	99	5				21	17		1		143	2.23
Aethomys kaiseri	Bush rat	2	74	3	5	4	0	1	8	14		111	1.73
Gerbillus	Pygmy gerbil	22					44	29				95	1.48
Saccostomus mearnsi	African pouched rat	60	3				7		1	10		81	1.26
Acomys	Spiny mouse		7			3			55	1		66	1.03
Graphiurus murinus	African dormouse	8	29		1	13			4		4	59	0.92
Thallomys paedulcus	Acacia rat	53										53	0.83
Dasymys incomptus	Shaggy swamp rat	7	6	6	3					16		38	0.59
Zelotomys hildegardeae	Broad-headed mouse	22	4			1	3			2		32	0.50
Pelomys fallax	Groove-toothed creek rat	0	8	2	8	2				3		23	0.36
Suncus	Shrew	13										13	0.20

Myomys yemeni	African groove-toothed rat		3	1						5		9	0.14
Praomys	Soft-furred rat	3	1						4			8	0.12
Grammomys	Thicket rat								8			8	0.12
Otomys angoniensis	African swamp rat	3	1	3								7	0.11
Mastomys pernanus	Multimammate rat	3	1		1				1			6	0.09
Thamnomys	Thicket rat		6									6	0.09
Elephantulus rufescens	Long-eared elephant shrew	5										5	0.08
Lemniscomys striatus	Striped grass mouse				3						2	5	0.08
Crocidura 2	Shrew	1				2						3	0.05
Grammomys dolichurus	Thicket rat				1						1	2	0.03
Tachyoryctes daemon	African mole rat						1	1				2	0.03
Aethomys nyikae	Bush rat					2						2	0.03
Rattus rattus	Black rat	1										1	0.02
Mus (Nannomys) tenellus	Mouse										1	1	0.02
Pedetes capensis	Spring hare							1				1	0.02
Gerbillus pusillus	Pygmy gerbil						1					1	0.02
Oenomys hypoxanthus	Rufous-nosed rat					1						1	0.02
Praomys jacksoni	Soft-furred rat				1							1	0.02
Lophuromys sikapusi	Brush-furred mouse		1									1	0.02
Total observations		2,966	767	519	40	62	409	306	159	1,167	9	6404	å 100

Sources: Includes published records from Misonne and Verschuren (1966), Senzota (1982, 1984), Packer (1983), and Magige and Senzota (2006), and unpublished data from A. R. E. Sinclair and D. Reed.

Note: Variation in sampling effort among habitat types could not be taken into account when compiling this table, so the total number of observations is only indicative of relative effort.

species: *Mastomys pernanus*, two species of *Aethomys*, and the African mole rat *Tachyoryctes daemon*, which was somewhat out of its normal montane habitat range. It is worth noting that many species of rare small mammals were likely to be recorded only during years when *Arvicanthus* and *Mastomys* were in outbreak mode.

Habitat Types Surveyed

Figure 12.1 illustrates the following habitat types that were surveyed: *Acacia woodland* (6425.4 km^2; 39% of total area inside Serengeti National Park). Much of the sampling effort to date has been expended in *Acacia* woodland, and this habitat is the most extensive in the ecosystem. There are many *Acacia* species that make up this habitat (Sinclair 1979) and most of them grow in more or less monospecific stands, one of the unusual aspects of the Serengeti compared with other parts of Africa. The different *Acacia* species are indicators representing several different communities of shrubs, herbaceous dicots, and monocots (see chapter 5 for a more complete description). Thus, the diversity of subhabitats within the larger *Acacia* woodland habitat must contribute to the high diversity of small mammals—some 21 species.

By far the most abundant species in *Acacia* habitat was the grass rat, *Arvicanthis niloticus*, with almost 30% of all records. It lives in termite hills as well as kopjes, and is most active during the day. It shows periodic outbreaks in abundance simultaneously with the multimammate rat *Mastomys natalensis*, which made up 16% of the records. However, these proportions are somewhat misleading because *Arvicanthis niloticus* was far more abundant in the 1970s than in the first decade of the twenty-first century. In the latter decade *Mastomys natalensis* was the dominant species. Third most abundant were shrews (*Crocidura* sp.) at 11%. This habitat also supported an abundance of *Steatomys* (fat mice), omnivores that occur in a broad range of vegetation types; *Gerbilliscus*, large gerbils that tunnel and feed on roots and tubers; *Dendromus* (tree mice), which prefer trees in moist grassland; *Saccostomus* (pouched rats) that are terrestrial granivores often living in termitaria; and *Mus*, probably several species in the African subgenus *Nannomys*. The acacia tree rat (*Thallomys paedulcus*), which makes its nest in *Acacia* trees and feeds on tree leaves, berries, and buds, was only recorded in the *Acacia* habitat. Although no specific surveys took place for mammals such as the Crawshays hare (*Lepus crawshayi*) and the lesser bush baby (*Galago senegalensis*), they were commonly observed in this habitat.

Terminalia woodland (339.8 km^2; 2% of total area). This vegetation type is a version of Miombo woodland, which is much more extensive in southern

Africa. It is characterized by the tall trees *Terminalia mollis* and *Combretum molle*, and several species of shrubs such as *Rhus* and *Heeria*. In general, this woodland is described as a broad-leaved form, in contrast to the fine-leaved acacias. A suite of grasses associated with these woody species are found on the granitic soils of the northwest Serengeti.

Despite the much smaller area of this habitat, it supported the same species richness of small mammals (21) as *Acacia* woodland. However, in *Terminalia* woodland the small mammal community was quite different: *Mastomys natalensis* took over from *Arvicanthis niloticus* (which was almost absent in this habitat) representing 47% of the records, and three others (*Steatomys parvus*, *Aethomys kaiseri*, and the striped mouse *Lemniscomys barbarus*) appeared in moderate numbers. In particular *Aethomys kaiseri* (bush rats) seemed to prefer this habitat, being very rare in *Acacia* woodlands. Bush rats live among rocks and termitaria and are partly arboreal granivores. Four of the rarer species recorded in *Terminalia* habitat (*Acomys* [the spiny mouse], *Myomys yemeni* [African groove-toothed rat], *Thamnomys* [the thicket rat], and *Lophuromus sikapusi* [the brush-furred mouse]) were not recorded in the *Acacia* habitat despite the much greater trapping effort in the latter.

Floodplain (898.3 km^2; 5% of total area). Speke Bay is a 50 km extension of Lake Victoria eastward. It is about 30 km wide and very shallow (about 20 m in depth). During dry episodes of prehistory the lake dried up, and the Serengeti would have extended further to the west by 50 km or more. During wet periods the lake was several meters deeper, and Speke Bay extended east some 35 km to the Nyakaromo hills near Kirawira. Today this area is flat, with a series of parallel sandy ridges that were ancient lake shores. In between the ridges lie silt- and clay-impeded drainage soils, which in the wet season become flooded for several weeks. We recognize this far-western area as a specialized habitat, the Ndabaka floodplain. Also on these plains are extensive *Acacia seyal* woodlands, a type of gall-acacia that can tolerate prolonged inundation.

These floodplains were dominated by *Mastomys natalensis* (80% of records). This species thrives in moist grassland, which may act as a refuge habitat during periods when the species is at extremely low abundance throughout the Serengeti. In contrast, *Arvicanthis niloticus* comprised only 10% of the records in floodplain habitat, and appeared to prefer better-drained grassland. Shrews were relatively abundant, and certain rarer flooded grassland rodents—*Dasymys*, *Pelomys*, and *Otomys*—were also present. Although species richness (10) was low in this unusual habitat, it was important for the more specialized species.

Riverine forest (212.2 km^2; 1% of total area). These forests occur along the major rivers of the Serengeti ecosystem, with fragments away from the Mara

River on ridges and hillsides in the north, and they are described at length in chapter 9. Here we document the small mammals found only in the Mara forests; little is known of the western rivers.

This habitat had a different group of species from that of the savanna, with moderate species richness (12). The creek rat (*Pelomys*) was the most common species (20% of records). Two species of striped mouse were found at greater frequency (26%) in the northern forests than in other habitats. Also recorded were *Mastomys natalensis* and *Aethomys kaiseri*, and *Dasymys sp.*, which prefer moist and dense habitats. In addition, three rare species were recorded in the riverine forest habitat—*Mastomys pernanus, Grammomys dolichurus* (thicket rat), and *Praomys jacksoni* (Jackson's soft-furred mouse)—which lives in forests and around rocks, and is arboreal.

The larger trees in the montane forests of the Mara River and its tributaries provide habitat for the tree hyrax (*Dendrohyrax*). In contrast, in the western lowland forests tree hyraxes are absent but greater galagos (*Otolemur crassicaudatus*) were observed to be present.

Riverine grassland (41.0 km^2; <1% of total area). This habitat is characterized by tall (2 m) coarse grasses found commonly as narrow strips along drainage channels and rivers throughout the woodlands. The habitat occurs in a variety of forms: one type occurs as temporary drainage lines at the base of the soil catena (chapter 3), usually 10–50 m wide; another form occurs adjacent to the riverine forest where the river floods in the wet season and thus can be as wide as 200 m, although is usually about 50 m in width. The most palatable grasses in this habitat are *Panicum maximum* and *Setaria sphacelata*, while *Imperata* sp. (elephant grass) is woody and bamboo-like.

Some 13 species of small mammals were recorded in riverine grasslands. They appeared to be another of the refuges for *Mastomys natalensis*, which was recorded in this habitat during the low phase between outbreaks. It is therefore possible that riverine grasslands act as a "source" for *M. natalensis* to repopulate other habitat types during outbreaks. Shrews were also recorded in this habitat and the nocturnal dormouse (*Graphiurus murinus*) was moderately common, probably because of the availability of large trees with hollows where dormice can den during the day. Two rare species of rodent were found in this habitat: *Aethomys nyikae* (Nyika rock rat) and *Oenomys hypoxanthus*. Both feed on a variety of seeds and green vegetation in moist areas.

Long-grass plains (1521.5 km^2; 9% of total area). Although the treeless plains can be divided into several different vegetation communities depending on the degree of resolution, in this chapter we need describe only two. The area northwest of the Serengeti plains has a similar grass layer to that of open grasslands within the *Acacia* woodlands. The dominant tall grass

(1 m) is *Themeda triandra*, with many other subsidiary species such as *Pennisetum mezianum* and *Sporobolus pyramidalis*. The plains differ from *Acacia* woodlands in the diversity of woody shrubs and forbs. The long-grass plains are demarcated from the woodlands by an abrupt boundary—to one side lie trees and shrubs, whereas on the other side, barely 10 m away, there are none. Forbs are much more abundant on the woodland side, these being associated with microhabitats under trees, or even sites from which trees have long ago died and disappeared.

The small mammal community was similar to that of the *Acacia* woodland although much less diverse (13 species). *Arvicanthis niloticus* was most common in this well-drained habitat, but shrews and fat mice (*Steatomys*) were also abundant and *Mus, Gerbillus, Gerbilliscus*, and *Dendromus melanotis* were found in moderate numbers. This last species, despite its name, does not necessarily require trees. Other common species of the woodlands were recorded in the long-grass habitat, but only in low numbers. The insectivorous *Zelotomys* was recorded here. In similar long-grass plains west of Nyamalumbwa, on the Kenya border, the African mole rat (*Tachyoryctes*) was recorded by Misonne and Verschuren (1966).

Short-grass plains (1981.5 km^2; 12% of total area). The short-grass plains occur only on the eastern extremity of the treeless Serengeti Plains. In the north they are adjacent to the long-grass plains, but the plant species complex differs radically from the long-grass habitat. First, there are many fewer grass species and they cover a smaller basal area. The dominant grasses, most of them less than 10 cm high, are *Sporobolus spicatus* and *Andropogon greenwayi*, with the sedge *Kyllinga* also common. Overall, grasses cover only 40% of the ground area and 20% is bare ground, leaving 40% for a speciose group of forbs. All are prostrate to mitigate the heavy impacts of grazing and trampling by large herbivores. In the south, the eastern boundary is more diffuse, being separated from long grass by a variety of specialized herbaceous and woody habitats that lie along Olduvai Gorge.

Twelve small mammal species were observed in short-grass habitats. A common feature of the well-drained short-grass sites on the northern plains were colonies of spring hares (*Pedetes capensis*). They live in burrows, digging entrances in clusters. The Cape hare (*Lepus capenis*) was also recorded on the plains in both long- and short-grass zones. This was also the main habitat for the gerbil species *Gerbilliscus robusta* and *Gerbillus*. As for the long-grass areas *Steatomys parvus* and *Dendromus melanotis*, along with shrews, were the most common species. Unexpectedly, the African mole rat (*Tachyoryctes*) was recorded by Misonne and Verschuren (1966) near Lake Masek.

Kopjes and rocky hills (7.4 km^2; <1% of total area). On both the Serengeti plains and in the *Acacia* woodland, small hillocks occur, comprised of large boulders up to 20 m high. These rock piles, known as kopjes, can vary from a few meters across to 100 m or more, especially at Moru kopjes on the western side of the long-grass plains. Shrubs and large broad-leaved trees, such as *Ficus*, grow among the rocks and form dense thickets. The steep rocky sides of the large hills in the Serengeti woodlands form a similar type of habitat.

This is a specialized habitat dominated by two cohabiting hyrax species, the rock hyrax (*Procavia*) and the bush hyrax (*Heterohyrax*). These occurred largely on the woodland kopjes or at the edge of the long-grass plains and were absent from kopjes far out on the plains. Kopjes are a specialized habitat for rodents, and relatively few species (10) were recorded here. The grass rat (*Arvicanthis niloticus*) was very common because it uses crevasses among the rocks to hide. It is possible that kopjes also act as a source for *M. natalensis* to repopulate other habitat types during outbreaks, as they were recorded in kopjes between outbreaks. Kopjes were also preferred habitat for the spiny mouse (*Acomys*), which was very common. The thicket rat (*Grammomys*), bush rat (*Aethomys*), and soft-furred rat (*Praomys*) were also often recorded.

Montane (582.5 km^2; 4% of total area). Misonne and Verschuren (1966) trapped on the sides of Kuka Mountain (near Klein's Camp, northeast of the park) in a patch of montane forest. The few species they recorded have all been found in montane forests along the Mara River (see riverine forest, above). On Kuka itself, at about 2000 m, several red rock hares (*Pronolagus rupestris*) were observed in 2005 (A. R. E. Sinclair, pers. obs.). Their extent throughout the rest of the Serengeti is unknown. The dormouse, spiny mouse, and *Mastomys natalensis* also occured on rocky hillsides.

Villages and cultivation. Cultivation around the western sides of the Serengeti ecosystem is usually in the form of maize, millet, and cassava fields with some rice fields in the far west near Lake Victoria, and small-scale fruit and vegetable farming. The fields are small, usually less than one hectare and bordered by dense sisal (*Agave sisalana*) hedgerows. Houses are built close to the fields and grain storage units are near the houses.

The small mammal community in villages was moderately species rich; less so than in *Acacia* woodland, but nevertheless 16 species were recorded during a relatively short period of trapping (2004–2010). Three species, *Myomys, Pelomys*, and *Acomys*, were recorded in villages similar to *Terminalia* woodland but dissimilar to *Acacia* woodland. In general, the two most common species of the natural woodlands are also the dominant species in cultivated areas, with *Gerbilliscus* also common. In addition, the exotic black rat (*Rattus rattus*) was found in cultivated habitats but almost never recorded in native habitats (Magige and Senzota 2006).

Small Mammal Diversity by Habitat

Prior to our systematic study, there was relatively little information on the diversity of small mammals in relation to the variety of habitat types found throughout Serengeti National Park (Reed 2007). Species richness, Shannon's H′, the exponential form of Shannon's H′ (N_1), and evenness (Krebs 1999; Magurran 2004) were used to describe the diversity of small mammals across the 10 habitat types surveyed in the greater Serengeti ecosystem. Species richness was higher than previously recorded: we described 37 species of small mammals in the greater Serengeti ecosystem, nine more than the 23 documented in owl pellets by Reed (2007), and our study was probably an underestimate because some genera (e.g., *Mus*) likely contained more than one species that were not distinguishable morphologically. Highest H′ and N_1 diversity was recorded in the forest, riverine, and long-grass habitat types, and lowest in the floodplains and in cultivated areas. All diversity measures highlighted several important habitat types for some of the rarer small mammals in the system, particularly montane, long grass, riverine, and forest habitat types.

Variability in community composition among habitat types was measured using the Bray-Curtis index of similarity (Bray and Curtis 1957; table 12.2). This measure highlighted the importance of several of the smaller-scale habitat patches in the ecosystem, particularly the kopjes, forest, and riverine areas, which shared few small mammal species with more extensive habitat types such as the *Acacia* and *Terminalia* woodlands. The Bray-Curtis measure also indicated that, whereas villages and cultivated areas shared a number of small mammal species with *Acacia* and *Terminalia* woodland, they had almost no species in common with these smaller, specialized habitat types inside the park (table 12.2). Thus, specialized habitat types comprising a relatively small proportion of the overall area inside the park are vital in maintaining high diversity of small mammals in the greater Serengeti ecosystem.

PART 2: FLUCTUATIONS IN SMALL MAMMAL ABUNDANCE, 1968–2010

Methods of Quantifying Small Mammal Abundance

Methods varied little among historical records (collected 1968–1999), in the Serengeti Biodiversity Program (1999–2010), and in studies published from 1966 to the present. In most cases, small mammals were captured in Sherman live traps on grids with trap spacing varying from 10 to 50 m, for

Table 12.2 Diversity of small mammals in 10 habitat types in the greater Serengeti ecosystem, 1962–2010

Diversity	*Acacia* **woodland**	*Terminalia* **woodland**	**Flood-plain**	**Riverine forest**	**Riverine grassland**	**Long-grass plains**	**Short-grass plains**	**Kopje and rocky hills**	**Villages and cultivation**	**Montane**	**All habitats**
Richness (S)	21	21	10	12	13	13	12	10	15	5	37
Shannon's H′	1.82	1.93	0.77	2.23	2.09	2.11	1.70	1.38	1.34	1.43	2.06
N_1 (= $e^{H'}$)	6.17	6.89	2.16	9.30	8.08	8.25	5.47	3.97	3.82	4.18	7.85
H evenness	0.60	0.63	0.33	0.90	0.82	0.82	0.68	0.60	0.49	0.89	0.57
Bray-Curtis similarity index											
Acacia woodland	1.00										
Terminalia woodland	0.33	1.00									
Floodplain	0.30	0.68	1.00								
Riverine forest	0.02	0.09	0.06	1.00							
Riverine grassland	0.03	0.14	0.13	0.33	1.00						
Long-grass plains	0.23	0.35	0.26	0.08	0.15	1.00					
Short-grass plains	0.18	0.30	0.09	0.08	0.11	0.59	1.00				
Kopjes and rocky hills	0.06	0.12	0.18	0.11	0.15	0.28	0.03	1.00			
Villages and cultivation	0.49	0.51	0.60	0.05	0.07	0.21	0.13	0.13	1.00		
Montane	0.00	0.01	0.00	0.20	0.14	0.01	0.01	0.05	0.00	1.00	

Note: Data sources as in table 12.1.

periods between 3 and 10 nights. Bait types included peanut butter, small dried fish, rolled oats, roasted coconut, rice, or maize flour. Depending on the study, traps were checked morning and evening to catch both diurnal and nocturnal species. Animals were identified in the field, their sex and breeding status determined, and in some studies body mass and head-body length measured. In some cases animals were marked using ear tags, ear clipping, or fur clipping. Beginning in 2000, tissue samples were occasionally taken for species identification by DNA analysis. Also, shrews (*Suncus, Elephantulus,* and *Crocidura* spp.) were systematically recorded both in live traps and pitfall traps in addition to rodents.

When quantifying the dynamics of rodents and shrews, capture data were converted to trap catch (proportion of traps containing an animal, corrected for nonfunctional traps). In most years from 1968 to 1998, trapping occurred annually. Since 2000, trapping was carried out at least three times per year in several habitat types (see below). Trap catches of rodents and shrews were analyzed separately, but species were combined within these two groupings. For many of the rodent species, there were too few data (both spatial and temporal) to analyze species-specific rates of increase or demographic rates separately. This approach is justified because (a) we expected similar mechanisms to be driving the dynamics of many of the species, (b) we expected changes in abundance during outbreaks to dwarf all other population changes, and (c) our aim was to infer general spatial and temporal patterns in one functional group (small mammals) in the greater Serengeti ecosystem rather than to make specific inferences about individual species. Such approaches have been used elsewhere (e.g., Jacsic et al. 1992; Krebs et al. 2003).

Rodents

In order to quantify fluctuations in rodent abundance inside Serengeti National Park, a natural ecosystem, available data on historical abundance were collated from Senzota (1978, 1982, 1984), Packer (1983), and unpublished data from D. Reed (1999) and A. R. E. Sinclair (1972–1973). These data suggest at least seven major peaks in abundance from 1968 to 1999 (1968, 1977, 1978, 1987, 1988, 1992, and 1998). There were either no records for intervening years, or those years were recorded in field notes as having no or few rodents present. No data were available for the "border closure" period from 1978 to 1985 (Sinclair 1995). The magnitude of the peaks varied from 47% trap catch in 1972 (quantified by Senzota 1982) to "very high rodent abundance" from field notes (all peaks except 1998; assigned 80% trap catch) to "plague" (1998; assigned 100% trap catch).

Systematic quantitative data on small mammal abundance were collected from 2000. Three further peaks in abundance occurred in 2001 (49% trap catch), 2002 (62%), and 2007 (>80%), with very low (<5%) or zero rodent abundance in intervening years. Data from 2000 onward were allocated to three time periods in a calendar year. Period 1 (January–March) was used to estimate rodent abundance prior to the breeding season. Period 2 (April–September) was used to estimate maximum abundance during the peak breeding phase after the wet season. Period 3 (October–December) was used to estimate dry-season abundance of rodents, and to determine how long (if at all) breeding continued into the dry season. The timing of each small mammal census varied within each period, so the average trap catch of rodents in periods 1 and 3, and the highest recorded trap catch in period 2 (to capture any peaks in abundance), were used for analysis. Data from the full 40-year time series were analyzed using R (version 2.10.1).

Putting the historical field data (1968–1999) and recent data (2000–2010) together we inferred that, over a period of more than 40 years, episodic outbreaks of rodents (defined as trap catch > 50%) occurred with multiannual periodicity in Serengeti National Park. There was no evidence of regular periodicity (cycles) when this 40-year time series was analyzed for cyclicity using the (ACF) autocorrelation function in R, so for the remainder of this chapter we focus on extrinsic driver(s) of rodent dynamics and the implications for the Serengeti ecosystem.

Rodent Abundance in Different Habitats

For five of the ten habitat types described in part 1 (*Acacia* woodlands, *Terminalia* woodlands, Ndabaka floodplain, riverine forest, and riverine grassland habitats), the temporal sequence of data was detailed enough from 2000 to 2010 to quantify relative changes in rodent abundance in each habitat type. Although relative trap catch varied between habitat types with more modest peaks observed in *Terminalia* and *Acacia* woodlands, without exception peaks in rodent abundance occurred simultaneously in all five habitat types (fig. 12.2)—further evidence of an extrinsic driver of rodent abundance across a range of habitat types. In further analyses (below) of small mammal dynamics, we combine abundance data from all habitat types except kopjes, as the latter were not surveyed consistently through time.

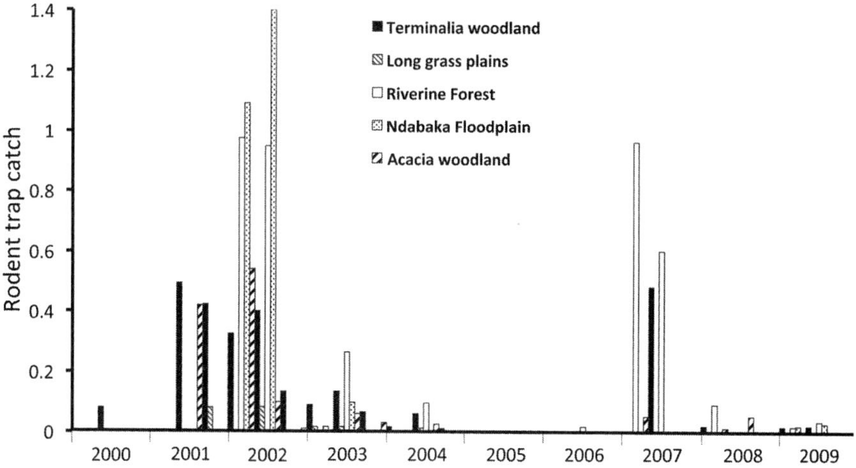

Fig. 12.2 Fluctuations in rodent abundance (proportional trap catch of live-captured rodents in Sherman traps) in five habitat types in the Serengeti, 2000–2010 (see text for description of habitats). Trap catch > 1.0 indicates occasions where two or more rodents were captured per trap.

Shrews

Quantitative data on shrew abundance were available from 1999 to 2010. Although this time series was too short to determine possible drivers of abundance (see below for rodents), shrew abundance (expressed as trap catch) showed fluctuations with moderate peaks in 2001, 2002, and 2007, and their abundance was strongly correlated with rodent abundance over the same time period (nonlinear regression $y = 0.2624*(1—e^{-(7.77*x)})$, $t_{20} = 2.14$, $r^2 = 0.49$, $P = 0.04$; fig. 12.3). Unlike rodents, shrews did not "outbreak" to very high abundance in peak years, but these data suggest that shrews respond to environmental cues (e.g., pulsed food availability) in the same way that rodents do.

WHAT DRIVES RODENT DYNAMICS IN THE SERENGETI?

The only consistently available explanatory variable for the 40-year time series of rodent outbreaks was rainfall, with monthly rainfall recorded in various parts of the park since 1938. In Africa and elsewhere, rainfall or other climatic factors have been postulated to drive system productivity and therefore food supply for rodents, both in agricultural areas (e.g., Stenseth et al. 2003) and natural ecosystems (e.g., Letnic and Dickman 2010). In

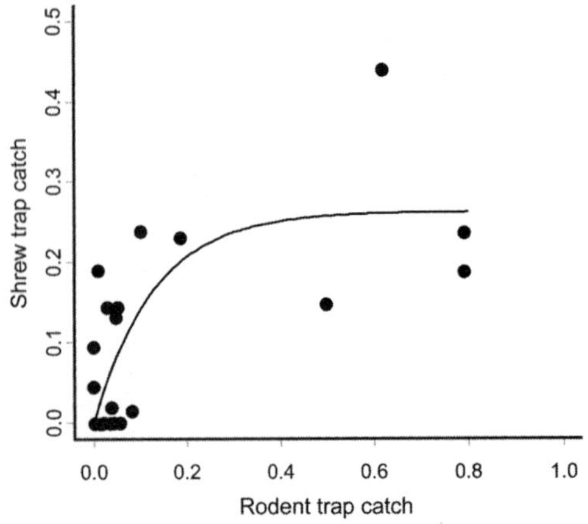

Fig. 12.3 Relationship between rodent abundance (all species) and shrew abundance (*Crocidura, Suncus,* and *Elephantulus* spp.), 1999–2010. Trap catch of shrews was derived from both pitfall and Sherman traps, both corrected for lost trap nights. Shrews = $a^*(1 - e^{-(b^*rodents)})$ where a and b are parameters of a two-parameter asymptotic nonlinear regression; a = 0.2624 ± 0.0450, b = 7.77 ± 3.6295; difference from null model $F_{1,20} = 18.84$, $P < 0.0001$.

Africa, however, this link has been made primarily for two species (*Mastomys* and *Arvicanthis*) in agricultural areas. *Mastomys* outbreaks are generally triggered by increased food availability during the dry season when rainfall exceeds a threshold amount in the immediately preceding wet season (Leirs et al. 1996, 1997; Julliard et al. 1999; Stenseth et al. 2001; Lima et al. 2003; Davis et al. 2004). For instance, Leirs et al. (1996) found that the short rains explained 69% of the variation in *Mastomys* outbreak occurrence in Morogoro (eastern Tanzania) between 1925 and 1990.

The wet season in the Serengeti lasts from October to May and is bimodal (Sinclair 1979) with short-season rains falling between October and December and long rains between March and May. Rainfall data were interpolated from weather stations in the Western Corridor and in the center of the park. We undertook a logistic regression analysis of rainfall and rodent outbreaks (1 = outbreak; 0 = no outbreak in a given year) in the Serengeti from 1968 to 2010. Shrews were not included in the analysis. There was a significant positive relationship between short-season rainfall in the Western Corridor and outbreak occurrence ($z = 2.13$, $P = 0.03$, 18% deviance explained; fig. 12.4). There was also a significant positive effect of both short-season rainfall from the center of the park and interpolated wet-season (short + long) rains on the probability of a rodent outbreak (center and long rains both $z = 2.02$, $P = 0.04$), although the predictive ability of both models (13% and 15% of

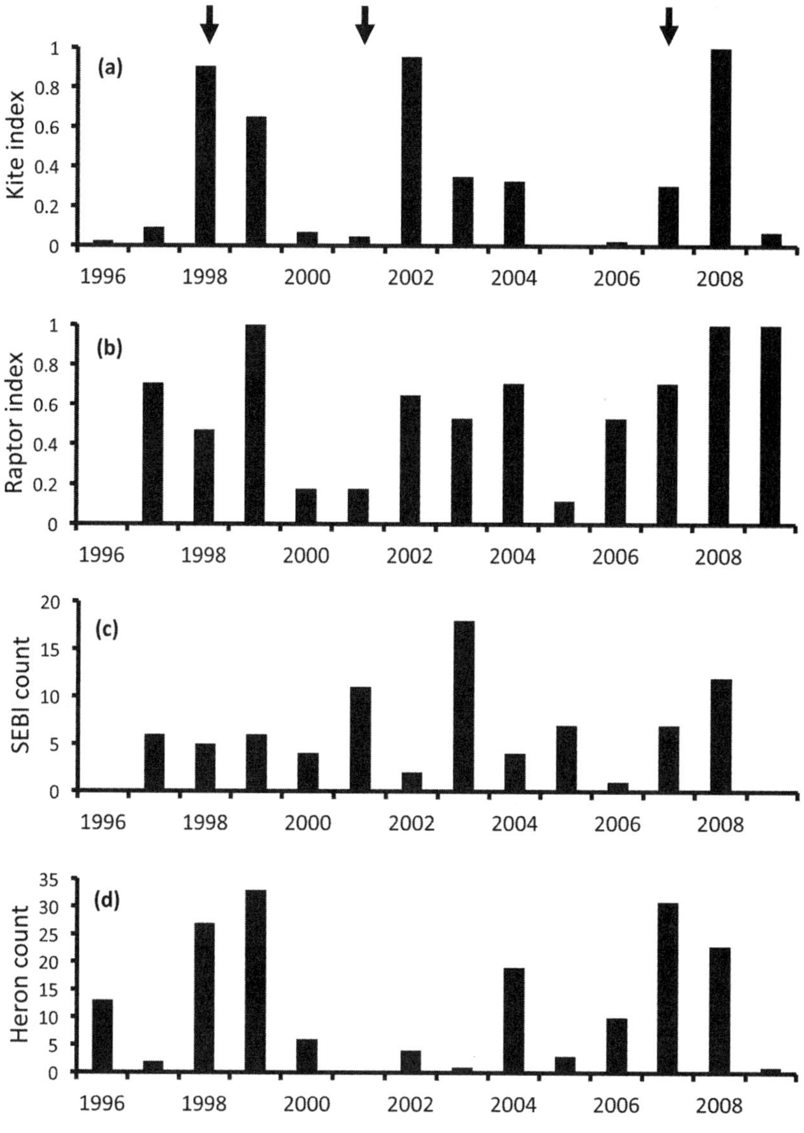

Fig. 12.4 Counts and/or indices of (a) black-shouldered kites; (b) five raptor species combined [black-chested snake eagle, brown snake eagle, long-crested hawk eagle, martial eagle, and tawny eagle]; (c) secretary birds (SEBI); and (d) black-headed herons, 1996–2010. Arrows indicate rodent outbreaks.

deviance explained respectively) was lower than the predictive ability of the model using rainfall data from the Western Corridor. We therefore concluded that rainfall is a major driver of rodent abundance in the Serengeti, a natural ecosystem, and that short rains >250 mm in western Serengeti will result in a rodent outbreak in the ecosystem.

From 1999 to 2010, more frequent trapping surveys of rodents took place, and three data points on rodent abundance were available in each calendar year. Using these recent data we also sought evidence of density-dependence in Serengeti rodent populations by examining the interaction between abundance and rainfall. Regression models were fitted to the abundance of rodents in period 2 (April–September) using rodent abundance in the previous two periods (October–December and January–March) and short rains as explanatory variables. Also, we attempted to explain rodent abundance in period 2 by (a) fitting a model with rainfall + abundance in period 1, and (b) by dropping rainfall from the model altogether and using only the number of rodents in the previous two time periods as explanatory variables. There was no significant interaction between rodent abundance in previous time periods and short rains ($t_{10} = 1.23$, $r^2 = 0.33$, $P = 0.25$), no effect when the interaction was dropped from the model ($t_{11} = 1.73$, $r^2 = 0.23$, $P = 0.11$), and no effect of rodent abundance alone in previous time periods (i.e., when rainfall was dropped from the model; $t_{14} = 0.67$, $r^2 = 0.03$, $P = 0.51$).

These models have some limitations. We were not able to determine direct drivers of rodent abundance because data on food availability such as green grass, forbs, seeds, and invertebrates were not available (although we know that rainfall influences grass growth on both the short- and long-grass plains; Sinclair 1975). In addition, the 40-year time series, while made up primarily of *Mastomys* and *Arvicanthis* outbreaks, contained a number of other rodent genera, but data on individual species were too sparse to analyze individually. Senzota (1982), Packer (1983), and unpublished field notes (A. R. E. Sinclair) all suggest that most outbreaks in the Serengeti up to 1998 were *Arvicanthis niloticus*, whereas the outbreaks in 2002 and 2007 included primarily *Mastomys natalensis*, but also *Arvicanthis niloticus* and several other genera. Elsewhere in Africa, different species have been observed to outbreak in different years; for example, Leirs et al. (1996) reported that rodent outbreaks in Morogoro were *Mastomys* in all years except for an *Arvicanthis* outbreak in 1977. Combining multiple rodent species likely substantially reduced the predictive ability of the models. Nevertheless, this is the first quantitative evidence for outbreaks of at least two known outbreak species (*Arvicanthis niloticus* and *Mastomys natalensis*) in a natural (cf. agricultural) ecosystem in Africa.

PREDATOR RESPONSES TO RODENT OUTBREAKS

Birds of Prey

Black-shouldered kites. The abundance of black-shouldered kites was recorded in field notes from 1968, and kites were counted systematically on transects inside and outside the park from 1997 (chapter 13). To analyze the response of kites to rodents, historical records of black-shouldered kite abundance inside the park (1968–1997) were standardized against the maximum number of black-shouldered kites recorded (in 1997) and scaled to 1. More recent systematic records (1998–2010) were standardized against the maximum number of kites recorded (in 2008) and also scaled to 1, resulting in a full 41-year index of kite abundance from 1968 to 2010. (Scaling the entire data set to one maximum value did not affect these analyses). Where possible, data on black-shouldered kite abundance from 1998 to 2010 were split into the same three time periods as for rodents within each calendar year in order to improve the statistical power of the model. Black-shouldered kites often peaked simultaneously with rodents, but lagged behind rodent abundance in their decline phase (fig. 12.4a). The relationship between rodent abundance and black-shouldered kite abundance was highly significant (linear regression model; $t_{31} = 4.12$, $r^2 = 0.35$, $P < 0.001$; fig. 12.5a). Despite the significant relationship, the model did not fit the data particularly well, so we also analyzed the relationship between kites and rodents using a nonparametric Spearman rank correlation. The

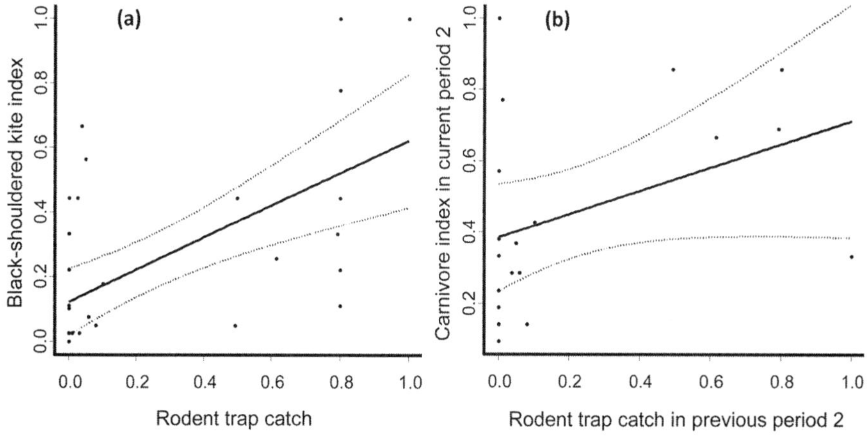

Fig. 12.5 Relationship between rodent abundance and (a) black-shouldered kite index in the same year, 1968–2010; and (b) small- to medium-sized carnivore index in subsequent year, 1993–2010. Dotted lines are 95% confidence intervals around linear regressions. See text for explanation of periods and carnivore species used in the analysis.

result confirmed the strong response of kites to rodents ($n = 33$, $r_s = 0.67$, $P < 0.001$).

We also explored a lagged response of kites to rodent abundance. Kites are known to be nomadic and move across large distances in response to food availability, but they also show a pronounced breeding response to rodent abundance and can produce three to four broods per year in a good year. In poor years high mortality of young due to starvation has been observed, but in years of high food availability their productivity is high for a raptor (Brown, Urban, and Newman 1982). We found a positive relationship between rodent abundance in one year, and kite abundance in the following year (linear regression model, $t_{29} = 2.25$, $r^2 = 0.15$, $P = 0.03$), although the lagged relationship was weaker than the direct relationship. The weak lagged response of black-shouldered kites to rodents was confirmed using a Spearman rank correlation that was significant at the 10% level ($n = 33$, $r_s = 0.68$, $P < 0.08$). The timing of when this slightly lagged response of kites was captured in our data depended on the timing of the rodent peak after the end of the rains (sensu Leirs et al. 1996), as well as the timing of both rodent and kite counts. We infer that the response of black-shouldered kites to rodents comprises both an immediate nomadic response of adults to food availability, followed by a breeding response in the six-month period following the rodent peak. Thus rodent abundance, driven by short-season rains, both influences the distribution and drives the population dynamics of a key bird of prey in the Serengeti ecosystem.

Other rodent-eating birds. Several other birds of prey (the black-chested snake eagle [*Circaetus pectoralis*], brown snake eagle [*Circaetus cinereus*], long-crested hawk eagle [*Lophaetus occipitalis*], martial eagle [*Polemaetus bellicosus*], and tawny eagle [*Aquila rapax*]), as well as the black-headed heron (*Ardea melanocephala*), owls, and secretary bird (*Sagittarius serpentarius*) all eat rodents, and some are rodent specialists (Brown, Urban, and Newman 1982). These species were recorded on transects throughout the park from 1997 to 2010 (chapter 13). The abundance of each of these groups or species also fluctuated in response to rodent outbreaks (fig. 12.4, panels b, c, and d). When combined into a single index of abundance, the five raptor species listed above (i.e., excluding kites, owls, herons, and secretary birds) showed a one-year lagged response to rodent abundance that was marginally significant (linear regression model, $t_{11} = 1.97$, $r^2 = 0.26$, $P = 0.07$); there was also marginal evidence of a response in secretary bird abundance to rodent abundance ($t_{11} = 2.00$, $r^2 = 0.33$, $P = 0.08$), suggesting delayed breeding responses to rodent outbreaks. Overall the dynamics of several birds of prey, in addition to black-shouldered kites, appear to be driven by rodent abundance in the Serengeti. There was no statistical evidence of an immediate or lagged response of black-headed herons to rodent abundance (immediate: $t_{12} = 1.23$, $r^2 = 0.11$, $P = 0.24$; lagged: $t_{11} = 1.13$, $r^2 = 0.10$, $P = 0.28$), and there were too few owl data to analyze statistically, although they did show a peak in abundance in 2007.

Mammalian Carnivores

Data on the abundance of small- to medium-sized carnivores (defined as body mass of 1–18 kg) were available from 1993 to 2010, a period spanning the four most recent rodent outbreaks. Daytime sightings of small carnivores were collected by S. Durant from 1993 to 2006 as part of the Serengeti Cheetah Project (Durant, Bashir et al. 2007). Small carnivores were also counted on transects at night by M. Craft et al. (details in chapter 15), both inside the park and in Mugumu Village and adjacent cultivated agricultural areas to the northwest of the park, from 2003 to 2010. Data on carnivore sightings inside the park were combined to form a single time series of carnivore abundance from 1993 to 2010. Species included in the time series were black-backed jackal (*Canis mesomelas*), golden jackal (*Canis aureus*), side-striped jackal (*Canis adustus*), caracal (*Caracal caracal*), wildcat (*Felis silvestris*), serval (*Leptailurus serval*), common and large-spotted genet (*Genetta genetta* and *Genetta maculata*), honey badger (*Mellivora capensis*), and white-tailed mongoose (*Ichneumia albicauda*). Only species that (a) had a known murid preference (Kingdon 1977; Geertsma 1985; Moehlman 1986), and (b) were recorded in both data sets were included in the time series. Each data set was standardized by scaling to 1 relative to maximum counts in 1996 (S. Durant data) and 2004 (M. Craft data). In addition to the above species, the abundance of domestic dogs (*Canis lupis*) and cats (*Felis catus*) was recorded in villages and cultivated areas (K. Hampson data). Where possible, data on carnivore abundance were split into the same three time periods as for rodents within each calendar year in order to improve the statistical power of the models.

As for black-shouldered kites, abundances of small carnivores inside the park often peaked simultaneously with rodents, but lagged behind rodents in the decline phase. With the exception of the honey badger, which has a gestation period of 153–162 days (Hancox 1992), most of the small carnivore species have gestation periods of 55–70 days (Kingdon 1977), so we expected time lags of 6–12 months in the numerical response of small rodent-eating carnivores to rodent outbreaks, taking into account the breeding response and subsequent growth of juveniles to "sightable" age. Indeed, there was no evidence of an immediate response to the abundance of rodent-eating carnivores (linear regression model, $t_{11} = 0.94$, $r^2 = 0.05$, $P = 0.36$), but there was weak evidence for a 12-month time lag in the carnivore response to rodent outbreaks ($t_{11} = 1.80$, $r^2 = 0.16$, $P = 0.09$; fig. 12.5b) (there were too few data to explore a six-month time lag). Thus rodent abundance, in addition to driving the population dynamics of birds of prey, also influences the dynamics of small- to medium-sized mammalian carnivore species inside the park.

RODENT AND PREDATOR DYNAMICS IN AGRICULTURAL AREAS OUTSIDE THE PARK

Rodents were censused in two cultivated agricultural areas/villages (Bwitengi and Robanda) growing maize and millet crops northwest of the park from 2004 to 2009. Mammalian carnivores and black-shouldered kites were also counted on transects (2003–2010 and 1997–2010, respectively; chapters 15 and 13) in Bwitengi, Robanda, and Mugumu. Although peaks in rodent abundance did occur in cultivated areas (in 2004 and 2007), trap catch of rodents was more consistent both within and between years (often >10%) in village areas, compared with trap catch inside the park boundary where rodent abundance declined to zero or near-zero between outbreak years. We emphasize the relatively short length of this time series; indeed, it was too short to determine drivers of rodent (and consequently carnivore) abundance in cultivated areas. Very few wild carnivores were observed in cultivated areas, and an index of their abundance was not correlated with rodent abundance ($t_5 = 0.80$, $r^2 = 0.11$, $P = 0.46$). Too few kites were observed (eight individual kites in a ten-year period) to statistically analyze their response to rodent abundance in cultivated areas. It is possible that the absence of carnivores and raptors in cultivated areas could contribute to the overall higher abundance of rodents in nonoutbreak years. Domestic cats and dogs were present in higher numbers in cultivated areas compared with kites and wild mammalian carnivores (chapter 15), but their abundance was not correlated with the abundance of rodents ($t_5 = 0.26$, $r^2 = 0.01$, $P = 0.80$).

ROLE OF SMALL MAMMALS IN THE GREATER SERENGETI ECOSYSTEM

Small Mammal Diversity and Distribution

Although there appear to be 37 species of small mammals in the Serengeti ecosystem, the spatial distribution of these species varies widely and in some instances the occurrence of individual species was quite localized. This suggests that observed fluctuations in climate, as well as heterogeneity in topography, soil productivity, and vegetation types, are essential for maintaining the diversity of small mammals in the Serengeti. Both rainfall and topography are important contributors to the distribution of woodlands, savanna grasslands, and open grasslands in the Serengeti ecosystem (Anderson et al. 2008; Reed et al. 2009). Coupled with our finding that specialized habitat types comprising a relatively small proportion of the overall area (e.g., floodplains, riverine forests, kopjes, and montane areas) are vital in maintaining the high diversity of small mammals in the Serengeti, we hypothesize that there are likely to be both large- and small-scale and short- and long-term drivers of small mammal diversity, distribution, productivity, and abundance in the ecosystem. This hypothesis is consistent with observations of climate, topography, and vegetation affecting other rodent species in Africa (e.g., Fitzherbert et al. 2006; Krystufek, Haberl, and Baxter 2007; Meyer et al. 2007, 2008; Crespin et al. 2008; Mulungu et al. 2008), but would require more detailed long-term data to properly test. The role of predators in maintaining diversity in the Serengeti ecosystem is also unknown at the present time.

Although by no means the least diverse habitat type, villages and cultivated areas had lower Shannon's H′ and evenness measures than many of the rarer habitat types in the ecosystem. This is because they were dominated by relatively common species such as *Mastomys, Arvicanthis*, and *Gerbilliscus*, which they shared with habitats inside the park boundary, so their overall contribution to small mammal diversity in the ecosystem was relatively low. This finding is consistent with Magige and Senzota (2006), who also observed greater diversity of small mammals inside the park boundary compared with either game reserves or unprotected areas. It suggests that human activity can impact negatively on small mammal diversity and again emphasizes the vital role of rare and/or heterogeneous habitats in maintaining overall diversity in the system, which has also recently been suggested for other parts of Tanzania (Mulungu et al. 2008; Makundi et al. 2010). Magige (2013) recorded more species in agriculture than within the Serengeti, but her overall number of species was small.

Gradients in rainfall and vegetation and variation in topography and soil productivity in the ecosystem not only influence diversity directly, but also indirectly affect the multiannual population dynamics of small mammals. Habitats such as floodplains and kopjes may serve as ecological refugia during periods of low abundance, enabling rapid spatial expansion, particularly of outbreak rodents, during favorable periods. This is akin to the "donor and recipient habitat" hypothesis postulated by Singleton, Tann, and Krebs (2007) to explain outbreaks of house mice (*Mus domesticus*) in southeastern Australia. The story is likely to be more complex than this, however, with ecological refugia, temporal and spatial pulses in food resources, and predation all playing a role in small mammal diversity and dynamics; similar "state and transition" processes are thought to occur in Australian desert ecosystems (Letnic and Dickman 2010). These processes are likely strong influences in outbreak dynamics, and we would predict a strong spatial component to outbreaks of small mammals in the Serengeti ecosystem driven by a complex interplay between differing responses of food quality and quantity to rainfall, the ability of individual species to respond to changes in food supply, habitat-mediated predation risk, and other spatial factors (Perrin and Kotler 2005; Bonnet et al. 2010; Letnic and Dickman 2010).

Trophic Importance of Small Mammals

The current working hypothesis is that small mammal dynamics in the Serengeti ecosystem are driven by bottom-up processes, with greater food availability in years of higher short-season rainfall resulting in outbreak conditions, particularly for species such as *Arvicanthis* and *Mastomys*. Years with high volumes of short and long rains in a pronounced bimodal pattern seem particularly susceptible to rodent outbreaks, and this was evident in our statistical models. It also fits with observations by Leirs et al. (1996). Strong bottom-up, rainfall, and food-driven dynamics have been observed previously in outbreak rodents in agricultural areas (Davis et al. 2004). However, the value of our findings is twofold. First, we have presented the first data on multiannual fluctuations in African rodents other than *Mastomys* in a variety of habitat types, and in a natural (as opposed to human-modified) ecosystem. Second, the observed links between rodents, small carnivores, and birds of prey suggest that the spatiotemporal dynamics of small mammals are extremely important to predators in the Serengeti ecosystem.

Small mammals are an important food source for both mammalian and avian predators elsewhere in Africa (Poulet 1974; McCauley et al. 2006; Sliwa 2006; Brown, Perrin, and Hoffman 2007; Granjon and Traore 2007; Roberts et al. 2007) and Europe (Balbontin et al. 2008). Our findings from the Serengeti support the notion that small mammals are essential prey items, particularly for black-shouldered kites. Other rodent-eating birds of prey and small- to medium-sized mammalian carnivores also appear to require small mammal prey. As such, small mammals can have direct effects on the abundance and possibly the distributions and persistence of these predators in the ecosystem. Although consistent time-series data were available for just nine mammalian carnivore species, rodents are known to be an important food source for a much wider variety of mammalian carnivores, as well as birds of prey, in the Serengeti ecosystem (Brown, Urban, and Newman 1982; Durant, De Luca et al. 2007). As such, understanding fluctuations in small mammal abundance in the greater Serengeti ecosystem may ultimately help in conservation of the rarer small- to medium-sized mammalian carnivores, particularly those with threatened or endangered conservation status (Durant, De Luca et al. 2007).

According to our data, the role of top-down control of small mammal populations by predators in the Serengeti is much less clear. Because we could not experimentally manipulate small mammal dynamics, we cannot conclusively rule out the hypothesis that predators have a role to play in maintaining rodents at virtually undetectable numbers during the "low" phase in intervening years. However, abundances of small carnivores and birds of prey often lagged behind rodent numbers in both their peak and decline phases (fig. 12.4). Such lagged declines in carnivores have been observed in numerous other studies (e.g., Jaksic 2001; Duncan, Swengel, and Swengel 2009) and they suggest that predators respond numerically to the dynamics of small mammals, rather than driving them. Conversely, rodent abundances in villages and cultivated areas usually remained above 10% trap catch in nonoutbreak years, whereas inside the park they were barely detectable. In cultivated areas there were fewer wild mammalian predators (Craft et al., chapter 15) and black-shouldered kites (Jankowski et al., chapter 13) than inside the park boundary. This may suggest that there were too few wild predators to suppress rodents to low levels in villages and cultivated areas, but this observation is confounded by the possibility that higher levels of background food for rodents probably exist in the form of stored and spilled grains in cultivated areas (Mohr et al. 2007). We also do not know whether domestic dogs and cats replace the functional role of wild predators in cultivated areas, but given that their abundance is regulated by humans (M. Craft and K. Hampson, pers. obs.), this seems unlikely. The bottom line is that the role of predators in controlling the abundance of small mammals in the Serengeti is yet to be established.

Some authors have suggested that rodent outbreaks inside the park provide source populations for outbreaks to occur in cultivated agricultural areas outside the park (e.g., Magige and Senzota 2006). This does not seem plausible due to the large spatial scale over which outbreaks operate. For example, Van Hooft et al. (2008) suggested that the spatial scale at which *Mastomys* dispersal and recolonization occurs is about 300 ha, much smaller than the 14,000 km^2 Serengeti National Park. While it is possible that in outbreak years there are high rates of rodent immigration into cultivated areas from the park, it is equally likely that villages and cultivated areas surrounding the park provide another important refuge for the more common rodent species such as *Mastomys, Arvicanthis*, and *Gerbilliscus* during low years. Evidence for this scenario exists for outbreaking house mice in Australian agricultural habitats (Singleton, Tann, and Krebs 2007).

Our data suggest that short-season rains >250 mm in the west Serengeti usually generate a rodent outbreak. This threshold is somewhat lower than the 380 mm predicted by Leirs et al. (1996) to generate *Mastomys* outbreaks in agricultural areas in central and southern Tanzania. Our interpolated rainfall data were collected relatively close (<30 km) to the rodent-trapping sites, so there is unlikely to be a spatial discrepancy between actual rainfall and observed rodent outbreaks in our model. It is more likely to be a site effect, and we propose three possible ecological explanations for the difference. First, Leirs et al. (1996) developed their model for just a single species (*Mastomys natalensis*), and the single-species model (also developed with over 70 years of time-series data) had greater predictive ability compared with our statistical model (69% cf. 18%), which combined several rodent species. However, for the rodent outbreaks we recorded in the Serengeti, *Mastomys* was the dominant outbreak species (as with Leirs et al. 1996), particularly in the most recent years, so this explanation is not entirely satisfactory.

Second, we would expect high primary productivity in response to rainfall to be more spatially heterogeneous in a natural ecosystem containing a variety of topographic features, soil types, and vegetation types (e.g., Bonnet et al. 2010) compared with the relatively monocultural agricultural areas studied by Leirs et al. (1996). Not surprisingly, the crop monoculture model (Leirs et al. 1996) showed a much steeper threshold for rodent outbreaks compared with the natural ecosystem (fig. 12.6). This fits with the notion of a smaller environmental window for growing crops in a good rainfall year compared with a poor year, and also with the idea that increasing complexity in the system leads to less predictable responses of herbivores to primary productivity in space and time (Bonnet et al. 2010).

Third, it is possible that a higher level of background food available in the form of spilled grains in agricultural areas (Mohr et al. 2007) could result in a higher starting density, altering the rapidity of the *Mastomys* outbreak response. Although we did not find any evidence for an effect of starting density on rodent outbreaks in the Serengeti, an interaction between rainfall and density dependence was proposed for *Mastomys* in a statistical model developed by Leirs et al. (1997) and it seems plausible that it would be detected for species about which detailed information on demographic rates is available. Because we found no evidence for density dependence in rodent outbreaks in the natural ecosystem, we fall back on the explanation that combining a number of habitat types and rodent species into

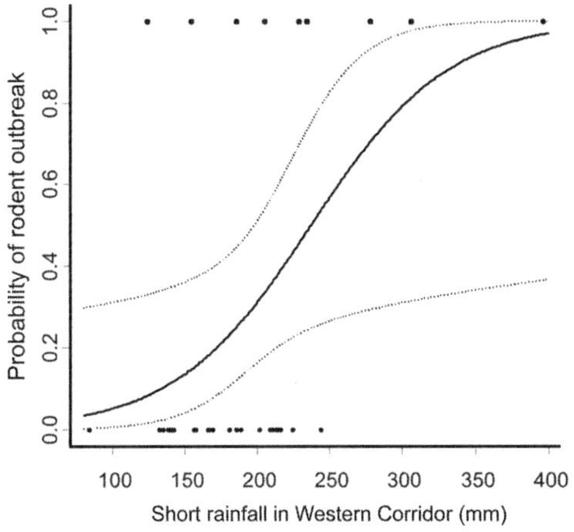

Fig. 12.6 Probability of rodent outbreaks in Serengeti National Park (1968–2010), modeled using a logistic regression analysis where 1 = outbreak and 0 = no outbreak. Dotted lines are 95% confidence intervals. Outbreaks are defined as >50% trap catch and include all species of rodents, but primarily *Mastomys* and *Arvicanthis*. Note that $z = 2.13$, $P = 0.03$; logistic regression explained 18% of the total deviance.

one model substantially reduced its predictive power, and we conclude that outbreaks inside the park are probably complicated by a number of factors including a rainfall gradient, spatial variation in habitat types and refugia, and a greater diversity of rodent species compared with agricultural areas outside the park and elsewhere in East Africa.

GLOBAL ENVIRONMENTAL CHANGE: HOW WILL SMALL MAMMALS RESPOND?

Climate and Land-Use Change and the Impact of Infectious Diseases

In Africa and internationally, there is increasing concern about the effects of global-change drivers on natural ecosystems (e.g., Tylianakis et al. 2008). Two drivers of particular interest in the greater Serengeti ecosystem are land-use change (i.e., increasing human pressure and intensification of agriculture in unprotected areas outside the park; Cleaveland et al. 2008; Sinclair 2008) and climate change (i.e., long-term changes in rainfall [Hulme et al. 2001; Ritchie 2008], which could affect the frequency and intensity of rodent outbreaks). How might small mammals respond to global environmental drivers? Global changes will have implications for both the diversity and dynamics of rodents in the greater Serengeti ecosystem as they do elsewhere in Africa (Blaum, Rossmanith, and Jeltsch 2006; Laurance et al. 2006, 2007; Mohr et al.

2007; Nicolas, Bryja et al. 2008; Nicholas, Mboumba et al. 2008) and internationally (Jaksic et al. 1992), so the answer to this question is important both for conservation of rodents and their predators and for understanding the economic and disease impacts of rodents on humans and wildlife.

Increased frequency of contacts between humans and wildlife (e.g., through land-use change and human encroachment) may exacerbate the problem of emerging infectious diseases (Ahmed et al. 2008; Jones et al. 2008). Rodents play a role in transmission of infectious diseases to humans in the case of plague (*Yersinia pestis*; Gratz 1997), toxoplasmosis (Elmore et al. 2010), leptospirosis (Levett 2001), hantavirus (Klein and Calisher 2007), and possibly hepatitis (Johne et al. 2010). They may also facilitate disease outbreaks in wild carnivore populations by increasing the abundance of peridomestic carnivores (such as white-tailed mongooses, genets, and jackals; Craft et al., chapter 15). In villages and agricultural areas, these species are then thought to come into contact with a broader range of wild carnivore species inside the park (Cleaveland et al. 2008). Canine distemper virus, rabies, and parvovirus are three such diseases; they can threaten the long-term survival of endangered carnivores such as wild dogs (*Lycaon pictus*), and can cause major perturbations in the abundance of lions (*Panthera leo*) (Cleaveland et al. 2008; Munson et al. 2008; Hampson et al. 2009). If small mammal outbreaks could be predicted, then disease could be controlled efficiently by implementing ecologically based rodent control (Massawe et al. 2007; Meerburg, Singleton, and Leirs 2008) when the risks to humans and wild carnivores are greatest. This would have important implications for the management and conservation of threatened and endangered carnivores. However, the dynamics of multiple diseases and hosts at an ecosystem level are complex and require a variety of empirical and theoretical approaches to determine cause and effect (Plowright et al. 2008; Tompkins et al. 2010), and we still know relatively little about the role of rodents in disease transmission in the greater Serengeti ecosystem.

Climate changes are known to have profound impacts on arid ecosystems worldwide (Holmgren et al. 2006); such changes may have both direct and indirect effects on small mammals in the Serengeti. For example, Ritchie (2008) observed a decreasing trend in both annual rainfall and variability in rainfall for the park from 1960 to the present day, although changes in rainfall variability for East Africa are difficult to predict (Hulme et al. 2001). If our bottom-up model is correct, we might predict fewer rodent outbreaks in the Serengeti with decreased variability in rainfall; on the other hand, increased variability may produce more frequent outbreaks. Given the importance of rodents as a prey source for numerous predators discussed above, a reduced prey base could potentially threaten the long-term persistence of some species of small- and medium-sized carnivores in the greater Serengeti ecosystem. Long-term climate trends might also alter community composition of rodent populations (e.g., Thiam, Ba, and Duplantier 2008), and influence the extent of savanna vegetation (Ritchie 2008), which may in turn influence the abundance, community composition and dynamics of rodents and their predators (Blaum et al. 2007). Better understanding of the link between climate and rodent outbreaks would also be extremely useful for the livelihoods of agriculturalists surrounding the park because of the economic impacts of rodents in those areas.

CONCLUSIONS

In this chapter we have summarized over 45 years of data on small mammal diversity and abundance collected in the greater Serengeti ecosystem. In doing so we have quantified the diversity of small mammals by habitat type and have increased the known number of species recorded in the system to a minimum of 37. We have recorded fluctuations in rodent abundance from 1968 to 2010 and demonstrated for the first time that outbreaks in rodent abundance occur in natural ecosystems in Africa, not just in agricultural systems. Short-season rainfall is a major driver of rodent outbreaks and, in turn, small mammals play a vital role in the food web of the greater Serengeti ecosystem because several species of predators show a strong response to rodent abundance.

In addition to the linkages between rainfall, primary productivity, small mammals, and predators described in this chapter, there is increasing evidence that small mammals play a wide range of subtle and indirect roles in the function of savanna ecosystems. For example, small mammals disperse seeds and prey on seeds and seedlings of tree species, thereby influencing tree establishment and woodland dynamics (Keesing 1998; Shaw, Keesing, and Ostfeld 2002; Walters et al. 2005; Goheen et al. 2010), and they respond numerically to ungulate grazing and burning (Senzota 1983; Keesing 1998, 2000; Yarnell et al. 2007; Hagenah, Prins, and Olff 2009), sometimes resulting in increased food supply for predators such as snakes (McCauley et al. 2006). It is likely that the influence of small mammals extends beyond these known effects to as-yet-unquantified effects on trophic interactions and ecosystem function. For example, the numerical responses of predators to small mammals may indirectly impact other trophic levels such as invertebrates, amphibians, and reptiles; other than the studies above, the roles of small mammals in nutrient cycling and ecosystem engineering have barely been quantified for African ecosystems. Our 45-year time series demonstrates the value of long-term ecological data as a vital baseline from which to predict the impact of future global environmental changes on the Serengeti ecosystem, and highlights the influential role of small mammals in maintaining the diversity and driving the population dynamics of a wide range of avian and mammalian predators. Further research on individual species' responses to environmental and habitat factors is surely warranted.

ACKNOWLEDGMENTS

This work was supported by the Canadian Natural Sciences and Engineering Research Council grants to ARES, and Frankfurt Zoological Society funding to the Serengeti Biodiversity Program. We thank Tanzania National Parks for permission for WAR and AEB to visit Serengeti National Park in 2006 and 2007. Work by AEB, WAR, and GJF was partially supported by the New Zealand Ministry for Science and Innovation Program C09X0909. Funding for night transects for carnivores were supported by NSF grant EF-0225453, Lincoln Park Zoo, and Canadian NSERC funding. We would like to thank the Serengeti Viral Transmission Dynamics Project (especially B. Chunde, C. Mentzel, and T. Lembo) and the Serengeti Biodiversity Program for the carnivore night-transect data. Anne Sinclair entered and organized some of the rodent-trapping data. Herwig Leirs and colleagues identified assorted small-mammal specimens from their DNA. Two anonymous reviewers made valuable comments on an earlier draft.

REFERENCES

Ahmed, E. H. A. R., J.-F. Ducroz, A. Mitchell, J. Lamb, G. Contrafatto, C. Denys, E. Lecompte, and P. J. Taylor. 2008. Phylogeny and historical demography of economically important rodents of the genus *Arvicanthis* (Mammalia: Muridae) from the Nile Valley: Of mice and men. *Biological Journal of the Linnean Society* 93:641–55.

Anderson, T. M., J. Dempewolf, K. L. Metzger, D. N. Reed, and S. Serneels. 2008. Generation and maintenance of heterogeneity in the Serengeti ecosystem. In *Serengeti III: Human impacts on ecosystem dynamics*, ed. A. R. E. Sinclair, C. Packer, S. A. R. Mduma, and J. M. Fryxell, 135–82. Chicago: University of Chicago Press.

Balbontin, J., J. J. Negro, J. H. Sarasola, J. J. Ferrero, and D. Rivera. 2008. Land-use change may explain the recent range expansion of the black-shouldered kite *Elanus caeruleus* in southern Europe. *Ibis* 150:707–16.

Blaum, N., E. Rossmanith, and F. Jeltsch. 2006. Land use affects rodent communities in Kalahari savannah rangelands. *African Journal of Ecology* 45:189–95.

Blaum, N., E. Rossmanith, A. Popp, and F. Jeltsch. 2007. Shrub encroachment affects mammalian carnivore abundance and species richness in semiarid rangelands. *Acta Oecologica* 31:86–92.

Bonnet, O., H. Fritz, J. Gignoux, and M. Meuret. 2010. Challenges of foraging on a high-quality but unpredictable food source: The dynamics of grass production and consumption in savanna grazing lawns. *Journal of Ecology* 98:908–16.

Bray, J. R., and J. T. Curtis. 1957. An ordination of the upland forest communities of southern Wisconsin. *Ecological Monographs* 27:326–49.

Brown, M., M. Perrin, and B. Hoffman. 2007. Reintroduction of captive-bred African grass-owls *Tyto capensis* into natural habitat. *Ostrich* 78:75–79.

Brown, L., E. K. Urban, and K. Newman. 1982. *Birds of Africa, vol. 1: Ostriches to birds of prey.* Paris: Lavoisier.

Cleaveland, S., C. Packer, K. Hampson, M. Kaare, R. Kock, M. Craft, T. Lembo, T. Mlengeya, and A. Dobson. 2008. The multiple roles of infectious diseases in the Serengeti ecosystem. In *Serengeti III: Human impacts on ecosystem dynamics*, ed. A. R. E. Sinclair, C. Packer, S. A. R. Mduma, and J. M. Fryxell, 209–39. Chicago: University of Chicago Press.

Crespin, L., Y. Papillon, D. Abdoulaye, L. Granjon, and B. Sicard. 2008. Annual flooding, survival, and recruitment in a rodent population from the Niger River plain in Mali. *Journal of Tropical Ecology* 24:375–86.

Davis, S. A., H. Leirs, R. Pech, Z. Zhang, and N. Chr. Stenseth. 2004. On the economic benefit of predicting rodent outbreaks in agricultural systems. *Crop Protection* 23: 305–14.

Duncan J. R., S. R. Swengel, and A. B. Swengel. 2009. Correlations of northern saw-whet owl *Aegolius acadicus* calling indices from surveys in southern Wisconsin, USA, with owl and small mammal surveys in Manitoba, Canada, 1986–2006. *Ardea* 97:489–96.

Durant, S. M., S. Bashir, T. Maddox, and K. Laurenson. 2007. Relating long-term studies to conservation practice: The case of the Serengeti Cheetah Project. *Conservation Biology* 21:602–11.

Durant, S. M., D. De Luca, T. R. B. Davenport, E. Konzo, and A. Lobora. 2007. Proceedings of the 1st Tanzania Small Carnivore Conservation Action Plan Workshop, April19–21, 2006. Arusha, Tanzania Wildlife Research Institute (TAWIRI). http://www.tanzaniacarnivores.org/information/publications.

Elmore, A. A., J. L. Jones, P. A. Conrad, S. Patton, D. S. Lindsay, and J. P. Dubey. 2010. *Toxoplasma gondii*: Epidemiology, feline clinical aspects, and prevention. *Trends in Parasitology* 26:190–96.

Fitzherbert, E., T. Gardner, T. Caro, and P. Jenkins. 2006. Habitat preferences of small mammals in the Katavi ecosystem of western Tanzania. *African Journal of Ecology* 45: 249–57.

Geertsma, A. A. 1985. Aspects of the ecology of the serval *Leptailurus serval* in the Ngorongoro Crater, Tanzania. *Netherlands Journal of Zoology* 35:527–610.

Goheen, J. R., T. M. Palmer, F. Keesing, C. Riginos, and T. P. Young. 2010. Large herbivores facilitate savanna tree establishment via diverse and indirect pathways. *Journal of Animal Ecology* 79:372–82.

Granjon, L., and M. Traore. 2007. Prey selection by barn owls in relation to small mammal community population structure in a Sahelian agro-ecosystem. *Journal of Tropical Ecology* 23:199–208.

Gratz, N. G. 1997. The burden of rodent-borne diseases in Africa south of the Sahara. *Belgian Journal of Zoology* 127:71–84.

Hagenah, N., H. H. T. Prins, and H. Olff. 2009. Effects of large herbivores on murid rodents in a South African savanna. *Journal of Tropical Ecology* 25:483–92.

Hampson, K., J. Dushoff, S. Cleaveland, D. T. Haydon, M. Kaare, C. Packer, and A. Dobson. 2009. Transmission dynamics and prospects for the elimination of canine rabies. *PLoS Biology* 7 (3): e1000053. doi:10.1371/journal.pbio.1000053.

Hancox, M. 1992. Some aspects of the distribution and breeding biology of honey badgers. *Small Carnivore Conservation* 6:19.

Holmgren, M., P. Stapp, C. R. Dickman, C. Gracia, S. Graham, J. R. Gutiérrez, C. Hice, F. Jaksic, D. A. Kelt, M. Letnic, et al. 2006. Extreme climatic events shape arid and semiarid ecosystems. *Frontiers in Ecology & the Environment* 4:87–95.

Hulme, M., R. Doherty, T. Ngara, M. New, and D. Lister. 2001. African climate change: 1900–2100. *Climate Research* 17:145–68.

Jaksic, F. M. 2001. Ecological effects of El Niño in terrestrial ecosystems of western South America. *Ecography* 24:241–50.

Jaksic, F. M., J. E. Jimenez, S. A. Castro, and P. Feinsinger. 1992. Numerical and functional response of predators to a long-term decline in mammalian prey at a semi-arid neotropical site. *Oecologia* 89:90–101.

Johne, R., G. Heckel, A. Plenge-Bönig, E. Kindler, C. Maresch, J. Reetz, A. Schielke, and R. G. Ulrich. 2010. Novel hepatitis E virus genotype in Norway rats, Germany. Emerging infectious diseases (serial on the Internet; accessed January 2012). http://dx.doi.org/10.3201/eid1609.100444.

Jones, K. E., N. G. Patel, M. A. Levy A. Storeygard, D. Balk, J. L. Gittleman, and P. Daszak. 2008. Global trends in emerging infectious diseases. *Nature* 451:990–93.

Julliard, R., H. Leirs, N. Chr. Stenseth, N. G. Yoccoz, A.-C. Prevot-Julliard, R. Verhagen, and W. Verheyen. 1999. Survival-variation within and between functional categories of the African multimammate rat. *Journal of Animal Ecology* 68:550–61.

Keesing, F. 1998. Impacts of ungulates on the demography and diversity of small mammals in central Kenya. *Oecologia* 116:381–89.

———. 2000. Cryptic consumers and the ecology of an African savanna. *BioScience* 50: 205–15.

Kingdon, J. 1977. *East African mammals: An atlas of evolution in Africa: Carnivores.* London: Academic Press.

Klein, S., and C. Calisher. 2007. Emergence and persistence of Hantaviruses. *Current Topics in Microbiology and Immunology* 315:217–52.

Krebs, C. J. 1999. *Ecological Methodology*. Menlo Park, CA: Addison Wesley Longman.

Krebs, C. J., K. Danell, A. Angerbjörn, J. Agrell, D. Berteaux, K. A. Bråthen, O. Danell, S. Erlinge, V. Fedorov, K. Fredga, et al. 2003. Terrestrial trophic dynamics in the Canadian arctic. *Canadian Journal of Zoology* 81:827–43.

Krystufek, B., W. Haberl, and R. M. Baxter. 2007. Rodent assemblage in a habitat mosaic within the valley thicket vegetation of the Eastern Cape province, South Africa. *African Journal of Ecology* 46:80–87.

Laurance, W. F., B. M. Croes, N. Guissouegou, R. Buij, M. Dethier, and A. Alonso. 2007. Impacts of roads, hunting, and habitat alteration on nocturnal mammals in African rainforests. *Conservation Biology* 22:721–32.

Laurance, W. F., B. M. Croes, L. Tchignoumba, S. A. Lahm, A. Alonso, M. E. Lee, P. Campbell, and C. Ondzeano. 2006. Impacts of roads and hunting on central African rainforest mammals. *Conservation Biology* 20:1251–61.

Laurie, W. A. 1971. The food of the barn owl in Serengeti National Park, Tanzania. *Journal of the East African Natural History Society* 28:1–4.

Leirs, H., N. Chr. Stenseth, J. D. Nichols, J. E. Hines, R. Verhagen, and W. Verheyen. 1997. Stochastic seasonality and nonlinear density-dependent factors regulate population size in an African rodent. *Nature* 389:176–80.

Leirs, H., R. Verhagen, W. Verheyen, P. Mwanjabe, and T. Mbise. 1996. Forecasting rodent outbreaks in Africa: An ecological basis for *Mastomys* control in Tanzania. *Journal of Applied Ecology* 33:937–43.

Letnic, M., and C. R. Dickman. 2010. Resource pulses and mammalian dynamics: Conceptual models for hummock grasslands and other Australian desert habitats. *Biological Reviews* 85:501–21.

Levett, P. N. 2001. Leptospirosis. *Clinical Microbiology Reviews* 14:296–326.

Lima, M., N. Chr. Stenseth, H. Leirs, and F. Jacsic. 2003. Population dynamics of small mammals in semi-arid regions: A comparative study of demographic variability in two rodent species. *Proceedings of the Royal Society of London B* 270:1997–2007.

Magige, F. J. 2013. Rodent species diversity in relation to altitudinal gradient in northern Serengeti, Tanzania. *African Journal of Ecology* 51 (4): 618–24. doi:10.1111/aje.12075.

Magige, F. J., and R. Senzota. 2006. Abundance and diversity of rodents at the human-wildlife interface in western Serengeti, Tanzania. *African Journal of Ecology* 44:371–78.

Magurran, A. E. 2004. *Measuring biological diversity*. Malden, MA: Blackwell.

Makundi, R. H., A. W. Massawe, L. S. Mulungu, and A. Katakweba. 2010. Species diversity and population dynamics of rodents in a farm-fallow field mosaic system in central Tanzania. *African Journal of Ecology* 48:313–20.

Massawe, A. W., W. Rwamugira, H. Leirs, R. H. Makundi, and L. S. Mulungu. 2007. Do farming practices influence population dynamics of rodents? A case study of the multimammate field rats, *Mastomys natalensis*, in Tanzania. *African Journal of Ecology* 45:293–301.

McCauley, D. J., F. Keesing, T. P. Young, B. F. Allan, and R. M. Pringle. 2006. Indirect effects of large herbivores on snakes in an African savanna. *Ecology* 87:2657–63.

Meerburg, B. G., G. R. Singleton, and H. Leirs. 2008. The year of the rat ends—Time to fight hunger! *Pest Management Science* 65:351–52.

Meyer, J., D. Raudnitschka, J. Steinhauser, F. Jeltsch, and R. Brandl. 2008. Biology and ecology of *Thallomys nigricauda* (Rodentia, Muridae) in the Thornveld savannah of South Africa. *Mammalian Biology* 73:111–18.

Meyer, J., J. Steinhauser, F. Jeltsch, and R. Brandl. 2007. Large trees, acacia shrubs, and the density of *Thallomys nigricauda* in the Thornveld savannah of South Africa. *Journal of Arid Environments* 68:363–70.

Misonne, X., and J. Verschuren. 1966. Les rongeurs et lagomorphes de la region du parc national du Serengeti (Tanzanie). *Mammalia* 30:517–37.

Moehlman, P. 1986. Ecology of cooperation in canids. In *Ecological aspects of social evolution*, ed. D. I. Rubenstein and R. W. Wrangham, 64–86. Princeton, NJ: University Press.

Mohr, K., H. Leirs, A. Katakweba, and R. Machang'u. 2007. Monitoring rodent movements with a biomarker around introduction and feeding foci in an urban environment in Tanzania. *African Zoology* 42:294–98.

Monadjem, A., and M. Perrin. 2003. Population fluctuations and community structure of small mammals in a Swaziland grassland over a three-year period. *African Zoology* 38: 127–37.

Mulungu, L. S., R. H. Makundi, A. W. Massawe, R. S. Machung'u, and N. E. Mbije. 2008.

Diversity and distribution of rodent and shrew species associated with variations in altitude on Mount Kilimanjaro, Tanzania. *Mammalia* 72:178–85.

Munson, L., K. A. Terio, R. Kock, T. Mlengeya, M. E. Roelke, E. Dubovi, B. Summers, A. R. E. Sinclair, and C. Packer. 2008. Climate extremes promote fatal co-infections during canine distemper epidemics in African lions. *PloS One* 3 (6): e2545.

Nicolas, V., J. Bryja, B. Akpatou, A. Konecny, E. Lecompte, M. Colyn, A. Lalis, A. Couloux, C. Denys, and L. Granjon. 2008. Comparative phylogeography of two sibling species of forest-dwelling rodent (*Praomys rostratus* and *P. tullbergi*) in west Africa: Different reactions to past forest fragmentation. *Molecular Ecology* 17:5118–34.

Nicolas, V., J.-F. Mboumba, E. Verheyen, C. Denys, E. Lecompte, A. Olayemi, A. D. Missoup, P. Katuala, and M. Colyn. 2008. Phylogeographic structure and regional history of *Lemniscomys striatus* (Rodentia: Muridae) in tropical Africa. *Journal of Biogeography* 35:2074–89.

Packer, C. 1983. Demographic changes in a colony of Nile grass rats (*Arvicanthis niloticus*) in Tanzania. *Journal of Mammalogy* 64:159–61.

Perrin, M. R., and B. P. Kotler. 2005. A test of five mechanisms of species coexistence between rodents in a southern African savanna. *African Zoology* 40:55–61.

Plowright, R. K., S. H. Sokolow, M. E. Gorman, P. Daszak, and J. E. Foley. 2008. Causal inference in disease ecology: Investigating ecological drivers of disease emergence. *Frontiers in Ecology & Environment* 6:420–29.

Poulet, A. R. 1974. Recherches ecologiques sur une savane sahelienne du ferlo septentrional, Sénégal: Quelques effets de la secheresse sur le peuplement mammalien. *La Terre et la Vie, Revue d'Ecologie Appliquée* 28:124–30.

Reed, D. N. 2007. Serengeti micromammals and their implications for Olduvai paleoenvironments. In *Hominin environments in the East African Pliocene: An assessment of the faunal evidence*, ed. R. Bobe, Z. Alemseged, and K. Behrensmeyer, 217–55. New York: Kluwer.

Reed, D. N., T. M. Anderson, J. Dempewolf, K. Metzger, and S. Serneels. 2009. The spatial distribution of vegetation types in the Serengeti ecosystem: The influence of rainfall and topographic relief on vegetation patch characteristics. *Journal of Biogeography* 36: 770–82.

Ritchie, M. 2008. Global environmental changes and their impact on the Serengeti. In *Serengeti III: Human impacts on ecosystem dynamics*, ed. A. R. E. Sinclair, C. Packer, S. A. R. Mduma, and J. M. Fryxell, 183–208. Chicago: University of Chicago Press.

Roberts, P. D., M. J. Somers, R. M. White, and J. A. J. Nel. 2007. Diet of the South African large-spotted genet *Genetta tigrina* (Carnivora, Viverridae) in a coastal dune forest. *Acta Theriologica* 52:45–53.

Senzota, R. B. M. 1978. Some aspects of the ecology of two dominant rodents in the Serengeti ecosystem. MS thesis, University of Dar es Salaam, Tanzania.

———. 1982. The habitat and food habits of the grass rats (*Arvicanthis niloticus*) in the Serengeti National park, Tanzania. *African Journal of Zoology* 20:241–52.

———. 1983. A case of rodent-ungulate resource partitioning. *Journal of Mammalogy* 64: 326–29.

———. 1984. The habitat, abundance and burrowing habits of the gerbil, *Tatera robusta*, in the Serengeti National Park, Tanzania. *Mammalia* 48:185–95.

Shaw, M. T., F. Keesing, and R. S. Ostfeld. 2002. Herbivory on *Acacia* seedlings in an east African savanna. *Oikos* 98:385–92.

Sinclair, A. R. E. 1975. The resource limitation of trophic levels in tropical grassland ecosystems. *Journal of Animal Ecology* 44:497–520.

———. 1979. The Serengeti environment. In *Serengeti: Dynamics of an ecosystem*, ed. A. R. E. Sinclair and M. Norton-Griffiths, 31–45. Chicago: University of Chicago Press.

———. 1995. Serengeti past and present. In *Serengeti II: Dynamics, management and conservation of an ecosystem*, ed. A. R. E. Sinclair and P. Arcese, 3–30. Chicago: University of Chicago Press.

———. 2008. Integrating conservation in human and natural systems. In *Serengeti III: Human impacts on ecosystem dynamics*, ed. A. R. E. Sinclair, Packer, C., S. A. R. Mduma, and J. M. Fryxell, 471–95. Chicago: University of Chicago Press.

Singleton, G. R., C. R. Tann, and C. J. Krebs. 2007. Landscape ecology of house mouse outbreaks in south-eastern Australia. *Journal of Applied Ecology* 44:644–52.

Sliwa, A. 2006. Seasonal and sex-specific prey composition of black-footed cats *Felis nigripes*. *Acta Theriologica* 51:195–204.

Stenseth, N. Chr., H. Leirs, S. Mercelis, and P. Mwanjabe. 2001. Comparing strategies for controlling an African pest rodent: An empirically-based theoretical study. *Journal of Applied Ecology* 38:1020–31.

Stenseth, N. Chr., H. Leirs, A. Skonhoft, S. A. Davis, R. P. Pech, H. P. Andreassen, G. R. Singleton, M. Lima, R. S. Machang'u, R. H. Makundi, et al. 2003. Mice, rats and people: The bio-economics of agricultural rodent pests. *Frontiers in Ecology and the Envrionment* 1:367–75.

Swynnerton, G. 1958. Fauna of the Serengeti National Park. *Mammalia* 22:435–50.

Thiam, M., K. Ba, and J. M. Duplantier. 2008. Impacts of climatic changes on small mammal communities in the sahel (west Africa) as evidenced by owl pellet analysis. *African Zoology* 43:135–43.

Timbuka, C. D., and J. Kabigumila. 2006. Diversity and abundance of small mammals in the Serengeti Kopjes, Tanzania. *Tanzania Journal of Science* 32:1–12.

Tompkins, D., A. M. Dunn, M. J. Smith, and S. Telfer. 2010. Wildlife diseases: From individuals to ecosystems. *Journal of Animal Ecology* 80 (1): 19–38. doi: 10.1111/j.1365--2656.2010.01742.x.

Tylianakis, J., R. K. Didham, J. Bascompte, and D. A. Wardle. 2008. Global change and species interactions in terrestrial ecosystems. *Ecology Letters* 11:1351–63.

Van Hooft, P., J. F. Cosson., S. Vibe-Peterson, and H. Leirs. 2008. Dispersal in *Mastomys natalensis* mice: Use of fine-scale genetic analyses for pest management. *Hereditas* 145:262–73.

Walters, M., S. J. Milton, M. J. Somers, and J. J. Midgley. 2005. Post-dispersal fate of *Acacia* seeds in an African savanna. *South African Journal of Wildlife Research* 35:191–99.

Yarnell, R. W., D. M. Scott, C. T. Chimimba, and D. J. Metcalfe. 2007. Untangling the roles of fire, grazing and rainfall on small mammal communities in grassland ecosystems. *Oecologia* 154:387–402.

THIRTEEN

Bird Diversity of the Greater Serengeti Ecosystem: Spatial Patterns of Taxonomic and Functional Richness and Turnover

Jill E. Jankowski, Anthony R. E. Sinclair, and Kristine L. Metzger

There is much scientific concern that the loss of biodiversity will reduce the capacity of our ecosystems to provide important services and will decrease the resiliency of natural systems to counteract the effects of anthropogenic change. Moreover, the consequences of species extinction on ecosystems may not be detected until conditions deteriorate beyond our ability to reverse the loss. Some ecosystems respond to disturbances by losing few species at first and many later on, producing multiple states that depend not only upon loss of species, but also loss of functional groups that affect ecosystem properties (Walker, Kinzig, and Langridge 1999; Díaz et al. 2007). Remedial measures are most likely to be effective during the period of initial slow change, if such change could be detected.

Recognizing when ecosystems are in danger of functional loss and degradation requires a baseline comparison to the composition and diversity of intact systems, which can be used as ecological benchmarks relatively free of human interference. Aspects of community structure vary through space and time, in both predictable and in stochastic ways. Through monitoring efforts, we can identify the major drivers (or at least environmental correlates) of variation in diversity and community structure within intact systems. Then it becomes possible to distinguish between long-term environmental trends and more direct impacts of human exploitation. Protected areas potentially provide a basic global network of reference sites (Arcese and Sinclair 1997; Sinclair 1998), assuming that the protected area is sufficient in area to support viable populations. To understand this problem we need to know what habitats species require, as well as whether habitats protect unique assemblages of species.

We can generally quantify aspects of community structure and composition through an analysis of species diversity. Specifically, alpha diversity, sometimes represented in simple form as the number of species occurring in a given area, indicates the richness of survey locations or designated habitats, whereas beta diversity quantifies the extent to which survey locations differ in species composition (Magurran 2004). When survey locations are positioned along environmental gradients or associated with distinct habitats, an analysis of beta diversity can reveal the effect of habitat transitions on species composition. This is useful to test whether recognizable habitat attributes that are expected to drive changes in structure and composition are actually relevant predictors. Such analysis is typically performed in terms of taxonomic richness and composition, but it is also revealing to consider the richness and composition of functional groups within and across habitats, as represented by the morphological, physiological, or ecological traits of species that make up communities (Petchy and Gaston 2006; Schmera, Erős, and Podani 2009). This "functional diversity" perspective may be more relevant in characterizing ecosystem function and trophic complexity, depending upon the attributes examined. Importantly, spatial patterns in functional richness and composition may not be congruent with diversity patterns quantified in taxonomic terms, and such discrepancies may allow environmental drivers of diversity and community assembly to be tested. Furthermore, prudent conservation strategies may be directed to preserve both taxonomic and functional diversity. Any discrepancy between these facets of diversity should therefore be examined (Devictor et al. 2010).

In this study we characterize the diversity and composition of bird communities across the Serengeti ecosystem. We ask a number of questions to understand how both richness and composition change across space, and we also determine the extent to which bird communities are differentiated by recognizable habitats that we have a priori identified. For example, do birds differentiate habitats as we have recorded them using indicator tree or grass species? Are similar habitat groups recognizable based on both taxonomic and functional composition? Answers to these questions should allow us to identify the essential habitats for bird conservation in protected areas and understand the changes in bird communities already described for habitats modified by agriculture (Sinclair, Mduma, and Arcese 2002; Sinclair et al., forthcoming).

Descriptions of the bird fauna in Serengeti are rudimentary. An early annotated but incomplete list was published by Schmidl (1982). Previous research has documented how migrant species from the Palearctic were accommodated in the Serengeti habitats as the intertropical convergence zone arrived with the short rains bringing with it swarms of migrating insects in October–November (Sinclair 1978). Studies of grassland birds were conducted by Folse (1982) and Gottschalk, Ekschmitt, and Bairlein (2007). A comparison of the bird communities in native savanna and agriculture has shown a marked decrease in species as agriculture has modified the habitats (Sinclair, Mduma, and Arcese 2002; Sinclair et al., forthcoming). Nkwabi et al. (2011) documented the impacts of two natural disturbances to the grass layer, fire and grazing by migrant ungulates, on bird communities of the long-grass Serengeti Plains. They found that burning had a marked effect by changing the species composition from those living in tall grass to those living in short grass. Similarly, Nkwabi et al. (chapter 14) have examined these disturbances on savanna bird communities, documenting an increase in insectivores immediately after a fire, most likely due to an abundance of easily detected burnt insects, and of granivores, possibly due to easily visible seeds. These effects, however, are short-lived, and disappeared after a month. Grazing, by comparison, had little effect on ground feeding guilds. Both disturbances reduced the abundance of tree and shrub feeders. Richness of species did not change with fire or grazing; instead, species that were sensitive to the disturbance were replaced by others that benefited from the disturbance. Thus, there was functional compensation within the communities.

Taken together, these studies suggest that habitat types that represent successional stages in recovery from disturbance are primary determinants of the composition of Serengeti communities. These observations may extend to a broader range of Serengeti habitats, and spatial variation in species composition and functional composition should correspond to transitions between habitat types. Here we use an extensive survey dataset of Serengeti bird communities collected in the years 2003–05 to assess the extent to which bird communities are distinctive across habitats, and to quantify the rate of change in composition moving across major environmental gradients in the Serengeti ecosystem. This dataset includes surveys performed across sites, representing nearly 60 habitat types typical of the Serengeti, as well as linear transects that encompass transitions from short- and long-grass plains to savanna woodlands and important shifts in precipitation regimes. We intend this study to be a comprehensive assessment of the Serengeti avifauna that additionally highlights the primary drivers of diversity and change in community structure, both in terms of taxonomic and functional composition. Our results can then serve as a benchmark for comparison to systems undergoing changes due to direct impacts of human disturbance.

STUDY AREA

The location, protected status, geography, geology, and climate of the Serengeti have been described in chapters 2 and 3. The Serengeti ecosystem (34–36° E, 1–4° S) in Tanzania, East Africa, has been protected as a national reserve or park since the 1920s. In general, from the point of view of birds, there is a wet season from approximately November to June and a dry season from July to October. Nearly all species breed in response to rain although the start of breeding varies with the group (Sinclair 1978; Nkwabi 2014). Weather responds to the effects of El Niño Southern Oscillation with either floods or droughts occurring at intervals of 4–6 years (Sinclair et al. 2013; Fryxell et al. chapter 7).

The predominant vegetation of the protected area in Serengeti is a savanna dominated by fine-leaved *Acacia* and broad-leaved *Terminalia* trees (see figure 12.1 in Byrom et al. chapter 12 for a map of the main habitats). The *Acacia* savanna is made up of different species that effectively form monospecific stands in patches about 200 m across and that separate along a soil gradient called the catena: at the top of low ridges the soil is shallow and rocky; these ridges support *A.tortilis*, *A. senegal*, and *A. hockii* together with two species of *Commiphora*. In midslope with deeper soils the dominant species *A.robusta* occurs, the most frequent large tree species in the ecosystem. On sandy washouts at the base of hills lives the wait-a-bit thorn, *A. mellifera*. Collectively, these areas make up the "upper catena." At the bottom of the slope, with deep silt soils that have impeded drainage, are the small gall Acacias, *A.drepanolobium* and *A.seyal*. Also in such soils is the tall tree, *Balanites aegyptiaca*. These areas collectively make up the "lower catena." There are also patches of open grassland in largely impeded drainage areas; most are small at 200–300 m across, but a few, such as Musabi and Ndoho in the west, and Togoro in the center, span 2–3 km. In general, the "Acacia savanna" is composed of a fine-scale mosaic of different monospecific stands of trees creating patches measuring 100–300 m across. The grass layer over most of the catena is dominated by red oats grass (*Themeda triandra*).

The *Terminalia* woodland, a type of broad-leaved miombo woodland characteristic of southern and central Africa, occurs in the far northwest of the ecosystem with a grass layer of tall (1 m) perennial *Hyparrhenia* species. It occurs on granitic rocky ridges which give way downslope to either open *Themeda* grassland or grassland with *A. gerrardii* trees; these trees replace *A. robusta* in the *Acacia* savanna.

A number of smaller but distinct habitats all occur within the savanna. Rocky hills that rise steeply 200–500 m above the surrounding savanna are found along the eastern boundary and through the middle of the western corridor. These hills support *Combretum* woodland on the lower stony slopes, a subset of the *Terminalia* woodlands further north, with similar grasses and herbs. The main rivers support dense riverine forests that depend on groundwater. These forests are of two sorts: that along the Mara River in the north is montane in origin from the Loita hills; whereas that along the western rivers, Grumeti and Mbalageti, is lowland Congo forest in origin (see Turkington et al., chapter 9). Numerous small seasonal streams dissect the savanna and support a thin strip of bushes and riverine trees; the larger streams have large Acacias, *A. xanthoploea* and *A. kirkii*. Smaller drainages occur as wet grasslands with small bushes and rushes. There are three lakes, Lagarja and Masek, which form the top end of Olduvai Gorge on the southeast plains, and Magadi on the western edge of the plains. All are very shallow and highly alkaline. Freshwater is confined to the rivers, a few springs seeping out of the hills and Lake Victoria at Speke Gulf in the far west.

The southeastern part of the ecosystem is open, treeless grassland with a gradient of long grass (similar to that of the *Acacia* savanna) in the northwest of the plains grading to intermediate grasslands of *Pennisetum* (20–50 cm) in the center, and short grasslands (5–15 cm) in the far east and south. Detailed descriptions of these habitats are given in chapter 3 and Sinclair et al. (2008).

In the past, similar natural savanna extended west of the present park borders covering about 2050 km^2 until agriculture, small holdings with cereal and root crops, took over in the mid-nineteenth century; these land cover types extended eastward in the 1950s and now abut the western border of the natural ecosystem. The present agricultural areas were originally similar in flora and fauna, geology, soil nutrients, and other ecological features to those of the native savanna (Sinclair 1995; Sinclair et al. 2008; Sinclair et al. chapter 2). Agriculture, which forms an abrupt boundary with the savanna on the western border, has removed most trees. Many small native shrubs surround crops of millet and cassava.

METHODS

Bird Surveys

Data on the bird communities were obtained by one of three methods: transects by vehicle along tracks and roads; sites located in selected habitats; and ad hoc points. These data have been collected over several decades, beginning in 1964 and continuing to the present. As the purpose of the current study is to assess spatial patterns in the Serengeti bird fauna, we selected a snapshot period of two years (2003–05) within this larger dataset for our analysis in order to control for changes in species composition and shifting habitat types over time. This time period was chosen because it represents years with a fairly typical precipitation regime (i.e., no major drought or extra heavy wet season) with a large number of observations that were systematically distributed throughout the wet season (November to June), when most detections were recorded.

Transects. Twenty eight transects of lengths varying from 15 km to 100 km were placed over the majority of the ecosystem along tracks. Transects were driven slowly (30–50 kph), stopping where necessary using two observers and a recorder. All species, their numerical abundance, and habitat were recorded up to 50 m either side of the track. Each transect was divided into 5-km segments and records for species, as well as information on occurrence of habitat types, were pooled for each segment. The bird groups counted were medium and large sized insect-eating species, and all seed and fruit eating species. These focal species were easily detectable in the open grassy habitats that we surveyed. Forests were not sampled by this method.

Three transects were run on a regular basis twice a year, at the end of the short rains in December–January and at the end of the long rains in May–June. Transect 1 crossed the plains from Seronera to Olduvai Gorge (75 km), and transect 4 ran from Seronera to Kirawira and sampled the savanna of the western corridor (100 km). Because transect 1 ends approximately where transect 4 begins, we examined bird communities across these transects as a single environmental gradient consisting of 35 segments of 5 km each (175 km total).

Sites. Sites were chosen so that they were homogeneous for a specific habitat (e.g., a stand of *A. robusta* trees, a wet grassland drainage line, or a thicket of *A. mellifera*). A total of 213 sites covering all habitats were established; a subset of these ($n = 76$) was surveyed in 2003–2005 and is used in our analysis of species and functional richness. A portion of the sites surveyed in 2003–2005 ($n = 41$), were also spatially matched to vegetation point-quarter plots (see vegetation surveys below) and are used for specific analyses of compositional differences. Surveys at sites include all birds seen or heard within a 50-m radius during a ten-minute count. Sites were surveyed between 7.00 and 10.00 hours, and site visits were rotated so that each site was surveyed at different times within this period. Habitats were represented across multiple sites, and habitat replicates were separated by at least 1 km.

Points. In addition to transect and site surveys, ad hoc sightings of birds were recorded by location and habitat whenever appropriate, usually when an unusual species appeared, or at a rare habitat such as a spring or rocky hillside. Point surveys do not represent a systematic means of data collection, so we have not included them in statistical analysis of bird diversity and composition. Point survey data are instead used to supplement the comprehensive species list of Serengeti birds.

Avian Functional Traits

Functional roles of species that identify the ecological niche of Serengeti birds are described using three trait categories: diet (five categories: frugivore, granivore, insectivore, omnivore, or vertebrate feeder); feeding location, namely the vertical stratum most commonly used for foraging (seven categories: ground, grass/herb layer, water, shrub layer, tree layer, or aerial); and mass. Mass data were obtained using the database of avian bird masses (Dunning 2008). For functional diversity indices, mass data were included as a continuous trait; however, for examining change in functional composition across space, mass was summarized as a categorical variable. Eleven categories were created, which represent a doubling of size classes (e.g., 0–25 g, 25–50 g, 50–100 g, 100–200 g, 200–400 g, etc., up to the largest categories of 6,400–12,800 g and >12,800 g). By assessing mass categorically, along with the other functional traits of diet and feeding location, we could combine species that were equivalent in all three functional trait categories and examine the species at a given locality in terms of functional composition (see data summary and analysis below).

Vegetation Surveys and Environmental Layers

Density and species composition of woody plants was measured by the point centered quarter (PCQ) method (Krebs 1989) at sites. Along a line that ran through a site, a random point was chosen at the start, and subsequent points were chosen randomly at distances not less than 50 m from the previous point along the line. This distance seemed to be sufficient so that the same individual plant was not measured in consecutive points. The area around each point was divided into four 90 degree quadrants and the distance was measured to the nearest tree in each quadrant. The tree species, diameter at 1 m, height of canopy, and canopy area were recorded. The density (D) of trees greater than 5 cm in diameter (at 1 m above ground) was estimated from the distances (r) across all the points (n) along the transect together with their standard errors.

$$D = 4(4n-1)/(\pi\sum(r^2))$$

Variance was calculated as

$$\text{Var}(D) = D^2/(4n-2)$$

And standard error as

$$\text{SE}(D) = \text{Sqrt}((\text{Var}(D)/4n)$$

Since the sites were chosen within the almost monospecific stands of trees, the densities represent those of the dominant tree species at a site.

Other environmental attributes for site and transect segment locations were extracted from raster images available for the Serengeti region. These variables include elevation, distance of a site or transect segment to permanent water, average precipitation for 2003–05, a habitat quality "greenness" index, and lat-long coordinates (for measurement of spatial proximity of survey locations). Elevation values were obtained from the NASA Shuttle Radar Topographic Mission Digital elevation model. The "greenness" index was developed using MODIS derived Normalized Difference Vegetation Index (NDVI) data as a surrogate for vegetation production (Carrol et al. 2004). Pixel values of monthly NDVI values greater than 0.5 were scored as high productivity, between 0.5 and 2.5 scored as medium productivity, and less than 0.25 scored as low productivity. Areas with high, medium, and low monthly productivity were summed to create composite maps of annual habitat quality. Precipitation information is continually collected across the Serengeti ecosystem over a network of rain gauges. Rainfall across the ecosystem is locally variable; therefore we refined our analysis to rainfall information collected at gauges that were within a 25 km distance from either a PCQ or sample transect location. See chapter 3 for further details on geographic information systems (GIS) raster layers.

Data Summary and Analysis

Our analysis focuses on a snapshot of the bird community using the years 2003–05 and considers data from the wet season only. Both of these restrictions are employed to control for yearly and seasonal variation in patterns of diversity and composition across the Serengeti ecosystem. Our primary survey methods, utilizing sites and transects, sample rather distinct but complementary areas within the Serengeti: transects span the open grasslands and savanna, occasionally passing through rocky hills and ridges supporting woodlands, whereas sites sample largely monotypic vegetation types within savanna and woodland, including the upper and lower catena, broad-leaved woodlands, swamp, and riverine areas. Due to the differences in spatial arrangement, extent and sampling protocols between sites and transects, we consider data from these two survey methods separately.

Species richness and functional diversity. The 617 bird species found in the Serengeti ecosystem are listed on the website [www.press.uchicago.edu/sites/serengeti/]. Names are those given in Gill and Donsker (2012). Species richness was determined for each site and for 5-km segments along transects using total number of species found in each location. Functional diversity was calculated using functional attribute diversity (FAD), an index that considers multiple traits simultaneously. Functional diversity indices were calculated using the program FDiversity, which operates from the R platform to calculate aspects of functional diversity from species' trait data (Casanoves et al. 2010).

Species richness and functional diversity data were analyzed for sites and transect segments in two ways. First, we summarized average species richness and functional diversity across primary habitats represented in site and transect surveys. Second, we examined the relationship between species richness and functional diversity to determine the degree to which communities become saturated in the functional roles of species with increasing species richness. We considered the discrepancy between species richness and functional diversity as an assessment of functional redundancy, with respect to the traits examined. Here functional redundancy is expressed as a saturation of FAD relative to species richness. It is important to remember that this saturation is influenced by the number of functional traits examined; if species are more finely distinguished in their functional roles, there will be a slower saturation of functional types with increasing species richness (also see discussion). Plots of species richness versus functional diversity identify habitat types of sites and transect segments for visual interpretation of habitat differences in functional diversity and/or species richness. Finally, we explored the extent to which species richness and functional diversity are explained by aspects of vegetation structure, using variables measured during vegetation PCQ surveys (using 41 sites with associated vegetation data).

Dissimilarity in species and functional composition. Change in species composition across space (i.e., beta diversity) was calculated using dissimilarity indices that compare pairs of localities, generated from a site by species presence-absence matrix (or, in the case of transect data, a 5-km transect segment by species matrix). Dissimilarity indices range from 0 to 1, where 0 indicates that two communities are completely distinct (i.e., sharing no species), and 1 indicates that all species are shared (Magurran 2004). To describe overall differences in species composition, we used Sørensen's dissimilarity index (β_{sor}), which expresses dissimilarity due to two processes: compositional change due to species loss or nestedness, and change due to species replacement (i.e., turnover). For select analyses, we used an approach by Baselga (2010) to partition total beta diversity (β_{sor}) into additive contributions of dissimilarity due to turnover (Simpson's dissimilarity β_{sim}; Lennon et al. 2001), which describes change in composition independent of differences in richness, and dissimilarity due to nestedness (β_{nes}), derived as the difference between β_{sor} and β_{sim}. To examine change in functional composition across space, we derived a list of all functional types by combining species with the same combination of categorical functional traits (i.e., diet, feeding location, and mass category) into a single functional species. We created a site by functional species matrix in order to calculate functional dissimilarity among sites and across transect segments using the same dissimilarity indices described above (β_{sor}, β_{sim}, and β_{nes}).

Pairwise dissimilarities were analyzed in four ways. First, for both site and transect data, we performed a cluster analysis using dissimilarity matrices for species composition and functional composition in order to describe major divisions among survey localities. The cluster analysis used dissimilarity matrices produced by β_{sor} (i.e., overall beta diversity) and an average linkage method. Cophenetic correlations were calculated to evaluate the degree to which resulting dendrograms represented actual distances among groups; this correlation was always ≥0.80. The cluster analysis for site data used the dissimilarity matrix from pairwise comparisons of the subset of 41 sites with environmental information, representing eight habitats: one of six dominant *Acacia* species (occurring in monotypic patches), rocky hills with *Combretum* woodlands, or broad-leaved *Terminalia* woodlands. A cluster analysis was also performed using data from all 76 sites, which represent a broader range of 15 habitats. In addition to those mentioned above, these habitats include dominant *Acacia* species occurring in riparian areas (*A. xanthoplea*, *A. polyacantha*), *Commiphora* species, as well as river, riverine forest, lake edge, and floodplain habitats. To reduce the number of clusters and represent these 15 habitats with a more comprehensive species list, species recorded at sites belonging to these habitats were pooled for analysis. The cluster analysis for transect data used the dissimilarity matrix from pairwise comparisons of 35 transect segments of 5 km each, which were identified as belonging to one of six habitats: short-, intermediate-, or long-grass plains, *Acacia* savanna, rocky hills with *Combretum* woodlands, or broad-leaved *Terminalia* woodlands. Second, for site and transect data, we used multiple regression on distance matrices (MRM) analysis (Legendre and Legendre 1998; Lichstein 2007) to assess how much variation in species and functional composition could be explained by space, vegetation and landscape structure, and local environmental conditions. Third, we used a Mantel analysis to examine the correlation between taxonomic and functional dissimilarity from pairwise comparisons within site and transect datasets (using site data both independently and pooled into habitats). Similar to our analysis of functional redundancy relative to species richness across survey localities, this correlation analysis of pairwise dissimilarity values permits an evaluation of the extent to which sites that are taxonomically distinct in composition are also functionally distinct. Significance for Mantel and MRM analysis is evaluated using permutation tests ($n = 999$ permutations). Finally, to examine major transitions in species and functional composition along environmental gradients, we plotted pairwise dissimilarity (using β_{sor}, β_{sim}, and β_{nes}) of adjacent 5-km transect segments that traversed a major precipitation gradient in the Serengeti ecosystem. For each dissimilarity index, we calculated the average and standard deviation of dissimilarity between 5-km segments, then identified peaks in dissimilarity along this gradient as transitions between segments with greater than one standard deviation above average dissimilarity. Qualitative habitat descriptions created for each 5-km segment were used to determine whether peaks of compositional change corresponded to shifts in habitat along this gradient.

Dissimilarity matrices for β_{sor}, β_{sim}, and β_{nes} were calculated in R v. 2.14.1 using functions provided by Baselga (2010) and by J. A. Di Rienzo (pers. comm.). Analyses used packages CLUSTER, ECODIST, STATS, and VEGAN (R Development Core Team 2010).

RESULTS

Regional Species Diversity, Functional Attributes, and Status of Serengeti Birds

To date, the Serengeti bird fauna is known to encompass a total of 617 species from 88 families and to represent over two-thirds of all avian orders. This estimate includes records of species from all survey methods dating back to initial surveys in 1964. Among the families represented by large numbers of species are Accipiters (Accipitridae: 52 species), Old World Flycatchers

Table 13.1 Avian guilds and feeding locations for the total 617 species of Serengeti birds

Feeding location	Fruit	Insects	Omnivore	Seeds	Vertebrates	Total (Feeding Location)
	Diet					
Tree	11	111	8	1	1	132
Shrub	11	40	6	5	--	62
Grass/Herb	--	18	8	35	--	61
Ground	--	79	16	37	5	137
Water	3	67	6	13	36	125
Aerial	--	26	--	--	--	26
Vertebrates	--	8	--	--	66	74
Status	**Fruit**	**Insects**	**Omnivore**	**Seeds**	**Vertebrates**	**Total (Status)**
Resident	20	220	37	72	62	411
Palearctic migrant	--	59	--	4	18	81
Intra-African migrant	1	19	--	2	3	25
Passage migrant	--	10	--	1	2	13
Seasonal mover	3	18	3	8	6	38
Occasional visitor	1	13	3	1	9	27
Vagrant	--	10	1	3	8	22
Total (Diet)	25	349	44	91	108	617

Notes: "Diet" indicates the dominant food type consumed, "feeding location" indicates the substrate on which the species is principally found, and "status" describes the residency or annual movements.

(Muscicapidae: 33 species), Weavers (Ploceidae: 33 species) and Cisticolas (Cisticolidae: 29 species). Of the 617 species recorded in the Serengeti ecosystem, by far the most abundant are the insectivores with 349 species, some 57% of the total (table 13.1). Omnivores are also largely insect feeders. Seed and fruit eaters make up 19%, and vertebrate feeders 17% of the total. Insectivores are approximately evenly divided between ground or grass/herb layer feeders (97 species) and tree or shrub layer feeders (151 species). Fruit eaters forage within the tree and shrub layer, while granivores generally forage on the ground or in the grass/herb layer. The majority of the bird fauna is made of resident species (411 species), and just over half of these are insectivores (table 13.1). Palearctic migrants (81 species) are effectively all insectivores or vertebrate feeders (the exceptions are four species of duck). Passage migrants (13 species), which are mostly from the Palearctic, are also largely insectivores, as are the intra-African migrants (25 species). The majority of granivores and frugivores (92 of 116) are residents. Generally, in this savanna biome, insectivores predominate, making up over half of the resident species and nearly all the migrant species.

Species Richness and Functional Diversity

The surveys analyzed in this study include 289 species, nearly half of the total bird species known to occur in the Serengeti. Of these, 92% (265 species) were detected at sites, and 52% (150 species) were detected along transects. Species richness varied considerably across sites and across 5-km transect segments. At a given site, richness could be as low as 2 or as high as 87 species, but averaged approximately 16 species per site. Transect segments exhibited higher average richness compared to sites (63.5 species per 5-km segment), ranging from 18 to 118 species per segment. Relative to mean species richness, sites were much more variable than transects in the number of species recorded: the coefficients of variation for sites and transect segments were 90% and 47% respectively. Functional diversity, as measured by FAD, was always substantially lower than species richness, reaching a maximum of 28 functional types at a single site (average of 10) and a maximum of 29 functional types at a transect segment (average of 18). Relative to mean functional diversity, the variability of site functional diversity was only slightly more than that of transect segments: the coefficient of variation in FAD for sites and transect segments was 50% and 32% respectively.

The extent to which habitats characterized species-rich and species-poor communities differed enormously between sites and transects (fig. 13.1). Across transects, the plains (i.e., short-grass, intermediate-grass, and long-grass plains) had consistently lower species richness and functional diversity compared to *Acacia* savanna, rocky hills (supporting *Combretum*), and broad-leaved *Terminalia* woodland (fig. 13.1b). Environmental attributes of transect segments, particularly elevation, distance from water and habitat type, explained 65% and 79% of the variation in species richness and functional diversity, respectively (table 13.2; species richness: $F_{6,29} = 9.1$, $p < 0.0001$; FAD: $F_{6,29} = 17.8$, $p < 0.0001$). Specifically, transect segments found at higher elevation and closer to a water source had higher species richness and functional diversity. By comparison, sites harboring a particular habitat could exhibit low or high species richness and functional diversity, and there was no strong trend for one habitat to have more species than another. Very generally swamps ($n = 3$ sites) had slightly higher species richness and functional diversity compared to upper and lower catena habitats (fig. 13.1a); however, vegetation structure and environmental attributes measured across sites (i.e., tree height, density, distance from a water

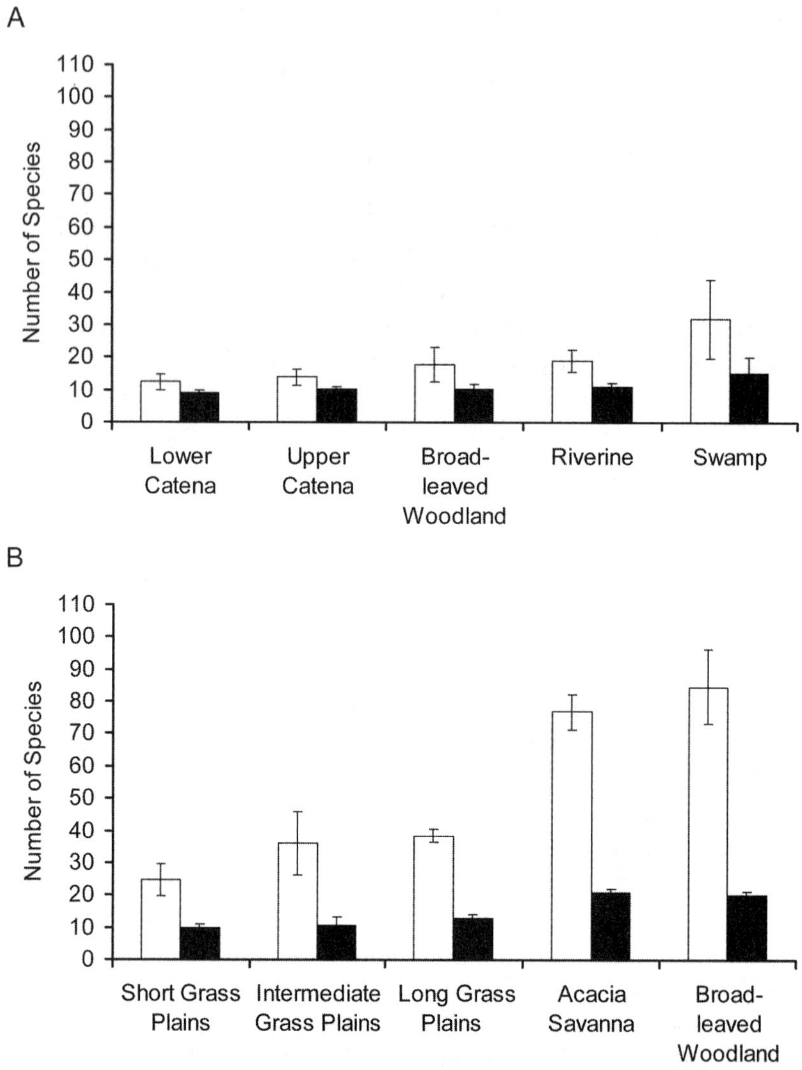

Fig. 13.1 Average number of taxonomic (white) and functional (black) species (with standard error bars) across macrohabitat types at sites (A) and 5-km transect segments (B). Macrohabitat groups differ slightly between sites and transects. The site macrohabitats of upper and lower catena are grouped as "*Acacia* Savanna" in transect surveys. Plains habitats are not sampled across sites, and riverine and swamp habitats are not explicitly sampled across transect segments.

Table 13.2 Linear models predicting species richness and functional diversity (FAD) across transect segments

Variables: Richness model	Parameter estimate	Standard error	*t*	*p*
(Intercept)	-55.0	51.7	-1.1	0.30
Elevation	0.10	0.04	2.64	**0.013**
Distance to water source	-1.82	0.58	-3.16	**0.004**
Habitat: Broad-leaved woodlands	14.16	11.09	1.28	0.212
Habitat: Short-grass plains	-28.16	17.81	-1.58	0.125
Habitat: Intermediate-grass plains	-13.42	2.13	-6.295	**<0.0001**
Habitat: Long-grass plains	-63.2	13.37	-4.73	**<0.0001**

Variables: FAD model	Parameter estimate	Standard error	*t*	*p*
(Intercept)	-13.11	7.93	-165	0.11
Elevation	0.026	0.0006	4.39	**0.0001**
Distance to water source	-0.34	0.09	-3.88	**0.0006**
Habitat: Broad-leaved woodlands	1.14	1.70	0.67	0.51
Habitat: Short-grass plains	-8.42	2.73	-3.08	**0.004**
Habitat: Intermediate-grass plains	-13.42	2.13	-6.29	**<0.0001**
Habitat: Long-grass plains	-14.34	2.05	-6.99	**<0.0001**

Notes: Habitat types use *Acacia* savanna as a basis of comparison. Predictors with *p*-values <0.05 are shown in bold.

source, elevation, precipitation, and habitat) were not significant predictors of variation in species richness or in functional diversity (species richness: $F_{12,28} = 0.76$, $p = 0.68$; FAD: $F_{13,27} = 1.55$, $p = 0.16$). An analysis with an expanded dataset of all 75 sites (including those without vegetation structure data) also showed that habitat type was not a significant predictor of species richness or functional diversity (species richness: $F_{4,70} = 1.31$, $p = 0.27$; FAD: $F_{4,70} = 0.60$, $p = 0.66$). Although sites dominated by larger trees species such as *Terminalia*, *Combretum*, and *A. tortilis* tended to support the highest number of species (>30) and functional types (>20), species richness and functional diversity was not related to tree size overall (species richness: $F_{2,72} = 0.30$, $p = 0.74$; FAD: $F_{2,72} = 0.20$, $p = 0.82$).

Functional diversity showed a saturating relationship with species richness for both sites and transects, never reaching more than 30 functional types at a given location (fig. 13.2). Perhaps because site richness was generally lower than richness of transect segments, there were many more instances of sites ($n = 16$) in which every species recorded exhibited a unique

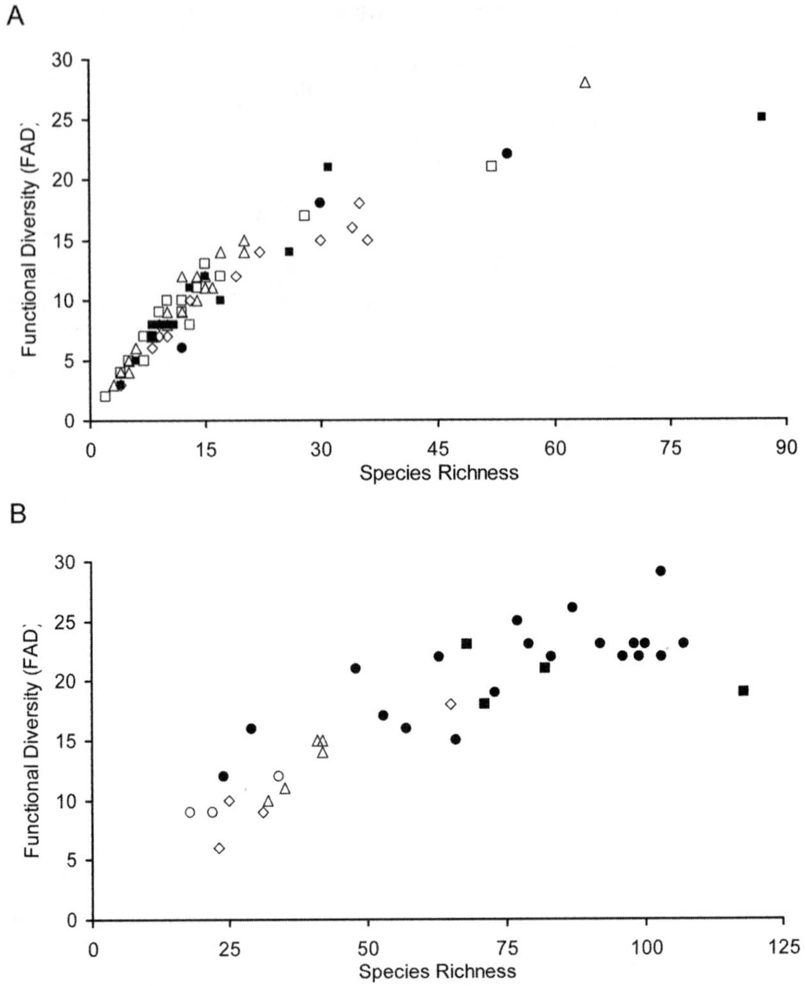

Fig. 13.2 Plot of functional diversity (FAD) with number of species at sites (A) and within 5-km transect segments (B) across macrohabitats. Site macrohabitats include lower catena (empty squares), upper catena (empty triangles), riverine (empty diamonds), swamp (filled circles), and broad-leaved woodland (filled squares). Transect macrohabitats include short- (empty circles), intermediate- (empty triangles), and long-grass plains (empty diamonds), *Acacia* savanna (filled circles), and broad-leaved woodland (filled squares). FAD shows the number of trait-attribute combinations across species in the sample.

combination of functional traits (i.e., no functional redundancy; species richness is equivalent to functional diversity). This only occurred at sites with 12 or fewer species. By contrast, there were no transect segments in which functional diversity was equivalent to species richness (i.e., all transect segments exhibited some functional redundancy).

Cluster analyses. Cluster analyses were performed for 5-km transect segments and for sites based on species and functional composition (figs. 13.3–13.6). The site-based cluster analysis showed no easily identifiable habitat divisions across *Acacia* savanna and broad-leaved woodlands. Many sites identified by a particular habitat were scattered across both taxonomic and functional dendrograms (e.g., rocky hills with *Combretum*, *A. senegal*, and *A. mellifera*). In other words, sites within these habitats are not as consistent in their species composition. In some cases, sites representing particular habitats were similar in bird species composition, indicating that there is more consistency from site to site within that habitat. These included sites dominated by *A. tortilis*, *A. robusta*, *A. seyal*, and, to a lesser extent, *A. drepanolobium* and broad-leaved *Terminalia* woodlands (fig. 13.3a). Of these, only sites dominated by *A. tortilis* and *Terminalia* (and *A. robusta*, to a lesser extent) were also grouped based on similarity in functional composition (fig. 13.3b).

The cluster analysis with pooled site data across 15 habitats provides a more comprehensive sample of the composition of each habitat and facilitates an assessment of habitat types most closely aligned in taxonomic and functional composition (fig. 13.4). A number of patterns are apparent. Taxonomically, habitats are largely grouped by dominant soil gradients (fig. 13.4a); *Acacia* species found in the upper catena (*A. robusta*, *A. tortilis*, *A. mellifera*, *A. senegal*) as well as *Terminalia* and *Combretum* woodland form a group at 0.6 dissimilarity. Some habitats within the lower catena (*A. seyal* and *A. drepanolobium*) and habitats surrounding river habitat (*A. polyacantha*, river and riverine forest) are also most similar to each other in species composition (although they group at a higher dissimilarity of 0.7). In some cases, habitats with shared structural elements are also grouped. For example, within the upper catena group, large *Acacia* species (*A. tortilis* and *A. robusta*) and small *Acacia* species (*A. senegal* and *A. mellifera*) are closest in species composition. Functionally, however, different habitat groups emerge (fig. 13.4b). There are a number of habitats from the upper and lower catena that are taxonomically distinct, but functionally similar (e.g., *Terminalia* and riverine forest; *A. senegal* and *A. drepanolobium*; *A. mellifera* and *A. polyacantha*). There is at least one example in which habitats closest in taxonomic composition are functionally distinct (*A. polyacantha*, river and riverine forest). This may represent a functional transition of species from areas dominated by early successional *A. polyacantha* to late successional riverine forests. The only habitat grouping consistent across taxonomic and functional dendrograms is that of large trees in the upper catena (*A. robusta*, *A. tortilis*, and *Combretum*).

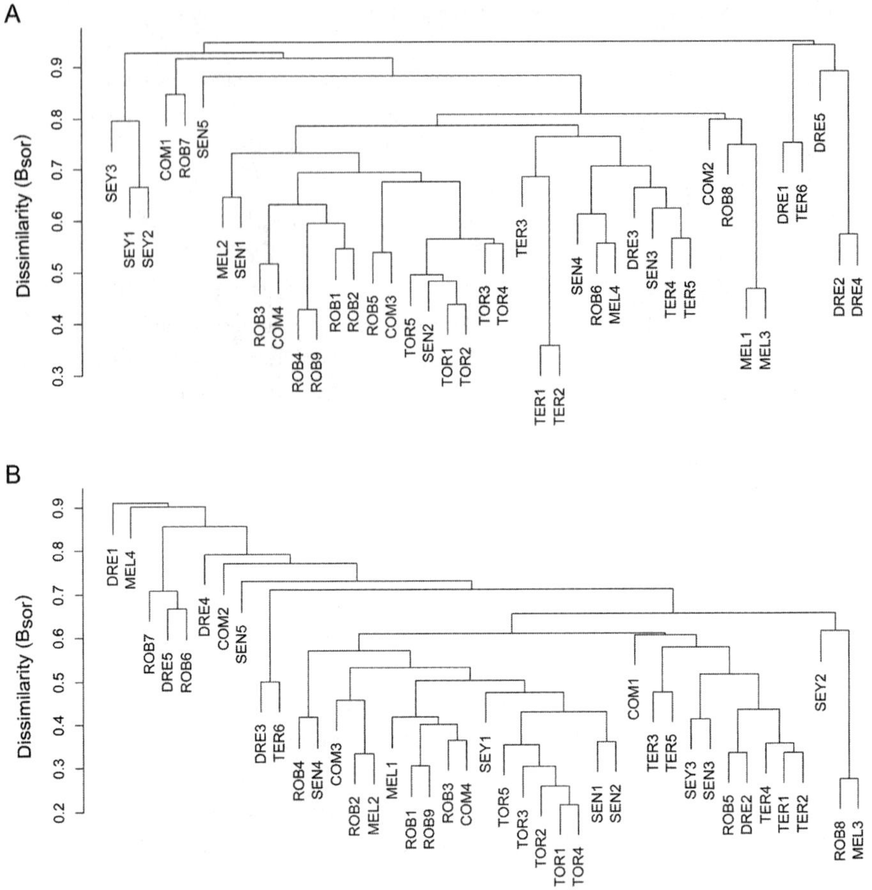

Fig. 13.3 Cluster analysis of 41 sites across *Acacia* savanna (with monotypic stands of *Acacia* species, broad-leaved *Terminalia* woodlands, and rocky hills with *Combretum* woodlands). Dendrograms group sites based on similarity in taxonomic composition (A) and functional composition (B) using β_{sor} and an average linkage method. The y-axis indicates the dissimilarity at which groups are joined (horizontal lines), where larger values indicate greater differences in composition. The dominant macrohabitat of each site is indicated by codes (*A. drepanolobium* = DRE; *A. mellifera* = MEL; *A. robusta* = ROB; *A. senegal* = SEN; *A. seyal* = SEY; *A. tortilis* = TOR; *Terminalia* = TER; *Combretum* = COM).

The cluster analysis of transect segments, based on similarity in taxonomic and functional composition, exhibited two major groups: one included the plains habitats (short-, intermediate-, and long-grass plains), and the other encompassed *Acacia* savanna, rocky hills, and woodland areas (fig. 13.5). These two groups were joined at a dissimilarity of 0.8 and 0.7 for taxonomic and functional dendrograms, respectively. Generally this grouping divides segments along transect 1 from those along transect 4, except for

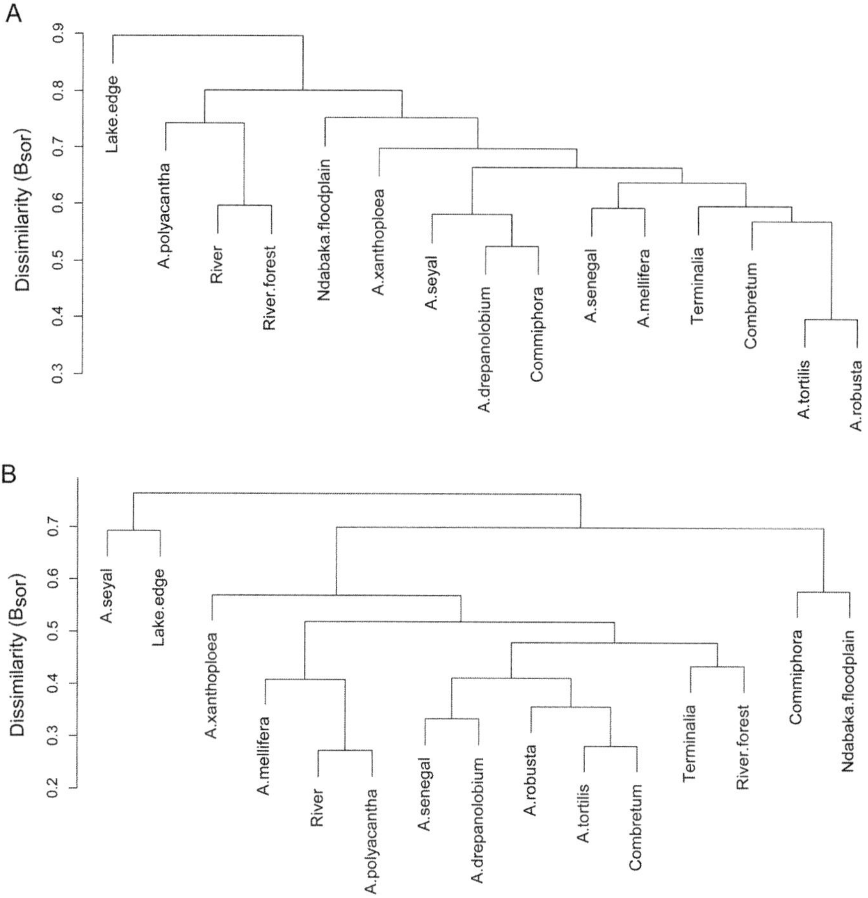

Fig. 13.4 Cluster analysis of 15 habitats found across the *Acacia* savanna, each of which includes data from all sites occurring within that habitat (76 total). Dendrograms group sites based on similarity in taxonomic composition (A) and functional composition (B) using β_{sor} and an average linkage method. The y-axis indicates the dissimilarity at which groups are joined (horizontal lines), where larger values indicate greater differences in composition.

segments along transect 1 which contain *Acacia* savanna (e.g., Naabi, Olduvai Gorge area). These were grouped with other segments supporting *Acacia* savanna, hills, and woodlands. Within the plains group, there was another recognizable subdivision between the short-grass plains and intermediate- and long-grass plains (with a dissimilarity of 0.7 and 0.55 for taxonomic and functional dendrograms, respectively). Transect segments within the intermediate- and long-grass plains group did not form separate subgroups in taxonomic composition (fig. 13.5a), but they largely formed subgroups in functional composition (dissimilarity of 0.5; fig. 13.5b).

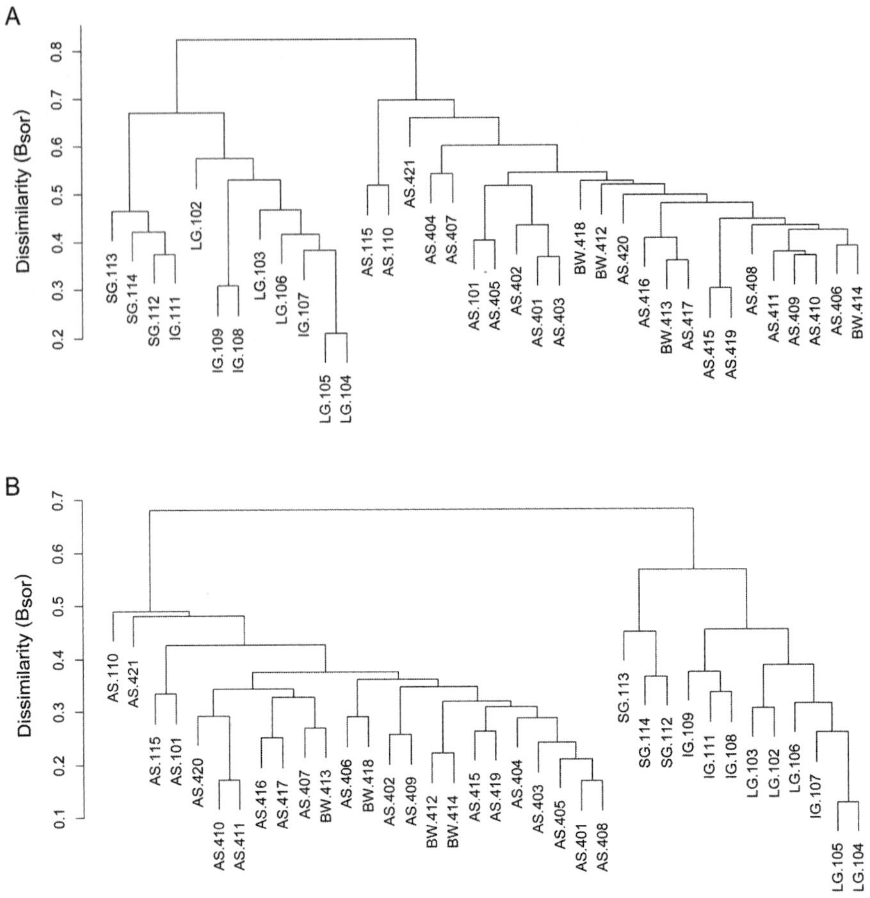

Fig. 13.5 Cluster analysis of 5-km transect segments along transects 1 and 4 during late wet-season surveys in 2003–05. Dendrograms group 5-km segments according to similarity in taxonomic composition (A) and functional composition (B), using β_{sor} and an average linkage method. The dominant macrohabitat of each 5-km segment is indicated by codes (short-grass plains = SG; intermediate-grass plains = IG; long-grass plains = LG; *Acacia* savanna = AS; broad-leaved woodland = BW).

Compositional variation explained by environment. We used MRM to determine whether differences in taxonomic and functional composition across sites and transect segments could be explained by differences in attributes of landscape, vegetation structure, and microclimate between survey locations (table 13.3). Predictors in these models included spatial proximity of survey locations, average precipitation in the 2003–05 period, elevation, habitat type, distance from a water source, and a greenness index. Our models for sites, though significant, explained little variation in dissimilarity, both in species composition ($R^2 = 0.10$, $p = 0.02$) and functional com-

Table 13.3 Significance of predictor variables in explaining bird species dissimilarity (β_{sor}) in a multiple regression on distance matrices (MRM)

	Sites		Transect segments	
Predictors	**Taxonomic**	**Functional**	**Taxonomic**	**Functional**
Spatial proximity	**0.02**	0.56	0.80	0.16
Avg precipitation	0.43	0.92	0.19	0.48
Habitat	**0.001**	**0.04**	**0.001**	**0.001**
Elevation	0.36	0.62	**0.001**	**0.001**
Distance to water	0.07	**0.005**	**0.001**	**0.001**
Greenness index	0.44	0.58	0.10	0.92

Notes: Taxonomic and functional composition were assessed for both sites and transect segments. Significance was evaluated based on 999 permutations. Predictors with *p*-values <0.05 are shown in bold.

position ($R^2 = 0.15$, $p = 0.001$). Predictors with significant effects included spatial proximity and habitat type for the species composition model and distance to water and habitat type for the functional composition model. Our models explained much more of the variation in dissimilarity across transect segments (taxonomic model: $R^2 = 0.62$, $p = 0.001$; functional model: $R^2 = 0.61$, $p = 0.001$); significant predictors, for both taxonomic and functional models, included elevation, distance to water, and habitat. The differences in functional composition across habitats appear to be driven by a number of diet and foraging substrate categories (table 13.4). The most apparent changes are in the number of ground-, shrub- and tree-foraging insectivores and ground- and herbaceous-foraging granivores, as well as the addition of frugivores (all foraging substrates) and tree-foraging species (all diet types) as one moves into habitats with trees.

Relationship between taxonomic and functional dissimilarity. We examined the correlation of taxonomic and functional dissimilarity matrices from sites, habitats (i.e., pooled site data), and transects using a Mantel analysis. If taxonomic dissimilarity increases between pairwise communities, this allows greater opportunity for communities to become more functionally divergent as well, generating a positive 1:1 relationship between taxonomic and functional dissimilarity. However, if functional redundancy is high across communities, we would expect this positive relationship to have a lower slope, with many examples of pairs of communities that are functionally similar despite being more unique in species composition. We find the latter expectation for each of the types of survey data examined (fig. 13.6). Functional dissimilarity increases with taxonomic dissimilarity but at a slower rate, and most pairwise comparisons of communities fall below the

Table 13.4 Number of species (indicated by shading) that belong to diet and foraging location categories (columns) for habitats surveyed across sites and transects (rows). Abbreviations below each diet group indicate foraging locations/substrates (T = tree; S = shrub; H = herbaceous cover; G = ground; W = water; A = aerial; V = vertebrates). Gray shading indicates one of five categories of species richness: white = no species; light gray = 1–6; medium gray = 7–12; dark gray = 13–18; black = 19 or more species.

Macrohabitats	Fruit			Insects							Omnivore				Seeds					Vertebrates		
Sites	T	S	W	T	S	H	G	W	A	V	T	S	H	G	T	S	H	G	W	G	W	V
Lower Catena																						
Upper Catena																						
Broad-leaved Woodlands																						
Riverine																						
Swamp																						
	Fruit			**Insects**							**Omnivore**				**Seeds**					**Vertebrates**		
Transects	T	S	W	T	S	H	G	W	A	V	T	S	H	G	T	S	H	G	W	G	W	V
Short-grass Plains																						
Intermediate-grass Plains																						
Long-grass Plains																						
Acacia Savanna																						
Broad-leaved Woodlands																						

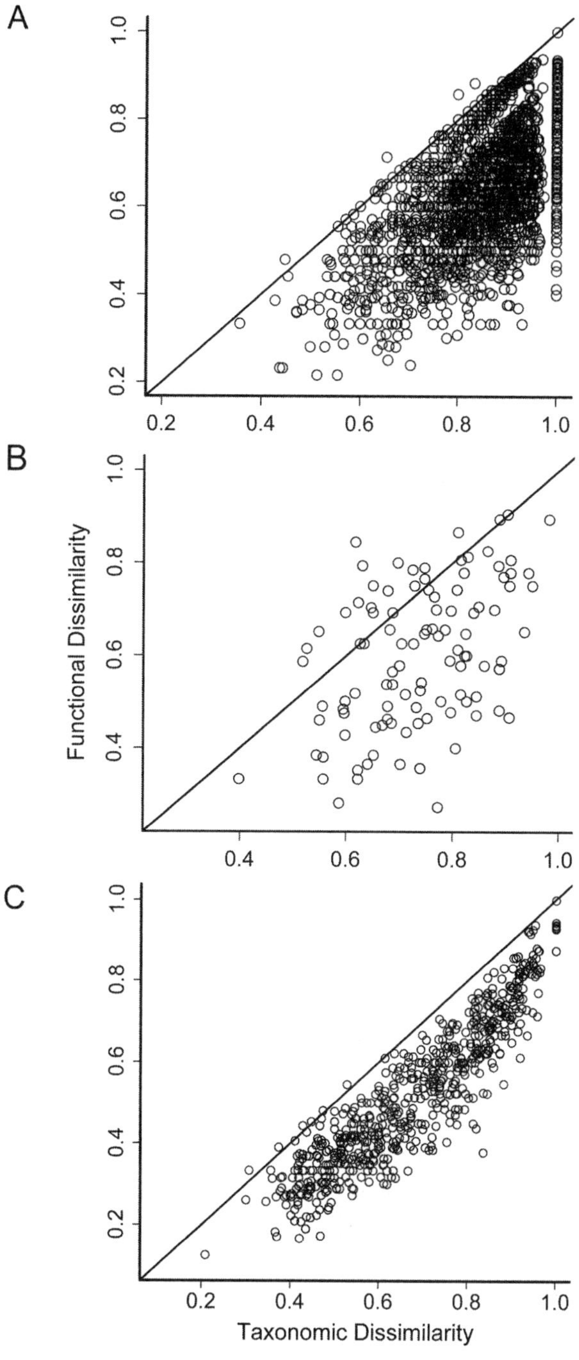

Fig. 13.6 Plot of functional dissimilarity with taxonomic dissimilarity, measured using β_{sor} for pairwise comparisons of sites (A), macrohabitats, using pooled site data (B), and transect segments (C).

1:1 line. Sites were more variable than transects in the relationship between taxonomic and functional dissimilarity (sites: $r = 0.64$, $p = 0.001$; habitats: $r = 0.47$, $p = 0.004$; transect segments: $r = 0.91$, $p = 0.001$) with a high number of pairwise communities showing high functional similarity despite large differences in species composition (fig. 13.6a).

Transitions along Environmental Gradients

Analysis of species dissimilarity and functional dissimilarity between adjacent transect segments showed that, on average, bird communities display moderate changes in composition within 5–10 kilometers (for taxonomic and functional dissimilarity, respectively: avg $\beta_{sor} = 0.48$ and 0.37, avg $\beta_{sim} = 0.38$ and 0.28, and avg $\beta_{nes} = 0.10$ and 0.09). Along the transect, β_{sor} between adjacent 5-km transect segments ranged from 0.21 to 0.76 for taxonomic dissimilarity, and from 0.13 to 0.68 for functional dissimilarity. A number of adjacent-segment comparisons showed dissimilarity values that were much higher than average dissimilarity—we refer to these as dissimilarity "peaks," identified by values greater than one standard deviation above mean dissimilarity of all adjacent-segment comparisons (figs. 13.7, 13.8). There were more peaks in taxonomic dissimilarity (13 peaks, for turnover and nestedness) compared to functional dissimilarity (9 peaks). The peaks in functional dissimilarity matched the location of peaks in taxonomic dissimilarity (with the exception of two peaks in functional turnover; fig. 13.8b), but there are several peaks in taxonomic dissimilarity that do not correspond to functional shifts in the bird community, indicating that some areas of transition in bird species composition are between functionally similar communities. Furthermore, each adjacent-segment comparison with turnover or nestedness greater than one standard deviation above the mean was associated with a noticeable habitat transition (figs. 13.7b, c, 13.8b, c). Generally, the patterns of dissimilarity peaks in the gradient analysis corroborated the groups defined by the cluster analysis: the major fauna transitions occur between plains habitats and *Acacia* savanna, *Acacia* savanna and broad-leaved woodlands (hills), and to a lesser extent, between short-, intermediate- and long-grass plains habitats. The nature of the dissimilarity among these groups, however, differed. Transitions between short- and intermediate-grass plains, intermediate- and long-grass plains, and long-grass plains and *Acacia* savanna exhibit nested patterns (figs. 13.8); the less diverse plains communities form subsets of the more diverse communities adjacent to them. By contrast, transitions between short-grass plains and *Acacia* savanna (at Olduvai Gorge), intermediate-

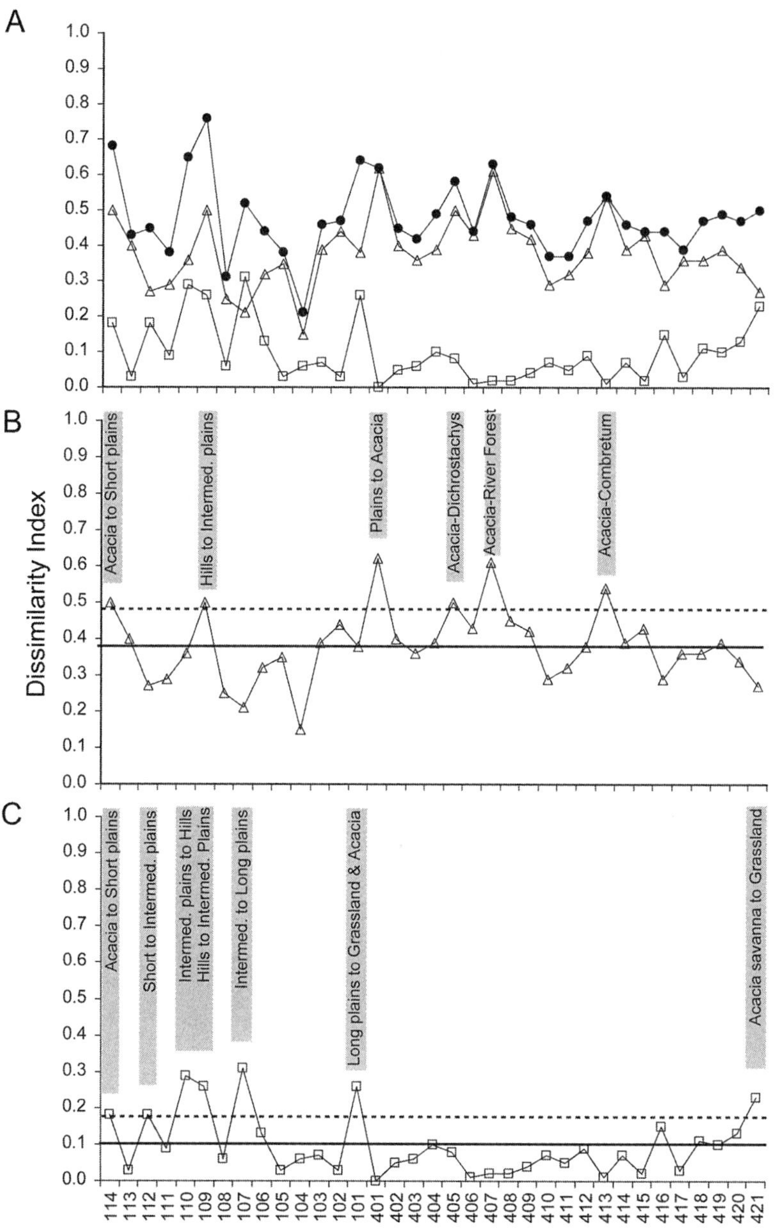

Fig. 13.7 Change in taxonomic composition between adjacent 5-km segments moving from west to east across transects 1 and 4 (x-axis shows numerical code for each segment). Each data point shows a dissimilarity value that compares that 5-km segment to the previous one. Sørensen's dissimilarity index (A) is partitioned into Simpson's (B) and nestedness-resultant (C) dissimilarity indices ($\beta_{sor} = \beta_{sim} + \beta_{nes}$) so that changes due to replacement of species (i.e., turnover; Simpson's index) can be assessed independently from changes due to loss or gain of species (i.e., nestedness-resultant index). For Simpson's and nestedness-resultant indices, the average dissimilarity between adjacent plots (solid line) and 1 standard deviation above this average (dotted line) are indicated. Peaks in compositional change above 1 sd of the mean are shown by grey shading. Shifts in macrohabitat type coincident with these peaks in compositional change, if any, are described by text in grey shaded area.

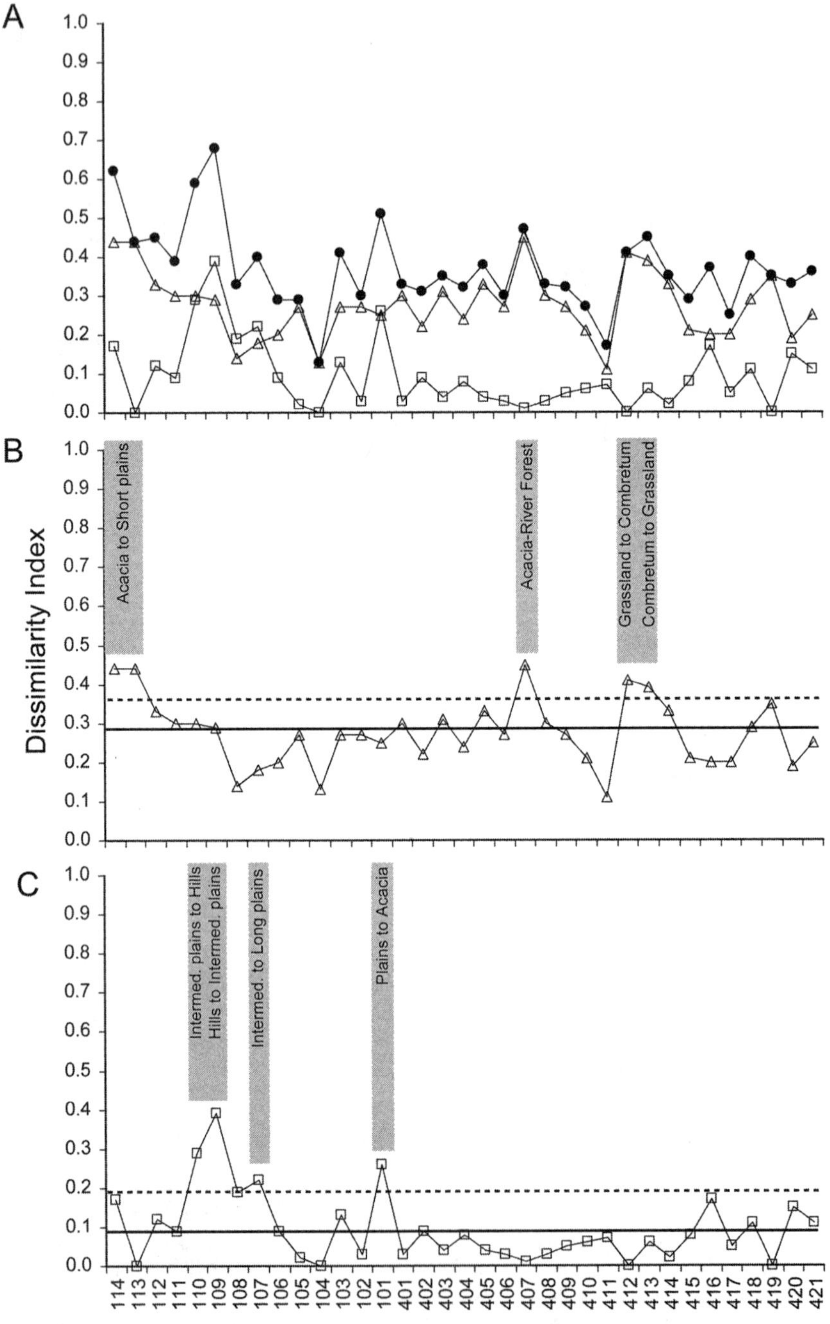

Fig. 13.8 Change in functional composition between adjacent 5-km segments moving from west to east across transects 1 and 4 (x-axis shows numerical code for each segment). See fig. 13.7 for details on panels A, B, and C.

grass plains and hills (Naabi Hill), as well as between *Acacia* savanna and riverine forest or broad-leaved woodlands, show peaks in turnover, such that one community is, to some extent, replaced by another. Some of these transitions show both peaks in turnover and nestedness, where communities also differ substantially in species richness, generally between plains and non-plains habitats (e.g., Olduvai and Naabi). This pattern was largely consistent both in taxonomic and functional compositional change.

DISCUSSION

The Serengeti supports a diverse regional bird fauna with a broad cross section of avian orders. Most locations with higher species richness (>15 species) also harbor bird communities with some level of functional redundancy in ecological roles (e.g., diet, foraging, and body size). Habitats defined in broad terms, with major differences in vegetation composition and structure, have fairly predictable bird communities, both in taxonomic composition and functional composition. These include short-, intermediate-, and long-grass plains (with generally lower species richness), and *Acacia* savanna in combination with broad-leaved woodlands (generally higher species richness). As a corollary, transitions between these major habitat types in the Serengeti landscape also pinpoint areas where bird communities experience the greatest change in functional and taxonomic composition. Notably, areas with the greatest change in functional composition match locations with greatest change in species composition, but the reverse is not always true; some locations show large shifts in species composition that occur between functionally similar communities. While broad habitat categories that we recognize also appear to be generally recognized by birds, there is a limit to which habitat subdivisions offer useful terms for identifying Serengeti bird communities. Monotypic stands of different *Acacia* species, for example, do not harbor (predictably) unique bird communities, and while there is substantial variation in species richness and functional diversity across sites within the *Acacia* savanna and broad-leaved woodlands, such richness and diversity are difficult to predict based on aspects of vegetation structure at finer scales. Thus, bird species of the Serengeti appear to be ecologically sorted in predictable ways, though this is visible only when using a larger lens of habitat differences. Within habitats, communities may be more variable (and less predictable) because of a major influence by stochastic processes on species composition and richness in any given season, by greater fluidity of bird species among structurally similar habitats as individuals track resources, few barriers to dispersal, or perhaps all of these. We now step through our analysis to further assess the ecological organization of Serengeti bird diversity and composition.

Species Richness and Functional Diversity

Differences in sampling intensity restrict our ability to compare richness and functional diversity patterns of survey sites and transects in absolute terms, though because some habitats were sampled using both methods, we can at least make conclusions about relative levels of species richness and functional diversity across Serengeti habitats. Across transects, we found that species richness and functional diversity were generally related to habitats that vary in vegetation complexity. Plains habitats have fewer species and less functional diversity compared to *Acacia* savanna and broad-leaved woodlands. In addition to habitat differences, elevation and distance to a water source were also important predictors of both species richness and functional diversity. Across sites, which focused only on habitats with more complex vegetation structure, variation in species richness was difficult to predict and was not explained by attributes of the vegetation or landscape. These patterns generally corroborate the long-held expected relationship between greater structural complexity of vegetation and higher species richness (MacArthur and MacArthur 1961; Cody 1985), at least when we consider major structural changes (e.g., from plains to *Acacia* savanna), but this relationship is not detectable when we consider fine-scale differences in vegetation complexity (e.g., variation in canopy height, vegetation density, and tree size within specific *Acacia* habitats).

Other avian studies in forest environments have shown that structural differences such as canopy height can have large and detectable effects on richness, particularly for communities dominated by insectivores, but such effects are measured over a large range of structural variation, from 3 to 30 meters, and result in the loss or gain of many foraging substrates (e.g., epiphytes and lianas; Jankowski et al. 2013). Perhaps, the magnitude of variation in structural variables in the Serengeti may not be sufficiently large (e.g., canopy height varies between three and eight meters) or may not create habitats that extend the resource spectrum, which prevents new guilds or additional foraging strategies within guilds from becoming established. It is also possible that some structural elements are critical for certain guilds of the Serengeti community, but specific relationships do not emerge in a community-wide analysis of species richness and functional diversity. For example, tree size should be a limiting habitat attribute for cavity-nesting birds, even though we find that habitats dominated by larger trees do not consistently support higher species richness or functional diversity. Thus, the magnitude and type of structural change is an important element of the structural complexity-richness relationship; in the Serengeti, increasing canopy height, stem density, or tree size appears to have little effect on the number of species (or functional types) compared to the effects of adding more complex grasses within plains or moving from plains to *Acacia* savanna and woodland vegetation.

Our analysis of functional diversity relative to species richness shows that functional types saturate quickly with increasing species richness, with a maximum of only 28–30 functional types, even in areas with over 100 species. This indicates that for most habitats, there are multiple representatives of any given functional type and there is redundancy of functional groups in all but the most species-poor sites. It is important to remember that assessments of absolute functional diversity are contingent upon the number of traits that are summarized across species. With our assessment of functional diversity, the recognition of additional traits would essentially allow more ways in which any two species could differ, and could increase our estimate of functional diversity of a community. This would also increase the level at which communities become functionally saturated and may influence our conclusions concerning the extent of functional redundancy. Qualitatively, however, our results would not change by recognizing additional functional traits. Areas of high species richness would still be likely to have higher functional diversity, and species-poor communities would still be more likely to have little to no functional redundancy compared to species-rich communities. Generally speaking, as long as habitats can maintain moderate levels of species richness (e.g., >50 species), the loss of any given species would not likely result in the loss of that species' ecological role. Further declines in species richness, however, could result in threshold effects, where the loss of a functional group (e.g., seed dispersers and seed predators) has measurable impacts across trophic levels.

Compositional Similarity and Transitions across Serengeti Habitats

Analysis of compositional differences across survey sites and transects indicates that bird communities are predictably sorted according to a number of Serengeti habitats and associated environmental attributes. As in our analysis of species richness, compositional differences are easily recognized across the most structurally different habitat types, principally between plains habitats and *Acacia* savanna/broad-leaved woodlands. Distinct bird communities are also found within short-, intermediate-, and long-grass plains habitats. Bird communities are not as easily recognized within *Acacia* savanna and broad-leaved woodlands, although species do tend to be associated with the catena soil gradient (particularly the upper catena), and species turnover can be high in transitional areas between *Acacia* savanna and riverine forest or *Combretum*, as shown in our transect analysis. These species-habitat associations are more difficult to predict, as even sites within the same habitat can vary substantially in composition. In addition to habitat type, environmental attributes such as elevation and distance to a water source are important predictors of change in composition. Across transect segments, these variables explained up to 62% of the variation in composition. Although these variables were also significant predictors across sites, they explained, at best, 15% of variation in composition. These patterns are consistent when considering species composition and functional composition (based on attributes of diet, foraging substrate, and body size), suggesting that bird communities are likely determined by the type of food resources and foraging opportunities supported by a given habitat and less by fine-scale partitioning of habitats with similar resources.

Our results on compositional differences are largely consistent with a recent study focused on Serengeti grassland birds (Gottschalk et al. 2007). This investigation conducted at ten 25-ha plots across Serengeti National Park similarly concluded that there were identifiable bird communities belonging to short-, intermediate-, and long-grass plains, and *Acacia* savanna; however, Gottschalk et al. (2007) found that intermediate- and long-grass plains habitats, together, exhibited bird communities more similar in composition to *Acacia* savanna, whereas we find a major division between the plains habitats, as one group, and *Acacia* savanna and broad-leaved woodlands, as another. At least two factors may contribute to this discrepancy. First, Gottschalk et al. (2007) consider abundances of species, whereas we only consider species presence-absence. Incorporating information on the dominance of shared species across habitats could greatly affect conclusions about similarity in species composition, especially with the large differences in density exhibited across Serengeti birds; Gottschalk et al. (2007) show that territory density varied from 1 to 40 across species and plots. In such an assessment, any rare species will have negligible effects on community similarity, whereas in our analysis, which weighs rare and common species equally, calculated differences will be influenced, to a greater extent, by changes in occurrence of rare species. Second, our study provides a much broader cross section of the Serengeti bird fauna, including data for 289 species (150 species along transects), whereas Gottschalk et al. (2007) base their analysis on 34 species, approximately two-thirds of which are small-bodied insectivorous passerines that lived in the herb or ground layers only. Given these differences in taxonomic and functional extent and in analysis of the Serengeti communities between studies, it is somewhat remarkable that we find such similar patterns.

By considering compositional patterns in both taxonomic and functional terms, our analysis reveals several notable aspects of functional redundancy across communities in the Serengeti. As we should expect, communities that are more dissimilar in species composition are also more divergent in the functional roles of species, and taxonomic dissimilarity sets an upper limit on how functionally distinct any two communities can be. However, there are many cases in which communities that differ greatly in species composition are quite functionally similar. Several examples of this functional redundancy across communities are shown between *Acacia* habitats belonging to the upper and lower catena soil gradient (fig. 13.4). Such a pattern could reflect spatial partitioning of species with similar functional roles across habitats, such that those species that are similar in body mass and share functional traits related to foraging behavior are more likely to occur in separate habitats than within the same habitat. This possibility warrants further investigation of the extent of functional overlap in co-occurring species, perhaps with explicit spatial and phylogenetic framework. In contrast to these discrepancies between functional and taxonomic similarity, there are a number of examples of Serengeti habitats that are similar in both taxonomic and functional composition, particularly the plains habitats, the *Acacia* savanna (collectively), and within the *Acacia* savanna, habitats within the upper catena dominated by large tree species (*A. tortilis*, *A. robusta*, and to a lesser extent, *Terminalia* and *Combretum*). These habitats may be important targets for conservation, as they appear to harbor unique communities both in their species composition and functional roles of species. Large-tree habitats, in particular, may be critical for some arboreal birds, and particularly cavity-nesting species (e.g., hornbills, woodpeckers, parrots).

Another aspect of the Serengeti bird fauna that we uncovered is that the identifiable bird communities represented by different habitats of the Serengeti can either form nested subsets of one another, a pattern driven by the loss or gain of species, or they can show patterns of turnover (i.e., replacement of one species group by another). Patterns of nestedness and turnover between habitats are largely consistent when considering changes in taxonomic or in functional composition. In particular, gradual increases in habitat structural complexity along the major precipitation gradient in the Serengeti, from short- to intermediate-grass plains, from intermediate- to long-grass plains, and from long-grass plains to *Acacia* savanna, show compositional change in the form of nested communities; as vegetation complexity increases, species and functional groups accumulate to yield more taxonomically and functionally diverse communities. These increases in species richness occur in approximately half of the diet and foraging substrate functional groups (table 13.4). Notably, the habitats at either end of this continuum—short-grass plains and *Acacia* savanna—when compared, predominantly represent distinct (rather than nested) communities, as noted by the high taxonomic and functional turnover in the southeastern end of the transect where a limited area of *Acacia* savanna gives way to the short-grass plains. Interestingly, our analysis of habitat transitions detects regions within the *Acacia* savanna (transect 4) that show high taxonomic and functional turnover, principally between *Acacia* savanna and either riverine forest or broad-leaved woodlands found in the hills region. These broad-leaved woodlands, however, have a less consistent taxonomic and functional composition. Broad-leaved woodland habitats were not closely grouped in the cluster analysis of transect segments, and likewise, analysis of sites revealed substantial variability in their composition. Thus, although these transitions exhibit changes in composition of bird communities, they do not do so in predictable ways. We should consider that the broad-leaved woodland sites in the hills region form a small subset of the data analyzed in this study, and future work may further investigate the composition of these habitats to determine if a larger sampling reveals distinctive features of the bird communities they harbor.

CONCLUSIONS AND FUTURE DIRECTIONS

Our analysis of the Serengeti bird fauna reveals communities that are invariably structured according to dominant habitats in this system, principally those that show major differences in structural complexity. Habitat subdivisions, although recognizable by human observers, are less useful in defining bird communities. Richness gradients are also evident across dominant habitats; the effect is that bird communities are nested across short-, intermediate-, long-grass plains, and *Acacia* savanna, as species representing multiple functional groups accumulate in more structurally complex communities.

It is important to remember that our assessment of the Serengeti bird fauna uses a three-year snapshot of diversity and composition; however, precipitation and natural disturbance regimes, which are major factors influencing Serengeti habitats, and standing vegetation structure can vary substantially seasonally and across years, such that patterns in diversity and composition observed in our analysis of "typical" wet seasons may be intensified or weakened in other years, as during El Niño climate oscillations (Sinclair et al. 2013; Fryxell et al. Chapter 7). An important next step will be to understand how the avian habitat relationships and levels of functional redundancy that we have uncovered in this analysis shift over these climate cycles.

ACKNOWLEDGMENTS

The directors of Tanzania National Parks and the Tanzania Wildlife Research Institute gave permission to work in the park. We thank especially Simon Mduma, Anne Sinclair, Ally Nkwabi, John Bukombe, John Mchetto, Stephen Makacha, and Joseph Masoy for their assistance in the fieldwork and Julio Di Rienzo for his assistance with data analysis. This work was supported by the Canadian Natural Sciences and Engineering Research Council grant to ARES.

REFERENCES

Arcese, P., and A. R. E. Sinclair. 1997. The role of protected areas as ecological baselines. *Journal of Wildlife Management* 61:587–602.

Baselga, A. 2010. Partitioning the turnover and nestedness components of beta diversity. *Global Ecology and Biogeography* 19:134–43.

Carroll, M. L., C. M. DiMiceli, R. A. Sohlberg, and J. R. G. Townshend. 2004. 250m MODIS normalized difference vegetation index, University of Maryland. http://glcf.umd.edu/data/ndvi/.

Casanoves, F., L. Pla, J. A. Di Rienzo, and S. Díaz. 2010. FDiversity: a software package for the integrated analysis of functional diversity. *Methods in Ecology and Evolution* doi: 10.1111/j.2041-210X.2010.00082.x.

Cody, M. L., ed. 1985. *Habitat selection in birds*. New York: Academic Press.

Devictor, V., D. Mouillot, C. Meynard, F. Jiguet, W. Thuiller, and N. Mouquet. 2010. Spatial mismatch and congruence between taxonomic, phylogenetic and functional diversity: The need for integrative conservation strategies in a changing world. *Ecology Letters* 13:1030–40.

Díaz, S., S. Lavorel, F. de Bello, F. Quétier, K. Grigulis, and T. M. Robson. 2007. Incorporating plant functional diversity effects in ecosystem service assessment. *Proceeding of the National Academy of Sciences* 104:20684–89.

Dunning, J. B. 2008. *CRC handbook of avian body masses*. Boca Raton, FL: CRC Press, Taylor & Francis Group, LLC.

Folse, L. J. 1982. An analysis of avifauna-resource relationships on the Serengeti plains. *Ecological Monographs* 52:111–27.

Gill, F. and D. Donsker, eds. 2012. IOC World Bird Names (v 3.3). Available at http://www.worldbirdnames.org/. [Accessed October 2012].

Gottschalk, T. K., K. Ekschmitt, and F. Bairlein. 2007. Relationships between vegetation and bird community composition in grasslands of the Serengeti. *African Journal of Ecology* 45:557–65.

Jankowski, J. E., C. L. Merkord, W. F. Rios, K. C. Cabrera, R. N. Salinas, and M. R. Silman. 2013. The relationship of tropical bird communities to tree species composition and vegetation structure along an Andean elevational gradient. *Journal of Biogeography* 40: 950–62.

Krebs, C. J. 1989. *Ecological methodology*. New York: Harper & Row.

Legendre, P., and L. Legendre. 1998. *Numerical ecology*, 2nd English ed. Amsterdam: Elsevier Science.

Lennon, J. J., P. Koleff, J. J. D. Greenwood, and K. J. Gaston. 2001. The geographical structure of British bird distributions: Diversity, spatial turnover and scale. *Journal of Animal Ecology* 70:966–79.

Lichstein, J. W. 2007. Multiple regression on distance matrices: A multivariate spatial analysis tool. *Plant Ecology* 188:117–31.

MacArthur, R. H., and J. MacArthur. 1961. On bird species diversity. *Ecology* 42:594–98.

Magurran, A. E. 2004. *Measuring biological diversity*. Oxford: Blackwell Science Ltd.

Nkwabi, A. K. 2014. Influence of habitat structure and seasonal variation on abundance, diversity and breeding of bird communities in selected parts of the Serengeti National Park, Tanzania. PhD diss., University of Dar es Salaam, Tanzania.

Nkwabi, A. K., A. R. E. Sinclair, K. L. Metzger, and S. A. R. Mduma. 2011. Disturbance, species loss and compensation: Wildfire and grazing effects on the avian community and its food supply in the Serengeti Ecosystem, Tanzania. *Austral Ecology* 36:403–12.

Petchy, O. L., and K. J. Gaston. 2006. Functional diversity: Back to basics and looking forward. *Ecology Letters* 9:741–58.

R Development Core Team. 2010. *R: A language and environment for statistical computing*. R Foundation for Statistical Computing, Vienna, Austria. Accessed November 20, 2011, http://www.R-project.org/.

Schmera, D., T. Erős, and J. Podani. 2009. A measure for assessing functional diversity in ecological communities. *Aquatic Ecology* 43:157–67.

Schmidl, D. 1982. *The birds of the Serengeti National Park, Tanzania*. London: BOU Check-list no. 5, British Ornithologists' Union.

Sinclair, A. R. E. 1978. Factors affecting the food supply and breeding season of resident birds and movements of palaearctic migrants in a tropical African savannah. *Ibis* 120: 480–97.

Sinclair, A. R. E. 1995. Equilibria in plant-herbivore interactions. In *Serengeti II: Dynamics, management and conservation of an ecosystem*, ed. A. R. E. Sinclair and P. Arcese, 91–113. Chicago: University of Chicago Press.

Sinclair, A. R. E. 1998. Natural regulation of ecosystems in Protected Areas as ecological baselines. *Wildlife Society Bulletin* 26:399–409.

Sinclair, A. R. E., S. A. R. Mduma, and P. Arcese. 2002. Protected areas as biodiversity benchmarks for human impact: Agriculture and the Serengeti avifauna. *Proceedings of the Royal Society of London B* 269:2401–5.

Sinclair, A. R. E., K. L. Metzger, J. M. Fryxell, C. Packer, A. E. Byrom, M. E. Craft, K. Hampson, T. Lembo, S. M. Durant, and G. J. Forrester, et al. 2013. Asynchronous food web pathways could buffer the response of Serengeti predators to El Niño southern oscillation. *Ecology* 94:1123–30.

Sinclair. A. R. E., A. Nkwabi, S. A. R. Mduma, and F. Magige. Forthcoming. Responses of the Serengeti avifauna to long-term change in the environment. *Ostrich. Journal of African Ornithology*.

Sinclair, A. R. E., C. Packer, S. A. R. Mduma, and J. M. Fryxell, eds. 2008. *Serengeti III: Human impacts on ecosystem dynamics*. Chicago: University of Chicago Press.

Walker, B. H., A. Kinzig, and J. L. Langridge. 1999. Plant attribute diversity, resilience, and ecosystem function: The nature and significance of dominant and minor species. *Ecosystems* 2:95–113.

FOURTEEN

The Effect of Natural Disturbances on the Avian Community of the Serengeti Woodlands

Ally K. Nkwabi, Anthony R. E. Sinclair, Kristine L. Metzger, and Simon A. R. Mduma

THE ROLE OF DISTURBANCE IN ECOLOGICAL COMMUNITIES

Heterogeneity of environment, landscapes, and habitat is thought to contribute to biodiversity, ecosystem function, and stability of the community by maintaining trophic structure (Terborgh et al. 2001; du Toit, Rogers, and Biggs 2003; Terborgh and Estes 2010). Such heterogeneity can be produced by both spatial and temporal variability of disturbances resulting in a change in both species composition and their abundances (Adler, Raff, and Lauenroth 2001; Fuhlendorf and Engle 2001, 2004; Fuhlendorf et al. 2006). The change in species richness or abundance can be due to either natural disturbances such as fluctuations in climate, wildfire, or changes in other species populations, change imposed by humans, or both combined. However, disturbance, particularly that imposed by humans from habitat modifications such as agriculture, can be so severe that there is a decline in biodiversity as seen in the bird fauna, as happened in Britain (Fuller et al. 1995; Robinson et al. 2004; Butler, Vickery, and Norris 2007). To distinguish between the alternative outcomes of stability or decline of biodiversity from disturbance, one needs an ecological benchmark that acts as a baseline. Protected areas such as national parks act as a basic global network of baseline sites in which to study disturbances in the absence of agricultural effects (Sinclair 1998; Arcese and Sinclair 1997; Sinclair and Byrom 2006; Newmark 2008).

The effect of fire and grazing has been examined by Fuhlendorf et al. (2008) on the tallgrass prairies of the Great Plains of North America. They emphasized that there is an interaction of these two disturbances because burning attracts grazers—they called this synergistic impact "pyric herbivory." In Africa a similar interaction of these disturbances has been studied in western Serengeti (Hassan et al. 2008). Gregory, Sensenig, and Wilcove (2010) have documented the separate effects of experimental fires and livestock grazing on bird communities in Kenya grasslands.

Disturbance in the Serengeti Savanna

Grass fires have dominated African savanna landscapes for hundreds of millennia (Bird and Cali 1998) and plants are fire tolerant. Hence, the vegetation in savanna Africa has become adapted and shaped by fire so that it can be considered as a natural disturbance (van Langevelde et al. 2003), and this applies in the Serengeti ecosystem (Norton-Griffiths 1979; Sinclair 1979; Stronach 1988; Sinclair et al. 2007). Equally, grazing by native ungulates has been impacting African savannas for millions of years so that the grass layer is adapted to these effects.

The Serengeti ecosystem supports some 613 species of birds, of which there are some 95 insectivores and 71 seed eaters in the ground/herb layer, and 150 insectivores versus 28 fruit/seed eaters in the tree or shrub layer. Impacts of fire and grazing on the avifauna have already been examined in the treeless long grasslands (Nkwabi et al. 2011). Savanna, however, supports trees with their additional complement of birds not found in grasslands. The importance of African savannas for birds is reflected in the alarming decline of Afro-Palearctic migrant bird species since 1970; it is not yet clear whether the cause of these declines lies in the African savannas or in the Palearctic (Sanderson et al. 2006).

Fire is frequent in the Serengeti savannas, usually occurring in July when the grassland dries out. Fires are set by humans; lightning strikes occur in October and November at the beginning of the rains but by that time, all combustible material has been removed (Norton-Griffiths 1979; Dempewolf et al. 2007). Grazing impacts also occur first in July as the vast migrating herds move off the plains and into the savanna. Grazing from hundreds of thousands of animals, predominantly wildebeest, reduces the grass layer by some 50% of its biomass within a few days, and so creates a rapid change in grass structure (Sinclair 1977); these animals do not return until the next wet season four to six months later.

AIM OF THIS STUDY

In this chapter we examine whether the impacts of fire and grazing disturbances of the grass layer result in a decline or increase in biodiversity of birds in the Serengeti savanna. In particular, we document changes in the abundance, richness, and composition of birds in the two major savanna components, the grass layer and the tree/shrub communities.

Both burning and grazing disturbances have their major effect by altering the grass structure from long grass (80–100 cm) to shorter grass (5–25 cm), and both occur at about the same time of year, the beginning of the dry season. Fire removes effectively all the biomass, which must then regrow as short grass. Because fires occur at the start of the dry season subsequent regrowth is in the form of leaves less than 15 cm high with no flower heads. Grass then remains in this structure until the rains begin in November. For the first month there is usually so little grass that grazers avoid the burns. In the second and third months there may be some regrowth which attracts wildebeest in particular. Thus, we examine two time periods subsequent to the fire, a postfire period of one month (period 1), and a later dry season period of two months (period 2).

Grazing removes some 50% of the biomass leaving stubble 10–25cm high (Sinclair 1977, Sinclair, Dublin, and Borner 1985). This stubble is usually too short to support fire so there is no interaction effect in this treatment. Fire is a more extreme disturbance than grazing in altering the grass structure. In addition, fire, unlike grazing, kills most of the insects living in the long grass. After fire, time is required both to regrow green vegetation and to recolonize the habitat with insects.

We predict, therefore, that burning would affect the bird community more than grazing, particularly in time period 1. Second, we predict that birds using the grass or ground layers would be more directly disturbed than those using the shrub or tree layers. Third, the reduction in grass biomass with both disturbances would reduce the food for birds, both insects and seeds, and so reduce bird abundance; alternatively the lower grass biomass allows greater visibility for searching the ground and increases access to both types of food, thus increasing bird abundance. Fourth, the change in grass structure also predicted a change in species by favoring those that were ground feeders rather than grass stem feeders. In other words there was compensation for species loss by inclusion of functionally equivalent species, as found by Nkwabi et al. (2011).

THE STUDY AREA

The Savannas

The Serengeti savanna is dominated by fine-leaved *Acacia* trees over most of the area. The dominant species in the area we studied were *A. robusta* and *A. tortilis* with shrubs of *Grewia bicolor* and *Cordia sp.* The far northwest supports a form of broad-leaved woodland dominated by *Terminalia* and *Combretum* species with many shrubs such a *Croton, Techlia, Euclea, Rhus,* and *Heeria* (see chapter 9 for a greater description of vegetation in this habitat). In these savannas deeper soils allow larger tussock grass species, the dominants being *Themeda triandra* and *Pennisetum mezianum* in *Acacia*, and *Hyparrhennia* grasses in *Terminalia* (Herlocker 1976). The *Acacia* savanna is generally drier with some 800 mm rain per year while the *Terminalia* receives 1,200 mm per year (Norton-Griffiths, Herlocker, and Pennycuick 1975).

The Bird Fauna

Descriptions of the avifauna in the Serengeti ecosystem are given by Jankowski et al. in chapter 13 and in Sinclair (1978), Folse (1982), Sinclair, Mduma, and Arcese (2002), and Nkwabi et al. (2011). The whole Serengeti ecosystem supports some 613 species of birds made up of 45 groups (families or subfamilies) of insectivorous birds comprising 279 species, and 16 groups of granivorous or fruit-eating birds made up of 113 species (see appendix 4 at www.press.uchicago.edu/sites/serengeti/). Not all of these species appeared in our studies because they used habitats that we were not sampling, such as forest, floodplain, rocks and hills, or short-grass plains. We consider in this chapter only those inhabiting the two major types of savanna woodlands of *Acacia* and *Terminalia*. For ease of documentation we use acronyms for species in the tables of results, the full names of which are given in table 14.1.

METHODS

Sites and Treatments

Sites for burning and grazing impacts were chosen each year immediately after these treatments had taken place in June or July for the years 2001–04 and 2009. The migratory herds of wildebeest move into the savannas each year in June and their routes and timing have been highly predict-

Table 14.1 Acronyms and common names used in the text

ACRONYM	NAME
astl	Athi short-toed lark
batf	Banded tit-flycatcher
bbwa	Buff-bellied warbler
bhbs	Brown-headed tchagra
blba	Black-lored babbler
bwpl	Black-winged plover
cofr	Coqui francolin
crci	Croaking cisticola
crpl	Crowned plover
csfl	Chin-spot flycatcher
daba	Usambiro barbet
fcla	Fawn-colored lark
flla	Flappet lark
gbca	Grey-backed camaroptera
gbfi	Grey-backed fiscal
gwho	Green wood-hoopoe
hegu	Helmeted guineafowl
hist	Hildebrandt's starling
mash	Magpie shrike
nobr	Northern brubru
nuwo	Nubian woodpecker
raci	Rattling cisticola
rbbw	Red-billed buffalo weaver
rbsr	Red-backed scrub-robin
rcla	Red-capped lark
rtti	Red-throated tit
rtwe	Rufous-tailed weaver
scbo	Slate-colored boubou
stki	Striped kingfisher
sust	Superb starling
wast	Wattled starling
wcsh	White-crowned shrike
whbw	White-headed buffalo weaver
ytlc	Yellow-throated longclaw
zici	Zitting cisticola

able for the past 40 years. Three sites were selected for the treatment plots after the wildebeest had arrived in the long grasslands and after burning commenced, the two treatments starting effectively simultaneously. The selection of sites had to await these events in order to place the treatments. Within a site there were three replicates for grazing and three for burning with paired undisturbed plots (ungrazed and unburnt). The exact locations of the grazed plots were chosen once the herds had moved through, which normally takes a few days. Burns were set by park rangers during the time that the wildebeest migration moved through. The burnt treatment plots were located within these burned areas. The matched undisturbed plots were then placed nearby where neither grazing nor burning had previously taken place. Each replicate plot was 70 m × 70 m in size (0.5 ha), located using a Global Positioning System (GPS), and marked by flagging tape at the corners of the site. The demarcation defined the boundaries for counting birds.

Censusing Birds

Six plots were surveyed per day (that is, all burn and undisturbed or grazed and undisturbed were counted on one day). These counts were repeated for three days and the numbers for each plot averaged over the three days. The number of days was determined from a pilot study where plots were counted for up to six days. Since all species present on a plot had been recorded after three days, this number of days was chosen. A 20-minute count period was used in each plot, and the time of day between 0630 and 1030h in the morning was rotated among the six plots to even out this possible bias. Birds were counted by walking slowly across the plot and back again. Birds were identified by both sight and call, and numbers recorded (Pomeroy and Tengecho 1986; Pomeroy 1992; Bibby et al. 2000). The open nature of the habitat allowed us to locate individual birds and prevent double counting; if a bird moved across the plot it would be seen and noted.

Birds were counted in two time periods. Period 1: Counts were conducted from shortly (<1 week) after burning or grazing affected the study plots until end of July, this period having little or no regrowth of grass. Period 2: Plots were monitored through August and September, this time being in the middle of the dry season but after short, green regrowth had taken place.

Data Analysis

In this analysis we included all birds that fed on insects or seeds in the tree, herb and ground layers. We excluded vertebrate feeders and aerial insectivores because their activities generally covered a much larger scale than that of our experiments. Birds were classified according to feeding guild, namely insectivores or granivores. Frugivores were included with granivores because there were too few species to treat them as a separate category. Birds were also classified into three groups according to where and how they fed, namely (1) ground plus herb layer feeders, (2) perching species that used trees to see insects in the grass layer, and (3) tree plus shrub feeders. The first of these had approximately equal numbers of species of insect and seed feeders, the second group was entirely insectivores, and the third group was largely insectivores.

The abundances of birds in each of the above categories in the different treatments and time periods were averaged across replicate sites and Standard Errors computed for each treatment and time period. Frequency distributions of species in the two savanna habitats and treatments were compared using the Bray-Curtis coefficient (*B*) as the similarity value (1–*B*) such that zero is complete dissimilarity and unity is complete similarity (Krebs 2001). Abundances of individual species were used in the comparisons.

RESULTS

Bird Species Abundances

The majority of birds observed were insectivores, which is to be expected since they predominate in the Serengeti bird fauna. In time period 1, within a few days after the fire, burned plots showed close to doubled numbers of insectivores (fig. 14.1a) for those that fed on the ground in Acacia savanna. The effect disappeared by period 2. An increase of insectivores was also recorded in *Terminalia* with a greater effect in period 2 (fig. 14.1b). This increase was also reflected in perching feeders although the small sample size required merging the two time periods (fig. 14.1c). This result was also seen on burned plots in the long-grass plains (Nkwabi et al. 2011). In contrast, the opposite result was seen for those feeding in trees and shrubs in both woodlands (fig. 14.1d, e). Grazed plots showed little effect of the treatment (fig. 14.1). In general there were slightly more individuals in the ungrazed plots than in the grazed plots in both time periods and for all feeding locations.

Ground feeding granivores showed a small but consistent increase in numbers on both burned and grazed plots in the first period but there was

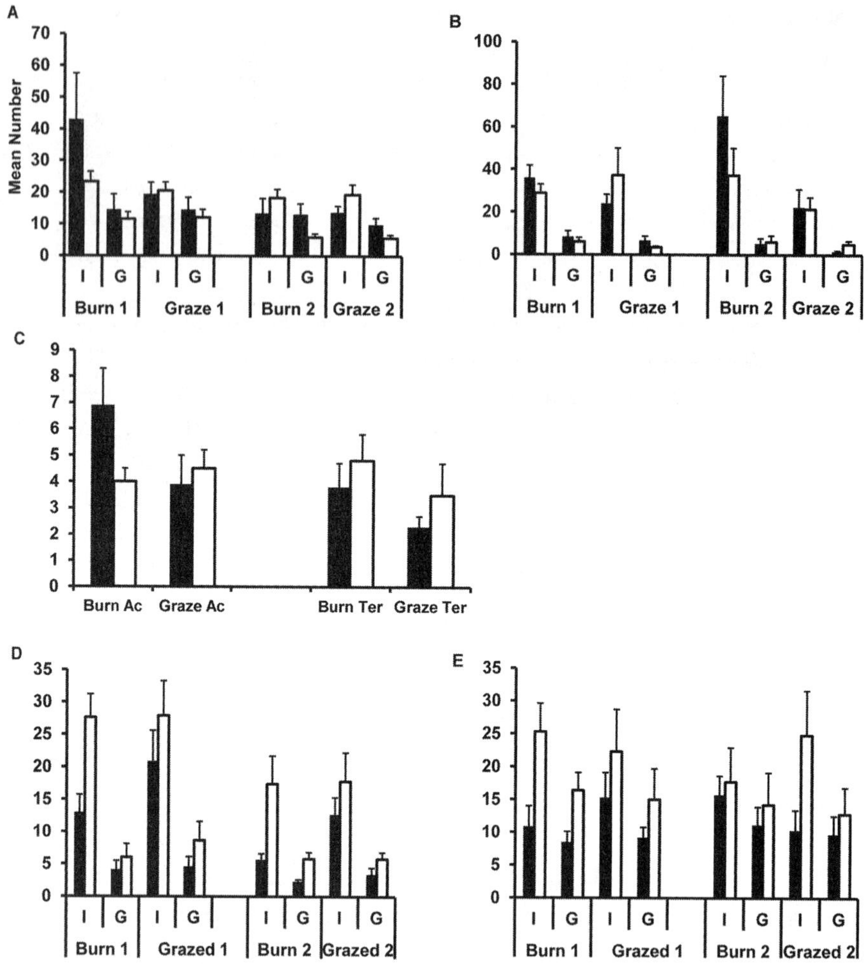

Fig. 14.1 The mean densities of insectivores (I) and granivores (G) on sites with treatments (solid) and controls (open). The first time period is indicated by Burn 1, Grazed 1, the second by Burn 2, Grazed 2. Panel A, ground feeding birds in Acacia. Panel B, ground feeding birds in Terminalia. Panel C, perching birds (insectivores only) for both time periods combined in Acacia (Ac) and Terminalia (Ter). Panel D, tree and shrub feeders in Acacia. Panel E, tree and shrub feeders in Terminalia. Mean densities calculated across replicates within habitats. Vertical bars are 1 SE.

less effect in the second period in *Terminalia* (fig. 14.1). Thus, the effect of removal of plant biomass was pronounced in the first month but less in the remaining two months of observation. In contrast, birds feeding on seeds and fruits in trees were consistently less abundant on both burned and grazed plots in both woodland types. The disturbances had a negative effect on tree feeders in general.

Bird Species Richness

The number of species on the plots showed little change with treatment (table 14.2). There were 15–19 insectivore species feeding on the ground in both burned and unburned plots. There were 13 species on grazed plots compared to 17–19 on ungrazed plots. There were 6–8 species of perching insectivores, and numbers did not show much response to either treatment. Similarly tree and shrub insectivores varied between 13 and 22 species but not in any consistent way with treatment.

Granivore species were also little affected by treatment. There were 8–12 ground feeding species and 3–10 species in trees, with more present in period 1 than in period 2. In general, species richness did not respond to the removal of grass biomass.

Bird Species Composition

Insectivores. The similarity of species composition between treatment and control was measured from their relative abundances by the Bray-Curtis coefficient (1–*B*) (Krebs 2001). In almost all cases there was a lower similarity in the species between burnt and unburnt plots than between grazed and ungrazed plots, indicating that burning had a greater impact on the bird community than did grazing (table 14.3). The only exception was for granivores in *Terminalia* where sample sizes were too small to say anything definite.

Counts for the two time periods in *Acacia* savanna were summed because there was little difference between the two periods, to compare the frequency distribution of species in treated and untreated plots. Figure 14.2a shows the top ten most abundant ground-feeding insectivores on unburnt plots and their equivalent counts on the burnt plots. The numbers and proportion of the community is given in table 14.4. The most abundant species, rattling cisticola, was reduced by some 80% on burnt plots because it normally inhabits tall grass and shrubs that are largely removed with burning. However, six species increased in density and frequency with burning. These include starlings, larks, buffalo-weavers, and game birds, all of which feed on the ground. In addition the rufous-tailed weaver and crowned plover appear on burnt ground when they are absent or rare in long-grass areas (table 14.4).

Figure 14.2b shows the ten most abundant ground-feeding insectivores on ungrazed plots and their equivalant counts on grazed plots. The numbers and proportion of the community are given in table 14.4. In contrast

Table 14.2 Species richness in Acacia and Terminalia savanna woodland sites with burning or grazing compared with their controls. Period 1 was one month after, and period 2 was two months after disturbance

	Acacia					Terminalia			
	Burnt	**Unburnt**	**Grazed**	**Ungrazed**		**Burnt**	**Unburnt**	**Grazed**	**Ungrazed**
Ground insectiv					Ground insectiv				
Period 1	16	15	13	17	Period 1	17	20	16	16
Period 2	18	19	13	19	Period 2	11	13	14	14
Ground granivores					Ground granivores				
Period 1	8	8	11	9	Period 1	5	4	4	2
Period 2	12	10	12	10	Period 2	4	4	3	5
Perching insectiv					Perching insectiv				
Period 1	6	8	6	6	Period 1+2	5	4	4	4
Period 2	6	7	4	7					
Tree insectiv					Tree insectiv				
Period 1	19	21	19	22	Period 1	15	19	16	14
Period 2	13	18	18	18	Period 2	17	15	11	15
Tree granivores					Tree granivores				
Period 1	6	7	6	10	Period 1	8	7	7	7
Period 2	3	5	5	5	Period 2	7	7	7	7

to the burn treatment, grazing produced a decline in eight of the ten species although the change was very small. In general, grazing had little effect on the community as indicated by the high similarity values (table 14.3). The *Terminalia* savanna showed similar changes in bird species composition to that of *Acacia*.

The effect of burning and grazing on the tree or shrub-feeding insectivore community is variable (table 14.5). With burning, half of the top ten species in untreated plots decreased and half increased. In addition, green wood-hoopoe appeared in burnt plots when they were absent in unburnt plots. With grazing there was little change in frequency compared to the ungrazed plots.

Seed and fruit feeders. The small increase in abundance of birds on burnt ground in *Acacia* woodland was due to the increase in a suit of small weavers, sparrows, and estrildines, and there was a similar but less noticeable effect with grazing. Sample sizes were, however, too small to detect significant changes.

In contrast to ground feeders, tree and shrub feeders declined in number with burning. In both woodland types, parrots, barbets, bulbuls, and turacos preferred unburnt plots over burnt plots. There were no species more abundant in either burnt or grazed plots in *Terminalia*, and only the blue-eared glossy starling was more abundant in *Acacia* burnt plots. In general, burning and grazing did not attract tree-feeding fruit or seed eaters.

Compensation

Nkwabi et al. (2011) found in the long grassland of the plains one lark species, the red-capped, replaced the athi short-toed lark with burning, thus providing evidence that species replace each other in functional roles—a process we call compensation (fig. 14.3). In the savanna the habitats are structurally more complex with concomitantly more species. With ground layer insectivores those using long grass, such as cisticolas and prinias, were replaced by starlings, larks, plovers, and gamebirds in both woodland types after fire (table 14.6).

In ground layer seedeaters the white-bellied canary was more abundant in unburnt plots and they were replaced by functionally equivalent yellow-rumped seedeaters (also a canary), grey-headed social weavers, and grey-headed sparrows.

Perching species also responded to burning, in that long grass was preferred by the grey-backed fiscal shrike, which was replaced by the white-

Table 14.3 The Bray-Curtis coefficients of similarity (1–*B*) for the bird communities in the Acacia and Terminalia savannas comparing burnt and grazed sites with their unburnt and ungrazed controls

Bray-Curtis similarity values					
Acacia			**Terminalia**		
	Burnt vs. unburnt	**Grazed vs. ungrazed**		**Burnt vs. unburnt**	**Grazed vs. ungrazed**
Ground insectiv			Ground insectiv		
Period 1	0.308	0.745	Period 1	0.371	0.592
Period 2	0.564	0.692	Period 2	0.473	0.704
Ground granivores			Ground granivores		
Period 1	0.462	0.539	Period 1	0.622	0.128
Period 2	0.388	0.455	Period 2	0.500	0.240
Perching insectiv			Perching insectiv		
Period 1	0.532	0.707	Period 1+2	0.540	0.550
Period 2	0.480	0.542			
Tree insectiv			Tree insectiv		
Period 1	0.552	0.760	Period 1	0.521	0.655
Period 2	0.473	0.722	Period 2	0.460	0.512
Tree granivores			Tree granivores		
Period 1	0.321	0.466	Period 1	0.658	0.648
Period 2	0.200	0.550	Period 2	0.671	0.803

Notes: Period 1 was one month after, and period 2 was two months after disturbance. Burning creates a larger dissimilarity than grazing among bird communities. Names for acronyms are given in table 14.1. Latin names are given in appendix 4 at www.press.uchicago.edu/sites/serengeti/. Totals and percentages are for the complete list of species.

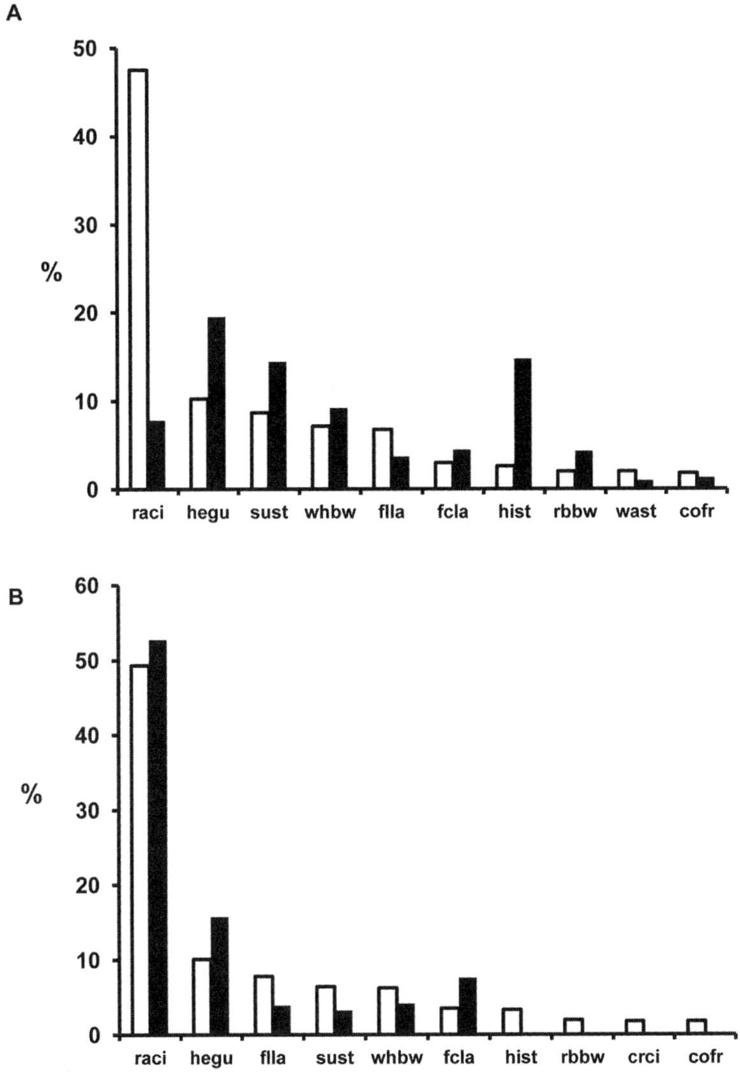

Fig. 14.2 (A) The proportion of the top 10 most abundant insectivore species in unburnt Acacia sites (open bars) compared with that of the same species in burnt sites (solid bars). (B) The proportion of the top 10 most abundant insectivore species in ungrazed Acacia sites (open bars) compared with that of the same species in grazed sites (solid bars). Names for acronyms are given in table 14.1, and Latin names are given in appendix 4 at www.press.uchicago.edu/sites/serengeti/.

Table 14.4 *Top panel*, the number and percentage of the top 10 ground feeding insectivores in *Acacia* savanna in the unburnt and ungrazed sites, compared to that of the same species in the burnt and grazed sites; *bottom panel*, the number and percentage of the top 10 ground feeding insectivores in *Acacia* savanna in the burnt and grazed sites, compared to that of the same species in the unburnt and ungrazed sites

Top 10 species in untreated sites									
Species	***N* unburnt**	**%**	***N* burnt**	**%**	**Species**	***N* ungrazed**	**%**	***N* grazed**	**%**
raci	241	47.5	49	7.8	raci	254	49.3	231	52.7
hegu	52	10.3	123	19.5	hegu	52	10.1	69	15.8
sust	44	8.7	91	14.4	flla	40	7.8	17	3.9
whbw	36	7.1	58	9.2	sust	33	6.4	14	3.2
flla	34	6.7	23	3.6	whbw	32	6.2	18	4.1
fcla	15	3.0	28	4.4	fcla	18	3.5	33	7.5
hist	13	2.6	93	14.7	hist	17	3.3	1	0.2
rbbw	10	2.0	27	4.3	rbbw	10	1.9	0	0.0
wast	10	2.0	6	1.0	crci	9	1.7	1	0.2
cofr	9	1.8	8	1.3	cofr	9	1.7	0	0.0
Total	507		631			515		438	

Top 10 species in treated sites

Species	*N* burnt	%	*N* unburnt	%	Species	*N* grazed	%	*N* ungrazed	%
hegu	123	19.5	52	10.3	raci	231	52.7	254	49.3
hist	93	14.7	13	2.6	hegu	69	15.8	52	10.1
sust	91	14.4	44	8.7	fcla	33	7.5	18	3.5
whbw	58	9.2	36	7.1	bwpl	24	5.5	0	0.0
crpl	52	8.2	6	1.2	whbw	18	4.1	32	6.2
raci	49	7.8	241	47.5	flla	17	3.9	40	7.8
rtwe	34	5.4	0	0.0	sust	14	3.2	33	6.4
fcla	28	4.4	15	3.0	ytlc	8	1.8	2	0.4
rbbw	27	4.3	10	2.0	crpl	6	1.4	6	1.2
flla	23	3.6	34	6.7	zici	3	0.7	6	1.2
Total	631		507			438		515	

Notes: Names for acronyms are given in table 14.1. Latin names are given in appendix 4 at www.press.uchicago.edu/sites/serengeti/. Totals and percentages are for the complete list of species.

Table 14.5 *Top panel*, the number and percentage of the top 10 tree and shrub feeding insectivores in *Acacia* savanna in the unburnt and ungrazed sites, compared to that of the same species in the burnt and grazed sites; *bottom panel*, the number and percentage of the top 10 tree and shrub feeding insectivores in *Acacia* savanna in the burnt and grazed sites, compared to that of the same species in the unburnt and ungrazed sites

Top 10 species in untreated sites									
Species	***N* unburnt**	%	***N* burnt**	%	**Species**	***N* ungrazed**	%	***N* grazed**	%
bbwa	94	18.7	10	4.9	bbwa	118	21.9	67	16.1
batf	89	17.7	48	23.5	batf	87	16.2	83	20.0
rtti	56	11.2	12	5.9	rtti	48	8.9	23	5.5
blba	40	8.0	4	2.0	scbo	43	8.0	6	1.4
nobr	37	7.4	28	13.7	nobr	41	7.6	37	8.9
scbo	37	7.4	10	4.9	csfl	39	7.2	25	6.0
csfl	32	6.4	27	13.2	blba	28	5.2	20	4.8
rbsr	25	5.0	6	2.9	rbsr	25	4.6	29	7.0
bhbs	21	4.2	2	1.0	bhbs	23	4.3	34	8.2
daba	11	2.2	12	5.9	daba	15	2.8	15	3.6
Total	502		204			538		416	

Top 10 species in treated sites

Species	*N* burnt	%	*N* unburnt	%	Species	*N* grazed	%	*N* ungrazed	%
batf	48	23.5	89	17.7	batf	83	20.0	87	16.2
nobr	28	13.7	37	7.4	bbwa	67	16.1	118	21.9
csfl	27	13.2	32	6.4	nobr	37	8.9	41	7.6
rtti	12	5.9	56	11.2	bhbs	34	8.2	23	4.3
daba	12	5.9	11	2.2	rbsr	29	7.0	25	4.6
nuwo	12	5.9	11	2.2	csfl	25	6.0	39	7.2
gwho	11	5.4	0	0.0	rtti	23	5.5	48	8.9
bbwa	10	4.9	94	18.7	gbca	21	5.0	10	1.9
scbo	10	4.9	37	7.4	blba	20	4.8	28	5.2
rbsr	6	2.9	25	5.0	nuwo	17	4.1	11	2.0
Total	204		502			416		538	

Notes: Names for acronyms are given in table 14.1. Totals and percentages are for the complete list of species.

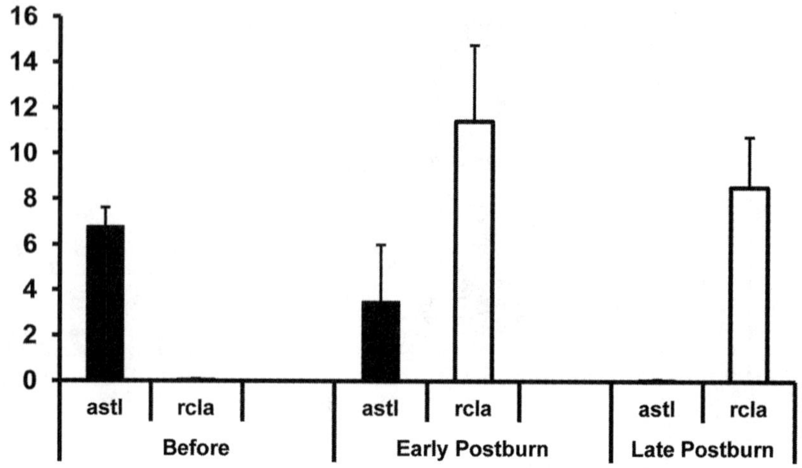

Fig. 14.3 Mean abundance per site of red-capped lark (rcla open bar) and athi short-toed lark (astl solid bar) on burned areas of the long-grass plains. Data from Nkwabi et al. (2011).

Table 14.6 Numbers of birds in different ground feeding insectivore groups

	Acacia ground insectivores		Terminalia ground insectivores	
	Burnt	Unburnt	Burnt	Unburnt
Cisticolas	49	258	58	220
Starlings	184	57	91	6
Plovers, Coursers	58	6	58	5
Gamebirds	123	52	33	15
Buffalo-weavers	85	46	7	1
Both woodlands				
Yellow-throated longclaw	0	16		
Grassland pipit	7	2		
Flappet lark	34	43		
Fawn-coloured lark	34	17		

Notes: Cisticolas that feed in long grass are replaced by ground feeding species. Longclaws (a pipit) and flappet larks in long grass are replaced by grassland pipits and fawn-colored larks on burnt ground. For Latin names, see appendix 4 at www.press.uchicago.edu/sites/serengeti/.

crowed shrike in burnt areas (fig. 14.4). These two species are functional equivalents. Grazing had no clear effect on species replacement.

In both woodland types there was a general negative effect of burning and grazing on tree feeding species. As a result was there were no clear replacements for those that declined with the treatments.

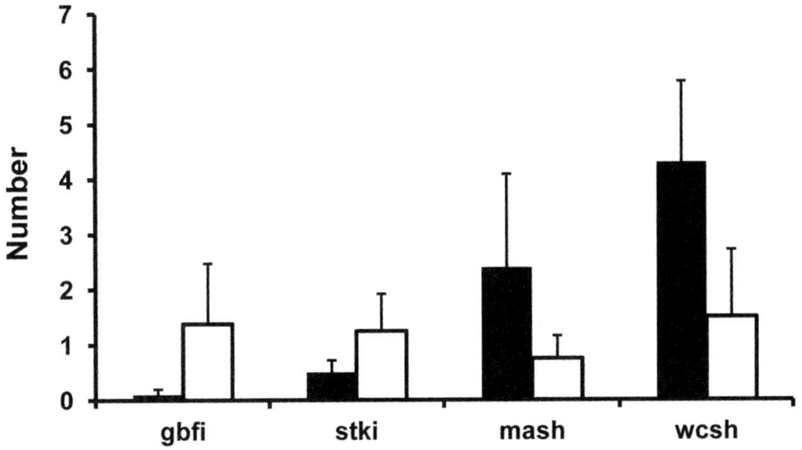

Fig. 14.4 Mean abundance per site of perching insectivorous species on burned (solid bars) and unburnt sites (open bars) in Acacia savanna. Grey-backed fiscal shrike (gbfi) and striped kingfisher (stki) were more frequent in unburnt sites, whereas magpie shrike (mash) and white-crowed shrike (wcsh) were more abundant on burnt sites.

DISCUSSION

This study investigated how the abundance and richness of bird species in savanna woodlands changed with two types of disturbance to the grass layer, burning and grazing. Our initial predictions on the effects of disturbance for the avifauna were fourfold, namely (1) that burning would affect the bird community more than grazing; (2) that birds using the grass or ground layers would be more directly disturbed than those using the shrub or tree layers; (3) that reduction in grass biomass would either reduce the food for birds and so reduce bird abundance, or increase visibility and access to food on the ground and thus increase bird abundance; and (4) the loss of some bird species would compensated for by the addition of other species, as found by Nkwabi et al. (2011).

Our results corroborate the first prediction: the effects of burning were more marked in terms of both the abundance of species and similarity of species composition than those of grazing. We found that the abundance of insectivores that fed on the ground (ground and perching feeders) doubled immediately after a fire in both *Acacia* and *Terminalia* savanna. This effect disappeared after a month. Ground-feeding granivores showed no response to fire. The increase of insectivores after fire was due to birds attracted to insects that had been killed by the fire. Grazing had no effect on either insectivores or granivores feeding on the ground. In contrast, both fire and grazing reduced insectivores and granivores feeding in trees and shrubs for both time periods. Clearly, these disturbances had a negative effect on the tree and shrub avifauna.

These results differed from those we recorded on the open long-grass plains where abundance increased with both disturbances. In those studies insect abundance was also measured and we found that the abundance of insects declined with both disturbances. On the open plains, therefore, bird communities changed in response to alterations in the structure of their habitat providing different conditions for new species to enter, and enabling them to access insect food and lowering its abundance. This suggests there is a change in ecosystem function through a change in the top-down predation cascade.

In general, tree and shrub feeders were less influenced than grass layer feeders, confirming the second prediction. However, the total richness did not change with disturbance but the relative abundances changed. There was higher abundance of birds on burned plots, both insect and seed feeders, than on grazed plots, particularly in the first time period. This result supports the third prediction if there was an increase in accessibility to ground food sources, and it is further confirmed by the fact that tree and shrub feeders declined in number instead; they would not be affected by increases in accessibility and, in contrast, would be negatively affected by destruction of food sources.

Finally, we found that there was compensation in grass layer feeders with ground feeding species replacing grass stem feeders after burning; such compensation did not occur in the tree and shrub feeders. On the plains, species that prefer long grass declined in number with the onset of disturbance. However some were replaced by functional equivalents. Thus, the red-capped lark and the athi short-toed lark, which are functionally equivalent, replaced each other when the grassland changed from long to short grass in both treatments. This replacement compensated for the loss of some species due to the disturbance, and contributed to maintaining both diversity and ecosystem function. In the savanna we also found that some functional groups replaced each other among the ground feeders. Thus, those living in long grass, such as small warblers (cisticolas and prinias), were replaced after fire by those living on the ground (larks and plovers). Two species of canaries (granivores) replaced each other as well. Perching insectivores also had preferences of grass structure so that one (grey-backed fiscal) using long grass was replaced by another (white-capped shrike) feeding on burnt ground. However, no functional replacements were found among tree feeders after disturbances.

Grazing had a lesser effect on bird abundance and diversity (as judged by the Bray-Curtis statistic) than burning. The aspect of grazing that we examined was the removal of grass over a few days at the beginning of the dry season. This comparatively low impact differs from the grazing of cattle, which can be heavy and continuous so that the grass sward is maintained in a short grass state over long periods. Because of this difference in impact our studies of native grazers showed a very different effect on the bird communities than did the studies of cattle grazing (Gregory, Sensenig, and Wilcove 2010).

The more important aspect is that the mosaic of patches—the heterogeneity of the habitat—results in an overall change in the diversity of species in that relative abundances change; less common species in tall grassland become more common in burnt areas. Thus, more species benefit from the combination of disturbed and undisturbed grassland. However, this conclusion does not apply to tree and shrub species which uniformly decline with burning, but not with grazing. Thus, a management policy to create heterogeneity in the grasslands of the Serengeti habitats through moderate application of burning, combined with the natural disturbances of grazing, would promote favorable conservation conditions for more bird species (Christensen 1997). This is particularly important as many bird species are losing habitat with the progressive conversion to agriculture outside of natural areas in Africa (Sinclair, Mduma, and Arcese 2002; Ogada and Keesing 2010).

In other studies Jansen, Little, and Crowe (1999) found that the richness of grassland birds and the density of the red-winged francolin (*Francolinus levaillantii*) were negatively correlated with grazing intensity and annual burning, whereas grey-winged francolin (*F. africanus*) density was positively correlated with grazing intensity in South Africa. We found that grazing by wild herbivores had less effect than burning. O'Reilly et al. (2006) have documented the effects of experimental fires on African savanna birds in Kenya. They recorded that burning increased bird diversity relative to controls but there were no significant effects on either richness or total abundance. Our studies showed little change in richness, but because species replaced each other, the combined set of species increased in number as a result of the disturbances. Woinarski (1990) found in northern Australia that short-term effects of fire involved a substantial increase in both diversity and density of birds. Thus, our results are similar in the early increase in abundance, but not in diversity from burning.

CONCLUSION

Natural disturbances such as fire and grazing create a mosaic of patches during the dry season. We found that fire, in particular, changed avian abundance. Although richness of species did not change with either disturbance, there was a turnover of species, with some species coming in to replace those whose populations were reduced by fire. Thus, there was an overall increase in diversity as a result of these natural disturbances. In contrast, major habitat modifications by humans due to agriculture have been found to significantly reduce both richness and abundance by as much as 50% (Sinclair, Mduma, and Arcese 2002). In essence, large disturbances to habitat such as agricultural modifications are detrimental to the conservation of savanna avifauna, whereas moderate disturbances such as burning enhance the overall diversity.

ACKNOWLEDGMENTS

Jill Jankowski provided constructive comments on the manuscript. The directors of Tanzania National Parks and the Tanzania Wildlife Research Institute gave permission to work in the park. We especially thank John Mchetto, Stephen Makacha, and Joseph Masoy for their assistance in the fieldwork. This work was supported by the Canadian Natural Sciences and Engineering Research Council grant to ARES.

REFERENCES

Adler, P. B., D. A. Raff, and W. K. Lauenroth. 2001. The effect of grazing on the spatial heterogeneity of vegetation. *Oecologia* 128:465–79.

Arcese, P., and A. R. E. Sinclair. 1997. The role of protected areas as ecological baselines. *Journal of Wildlife Management* 61:587–602.

Bibby, C. J., N. D. Burgess, D. A. Hill, and S. Mustoe. 2000. *Bird census techniques,* 2nd ed. London: Academic Press.

Bird, M. I., and J. A. Cali. 1998. A million-year record of fire in sub-Saharan Africa. *Nature* 394:767–69.

Butler, S. J., K. A. Vickery, and K. Norris. 2007. Farmland biodiversity and the footprint of agriculture. *Science* 315:381–84.

Christensen, N. L. 1997. Managing for heterogeneity and complexity on dynamic landscapes. In *The ecological basis for conservation: Heterogeneity, ecosystems, and biodiversity,* ed. S. T. A. Pickett, R. S. Ostfeld, M. Shachak, and G. E. Likens, 167–86. New York: Chapman and Hall.

Dempewolf, J., S. Trigg, R. S. DeFries, and S. Eby. 2007. Burned area mapping of the

Serengeti-Mara region using MODIS reflectance data. *Geoscience and Remote Sensing Letters, IEEE* 4:312–16.

du Toit, J. T., K. H. Rogers, and H. C. Biggs, eds. 2003. *The Kruger experience*. Washington, DC: Island Press.

Folse, J. L. 1982. An analysis of avifauna-resource relationships on the Serengeti Plains. *Ecological Monographs* 52:111–27.

Fuhlendorf, S. D., and D. M. Engle. 2001. Restoring heterogeneity on rangelands: Ecosystem management based on evolutionary grazing patterns. *BioScience* 51:625–32.

Fuhlendorf, S. D., and D. M. Engle. 2004. Application of the fire-grazing interaction to restore a shifting mosaic on tallgrass prairie. *Journal of Applied Ecology* 41:604–14.

Fuhlendorf, S. D., D. M. Engle, J. Kerby, and R. Hamilton. 2008. Pyric herbivory: Rewilding landscapes through the recoupling of fire and grazing. *Conservation Biology* 23:588–98.

Fuhlendorf, S. D., W. C. Harrell, D. M. Engle, R. G. Hamilton, C. A. Davis, and D. M. Leslie. 2006. Should heterogeneity be the basis for conservation? Grassland bird response to fire and grazing. *Ecological Applications* 16:1706–16.

Fuller, R. J., R. D. Gregory, D. W. Gibbons, J. H. Marchant, J. D. Wilson, S. R. Baillie, and N. Carter. 1995. Population declines and range contractions among lowland farmland birds in Britain. *Conservation Biology* 9:1425–41.

Gregory, N. C., R. L. Sensenig, and D. S. Wilcove. 2010. Effects of controlled fire and livestock grazing on bird communities in East African savannas. *Conservation Biology* 24:1608–16.

Hassan, S. N., G. M. Rusch, H. Hytteborn, C. Skarpe, and I. Kikula. 2008. Effects of fire on sward structure and grazing in western Serengeti, Tanzania. *African Journal of Ecology* 46:174–85.

Herlocker, D. J. 1976. Structure, composition, and environment of some woodland vegetation types of the Serengeti National Park, Tanzania. PhD diss., Texas A&M University.

Jansen, R., R. M. Little, and T. M. Crowe. 1999. Implications of grazing and burning of grasslands on the sustainable use of francolins (*Francolinus* spp.) and on overall bird conservation in the highlands of Mpumalanga province, South Africa. *Biodiversity and Conservation* 8:587–602.

Krebs, C. J. 2001. *Ecological methodology*. New York: Harper & Row.

Newmark, W. D. 2008. Isolation of African protected areas. *Frontiers of Ecology and Environment* 6:321–28.

Nkwabi. A. K., A. R. E. Sinclair, K. L. Metzger, and S. A. R. Mduma. 2011. Disturbance, ecosystem function and compensation: The influence of wildfire and grazing on the avian community in the Serengeti ecosystem, Tanzania. *Austral Ecology* 36:403–12.

Norton-Griffiths, M. 1979. The influence of grazing, browsing, and fire on the vegetation dynamics of the Serengeti. In *Serengeti: Dynamics of an ecosystem*, ed. A. R. E. Sinclair and M. Norton-Griffiths, 310–52. Chicago: University of Chicago Press.

Norton-Griffiths, M., D. Herlocker, and L. Pennycuick. 1975. The patterns of rainfall in the Serengeti ecosystem, Tanzania. *East African Wildlife Journal* 13:347–74.

Ogada, D. L., and F. Keesing. 2010. Decline of raptors over a three-year period in Laikipia, central Kenya. *Journal of Raptor Research* 44:129–35.

O'Reilly, L., D. Ogada, T. M. Palmer, and F. Keesing. 2006. Effects of fire on bird diversity and abundance in an East African savanna. *African Journal of Ecology* 44:165–70.

Pomeroy, D. E. 1992. *Counting birds*. Nairobi: African Wildlife Foundation.

Pomeroy, D. E., and B. Tengecho. 1986. A method of analysing bird distributions. *African Journal of Ecology* 24:243–53.

Robinson, R. A., R. E. Green, S. R. Baillie, W. J. Peach,. and D. L. Thomson. 2004. Demographic mechanisms of the population decline of the song thrush *Turdus philomelos* in Britain. *Journal of Animal Ecology* 73:670–82.

Sanderson, F. J., P. F. Donald, D. J. Pain, I. J. Burfield, and F. P. J. van Bommel. 2006. Long-term population declines in Afro-Palearctic migrant birds. *Biological Conservation* 131:93–105.

Sinclair, A. R. E. 1977. *The African buffalo. A study of resource limitation of populations.* Chicago: University of Chicago Press.

———. 1978. Factors affecting the food supply and breeding season of resident birds and movements of palaearctic migrants in a tropical African savannah. *Ibis* 120:480–97.

———. 1979. Dynamics of the Serengeti ecosystem: Process and pattern. In: *Serengeti: Dynamics of an ecosystem*, ed. A. R. E. Sinclair and M. Norton-Griffiths, 1–30. Chicago: University of Chicago Press.

———. 1998. Natural regulation of ecosystems in protected areas as ecological baselines. *Wildlife Society Bulletin* 26:399–409.

Sinclair, A. R. E., and A. Byrom. 2006. Understanding ecosystems for the conservation of biota. *Journal of Animal Ecology* 75:64–79.

Sinclair, A. R. E., H. Dublin, and M. Borner. 1985. Population regulation of Serengeti wildebeest: A test of the food hypothesis. *Oecologia* 65:266–68.

Sinclair, A. R. E, S. A. R. Mduma, and P. Arcese. 2002. Protected areas as biodiversity benchmarks for human impacts: Agriculture and the Serengeti avifauna. *Proceedings of the Royal Society, London B* 269:2401–5.

Sinclair, A. R. E, S. A. R. Mduma, J. G. C. Hopcraft, J. M. Fryxell, R. Hilborn, and S. Thirgood. 2007. Long term ecosystem dynamics in the Serengeti: Lessons for conservation. *Conservation Biology* 21:580–90.

Stronach, N. R. H. 1988. Grass fires in Serengeti National Park, Tanzania: Characteristics, behavior and some effects on young trees. PhD diss., University of Cambridge.

Terborgh, J., and J. A. Estes, eds. 2010. *Trophic cascades: Predators, prey, and the changing dynamics of nature.* Washington, DC: Island Press.

Terborgh, J., L. Lopez, P. V. Nunez, M. Rao, G. Shahabuddin, G. Orihuela, M. Riveros, R. Ascanio, G. H. Adler, and T. D. Lambert, et al. 2001. Ecological meltdown in predator-free forest fragments. *Science* 294:1923–26.

van Langevelde, F., C. A. D. M. van de Vijver, L. Kumar, J. van de Koppel, N. de Ridder, J. van Andel, A. K. Skidmore, J. W. Hearne, L. Stroosnijder, and W. J. Bond, et al. 2003. Effects of fire and herbivory on the stability of savanna ecosystems. *Ecology* 84: 337–50.

Woinarski, J. C. Z. 1990. Effects of fire on the bird communities of tropical woodlands and open forests in northern Australia. *Australian Journal of Ecology* 15:1–22.

FIFTEEN

Carnivore Communities in the Greater Serengeti Ecosystem

Meggan E. Craft, Katie Hampson, Joseph O. Ogutu, and Sarah M. Durant

Carnivores in the Serengeti are abundant and diverse; there are at least 30 species of carnivores in the greater Serengeti ecosystem, ranging in average body size from 0.35 kg (common dwarf mongoose, *Helogale parvula*) to 170 kg (male lion, *Panthera leo*) (Schaller 1972; Waser et al. 1995; Mduma and Hopcraft 2008; Durant, Craft et al. 2010). In the global context, this number of carnivore species in a single system is impressive, putting the Serengeti (and more broadly, central and southeast Africa) in the "especially high" category of carnivore numbers, along with carnivores in southeast Asia and the Philippines (Loyola et al. 2009). Although the Serengeti carnivore guild is species rich, the distribution of these carnivore populations varies widely throughout the ecosystem. The composition of the carnivore community within the ecosystem is affected by both biotic factors, such as interspecific competition or prey abundance, and abiotic factors, such as spatial gradients in soil type or rainfall. Furthermore, interactions with human populations exert a strong influence on carnivores at the edges of the ecosystem.

Carnivores are a taxon of conservation importance, as 32% of all carnivore species are threatened (Sechrest et al. 2002); moreover, their position at the top of the food chain means that they may have important impacts on the functioning of ecosystems (Ginsberg 2001; Estes 2011). Ecosystem function and stability can be influenced by both *top-down* processes, such as predation, and by *bottom-up* processes such as primary production and food availability (Pace et al. 1999). Large carnivores in the Serengeti exert a top-down effect on ungulates; an example of this is the decline of resident prey numbers due to increases in lion and hyena numbers post rinderpest (Sinclair et al. 2008). Carnivores themselves are likely subject to both top-down and bottom-up pressures (Kissui and Packer 2004; Ritchie and Johnson 2009).

However, apart from a handful of species, information on carnivore ecology and abundance is rarely available. This is because there are numerous logistical and theoretical challenges in gathering information about carnivores. Carnivores are notoriously difficult to census; they are cryptic, found in low densities, can have extensive ranging patterns, are often nocturnal, and are shy by nature. Most carnivore studies have focused on single species, especially those that are large and charismatic. Serengeti examples include studies of jackals, *Canis aureus* and *C. mesomelas*; lions; cheetahs, *Acinonyx jubatus*; and spotted hyenas *Crocuta crocuta* (Moehlman et al. 1983; Packer et al. 2005; Durant et al. 2007; Hofer and East 2008). Thus few quantitative descriptions of entire carnivore communities are available, especially beyond the boundaries of protected areas.

In this chapter we will review available data to develop an understanding of carnivore communities in the Serengeti. We start with some definitions used specifically in this chapter. Here the Serengeti ecosystem is synonymous with the greater Serengeti-Mara ecosystem and is defined by those areas in Kenya and Tanzania where migrating herbivores frequent (see chapters 2 and 6). The northernmost area of the Serengeti ecosystem, the Mara Region, is in Kenya. The Mara Region consists of a protected area (where only wildlife conservation and tourism are permitted) called the Maasai Mara National Reserve (MMNR), and pastoral regions to the north and east of the protected area, or reserve. These pastoral areas can be subdivided into low-density and high-density human pastoral landscapes. For the purposes of this chapter, the Tanzanian portion of the Serengeti ecosystem consists of the Serengeti National Park (SNP), a protected area intended for wildlife conservation and tourism, and areas surrounding the park: agropastoralist communities to the west and pastoralists to the east. The vegetation of the Serengeti National Park broadly consists of woodlands to the north and west and plains to the southeast. Long-grass plains (LGP) are found in the northwest section of the plains while short-grass plains (SGP) are in the southeast section of the plains, extending eastward out of the Serengeti National Park into the pastoralist landscapes. (For more description of vegetation and human ethnic composition of these areas, see chapters 2, 3, and 16). When appropriate, we try to reduce analyses down to the smallest area possible in order to maximize spatial resolution for specific questions; hence, there are myriad nested, and sometimes overlapping, regional definitions.

We ask how the distribution of carnivores is influenced by (a) seasonal prey migrations, (b) landscape variables, (c) interspecific competition, and (d) the presence of human populations. We draw on the wide variety of published data on carnivores in and around the Serengeti ecosystem, which we supplement with new data from nighttime transect surveys in the Serengeti-Mara ecosystem. These surveys provide additional information to aid our understanding of the abundance and distribution of nocturnal species in the ecosystem. With these new data sources and analyses, we see that season, habitat, interspecific competition, and humans all influence carnivore numbers in ways not previously explored in this book series. We review long-term trends and throughout discuss some of the challenges facing in-depth spatial and temporal studies of carnivore communities. To our knowledge this is the first formal synthesis of carnivore communities from the Mara Region and the Tanzanian portion of the Serengeti ecosystem.

STUDYING CARNIVORES IN THE SERENGETI ECOSYSTEM

Here we synthesize data from a variety of sources. Some of these sources are derived from studies designed specifically for studying carnivores, while others are derived from secondary activities, such as opportunistic sightings of small carnivores while conducting other research duties. Accurate, long-term monitoring is expensive, time consuming, and requires standardized protocols to ensure comparability of counts. This gets more challenging as one extends the scale of monitoring across spatial areas and across multispecies communities. Monitoring populations of diverse species may require the use of a variety of different methods—consider diurnal versus nocturnal carnivores—which makes the selection and implementation of a universal and accurate survey method additionally problematic. In this chapter we review existing information and use new analytic tools to glean useful and relevant information on the whole carnivore community from varying data sources over multiple time periods in different spatial areas.

There have been a myriad of approaches to carnivore research in the Serengeti ecosystem. Eight species have been the subject of in-depth study: lion from the 1960s to the present (Packer et al. 2005); cheetah from 1974 to the present (Caro 1994; Durant, Craft et al. 2010; Durant, Dickman et al. 2010); spotted hyena in the Serengeti National Park from 1987 to the present (Hofer and East 2008) and in the Maasai Mara National Reserve (MMNR) from 1979 to present (Frank 1986; Holekamp and Dloniak 2010); African wild dogs *Lycaon pictus* periodically from 1965 to their presumed disappearance in the early 1990s—and their consequent reappearance in the first decade of the twenty-first century (Burrows 1995; Creel, Creel, and Montfort 1997; Marsden, Wayne, and Mable 2012); golden and black-backed jackal from the mid-1970s to the present (Moehlman et al. 1983); bat-eared fox from 1986 to 1990 (Maas and MacDonald 2004); and common dwarf mongoose from 1974 to 1987 (Rood 1990). Most of this research has focused on the southern part of the Serengeti ecosystem, where habitat is more open and carnivores are easier to observe and are habituated by tourists. However, single-species studies are limited in their ability to understand the interactions between the suite of different carnivore species in the ecosystem. The fact that new species are still being discovered in the ecosystem (Durant, Craft et al. 2010) is a testament to the lack of attention given to the carnivore community as a whole, as well as the overall difficulty of studying these taxa.

At the community level, the available information shows that community interactions are complex. For example, although large carnivores such as lions, spotted hyenas, and cheetahs all generally use the same type of habitat, research on cheetahs shows that there is a fine-scale temporal-spatial system of partitioning that enables cheetah to avoid lions and hyenas in the south of the Serengeti (Durant et al. 1988; Durant, Dickman et al. 2010). In addition, a series of carnivore counts, started in 1977 and 1986 and continued again in 2002/3 and 2005, demonstrated declines in golden jackals, black-backed jackals, and bat-eared foxes. While the numbers of other abundant carnivores remained relatively stable, the cause of the decline in jackals and foxes is not yet known (Durant et al. 2011). Additional information about the wider carnivore community has only become available recently.

There have been three main methods used for surveying carnivores in the ecosystem: (a) transect surveys, both fixed width (strip transects) and distance based (line transects) (Buckland 1993; Thomas et al. 2010), with the added aid of spotlights at night; (b) camera trapping; and (c) opportunistic sightings. We cover each data-collection method briefly below and recommend other references where appropriate for more details. Care was taken to only compare similar types of data; information on analyses will follow in later sections.

Transect Surveys

Daytime surveys. Fixed-width transect surveys, and, more recently, distance-based surveys, have been used to estimate carnivore density on the Serengeti plains (Serengeti Research Institute 1977a, 1977b; Campbell and

Borner 1986; Durant et al. 2011). Here we will discuss results from Durant et al. (2011) in which, (a) the previous fixed-width surveys were adjusted to account for decreasing detection of carnivores at increasing distance from the transect line, and (b), new distance-based surveys for 2002/3 and 2005 performed along the same routes as the previous surveys were analyzed (for more survey details, see Durant et al. [2011]). Data from distance sampling surveys conducted in and around the Maasai Mara National Reserve are also analyzed here. In the MMNR, for the wet and dry seasons of 2005 and 2006, 42 transects, spaced at least 1.5 km apart, were distributed over four landscapes: *reserve* (inside the MMNR), *boundary* (on the boundary of the MMNR and pastoral areas), *low-density* pastoralist (areas with few pastoralists), and *high-density* pastoralist (areas with large populations of pastoralists) (MMNR transects are shown in fig. 15.1, inset). Twelve transects that were 7 km long were each driven in the reserve boundary and high-density pastoral landscapes. Only 6 transects could be established in the low-density pastoral landscape: 4 of 7 km length and 2 of 6 km length. Between 90 and 95% of transects were established entirely off-road to minimize biases due to the road location (Buckland 1993); however, parts of some transects utilized roads in order to pass through patches of dense vegetation, avoid rocks or rivers in the path, or to avoid the vehicle getting stuck in heavy clay soils. An observation platform was mounted on the top of each of four observation vehicles to enhance the ability to detect animals; the vehicle was driven at a maximum speed of 10 km per hour.

Nighttime surveys. Transects targeting nocturnal carnivores were conducted at night using spotlights throughout the main road network across the Tanzanian portion of the Serengeti ecosystem by the Serengeti Viral Transmission Dynamics Team and the Serengeti Biodiversity Project. Transects were driven along roads within the Serengeti National Park, outside the national park in both the agropastoralist areas to the west and the pastoralist areas to the east with two observers inside the vehicle, each holding a spotlight directed to either side (Tanzanian transects are shown in fig. 15.1, inset). Transects were repeated throughout the year at monthly or quarterly intervals from 2003 to 2011 (both in and around the Serengeti National Park). The transects in Tanzania were restricted to roads, covering 20 km stretches in areas outside of the national park and 40 km stretches of road within the national park. Surveying from roads introduces inherent biases, especially for species with tendencies to follow or avoid roads, which will likely be over- or undercounted, respectively. We therefore restrict our comparisons to indices using these data, which does enable estimates of trends. Because of the difficulty in identifying some species at night, all species of jackals and genets in this data set have been lumped together as "jackal

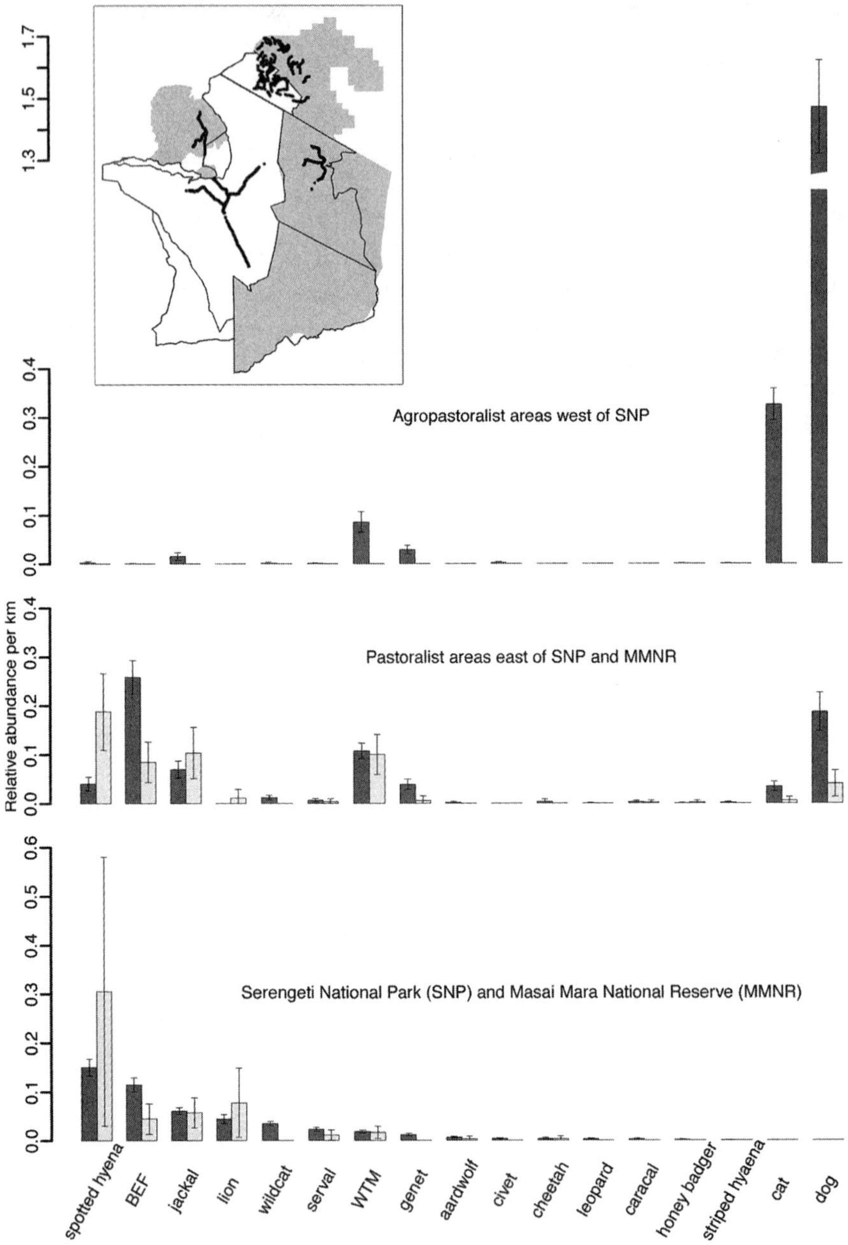

Fig. 15.1 Relative abundances of carnivore species, with 95% confidence intervals, estimated from night-time transects. Dark gray bars indicate abundances estimated from transects in Tanzania (located in and around Serengeti National Park) and light gray bars indicate the abundances estimated from Kenya (in and around the Maasai Mara National Reserve). Encounters are summarized by survey area, as indicated by the inset showing of the location of transect routes as a series of black dots; the gray-shaded zones represent areas where humans live, while white zones include MMNR to the north and SNP to the south and are protected areas for conservation and tourism. The top bar chart is for agropastoralist areas west of SNP, the middle bar chart is for pastoralist areas east of SNP and MMNR, and the bottom chart is for SNP and MMNR. In and around SNP, transects were repeated monthly from 2003 to 2009, but sampling intensity was reduced to quarterly from 2007 to 2011. In and around the MMNR transects were repeated in the wet and dry seasons of 2005 and 2006. BEF = bat-eared fox and WTM = white-tailed mongoose.

spp" and "genet spp." In the Mara, nighttime surveys were performed by the same counting teams on the same transects as daytime transects, although at night each vehicle used two spotlights mounted on the roof rack.

Camera Trapping

Camera trapping is a noninvasive tool for documenting carnivore presence throughout the day and night. This chapter makes use of published analyses of camera-trapping data collected by the Tanzania Carnivore Program (TAWIRI) in the Maswa Game Reserve (79 locations, 1,260 camera trap days), Ngorongoro highlands (52 locations, 1,008 camera trap days), and the Serengeti riverine forest (40 locations, 1,219 camera trap days) between April 2005 and June 2008 (Pettorelli et al. 2010; Durant, Craft et al. 2010). Cameras were set up in fixed locations at least 1 km apart, were triggered by infrared sensors, and no bait was used to lure carnivores to the cameras. More details about this survey methodology can be found in Pettorelli et al. (2010).

Opportunistic Sightings

Opportunistic sightings of carnivores were collected by various sources in Tanzania and have been summarized in Durant, Craft et al. (2010). This data includes georeferenced locations from the methods used above, as well as locations of all carnivores seen by the Serengeti Lion Project from 2004 to 2008 while radio-tracking for lions; on rarely sighted carnivores (excluding lions, spotted hyenas, golden jackals, black-backed jackals, and bat-eared fox) seen by SMD from 1993 to 2008 while searching for cheetah; and data on all carnivore species submitted to the Tanzania Carnivore Program database from 2002 to 2008 (see http://www.tanzaniacarnivores.org.).

The Species

A total of 30 species have been documented within the ecosystem (Mduma and Hopcraft 2008; Durant, Craft et al. 2010; see also appendix 1 at www.press.uchicago.edu/sites/serengeti/): lion; leopard, *Panthera pardus*; cheetah; serval, *Leptailurus serval*; caracal, *Caracal caracal*; African wildcat, *Felis silvestris*; golden jackal; black-backed jackal; side-striped jackal, *Canis audustus*; African wild dog; bat-eared fox; zorilla, *Ictonyx striatus*; African striped weasel, *Poecilogale albinucha*; African honey badger, *Mellivora capensis*; African civet, *Civettictis civetta*; spotted-necked otter, *Hydrictis maculicollis*; cape clawless otter, *Aonyx capensis*; African palm civet, *Nandinia binotata*; common genet, *Genetta genetta*; Egyptian mongoose, *Herpestes ichneumon*; slender mongoose, *Galerella sanguinea*; common dwarf mongoose; marsh mongoose, *Atilax paludinosus*; banded mongoose, *Mungos mungo*; white-tailed mongoose, *Ichneumia albicauda*; aardwolf, *Proteles cristata*; spotted hyena and striped hyena, *Hyaena hyaena*; large spotted genet, *Genetta maculate*; and the bushy-tailed mongoose, *Bdeogale crassicauda*. The latter species has only been recorded at six sites in the Ngorongoro highlands during a camera-trapping survey (Durant, Craft et al. 2010), but the fact that it went undetected for so long makes it possible that it occurs within the SNP as well. Not only does the Serengeti ecosystem contain unusually high carnivore species richness (for example, Kruger National Park has documented 27 species while the whole of Botswana and Namibia harbor 23 and 18 species of carnivore, respectively), but it is also a stronghold for some of the highest densities of large carnivores recorded in Africa (Caro 1994).

THE IMPORTANCE OF SEASONAL PREY MIGRATION

The Serengeti-Mara ecosystem is characterized by wet and dry seasons, with the wet season from November until May (until June in the Mara), and the long dry season from June to October (from July to October in the Mara). There is often a short dry season between January and February before the start of the long rains in March. As we have seen in previous books in this series, this seasonality drives major changes in vegetation phenology and productivity and herbivore distribution. Inevitably, it also has important impacts on carnivores. For the large carnivores, the annual migration of ungulates from the short-grass plains to the woodlands in the north and west and into the Mara Reserve has a major impact on food availability. The Serengeti wildebeest, zebra, and eland migrate to and stay in the Maasai Mara Region from July to at least October (Hopcraft et al., chapter 6). In addition, the ungulates bring with them a great abundance of flies and dung beetles, providing food that may be directly eaten by some of the smaller carnivores (Waser et al. 1995). Rodent and bird abundances also change by season (Byrom et al., chapter 12). Food availability is therefore expected to fluctuate seasonally to some degree for all carnivore species. In the Mara Region, tall and dense grass cover in the late wet season drives a more local migration, and many herbivores disperse to pastoral ranches surrounding the reserve that have short, actively growing, and nutritious grasses. High predation risk and reduced quality of mature grasses in the reserve have been reported as the likely reason for this movement (Stelfox et al. 1986; Ogutu et al. 2008; Bhola, Ogutu, Piepho et al. 2012; Bhola, Ogutu, Said et al. 2012). Within the Mara Region, there is an additional migration involving movements of wildebeest, zebra, and Thomson's gazelles to the Mara Reserve from the Loita plains in the northeast in the dry season and back again in the wet season.

Most carnivores in the Serengeti ecosystem are territorial, and hence have to find sufficient resources in their territories throughout the year despite the fluctuations of the seasons. However, a few have developed interesting mechanisms to adapt to the variable prey availability. Both wild dog and cheetah are able to follow their prey over their migration, moving from the far southeast of the ecosystem in the wet season up into the woodlands in the dry season (Durant et al. 1988). Lions, although territorial, shift their territories southward and eastward with the rains, allowing them to make brief forays even further south to access migratory prey (Schaller 1972). This is aided by the fact that the southernmost and easternmost areas of the Serengeti short-grass plains (in the Ngorongoro Conservation Area) are largely unoccupied by lions. Perhaps the most dramatic adaptation, though, comes from the spotted hyena, which has adopted a commuting system whereby adults will leave their den site for days at a time commuting to the migration throughout the ecosystem. Hyena cubs can go without milk for many days, while their mother forages at a mean distance of 40 km away (Hofer and East 1993; Hofer and East 1995).

The seasons, therefore, present some challenges in estimating carnivore densities in the Serengeti ecosystem. We expect the abundance of some species to increase in an area when their herbivore prey are present—for example, spotted hyena densities in the south of the Serengeti Park are substantially supplemented by commuting hyenas during the wildebeest and zebra migration in the wet season and these hyenas are correspondingly depleted due to the absence of resident commuters during the dry season; cheetah densities should broadly drift in response to Thomson's gazelle movements; and lions shift south with the wildebeest and zebra, appearing in areas where they are absent in the dry season.

The distance-based surveys shed some insight into seasonal density estimates; however, we have to interpret the estimates cautiously. We calculated a seasonal density estimate per species by averaging the 2002/3 and 2005 densities from figure 3 in Durant, Craft et al. (2010; fig. 15.2). There are three potential explanations for seasonal changes in density using survey techniques (versus individual behavioral studies): (a) the overall population is actually changing between seasons (this is very unlikely given that demographic studies do not support this assumption); (b) detection rates

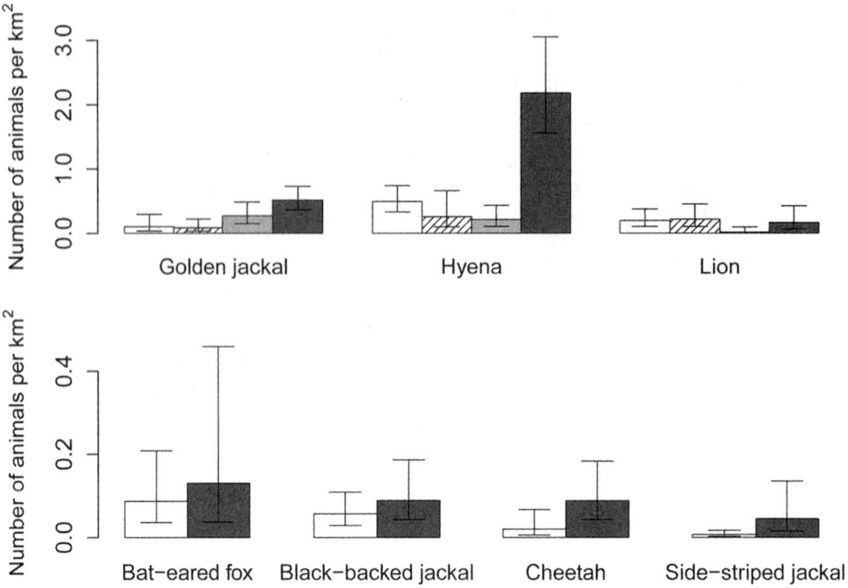

Fig. 15.2 Seasonal changes in Serengeti National Park carnivore densities. In the top panel, for the more abundant species, we have shown densities for the short-grass plains (SGP) and long-grass plains (LGP) where bar colors represent the following: white = LGP dry; diagonal hatching = LGP wet; light gray = SGP dry; dark gray = SGP wet. Error bars are 95% confidence limits. In the bottom panel, average densities are shown for dry season (white) and wet season (dark gray). Adapted from (Durant et al. 2011).

are confounded with season (but this was not true); and (c) there is movement between areas. Thus, we believe that changes in density detected by this survey method between areas are due to movement. Specifically, on the Serengeti short-grass plains (SGP), the commuting system of the hyenas can be easily detected from distance-based surveys, as seen by a tenfold increase in dry season abundance relative to the wet season, when the plains are flooded with commuters (fig. 15.2; Durant et al. 2011). In the dry season, hyena numbers are depleted by commuters who have left their territories to forage in the north and west of the park (including the long-grass plains [LGP]). Seasonal changes in abundance for other species were less marked, or absent. There was evidence that lions shifted south and east onto the short-grass plains in the wet season as expected, and there were slightly higher numbers of cheetah in the wet season on the plains (both SGP and LGP); however, no seasonal changes are apparent for golden, black-backed, and side-striped jackals or bat-eared fox, as might be expected for small carnivores with territorial behavior (fig. 15.2). Outside the plains, data are limited; however, we expect to see the changes in large carnivore abundance mirrored in the woodlands to the north and the west and there is no reason to suppose that those carnivores showing constant population size across the seasons on the plains should vary in the woodlands.

Based on this knowledge of carnivore ecology in the SNP, we might expect that carnivore density might decrease in the wet season in the MMNR and increase in the dry season when the herds are present (and SNP carnivores might appear with the herds). However, some unexpected differences emerged when we compared daytime transect density estimates from inside the MMNR and the SNP plains (fig. 15.3A). Spotted hyenas were more abundant in the SNP survey area than in the MMNR for both the wet and dry seasons. And in contrast to the SNP, spotted hyena densities in the MMNR were relatively stable throughout the year, consistent with observations from a largely residential group of well-studied hyenas that do not commute (Holekamp and Dloniak 2010). This is consistent with another study from the same Mara hyena study population showing no significant correlations between hyena movement and rainfall patterns (Kolowski and Holekamp 2006).

From surveys performed during the day, lions appear to have similar densities in the MMNR and in the SNP. Nomadic lions are thought to follow the migration (Schaller 1972; Craft et al. 2011), so it is expected that lion numbers might rise in the dry season in the MMNR and rise during the wet season on the SNP short-grass plains (Ogutu and Dublin 2002). While this trend was supported in the SNP data, the trend in the Mara might indicate that lions are more abundant in the wet season—however, because of overlapping confidence limits, we are limited in our interpretation of these numbers.

Black-backed jackal numbers are similar in the MMNR and the SNP, except in the Mara Reserve in the wet season, when very few jackals were detected (fig. 15.3A). In the SNP we did not detect any seasonal differences between estimated densities, and this is consistent with the observation that Serengeti black-backed jackals are strongly territorial (Moehlman 1983). Mara black-backed jackals, however, are at higher densities in the pastoral lands in the wet season and in higher densities in the reserve in the dry season (fig. 15.3B). Because jackals are extremely flexible in their behavior (Loveridge and Macdonald 2003), and there are larger numbers of small prey (such as gazelle) outside the reserve than inside the reserve (Bhola, Ogutu, Piepho et al. 2012; Bhola, Ogutu, Said et al. 2012), jackals likely shift territories seasonally in response to resource availability.

Season can also drive demographic events, such as breeding, particularly in the case of the smaller carnivores, many of which give birth during the wet season. Data from nighttime transects hints at seasonal fluctuations in numbers of bat-eared fox, jackal, and wildcat, probably linked to variation

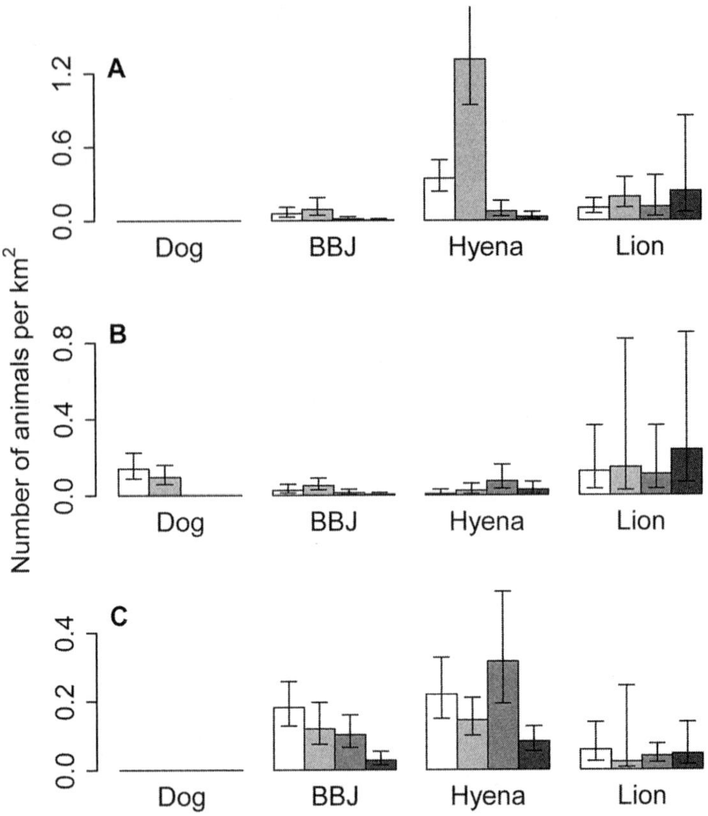

Fig. 15.3 Seasonal carnivore densities (number of animals per square kilometer) with comparisons for (A) daytime densities inside the protected areas for both Maasai Mara National Reserve (reserve) and Serengeti National Park (SNP, averaged across both long- and short-grass plains for the 2005 surveys) where white = SNP dry; light gray = SNP wet; gray = reserve dry; and dark gray = reserve wet; (B) daytime comparisons from inside Maasai Mara National Reserve (reserve) and outside the reserve in pastoralist areas (pastoral); and (C) night estimates inside (reserve) and outside (pastoral) the Maasai Mara National Reserve where for both B and C white = pastoral dry; light gray = pastoral wet; gray = reserve dry; and dark gray = reserve wet. Dog = domestic dog; BBJ = black-backed jackal; hyena = spotted hyena.

in food supply or demographic events; however, seasonal fluctuations are swamped by year-to-year variation (fig. 15.4). Byrom et al. (chapter 12) also discuss variation in small carnivore food supply.

From the perspective of the ungulate, migration has been proposed as a mechanism to escape predation (Fryxell, Greever, and Sinclair 1988). However, the research on hyenas, lions, and jackals presented here indicates that large, territorial predators can also adapt to this changing food source. While it is likely that the percentage of migratory prey in the diet of an individual territorial predator does vary seasonally (Fryxell et al. 2007), it

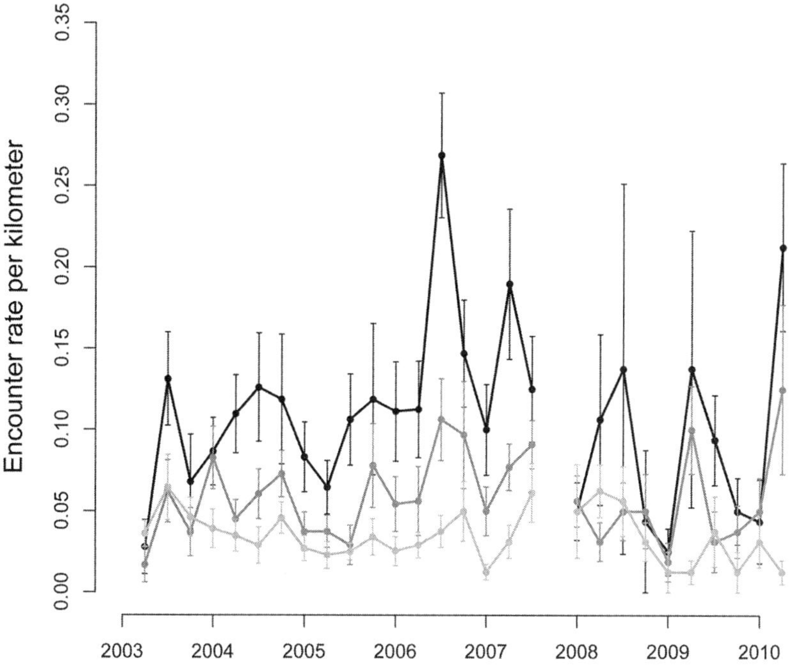

Fig. 15.4 Comparisons of trends in nocturnal abundance for selected species in Serengeti National Park (summarized over all four transects) aggregated at quarterly intervals with standard error bars. Black = bat-eared fox; gray = jackal; and light gray = wildcat. "Jackal" includes all jackal species.

remains untested whether the feeding rate of migratory predators remains similar in the dry versus wet season; hence, it is still possible that ungulates do manage a partial relief from predation.

THE IMPORTANCE OF HABITAT

Like all taxa, carnivora contains specialist species, such as the spotted-necked otter, whose distribution is completely tied to rivers and lakes, and species traditionally thought as being more generalist, such as the spotted hyena, which can be found in a variety of habitats, provided they harbor sufficient prey. However, quantitative tests of these assumptions are rare, and for most carnivore species living within the Serengeti ecosystem, information on habitat selectivity was completely absent until recently. Broadly, cheetah, lion, and spotted hyena are found throughout the ecosystem, and Durant, Craft et al. (2010) suggest that this is also the case for slender mongoose, wildcat, and serval. The plains are the haunt of the golden jackal;

the woodlands for leopard, black-backed, and side-striped jackal. Of the remaining species, habitat selectivity cannot be conclusively determined. Evidence from camera traps suggests that the large spotted genet is concentrated around the river valleys, while the common genet is more frequently seen outside these areas. Aardwolf and bat-eared fox are dependent on termites, and hence their distributions are expected to be tied to the distribution of the *Hodotermitid* termites on which they depend—but even less is known about these termites than the carnivores themselves.

An analysis of habitat selectivity for 13 ecogeographical habitat variables, using all the georeferenced data available from opportunistic sightings, transects, and camera-trap surveys, provided a recent advance in our understanding of the distribution of carnivores in the Serengeti (Durant, Craft et al. 2010). This analysis was undertaken for 18 of the most abundant 30 species in the Serengeti ecosystem. As defined by Hirzel et al. (2002), a species with high habitat selectivity is expected to have a high *marginality*, indicating that it is selecting habitats that differ from the global average, and/or a low *tolerance*, indicating that it is selecting habitats with a narrow range over the ecogeographical variables. This study demonstrated clear habitat selection by some species, such as golden jackal, but found little evidence of selection by others; for example, slender mongoose was found across a large range of habitats in the ecosystem (table 15.1). Other species, such as lion and cheetah, are likely to range widely across the ecosystem and may not be greatly influenced by the habitat variables used in this analysis; for still other species, such as the bat-eared fox, it is possible that the habitat variables used do not correlate well with the termites on which this species depends. Interestingly, this study did not find any evidence of a relationship between habitat selectivity and body size, despite predictions that species should become less selective with increasing size (Brown and Maurer 1989).

INTERSPECIFIC COMPETITION

Habitat may not be the only determinant of carnivore distribution. Intraguild relationships could have strong impacts on the populations and distribution of individual species, because carnivores are better equipped than many other taxa to kill each other (Palomares and Caro 1999; Donadio and Buskirk 2006). Impacts from interspecific killings have been seen at the population level for cheetah—lions and hyenas heavily predated cheetah cubs (Laurenson 1995; Durant, Kelly, and Caro 2004). Interference competition has also been seen in the Serengeti ecosystem as both wild dog (Creel and Creel 1996) and cheetah lose kills to lions and hyenas (Durant

Table 15.1 Coefficients of marginality and tolerance for 18 species of Serengeti carnivores

Species	**Marginality**	**Tolerance**	***N***	**% explained by marginality**	**% explained by first axis of specialization**
Aardwolf	0.861	0.547	107	0.176	0.269
Banded mongoose	1.278	0.517	274	0.292	0.295
Black-backed jackal	1.013	0.687	1,858	0.117	0.321
Bat-eared fox	0.815	0.723	1,298	0.151	0.254
Golden jackal	1.604	0.413	1,001	0.175	0.248
Caracal	0.952	0.573	77	0.122	0.322
Cheetah	1.308	0.564	1,273	0.184	0.247
Civet	0.794	0.577	58	0.192	0.249
Dwarf mongoose	1.101	0.348	57	0.380	0.233
Honey badger	1.140	0.625	147	0.075	0.260
Leopard	0.677	0.603	248	0.219	0.307
Lion	1.315	0.509	6,699	0.352	0.217
Serval	0.600	0.660	365	0.174	0.225
Slender mongoose	0.703	0.699	86	0.097	0.221
Spotted hyena	1.203	0.622	5,066	0.152	0.241
Side-striped jackal	1.268	0.440	126	0.352	0.198
White-tailed mongoose	0.522	0.523	503	0.231	0.323
Wildcat	0.571	0.639	397	0.179	0.202

Notes: A global marginality factor of more than 1 means that the species lives in a very particular habitat relative to the reference set, while a tolerance of much less than 1 indicates some form of specialization. The *N* indicates the number of sightings per species (Durant, Craft et al. 2010).

1998; Durant 2000a, 2000b). It is highly likely that such interactions do not stop at the large carnivore guild, as Serengeti ecosystem intraguild killings have been observed for lion on cheetah and spotted hyena; for spotted hyena on cheetah; for leopard on cheetah; for cheetah on black-backed jackal and bat-eared fox; for serval on dwarf mongoose, and so forth. These observations of larger carnivores killing the next smallest carnivore are consistent with predictions that interspecific killings are more frequent at intermediate differences in body size (as opposed to similar body size or very large differences in body size; Donadio and Buskirk [2006]). As yet, we have only sporadic glimpses as to how these intraguild relationships might impact interrelationships between different species and modify their distribution through space and time. We agree with a recent review calling for "better understanding of the complexity of species interactions in multi-predator communities" (Ritchie and Johnson 2009).

THE INFLUENCE OF HUMANS

Carnivores, from their position at the top of the food chain, are particularly vulnerable to the negative impacts of people. Carnivores experience direct conflicts with humans (e.g., ritual lion killing and retaliatory killings caused by livestock depredation) as well as other indirect conflicts such as exclusion from agricultural environments that lack natural prey, and multihost disease spillover from domestic animals (for more details see Hampson et al., chapter 21). This is especially true for the largest and most wide ranging of carnivores, which do not recognize the boundaries of protected areas, and become particularly vulnerable to impacts on their borders (Woodroffe and Ginsberg 1998). In a few cases, however, opportunistic wild carnivore species might actually thrive in "disturbed" environments (Macdonald 1979), further complicating efforts to understand carnivore communities in human-occupied areas of the ecosystem. An understanding of the relationship between carnivores and the increasing impacts of people is therefore crucial to inform the sound management of an ecosystem, to maintain viable populations of large carnivores, and to minimize the disruption of smaller ones.

Here, we examine the quantitative relationships between carnivore abundance and human settlement. We expect differences in carnivore composition inside versus outside protected areas due to human presence. We would also expect differences in carnivore community composition in accordance with human land-use type; for example, markedly higher carnivore species richness is found in pastoralist areas than in agricultural areas (Msuha et al. 2012). When pastoral lifestyles are followed without hunting, savanna ecosystems can retain their biodiversity and much of their function (Homewood et al. 2001). In contrast, widespread land-use change to support crops and increased population growth, or woodland clearance for charcoal burning, generally results in habitats that are much less hospitable to wildlife. Hence, more biodiverse carnivore communities should exist in pastoral than in agropastoralist communities. To quantify any differences between carnivore communities with respect to humans, we explore a suite of metrics: abundance, richness, diversity, and species-specific densities.

Carnivore Abundance per Land-Use Type

Over 10,000 km of spotlight night transects were driven within the Serengeti ecosystem; more than 8,000 km in SNP and 1,800 km in areas to the pastoralist east and agropastoralist west of SNP, amassing over 6,000 observations of wild and domestic carnivores, of which over half were outside SNP. In the Mara Region, over 3,900 km were driven during both day and night. Relative abundance estimates and 95% confidence intervals for each species were calculated from the number of encounters per kilometer driven. Species abundances and assemblage composition differed markedly between areas of human influence: agropastoralist areas to the west of SNP, pastoralist areas to the east of SNP and outside of MMNR, and within the two protected areas (SNP and MMNR; fig. 15.1). Domestic dogs and cats predominated in the agropastoralist zone to the west and, although they were still among the most abundant carnivore species, they were found less frequently in the pastoralist zone. Dogs and cats were entirely absent inside the SNP and MMNR protected areas. A range of wild carnivore species occurred throughout the Serengeti ecosystem, but only three were regularly seen in the agropastoralist area: white-tailed mongoose, genets, and jackals. Surprisingly, genets and white-tailed mongooses were more abundant in the agropastoralist areas than inside the protected areas. Inside SNP the carnivore assemblages were the most diverse, and included all the top predators and several rare and cryptic species such as caracal, aardwolf, leopard, and striped hyena. Similar diversity was seen in the MMNR, but because fewer transects were driven, the species abundance estimates have large confidence intervals, and some of the rarer, more cryptic species were not seen. In the pastoralist areas outside both the SNP and MMNR, opportunistic species such as genet, white-tailed mongoose, and jackal were common, as were bat-eared foxes and spotted hyenas. In addition to white-tailed mongooses and jackals, bat-eared foxes were more abundant in the pastoralist areas than inside the protected areas. As intensification of human settlements increase across the gradient from protected area to pastoralist to agropastoralist zones, larger wild carnivores were excluded, although a select few small wild species seem to thrive in the human-dominated landscapes. These wild opportunists could play an important role in transmission of multihost pathogens between domestic dogs to large, or rare, wild carnivores that do not exist in these human-dominated landscapes (Cleaveland et al. 2008).

Wild Carnivore Species Richness and Diversity

Species *richness* is simply the number of different species in a given area, while species *diversity* quantifies the relative abundance of each species. The Serengeti night transects were divided into 25 km^2 grids, the length of the transect through the grid was calculated, domestic cats and dogs were excluded, and only data where each of the 10 transects were completed in

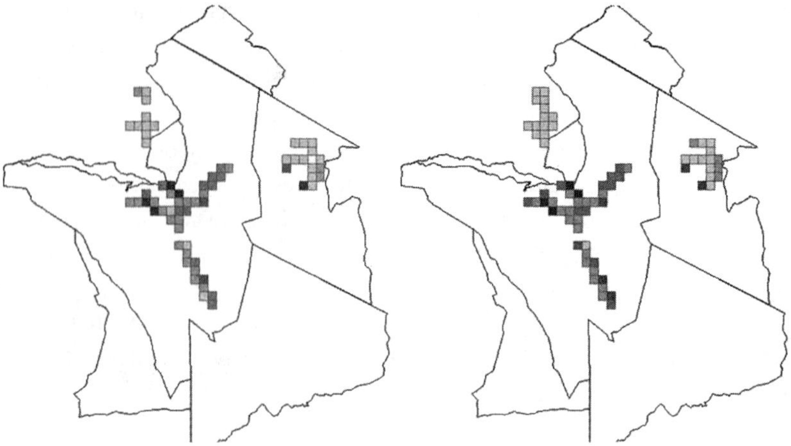

Fig. 15.5 An index of wild carnivore richness (left panel) and diversity (right panel) per 25 km² estimated from transects driven at night. Darker shades represent higher values.

a certain month were used, leaving 21 months of complete data. Species diversity was calculated using the Shannon-Weaver diversity index (H) (Shannon and Weaver 1949) and was divided by the area traveled in each grid cell. Because jackals and genets were not separated by species, and these transects only captured larger carnivores active at night, richness and diversity are underestimated; however, we can use these measures as a comparative index.

The resulting values for richness and diversity tell a similar story to the night transect abundance data, as nighttime richness and diversity were both highest inside the Serengeti National Park (fig. 15.5). Within the SNP, richness and diversity are highest in the woodlands (specifically, the three northern-most transects inside the park) and lowest on the plains (the southern-most transect). However, even the relatively low species richness/diversity found on the plains was generally greater than in the pastoralist area to the east of the park, and much greater than in the agricultural area to the west, which in particular had very low species richness and wild carnivore diversity. These metrics also confirm that humans, with their various land-use types, impact carnivore richness and diversity.

Carnivore Density in Relation to Humans

The estimation of densities is a gold standard for monitoring and can be accurately carried out using distance-sampling methods. However, this method is data hungry and comes with a cost, as at least 60–80 observations are needed to reliably estimate densities for each species, region, and time unit of interest (Buckland et al. 2001). For the carnivore species with sufficient numbers of observations, we calculated densities using the software Distance (Thomas et al. 2010). In order to quantify the likely impact of humans on carnivore densities we compared the daytime carnivore estimates for domestic dogs, black-backed jackals, hyenas, and lions found inside the MMNR protected area, with those found in the pastoralist areas (pastoral) outside the reserve (fig. 15.3A). Spotted hyenas were significantly more abundant inside the protected area than outside in the pastoral areas during the dry season, while the wet season had similar densities in both locales. Interestingly, humans seem to have less of an impact on smaller species such as jackals; during the wet season, black-backed jackals seem to be more common in the pastoral areas than inside the protected area, although similar numbers were detected during the dry season. Due to the low number of lion groups detected, (16 and 18 groups of lions spotted in two years in the dry season and wet seasons, respectively), the confidence limits on the density estimates were too wide to statistically differentiate between protected and pastoralist areas, regardless of season. Nevertheless, the surveys indicate a surprising number of lions outside the reserve, which was more than expected from previous surveys (Ogutu, Bhola, and Reid 2005).

Why Are There Differences in Assemblages?

Human-modified landscapes negatively affected the number of wild carnivore species. Domestic carnivores dominated the carnivore community in agropastoralist areas, although a few small opportunistic carnivores were present in relatively high numbers. In pastoralist areas, more wild carnivores coexisted with domestic carnivores than in the agropastoralist areas. Domestic carnivores were not found inside the protected areas, while the highest abundance and numbers of species of wild carnivores were found here.

The impact of human populations is likely to be the most important factor in shaping the large carnivore guild outside protected areas. One possible explanation has to do with prey base, as very few resident wild ungulates survive in agropastoralist areas except for the occasional dik-dik. During the annual migration, wildebeest do pass through agropastoralist villages and often suffer heavy mortality as a result. In general agropastoralist areas are hostile for ungulates and thus can support only very few large predators and, secondly, those carnivores caught venturing into these populated areas are rapidly killed or harassed (see Hampson et al., chapter 21). Edges of protected areas are known to be sinks for large carnivores because of high levels of human persecution (Woodroffe and Ginsberg 1998), and this is likely the case in the Serengeti, especially along the western park boundary, where human densities are highest. In contrast, in pastoralist areas there is still a relatively large prey base, which supports substantive populations of large carnivores (Maddox 2003; Ogutu et al. 2011). Although few lions were observed on transects in pastoralist areas (and none in pastoral areas in Tanzania), this may be a direct consequence of lower densities, behavioral avoidance due to the considerable persecution of lions by people, and more generally, harassment of carnivores by people, livestock, and dogs (Ikanda and Packer 2008). Hyenas are quite abundant, especially in the pastoral areas around the MMNR. Although both lions and hyenas exhibit behavioral plasticity to cope with living alongside humans (Boydston et al. 2003; Pangle and Holekamp 2010; Mogensen, Ogutu, and Dabelsteen 2011), lions are more susceptible to persecution than hyenas (Kissui 2008).

In agropastoralist villages, the small wild carnivore guild is dominated by opportunists, which are commonly reported as pest species that predate on poultry and small stock. The omnivorous nature of these species (e.g., white-tailed mongoose) enables them to thrive in peridomestic environments, but in more pristine environments they are less abundant, possibly due to greater interspecific competition and predation by other carnivore species, such as mesocarnivore or mesopredator release (Terborgh and Winter 1980; Ritchie and Johnson 2009; fig. 15.1). Widespread cultivation may degrade habitat quality for more specialized species such as bat-eared foxes, but deliberate killing of wild carnivores by people may confound such effects. In pastoral landscapes, small carnivores are generally more abundant than in the agropastoralist areas (fig. 15.1). Solitary or cryptic carnivores such as caracal, serval, aardwolf, and striped hyena were only found in extremely small numbers in all zones, making rigorous quantitative comparison difficult. Nonetheless, all these rare species appear to be surviving better in pastoralist areas than in agropastoralist areas, and are at similar levels as found in uninhabited protected areas (SNP and MMNR).

Our methods support the idea that domestic dog, and probably cat, abundances are directly correlated with human densities. Domestic dogs accompany herders or pedestrians in the daytime and stay at home at night, although less is known about the activity and movement patterns of domestic cats. Because domestic animals in this ecosystem are owned, and there are few, if any, wandering or ownerless dogs, household surveys are an accurate way to survey domestic dog abundance. From household surveys, the ratio of humans to dogs is approximately 6.7:1 in both agropastoralist and pastoralist areas (Lembo et al. 2008). Less is known about ownership patterns of domestic cats, but because they (a) hybridize with the African wildcat (Nowell and Jackson 1996) and (b) villagers have difficulties visually distinguishing a wildcat from a domestic cat (Andrew Ferdinands, pers. comm.), it is possible that not all domestic cats are owned.

Despite the time and financial efforts dedicated to nighttime transects and distance-sampling methods, some carnivore species are still not represented. For example, African wild dog packs have been steadily growing in number in pastoralist areas to the east of SNP since 2004 (Marsden et al. 2012). However, given their low absolute densities wild dogs are unlikely to be detected by transect sampling. Another reason that some carnivore species might not be well represented is that night transects do not capture diurnal species, such as slender, dwarf, and marsh mongoose, which have been directly observed and confirmed in the ecosystem by camera trapping.

As a cautionary tale, the time of day that a survey was done can affect density estimates. As a simple example of this phenomenon, we looked at density estimates from day and night distance-sampling surveys in Kenya. When we compared density estimates for three common species, there was a bias due to differing activity patterns in carnivores. For example, by comparing fig. 15.3C with fig. 15.3B, mean hyena and jackal densities appear higher at night than during the day, whereas densities of lions appear to be higher during the daytime than at night, although there are overlapping confidence intervals. Because hyenas and jackals are resting and often not spotted during the daytime, these animals would be undercounted during daytime counts. Therefore, we must be careful to only compare estimates from equivalent time periods (e.g., same time of day). This also applies to making sure comparisons are made between similar study designs and spatial areas. In our case, distance sampling and spotlight transects are useful for detecting long-term trends because the same method is applied under consistent conditions. Because of our diverse sources of data, it is difficult to calculate aggregate measures of community structure; for example, our species richness and diversity figure only indicates comparative richness and diversity for carnivores that can easily be seen at night by a spotlight.

LONG-TERM TRENDS

As we have seen in earlier books in this series, the Serengeti ecosystem is dynamic and is continually changing. It has undergone the impacts of disease, changes in climate, varying management practices, and the increasing impacts of people on its borders. All these factors will have complex and interacting impacts on its biodiversity, including its 30 species of carnivores. A classic example is the increase in large predator numbers following the increase in ungulates after the eradication of rinderpest from the Serengeti-Mara ecosystem (Sinclair et al. 2008). Perturbations to vegetation such as changes in burning patterns, or heavy and sustained livestock grazing, could also cause long-term changes. These human-induced changes can cause a shift in the vegetation state from grassland to woodland or the length of grass, thereby influencing prey and predator distribution (for example, because golden jackals are found in open grasslands, if an area changes from grassland to woodland we might expect golden jackal numbers to decrease). Numbers of carnivores could also change due to direct conflict with humans (see Hampson et al., chapter 21). It is important to be able to detect, determine the direction and magnitude of long-term changes, and ultimately to determine if the change is caused by "natural" ecosystem processes, or is caused by negative consequences of humans, such as poaching, loss of habitat, or human-wildlife conflict (Sinclair et al. 2007). Monitoring and documenting these changes is necessary to inform management practices, and to test the efficacy of management interventions. Understanding the interrelationships between these species, and the long-term forces on the ecosystem, is critical for ensuring the Serengeti continues to support a full suite of carnivores for years to come.

We looked for long-term changes in carnivore abundance in the Serengeti National Park by comparing new distance-based carnivore surveys with the 1977 and 1986 fixed-width surveys (Durant et al. 2011). We looked for long-term increases or declines; golden jackal, black-backed jackal, and bat-eared fox densities were detected as declining (fig. 15.6). In the other trend analysis we found a rapid decline in golden jackal numbers in the most recent survey compared to those in all previous surveys. No significant changes were found in densities of spotted hyenas, lions, cheetah, and side-striped jackal. Methods for these trend analyses are described further in Durant et al. (2011).

Although we do not have comparable longer-term systematic carnivore surveys inside and outside the MMNR of Kenya, ungulate numbers there are dropping. This is likely due to human influences including drought, increased cultivation, growing human settlement, illicit hunting, and livestock incursions into protected areas (Ogutu et al. 2011). These influences will undoubtedly have long-term effects on carnivore populations. However, the ongoing formation of conservancies in the pastoral lands of the Mara that began in 2006 is promoting expansion of the ungulate and carnivore populations on these landscapes.

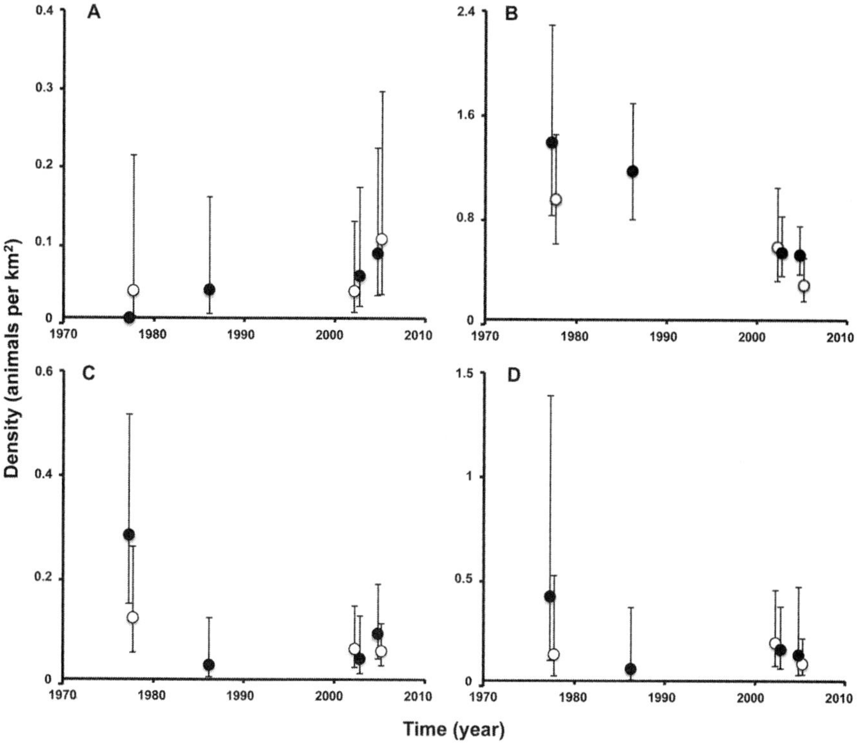

Fig. 15.6 Density and 95% confidence intervals of carnivore species that show a long-term decline. (A) golden jackal on the long-grass plains, (B) golden jackal on the short-grass plains, (C) black-backed jackal on both the long- and short-grass plains, and (D) bat-eared fox on both the long- and short-grass plains. Open circles represent dry season, whereas solid circles represent wet season estimates. Figure adapted from Durant et al. (2011).

CONCLUSION

The Serengeti-Mara ecosystem harbors a rich community of carnivores, and we are making gains to understand their ecology and distribution. In this system seasonality is important in driving many of the large carnivores' movements and abundance patterns, while the distribution of all carnivores is also influenced by the availability of their preferred habitats in the ecosystem. However, even in an ecosystem the size of the Serengeti, carnivores are still influenced by the impacts of humans. Human activities, particularly from higher-density agropastoralist populations, severely impact the structure and composition of carnivore assemblages, with far- reaching consequences. Agropastoralist areas likely act as sinks for wide-ranging, larger-bodied carnivore species and potentially as sources for opportunistic omnivorous species, whereas carnivore guilds are more complete and diverse in pastoralist areas—further confirming their conservation value. Long-term trends in abundance can be detected using the methods we present here, and future cross-disciplinary research needs to be implemented to determine the drivers of these trends (for example, in our case, declines in golden jackal abundance).

It is challenging to come up with a universal survey method that is accurate for a suite of carnivores, yet we highlight the importance of long-term monitoring. Currently, small animals are often missed during our sampling and, in general, we lack universal sampling protocols. However, technical advances are being made that should remove these biases (Pettorelli et al. 2010). At the same time, methodological advances are also being made, as seen with a new theory that is refining variance estimation for grouped species (Fewster 2011). We remain optimistic that despite these current challenges, attention to long-term monitoring is crucial for the management of carnivores in the Serengeti-Mara ecosystem.

ACKNOWLEDGMENTS

MC was supported by the US National Science Foundation under grant no. OISE-0804186. We would like to thank the Serengeti Viral Transmission Dynamics Project (especially B. Chunde, C. Mentzel, and T. Lembo) and the Serengeti Biodiversity Project (A. R. E. Sinclair, J. Fryxell, S. Mduma, and J. Masoy) for the night transect data; these transects were supported by NSF grant EF-0225453, Lincoln Park Zoo, and Canadian NSERC funding. The Mara distance sampling study was funded by ILRI's (International Livestock Research Institute) 37 donors and grants from the National Science Foundation (DEB 0342820 and BCS 0709671). Point data, carnivore transects, and camera-trap data in SNP were supported by the Darwin Initiative, the Wildlife Conservation Society, Zoological Society of London, St. Louis Zoo WildCare Institute, and the Frankfurt Zoological Society. Thanks to K. Metzger for plotting the richness and diversity figure and to two anonymous reviewers for helpful comments to improve this manuscript. Finally we thank TANAPA, Ngorongoro Conservation Area Authority, and TAWIRI for their support.

REFERENCES

Bhola, N., J. O. Ogutu, H. Piepho, M. Said, R. Reid, T. Hobbs, and H. Olff. 2012. Comparative changes in density and demography of large herbivores in the Masai Mara Reserve and its surrounding human-dominated pastoral ranches in Kenya. *Biodiversity and Conservation* 21:1509–30.

Bhola, N., J. O. Ogutu, M. Y. Said, H.-P. Piepho, and H. Olff. 2012. The distribution of large herbivore hotspots in relation to environmental and anthropogenic correlates in the Mara region of Kenya. *Journal of Animal Ecology* 81 (6): 1268–87. doi: 10.1111/j.1365-2656.2012.02000.x

Boydston, E. E., K. M. Kapheim, M. Szykman, and K. E. Holekamp. 2003. Individual variation in space use by female spotted hyenas. *Journal of Mammalogy* 84:1006–18.

Brown, J. H., and B. A. Maurer. 1989. Macroecology: The division of food and space among species on continents. *Science* 243:1145–50.

Buckland, S. T. 1993. *Distance sampling: Estimating abundance of biological populations.* London: Chapman and Hall.

Buckland, S. T., D. R. Anderson, K. P. Burnham, J. L. Laake, D. L. Borchers, and L. Thomas. 2001. *Introduction to distance sampling: Estimating abundance of biological populations.* Oxford: Oxford University Press.

Burrows, R. 1995. Demographic changes and social consequences in wild dogs, 1964–1992. In *Serengeti II: Dynamics, management, and conservation of an ecosystem*, ed. A. R. E. Sinclair and P. Arcese, 400–20. Chicago: University of Chicago Press.

Campbell, K. L. I., and M. Borner, eds. 1986. Census of predators on the Serengeti plains, May 1986. Unpublished report, Serengeti Ecological Monitoring Programme, Frankfurt Zoological Society.

Caro, T. M. 1994. *Cheetahs of the Serengeti Plains: Group living in an asocial species*. Chicago: University of Chicago Press.

Cleaveland, S., C. Packer, K. Hampson, M. Kaare, R. Kock, M. Craft, T. Lembo, T. Mlengeya, and A. Dobson. 2008. The multiple roles of infectious diseases in the Serengeti ecosystem. *In Serengeti III: Human impacts on ecosystem dynamics*, ed. A. R. E. Sinclair, S. Mduma, and J. Fryxell, 209–39. Chicago: University of Chicago Press.

Craft, M. E., E. Volz, C. Packer, and L. A. Meyers. 2011. Disease transmission in territorial populations: The small-world network of Serengeti lions. *Journal of the Royal Society Interface* 8:776–86.

Creel, S., and N. M. Creel. 1996. Limitation of African wild dogs by competition with larger carnivores. *Conservation Biology* 10:526–38.

Creel, S., N. M. Creel, and S. L. Monfort. 1997. Radio collaring and stress hormones in African wild dogs. *Conservation Biology* 11:544–48.

Donadio, E., and S. Buskirk. 2006. Diet, morphology, and interspecific killing in Carnivora. *American Naturalist* 167:524–36.

Durant, S. M. 1998. Competition refuges and coexistence: An example from Serengeti carnivores. *Journal of Animal Ecology* 67:370–86.

———. 2000a. Living with the enemy: Avoidance of hyenas and lions by cheetahs in the Serengeti. *Behavioral Ecology* 11:624–32.

———. 2000b. Predator avoidance, breeding experience and reproductive success in endangered cheetahs, *Acinonyx jubatus*. *Animal Behaviour* 60:121–30.

Durant, S. M., S. Bashir, T. Maddox, and M. K. Laurenson. 2007. Relating long-term studies to conservation practice: The case of the Serengeti Cheetah Project. *Conservation Biology* 21:602–11.

Durant, S. M., T. M. Caro, D. A. Collins, R. M. Alawi, and C. D. Fitzgibbon. 1988. Migration patterns of Thomson's gazelles and cheetahs on the Serengeti Plains. *African Journal of Ecology* 26:257–68.

Durant, S. M., M. E. Craft, C. Foley, K. Hampson, A. L. Lobora, M. Msuha, E. Eblate, J. Bukombe, J. McHetto, and N. Pettorelli. 2010. Does size matter? An investigation of habitat use across a carnivore assemblage in the Serengeti, Tanzania. *Journal of Animal Ecology* 79:1012–22.

Durant, S. M., M. E. Craft, R. Hilborn, S. Bashir, J. Hando, and L. Thomas. 2011. Long-term trends in carnivore abundance using distance sampling in Serengeti National Park, Tanzania. *Journal of Applied Ecology* 48:1490–500.

Durant, S. M., A. J. Dickman, T. Maddox, M. Waweru, and N. Pettorelli. 2010. Past, present and future of cheetah in Tanzania: From long term study to conservation strategy. In *Biology and conservation of wild felids*, ed. D. MacDonald and. A. L. Macdonald, 383–82. Oxford: Oxford University Press.

Durant, S. M., M. Kelly, and T. M. Caro. 2004. Factors affecting life and death in Serengeti cheetahs: Environment, age, and sociality. *Behavioral Ecology* 15:11–22.

Estes, J. A., J. Terborgh, J. S. Brashares, M. E. Power, J. Berger, W. J. Bond, S. R. Carpenter, T. E. Essington, R. D. Holt, J. B. C. Jackson, et al. 2011. Trophic downgrading of planet Earth. *Science* 333:301–6.

Fewster, R. M. 2011. Variance estimation for systematic designs in spatial surveys. *Biometrics* 67:1518–31.

Frank, L. G. 1986. Social organization of the spotted hyaena (*Crocuta crocuta*). I. Demography. *Animal Behaviour* 34:1500–09.

Fryxell, J. M., J. Greever, and A. R. E. Sinclair. 1988. Why are migratory ungulates so abundant? *American Naturalist* 131:781–98.

Fryxell, J. M., A. Mosser, A. R. E. Sinclair, and C. Packer. 2007. Group formation stabilizes predator-prey dynamics. *Nature* 449:1041–43.

Ginsberg, J. R. 2001. Setting priorities for carnivore conservation: What makes carnivores different? In *Carnivore conservation*, ed. J. L. Gittleman, S. M. Funk, D. W. Macdonald, and R. K. Wayne, 498–523. Cambridge: Cambridge University Press.

Hirzel, A. H., J. Hausser, D. Chessel, and N. Perrin. 2002. Ecological-niche factor analysis: How to compute habitat-suitability maps without absence data? *Ecology* 83:2027–36.

Hofer, H., and M. L. East. 1993. The commuting system of Serengeti spotted hyaenas: How a predator copes with migratory prey. III. Attendance and maternal care. *Animal Behaviour* 46:575–89.

———. 1995. Population dynamics, population size, and the commuting system of Serengeti spotted hyenas. In *Serengeti II: Dynamics, management, and conservation of an ecosystem*, ed. A. R. E. Sinclair and P. Arcese, 332–63. Chicago: University of Chicago Press.

———. 2008. Siblicide in Serengeti spotted hyenas: A long-term study of maternal input and cub survival. *Behavioral Ecology and Sociobiology* 62:341–51.

Holekamp, K. E., and S. M. Dloniak. 2010. Intraspecific variation in the behavioral ecology of a tropical carnivore, the spotted hyena. In *Advances in the study of behavior*, ed. R. Macedo, 189–229. New York: Academic Press.

Homewood, K., E. F. Lambin, E. Coast, A. Karlukl, I. Kikula, J. Kivella, M. Y. Said, S. Serneels, and M. Thompson. 2001. Long-term changes in Serengeti-Mara wildebeest and land cover: Pastoralism, population, or policies? *Proceedings of the National Academy of Sciences* 98:12544–49.

Ikanda, D., and C. Packer. 2008. Ritual vs. retaliatory killing of African lions in the Ngorongoro Conservation Area, Tanzania. *Endangered Species Research* 6:67–74.

Kissui, B. M. 2008. Livestock predation by lions, leopards, spotted hyenas, and their vulnerability to retaliatory killing in the Maasai steppe, Tanzania. *Animal Conservation* 11:422–32.

Kissui, B. M., and C. Packer. 2004. Top-down population regulation of a top predator: Lions in the Ngorongoro Crater. *Proceedings of the Royal Society, B* 271:1867–74.

Kolowski, J. M., and K. E. Holekamp. 2006. Spatial, temporal, and physical characteristics of livestock depredations by large carnivores along a Kenyan reserve border. *Biological Conservation* 128:529–41.

Laurenson, M. K. 1995. Implications of high offspring mortality for cheetah population dynamics. In *Serengeti II: Dynamics, management and conservation of an ecosystem*, ed. A. R. E. Sinclair and P. Arcese, 385–99. Chicago: University of Chicago Press.

Lembo, T., K. Hampson, D. T. Haydon, M. Craft, A. Dobson, J. Dushoff, E. Ernest, R. Hoare, M. Kaare, T. Mlengeya, et al. 2008. Exploring reservoir dynamics: A case study of rabies in the Serengeti ecosystem. *Journal of Applied Ecology* 45:1246–57.

Loveridge, A. J., and D. W. Macdonald. 2003. Niche separation in sympatric jackals (*Canis mesomelas* and *Canis adustus*). *Journal of Zoology* 259:143–53.

Loyola, R. D., L. G. R. Oliveira-Santos, M. Almeida-Neto, D. M. Nogueira, U. Kubota, J. A. F. Diniz-Filho, and T. M. Lewinsohn. 2009. Integrating economic costs and biological traits into global conservation priorities for carnivores. *PLoS ONE* 4:e6807.

Maas, B., and D. MacDonald. 2004. Bat-eared foxes. In *Biology and conservation of wild canids*, ed. D. MacDonald and C. Sillero-Zubiri, 227–42. Oxford: Oxford University Press.

Macdonald, D. W. 1979. The flexible social system of the golden jackal, *Canis aureus*. *Behavioral Ecology and Sociobiology* 5:17–38.

Maddox, T. 2003. *The ecology of cheetahs and other larger carnivores in a pastoralist-dominated buffer zone*. PhD diss., University College and Institute of Zoology, London.

Marsden, C. D., R. K. Wayne, and B. K. Mable. 2012. Inferring the ancestry of African wild dogs that returned to the Serengeti-Mara. *Conservation Genetics*. 13:525–33.

Mduma, S. A. R., and J. G. C. Hopcraft. 2008. The main herbiverous mammals and crocodile in the greater Serengeti ecosystem. In *Serengeti III: Human impacts on ecosystem dynamics*, ed. A. R. E. Sinclair, C. Packer, S. A R. Mduma, and J. M. Fryxell, 497–505. Chicago: University of Chicago Press.

Moehlman, P. 1983. Socioecology of silver-backed and golden jackals (*Canis mesomelas* and *Canis aureus*). In *Advances in the study of mammalian behavior*, ed. J. Eisenberg and D. Kleinman, 423–53. Lawrence, Kansas: American Society of Mammalogists.

Mogensen, N. L., J. O. Ogutu, and T. Dabelsteen. 2011. The effects of pastoralism and protection on lion behaviour, demography and space use in the Mara Region of Kenya. *African Zoology* 46:78–87.

Msuha, M. J., C. Carbone, N. Pettorelli, and S. M. Durant. 2012. Conserving biodiversity in a changing world: Land-use change and species richness in northern Tanzania. *Biodiversity and Conservation* 21: 2747–59.

Nowell, K., and P. Jackson. 1996. *Wild cats: Status survey and conservation action plan.* IUCN/SSC Cat Specialist Group. Gland, Switzerland, World Conservation Union.

Ogutu, J. O., N. Bhola, and R. Reid. 2005. The effects of pastoralism and protection on the density and distribution of carnivores and their prey in the Mara ecosystem of Kenya. *Journal of Zoology* 265:281–93.

Ogutu, J. O., and H. T. Dublin. 2002. Demography of lions in relation to prey and habitat in the Maasai Mara National Reserve, Kenya. *African Journal of Ecology* 40:120–29.

Ogutu, J. O., N. Owen-Smith, H. P. Piepho, and M. Said. 2011. Continuing wildlife population declines and range contraction in the Mara region of Kenya during 1977–2009. *Journal of Zoology* 285:99–109.

Ogutu, J. O., H. P. Piepho, H. T. Dublin, N. Bhola, and R. S. Reid. 2008. Rainfall influences on ungulate population abundance in the Mara-Serengeti ecosystem. *Journal of Animal Ecology* 77:814–29.

Pace, M. L., J. J. Cole, S. R. Carpenter, and J. F. Kitchell. 1999. Trophic cascades revealed in diverse ecosystems. *Trends in Ecology and Evolution* 14:483–88.

Packer, C., R. Hilborn, A. Mosser, B. Kissui, M. Borner, G. Hopcraft, J. Wilmshurst, S. Mduma, and A. R. E. Sinclair. 2005. Ecological change, group territoriality, and population dynamics in Serengeti lions. *Science* 307:390–93.

Palomares, F., and T. M. Caro. 1999. Interspecific killing among mammalian carnivores. *American Naturalist* 153:492–508.

Pangle, W. M., and K. E. Holekamp. 2010. Lethal and nonlethal anthropogenic effects on spotted hyenas in the Masai Mara National Reserve. *Journal of Mammalogy* 91:154–64.

Pettorelli, N., A. L. Lobora, M. J. Msuha, C. Foley, and S. M. Durant. 2010. Carnivore biodiversity in Tanzania: Revealing the distribution patterns of secretive mammals using camera traps. *Animal Conservation* 13:131–39.

Ritchie, E. G., and C. N. Johnson. 2009. Predator interactions, mesopredator release and biodiversity conservation. *Ecology Letters* 12:982–98.

Rood, J. P. 1990. Group size, survival, reproduction, and routes to breeding in dwarf mongooses. *Animal Behaviour* 39:566–72.

Schaller, G. B. 1972. *The Serengeti lion: A study of predator-prey relations.* Chicago: University of Chicago Press.

Sechrest, W., T. M. Brooks, G. A. B. da Fonseca, W. R. Konstant, R. A. Mittermeier, A. Purvis, A. B. Rylands, and J. L. Gittleman. 2002. Hotspots and the conservation of evolutionary history. *Proceedings of the National Academy of Sciences* 99:2067–71.

Serengeti Research Institute. 1977a. Census of predators and other animals on the Serengeti plains, May 1977. Unpublished report, Serengeti Wildlife Research Center.

———. 1977b. Census of predators and other animals on the Serengeti plains, October 1977. Unpublished report, Serengeti Wildlife Research Center.

Shannon, C. E., and W. Weaver. 1949. *The mathematical theory of communication.* Urbana: University of Illinois Press.

Sinclair, A. R. E., J. G. C. Hopcraft, H. Olff, S. A. R. Mduma, K. A. Galvin, and G. J. Sharam. 2008. Historical and future changes to the Serengeti ecosystem. In *Serengeti III: Human impacts on ecosystem dynamics*, ed. A. R. E. Sinclair, C. Packer, S. A. R. Mduma, and J. M. Fryxell, 7–46. Chicago: University of Chicago Press.

Sinclair, A. R. E., S. A. R. Mduma, J. G. C. Hopcraft, J. M. Fryxell, R. Hilborn, and S. Thirgood. 2007. Long-term ecosystem dynamics in the Serengeti: Lessons for conservation. *Conservation Biology* 21:580–90.

Stelfox, J. G., D. G. Peden, H. Epp, R. J. Hudson, S. W. Mbugua, J. L. Agatsiva, and C. L. Amuyunzu. 1986. Herbivore dynamics in southern Narok, Kenya. *Journal of Wildlife Management* 50:339–47.

Terborgh, J., and B. Winter. 1980. Some causes of extinction. In *Conservation biology: An evolutionary-ecological perspective,* ed. M. E. Soule and B. A. Wilcox, 119–33. Sunderland, Mass: Sinauer.

Thomas, L., S. T. Buckland, E. A. Rexstad, J. L. Laake, S. Strindberg, S. L. Hedley, J. R. B. Bishop, T. A. Marques, and K. P. Burnham. 2010. Distance software: Design and analysis of distance sampling surveys for estimating population size. *Journal of Applied Ecology* 47:5–14.

Waser, P. M., L. F. Elliot, N. M. Creel, and S. R. Creel. 1995. Habitat variation and mongoose demography. In *Serengeti II: Dynamics, management, and conservation of an ecosystem,* ed. A. R. E. Sinclair and P. Arcese, 421–47. Chicago: University of Chicago Press.

Woodroffe, R., and J. R. Ginsberg. 1998. Edge effects and the extinction of populations inside protected areas. *Science* 280:2126–28.

PART III

The Human Ecosystem and Its Response to Disturbance

SIXTEEN

The Plight of the People: Understanding the Social-Ecological Context of People Living on the Western Edge of Serengeti National Park

Eli J. Knapp, Dennis Rentsch, Jennifer Schmitt, and Linda M. Knapp

Serengeti is perhaps best known for its status as a national park. In reality, however, the national park forms one piece of a complex social-ecological system with more than five types of land-use designations with varying governmental restrictions. This complex system is often referred to as the Greater Serengeti Ecosystem (GSE). Although people are prohibited from living within the park, hundreds of thousands reside within the ecosystem and have profound impacts on its wildlife and natural resources. At the same time, wildlife affects the people living nearby in ways that are both beneficial and harmful. This reciprocal relationship creates a complex milieu with profound implications for both wildlife management and socioeconomic development.

Until recently much ecological and anthropological research attention focused on Maasai populations on the eastern side of Serengeti National Park (hereafter, SNP). While human populations have lived for centuries west of the park, they have been geographically more isolated and relatively small (Shetler 2007). As a result, local interactions with the resource base west of the area now known as SNP were comparatively insignificant. Other factors that isolated western populations were poorly maintained roads, difficult river crossings, unpredictable rainfall, abundance of tsetse fly, and periodic border closures with Kenya. In the last several decades, however, human population size has steadily increased in the west (Kaltenborn et al. 2008). As a result, local resource use has markedly increased and human impacts have more directly and severely affected the park (Sinclair et al. 2007).

Since human population growth and resource use is unlikely to attenuate quickly on this western side, it is important to understand the nature of these escalating human-ecological interactions. While human-ecological interactions occur throughout western Serengeti extending from the park's border all the way to Lake Victoria, the term "western Serengeti" is used more narrowly here, including only those households lying within 18 km of a protected area boundary. After 18 kilometers, park-related human-wildlife interactions decrease considerably (Knapp 2009).

Human presence to the west of the GSE is not a recent phenomenon. The system has been an active zone of interaction for millennia as people have used it for dwelling, cultural ceremonies, hunting, and resource collection (Shetler 2007). Throughout this time, humans and wildlife have coexisted favorably as human population densities were low and wildlife and natural resources were abundant. Such peaceful coexistence, however, has been challenged due to conflicts spurred largely by poverty and population growth (Kaltenborn et al. 2008). Various agencies have responded by enforcing restrictions on illegal resource use with more regularity and extent (Hilborn et al. 2006; Knapp 2009). More attention has been paid to Serengeti's escalating conflicts which most often take the form of illegal hunting, grazing, and tree cutting, as well as crop destruction and livestock depredation by wildlife.

Underlying these conflicts are characteristics of the local people themselves—the knowledge, awareness, and attitudes they have toward the national park and its affiliated protected areas. Without a general understanding of local sentiment, it is difficult to predict the response—or behavioral adjustments—that people will make toward either of the two main conservation strategies used in the GSE: community-based conservation (CBC) which is an incentive-based approach, and antipoaching ("fences and fines") enforcement with subsequent penalties for arrested parties. Ideally, CBC is implemented by the communities themselves in an attempt to conserve wildlife and natural resources by increasing the benefits people receive from having conservation in their area (e.g., direct monetary payments or meat distribution, village infrastructural improvements) to overcome the costs of living with wildlife (e.g., crop damage, depredation on livestock). Regulatory or enforcement efforts, on the other hand, seek to increase the costs people incur from illegal natural resource use (e.g., fines, arrests) to overcome the benefits people receive from illegal resource use (e.g., bushmeat, fuelwood). If long-term conservation is ultimately to succeed, we propose that the net benefit (benefits minus the costs) a household or individual receives from local conservation need to exceed the net benefit that may be received through illegal natural resource use.

The livelihood strategies that households choose are a dual product of their values and socioecological histories. This chapter limits its focus to household livelihoods that are most impacted by the GSE due to living closely to protected area boundaries. It begins with a look at the people themselves—their ethnic backgrounds and the land that they occupy. It proceeds with an examination of the knowledge, awareness, and attitudes that the local people have toward SNP and conservation. Focus then turns to four primary interactions that households have with the park—crop destruction, wildlife depredation on livestock, park-related employment, and illegal hunting. The chapter concludes with a brief discussion of park-related employment, which is recommended here as a possible means of mitigating human-wildlife conflict in the region.

METHODS

This chapter features the work of two separate studies conducted in western Serengeti. One study (Knapp 2009) was composed of semistructured household interviews which were conducted in 15 villages located 0–18 km away from Serengeti National Park, or from one of the park's adjoining game reserves (Grumeti, Ikorongo, or Maswa). Village selection was predicated on proximity to Serengeti National Park and the verified presence of human-wildlife conflict. Human-wildlife conflict was verified through the use of archival district reports and open-ended structured interviews with district officials. A total of 722 households were selected using a stratified-random format. Households represented 46 subvillages in the Mara and Shinyanga Region which included the districts Serengeti, Bunda, and Meatu. These districts were chosen with nonprobability judgment sampling (Bernard 2006). One to three subvillages were selected from each village and 20–25% of selected subvillages were sampled. Due to existing patriarchal cultural mores, all interviews were administered in Swahili to the male head of the household and done with the aid of a local guide from the respective village as assigned by the local village chairman. If the head of the household was unavailable, interviews were carried out with the spouse or oldest child, provided the individual was over the age of 18 and demonstrated an adequate knowledge of household affairs. This group of 722 interviews is designated here as the primary sample.

As part of this study (Knapp 2009), three subset surveys were conducted with strategically chosen respondents based upon their participation in, or knowledge of, sensitive issues relating to Serengeti National Park. The three groups chosen included 104 acknowledged illegal hunters, 50 individuals with current park-related employment, and 14 key informants who demonstrated in-depth knowledge about specified issues. Methods for identifying and interviewing illegal hunters varied from those used for the primary sample. Since poaching is a sensitive issue that carries penalties if arrested, respondents were selected with a snowball sampling technique (Bernard 2006). Interviews were organized with the help of trust-based relationships developed with three key informants from Serengeti, Bunda, and Meatu Districts. Respondents remained anonymous and interviews were conducted in remote areas to ensure full confidentiality. This group is referred to here as the poaching subset sample. The subset sample of individuals with park-related employment mirrored those used for illegal hunting. Respondents were chosen using trust-based relationships and snowball sampling techniques. Park-related employment referred to all forms of employment that are in some way connected to, or dependent upon, Serengeti National Park, or one of its affiliated protected areas. An independently owned tourism camp, for example, qualified because it utilizes the park for clients.

In an effort to obtain information on sensitive issues (e.g., income level), interviews in this subset survey were conducted with individuals with whom the author had previously developed a trust-based relationship. Other respondents were selected through recommendations given by respondents or friends of the author made through weekly field contact. Full anonymity was assured to all respondents. This group is referred to as the employment subset sample.

A key informant subset sample of 14 individuals was conducted in a semistructured format with built-in open-ended questions. This was done for the purposes of providing context to larger issues pertinent to an understanding of the social-ecological system of western Serengeti. Key informant interviews were conducted with seven subvillage chairmen, two village executive officers, one head village chairman, one district livestock officer, one district agricultural officer, one zone commander of antipoaching rangers, and one foreign nongovernmental worker. Key informants were selected based on the criteria that they were currently involved in official affairs long enough to be well informed on local issues. All interviews for the three subset samples were conducted in the same time frame as the primary sample.

A second separate study was conducted in eight villages all located to the northwest of SNP (Schmitt 2010). This study similarly featured nonprobability judgment sampling and semistructured interviews conducted through a local nongovernmental research organization, Savannas Forever Tanzania. For each of the eight villages, five subvillages were randomly selected.

Within each subvillage, one to nine households were selected at random based upon official village rosters. This ensured that over 20 households per village were chosen. Interviews were conducted in Swahili by the Tanzanian researchers and administered with the household head. If unavailable, the next most knowledgeable respondent from the household was chosen provided this person was over the age of 18. Villages from this study were located in or within 5 km of a protected area that borders SNP. Data was collected from August 2006 through January 2007. While the conclusions and recommendations from this chapter are drawn from both studies detailed here, the results of the second study (Schmitt 2010) are limited to the forthcoming section entitled: "Perceived Costs and Benefits."

ETHNICITY AND HOUSEHOLD CHARACTERISTICS

In the most general sense, the people comprising western Serengeti are similar in that most descend from east Nyanza Bantu-speaking peoples. A closer look, however, reveals considerable ethnic diversity. Respondents from the primary sample represented 29 different ethnic groups. People of Kuria descent dominated the northwestern part of the system while Sukuma people (the largest ethnic group in Tanzania) dominated the southwestern portion. Ensconced within these two majorities were many smaller tribes including Ikoma, Natta, Issenye, and Ngurime. Many of these smaller ethnic groups emerged in the early 1800s and were officially codified during the years of British and German colonialism in the early 1900s (Shetler 2007). For the most part, the varying ethnic groups bordering Serengeti's western side have enjoyed a peaceful coexistence that has stemmed from a need for mutual support (Shetler 2007).

Weather irregularities, ongoing changes in colonialism and government, and pressures exerted by the Maasai historically resulted in a cross-regional awareness and willingness to band together in loose coalitions and communities. Individual family members were intentionally separated among several communities, for example, to prevent the death of entire households in the event of a Maasai raid from the east (Shetler 2007). Likewise, cohorts of similarly aged males of varying ethnic backgrounds routinely banded together to repel further Maasai incursions, largely coming from the north and east. Mixed settlements sprang out of such initiatives and ethnic descent took on less importance. This inclusive legacy may be evidenced by the fact that 29% of respondents had a spouse of another ethnic descent. While it was not uncommon for households to exceed 20 members, the average western Serengeti household contained about eight people. Larger households had greater resource requirements but were frequently perceived by respondents to be advantageous in rural areas because a larger domestic labor force allowed more livelihood diversification.

LAND AVAILABILITY IN WESTERN SERENGETI

In Tanzania, every piece of land is designated by one of four categories which include private land, conservation land (protected areas), village land, and/or communal land. The rural lands that comprise the social-ecological system of western Serengeti are a mosaic of many properties which generally lack fences, boundaries, signs, and even deeds. Even so, ownership and status of each parcel is widely known—and generally respected—by local peoples. Failure to respect ownership, such as allowing cattle to graze in another person's maize field, often results in sanctions levied by a locally elected village land committee. Land is not, in fact, owned in Tanzania. Rather, all land in Tanzania is under the current president's control and is the property of the nation, governed by the department of land (Shetler 2007). With the exception of protected areas, private and village holdings form a communal type of tenure, locally recognized as the "deemed right of occupancy" (Kideghesho et al. 2006).

The availability of land has decreased considerably in recent years. A generation ago, contiguous tracts of land were readily available to local residents who wished to increase their holdings under a deemed right of occupancy. All that was required of an individual, or household, was to present a letter/application to the locally elected village council. After perfunctory background checks by village officials, parcels were distributed indefinitely and without charge. At the time of this research, this process was still in operation in all of the sampled villages. Although the process is unchanged, local officials from each village stated that there are fewer—if any—land parcels available to disburse. Most of the remaining tracts of land that were available at the time of research were of poor quality for agriculture and/or livestock. Key informant interviews revealed that just 38% of villages in Serengeti district still had available land. If current population trends continue as they are projected to (Kaltenborn et al. 2008), it is not difficult to anticipate complete elimination of available land within a few generations (Knapp 2009).

Western Serengeti has also seen a significant increase in land subdivision. This is reflected in the trends observed over the span of one generation. Current households owned an average of 8.9 acres. The parents of these respondents, however, owned 43.6 acres. Like much of Tanzania, land is passed down in western Serengeti through the paternal line and subdivided among a father's sons. Although households own just one-fourth of the land that their parents did, land ownership in western Serengeti ranks higher than in other areas. A study by Bahiigwa, Mdoe, and Ellis (2005) in the Kilosa and Morogoro rural districts of Tanzania revealed that average household land ownership amounts to 3.8 acres. Households within 18 km of a protected area boundary in Serengeti owned nearly twice this amount. This relative land "abundance" in western Serengeti may account for some of the encroachment in recent decades on SNP's boundary. Land availability was the primary lure (41%) for immigrants that had moved into the western Serengeti study area within the last ten years. This underscores the fact that while land availability on the whole may be decreasing closer to the park, it has still drawn more households in, likely exacerbating existing human-wildlife conflicts.

Somewhat surprisingly, households actively used only 4.9 acres (55%) of their total holdings over the course of a year. In most cases, households used the remaining 45% of their land as grazing for livestock or as rental parcels for households lacking adequate holdings. Other unused parcels were said to be "resting" or lying fallow for the purposes of soil regeneration. What is clear from these findings is that extreme agricultural intensification has not yet taken hold and that households in western Serengeti are land-rich compared to other areas of Tanzania.

Somewhat contradictory, however, was that over half (62%) of the sample reported that their current landholdings were insufficient in meeting their basic needs. When key informants were pressed to explain this discrepancy, several explained that they viewed their holdings as inadequate because they would not be enough to divide among their male heirs. Although land would be passed down, respondents recognized that the subdivided plots would not be enough to sustain each respective household. For whatever reason, increasing the productivity of land was not mentioned, be it through improved seeds, fertilizer application, or irrigation channels. While agricultural intensification is not recommended as a solution here, it is likely that land productivity could be increased if additional labor and capital were available.

A little over 10% of western Serengeti households reported that they did not own land, but just 2% claimed not to use any land for agriculture or livestock whatsoever. This suggests the presence of an active rental—or informal—market for land. Several respondents stated they use land from friends or relatives free of charge. Other land-poor households "rented" land in exchange for a fraction of the harvest they obtained in a given growing season. This suggests the existence of active social networking and reciprocity. The fact that 98% of households utilize land for agriculture or livestock to some degree underscores considerable dependence upon land in current livelihood strategies.

The villages next to Serengeti's western boundary have varying amounts of communally held lands as well. According to village officials, these ranged from 50 to 100 acres per village and adjoined existing rivers or watershed tributaries. The widely acknowledged purpose behind the designation of such areas is for the provision of grazing lands and water access for livestock. In addition, these areas have been heavily utilized for firewood collection for cooking fuel. The reliance on communal lands varies with the household. As might be expected, land-poor households cannot afford to graze livestock on limited landholdings. What is surprising, however, is that middle-tier landholders chose not to use their private holdings for regular livestock grazing. Optimistically, this implies that suitable pasture exists in communal lands. Also plausible, however, is the notion that current practices may be foreshadowing a tragedy of the commons scenario in which communal lands are degraded to a point of limited, or diminishing, returns (Hardin 1968). Coupled with ongoing subdivision, overuse of the commons may eventually force more households to graze in protected areas if they hope to maintain current livestock numbers.

PERCEIVED COSTS AND BENEFITS

As will be discussed shortly, households west of the GSE have many interactions—both harmful and beneficial—with the park and its protected areas. Despite this, an alarming percentage of households lack even the most basic knowledge and awareness about SNP's existence, boundaries, and regulations. Amid the people living northwest of the park, for example, just 40.2% of surveyed households had any awareness of the existence of Serengeti National Park and only half (49.7%) knew the location or name of their nearest protected area. This raises three fundamental issues, or "problems," that need to be addressed if improved livelihoods and more successful conservation are to be achieved in the GSE.

The first issue is that local people with limited knowledge of the GSE are unlikely to know how natural resource use is restricted or regulated, be it hunting, grazing, or fuelwood collection. Nor is it likely that people thoroughly or even partially understand the differences in resource utilization between the five land-use designations attached to SNP's adjoining protected areas (see chapter 2, fig. 2.1 for the different administrative areas).

In short, SNP (14,763 km^2) forms the heart of the GSE and is the most restrictive among Tanzania's protected areas. Legal uses in SNP are limited to nonconsumptive utilization (game-viewing, research, and photographic tourism). Access and resource use by local peoples is strictly prohibited and routinely enforced (Hilborn et al. 2006). Due to restrictions from Wildlife Conservation Act No. 12 in 1974, licensed hunting is allowed but all other forms of access, cultivation, and livestock grazing are prohibited in Ikorongo (563 km^2) and Grumeti (1900 km^2) Game Reserves. Local confusion likely exists because these uses are legal in the Loliondo Game Controlled Area, east of the park. Moreover, another land-use designation called the Ikona open area is located between the Ikorongo and Grumeti reserves. In the open area, restrictions on use are determined and enforced locally. These various jurisdictions receive periodic support by nongovernmental organizations. The added role of nongovernmental organizations may inadvertently add to local confusion because ephemeral funding cycles within these organizations cause relatively sudden increases in enforcement and conservation initiatives which may be abruptly truncated when outside financial support dries up.

In the northwest section of the GSE, just 26% of surveyed households possessed any knowledge about both SNP and the nearest protected area (Ikorongo or Grumeti Game Reserves). This relatively small portion of households with awareness of this suggests two things. One, it means that about three-quarters of all households are lumping protected areas and their associated—and varying—restrictions together. And two, it shows that most households do not know where one protected area ends and another begins. In other words, even those households that have sufficient understanding of the varying restrictions are challenged by not knowing boundaries between the reserves and the park.

If benefits of conservation are to be recognized at all in western Serengeti, of course, general awareness of SNP and its protected areas must be increased first. This process should start with local village leaders. Three-quarters of local leaders were knowledgeable about SNP and 62.5% had awareness of their closest protected area. While this was higher than other village households, it underscores the fact that general knowledge is lacking even at the highest level of village governance. While it is vital for establishing the link between benefits, conservation, and protected areas, it may also make local people more negative toward protected areas. For example, several respondents in the northwest directly blamed SNP for crop damage that they suffered from elephants. This potentially negative outcome is likely an unavoidable reality of increasing local knowledge of protected areas.

The second issue is that local people are unlikely to link benefits (and costs) with conservation initiatives put forward in their respective areas. While it may actually be good for conservation managers if people fail to link costs (e.g., crop damage caused by wildlife) to the GSE, it is decidedly inauspicious if the benefits of conservation initiatives are not recognized by local people. If benefits are not linked to protected areas, then any incentive for conserving nearby protected areas is greatly diminished. To better understand this relationship in western Serengeti, households in 15 villages in western Serengeti were asked if they received benefits from SNP. Reported benefits were strongly correlated with park-related employment (Pearson $R = 0.38$, $p < 0.001$). In other words, households with park-related employment were more likely to recognize, or link, this benefit to the park's presence than households without such employment (fig. 16.1). Most importantly, all households with park-related employment recognized that their employment was a benefit that was directly linked to SNP (Pearson $R = 1.00$, $p < .001$). Basic ecotourism theory predicts that households that know the source of their benefits have a greater likelihood to protect the resource upon which their livelihood depends than households who lack such information (Charnley 2005). In this way, the value of park-related employment is that it not only directly helps households through income, but it also engenders households with positive feelings toward SNP. The temptation here is to recommend more park-related employment opportunities for local households. But such recommendations would be premature until more in-depth analyses are done that test SNP's capacity for absorbing a significant number of additional employees from local communities. Another potential pitfall would be drawing more households closer to the park to take advantage of new opportunities. Lastly, more research is needed comparing the income gains from park-related employment relative to the opportunity costs of not farming. For these and several other reasons, increasing park-related employment opportunities to mitigate human-wildlife conflict should be done cautiously and judiciously.

Thirdly, conservation benefits that are distributed equally to households that practice illegal resource utilization (e.g., illegal hunting) and those that do not are unlikely to curb resource-destructive behaviors even if households recognize that such benefits come from the GSE. If all households receive benefits equally, there is no incentive for households to refrain from additional income-generating opportunities. Of people who were knowledgeable about SNP and its protected areas, just over one-third (34.2%) reported that they received a benefit of infrastructure by living close to the park. This included the creation of schools and dispensaries, and an improvement in local roads. This was much larger than employment which

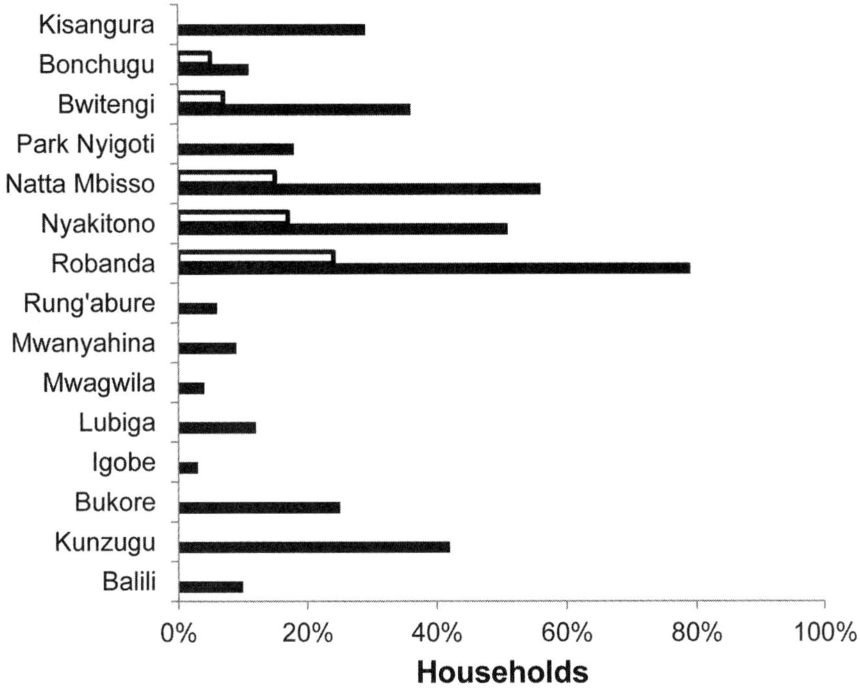

Fig. 16.1 Comparison of park-related employment and perceived park-related benefits among 722 households in 15 villages in western Serengeti, Tanzania. Black bar indicates percentage of households with a perception of park-related benefits. White bar indicates percentage of households with park-related employment.

was the second most mentioned benefit (6.8%). Infrastructure is of interest here because it is a clear example of a benefit from which all local households benefit equally. While benefits like infrastructure may truly help local communities and be good for improving public relations for conservation organizations, they fall short in that they equally reward all households, those that are involved in illegal activities and those that are not. As such, infrastructural improvements may be merely exacerbating the problem they are attempting to resolve. These benefits may be acting as a lure for households to immigrate into western Serengeti from outside areas or for current households to relocate nearer park boundaries to take better advantage of new services and improvements. For these reasons, governmental and outside organizations must find a way to make benefits conditional on positive conservation outcomes.

As stated previously, many households lack any knowledge whatsoever about the existence of SNP or the nearest protected area. For those that were knowledgeable, the percentage of local households that have positive feelings (49.4%) toward SNP was roughly equivalent to those with negative feelings (47.2%). An even higher percentage (63%) had negative feelings toward their closest protected areas while only 32% had positive feelings. While theory predicts that individuals with positive feelings toward a resource are more likely to protect it (Charnley 2005), there remains a gap between attitudes and behavior. For example, while infrastructural improvements may improve local attitudes toward conservation, they may not actually deter an individual from going out to poach wildlife. Due to this gap, and that attitudinal change often occurs gradually, conservation initiatives that occupy a household's time and alter behavior may be better alternatives over the short term (Knapp, 2007). While it is true that households with park employment can still hunt illegally and potentially facilitate "insider" hunting, the time to actually do so may be greatly reduced by the requirements of employment.

LIVELIHOOD STRATEGIES

Generally speaking, most of the people that live on the western margin of Serengeti National Park have experienced decreased farm yields and a reduced number and size of farm plots. They also have limited access to markets. Most practice horticulture or small scale, low intensity farming. Households generally farm small plots scattered throughout a village and have modest livestock holdings (chickens, goats, and cattle). While most produce is used for subsistence, some households succeed in growing a surplus which is often sold for use in other household pursuits, such as the acquisition of more land. Due to ever-increasing land subdivision, however, sole reliance on crops and livestock is no longer viewed as a viable livelihood strategy (Himmelfarb 2006). Households devoted most (3.3 acres) of their private landholdings (8.9 total acres) to subsistence farming and used a smaller portion (1.7 acres) for cash crops. Not surprisingly, households with larger landholdings devoted more acreage to both subsistence ($R^2 = 0.56$, $p < 0.001$, $N = 630$) and cash crops ($R^2 = 0.49$, $p < 0.001$, $N = 631$). The fact that virtually all households (98%) farmed underscores two critical points. One, farming plays a pivotal role in the livelihood strategies of the social-ecological system, and two, it suggests that most households are at least partially dependent upon rainfall and weather regimes. Dependency on stochastic factors such as weather is important for understanding human-wildlife conflicts in the GSE. If a severe drought occurs, for example, wildlife may be forced to leave protected areas in search of other food sources such as crop fields, or a persistent drought that destroys crops may cause local people to make up these losses by illegally hunting wildlife. This simple fact makes it difficult to predict how households may interact with the park from one year to next. It also makes it difficult to predict how wildlife will respond to the presence of accessible cultivation. Clearly, longer term multi-year studies are needed to better understand human-wildlife interactions in response to changes in rainfall and climate.

PARK EFFECTS ON LIVELIHOODS

Agriculture

Over generations, households in western Serengeti have adapted their agricultural choices to reflect the rainfall patterns. As such, the assortment of crops that most households use renders agriculture as a relatively low-input and risk-averse livelihood strategy (Ayalew et al. 2003). It is the primary form of income for households and allows them to spread risk through the selection of several crops and the utilization of heterogeneous plots. Although all crops were sold locally, cotton (*Gossypium sp.*) was the primary cash crop. In the year of this study, cotton earned households US $154 compared to US $14 for all the other crops combined. A total of 19 different crops were used in varying proportions in western Serengeti. According to key informants, decisions concerning what plots to use and which crops to plant depended on local market conditions, rainfall, available labor, assets, and soil conditions. For most surveyed households—especially those closer to park boundaries—crops were chosen according to likelihood, intensity, and duration of crop raiding by wildlife. The severity of crop raiding was evidenced by the fact that households lost an average of US $110 of potential income over a 12-month period. Many factors were involved in the amount of crop destruction a household suffered. As other studies have suggested, the proximity of agricultural fields to protected areas was strongly related to crop damage (Rao et al. 2002; Linkie et al. 2007; Cai et al. 2008). Pearson bivariate correlations between damage levels to seven common subsistence crops and distance from a protected area boundary were negatively correlated (table 16.1). As expected, the closer a household and its farm plots was to a boundary, the more crop damage the household incurred from wildlife (fig. 16.2).

The potential tradeoff from a household's standpoint, however, is that the closer it is to a protected area, the less time is needed to illegally extract

Table 16.1 Pearson correlation coefficients between distance to a protected area and crop damage

Distance to protected area	Pearson *R*	*p*-value	St. Dev.
Sacks destroyed			
Maize	–0.14	0.01	13.1
Millet	–0.24	<0.001	5.4
Sorghum	–0.20	<0.001	4.2
Potatoes	–0.13	0.01	3.9
Beans	–0.11	0.02	3.4
Cotton*	–0.12	0.02	163.8
Cassava**	–0.20	<0.001	1.0

Notes: $N = 422$, *Cotton measured in kilograms. **Cassava measured in hectares.

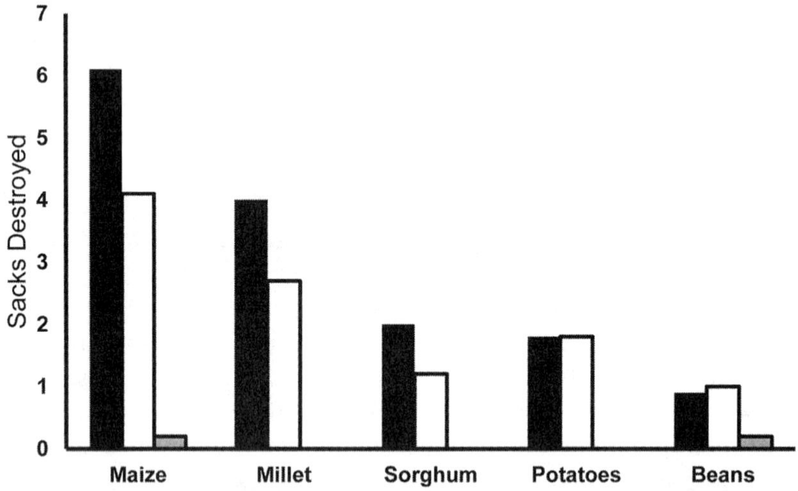

Fig. 16.2 Relationship between damage to five subsistence crops (in terms of average sacks destroyed per household per year) and the distance of 422 households to the nearest protected area boundary of Serengeti National Park, Tanzania. Dark bar represents households living within 5 km, white bar those within 10 km, and grey bar those beyond 10 km of the nearest protected area boundary.

resources. The other important distance measure was the distance that agricultural plots were from a household. On average a household's nearest agricultural plot was 1.4 km away. This required a 49 minute one-way traverse by foot, which was the most common form of travel for farmers. In addition to being an opportunity cost, such large distances likely account for why households are largely unable to defend their agricultural fields from wildlife.

Livestock

Livestock are an important complimentary asset for farming households in western Serengeti. While households may record low annual incomes, this does not necessarily mean they are poor. A household, for example, may have little annual income but instead own 100 cattle, which keeps the household out of poverty. When crop failures occur due to pest outbreaks, drought, or wildlife destruction, households with livestock can, and do, sell or slaughter their holdings to subsist through difficult periods. Annual sales of livestock provided households with an average of US $71. Exactly half of the sample, for example, reported that they sold livestock as a survival strategy during a year of major crop failure (fig. 16.3).

More importantly, however, livestock feature as a bank account as they are indivisible assets that are easier to retain than cash. Culturally, it is more difficult for households with extra cash to save it for future hardships when extended networks may ask to "borrow" some. Households are not dutifully bound to lend or give cattle to others who may come asking, as one would for cash.

The most commonly owned domestic animals were cattle, goats, and chickens. These animals play a vital role in asset accumulation and are in effect "traded up" in sequence (Ellis and Mdoe 2003). When enough chickens are obtained, for example, many may be sold off and money used to obtain a goat, which diversifies holdings and spreads risk among varying species. Better-off households, however, tend to maintain adequate numbers of

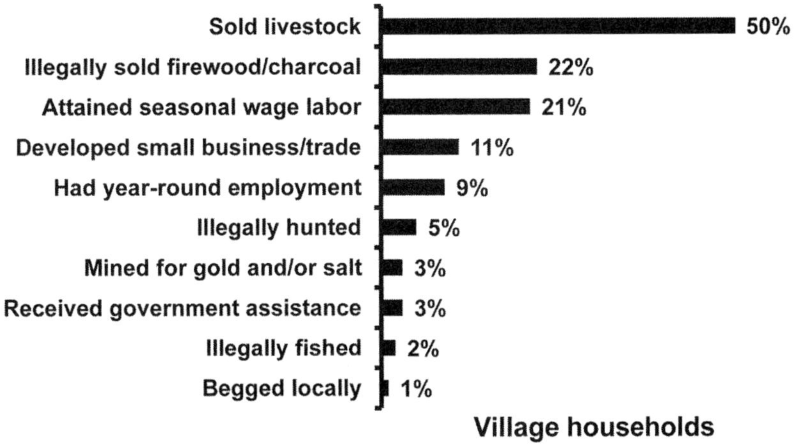

Fig. 16.3 Alternative (survival) strategies adopted by 722 households in 15 villages during years of major crop failure in western Serengeti. Dark bars indicate the percentage of households from all villages who adopted the various alternative survival strategies depicted on the y-axis.

Table 16.2 Correlation analysis of common livestock owned by households in western Serengeti

	Cattle	Goats	Sheep	Chickens
Cattle	1.0			
	(.)			
Goats	0.33**	1.0		
	(0.0)	(.)		
Sheep	0.27**	0.24**	1.0	
	(0.0)	(0.0)	(.)	
Chickens	0.38**	0.26**	0.12**	1.0
	(0.0)	(0.0)	(0.002)	(.)

Notes: Numbers in parentheses indicate significance levels.

**Correlation is significant at 0.01 level (two-tailed).

(.) Correlation cannot be calculated.

each livestock species (table 16.2). This strategy may also help households better utilize the mixed grassland-shrubland-woodland mosaic (Hella, van Huylenbroeck, and Mlambiti 2001) that comprises many of the village and grazing areas west of the park. Cows and sheep may graze, for example, while goats browse on woody growth, and chickens forage on insects.

Loss of livestock to wildlife and disease were common for households in western Serengeti. Disease claimed a larger percentage of cattle (58%) than did wildlife (40%) but this trend did not apply for sheep, goats, and chickens (fig. 16.4). Each household lost 0.4 cattle from an average herd consisting of 7.1 animals. Despite this loss, households still accrue 1.8 cows per year through natural herd increase as an average of 2.2 calves were born over the course of a year. In terms of depredation and/or disease, 7% of the total sample lost at least one cow in the last year. This is more significant than it may appear, as under half (43%) of households owned cattle at the time of the survey.

A higher percentage of households owned goats or sheep (49%) than cattle. This is likely because small stock were less expensive to obtain than cattle. Moreover, nearly one-third (29%) of households reported the loss of at least one goat or sheep in the last year to wildlife or disease. These higher losses are not a function of households owning more goats and sheep than cattle because households with cattle, along with goats and sheep, actually owned more cattle (7.1) than goats and sheep (4.7). A more likely scenario is that large carnivores preferentially prey on smaller stock (Mills and Hofer 1998; Nyahongo et al. 2007). Other possible scenarios are that

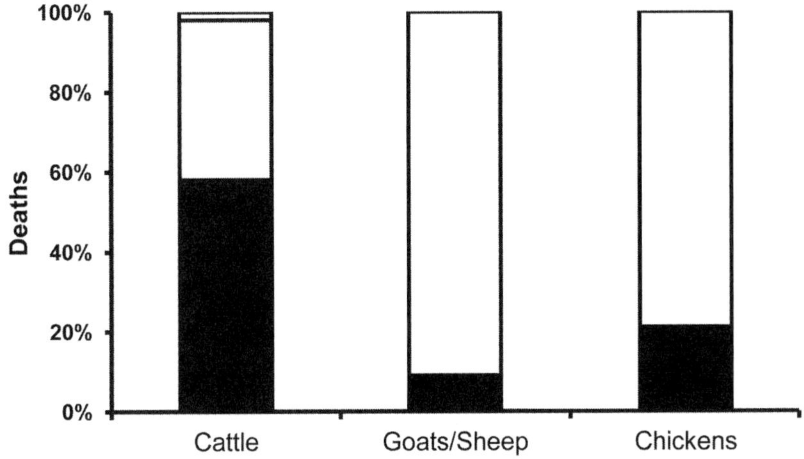

Fig. 16.4 Percentage of livestock animals kept by 722 households in 15 villages that died in a one-year period from disease, wildlife, or other causes in western Serengeti, Tanzania. White bar represents death from wildlife; black bar represents death from disease; gray bar represents death from other factors.

households with small stock practice less vigilant custodianship due to their lower values of worth.

The average household reported a loss of over half (55%, or 3.9 individuals) of their chicken holdings in the span of one year. Compared with cattle, goats, and sheep, this represents a significant loss of potential protein as households slaughtered 3.4 chickens in the previous year. In contrast, the average household slaughtered just 0.6 goats, 0.1 sheep, and 0.1 cows. One chicken offers approximately two meals for an average size western Serengeti family of 8 members. Although such losses may appear insignificant, any loss of protein is critical for households living at the margin. Moreover, diets severely lacking in protein may prompt individuals to illegally hunt wildlife (Knapp 2010). Chicken holdings differed from the other livestock based on the number of wildlife species that predated on them. Chickens were prey to 12 different species compared to goats and sheep being predated by two species (leopard, hyena) and cattle by one (hyena). The larger array of predators that prey on chickens may make custodianship more difficult than that of the other livestock species, or again, it may be that households are not as careful with chicken holdings because of their smaller respective value.

In addition to the loss of protein, losses of livestock directly affected household economies by limiting the potential income they might earn. Summed together, households reported a loss of US $51 in the last year compared to earnings of US $71. If the year that this research was collected is representative, this means that households are losing well over half of what they earn in one year.

Employment

Employment of any form is a highly sought after livelihood diversification strategy in western Serengeti. Year-round employment is relatively rare (12% of households) in the region while part-time, seasonal, or noncontractual employment is much more common. Much of the seasonal employment in western Serengeti is farm-related while a smaller percentage is nonfarm, or off-farm. Perhaps unsurprisingly, the most common form of year-round employment is related in some way to Serengeti National Park or one of the protected areas (6% of households). The direct fiscal impact that year-round employment can have is evidenced by the fact that on average, it contributed the third highest income source to households (US $62), ranking only behind agricultural sales (US $168) and livestock sales (US $71) despite far fewer households having obtained it. In other words, year-round employment provides far higher incomes than households can achieve through other more conventional avenues.

Park Effects through Employment

Theory suggests that local persons (or households) with park-related employment will have a greater incentive for protecting the system that provides their livelihood (DeFries et al. 2007). The implicit idea is that households with park-related employment benefit more—and feel they benefit more—than households that lack employment. This was generally supported in western Serengeti (fig. 16.1). Interestingly, however, park-related employment was only reported in Serengeti district, where it comprised 10% of all households ($X^2 = 32.69$, df = 2, $p < 0.001$). This may account for why a greater percentage of households reported benefits from the park in Serengeti district than in Bunda or Meatu districts ($X^2 = 77.94$, df = 2, $p < 0.001$).

The subset sample of 50 individuals with park-related employment revealed that the highest income went to individuals with government jobs associated with Tanzania National Parks (TANAPA). In contrast, the lowest incomes were those of respondents working with private safari companies (fig. 16.5). The average income from households with park-related employment was US $1,772, which was over one thousand dollars more than households without park-related income. When total income of the

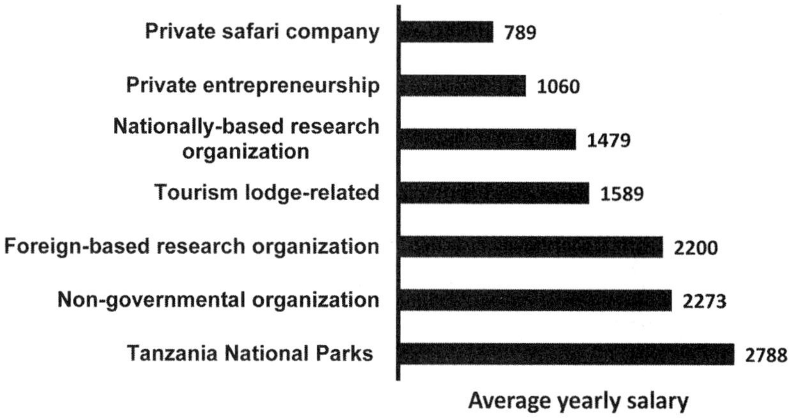

Fig. 16.5 Reported yearly salaries for 50 respondents with park-related employment in Serengeti National Park. Yearly salaries are given in United States dollar currency with the currency exchange at the time the interviews were conducted.

unemployed households is divided by the average adult equivalent value (6.5) of people living in an average size household, the amount per person was US $57, which is below Tanzania's national poverty line for rural areas in 2001 (US $61). When the same calculation is performed for households with employment, however, the amount per person is US $267, which is over four times greater than the poverty line. This demonstrates how park-related employment for only one individual in a household can lift all the occupants out of poverty.

Households with park-related employment may help keep relatives and friends out of poverty too. The subset sample revealed that 90% reported giving an undisclosed portion of their income to others whether or not such funds were solicited. Noting that the questionnaire categories were not mutually exclusive, most of the respondents gave remittances to parents (70%), while a smaller percentage gave money to siblings (34%), extended family (28%), and unrelated persons (14%). Monies were given for basic necessities, health needs, and education. Remittances are critical in rural Tanzania, as social security or related governmental programs are nonexistent. Elderly parents and grandparents who no longer have the physical strength to farm or herd livestock may be entirely dependent upon the remittances provided from their offspring.

Perhaps the most important aspect of park-related employment for conservation may be the reduced dependence that such households have upon land-intensive activities and SNP's natural resources. This study in

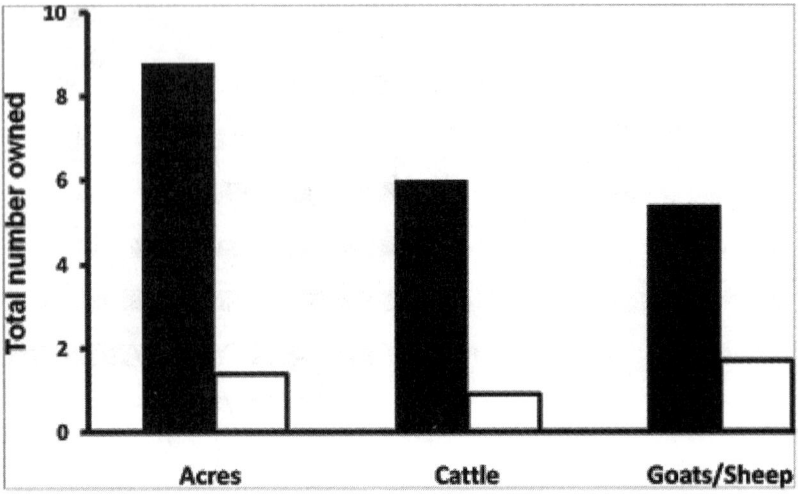

Fig. 16.6 Assets owned by households in the primary sample without park-related employment and the subset sample of households with park-related employment. Black bar represents 722 households living within 18 kilometers of a protected area boundary in western Serengeti. White bar represents the subset sample of 50 individuals with park-related employment in Serengeti National Park.

western Serengeti aligned with several others around the world that show this trend (DeFries et al. 2007; Vina et al. 2007). For example, households with park-related employment in western Serengeti utilized just 1.4 acres while households without employment used 8.8 acres ($F = 11.18$, df = 1, $p = 0.001$) (fig. 16.6).

Households with year-round or park-related employment also enjoy the advantage of having a smooth income source, or one that is decoupled from environmental stochasticity. While other households may need to turn to poaching during severe droughts, employed households need not change their livelihood strategy. Employed households earned less income from crop sales (US $64) than did unemployed households (US $150) ($F = 3.91$, df = 1, $p = 0.04$). The trend was also observed for livestock holdings. Households with park-related employment owned just 0.9 cattle and 1.7 goats and sheep compared to households without employment that owned 6.0 cattle ($F = 5.25$, df = 1, $p = 0.02$) and 5.4 goats and sheep ($F = 8.147$, df = 1, $p = 0.004$). The explanation here may be that households with employment have less opportunity or time to farm, and less of a need for alternative income sources.

Park Effects through Illegal Hunting

Illegal hunting—or "poaching"—has been called the primary threat to the Greater Serengeti Ecosystem (Sinclair 1995). One of the most abundantly harvested animals by poachers is wildebeest (fig. 16.7). This is critical because the migration of the wildebeest determines the structure and function of the ecosystem. Since the social-ecological system is linked, damage to any one part of it can have significant impacts on livelihoods and household economies, which is of interest here. While illegal hunting is widespread over the entire GSE, recent studies have suggested that it is concentrated on the western boundary of the ecosystem where human population densities are higher (Arcese, Hando, and Campbell 1995; Packer 1996; Loibooki et al. 2002; Holmern et al. 2002; Thirgood et al. 2004; Rentsch et al. chapter 22).

Without a purchased permit, all hunting is illegal in Tanzania. For the purposes here, illegal hunting and poaching are treated synonymously and refer to all hunting done inside or outside Serengeti National Park or any of its affiliated protected areas. While this may seem straightforward, the context of illegal hunting in the GSE is complicated by the fact that hunters have not always been poachers. Before the German and British colonial governments, for example, the GSE was completely unregulated and provided western Serengeti households with a reliable source of protein. Even after colonial governments began operating, antihunting regulations were rarely and irregularly enforced because human populations were low and impacts of hunting were minimal. This period of inchoate and unenforced

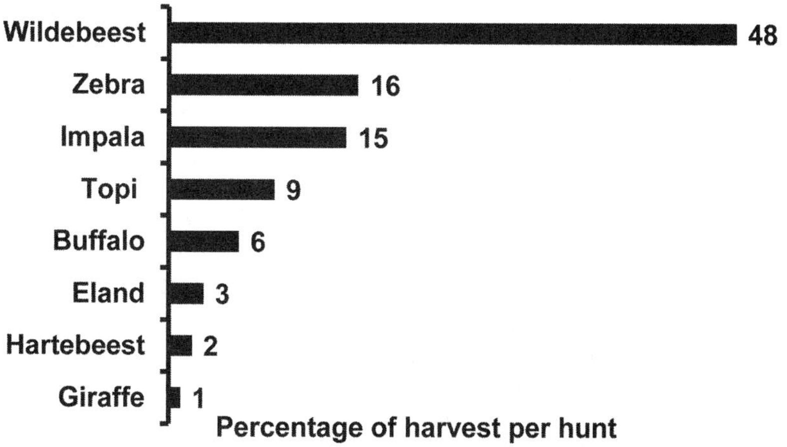

Fig. 16.7 Percentage of the catch in a day's hunt comprised by each wildlife species in western Serengeti, Tanzania.

regulations changed shortly after World War II. According to the research of Shetler (2007), it is from this time to present when hunting changed from a local strategy to diversify local economies into the current phenomenon that has transformed bushmeat into a widespread commodity. Viewed within this historical context, ascribing the pejorative label of poacher to a young western Serengeti male arrested with dried impala meat may be technically accurate but culturally misleading. In the colonial context, hunters were viewed as cruel and hardened indiscriminate killers (Shetler 2007). Today, the reality is that the majority of local hunters are poor and uneducated young farmers (Loibooki et al. 2002) who are seeking to supplement income shortfalls and household protein sources (Noss 2000; Bennett et al. 2007). Results from the subset sample of households with at least one acknowledged hunter bears this out.

Illegal hunters were all relatively young males (34.8 years old) which was lower than the average 44.9 years for males from the primary sample ($F = 42.00$, df = 1, $p < 0.001$). Over three-fourths (84%) of the sample were married and 17% were married to more than one wife. Eight tribes were represented among the 104 respondents with the largest being Ikoma (52%), Kuria (24%), and Sukuma (15%). Since interviews were necessarily predicated on trust generated through relationships built with key informants, the ethnic composition shown here is only significant in that it suggests that illegal hunting in western Serengeti is not limited to only one or two ethnic groups.

Analyses from education revealed that the subset sample of acknowledged poachers had actually attained a higher level of primary schooling ($F = 11.23$, df = 1, $p = 0.001$) but a lower level of secondary schooling than the primary sample ($F = 5.21$, df = 1, $p = 0.02$) (fig. 16.8). To generalize that poachers are uneducated in comparison to the general population, therefore, is technically incorrect. The generalization is fair, however, in regards to the utility of a primary versus a secondary education. In Tanzania, seven years of primary schooling are provided free by the government to all children. Except in various cases in which children are kept home to help with household duties, most children—even in rural locations—attend and finish primary school. As a result, the completion of primary school falls short in distinguishing individuals, as it affords only remedial skills comparable to an elementary education in the United States. Completion of secondary school, however, does appear to distinguish students in rural parts of Tanzania making them more employable. The limiting factor for many potential students is the annual fees needed for secondary school which many rural households cannot afford. This may account for why just 1% of admitted poachers had finished secondary school compared to 7% of male respon-

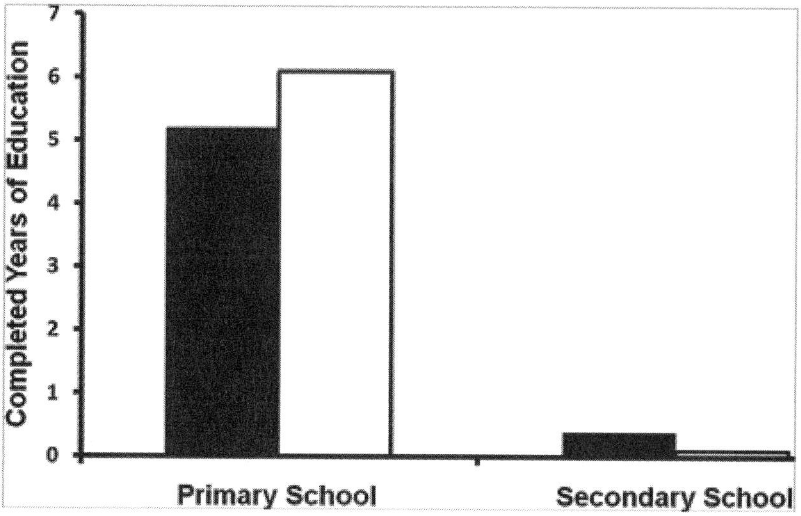

Fig. 16.8 Education comparison between the primary sample of 722 households and the subset sample of 104 acknowledged illegal hunters in western Serengeti, Tanzania. Black bar indicates primary sample; white bar indicates acknowledged illegal hunters.

dents from the primary sample ($X^2 = 5.11$, df = 1, $p = 0.02$). While this may not appear to be a sizable difference, it is argued here that the line between a poaching household and a nonpoaching household is thin, and seemingly insignificant differences between educational levels, income, opportunity time (Knapp 2007), household size, health, and nutrition may aggregate in such a way that prompts one household to poach while keeping another from it.

In addition to educational deficits, lower income from crop and livestock sales also appears to play a role in illegal hunting. Although poaching households earned slightly more from illegal home-brewed beer sales and seasonal employment than nonpoaching households ($F = 5.97$, df = 1, $p = 0.02$), they earned far less from livestock sales ($F = 7.48$, df = 1, $p = 0.01$), agricultural sales ($F = 6.62$, df = 1, $p = 0.01$), and year-round employment ($F = 3.81$, df = 1, $p = 0.05$) (fig. 16.9).

Several important points emerge from these findings. One, both poaching households and the primary sample's households attempt to diversify their livelihoods. This is demonstrated by the fact that both groups drew income from at least five different sources. In other words, it is incorrect to assume that poaching households hunt exclusively and forego other income-generating activities. Two, income generated from poaching may be an attempt by poaching households to recoup some of their shortfalls

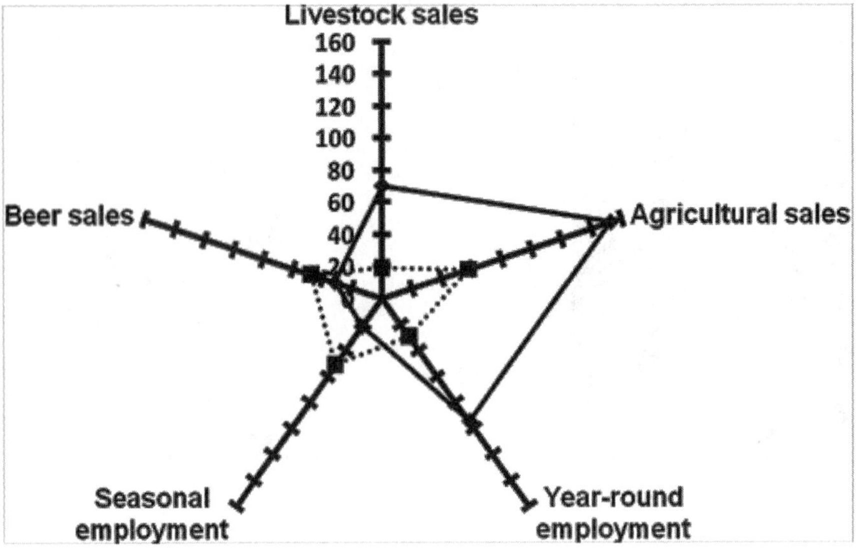

Fig. 16.9 Comparison of income generated over a yearlong period from five common sources between the primary sample of 722 households and the subset sample of 104 acknowledged illegal hunting households in western Serengeti, Tanzania. Solid line represents primary sample; dotted line represents households that engage in illegal hunting. Numbers on the bars refer to the United States dollar amount at the time the interviews were administered.

in livestock sales, agricultural sales, and year-round employment. The observed shortfalls in these three venues appear to be adequately covered by income generated through bushmeat sales. Bushmeat sales earned poaching households US $482 in a one-year period which amounted to nearly three-quarters (73%) of total income. In contrast, livestock sales contributed US $19 to the average poacher's household economy. An analysis of total income revealed that a poaching household generated US $243 more income over a year than did a household from the primary sample ($F = 15.22$, df = 1, $p < 0.001$). At the very least, this finding shows that illegal hunting is capable of making significant economic contributions to household economies in western Serengeti. What is less clear is causation. Are deficiencies in land, crop production, and livestock causing people to poach, or are the increased financial benefits that an individual may accrue through poaching—and the time spent doing it—causing them not to farm and herd livestock?

Although poaching households earned more income over a yearlong period than households from the primary sample, it is unlikely that they are more wealthy, with wealth being defined as having more liquid assets. This is demonstrated by the fact that households from the primary sample owned more cattle ($F = 10.79$, df = 1, $p < 0.001$), goats and sheep ($F = 14.28$,

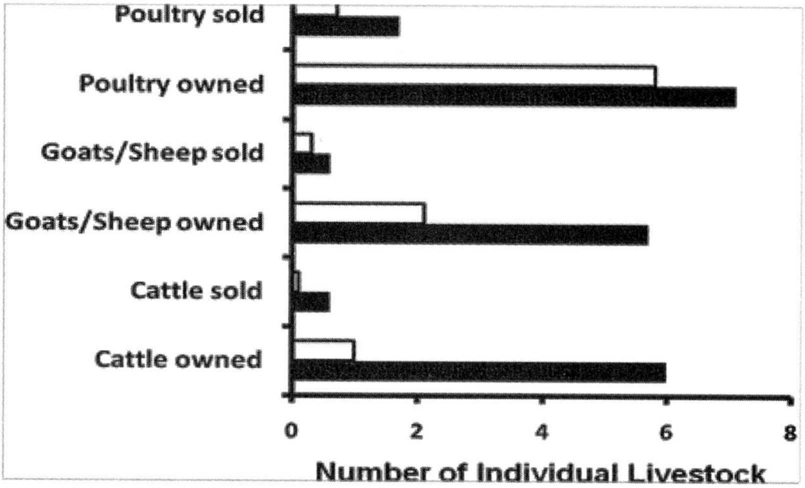

Fig. 16.10 Comparison of numbers of livestock owned and sold between the primary sample of 722 households and the subset sample of 104 acknowledged households engaged in illegal hunting in western Serengeti, Tanzania. Black bar represents primary sample; white bar represents households that engage in illegal hunting.

df = 1, $p < 0.01$) and poultry than poaching households (fig. 16.10). As a result, households from the primary sample also sold more cattle ($F = 6.6$, df = 1, $p = 0.002$), goats and sheep, and poultry than poaching households. Although assets and income have important differences, what is clear here is that households are poaching primarily for income-generating reasons. Over half (60%) of acknowledged poachers directly listed monetary factors as their primary motivation to engage in illegal hunting activities. Furthermore, 8% listed food shortages and 7% listed a lack of employment as a motivation to poach. When the question was posed in terms of benefits rather than motivation, income generation still surfaced as the leading benefit that poaching households listed as having received from Serengeti National Park (fig. 16.11). Ironically, this may account for why a greater percentage of poaching households reported they received benefits from SNP (55%) than did households from the primary sample (31%) ($X^2 = 22.62$, df = 1, $p < 0.001$).

A SYNTHESIS OF PARK EFFECTS ON HOUSEHOLD ECONOMIES

Through the four primary human-wildlife interactions of crop destruction, livestock depredation, park-related employment, and illegal hunting, it is clear that Serengeti National Park has considerable effects on the house-

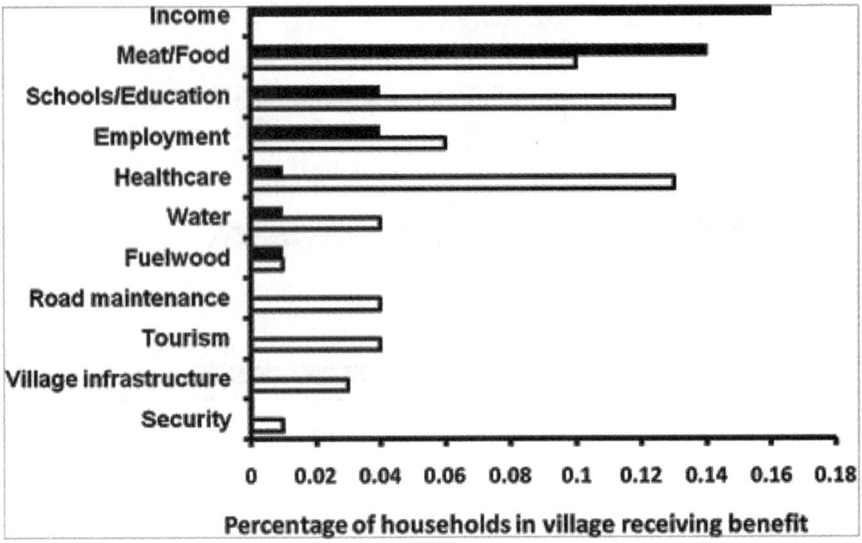

Fig. 16.11 Comparison of reported benefits received from Serengeti National Park between 722 households from the primary sample and acknowledged households engaged in illegal hunting in western Serengeti, Tanzania. Black bar represents subset sample of households that engage in illegal hunting; white bar indicates households from the primary sample. Interviewer asked respondents to list the most significant benefit received from the park.

hold economy. Two of these interactions (crop destruction and livestock depredation) fiscally hinder livelihoods, while two of them (park-related employment and poaching) augment them. Although it is challenging to separate these effects in the linked social-ecological system of western Serengeti, it is done so here merely to highlight how the park contributes to household income and the general household economy. The average revenue of a household in western Serengeti amounted to US $371 when considered apart from the effects of Serengeti National Park (fig. 16.12). As shown, the greatest contribution came from agricultural sales (US $168). This was followed by livestock sales (US $71), trade/small business (US $50), year-round contractual employment (US $37), beer sales (US $23), seasonal noncontract wage labor (US $22), and remittances (US $18).

When the park's effects are layered in, an interesting picture emerges. The most significant negative contributions—or losses of potential revenue—came from crop raiding by wildlife (US $110) (fig. 16.13). We did not consider injuries and death to humans because of their rarity and the difficulty of attaching monetary value to such phenomena. One household (out of 722) had experienced a death in the previous year (a hyena had killed a young girl in Robanda village).

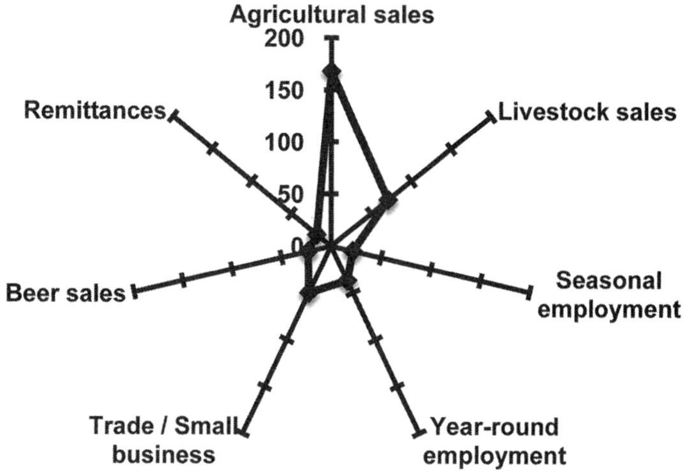

Fig. 16.12 Income generation from the seven most common sources for 722 households in 15 villages from the primary sample in western Serengeti, Tanzania. Units on the bars indicate United States dollar amounts at the time the interviews were administered.

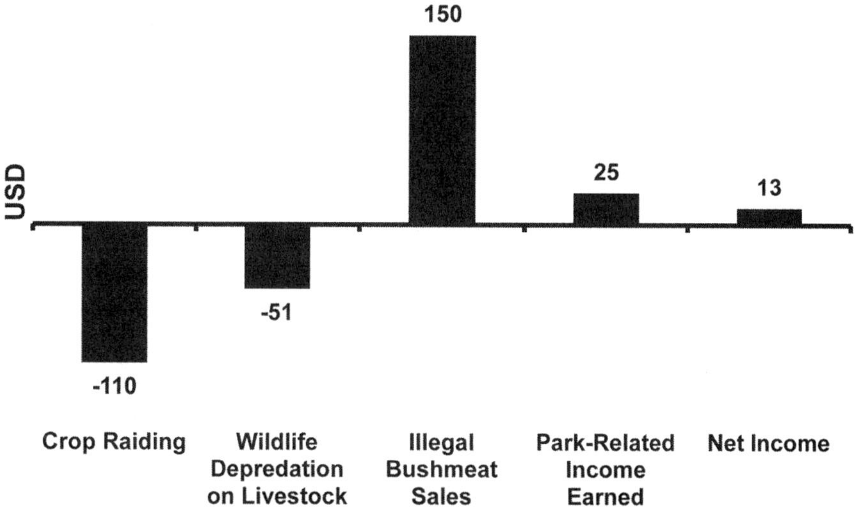

Fig. 16.13 Positive economic contributions (illegal bushmeat sales and park- related employment) and negative economic contributions (crop raiding by wildlife and wildlife depredation on livestock) to 722 household economies in western Serengeti, Tanzania. USD is US dollars per year.

Wildlife depredation on livestock was also important, accounting for an average loss of US $51. Positive contributions came from illegal bushmeat sales which netted households an average of US $150. This category was followed by park-related employment which brought in US $25. If there was such a thing as an average household that experienced all four of the major park-related effects, it would experience a net gain of US $13. Obviously, this is not the case, as some households benefit considerably and experience few losses while others suffer with few benefits. But what this does reveal is that the direct financial benefits of SNP—both legal and illegal—can successfully offset the losses from that a park's wildlife can inflict. This contradicts several studies which assert that protected areas generally have negative impacts on human well-being (Ghimire and Pimbert 2000; Colchester 2004).

Another possible explanation for these findings may be that households that live near the park in western Serengeti are compensating losses (e.g., crop raiding) through legal (employment) or illegal (bushmeat sales) means. This may help explain why in-migration continues to occur at a high rate and why households continue to move closer to the park. In terms of conservation, if losses of living near a protected area can be recouped, or avoided altogether, then households will continue to multiply and expand, and human-wildlife conflicts may increase. The ultimate cost of living near a protected area, of course, is wildlife-caused human injury or death. Just how such monumental losses factor in to people-park interactions merits further study.

As this chapter has shown, living next to large protected areas like the GSE is fraught with a complex milieu of interactions that may be harmful, beneficial, or both. Moreover, interactions can be quick and direct, like losing a maize crop overnight to an elephant, or slow and indirect, like opportunity time that is lost in guarding one's crops over several years. If the goal of conservation is to decrease human-wildlife conflict and increase biodiversity, there presently is no one-size-fits-all solution for doing this. Although park-related employment has been offered here as a workable initiative, it currently has many unanswered questions that need to be researched further. For one thing, there are simply too many people in western Serengeti and employment opportunities are too few. Secondly, any employment that is not year-round may allow employees to continue illegal hunting or grazing activities (Knapp 2007). Thirdly, advertising employment opportunities may lure households into the area which may increase local population size and exacerbate the issue. Finally, poachers are the greatest threat to the system but they lack the necessary amount of secondary education to procure employment opportunities. In lieu of these obstacles, park-related employment opportunities must be handled carefully and strategically—if at all.

If opportunities are generated by park-related agencies then three steps are recommended here. First, job opportunities should be refocused on households living nearer the park. Secondly, training should be given to those lacking necessary education credentials with special opportunities reserved for previously arrested poachers, and thirdly, in-migration into the area must be curbed by local zoning laws. While park-related employment may have drawbacks that require further study, its strengths are noteworthy, namely: decoupling households from unpredictable farming livelihoods, smoothing annual income, lifting households out of poverty, occupying time and altering behavior, decreasing the ecological footprint of employed households, and better linking conservation to protected areas. So while park-related employment should not—and cannot—be expected to be the sole driver of conservation and development (du Toit, Walker, and Campbell 2004), its prospects in western Serengeti should be thoroughly explored.

ACKNOWLEDGMENTS

We are grateful to the people living in the communities west of Serengeti National Park who invited us into their homes and told about their lives. We are also thankful for the supervision of TAWIRI, TANAPA, and COSTECH who allowed us to carry out this research. Funding was provided by the National Science Foundation (Grant #DEB-0308486).

REFERENCES

Arcese, P., J. Hando, and K. Campbell, K. 1995. Historical and present-day antipoaching efforts in Serengeti. In *Serengeti II: Dynamics, management, and conservation of an ecosystem*, ed. A. R. E. Sinclair and P. Arcese, 506–33. Chicago: University of Chicago Press.

Ayalew, W., J. M. King, E. Bruns, and B. Rischkowsky. 2003. Economic evaluation of small-holder subsistence livestock production: Lessons from an Ethiopian goat development program. *Ecological Economics* 45:473–85.

Bahiigwa, G., N. Mdoe, and F. Ellis. 2005. Livelihoods research findings and agriculture-led growth. *Institute of Development Studies Bulletin* 36:115–20.

Bennett, E. L., E. Blencowe, K. Brandon, D. Brown, R. W. Burn, G. Cowlishaw, G. Davies, H. Dublin, M. Rowcliffe, and F. M. Underwood, et al. 2007. Hunting for consensus: Reconciling bushmeat harvest, conservation, and development policy in west and central Africa. *Conservation Biology* 21:884–87.

Bernard, H. R. 2006. *Research methods in anthropology: Qualitative and quantitative approaches, fourth ed.* Lanham, MD: Rowman & Littlefield, AltaMira.

Cai, J., Z. Jiang, Y. Zeng, C. Li, and B. D. Bravery. 2008. Factors affecting crop damage by wild boar and methods of mitigation in a giant panda reserve. *European Journal of Wildlife Research* 54:723–28.

Charnley, S. 2005. From nature to ecotourism? The case of the Ngorongoro Conservation Area, Tanzania. *Human Organization* 64:75–88.

Colchester, M. 2004. Conservation policy and indigenous peoples. *Environmental Science and Policy* 7:145–53.

DeFries, R., A. Hansen, B. L. Turner, R. Reid, and J. Liu. 2007. Land use change around protected areas: Management to balance human needs and ecological function. *Ecological Applications* 17:1031–38.

du Toit, J. T., B. H. Walker, and B. M. Campbell. 2004. Conserving tropical nature: Current challenges for ecologists. *Trends in Ecology and Evolution* 19:12–17.

Ellis, F., and N. Mdoe. 2003. Livelihoods and rural poverty reduction in Tanzania. *World Development* 31:1367–84.

Ghimire, K. B., and M. P. Pimbert. 2000. *Social change and conservation: An overview of issues and concepts.* London: Earthscan Publications Limited.

Hardin, G. 1968. The tragedy of the commons. *Science* 162:1243–48.

Hella, J. P., G. van Huylenbroeck, and M. E. Mlambiti. 2001. Small farmers' adaptive efforts to rainfall variability and soil erosion problems in semiarid Tanzania. *Journal of Sustainable Agriculture* 22:19–38.

Hilborn, R. A., M. Borner, J. Hando, G. Hopcraft, M. Loibooki, S. Mduma, and A. R. E. Sinclair. 2006. Effective enforcement in a conservation area. *Science* 314:1266.

Himmelfarb, D. 2006. Moving people, moving boundaries: The socioeconomic effects of protectionist conservation, involuntary resettlement and tenure insecurity on the edge of Mt. Elgon National Park, Uganda. Agroforestry in Landscape Mosaics Working Paper Series. World Agroforestry Center, Tropical Resources Institute of Yale University, and the University of Georgia.

Holmern, T., E. Roskaft, J. Mbaruka, S. Y. Mkama, and J. Muya. 2002. Uneconomical game cropping in a community-based conservation project outside the Serengeti National Park, Tanzania. *Oryx* 36:364–72.

Kaltenborn, B. P., J. W. Nyahongo, J. R. Kidegesho, and H. Haaland. 2008. Serengeti National Park and its neighbours—Do they interact? *Journal for Nature Conservation* 16:96–108.

Kideghesho, J. R., J. W. Nyahongo, S. W. Hassan, T. C. Tarimo, and N. E. Mbije. 2006. Factors and ecological impacts of wildlife habitat destruction in the Serengeti ecosystem in northern Tanzania. *African Journal of Environmental Assessment and Management* 11:17–32.

Knapp, E. J. 2007. Who poaches? Household economies of illegal hunters in western Serengeti, Tanzania. *Human Dimensions of Wildlife* 12:195–96.

Knapp, E. J. 2009. Western Serengeti people shall not die: The effects of Serengeti National Park on rural household economies in Tanzania. PhD diss., Colorado State University, Fort Collins, CO.

Knapp, L. M. 2010. Human health in western Serengeti: Using three methodologies to better understand the interactions and impacts of conservation, culture, and poverty. MA thesis, Colorado State University, Fort Collins, CO.

Linkie, M., Y. Dinata, A. Nofrianto, N. and Leader-Williams. 2007. Patterns and perceptions of wildlife crop raiding in and around Kerinci Seblat National Park, Sumatra. *Animal Conservation* 10:127–35.

Loibooki, M., H. Hofer, K. L. I. Campbell, and M. L. East. 2002. Bushmeat hunting by communities adjacent to the Serengeti National Park, Tanzania: The importance of livestock ownership and alternative sources of protein and income. *Environmental Conservation* 29:391–98.

Mills, M. G. L., and H. Hofer. 1998. Hyaenas: Status survey and conservation action plan. International Union for Conservation of Nature and Natural Resources. Hyaena Specialist Group. Gland, Switzerland, and Cambridge, United Kingdom.

Noss, A. 2000. Cable snares and nets in the Central African Republic. In *Hunting for Sustainability in Tropical Forest*, ed. J. G. Robinson and E. L. Bennett, 282–304. New York: Columbia University Press.

Nyahongo, J. W. 2007. Depredation of livestock by wild carnivores and illegal utilization of natural resources by humans in the western Serengeti, Tanzania. PhD diss., Norwegian University of Science and Technology, Trondheim, Norway.

Packer, C. 1996. Who rules the park? *Wildlife Conservation* 99:36–39.

Rao, K. S., R. K. Maikhuri, S. Nautiyal, and K. G. Saxena. 2002. Crop damage and livestock depredation by wildlife: A case study from Nanda Devi Biosphere Reserve, India. *Journal of Environmental Management* 66:317–27.

Schmitt, J. 2010. Improving conservation efforts in the Serengeti ecosystem, Tanzania: An examination of knowledge, benefits, costs, and attitudes. PhD diss., University of Minnesota.

Shetler, J. B. 2007. *Imagining Serengeti: A history of landscape memory in Tanzania from earliest times to the present*. Athens: Ohio State University Press.

Sinclair, A. R. E. 1995. Population limitation of residents herbivores. In *Serengeti II: Dynamics, management, and conservation of an ecosystem*, ed. A. R. E. Sinclair and P. Arcese, 194–219. Chicago: University of Chicago Press.

Sinclair, A. R. E., S. A. R. Mduma, J. G. C. Hopcraft, J. M. Fryxell, R. Hilborn, and S. Thirgood. 2007. Long-term ecosystem dynamics in the Serengeti: Lessons for conservation. *Conservation Biology* 21:580–90.

Thirgood, S., N. Mosser, S. Tham G. Hopcraft, E. Mwangomo, T. Mlengeya, M. Kilewo, J. Fryxell, A. R. E. Sinclair, and M. Borner. 2004. Can parks protect migratory ungulates? The case of the Serengeti wildebeest. *Animal Conservation* 7:113–20.

Vina, A., S. Bearer, X. Chen, G. He, M. Linderman, L. An, H. Zhang, Z. Ouyang, and J. Liu. 2007. Temporal changes in giant panda habitat connectivity across boundaries of Wolong Nature Reserve, China. *Ecological Applications* 17:1019–30.

SEVENTEEN

Transitions in the Ngorongoro Conservation Area: The Story of Land Use, Human Well-Being, and Conservation

Kathleen A. Galvin, Randall B. Boone, J. Terrence McCabe, Ann L. Magennis, and Tyler A. Beeton

Arriving in the Ngorongoro Conservation Area (NCA) and at Ngorongoro Crater invokes images of pristine Africa full of exotic animals with a past that travels into the origins of *Homo sapiens*. The NCA is, indeed, unique because it is a multiple-use area, conserving wildlife and preserving archaeological sites from the dawn of human evolution, while at the same time being home to resident Maasai pastoralists. As such, the NCA is a place where multiple goals prevail including wildlife conservation, sustainable human livelihoods, Maasai culture, and tourism (Arhem 1985). The NCA is distinctive to all of the country's other protected areas as they are designated as game reserves, wildlife management areas, forest reserves, or national parks. The NCA is currently governed by the Ngorongoro Conservation Area Authority (NCAA) under the auspices of the Wildlife Division for the Ministry of Natural Resources and Tourism (MNRT). Pressures that affect other pastoral regions such as land fragmentation, privatization, diversification, and intensification of income-related activities are not common in the NCA. Rather, a complex history of conservation policies in conjunction with human population dynamics, food insecurity, disease interactions, and land-use changes through time informs our understanding of the current conditions of the NCA. We explore components of this system with an emphasis on human population changes, food security, wildlife-livestock dynamics, and land-use changes within the context of conservation policy. These factors account, in large part, for the current state of conservation and human well-being in the NCA. We start with a history of conservation

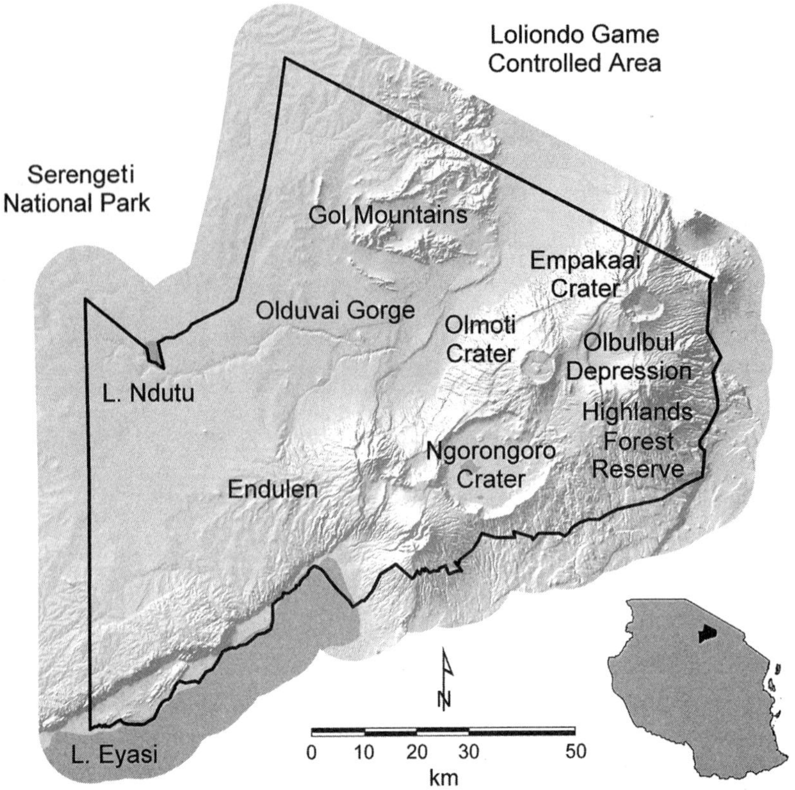

Fig. 17.1 Map of the Ngorongoro Conservation Area. Topography is shown in gray shading, derived from Shuttle Radar Topography Mission data from 2000, and water is shaded a darker gray. Selected locations are labeled, and an inset shows the location of NCA within Tanzania.

policy. We then explore human well-being as a function of food security, then look at land use by first considering some aspects of livestock-wildlife interactions, and finally, looking at cultivation. The chapter is a synthesis of scholarly literature, while also providing primary data on human health and on cultivation. Fig. 17.1 is a map of the NCA with its prominent features.

HISTORY OF LAND USE

European knowledge of the spectacular beauty and abundant wildlife populations of the Serengeti plains and the Ngorongoro highlands can be traced back to the German colonial period. European views of this type of nature, however, emerged much earlier. Individualism, ideals of progress and capitalism, begun in the Enlightenment, carried over into the nineteenth century. With a rising middle class due to the processes of industrialization in Europe, accumulation of wealth and privatization of land were desired goals. The enclosure movement, especially the privatization of land, effectively developed the notion of the commodification of nature in the name of individual land rights and particular land uses. The land was transformed into a commodity protected by contract. Thus, institutional histories associated with colonialism and political interests have driven African land use, resource rights and ownership since colonial times (Nelson 2010).

The first European to visit what is now known as the Ngorongoro Conservation Area (NCA) was the Austrian Oscar Baumann. He camped in the Ngorongoro Crater, (henceforth, the Crater), in March of 1892 while on his famous "Maasai Expedition of 1891/1893" (see Sinclair et al., chapter 2). He wrote of the abundant and "magnificent" (Baumann 1894) wildlife in the Crater, and also provided some of the first accounts of the devastating effects on the Maasai of the period referred to as Emuatai (Waller 1988). During this period, which lasted from approximately 1883 to 1902, Maasai livestock were weakened due to the spread of contagious bovine pleuropneumonia, followed by the introduction of rinderpest. Rinderpest was unknown in East Africa until this time and was extremely virulent; reports of whole herds dying in the course of a day or two were not uncommon. For the Maasai, this was followed by outbreaks of smallpox in 1892, and eventually led to internecine warfare among Maasai territorial groups (Ofcansky 1981; Waller 1988).

Following Baumann, numerous German expeditions visited the Crater Highlands and sometime around 1899 two German brothers, Adolph and Friedrich Wilhelm Siedentoph, established farms on the Crater floor (Fosbrooke 1972). The Siedentophs remained there farming, ranching, and hunting in the Crater until the German defeat in World War I. During the German colonial period 18 game reserves were created, but no reserves were appointed for the NCA or the Serengeti. In 1891 the Germans instituted control over hunting and by 1896 the first hunting reserves were established (Nelson et al. 2009). British and American hunting expeditions continued to visit the area and the impact on wildlife was becoming more and more apparent. In 1919 Tanzania became a British possession (Neumann 2001). In 1921 the Game Preservation Act regazetted all game reserves, but maintained that indigenous populations should still retain the right of subsistence hunting. In 1928, the British colonial government declared the Crater a "closed reserve," meaning that all hunting and photography was prohibited unless permitted (Lenhart and Casimir 2001). The government of Tanganyika later signed into law the Game Ordinance of 1940. Although the new law was not seen as being particularly effective at limiting indigenous hunting (as it was originally intended), it did create a new category of protected area—the national park. The Serengeti, including parts of Ngorongoro, was now officially Tanzania's first national park. In 1948 the legal protection for Serengeti National Park was signed into law, but the park was not exclusionary—local people could still live there and their land-use rights, except for hunting, were not restricted.

By the mid-1950s conservationists began to be concerned about the compatibility of conservation and cultivation (Nelson et al. 2009). Perkin (1997) suggests that an estimated 10,000 people lived in Serengeti National Park (SNP) and approximately 1,000 were non-Maasai cultivators; in 1954 all cultivation was banned within the park. Because it was believed that the Maasai subsisted on milk, meat, and blood, it was assumed the eviction of the cultivators should pose no problems to the pastoral community. However, unlike some pastoral peoples at that time in East Africa, the Maasai incorporated a significant amount of grain in their diet that was acquired by cultivating it themselves or by exchange relationships with local cultivators. The result was conflict among the residents of SNP and the SNP authorities. This lead to a government-sanctioned study and the publication of a "white paper" in 1956, which advocated setting aside the western Serengeti, Ngorongoro Crater, Empakaai Crater, and the Northern Highland Forest for wildlife and conservation, while the rest of the park would be open to unrestricted pastoralism and conservation (Perkin 1997). This proposal was strongly opposed by international conservation organizations and a committee of enquiry was appointed to propose a new plan. The recommendations of the committee of enquiry form what became the Ngorongoro Conservation Ordinance of 1959. The western section of SNP and a new northern extension were to be managed as a national park based on the Yellowstone model (no human inhabitants except for administration, researchers, and tourists); the Ngorongoro Conservation Area, including half of the western plains, the whole of the highlands, and an expansion to include the Endulen Game Controlled Area. In the NCA the pastoralists retained the rights of "habitation, cultivation, and socioeconomic development" (Ministry of Lands, Natural Resources and Tourism [MLNRT] 1990, cited in Perkin 1997, 22). However, Grant ([1957], in Sinclair et al., chapter 2) suggests that in 1954 there were 194 Maasai living in the Gol Mountains, which is today on the northern edge of the NCA. Rather than moving people out of SNP, the park boundaries were designed to work with Maasai groups living in the Gol mountains and along Olduvai Gorge. This historical account is very different than that suggested above. The result, in any case, was the establishment of the NCA and SNP. The original

NCA Authority was made up of conservation officials from the forest, game, veterinary, and water development departments along with five Maasai representatives plus the district officer acting as chairman (Shivji and Kapinga 1998). This was changed in 1961, with a new position of conservator as chairman, the regional heads of the departments mentioned above, and included the district commissioner, but only one Maasai representative. Although this administrative structure remained in place for the next 14 years, international conservation organizations were advocating for the NCA to be managed as a national park. The Tanzanian government responded with the signing of the Wildlife Conservation Act of 1974, which eliminated local land rights and resulted in more state control over resources. This was followed by the Game Parks Laws Act of 1975. The act overhauled the 1959 NCA ordinance and reconstituted the management structure. The new authority created by the Game Parks Laws Act consisted of a chairman appointed by the president of the republic, a board of directors appointed by the relevant minister responsible for the conservation of natural resources, and under the chairman and the board, the conservator, also appointed by the president. This established the Ngorongoro Conservation Area Authority (NCAA) as a corporate body rather than a division of a government ministry (Shivji and Kapinga 1998). One major impact on the Maasai was a provision that banned all cultivation and required the removal of all inhabitants of Ngorongoro, Empakaai, and Olmoti Craters as well as restricting use of water resources in the Olduvai River, due in part to the belief that a human presence in these areas would hinder wildlife conservation. There were concerns of increasing livestock pressure and soil and vegetation degradation (Perkin 1997). Farms were expanding at the periphery of the NCA impinging on wildlife migratory routes, clustered around permanent water sources, and animals were killed to protect crops (Runyoro 1994). The abolishment of cultivation in 1975 was accompanied with two promises. One was to open a branch of the regional trading company to allow pastoralists to obtain grain and the second was the development of a productive livestock economy (Parkipuny 1997). However, there was a subsequent deterioration of all of these services despite growing NCAA income from tourism. The ban on cultivation not only resulted in significant hardships for many Maasai families, it also created distrust and resentment among many of the residents of the NCA toward the NCAA. Many people continued to cultivate on small plots far into the forest, but if found these individuals faced severe fines and possible imprisonment (Runyoro 1994; Perkin 1993).

Restrictions on herd movement also occurred as the Northern Highland Forest, Ngorongoro Crater, Empakaai Crater, and Olduvai Gorge were banned from grazing. Then as now, many NCA residents feel that their rights have been neglected and trampled upon while conservation efforts have been promoted, which is contrary to the promises that Maasai livelihoods would be part of the NCA as a community-based and multiple-use conservation area.

The ban on cultivation remained in effect until 1992, when it was "temporarily" lifted by the prime minister of Tanzania following phase 1 of the Ngorongoro Conservation and Development Project which, in 1987, concluded that static livestock populations were failing to keep pace with human population growth and that human development had deteriorated (Malapas and Perkin 1997). Under the September 1992 plan each household of a woman could cultivate one acre using hand tools. Since that time the issue of cultivation has been debated and contested, with conservationists advocating that the ban be reinstated while residents and some nongovernmental organizations (NGOs) arguing that the NCA residents should be able to continue to engage in small-scale cultivation. The issue has remained contentious, with the prime minister of Tanzania, Frederik Simaye, making a speech in 2001 reinstating the ban and president Benjamin William Mkapa making a speech two weeks later saying that small-scale cultivation would be permitted into the foreseeable future (McCabe 2003). In an attempt to find a solution to the cultivation issue, the NCAA purchased land in the vicinity of Ol Donyo Sambu, near the Kenya border, for those families who wish to cultivate, but this area is considered inadequate by many residents as well as too remote and lacking infrastructure.

In 1994 the Maasai Pastoral Council was established to involve the Maasai community in the planning and management decisions, especially with regard to food security and community development (Kipuri and Sorensen 2008). The input of this group remains limited and still lacks a voice in the NCAA. Revenue to the pastoral community from gate fees has actually declined since its establishment (UNESCO/IUCN 2007; Thirgood et al. 2008). A Danida (Danish International Development Agency) funded nongovernmental organization (NGO), the ERETO Ngorongoro Pastoralist Project (ERETO-NPP), was established in 2002 with help from the Dutch government to support local communities with development in livestock care and food security (Kangera 2002; Kipuri and Sorenson 2008).

The controversy over a rapidly increasing human population, expanding cultivation, the depletion of water resources by the lodges on the Ngorongoro Crater rim, and the high number of vehicles carrying tourists into the Crater has resulted in the United Nations Educational, Scientific, and Cultural Organization (UNESCO) questioning the World Heritage status of the NCA (UNESCO/IUCN 2007). The NCA is both a cultural and natural World Heritage Site (WHS) because it encompasses the world's largest unflooded caldera, has diverse fauna such as a relict population of the black rhino and the largest population of African cats, and evidence of hominin occupation dating back 3.5 million years (United Nations Environment Program [UNEP] 2011). A study was commissioned by UNESCO and a series of recommendations were made to the Tanzanian government including reducing the human population to 25,000 people, banning cultivation, restricting vehicle use in the Crater, and implementing a moratorium on the building of new lodges, especially on the Crater rim (UNESCO/IUCN 2007). In 2009 cultivation was banned in the villages of Kapanjiro and Nyobi, on the escarpment overlooking Ol Doinyo Lengai (*the mountain of God* in the Maasai language), and it has now been extended to all residents of the NCA (*Arusha Times* 2009; UNEP 2011). The NCAA staff families of 3,000 people and lodge staff of about 2,000 near the Crater rim were to be moved to Kamyn estate, 5 km from Lodoare gate (Ihucha 2009). However, as we write, the World Heritage status of the NCA remains in doubt as does the future of the resident Maasai.

How has the policy history created the situation of today? We address three major issues: human well-being, land use in the form of livestock and their interactions with wildlife, and cultivation. These issues emerge as being important now as a result of the history of conservation policy. They are closely connected to the goal of the NCA, which is to be a multiple-use area supporting wildlife, people, and livestock. The current state of these issues has major implications for the future of the NCA. A mix of literature sources, our own data, and new analyses of satellite data are used to tell the story of transition in the NCA.

ISSUES IN THE NGORONGORO CONSERVATION AREA

Human Population and Well-Being

The human population has grown steadily since the creation of the NCA in 1959 due to both endogenous growth and to immigration. The growth rate from 1954 to 1994 was about 3.5% per year. The only time it dropped was from 1978 to 1980, probably due to the ban in cultivation in 1975 (Kijazi, Mkumbo, and Thompson 1997; table 17.1). Access to health care has improved maternal and child health and increased infant survival rates, but only up to a point (Maro 1997; McCabe 2003). However, what has been particularly problematic for the NCA Authority has been attempts to control immigration. Despite the restrictions on land use, the fertile volcanic soils have attracted cultivators; some are agropastoral Maasai, but many are from the Arusha and Meru ethnic groups (Sinclair et al. 2008). These groups have historical ties with the Maasai, and many of these in-migrants have

Table 17.1 Human-to-livestock demographics in the Ngorongoro Conservation Area (1954–2009)

Year	Human pop.	Cattle	Small stock	TLU	TLU/pers.
1954	10,633				
1960		161,034	100,689	133,061	*12.51
1962		142,230	83,120	116,536	
1963		116,870	66,320	95,420	
1964		132,490	82,980	109,500	
1966	7,387	94,580	68,590	79,758	10.80
1970	5,435	64,766	41,866	53,749	9.89
1974	12,665	123,609	157,568	115,785	9.14
1977	16,705	110,584	244,831	121,241	7.26
1978	17,982	107,838	186,985	109,430	6.09
1980	14,645	118,358	144,675	109,813	7.50
1984		109,724	100,948	96,162	
1987	22,637	137,398	137,389	122,282	5.40
1988	26,743	122,513	152,240	114,090	4.30
1993	37,352	77,243	148,288	80,824	2.16
1994	42,508	115,468	193,294	115,997	2.73
1998**	52,000	120,000	195,000	119,500	2.30
2009***	64,842	136,500	193,056	131,099	2.02

*The 1954 human population estimate was used for the 1960 TLU per person calculation.

** This is an estimation based on a graph in McCabe (2003).

*** This figure is based on the most recent census.

intermarried with Maasai residents of the NCA. Before the formation of the NCA, the villages of Kapanjiro, Nyobi, and to some extent Endulen were agricultural areas with mixed ethnic populations and many of those who have migrated into the NCA have settled in these villages.

In addition to cultivators, nonresident pastoralists have historically brought their livestock into the NCA during times of stress (Perkin 1997; Homewood and Rodgers 1991). There is a regular migration of pastoralists living north of the NCA to the Ndutu area in the NCA during the dry season of most years. In addition, the highlands of the NCA have been important drought refugia for pastoral Maasai from throughout northern Tanzania and sometimes southern Kenya. This seasonal and occasional episodic influx of people and livestock can significantly boost the number of people and livestock in the NCA. Although the NCA Authority forbids this in-migration, Maasai social norms require that all Maasai be allowed to temporarily migrate into areas where resources for their animals are available during times of stress (Galvin 2009). During the 2009 drought there were large numbers of temporary migrants and their cattle living throughout the NCA highland region (Ihucha 2009).

While human populations have grown, livestock numbers remain variable (see livestock and wildlife section) so that there has been a decline in the ratio of human to livestock populations. Maasai have become dependent on cultivation for their livelihoods and well-being.

One way of understanding human well-being is through assessment of food security as measured through diet and nutrition studies (McCabe, Nadtali, and Tumaini 1997; Galvin et al. 2002; Fratkin and Mearns 2003). One of the earliest looks at human well-being in the NCA as measured through diet and nutrition can be found in Homewood's comparative study of Maasai in Kenya and Tanzania, each under different development programs (Homewood and Rodgers 1991; Homewood 1992). This work was conducted in the early 1980s and Homewood (1992) found no nutritional differences between Maasai in the NCA compared to those in Kajiado district, Kenya. For the NCA Maasai, protein intake—primarily from milk—was adequate, but total food energy intake was less than 70% of the international standards. Grains comprised 50% of dietary energy intake while milk products and meat comprised about 45% of the diet.

The study by Homewood (1992) was followed by McCabe and colleagues who conducted two studies in the NCA to determine the economic and nutritional status of Maasai residents (McCabe, Perkin, and Schofield 1992; McCabe, Nadtali, and Tumaini1997; McCabe 2003). In 1989, and before the ban was lifted on cultivation, McCabe, Nadtali, and Tumaini (1997) looked at Maasai food security and found that pastoralists relied on grain, which provided 65% of caloric intake. Nevertheless, over 50% of children were under- or malnourished. Low weight-for-height in children showed little significant differences among wealthy versus poor households. More livestock were being sold to purchase grain than could be replaced through reproduction. Even worse, almost half of the cattle sold were females of reproductive age, which meant that the resident Maasai were mortgaging their future to meet immediate demands. The 1989 study revealed an impoverished population enmeshed in a poverty trap. The authors recommended that grain supplies should increase and grain storage options be made available.

The following study was conducted in 1994, two years after small-scale cultivation was permitted (McCabe 2003). The difference between the two studies is striking. After the ban was lifted, most families began to cultivate. Farm sizes were small with each household (defined here as a woman and her children—*enkaji* in Maa) cultivating slightly less than one acre. The majority of households cultivated maize, although people living in the highlands around Nainokanoka village cultivated potatoes. McCabe (2003) was able to convert the harvest in potatoes into a maize equivalent. For the sample of 104 households, the average yield was 4.94 bags (each bag is about 90–100 kg) of maize. This ranged from a high of 7 bags per acre in Endulen to 2.6 bags in Olbalbal. Although these yields may seem small, the harvest contributed 50% of all grains consumed. For Nainokanoka the harvest contributed 72% of grain consumed, for Endulen 57%, and for Olbalbal 38% consumed. The nutritional status of 306 children under the age of five was compared to McCabe, Nadtali, and Tumaini's (1997) study conducted during the ban on cultivation. The percentage of children classified as undernourished dropped from 38% to 35%, and more importantly the percentage of children classified as malnourished dropped from 19% to 3%. The number of livestock sold also declined significantly and no reproductive females were sold in the 1994 study. It was also clear that those interviewed felt that conditions had improved and that they had more control over their lives (McCabe 2003).

Results of a comparative study of well-being between Maasai living in the Loliondo Game Controlled Area (LGCA or Loliondo) and in the NCA suggested that in several measures of well-being (livestock to human ratios, acreage cultivated, and nutritional status) the Maasai of Loliondo were better off (Galvin et al. 2002; Lynn 2000; Smith 1999). Although the climatic patterns and general ecology of the two regions are virtually identical, land-use conservation policies within Loliondo were much less restrictive than those imposed on the Maasai in NCA. As a result, most households in Loliondo practiced cultivation that adhered only to minimal restrictions on grazing and agriculture, which may explain the differences in well-being between the two regions. Tropical livestock units ([TLUs], a measure of total animal biomass) per person were less than 3 in 1999 among livestock owners in the NCA, while in the LGCA Maasai still had greater than 10 TLUs per person. The TLUs decreased even more in the NCA when by 2004 TLUs owned were down to 2.1 (Runyoro 2007). Due to the limitations put on agriculture, the NCA Maasai cultivated a fraction (0.12 ha per person) of the acreage of Maasai in the LGCA (0.30 ha per person). There was no statistically significant effect of region on childrens' weights, but in general girls and boys in Loliando tended to weigh more (15% among two- to five-year-old boys; 17% among those six to thirteen; 4% among adolescents) than those in the NCA (Galvin et al. 2002). Triceps skinfold (TSF) measures, indicators of fat stores, the principal energy reserve of the body (Jelliffe 1966; Galvin 1992; Galvin and Little 1999; Galvin, Coppock, and Leslie 1999), were significantly different among women and men in the two areas, showing that NCA adults were leaner than those in the Loliondo area. The other adult measure of

Table 17.2 Mean body measures of pastoral mothers in Loliondo Game Controlled Area and Ngorongoro Conservation Area, northern Tanzania

	Weight	Height	TSF	UAC	BMI
Loliondo	50.3 (5.87)	160.5 (6.24)	17.5 (5.87)	25.2 (2.17)	19.5 (1.99)
NCA	46.9 (6.29)	158.2 (5.63)	13.8 (4.86)	25.0 (2.34)	18.7 (2.11)
p	0.001	<0.02	<0.001	Not significant	<0.02

Note: Measures include weight (kg), height (cm), triceps skinfold (TSF; mm), upper-arm circumference (UAC; mm), and body mass index (BMI; wt/ht^2). Values are followed by standard deviations in parentheses. The last row reports significance of comparisons of means between the study areas.

nutritional status, body mass index (BMI), a measure of leanness, was not significantly different among men in the two sites or among women, except for the18–29.9-year-old women from the NCA who were leaner ($p < 0.01$) (Galvin et al. 2002).

We also compared infant and maternal nutritional status in 1999 between the NCA and Loliondo, and though discussed elsewhere (e.g., Magennis and Galvin 1999), we report them here for the first time. The two samples, chosen randomly, were almost identical in maternal age distribution with Loliondo mothers showing a mean age of 25.2 years while those in the NCA were 25. Ninety-five mother-infant pairs were sampled in Loliondo and 65 in the NCA. Results show that in all measures except upper-arm circumference ([UAC], a rough measure of muscle mass), Loliondo women were larger and heavier than their NCA counterparts (table 17.2). On average, women in Loliondo were slightly more than 2 cm taller and weighed 3.5 kg more than mothers in the NCA.

For infants younger than two years of age (males and females combined), their lengths were not different between the two regions. There was also virtually no difference in mean weight among those infants less than six months of age. However, by the time infants reached 1.5 and 2 years of age their weights were significantly different ($p < 0.001$), with infants in the NCA lagging behind those in Loliondo by at least 1.5 kg (fig. 17.2). Maternal nutritional status was related to the nutritional status of her infant (controlling for infant length and age) as reflected in the mother's BMI scores and region. Table 17.3 shows that both maternal BMI and babies' length are significant predictors of a baby's weight. Region alone, however, is the most powerful predictor of an infant's weight. We did not explore whether or not this nutritional difference has a demonstrable effect on infant morbidity or mortality. Nevertheless, the data suggest that mothers' and infants' nutritional status was compromised in the NCA compared to Loliondo.

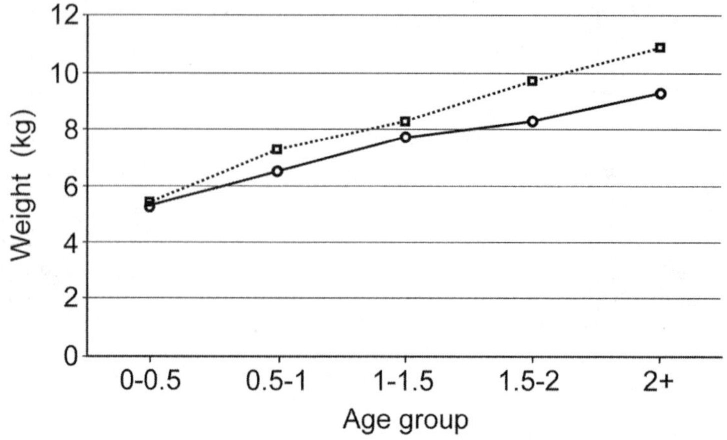

Fig. 17.2 Infant weight (kg) by region, comparing infants in Ngorongoro Conservation Area (open circles, solid line) and Loliondo Game Controlled Area (open squares, dotted line), northern Tanzania.

Table 17.3 Regression of infant weight with infant length, maternal body mass index (BMI), and the region in which they live (Loliondo Game Controlled Area or Ngorongoro Conservation Area)

	b (SEb)	**Standardized b**	p^a
Constant	–8.932 (0.869)		< 0.001
Maternal BMI	0.138 (0.034)	0.141	< 0.001
Infant length (cm)	0.195 (0.007)	0.934	< 0.001
Region	0.627 (0.147)	0.153	< 0.001

[a]–$r^2 = 0.819$, $p < 0.001$. Standard errors on *b* value follow in parentheses

With the NCA household data and information from the literature, we developed a pastoral household model and linked it to an ecological model to explore different scenarios and to understand their impacts on households and the ecosystem (Thornton, Galvin, and Boone 2003). We found that well-being in the NCA was under severe pressure even though small-scale agriculture was allowed. Even at the population levels in the late 1990s, all households were food insecure. When drought was introduced into the model, the impact on the environment was slight but food security was again the main concern (Galvin et al. 2004).

Another measure of pastoral well-being is the human-to-livestock ratio and the trends in the numbers of livestock to people over time. Across East Africa the ratio of livestock to people has been dropping and Maasai pastoralists in the NCA provide an extreme example of this declining trend

(Runyoro 2007; Homewood, Kristjanson, and Chenevix-Trench 2009). A comparison of human and livestock demographics demonstrates that livestock numbers have fluctuated around a long-term mean, while the human population has been steadily growing (table 17.1). There has been some growth in small stock numbers and a slight decrease in cattle population, but the total TLUs are nearly the same in the most recent census as they were in 1954. The net result has been progressive impoverishment of the Maasai living in the NCA.

These measures of human well-being throughout this 20-year period show that the Maasai have had an uneasy and difficult existence living within a conservation area. Livestock keeping alone is not enough to sustain the human population, in part because livestock populations have remained somewhat stable through time. The next section explores the interactions between livestock and wildlife.

Livestock and Wildlife

When debates concerning conservation and livestock keep occurring, it is often stated that rising livestock numbers place pressures on vegetation resources and water, and are thus incompatible with wildlife (Sandford 1983; Brockington and Homewood 2001). However, the data suggest that for East Africa generally, livestock populations have not grown and are often limited due to recurring drought events that kill livestock, and from disease interactions with wildlife (Campbell 1984; Brockington and Homewood 2001; Homewood 2004; Vetter 2005). In the NCA, cattle numbers have been reduced through the direct and indirect impact of disease, droughts, and the sale of animals to purchase grain.

Although a wide range of livestock diseases affect cattle in Ngorongoro, the most important disease in terms of mortality, especially calf mortality, is east coast fever (ECF). This may change as a new vaccine is available and mortality rates have dropped from as high as 70% for noninoculated calves to 4–5% for calves that have received the vaccine (Homewood et al. 2006). Another major cause of mortality is cerebral bovine theileriarosis (*ormilo* in the Maasai language of Maa), which affects adult animals. Both ECF and *ormilo* are transmitted by infected ticks. Ticks thrive in the high, relatively wet environment of the NCA, and cattle living in the highlands are exposed to ticks nearly year round (Fyumagwa et al. 2007). Until recently, limited access to both acaricide and veterinary services has made controlling tick-borne diseases difficult.

Although few cattle die from contracting malignant catarrhal fever

(MCF), the possibility of infection has had an important impact as herders limit their cattle's access to the short-grass plains during the months of January, February, and March. The MCF is a virus transmitted through the ocular and nasal secretions of wildebeest calves (Cleaveland et al. 2008; Russell, Stewart, and Haig 2009). Nearly all wildebeest calves are either born with the virus or become infected within a few days following their birth. Although the virus is benign in wildebeest, it is 100% fatal to cattle, which come into contact with the virus when grazing in areas where infected wildebeest calves have also grazed and shed the virus. There is no preventative vaccine to protect cattle or a cure once they have become infected. The only way to completely prevent cattle from becoming infected is to separate the two populations, and the most effective way to do this is to keep cattle from the Ngorongoro plains during the period when the wildebeest are calving. Boone et al. (2006) have estimated that cattle numbers could be increased by 35% if they were able to make use of the forage resources in the plains during the calving period of the wildebeest.

Drought has also had significant impacts on livestock numbers. One of the most severe droughts in recent memory ended in December 2009. According to a report presented by the district commissioner of Ngorongoro district to the deputy prime minister, 35–40% of all cattle in Ngorongoro district died due to drought-related causes between September and December of 2009, after livestock numbers were estimated for the district (see table 17.1).

Finally, previous studies (McCabe, Perkin, and Schofield 1992; McCabe 2003) have shown that during the ban on cultivation and up to the 1992 release of the ban, the Maasai in the NCA were selling more cattle than could be replaced through reproduction in order to purchase grain. This downward spiral was interrupted when residents of the NCA began to cultivate in 1992, but with the ban reinstated in 2009, NCA residents have to find new ways to supplement their livestock based economy. One option is for some NCA residents to migrate out of the NCA to find jobs. This has been occurring. This type of labor migration began in the Endulen, Ossinoni, and Kakesio areas in the mid- to late-1990s and is now common throughout the NCA (Fratkin and Mearns 2003; May and McCabe 2004). Unfortunately, most Maasai migrate into urban areas to work as guards—a job that is dangerous and low paying. However, many Maasai have been able to save enough money to purchase livestock and pay for veterinary drugs, food, and clothes. The vaccine for ECF may lead to a rebound of cattle in the NCA; the NCAA, however, already feels that the conservation area is overstocked with domestic livestock (McCabe 2002, 2003).

With livestock populations holding steady and an expanding human population, many people feel that the only means to increase food security is to cultivate. We conducted a study on cultivation early in the first decade of the twenty-first century with the use of satellite imagery and we repeat it here to assess changes in cultivation in the NCA. These changes necessarily have implications for human well-being and for conservation.

Cultivation

In 2000 we assessed responses to various management questions for the NCA using field data and simulation modeling (Boone et al. 2002, 2006). One of the questions was the extent of cultivation in the NCA. Using a Landsat 7 ETM+ satellite image from 2000, 3,967 ha (9,803 ac) of cultivation within the boundary of Ngorongoro Conservation Area was mapped (Boone et al. 2006; fig. 17.3). It was determined then that less than 1% of the NCA was cultivated; this cultivation was deemed not detrimental to wildlife or livestock and was important to promote Maasai food security. There was little evidence of soil erosion (assessed from FAO soil data) even though 85% of Maasai households in the NCA had adopted cultivation (Kijazi 1997). There were few ecosystem effects when cultivation grew in simulations to cover approximately 5% of the NCA. Maasai cultivation was generally small scale, though aerial surveys in 1993 showed that more than 50% of cultivation was being done by non-Maasai cultivators who were sometimes cultivating very large fields.

We updated that earlier map to 2010 using the same techniques. Unfortunately, in May 2003, the Landsat 7 hardware malfunctioned. The failure of its scan-line corrector has left wedge-shaped gaps in the data, with the center of the image running north-south complete, and gaps growing larger nearer the eastern and western edges of images. At the edges of the images (much of the area of interest here), a strip about 500–700 m wide is present, with the area of Earth being imaged in the stripes varying across images. Colleagues searched for alternative image sources that fit our needs, but none were available (see acknowledgments).

Although we could not provide an updated area in cultivation within the Ngorongoro Conservation Area, we could portray changes in spectral reflectance through time for select areas (fig. 17.4). We acquired from the US Geological Survey (USGS 2010) nine Landsat Satellite ETM+ images for path 169, row 62. After exploring the images and confirming that strips with data did not overlap well between images, we selected three images from the same time of year to portray the most recent at the time of analyses (March 11, 2010), the last image before the sensor hardware failed (February 4, 2003), and the original image (February 12, 2000). From the panchro-

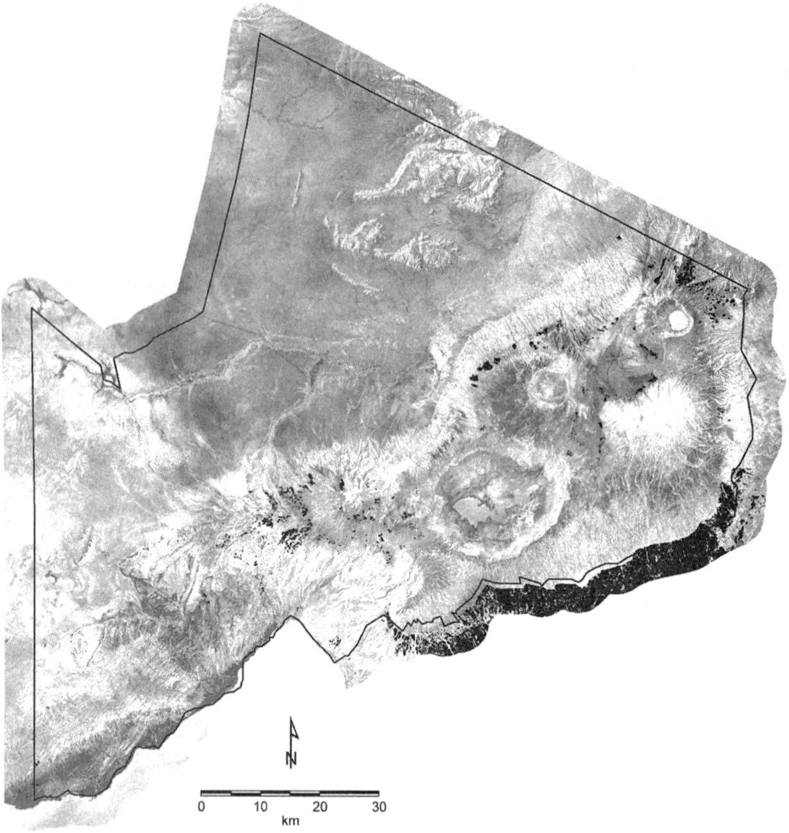

Fig. 17.3 Cultivation in the NCA in February 2000 is shown in black. Gray tones provide perspective, and are from the Landsat image used to map cultivation. Reproduced from Boone et al. (2006) with kind permission provided by Springer Publications. Shown with permission from Springer publications.

matic band in the images (band 8, 0.52 to 0.90 μm, pixels representing 15 m × 15 m patches of land), we selected strips representing 500 m × 3,000 m of land, inclined ca. 8 from north to align with the stripes in the 2010 image. These were drawn using grays stretched using twice the standard deviation of the pixel values (–2*SD = black, +2*SD = white).

Fig. 17.4 shows dark areas of cultivation and other disturbance in Ngorongoro landscapes. The images suggest that efforts to eliminate large-scale cultivation within the conservation area have been successful. Areas that were in cultivation are no longer so in 2010 (fig. 17.4a-c, g, l-q). Most of the large-scale cultivation northeast of Empakaai has revegetated (fig. 17.4a, b), along with extensive areas on the slopes of Olmoti (fig. 17.4g, h), and in Endulen (fig. 17.4l, n-q). Small-scale cultivation appears to be intact, for the

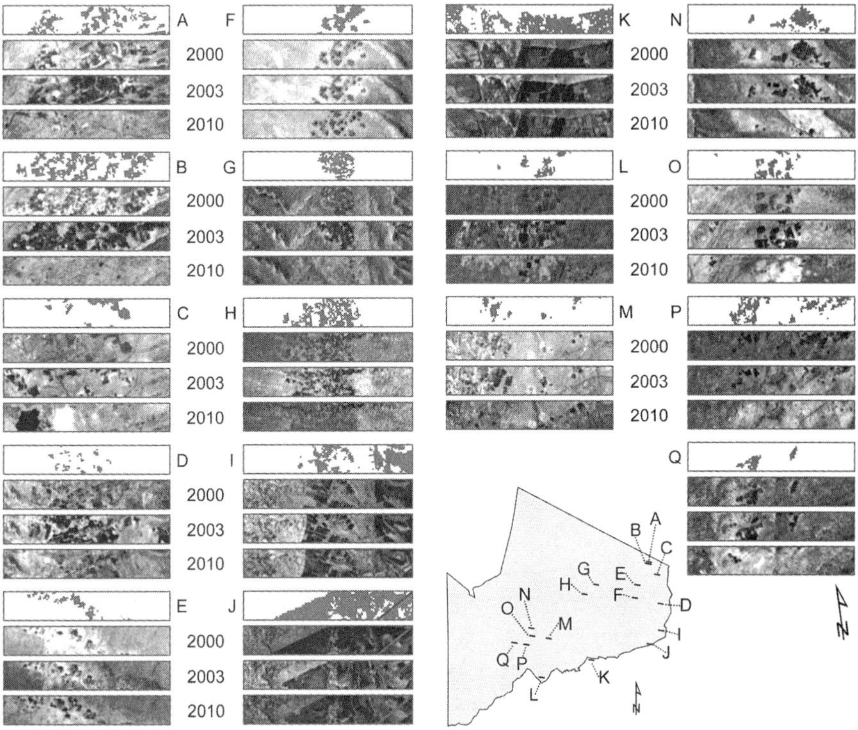

Fig. 17.4 Transects in NCA represent selected areas that were in cultivation in 2000. Transects A through Q each includes four panels. Changes in the distribution of cultivation may be inferred by comparing these panels. The first panel shows what was mapped as cultivation in February 2000 (i.e., fig. 17.3). The second panel below the first shows the Landsat ETM+ band 8 image from which cultivation was mapped (see Boone et al. 2006). An analogous image from February 2003 is shown in the third panel, and the fourth panel is from March 2010. The location of each transect is shown in the inset of NCA. For scale, the transects are each 500 m × 3000 m.

most part (fig. 17.4e, f). Encroachment on the southeast boundary of the conservation area continues (fig. 17.4d, i), and areas of extensive cultivation to the south of the conservation area remain (fig. 17.4j, k).

The accuracy of the cultivation shown in fig. 17.4 has not been accessed, although the patterns in most of the year 2000 image were readily identified (Boone et al. 2006). More importantly, what appears to be revegetated ground in 2010 may be short-term abandonment due to the recent 2009 severe drought in the region. However, several lines of evidence suggest that is not the case. First, Ngorongoro Conservation Area is a drought refuge for the people in the region. They move into the conservation area during drought because it is less affected. Many areas that show as cultivation in 2000 and 2003 remain in cultivation in 2010, which would not be the case if

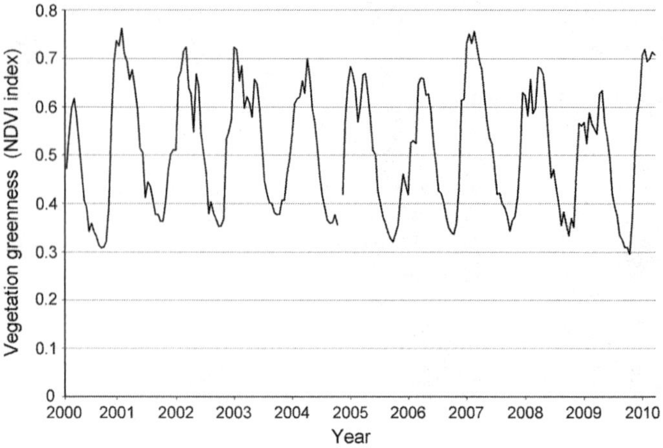

Fig. 17.5 The NDVI measures of vegetation greenness through time in the Ngorongoro Conservation Area, as derived from MODIS satellite images.

it were too dry to cultivate. Also, extensive areas of cultivation near the conservation area continue to be cultivated. Lastly, we graphed changes in Normalized Difference Vegetation Indices, measures of greenness as contained in MODIS satellite images (MOD13Q1; ORNL 2010) for Ngorongoro Conservation Area management blocks containing cultivation (fig. 17.5). The drought led to a decline in greenness in 2009, which is correlated closely with vegetation production in semiarid areas. However, by 2010, greenness was above the long-term greenness (fig. 17.4). Given food insecurity in the region, areas where residents were allowed to cultivate would presumably be in cultivation in 2010.

This analysis suggests that Maasai small-scale cultivation was intact in the decade from 2000 to 2010, that it did not increase and, in fact, decreased in extent. From a standpoint of human food security it seems vitally important, especially since the livestock-to-human ratio has so drastically decreased. However, what are the implications for wildlife conservation? Are there necessary trade-offs for human well-being, livestock production, wildlife habitat, and conservation for tourism? We address these questions in the following sections.

Conservation and Tourism

The Ngorongoro Conservation Area is one of the world's premier wildlife sanctuaries on the planet, and as such, attracts high-end tourism whose proceeds are important to Tanzania. The NCA has one of Africa's largest concentrations of wildlife. The spectacular Ngorongoro Crater is the largest intact, inactive, and unflooded caldera in the world (UNEP 2011). The Crater is home to a very high density of mammalian predators (especially lions) and large ungulates, including black rhinoceros and hippopotamus (Moehlman, Amato, and Runyoro 1996; Kangera 2002). Over a million wildebeest, with lesser numbers of zebra and gazelles, pass through the plains of the NCA every year (Leader-Williams, Kayera, and Overton 1996). On the plains is Olduvai Gorge, and the surrounding areas are the locations of world-renowned research on the evolution of *Homo sapiens* (Leakey 1966). Over 500 species of birds have been recorded within the NCA, which is in one of the world's endemic bird areas (Fishpool and Evans 2001). But people and their livestock also move through this land and cultivate when possible. Following the successful eradication of rinderpest, wildebeest populations have significantly increased, resulting in higher probability of affecting livestock with MCF. Maasai thus keep their livestock in the highlands during times that the wildebeest are on the plains (Machange 1997). However, with abundant livestock in the highlands and seasonal cultivation, the fear has grown that cultivation may block wildlife corridors that link the NCA highland forests with the Serengeti plains and Lake Eyasi (Kijazi 1997; Runyoro 2007). This may affect tourist satisfaction (DeLuca 2002; Boone et al. 2006; Honey 2008).

Further, the dramatically rising human population in the NCA is linked to growing villages in the NCA. As villages within the NCA grow, there is concern that increased detrimental stress on the ecosystem will affect vegetation and wildlife numbers, especially near the villages of Nayobi, Olpiro, and Edulen (Misana 1997). The irony of this is that tourism has also increased, and Ngorongoro Village, where NCAA personnel resided and where a number of people who work in the tourist lodges' frequent, has had the largest human population in the NCA. In 1988, out of 280 employees at the NCA, only seven were Maasai (Parkipuny 1997) and even in 2002 only a small fraction of Maasai men over the age of 15 were employed full time in the tourism industry (Coast 2002). Maasai are not hired because they lack basic skills necessary for the tourism industry. With most of the Maasai having very little education, this is not surprising. Women have even less access to economic opportunities than do men, most likely as a direct reflection of their lack of education (Coast 2002). The result is a very large workforce of non-Maasai Tanzanians living in the NCA.

The NCA, as one of the premier wildlife viewing areas of the world, draws the highest number of tourist visits in Tanzania (Charnley 2005). There are five lodges along the Ngorongoro Crater rim and one at Ndutu

Lake. Some lodges built around the rim of Ngorongoro Crater were built on local villages' land without the consent or even input from local populations (Honey 1999). There is also concern by biologists that water is being diverted to tourist lodges that used to be available to wildlife (Estes, Atwood, and Estes 2006). Nevertheless, the lodges and tourism have generated abundant revenues for the NCAA. Most tourists usually spend a day in the NCA as part of a week-long driving safari of surrounding game reserves. Though there are 16 villages in the NCA tourists rarely stop at these, although they typically visit one of four cultural bomas (homes) of traditional Maasai construction, and it is here that handicraft selling is managed. Runyoro (2007) found in his survey that 11% of households engaged in handicraft making and 40% of those households were in Endulen. However, craft selling accounted for only 2% of household income though, importantly, it is in control of women (Deluca 2002). The cultural bomas are also used for starting points for walking safaris. In the Walking Safari Management Plan, young Maasai men with some education work as guides and the number of safaris has increased since its inception (Deluca 2002). It is not known how successful these have been in generating income, but Maasai men guide tourists, as well as offer porter services and donkeys for hikes. The boma entrance fee is traditionally split between boma members and ward governments that use the funds for community development projects, while charges for donkeys, porters, and guides go to individuals (Charnley 2005). Because the program is limited, only a small proportion of the Maasai population in the NCA benefit from it (Deluca 2002).

The NCA seems to be at a defining moment. Its multiple goals to maintain premier status as a site for conservation, continue its tourism, and provide for the socioeconomic development of the resident Maasai seem at a crisis point. We explore this moment in time below.

CONCLUSION: THE NCA AS A MULTIPLE-USE PROTECTED AREA

Today the NCA appears to be a very successful conservation area and a major tourist destination. Conservation of wildlife is the theme that has encompassed NCAA policy since its inception and has been the focus of biologists, conservationists, policymakers, resident populations, human rights observers, local, national, and international NGOs, and others (Neumann 2001; Polasky et al. 2008; Homewood, Kristjanson, and Chenevix-Trench 2009; Nelson et al. 2009). People, along with their livestock and cultivation, have been seen by many as a disturbance on an otherwise pristine wildlife system. There has been conflict around cultivation since the 1950s, again in the1970s, and 1990s to the present. Human populations have risen, and at the same time so has food insecurity, because livestock biomass basically has not changed in decades and it alone can no longer support sustainable livelihoods. The number of tourists has greatly increased, as well as an associated increase in vehicles. All these changes may have negative effects on the ability of the NCA to conserve wildlife, especially in Ngorongoro Crater and along the Crater rim (Perkin 1997; Kijazi, Mkumbo, and Thompson 1997; Lynn 2000; Galvin et al. 2002; Estes, Atwood, and Estes 2006; Kipuri and Sorenson 2008).

The resident Maasai claim that conservation policies are responsible for a downward spiral of economic deprivation. Clearly the data here on human well-being suggest that Maasai are suffering from severe food insecurity and have done so for a long time. However, there are multiple factors that contribute to the state of Maasai well-being. Policy has certainly favored wildlife conservation as the primary commitment of the NCAA. Speaking on behalf of the Maasai, Fosbrooke, an early conservator of the NCA (in Homewood and Rodgers 1991), stated that Maasai were active deterrents to poaching and they limited agricultural encroachment from the outside. Clearly, Maasai ability to deter outsiders changed after the government allowed agriculture in the NCA in 1991 with half of the cultivators estimated to be nonresidents. The Maasai were not able to limit people moving into the NCA to cultivate, but neither has the NCAA been able to stem the tide of newcomers until recently. Non-Maasai are now being asked to move out of the NCA to cultivate elsewhere.

Reading the NCA from this historical perspective we see a system that has had ad hoc policies that have clearly focused on conservation and have largely ignored the other major component of the system—people—at least until there is a crisis. Is this community-based conservation? Hardly. The conservation value of the NCA currently is in jeopardy with the possibility of its World Heritage status being revoked. The Maasai's state of poverty is low. There seem to be no easy solutions at this point. It is not just one factor but a set of many reasons including cultivation, increasing human populations both local and external, increased tourism and the growth in lodges, lack of education for residents while providing increased health care, and wildlife and livestock disease incidence as well as its eradication that altogether have changed the NCA.

Some suggest that now that cultivation is banned, diversification in livelihood strategies outside of the NCA is one of the only alternatives to food insecurity (Philemon 2011). However, the data presented here do not support assumptions about cultivation undermining the conservation value of the NCA. As recently as July 2011, the minister from the Ministry of Natural Resources and Tourism warned people not to engage in farming along the edge of the NCA (Philemon 2011). He suggested that people interested in farming go elsewhere, stating there is room for citizens to farm in other places. Where is there open space for farming in Tanzania? How can an impoverished and poorly educated population purchase land? Incorporating residents in tourism is another means to help diversify livelihoods. How can this happen with such low levels of education, whereby almost 70% of NCA household heads have not attended school (Runyoro 2007)? This limits the options for pastoralists. Further, only 58% of children eligible to be in school actually are. Education level is directly correlated with benefitting from tourism.

The conservator of the NCAA stated in July 2011 that food insecurity is currently the key challenge facing communities in the NCA and that relief food would need to be increased (Philemon 2011). Economic development is perceived by some as the answer to food insecurity (e.g., Timmer 2000; Devereux 2001). In the 1980s the NCAA called for the evolution of pastoralism into commercial ranching and dairy enterprises (Homewood and Rodgers 1991). The 1996 NCA general management plan had as its food security goals to promote the livestock sector, phase out cultivation, and develop a program to move people out of the NCA to cultivate elsewhere (NCAA 1996). Today there is the call to make the Maasai "modern" by developing a cattle ranch in Kakesio outside the southwestern rim of the NCA (*Arusha Times* 2011). Maasai are to take their cattle to the ranch and leave them to be cared for by others and at some point collect money for their cattle. However, the traditional system of livestock movement associated with pastoralism is not the same as a ranching livelihood strategy (deRidder and Wagenaar 1986; Ellis and Swift 1988), and Campbell et al. (2006) argue that stocking regimes are entirely context specific. Livestock are associated with personal and collective identity and a livelihood where labor, norms of movement patterns, and ownership of livestock are embedded in pastoral families and culture. Ranching may be said to be primarily a commercial occupation, though it can be said to be embedded in a ranching culture, too.

How should decisions be made regarding the future of the NCA? Are there ways to keep people and wildlife in the NCA? Isn't that what was intended in the beginning? Should this be the future? Whatever the future should be, a comprehensive approach to conservation and human well-being that looks at the NCA as a coupled social-ecological system is a necessary first step (Chapin, Folke, and Kofinas 2009). By understanding the social, political, economic, and ecological processes, past and present, and being able to imagine as well as project scenarios of the future, one can assess the factors necessary to help the system be sustainable. This view corresponds to the International Council for Science's (ICSU) grand challenges in sustainability research; that is, looking at innovation and imagination to support sustainable development in the context of global change. The authors state, "It is increasingly clear that pathways to address rapid global change can only be found through inquiries that integrate the full range of sciences and humanities in ways that may lead to significant transformations in these disciplines as they are currently understood. It also requires the inclusion of local, traditional and indigenous knowledge" (ICSU 2010, Reid et al. 2009).

Respect for different knowledge bases and skills are good starting points for bringing different stakeholders to a discussion of the NCA's future. Having all stakeholders at the table can result in decisions for an integrated management for both ecological and economic reasons (Folke et al. 2005; Galvin et al. 2006; Berkes 2009; Roque de Pinho 2009). New institutions will need to emerge that facilitate adapting to risks from globalized markets such as tourism and the price of livestock products, climate change, maintaining ecosystem services, and accommodating local people's interests. The carbon market is an emerging source of income for some local people around the world, for example. It could provide the NCA Maasai increased agency and control over their livelihoods and further sustainable natural resource management for wildlife at the same time (Gomera, Rihoy, and Nelson 2010). Unless a broad view of the problem is taken into account, the NCA will likely have lower resilience to change in the future.

ACKNOWLEDGMENTS

Our thanks to Dr. C. J. Tucker and J. D. Nigro of the US National Aeronautics and Space Administration Goddard Space Flight Center for their help in identifying the best satellite imagery for our use. The MODIS data were provided by the Land Processes Distributed Active Archive Center (LP DAAC), located at the US Geological Survey (USGS) Earth Resources Observation and Science (EROS) Center (lpdaac.usgs.gov) and acquired through the Oak Ridge National Laboratory DAAC. Our thanks also go to Patrick Dorian for his help in reviewing the literature. We gratefully acknowledge the comments from two anonymous reviewers.

REFERENCES

Arhem, K. 1985. Pastoral man in the Garden of Eden. The Maasai of Ngorongoro Conservation Area, Tanzania. Uppsala Research Report in Cultural Anthropology, Uppsala, Sweden.

Arusha Times. 2009. New census for Ngorongoro likely. Issue 00577, July 25–31. http://www.arushatimes.co.tz/2009/29/front_page_4.htm.

———. 2011. Ranch to spur Maasai's pastoralists to modernity. Issue 00660, April 9–15. http://www.arushatimes.co.tz/2011/13/Local%20News_3.htm.

Baumann, O. 1894. *Durch Massailand zur niquelle: Reisen und forshungen der Massai-expedition des deutschen antisklaverei-Komite in den jahren 1891–1892*. Berlin: D. Reimer.

Berkes, F. 2009. Evolution of co-management: Role of knowledge generation, bridging organizations and social learning. *Journal of Environmental Management* 90:1692–1702.

Boone, R. B., M. B. Coughenour, K. A. Galvin, and J. E. Ellis. 2002. Addressing management questions for Ngorongoro Conservation Area, Tanzania, using SAVANNA modeling system. *African Journal of Ecology* 40:138–50.

Boone, R. B., K. A. Galvin, P. K. Thornton, D. M. Swift, and M. Coughenour. 2006. Cultivation and conservation in Ngorongoro Conservation Area, Tanzania. *Human Ecology* 34:809–28.

Brockington, D., and K. M. Homewood. 2001. Degradation debates and data deficiencies: The Mkomazi Game Reserve, Tanzania. *Africa: Journal of the International African Institute* 71:449–80.

Campbell, B. M., I. J. Gordon, M. K. Luckert, L. Petheram, and S. Vetter. 2006. In search of optimal stocking regimes in semi-arid grazing lands: One size does not fit all. *Ecological Economics* 60:75–85.

Campbell, D. J. 1984. Response to drought among farmers and herders in southern Kajiado District, Kenya. *Human Ecology* 12:35–64.

Chapin, F. S., III, C. Folke, and G. P. Kofinas. 2009. A framework for understanding change. In *Principles of ecosystem stewardship: Resilience based natural resource management in a changing world*, ed. F. S. Chapin III, G. P. Kofinas, and C. Folke, 3–28. New York: Springer.

Charnley, S. 2005. From nature tourism to ecotourism? The case of the Ngorongoro Conservation Area, Tanzania. *Human Organization* 64:75–88.

Cleaveland, S., C. Packer, K. Hampson, M. Kaare, R. Kock, M. Craft, T. Lembo, T. Mlengeya, and A. Dobson. 2008. The multiple roles of infectious diseases in the Serengeti ecosystem. In *Serengeti III: Human impacts on ecosystem dynamics*, ed. A. R. E. Sinclair, C. Packer, S. A. R. Mduma, and J. M. Fryxell, 209–39. Chicago: University of Chicago Press.

Coast, E. 2002. Maasai socioeconomic conditions: A cross-border comparison. *Human Ecology* 30:79–105.

Deluca, L. 2002. Tourism, conservation and development among the Maasai of Ngorongoro District, Tanzania: Implications for political ecology and sustainable livelihoods. PhD diss., University of Colorado at Boulder.

de Ridder, N., and K. T. Wagenaar. 1986. Energy and protein balances in traditional livestock systems and ranching in eastern Botswana. *Agricultural Systems* 20:1–16.

Devereux, S. 2001. Livelihood insecurity and social protection: A re-emerging issue in rural development. *Development Policy Reviews* 19:507–19.

Ellis, J. E., and D. M. Swift. 1988. Stability of African pastoral ecosystems: Alternate paradigms and implications for development. *Journal of Range Management* 41:450–59.

Estes, R. D., J. L. Atwood, and A. B. Estes. 2006. Downward trends in Ngorongoro Crater ungulate populations 1986–2005: Conservation concerns and the need for ecological research. *Biological Conservation* 131:106–20.

Fishpool, L., and M. Evans, eds. 2001. *Important bird areas for Africa and associated islands: Priority sites for conservation*. Birdlife International Conservation Series no. 11. Cambridge: Pisces Publications.

Folke, C., T. Hahn, P. Olsson, and J. Norberg. 2005. Adaptive governance of social-ecological systems. *Annual Review of Environmental Resources* 30:441–73.

Fosbrooke, H. 1972. *Ngorongoro: The eighth wonder*. London: Deutsch.

Fratkin, E., and R. Mearns. 2003. Sustainability and pastoral livelihoods: Lessons from East African Maasai and Mongolia. *Human Organization* 62:112–22.

Fyumagwa, R. D., V. Runyoro, I. G. Horak, and R. Hoare. 2007. Ecology and control of ticks as disease vectors in wildlife of the Ngorongoro Crater, Tanzania. *South African Journal of Wildlife Research* 37:79–90.

Galvin, K. A. 1992. Nutritional ecology of pastoralists in dry tropical Africa. *American Journal of Human Biology* 4:209–21.

———. 2009. Transitions: Pastoralists living with change. *Annual Review of Anthropology* 38:185–98.

Galvin, K. A., D. L. Coppock, and P. W. Leslie. 1999. Diet, nutrition and the pastoral strategy. In *Nutritional anthropology: Biocultural perspectives on food and nutrition*, ed. A. H. Goodman, D. L. Dufour, and G. H. Pelto, 86–96. Mountain View, CA: Mayfield Publishing Company.

Galvin, K. A., J. Ellis, R. B. Boone, A. L. Magennis, N. M. Smith, S. Lynn, and P. Thornton. 2002. Compatibility of pastoralism and conservation? A test case using integrated assessment in the Ngorongoro Conservation Area, Tanzania. In *Displacement, forced settlement and conservation*, ed. D. Chatty and M. Colchester, 36–60. Oxford: Berghahn Books.

Galvin, K. A., and M. A. Little. 1999. Dietary intake and nutritional status. In *Turkana herders of the dry savanna: Ecology and biobehavioral response of nomads to an uncertain environment*, ed. M. A. Little and P. W. Leslie, 125–145. Oxford: Oxford University Press.

Galvin, K. A., P. K. Thornton, R. B. Boone, and J. Sunderland. 2004. Climate variability and impacts on East African livestock herders: The Maasai of Ngorongoro Conservation Area, Tanzania. *African Journal of Range and Forage Science* 21:183–89.

Galvin, K. A., P. K. Thornton, J. Roque de Pinho, J. Sunderland, and R. B. Boone. 2006. Integrated modeling and its potential for resolving conflicts between conservation and people in the rangelands of East Africa. *Human Ecology* 34:155–83.

Gomera, M., L. Rihoy, and F. Nelson. 2010. A changing climate for community resource governance: Threats and opportunities for climate change and the emerging carbon market. In *Community rights, conservation and contested land*, ed. F. Nelson, 293–309. London: Earthscan.

Grant, H. St. J. 1957. *A report on human habitation in the Serengeti National Park*. Dar es Salaam, Tanzania: Government Printer.

Homewood, K. M. 1992. Development and the ecology of Maasai pastoralist food and nutrition. *Ecology of Food and Nutrition* 29:61–80.

———. 2004. Policy, environment and development in African rangelands. *Environmental Science and Policy* 7:125–43.

Homewood, K. M., P. Kristjanson, and P. Chenevix-Trench, eds. 2009. *Staying Maasai? Livelihoods, conservation and development in East African rangelands.* New York: Springer Science+Business Media.

Homewood, K. M., and W. A. Rodgers. 1991. *Maasailand ecology: Pastoral development and wildlife conservation in Ngorongoro, Tanzania.* Cambridge: Cambridge University Press.

Homewood, K. M., P. Trench, S. Randall, G. Lynen, and B. Bishop. 2006. Livestock health and socio-economic impacts of a veterinary intervention in Maasailand: Infection-and-treatment vaccine against east coast fever. *Agricultural Systems* 89:248–71.

Honey, M. S. 1999. Treading lightly? Ecotourism's impact on the environment. *Environment* 41:139–57.

———. 2008. *Ecotourism and sustainable development: Who owns paradise?*, 2nd ed. Washington, DC: Island Press.

Ihucha, A. 2009. Mwangunga sends SOS message to frustrated Maasai in Ngorongoro. *Eturbo News.* http://www.eturbonews.com/10676/mwangunga-sends-sos-message-frustrated-maasai-ngorongoro.

International Council for Science. 2010. Earth system science for global sustainability: The grand challenges. Paris, ICSU.

Jelliffe, D. B. 1966. *The assessment of the nutritional status of the community.* World Health Organization Monograph Series, no. 53. Geneva: WHO.

Kangera, R. 2002. Time, life and tides in Ngorongoro Park. *ArushaTimes*, 31 May, Arusha, Tanzania.

Kijazi, A. 1997. Principal management issues in the Ngorongoro Conservation Area. In *Multiple land use: The experience of the Ngorongoro Conservation Area, Tanzania,* ed. D. M. Thompson, 33–44. Cambridge: International Union for Conservation of Nature.

Kijazi, A., S. Mkumbo, and D. M. Thompson. 1997. Human and livestock population trends. In *Multiple land use: The experience of the Ngorongoro Conservation Area, Tanzania,* ed. D. M. Thompson, 169–80. Cambridge: International Union for Conservation of Nature.

Kipuri, N., and C. Sorensen. 2008. *Poverty, pastoralism and policy in Ngorongoro: Lessons learned from the Ereto/Ngorongoro pastoralist project with implications for pastoral development and the policy debate.* London: Ereto/International Institute for Environment and Development.

Leader-Williams, Kayera, N. J., and G. Overton, eds. 1996. *Community-based conservation in Tanzania.* Cambridge: International Union for Conservation of Nature.

Leakey, M. D. 1966. A review of the Oldowan culture from Olduvai Gorge, Tanzania. *Nature* 210:462–66.

Lenhart, L., and M. Casimir. 2001. Environment, property resources and the state: An introduction. *Nomadic Peoples* 5:6–20.

Lynn, S. J. 2000. Conservation policy and local ecology: Effects on Maasai land use patterns and human welfare in Northern Tanzania. MS thesis, Colorado State University.

Machange, J. 1997. Livestock and wildlife interactions. In *Multiple land use: The experience of the Ngorongoro Conservation Area, Tanzania,* ed. D. M. Thompson, 127–42. Cambridge: International Union for Conservation of Nature.

Magennis, A. L., and K. A. Galvin. 1999. Maternal-child nutrition among Maasai pastoralists, Loliondo district. Paper presented (by ALM) at the annual meeting of the American Anthropological Association, November 1999.

Malapas, R. C., and S. L. Perkin. 1997. The Ngorongoro conservation and development project: Background, objectives and activities. In *Multiple land use: The experience of the Ngorongoro Conservation Area, Tanzania*, ed. D. M. Thompson, 45–58. Cambridge: International Union for Conservation of Nature.

Maro, W. E. 1997. Education and health services. In *Multiple land use: The experience of the Ngorongoro Conservation Area, Tanzania*, ed. D. M. Thompson, 271–85. Cambridge: International Union for Conservation of Nature.

May, A., and J. T. McCabe. 2004. City work in a time of AIDS: Maasai labor migration in Tanzania. *Africa Today* 51:3–32.

McCabe, J. T. 2002. Giving conservation a human face? Lessons from forty years of combining conservation and development in the Ngorongoro Conservation Area, Tanzania. In *Conservation and mobile indigenous peoples: Dispacement, forced settlement, and sustainable development*, ed. D. Chatty and M. Colchester, 61–76. Oxford: Berghahn Books.

———. 2003. Sustainability and livelihood diversification among the Maasai of northern Tanzania. *Human Organization* 62:100–11.

McCabe, J. T., M. Nadtali, and A. Tumaini. 1997. Food security and the role of conservation. In *Multiple land use: The experience of the Ngorongoro Conservation Area, Tanzania*, ed. D. M. Thompson, 285–301. Cambridge: International Union for Conservation of Nature.

McCabe, J. T., S. Perkin, and C. Schofield. 1992. Can conservation and development be coupled among a pastoral people? The Maasai of the Ngorongoro Conservation Area, Tanzania. *Human Organization* 51:353–66.

Misana, S. B. 1997. Vegetation change. In *Multiple land use: The experience of the Ngorongoro Conservation Area, Tanzania*, ed. D. M. Thompson, 97–110. Cambridge: International Union for Conservation of Nature.

Moehlman, P., G. Amato, and V. Runyoro. 1996. Genetic and demographic threats to the black rhinoceros population in the Ngorongoro Crater. *Conservation Biology* 10: 1107–14.

NCAA. 1996. The Ngorongoro Conservation Area general management plan. In *Multiple land use: The experience of the Ngorongoro Conservation Area, Tanzania*, ed. D. M. Thompson, 417–40. Cambridge: International Union for Conservation of Nature.

Nelson, F. 2010. Introduction: The politics of natural resource governance in Africa. In *Community rights, conservation and contested land*, ed. F. Nelson, 3–31. London: Earthscan.

Nelson, F., B. Gardner, J. Igoe, and A. Williams. 2009. Community-based conservation and Maasai livelihoods in Tanzania. In *Staying Maasai? Livelihoods, conservation and development in East African rangelands*, ed. K. Homewood, P. Kristjanson, and P. C. Trench, 299–333. New York: Springer Science + Business Media.

Neumann, R. P. 2001. Africa's "Last Wilderness": Reordering space for political and economic control in colonial Tanzania. *Africa: Journal of the International African Institute* 71:641–65.

Oak Ridge National Laboratory (ORNL). 2010. *MODIS land product subsets*. Oak Ridge National Laboratory, Oak Ridge, Tennessee, USA [online]. http://daac.ornl.gov/MODIS (last updated August 4, 2009; last accessed April 20, 2010).

Ofcansky, T. P. 1981. The 1889–1897 rinderpest epidemic and the rise of British and German colonialism in eastern and southern Africa. *Journal of African Studies* 8:31–38.

Parkipuny, M. S. 1997. Pastoralism, conservation and development in the greater Serengeti region. In *Multiple land use: The experience of the Ngorongoro Conservation Area, Tanzania*, ed. D. M. Thompson, 143–68. Cambridge: International Union for Conservation of Nature.

Perkin, S. L. 1993. Integrating conservation and development: An evaluation of multiple land-use in the Ngorongoro Conservation Area, Tanzania. PhD diss., University of East Anglia, UK.

———. 1997. The Ngorongoro Conservation Area: Values, history and land use conflicts. In *Multiple land use: The experience of the Ngorongoro Conservation Area, Tanzania*, ed. D. M. Thompson, 19–32. Cambridge: International Union for Conservation of Nature.

Philemon, L. 2011. Ban on farming in Ngorongoro crater rim intact, says minister. *The Guardian*, July 12. http://www.ippmedia.com/frontend/index.php?l=31097.

Polasky, S., J. Schmitt, C. Costello, and L. Tajibaeva. 2008. Large-scale influences on the Serengeti ecosystem: National and international polity, economics, and human demography. In *Serengeti III: Human impacts on ecosystem dynamics*, ed. A. R. E. Sinclair, C. Packer, S. A. R. Mduma, and J. M. Fryxell, 347–78. Chicago: University of Chicago Press.

Reid, R. S., D. Nkedianye, M. Y. Said, D. Kaelo, M. Neselle, O. Makui, L. Onetu, S. Kiruswa, N. Ole Kamuaro, P. Kristjanson, et al. 2009. Evolving models to support communities and policy makers with science: Balancing pastoralism and wildlife conservation in East Africa. *Proceedings of the National Academy of Sciences, USA*. Early edition, online only. www.pnas.org/cgi/doi/10.1073/pnas.0900313106.

Roque de Pinho, J. 2009. "Staying together": People-wildlife relationships in a pastoral society in transition, Amboseli ecosystem, southern Kenya. PhD diss., Colorado State University.

Runyoro, V. A. 1994. The socio-economic and ecological implications of increasing sedentarisation in the Ngorongoro Conservation Area, Tanzania. MS thesis, Agricultural University of Norway.

———. 2007. Analysis of alternative livelihood strategies for the pastoralists of Ngorongoro Conservation Area, Tanzania. PhD diss., University of Morogoro, Tanzania.

Russell, G. C., J. P. Stewart, and D. M. Haig. 2009. Malignant catarrhal fever: A review. *Veterinary Journal* 179:324–35.

Sandford, S. 1983. *Management of pastoral development in the third world*. New York: John Wiley and Sons.

Shivji, I. G., and W. B. Kapinga. 1998. Maasai rights in Ngorongoro, Tanzania. London: International Institute for Environment and Development and Land Rights Research and Resources Institute.

Sinclair, A. R. E., J. Grant, C. Hopcraft, H. Olf, S. A. R. Mduna, K. A. Galvin, and G. J. Sharam. 2008. Historical and future changes to the Serengeti ecysystem. In *Serengeti III: Human impacts on ecosystem dynamics*, ed. A. R. E. Sinclair, C. Packer, S. A. R. Mduma, and J. M. Fryxell, 7–46. Chicago: University of Chicago Press.

Smith, N. M. 1999. Maasai household economy: A comparison between the Loliondo Game Controlled Area and the Ngorongoro Conservation Area, Northern Tanzania. MA thesis, Colorado State University.

Thirgood, S., C. Mlingwa, E. Gereta, V. Runyoro, R. Malpas, K. Laurenson, and M. Borner. 2008. Who pays for conservation? Current and future financing scenarios for the Serengeti ecosystem. In *Serengeti III: Human impacts on ecosystem dynamics*, ed. A. R. E. Sinclair, C. Packer, S. A. R. Mduma, and J. M. Fryxell, 443–70. Chicago: University of Chicago Press.

Thornton, P. K., K. A. Galvin, and R. B. Boone. 2003. An agro-pastoral household model for the rangelands of East Africa. *Agricultural Systems* 76:601–22.

Timmer, C. P. 2000. The macro dimensions of food security: Economic growth, equitable distribution, and food price stability. *Food Policy* 25:283–95.

United Nations Environment Programme (UNEP). 2011. *Ngorongoro Conservation Area, Tanzania*. World Conservation Monitoring Centre. (http://www.unep-wcmc-apps.org/sites/wh/pdf/Ngorongoro.pdf).

UNESCO/IUCN. 2007. Ngorongoro Conservation Area (United Republic of Tanzania). Report of the WHC/IUCN Reactive Monitoring Mission, May 2007.

United States Geological Survey (USGS). 2010. *USGS global visualization viewer*. USGS Earth Resources Observation and Science Center (EROS). Reston, Virginia, USA [online]. http://glovis.usgs.gov/BrowseBrowser.shtml (last updated April 7, 2010; last accessed April 23, 2010).

Vetter, S. 2005. Rangelands at equilibrium and non-equilibrium: Recent developments in the debate. *Journal of Arid Environments* 62:321–41.

Waller, R. D. 1988. Emutai: Crisis and response in Maasailand 1883–1902. In *The ecology of survival: Case studies from northeast African history*, ed. D. Johnson and D. Anderson, 73–114. Boulder, Colorado: Lester Crook Academic Publishing/Westview Press.

EIGHTEEN

Agricultural Expansion and Human Population Trends in the Greater Serengeti Ecosystem from 1984 to 2003

Anna B. Estes, Tobias Kuemmerle, Hadas Kushnir, V. C. Radeloff, and H. H. Shugart

Protected areas are impacted by what takes place around their borders, and the nature and severity of these impacts is influenced by reserve size, land use, and human populations. In particular, increasing human populations and land conversion around protected areas can cut off corridors, erode buffer areas, decrease effective reserve size, and lead to loss of biodiversity and ecosystem function (Cantú-Salazar and Gaston 2010; DeFries et al. 2005; Hansen and DeFries 2007). While these impacts are typically more pronounced in smaller reserves, even protected area complexes as large as the ~25,000 km^2 greater Serengeti ecosystem are not immune. For example, wildebeest (*Connocheates taurinus*) rely on areas outside of the protected area during critical dry season parts of their migration (Thirgood et al. 2004), human-wildlife conflicts create antagonism between local people and the conservation objectives of the protected area (Walpole et al. 2004; Hampson et al., chapter 21), local communities rely heavily on bushmeat from the protected areas (Campbell and Hofer 1995; Mfunda and Roskaft 2010; Mduma, Hilborn, and Sinclair 1998; Ogutu et al. 2009; Sinclair et al. 2007; Rentsch et al., chapter 22), and domestic dog populations act as reservoirs of diseases that spread to wild carnivores (Cleaveland et al. 2000; Hampson et al., chapter 21). These interactions between the park and surrounding communities underscore the need for protected area managers and community planners to understand both patterns of human population growth and the forces that cause population growth in adjacent lands.

Immigration is a common cause of elevated human population growth rates around protected areas, a phenomenon that has been identified on multiple continents (Wittemyer et al. 2008). However, the potential drivers of immigration around protected areas, while directly relevant to conservation interventions, are a source of debate (Joppa, Loarie, and Nelson 2010; Scholte and de Groot 2010). These drivers can be broadly grouped into push and pull factors.

Push factors cause people to leave their areas of origin and include lack of access to land and natural resources, declining soil productivity, and high population pressure. Pull factors that draw people into new areas can be greater availability of natural resources including land, employment, access to markets and social services, and reunification with family (Oglethorpe et al. 2007). These mechanisms have similarly been referred to as frontier engulfment (i.e., when agriculture expands and "bumps into" a protected area boundary) and attraction (pull factors) (Scholte and de Groot 2010). Prior studies assume that elevated immigration rates near protected areas result from pull factors, in particular perceived benefits associated with the protected areas themselves (Wittemyer et al. 2008), but empirical evidence for this assumption is sparse (Scholte and de Groot 2010). A better understanding of the importance of pull versus push factors is important given the strong conservation implications of human population growth surrounding protected areas. Mapping the spread of agriculture is one way to elucidate mechanisms behind population growth (Scholte and de Groot 2010; Joppa, Loarie, and Nelson 2010).

The general goal of this study was to discover why and how human populations and land cover around Serengeti are changing, and specifically to (a) quantify trends in human population growth and density that might be related to the presence of the park, (b) map land-cover change around the park, and (c) examine relationships between the observed trends in land-cover and human population change.

METHODS

Study Area

The 14,763 km^2 Serengeti National Park comprises the core of the greater Serengeti ecosystem. The park is bordered in Tanzania by game reserves, game controlled areas, wildlife management areas, the Ngorongoro Conservation Area, the Maasai Mara National Reserve, and surrounding Maasai areas in the Narok region in Kenya, all of which together make up the greater Serengeti ecosystem. In addition to the park, game reserves and the Maasai Mara exclude human habitation, while game controlled areas, parts of the wildlife management area, and village lands allow for habitation, farming, and livestock husbandry. Hunting occurs in game reserves, the game controlled area, and wildlife management areas. The Ngorongoro Conservation Area permits habitation and livestock, but seeks to limit farming, although some farming occurs. For simplicity, the complex of game reserves and the park, which exclude habitation, will henceforth be referred to as "the park."

Rainfall is highly variable in the greater Serengeti ecosystem, but typically peaks in December, and March–May. Rainfall is lower in the southeast (500 mm/yr) of the ecosystem than in the northwest (1,200 mm/yr) (Sinclair 1995). Due in part to the rainfall patterns, areas northeast, east, and southeast of the park are occupied by the primarily pastoral Maasai, while agriculturalists and agropastoralists populate the areas northwest, west, and southwest of the park, where climate is more conducive to agriculture. Agriculture is dominated by smallholder farmers with average farm sizes from 0.9 to 3 ha, the majority of which are cultivated by hand hoe (70% nationwide) or ox plough (20%) (www.tanzania.go.tz/agriculture). The highest human densities in the region occur near Lake Victoria, with an average population growth rate of ~3.1% between 1988 and 2002 (Polasky et al. 2008).

Demographic Data

Analyses were based on ward-level demographic data for Tanzania, digitized by the International Livestock Research Institute (ILRI) in Nairobi, and the 2002 population and housing census. Wards are collections of several villages, and their size is inversely correlated with human densities. Because human densities in the pastoralist-dominated areas east of the park are very low, the wards are much larger, making east-west comparisons difficult (fig. 18.1). We therefore used only the wards in the agropastoral areas west of the park to look at rates of land cover change and human demographic parameters at increasing distances from the park. Rates of population change by ward were calculated using the 1988 and 2002 census data. We separated the wards into the following zones for this analysis: wards with borders adjacent to the park (listed as 0 km distance in following results), and those with centroids (the geographical center of the ward) within 20, 40, 60, and 80 km of the park (fig. 18.1). Land-cover change was analyzed according to these same zones to examine relationships between human population density, growth rate, and agricultural expansion relative to proximity to the park. Urban wards (1 ward in the 40 km zone and 11 in the 80 km zone) in the 2002 census were excluded.

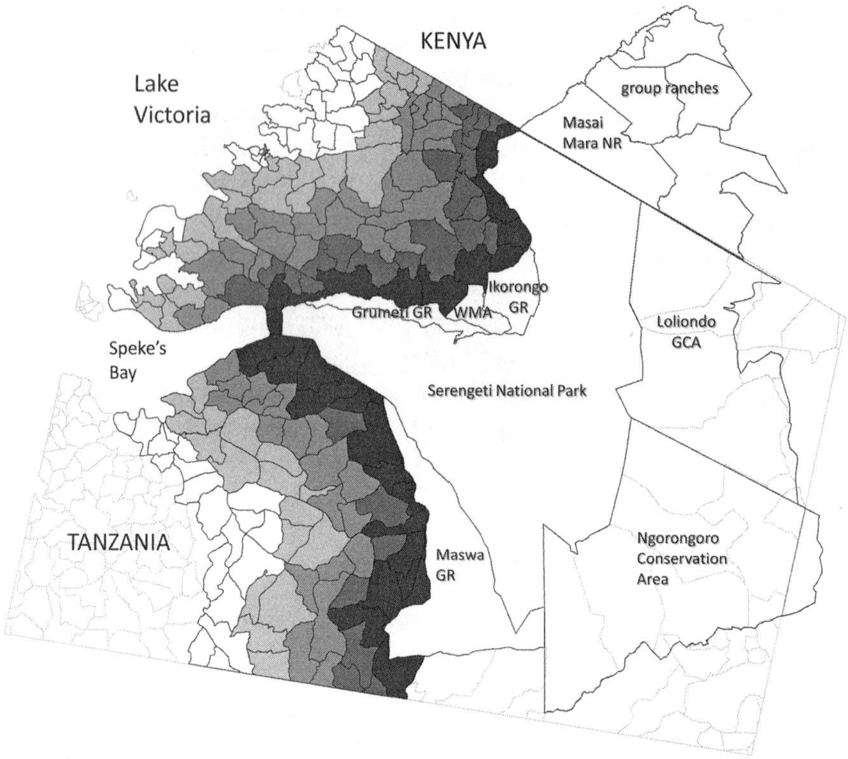

Fig. 18.1 Serengeti National Park and surrounding protected areas, with ward zones used in the analysis in western Serengeti. Wards adjacent to the park are indicated in dark grey, and those with centroids falling within 20, 40, 60, and 80 km from the park become successively lighter. Larger wards east of the park are shown in very light grey. GR = game reserve, WMA = wildlife management area, GCA = game controlled area, NR = national reserve. Reproduced from Estes et al. (2012).

Land-Cover Change Mapping

We mapped land-cover change in the greater Serengeti ecosystem using 30-m resolution Landsat TM and ETM+ imagery from the USGS archive (www.glovis.usgs.gov). The greater Serengeti ecosystem falls at the intersection of 4 Landsat footprints (path/row 169/061, 169/062, 170/061 and 170/062). We acquired images representing different phenological states for the time periods 1984–87 and 2002–03 (dates were partially determined by image availability), and used these in a support vector machine (SVM) classification with the following six classes: stable agriculture (unchanged between 1984 and 2003), stable savanna (including grassland, savanna, and woodland), stable forest, savanna and forest that converted to agriculture during the study period (henceforth referred to as "agricultural conversion"), water, and cloud. SVM are nonparametric classifiers capable of handling complex, nonlinear class boundaries that are common for change detection problems and frequently outperform traditional statistical classifiers (Huang, Townshend, and Davis 2002). The use of SVM and images of different phenological states greatly enhanced our ability to distinguish spectrally-similar classes, such as cropland and natural savanna (Kuemmerle et al. 2008). We parameterized the classification using high resultion imagery in Google Earth, and validated it using a standard confusion matrix (Congalton 1991). For more detail on remote sensing methods, please see (Estes et al. 2012).

We also used Landsat images from 1989–90, 1994–95 and 1999–2000 to examine rates of change in ~5-year time intervals. We assigned 218 random points (200 points stratified by the amount of converted land in each zone, with additional points added to get a minimum of 20 points in the outermost zones) in western Serengeti and checked in which period conversion had occurred. We calculated conversion rates based on the percent of the land area in each ward zone, and normalized by the number of years in each time period to yield annual conversion rates by time period and zone.

RESULTS

Human Population Trends

Human populations surrounding Serengeti changed rapidly in recent decades. In the agricultural areas west of Serengeti, population growth rates were highest where densities were lowest ($R^2 = 0.75$), and varied dramatically with distance from the park. Wards closest to the park had the lowest human densities (98 people/km^2) and the highest rates of human population growth (3.5% per year) over the study period from 1988 to 2002, while those farthest from the park and closest to Lake Victoria had the highest densities (160 people/km^2) and lowest growth rates (2.5% per year) (fig. 18.2).

Land-Cover Classification

The land-cover change analysis of the greater Serengeti ecosystem highlighted the stark contrast between protected and unprotected areas along the western boundary of the ecosystem (fig. 18.3), and regional and national

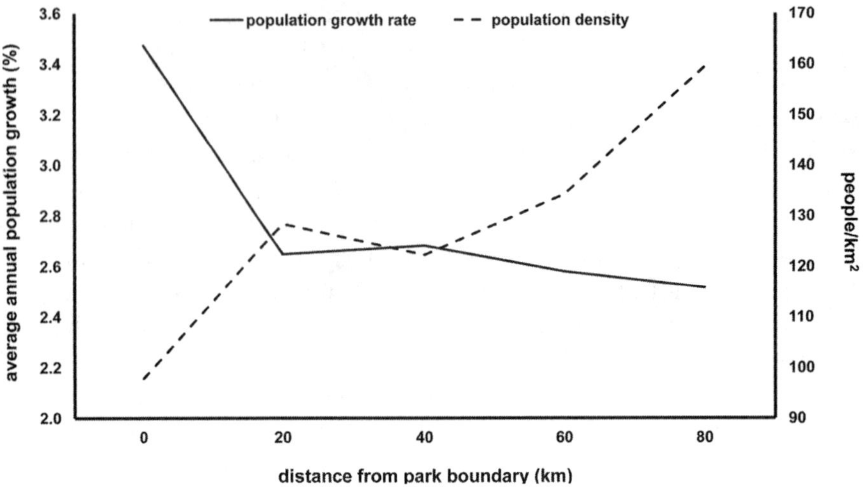

Fig. 18.2 The average annual rate of human population increase in western Serengeti from 1998 to 2002 was greatest in wards adjacent to the park boundary, while human density in 2002 remained low in these same areas, and increased with distance from the boundary. Reproduced from Estes et al. (2012).

differences in land use that reflect differences in culture, livelihood strategies, and land tenure policy. Areas east of the park are primarily pastoralist, and although agriculture is increasing in these areas (at least in Loliondo if not Ngorongoro), by 2003 agriculture was still minimal and found close to human settlements. In contrast, agriculture was by far the dominant land cover west of the park (fig. 18.4). Savanna was found primarily within the protected areas and parts of the adjacent Kenyan and Tanzanian Maasailand, and agricultural conversion was concentrated near the western boundary of the park in Tanzania. The park therefore remained more linked to its surroundings in the east than the west, where a sharpe edge had developed between the park and adjacent agriculture. In Kenya, however, most of the agricultural conversion was of isolated forest patches, particularly in the southern Mau forest area, and parts of the Narok region near the large wheat cropping schemes (fig. 18.3).

The overall classification accuracy of the land cover map was 83% (kappa = 0.79), with the agricultural conversion change class proving the most difficult to classify (producer's accuracy = 73.5%), perhaps partly because the lack of high resolution imagery in the earlier time periods necessitated visual interpretation of the coarser Landsat imagery, and agricultural classes appear fairly similar in those images. The stable classes of agriculture, forest, and savanna yielded producer's accuracies between 92–94% (table 18.1).

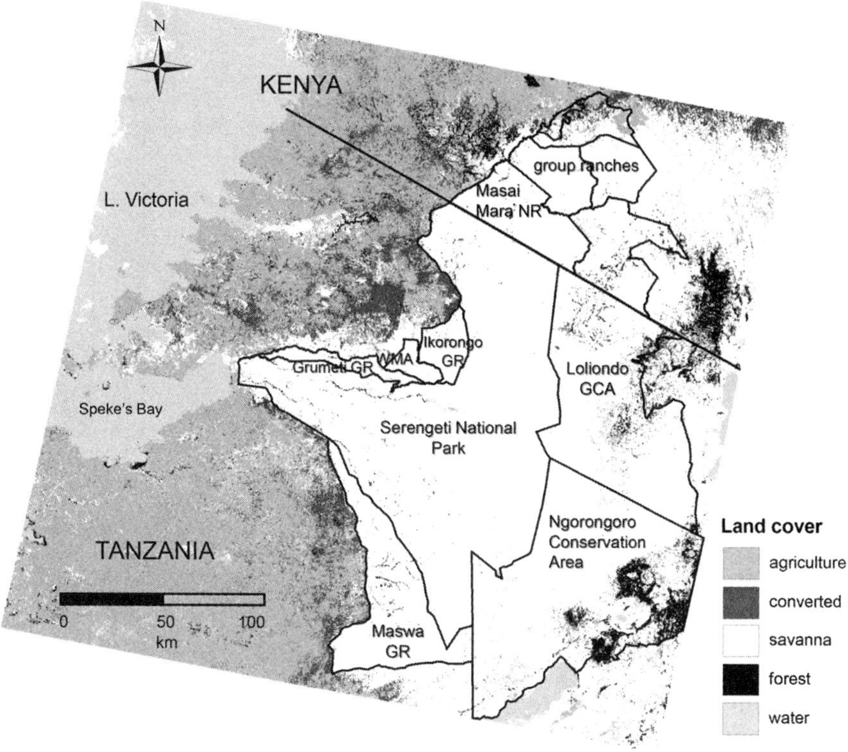

Fig. 18.3 Land cover in the greater Serengeti ecosystem showing stable savanna, agriculture and forest, and agricultural conversion from 1984 to 2003. GR = game reserve, WMA = wildlife management area, GCA = game controlled area, NR = national reserve. Modified from Estes et al. (2012).

Table 18.1 Class accuracies of the land cover classification*

	Producer's	User's	Omission	Commission
Stable agriculture	92.0%	74.2%	8.0%	25.8%
Agricultural conversion	73.5%	94.8%	26.5%	5.2%
Forest	91.9%	96.8%	8.1%	3.2%
Savanna	93.9%	74.4%	6.1%	25.6%

Source: Reproduced from Estes et al. (2012).

Note: * Producer's accuracy is the probability of a reference pixel being correctly classified, a measure of omission error, and user's accuracy is the probability that a pixel classified on the map actually belongs to that class on the ground, which reflects commission error.

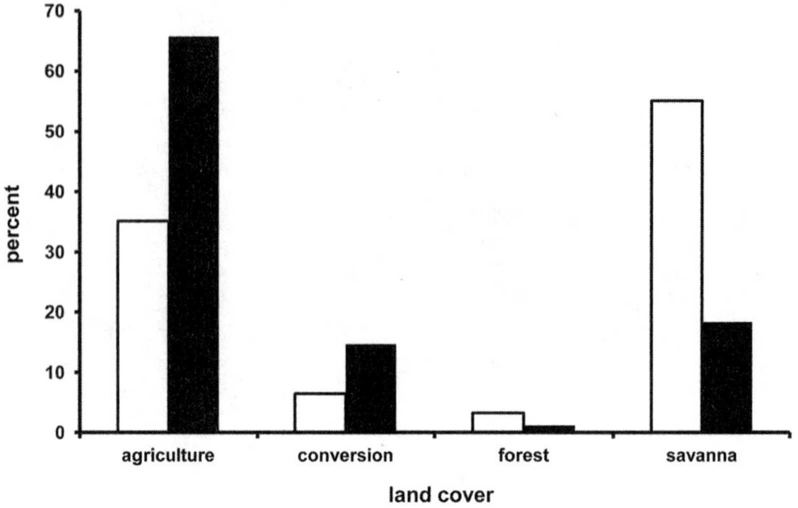

Fig. 18.4 Relative class proportions of the entire land cover classification (white bars), and the western Serengeti study area (black bars) defined by wards with centroids within 80 km of the park boundary. (Modified from Estes et al. 2012).

Land-Cover Change

In the western Serengeti, agricultural conversion was greatest close to park boundaries where there was the most remaining arable land, and conversion dwindled to almost nothing in the zones farthest from the park where agriculture was already well established by 1984, and there was almost no remaining arable land (fig. 18.5). Over the full study period from 1984–2003, zones closest to the park exhibited conversion rates between 1.6–2.0% of the land area in the zone per year, while the zones farthest from the park showed steadily decreasing annual conversion rates from ~1.0 to 0.1% per year (fig. 18.6). The high rates of conversion in the zones closest to the park drove the overall trends in conversion for all of western Serengeti, discussed below (fig. 18.7). Indeed, observed conversion rates would likely have been even higher if they reflected the true amount of remaining arable land. Some lands on steeper and rocky slopes are not suitable for agriculture but were included as potentially arable land in the overall calculations because it was not possible to accurately predict the occurrence of the unsuitable areas.

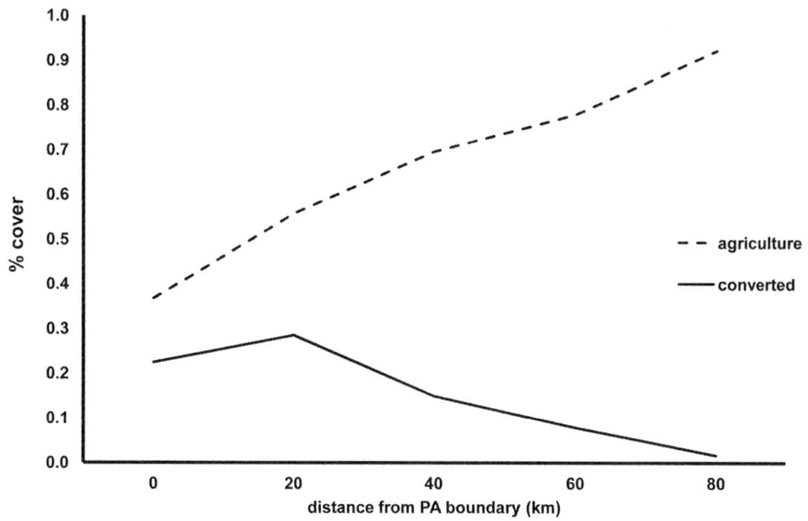

Fig. 18.5 Percent of each ward zone in western Serengeti under stable or converted agriculture from 1984 to 2003. Wards closest to the park boundary began with less agriculture, but showed the highest rates of conversion to new agriculture in the study period. (Reproduced from Estes et al. 2012).

Rates of Change by Period

The greatest overall conversion to new agriculture (41.7%) occurred during the earliest part of the study period, from 1984 to 1990. Total conversion then declined slowly with 30.7% converted to agriculture between 1990 and 1995, 12.4% between 1995 and 2000, and 15.1% between 2000 and 2003. Normalized annual conversion rates for the entire western Serengeti study area exhibited a similar trend, with the greatest annual change rates occurring in the earliest time period (1.02% per year), then declining to the lowest annual conversion rate (0.36% per year) between 1995–2000 (fig. 18.7). An increase in conversion rates to 0.74% per year between 2000 and 2003 appeared to be driven by conversion in the 20-km zone where conversion rates reached 2.3% of the zone per year, the highest observed rate for all zones and time periods (figs. 18.6, 18.7).

Rates of Change by Distance from the Park

The zone farthest from the park and closest to Lake Victoria (80-km zone) experienced the lowest rates of conversion in all time periods, and steadily decreasing rates (from 0.14–0.03% per year) from the beginning to the end

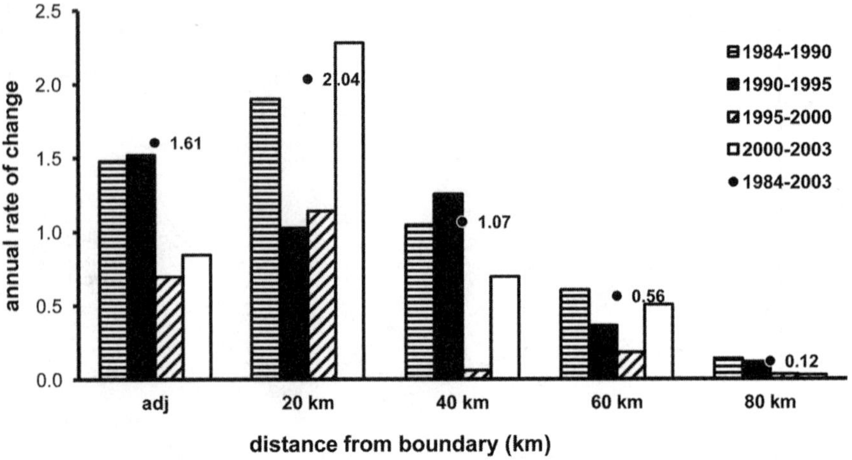

Fig. 18.6 Annual rates of conversion to cropland for each ward zone, by period of conversion. Dots represent annual rates of conversion within each zone for the entire study period. (Reproduced from Estes et al. 2012).

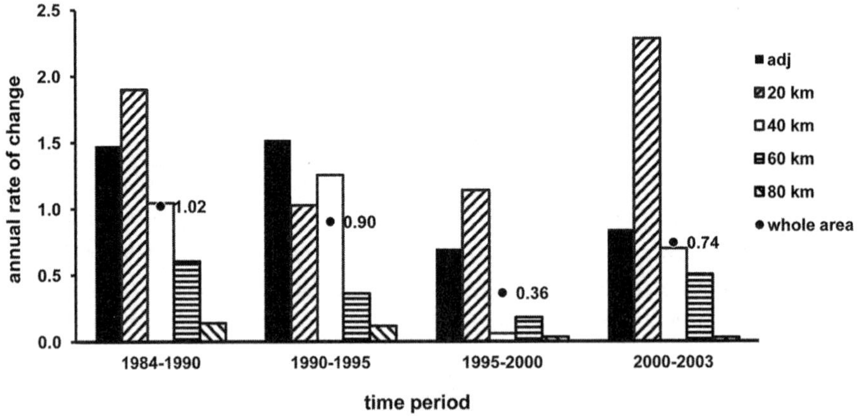

Fig. 18.7 Annual rates of conversion to cropland for each time period, by ward zone. The dot represents the annual rate of change for each time period estimated for the entire study area, defined by all wards with centroids 80 km or less from the park boundary. (Reproduced from Estes et al. 2012).

of the study period. Rates of conversion in the 60-km zone showed a similar decreasing trend with time, except for an increase in the most recent period. In fact, all zones, except for the 80-km zone, showed an increase in annual conversion rates in the most recent time period, relative to the period before it. Zones adjacent to the park and in the 40-km zone had the highest annual conversion rates in 1990–95, with rates between 1984 and 1990 a close second. The 20-km zone had greater annual rates of conversion than any other zone, and these were in the most recent and earliest time periods (2.3 and 1.9% per year, respectively), with wards adjacent to the park showing the next highest rates (~1.5% per year) in the first two time periods.

Land Cover and Human Population Trends

Taken together, analysis of changes in human populations and land cover in western Serengeti revealed some striking trends. Zones with the highest growth rates also had the lowest densities, and these were the same zones that exhibited the least existing agriculture at the beginning of the study period, the greatest conversion to new agriculture by the end of the study period, and were closest to the park. Conversely, both growth and conversion rates were lowest in those areas with the highest human densities and the greatest amount of stable agriculture, which were farthest from the park, and close to Lake Victoria (figs. 18.2, 18.5).

In contrast, in Loliondo in eastern Serengeti, different patterns of land use and human population trends prevailed, likely driven by different livelihood strategies and environmental conditions. However, because ward sizes in Loliondo were large, with boundaries of wards adjacent to the park extending 40–50 km from the park (fig. 18.1), it was not possible to compare growth rates in Loliondo at different distances from the park, as we were able to do in the more densely-settled western Serengeti. On average, though, population growth rates in Loliondo wards were around 2.8%, with densities ranging from only 4–20 (average of 10) people/km^2. Population growth rates west of the park were similar over the same distance, at around 2.9%, but densities were much higher, from 100–135 (average of 117) people/km^2.

DISCUSSION

Our analyses revealed dramatic differences in land-cover change and human population growth and density across the greater Serengeti ecosystem that can be linked to variations in land use, culture and sociopolitical conditions. We found the greatest conversion of natural areas to agriculture in the agropastoralist areas along the western border of the park in Tanzania, where population growth rates were also highest, and human densities lowest. Agricultural conversion in the pastoralist areas east of the park was minimal, and showed no patterns relative to the presence of the park. West of the park, agricultural conversion was inversely related to the amount of stable agriculture. Zones closest to the population centers near Lake Victoria and farthest from the park had the most agriculture at the start of the study period, and the least conversion to new agriculture by 2003. Conversely, zones closest to the park had very little agriculture at the beginning of the study period, but showed the greatest conversion to new agriculture.

The patterns that we found are typical of frontier engulfment and an indication of push factors causing human population spread (Oglethorpe et al. 2008; Scholte and de Groot 2010), and do not support the explanation that populations near the park increased because of economic opportunities provided by the park. Instead, people likely moved away from areas where resources have become scarce (i.e., near the highest population centers near the lake), and to places where resources are still available, which happen to be close to the park. Indeed, surveys in villages adjacent to the park in northwest and southwest Serengeti found that two-thirds to four-fifths, respectively, of the villages' populations are immigrants whose primary reason for moving to these areas was for grazing land or the opportunity to farm (Schmitt 2010). Similar results have been found in villages around Budongo Forest Reserve in Uganda, where the majority of regional migrants to the area came for access to land, likewise supporting the frontier engulfment model of immigration around protected areas (Zommers and MacDonald, 2012).

In pastoralist-dominated Loliondo east of Serengeti National Park, conversion of land to agriculture until 2003 was minimal, and mostly limited to areas immediately surrounding human dwellings, consistent with the typical establishment of Maasai farming in this area (McCabe, Leslie, and DeLuca 2010). Almost no agricultural expansion was detected near the park boundary, which may be partly the result of agreements between tour operators and villages in Loliondo that generated income for the villages while also limiting agriculture (Nelson, Gardner, and Igoe 2009). In contrast to western Serengeti, only one-quarter of the residents in villages adjacent to the park in Loliondo were immigrants, and their primary reason for immigrating was marriage (Schmitt 2010). In Loliondo, where cultural and environmental differences have led to slower adoption of agriculture, there is still abundant arable land, and lack of this resource therefore does not yet appear to be influencing people's movements, although competition for grazing land is increasing.

Our analysis of human population data in western Serengeti further upholds the argument that people were being pushed from densely-settled areas in search of resources. Rural population densities were highest close to Lake Victoria and farther from the park, which mirrors the trend in stable agriculture. Annual population growth rates were greatest in the areas with the lowest human densities, and these same areas showed the greatest extent of conversion to new agriculture. Visual inspection of the points used to assess conversion rates in different time periods indicated that more recent conversion in areas farthest from the park was often the result of converting remaining wetlands, highlighting limits to expanding agriculture there. Likewise, remaining patches of natural habitat in the more densely-farmed zones typically coincided with hills and rocky outcrops which are often less suitable for farming, indicating that conversion rates of remaining arable lands are actually even higher than reflected here. In contrast, new agriculture in zones closer to the park was typically the result of conversion from savanna, which may previously have been used for grazing land. Taken together, these trends indicate a migration of people away from heavily-farmed, densely-populated areas, and toward low-density areas with available land closer to the park boundary. Similar trends have been observed around a number of other protected areas, including Kafue National Park in Zambia (Joppa, Loarie, and Pimm 2009).

Our analysis of rates of agricultural conversion throughout the study period further supported the push mechanism behind immigration in areas near the park. In general, annual rates of conversion declined throughout the study period, increasing slightly in the most recent period from 2000 to 2003. The zone farthest from the park, which had the greatest amount of cultivated land at the beginning of the study period, exhibited the lowest annual rates of conversion to new agriculture in all time intervals. These rates declined in the earlier half of the study period and remained constant at just 0.03% per year from 1995 to 2003, whereas conversion in zones closer to the park was occurring at 0.7 to 2.3% of the land in those zones per year over the same period. This underscores the fact that even at the start of the study period, the zones closer to the population centers around Lake Victoria were limited in the amount of land not yet converted to agriculture, and that any further opportunities to expand agriculture in those areas were exhausted by the end of the study period.

Economic data from village surveys also suggest that pull factors are not the most probable cause for high population growth near the park, where poverty rates were higher. About three-quarters of households in villages directly adjacent to the park in northwest (75.1%) and southwest (71%) Serengeti are below the poverty line (Schmitt 2010), as compared to 42% and 46%, respectively, of the larger regions in western Serengeti to which these communities belong (Polasky et al. 2008), and 39% of rural populations in Tanzania as a whole (Tanzania National Bureau of Statistics 2002). Likewise, the majority of people in villages surrounding the park, the same villages in which the majority of people were immigrants, report no benefits from conservation or the park (Schmitt 2010). The fact that people are moving into very poor areas argues against an economic benefit, both real and perceived, to those living adjacent to the park.

Furthermore, areas closer to the park may actually be less desirable to agropastoralists due to high rates of human-wildlife conflict, which results in loss of crops and livestock, and sometimes human lives, and can substantially impact the livelihoods of subsistence farmers (Naughton-Treves and Treves 2005; Thirgood, Woodroffe, and Rabinowitz 2005). As a result, the newest immigrants to an area typically end up with undesirable plots on the periphery of the human-wildlife interface (Naughton-Treves 1997), a trend which has also been observed along the western Serengeti park boundary (J. Schmitt pers. comm.) where human-wildlife conflict is the most severe. In fact, 85% of people in villages adjacent to the park report a cost from wildlife, with 64% reporting the cost to be crop destruction (Schmitt 2010; Hampson et al. chapter 21).

The last decade has seen a particular increase of human-elephant conflict in villages in western Serengeti (Malugu 2010; Walpole et al. 2004). Farmers greatly fear elephants because they threaten their lives and destroy their crops. Decimated in the 1970s and 1980s, the Serengeti elephant population recovered and expanded its range westward following the international ban on the ivory trade in 1989. Increasing agricultural expansion and human populations adjacent to the park, in conjunction with the westward-expanding elephant population, has created a fertile ground for sometimes intense human-elephant conflicts, leading to human and elephant deaths, crop loss (Malugu 2010; Walpole et al. 2004), and likely antagonism toward the protected areas who are seen as the "owners" of the problem animals (Gadd 2005; Naughton-Treves, Holland, and Brandon 2005).

Similarly, the conversion of savanna to agriculture in western Serengeti, accompanied by increasing human populations, impacts the migratory wildebeest (Thirgood et al. 2004) which are at greater risk of hunting when closer to local communities. In fact, in addition to the push factors outlined above, there may be one pull factor of strong conservation concern, and that is the wildlife itself. Many people in western Serengeti consume bushmeat as a source of protein, and sell it to supplement their income (Mfunda and Roskaft 2010; Mduma, Hilborn, and Sinclair 1998; Rentsch et al. chapter 22). Since one of the key predictors of bushmeat hunting is the distance between one's home and the wildlife resource (Hofer et al. 2000), increasing human populations near the park likely resulted in increased bushmeat hunting.

Looking ahead, it is likely that human populations adjacent to the park in western Serengeti will continue to increase and convert remaining natural land to agriculture, unless broad-reaching conservation plans are enacted that also incorporate community engagement and district-level land-use planning. As Maasai continue to adopt agriculture (McCabe, Leslie, and DeLuca 2010), we may also expect to see increasing land pressure in Loliondo and in the Narok region in Kenya.

Although plans for a controversial paved road connecting Lake Victoria to the coast of Tanzania that would have bisected Serengeti National Park have been tabled, there are still plans to build paved roads into the communities east and west of the park (written comm. TZ Minister of Natural Resources 2011). This has the potential to dramatically increase the pace of agricultural conversion in Loliondo, and the competition for land. The roads, while likely beneficial to development in some respects (Escobal and Ponce 2001), also increase access to the area by agriculturalists from outside the region, where there are shortages of arable land. This could lead to a land grab by wealthy farmers from Arusha and other densely-settled areas, unless urgent action is taken to secure land rights in Loliondo. Only weeks after the announcement of the road, there were reports of prospecting for land in Loliondo by wealthy and powerful individuals from Arusha.

The conservation of the greater Serengeti ecosystem is clearly a challenge that extends well beyond the boundaries of the protected areas. Land shortages near Lake Victoria and even farther afield will continue to put pressure on remaining arable land around Serengeti, and national and global economic interests can exacerbate this process. Designing appropriate conservation strategies to preserve Serengeti will therefore necessitate collaboration between biological and social scientists, economists, multilevel governments, protected area managers, NGOs, and community organizations. In Serengeti, there are important collaborations being formed to enact a broader view of ecosystem management which can help achieve this goal (see www.serengetiecoforum.org). Participatory wildlife management in the form of a quite successful wildlife management area in western Serengeti is contributing income from conservation to local communities, which may help these areas stall the advance of agriculture, as similar collaborations appear to have done in Loliondo. Nevertheless, these efforts can collapse under the strain of land and population pressures, especially with immigration from outside their area of influence, and every effort must be made to try to prevent this from happening. Furthermore, when the assumption is made that immigration around protected areas is benefit-driven, a common response is to suggest the disbursement of benefits farther away from the protected area, to draw people out of the buffer areas

(DeFries et al. 2007; Wittemyer et al. 2008). Management plans are sometimes based on this assumption, and suggest building schools and health clinics in communities more removed from the protected areas. However, if immigration is mainly driven by push factors, such an intervention may be ineffective if not coupled with clear livelihood alternatives that are not based on the diminished resource, in this case, arable land. This speaks to the need for management interventions to be multifaceted, incorporating both land use planning and development of alternative incomes, perhaps in concert with efforts to increase agricultural yields in already developed lands, which can lead to "land sparing" in areas closer to the park (Baudron et al. 2011). Analysis of land-cover change and population trends can inform conservation planners of the spatial scales they should be targeting, and help inform zoning efforts that incorporate both critical species habitat and information about human development (Abbitt, Scott, and Wilcove 2000).

CONCLUSION

Contrary to the common assumption that immigration is the result of attractive forces associated with the protected area itself, our study suggests that people in western Serengeti are being pushed from their areas of origin by lack of arable land, which they find near the protected area. Effective, targeted conservation interventions require that we move beyond broad generalizations to correctly identify both the patterns and the drivers of human population trends and their relationship to both land-cover change and protected areas on a case-by-case basis.

ACKNOWLEDGMENTS

We gratefully acknowledge the support of TAWIRI, TANAPA, the Wildlife Division and COSTECH for permission to work in Serengeti, and Grumeti Reserves for access to their concessions. A. B. E. was supported by the USFWS, NSF Geography and Regional Sciences and Office of International Science and Engineering, a NASA Earth and Space Science Fellowship, the Jefferson Scholars Foundation, Cleveland Metroparks Zoo, and the Explorer's Club Washington. A. B. E. would also like to thank L. Gemmill and D. Lowry for their generous support of the study. T. K. gratefully acknowledges support by the Alexander von Humboldt Foundation and the European Commission (integrated project VOLANTE, FP7-ENV-2010-265104), and V. C. R. by the NASA Land-Cover and Land-Use Change Program and the NASA Biodiversity Program.

REFERENCES

Abbitt, R. J. F., J. M. Scott, and D. S. Wilcove. 2000. The geography of vulnerability: Incorporating species geography and human development patterns into conservation planning. *Biological Conservation* 96:169–75.

Baudron, F., M. Corbeels, J. A. Andersson, M. Sibanda, and K. E. Giller. 2011. Delineating the drivers of waning wildlife habitat: The predominance of cotton farming on the fringe of protected areas in the Mid-Zambezi Valley, Zimbabwe. *Biological Conservation* 144:1481–93.

Campbell, K., and H. Hofer. 1995. People and wildlife: Spatial dynamics and zones of interaction. In *Serengeti II: Dynamics, management, and conservation of an ecosystem*, ed. A. R. E. Sinclair and P. Arcese, 534–70. Chicago: University of Chicago Press.

Cantú-Salazar, L., and K. J. Gaston. 2010. Very large protected areas and their contribution to terrestrial biological conservation. *BioScience* 60:808–18.

Cleaveland, S., M. G. J. Appel, W. S. K. Chalmers, C. Chillingworth, M. Kaare, and C. Dye. 2000. Serological and demographic evidence for domestic dogs as a source of canine distemper virus infection for Serengeti wildlife. *Veterinary Microbiology* 72: 217–27.

Congalton, R. G. 1991. A review of assessing the accuracy of classifications of remotely sensed data. *Remote Sensing of Environment* 37:35–46.

DeFries, R., A. Hansen, A. C. Newton, and M. C. Hansen. 2005. Increasing isolation of protected areas in tropical forests over the past twenty years. *Ecological Applications* 15:19–26.

DeFries, R., A. Hansen, B. L. Turner, R. Reid, and J. Liu. 2007. Land use change around protected areas: Management to balance human needs and ecological function. *Ecological Applications* 17:1031–38.

Escobal, J., and C. Ponce. 2001. The benefits of rural roads: Enhancing income opportunities for the rural poor. *World Development* 29:497–508.

Estes, A. B., T. Kuemmerle, H. Kushnir, V. C. Radeloff, and H. H. Shugart. 2012. Land-cover change and human population trends in the greater Serengeti ecosystem from 1984 to 2003. *Biological Conservation* 147:255–63.

Gadd, M. E. 2005. Conservation outside of parks: Attitudes of local people in Laikipia, Kenya. *Environmental Conservation* 32:50–63.

Hansen, A. J., and R. DeFries. 2007. Ecological mechanisms linking protected areas to surrounding lands. *Ecological Applications* 17:974–88.

Hofer, H., K. L. Campbell, M. L. East, and S. A. Huish. 2000. Modeling the spatial distribution of the economic costs and benefits of illegal game meat hunting in the Serengeti. *Natural Resource Modeling* 13:151–77.

Huang, C., J. R. G. Townshend, and L. S. Davis. 2002. An assessment of support vector machines for land cover classification. *International Journal of Remote Sensing* 23: 725–49.

Joppa, L. N., S. R. Loarie, and A. Nelson. 2010. Measuring population growth around tropical protected areas: Current issues and solutions. *Tropical Conservation Science* 3: 117–21.

Joppa, L. N., S. R. Loarie, and S. L. Pimm. 2009. On population growth near protected areas. *PLoS ONE* 4 (1): e4279.

Kuemmerle, T., P. Hostert, V. C. Radeloff, S. Linden, K. Perzanowski, and I. Kruhlov. 2008.

Cross-border comparison of post-socialist farmland abandonment in the Carpathians. *Ecosystems* 11:614–28.

Malugu, L. T. 2010. Assessment of human-elephant conflicts in areas adjacent to Grumeti-Ikorongo Game Reserves, northern Tanzania. MS thesis, Sokoine University of Agriculture, Morogoro.

McCabe, J. T., P. W. Leslie, and L. DeLuca. 2010. Adopting cultivation to remain pastoralists: The diversification of Maasai livelihoods in northern Tanzania. *Human Ecology* 38:321–34.

Mduma, S. A. R., R. Hilborn, and A. R. E. Sinclair. 1998. Limits to exploitation of Serengeti wildebeest and implications for its management. In *Dynamics of tropical communities*, ed. D. M. Newbery, H. H. T. Prins, and N. D. Brown, 243–65. Oxford: Blackwell Science.

Mfunda, I. M., and E. Roskaft. 2010. Bushmeat hunting in Serengeti, Tanzania: An important economic activity to local people. *International Journal of Biodiversity and Conservation* 2:263–72.

Naughton-Treves, L. 1997. Farming the forest edge: Vulnerable places and people around Kibale National Park, Uganda. *Geographical Review* 87:27–46.

Naughton-Treves, L., M. B. Holland, and K. Brandon. 2005. The role of protected areas in conserving biodiversity and sustaining local livelihoods. *Annual Review of Environment and Resources* 30:219–52.

Naughton-Treves, L., and A. Treves. 2005. Socio-ecological factors shaping local support for wildlife: Crop-raiding by elephants and other wildlife in Africa. In *People and wildlife: Conflict or coexistence?*, ed. R. Woodroffe, S. Thirgood, and A. Rabinowitz, 252–77. Cambridge: Cambridge University Press.

Nelson, F., B. Gardner, and J. Igoe. 2009. Community-based conservation and Maasai livelihoods in Tanzania. In *Staying Maasai? Livelihoods, conservation, and development in East African rangelands*, ed. K. Homewood, P. Kristjanson, and P. C. Trench, 299–333. New York: Springer.

Oglethorpe, J., J. Ericson, R. E. Bilsborrow, and J. Edmond. 2007. People on the move: Reducing the impact of human migration on biodiversity. World Wildlife Fund and Conservation International Foundation, Washington, DC.

Ogutu, J. O., H.-P. Piepho, H. T. Dublin, N. Bhola, R. S. Reid. 2009. Dynamics of Mara–Serengeti ungulates in relation to land use changes. *Journal of Zoology* 278:1–14.

Polasky, S., J. Schmitt, C. Costello, and L. Tajibaeva. 2008. Larger-scale influences on the Serengeti ecosystem: National and international policy, economics and demography. In *Serengeti III: Human impacts on ecosystem dynamics*, ed. A. R. E. Sinclair, C. Packer, S. A. R. Mduma, and J. M. Fryxell, 347–78. Chicago: University of Chicago Press.

Schmitt, J. A. 2010. Improving conservation efforts in the Serengeti ecosystem, Tanzania: An examination of knowledge, benefits, costs, and attitudes. PhD diss., University of Minnesota, St. Paul.

Scholte, P., and W. T. de Groot. 2010. From debate to insight: Three models of immigration to protected areas. *Conservation Biology* 24:630–32.

Sinclair, A. R. E. 1995. Serengeti past and present. In *Serengeti II: Dynamics, management, and conservation of an ecosystem*, ed. A. R. E. Sinclair and P. Arcese, 3–30. Chicago: University of Chicago Press.

Sinclair, A. R. E., S. A. R. Mduma, J. G. C. Hopcraft, J. M. Fryxell, R. Hilborn, and S. Thirgood. 2007. Long-term ecosystem dynamics in the Serengeti: Lessons for conservation. *Conservation Biology* 21:580–90.

Tanzania National Bureau of Statistics. 2002. The National Statistics Office of Tanzania. Accessed at www.nbs.go.tz.

Thirgood, S., A. Mosser, S. Tham, G. Hopcraft, E. Mwangomo, T. Mlengeya, M. Kilewo, J. Fryxell, A. R. E. Sinclair, and M. Borner. 2004. Can parks protect migratory ungulates? The case of the Serengeti wildebeest. *Animal Conservation* 7:113–20.

Thirgood, S., R. Woodroffe, and A. Rabinowitz. 2005. The impact of human-wildlife conflict on human lives and livelihoods. In *People and wildlife: Conflict or coexistence?*, ed. R. Woodroffe, S. Thirgood, and A. Rabinowitz, 13–26. Cambridge: Cambridge University Press.

Walpole, M. J., Y. Ndoinyo, R. Kibasa, C. Masanja, M. Somba, and B. Sungura. 2004. An assessment of human-elephant conflict in the western Serengeti. Unpublished Report, Frankfurt Zoological Society, Arusha, Tanzania.

Wittemyer, G., P. Elsen, W. T. Bean, A. C. O. Burton, and J. S. Brashares. 2008. Accelerated human population growth at protected area edges. *Science* 321:123–26.

Zommers, Z., and D. W. MacDonald. 2012. Protected areas as frontiers for human migration. *Conservation Biology* 26:547–56.

NINETEEN

Infectious Diseases in the Serengeti: What We Know and How We Know It

Tiziana Lembo, Harriet Auty, Katie Hampson, Meggan E. Craft, Andy Dobson, Robert Fyumagwa, Eblate Ernest, Dan Haydon, Richard Hoare, Magai Kaare, Felix Lankester, Titus Mlengeya, Dominic Travis, and Sarah Cleaveland

The last twenty-five years have seen significant advances in our understanding of the role that pathogens play in the dynamics of natural communities and ecosystems. The case of rinderpest in the Serengeti provides a textbook example of how a single pathogen can dramatically modify the abundance of many other species in an ecosystem through a classical trophic cascade (Sinclair 1973; Sinclair 1977; Holdo et al. 2009). Detailed studies of canine distemper outbreaks in lions (Roelke-Parker et al. 1996; Kissui and Packer 2004; Munson et al. 2008; Craft et al. 2009), and rabies in domestic dogs (*Canis familiaris*) and wild carnivores (Gascoyne et al. 1993; Cleaveland and Dye 1995; Lembo et al. 2008; Hampson et al. 2009) have illustrated the impact and complex interactions of pathogens in Serengeti carnivores. Thanks also to advances in the development of mathematical models for host-parasite dynamics (Grenfell and Dobson 1995; Hudson et al. 2002), there is now collective interest among ecologists in the role that pathogens play in natural communities. Disease outbreaks are no longer regarded as random sources of mortality that disrupt the more traditionally studied predator-prey and herbivore-plant relationships; instead, pathogens are perceived as ubiquitous and diverse natural enemies that take a daily energetic toll from almost all free-living species. Their profound effects on birth and death rates and behavior significantly modify the way in which free-living species interact and how nutrients and trace elements move through ecosystems. A deeper understanding of the ecological role of disease has also led to more synergistic interactions between ecologists, veterinarians, and human health workers. Combined with concurrent progress in diagnostics and techniques for analysis of complex data, this shift is changing the way in which ecological and human/animal health data are collected and analyzed. In previous chapters we described impacts of infectious diseases on ecosystem health in the Serengeti (Cleaveland et al. 2008). Here we outline a pathway, based on this new perspective, which summarizes the available information on Serengeti pathogen dynamics and approaches used to elucidate them; specifically, we explore what we know and how we know it. A glossary of terms is given in box 19.1 at the end of the chapter, and Latin names for large mammals are given in appendix 1 at www.press.uchicago.edu/sites/serengeti/.

WHAT WE KNOW ABOUT SERENGETI PATHOGENS AND HOW WE KNOW IT

Epidemiological studies of Serengeti pathogens have illustrated that infectious diseases are important ecological perturbations that substantially affect the dynamics of ecosystems and their communities (Cleaveland et al. 2008). Impacts on human communities are direct (through morbidity and mortality) and indirect (affecting livestock production, food security, nutrition, and, as a result, socioeconomic status), with changing patterns of land use also arising as a consequence of decreased livestock production. In addition, diseases directly threaten the viability of Serengeti wildlife populations, and more broadly, the conservation of global biodiversity. With increased human and animal movements, there is also potential for many of these pathogens to disseminate beyond local communities or for new pathogens to be introduced.

Many of the important veterinary and human diseases in the Serengeti infect a range of species, but some of these may act as incidental or dead-end hosts. Hence, determining which species are critical for maintaining infection is a key research priority. Classifications of pathogens often focus on whether infection is harbored by domestic animals, humans, or wildlife. However, it is unclear precisely where many pathogens in the Serengeti fit in this context (fig. 19.1), since the interactions of a pathogen within a complex ecosystem can be hard to disentangle. Another important consideration is the pathogen transmission route which also varies substantially. As an example, it can range from environmental contamination to respiratory and venereal transmission (fig. 19.2), influencing the species affected, rate of spread, scale of persistence, and effects of external forces such as seasonal/climatic factors, human encroachment, poaching, and fire. Much research in the Serengeti has therefore focused on exploring how pathogens are

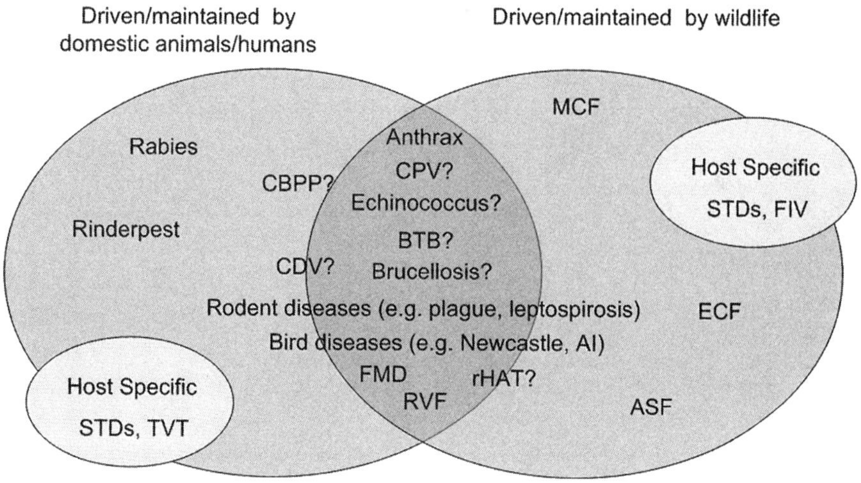

Fig. 19.1 Classification of Serengeti pathogens and diseases based on maintenance hosts (domestic animals/humans or wildlife). AI = avian influenza; ASF = African swine fever; BTB = bovine tuberculosis; CBPP = contagious bovine pleuropneumonia; CDV = canine distemper virus; CPV = canine parvovirus; ECF = east coast fever; FIV = feline immunodeficiency virus; FMD = foot-and-mouth disease; MCF = malignant catarrhal fever; rHAT = Rhodesian human African trypanosomiasis; RVF = Rift Valley fever; STD = sexually transmitted disease; TVT = transmissible venereal tumor.

maintained and transmitted among different host populations, which is critical for predicting, preventing, and controlling damaging disease outbreaks.

A number of diagnostic methodologies have been deployed to detect the presence and prevalence of diseases and infections in the Serengeti, involving field-based approaches to case detection and laboratory diagnostic techniques. A wide range of approaches have subsequently been adopted to develop effective surveillance systems for early detection of diseases and to analyze disease surveillance and ecological data to understand infection dynamics. The combination of tools has been key to generating a substantial body of epidemiological data (table 19.1) and drawing important inferences on the dynamics of many Serengeti pathogens (table 19.2).

TOOLS FOR INFECTIOUS DISEASE RESEARCH IN THE SERENGETI

The toolbox that we have used for infectious disease research in the Serengeti (fig. 19.3) includes three parts: (a) generating surveillance data (i.e. for detection of disease and infection) and developing surveillance systems using these data; (b) obtaining and using ecological data for disease investigation; and (c) quantitatively analyzing surveillance and ecological data to

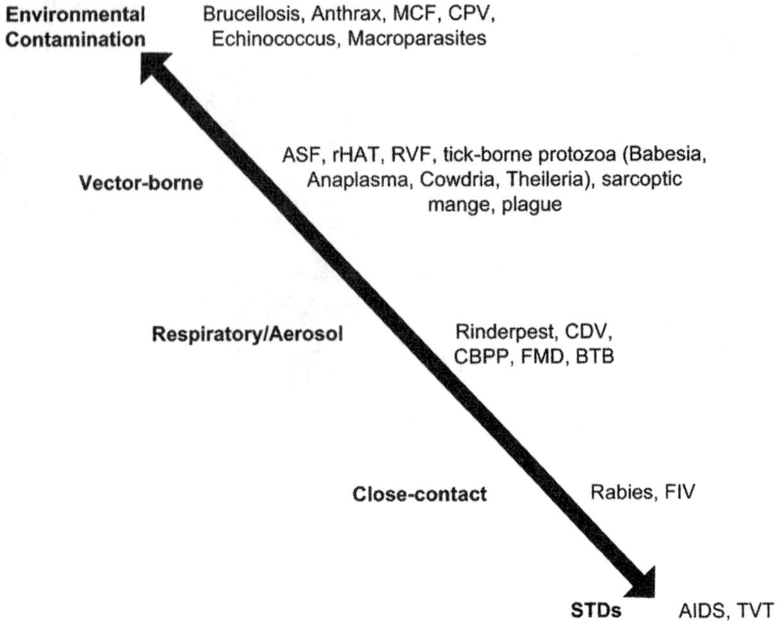

Fig. 19.2 Transmission routes of Serengeti pathogens. AIDS = acquired immune deficiency syndrome; ASF = African swine fever; BTB = bovine tuberculosis; CBPP = contagious bovine pleuropneumonia; CDV = canine distemper virus; CPV = canine parvovirus; FIV = feline immunodeficiency virus; FMD = foot-and-mouth disease; MCF = malignant catarrhal fever; rHAT = Rhodesian human African trypanosomiasis; RVF = Rift Valley fever; STD = sexually transmitted disease; TVT = transmissible venereal tumor.

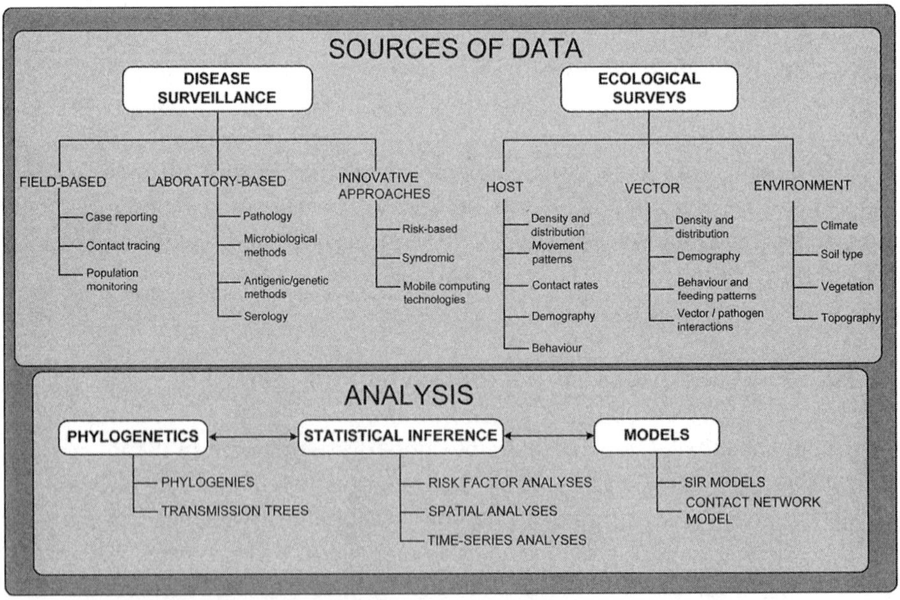

Fig. 19.3 Schema showing the tools used for infectious disease research in the Serengeti. SIR = susceptible-infected-recovered.

Table 19.1 Tools used for infectious disease research in the Serengeti and data generated deploying these tools

	Indicator data															
Tool/Activity	**Presence/absence**	**Incidence**	**Prevalence**	**Age-prevalence**	**Seroprevalence**	**Age-seroprevalence**	**Mobidity/mortality/case-fatality rates**	**Susceptibility**	**Indices of host abundance/density**	**Host demography**	**Indices of vector abundance/density**	**Vector infection prevalence**	**Risk factors**	**Transmission/transmission links**	**Pathogen origin/evolution time**	**Temporal/spatial patterns**
Existing records (veterinary/hospital)	x	x	x	x			x	x					x			x
Participatory techniques/contact tracing	x	x	x	x			x	x					x	x		x
Total counts/presence-absence/ transect surveys (host)	x						x	x	x							
Gross pathology	x		x	x				x								
Diagnostic pathology	x	x	x	x			x	x								x
Microscopy	x	x	x	x								x				
Isolation	x	x	x	x								x				
Antigenic/genetic methods	x	x	x	x								x				
Serology	x				x	x	x	x								x
Observational/behavioral studies	x						x	x		x						x
Host movement monitoring														x		
Counts (vector)											x			x		
Environmental monitoring (climate, soil, vegetation, rainfall)													x			x
Phylogenetics														x	x	
Transmission tree construction														x		
Time-series analyses													x			x
Spatial analyses													x			x
Network models														x		x
Other models														x		x

Table 19.2 Inferences on the dynamics of Serengeti pathogens drawn from a range of indicator data

		Inferences					
Indicator data	**Pathogen**	**Pathogenicity**	**Timing of infection and outbreaks**	**Spatial dynamics**	**Reservoir dynamics**	**Population impacts**	**Other (conservation, public health, livelihoods)**
Presence/ absence	Rabies			Transmission within or from neighboring districts (not from wildlife protected areas) implicated in persistence in dogs	Infection in dogs is not driven by wildlife in protected areas		
	Bovine TB	Variable pathogenicity in wildlife (compared to other African populations)					*Mycobacterium bovis* and atypical mycobacterial species potentially an important cause of human and animal disease
Incidence	Rabies		Dog outbreaks precede wildlife outbreaks	Clusters of wildlife cases are spatially linked to cases in domestic dogs		Potentially dramatic impacts on threatened carnivores	High incidence of human exposures to dog bites; substantial impacts on wildlife populations (particularly threatened canids) and livestock production

	CDV	Temporal patterns of cases in dogs, lions, and hyenas			Impacts on lion population variable	
	Anthrax	Cases/outbreaks in livestock and wildlife predicted by prolonged climate extremes	Outbreaks in wildlife clustered around endemic foci		No obvious impacts on wildlife species affected	
	Rinderpest			Cattle reservoir, not wildlife	Considerable impacts on wildebeest and buffalo	
Prevalence	rHAT			Multihost reservoir system potentially responsible for persistence		Identification of wildlife species harboring subspecies pathogenic for humans
	Echinococcus					Very high prevalence of cysts in human and livestock in pastoral systems

continued

Table 19.2 continued

		Inferences					
Indicator data	**Pathogen**	**Pathogenicity**	**Timing of infection and outbreaks**	**Spatial dynamics**	**Reservoir dynamics**	**Population impacts**	**Other (conservation, public health, livelihoods)**
Age-prevalence	rHAT	Decreasing prevalence with age may indicate development of immunity to *Trypanosoma brucei* in lions				No impact on lion survival	
Seroprevalence	Anthrax			Spatial heterogeneity and discovery of high-risk areas, linked to alkaline soils		Nonfatal infection common in carnivores, buffalo, and wildebeest. Invariably fatal infection in zebra	

	CDV	Variable pathogenicity in lions and dogs	Higher morbidity/ mortality due to infection when associated with cofactors			Effects on lion host fitness
Age-seroprevalence	CDV		Approximate timing of infection/ exposure in lions and dogs		Dogs source of infection, but not wildlife	
	Anthrax		Timing of outbreaks in villages using dog age-seroprevalence data			
Indices of host abundance/ density	Rabies		Potential for signalling outbreaks with high mortality		Transmission not proportional to wildlife species abundance, but a consequence of stochastic spillover events	Potential to show population declines when susceptible species are affected

continued

Table 19.2 continued

		Inferences					
Indicator data	**Pathogen**	**Pathogenicity**	**Timing of infection and outbreaks**	**Spatial dynamics**	**Reservoir dynamics**	**Population impacts**	**Other (conservation, public health, livelihoods)**
Host demography	CDV					Impacts on lion population variable	
	Rabies				Very high turnover in domestic dogs important for maintaining susceptibles for sustained transmission		Essential parameter for designing and evaluating control measures suggests 70% coverage necessary to prevent outbreaks in between annual vaccination campaigns
Indices of vector abundance/ density/ infection prevalence	rHAT			Spatial heterogeneity in disease risk due to variation in both vector density and infection prevalence			Quantification of human disease risk and identification of high risk areas for native wildlife, e.g., black rhino reintroductions
Risk factors	Bovine TB				Presence of wildlife a risk factor for cattle		

	Anthrax	Climatic associations	Soil associations explaining observed spatial heterogeneity	
Transmission links	Rabies			Most transmission dog-to-dog or dog-to-wildlife; high rates of between-species transmission
	CDV			Multihost transmission required for maintaining infection

Notes: TB = tuberculosis; CDV = canine distemper virus; rHAT = Rhodesian human African trypanosomiasis.

determine disease impacts and elucidate infection dynamics. Here we describe these tools and then illustrate how they can be effectively integrated to achieve a better understanding of complex disease systems, using rabies and trypanosomiasis as case studies.

Generating Disease Surveillance Data

Pathogen surveillance involves the detection of infection (presence of a pathogen) and/or disease (occurrence of cases and outbreaks) in a population to: (1) assess prespecified diseases and infections known to be continuously present (endemic) in the population; (2) detect diseases that are reemerging (i.e., known conditions reappearing or increasing in prevalence), emerging (i.e., previously unknown conditions), or exotic (i.e., known conditions occurring in areas where they had not been recorded) and/or demonstrate freedom from disease; and (3) identify risk factors for disease occurrence. This information is critical for planning, implementation, and evaluation of control measures, and prioritizing future health care needs that will ultimately have beneficial consequences for human and animal health, livestock economies, human livelihoods, and wildlife conservation.

Traditional definitions differentiate between monitoring, used to detect changes in the occurrence of endemic diseases (without necessarily eliciting a response), and surveillance, aimed at detecting incursions of nonendemic (reemerging, emerging, or exotic) diseases and eliciting a response to control them (OIE 2010). It has been argued that this distinction may not be required. Rarely would data be collected if no response or purpose is envisaged, and activities are better distinguished according to "surveillance purpose" (Hoinville et al. 2009).

Surveillance approaches can be described based on the origin and initiator of the surveillance information and have been distinguished as "passive" and "active," although the precise definition of these terms is widely debated. Broadly, active (targeted) surveillance describes data collection activities that are initiated by the investigator (i.e., who decides which specific information should be collected by the observer). In the case of wildlife disease surveillance, these activities commonly comprise cross-sectional surveys of the wildlife species of interest, but may also include regular screening of sentinel populations, serosurveys, vector trapping, rodent trapping, and scat analyses. In contrast, passive surveillance describes systems where data are gathered from records generated by the observer (e.g., clinician, farmer, veterinarian, ranger), often for other purposes, and from which surveillance information is extracted by the investigator. For example, for wildlife diseases, routine data collection by park authorities, as well as opportunistic sampling or necropsies during daily activities such as capture operations, problem animal removals, investigations of morbidity/mortality events, and examinations of road-killed, hunted, or poached animals can yield useful passive surveillance information. In general, active methods provide a more accurate assessment of disease occurrence, as they usually involve random or representative sampling of the population of interest, but they also tend to be more costly.

For some pathogens, infection always results in disease with detectable clinical or postmortem signs. However, identification of a pathogen in a host does not always indicate disease. Surveillance for pathogens in hosts that do not show concomitant signs of disease is important for understanding infection dynamics, identifying alternate (and possible reservoir) hosts, and detecting potential sources of emerging infections. Pathogens may also be detected as coinfections and exacerbating cofactors in diseases caused by other agents. For example, a high intensity of *Babesia* infection was thought to be a cofactor causing high mortality from canine distemper virus (CDV) infection in Serengeti lions in 1994 and conversely, recent CDV infection was thought to have increased the severity of babesiosis in Ngorongoro lions in 2001 (Munson et al. 2008). Indeed, the detection of pathogens and the quantification of parasite burdens can provide useful insights into a variety of fitness effects and their ecological ramifications (Jolles, Cooper, and Levin 2005; Jolles et al. 2008).

While collection and analysis of surveillance data will ideally lead to improved disease management, a discussion of disease control options is not an objective of this chapter and has been extensively covered elsewhere (Breed et al. 2009). It should however be emphasized that designing effective strategies to control multihost disease concerns poses considerable challenges and requires a clear understanding of pathogen dynamics to ensure that control programs are targeted at the species mostly responsible for maintenance (if disease elimination is the ultimate objective).

Field-Based Approaches to Case Detection

Collection of basic field data on disease occurrence is the first step in any epidemiological investigation. However, numerous practical, economic, and technical constraints render the collection of reliable data challenging, especially for wildlife diseases (Breed et al. 2009). Similarly, underreporting of most livestock and human diseases is a common problem throughout sub-Saharan Africa, particularly for endemic zoonoses which have been widely neglected (WHO/DFID 2006). Yet, low-technology disease surveillance options exist that have the potential to improve disease detection rates at low cost. In particular, training local field staff in recognizing visible diseases/abnormalities (e.g., sarcoptic/demodectic mange, rabies, papillomatosis, dermatomycosis, verminous dermatitis or elbow hygromas/parotid swellings in lions, squamous cell carcinomas in lions, tuberculosis in greater kudus, and carpal hygromas in ruminants with brucellosis) is likely to be a cost-effective means of improving the quality of disease surveillance. Additional activities that could be conducted by field staff may include investigating vulture activity, condition scoring of animals, recording simple data on the timing and location of animal deaths, photographing of lesions/conditions and basic sample collection (e.g., blood smears for anthrax, *Babesia*, or trypanosome detection—see also below). In addition, the lack of effective surveillance systems provides a powerful stimulus for developing innovative approaches that can rapidly and efficiently exploit recent technological advances ("leap-frogging" technology) such as mobile phone technology. Furthermore, there is no doubt that existing data could be utilized more effectively with the application of appropriate analytical tools.

Case reporting. Community knowledge of animal health issues (so-called existing veterinary knowledge) has long been recognized as an important source of infectious disease information: the Maasai have long emphasized the association between wildebeest (particularly the wildebeest calving season) and fatal malignant catarrhal fever (MCF) in cattle, which has considerably influenced land-use decisions by these communities (Cleaveland et al. 2001; Bedelian, Nkedianye, and Herrero 2007). In remote areas, and in the absence of laboratory support, data generated from reporting of cases provide a potential source of information often available through existing systems (e.g., records from government veterinary/livestock offices, abattoirs, hospitals). Such records, commonly generated by a number of different observers, may contain information on clinical signs/symptoms as well as mortality, which, for certain diseases, can be used for a presumptive diagnosis. Negative reports may indicate the absence of specific diseases.

For rabies, surveillance and diagnosis of human and animal cases is severely constrained in much of the developing world, but animal-bite injury records from local hospitals have been used as primary sources of data for long-term epidemiological studies in the Serengeti ecosystem. These records have proven reliable indicators of animal rabies incidence and human exposures (Hampson et al. 2008), and have provided valuable data to inform models for estimating national disease burden (Cleaveland et al. 2002), evaluating the efficacy of control measures (Cleaveland et al. 2003; Beyer et al. 2011), and making inferences about spatial dynamics (Beyer et al. 2011).

They have also been used to initiate detailed contact tracing studies (Lembo et al. 2008; Hampson et al. 2009; see also below).

In wildlife protected areas, the abundance of scavengers, thick bush, and high ambient temperatures hinder the detection of sick and dead animals and the collection of fresh diagnostic samples. As a result, surveillance often relies upon unconfirmed reports of suspected cases. But even with only limited opportunities for laboratory confirmation, case-reporting through a network of veterinarians from Tanzania National Parks (TANAPA) and Tanzania Wildlife Research Institute (TAWIRI), government livestock offices, rangers, scientists, tourists, and tour operators has provided the principal data source for the monitoring of rabies (Lembo et al. 2008) and anthrax (Hampson et al. 2011; Lembo et al. 2011) in the Serengeti. Uncertainty associated with the lack of diagnostic confirmation has been addressed both through small-scale intensive studies to quantify disease recognition probability, such as for rabies (Lembo et al. 2008), and application of appropriate analytical techniques, as in anthrax (Hampson et al. 2011; Lembo et al. 2011).

It has been suggested that investments in enhancing passive case reporting through increased awareness and improved information flow frameworks could be the single most cost-effective way to improve disease detection in areas with limited surveillance infrastructure (Hoinville et al. 2009; Halliday et al. 2012). Key to the success of this approach is the prompt follow-up of reported cases for retrieval of viable diagnostic material and feedback of results, both of which provide critical motivation for continued reporting.

Contact tracing. Oral testimony from local communities through key informants or household questionnaire surveys and whenever possible, combined with testing/sampling of individuals, has been an important active surveillance tool in epidemiological studies in the Serengeti. In particular, community-based contact tracing measures have proven enormously effective in increasing rabies detection rates in selected areas (Lembo et al. 2008; Hampson et al. 2009). This approach, typically used for the control of infectious diseases of humans, as in sexually transmitted diseases (Potterat, Dukes, and Rothenberg 1987) or emerging diseases such as severe acute respiratory syndrome (SARS) (Riley et al. 2003; Fraser et al. 2004), involves tracing individuals who have had a potentially infectious contact with an infected person. Contact tracing is however not commonly employed for the study and control of infectious diseases of animals. Contact tracing is particularly applicable for rabies investigations because transmission events are discrete, memorable and are often observed. In the Serengeti it has been possible to take advantage of the characteristic nature of transmission to track the spread of infection and simultaneously generate detailed data on key epidemiological parameters (Hampson et al. 2009). The spatiotemporal detail recovered largely from cases in domestic animals has allowed statistical inferences to be drawn about transmission events in wildlife (Lembo et al. 2008; see also case study). From a practical perspective, contact tracing has led to the discovery of cases in wildlife and has enabled the collection of valuable samples that would otherwise have remained undetected. Nonetheless, contact tracing has its drawbacks: the proportion of wildlife cases detected is much lower than that of domestic animals, and there is no means to continue a line of investigation about a wildlife case in comparison to a domestic animal case where the owners and neighbors can be interviewed. Contact tracing is therefore unlikely to be of great utility for pathogens where wildlife hosts are responsible for the majority of transmission or where transmission is less visible.

Monitoring of population abundance and density. Given the difficulties of detecting mortality in free ranging wildlife, a decline in population size may be the only indicator of a disease outbreak. While decreasing population size may be caused by factors other than disease, such as changes in predation or food availability, observing a decrease in population size can provide a warning of potential disease threats and enable targeted surveillance to establish if a disease outbreak is the cause. Monitoring of animal numbers through time can therefore be a valuable part of the disease surveillance toolkit, with additional benefits of providing population data for ecological and management purposes. However, indices have to be estimated with sufficient precision and at a frequency appropriate to the host species and pathogen of interest, such that statistically and biologically significant changes (from baseline "normal" fluctuations) can be detected. Furthermore, while valuable for monitoring purposes, data can rarely be generated and analyzed at a sufficiently high frequency to allow for a timely management response, particularly for acute infections or short-lived epidemics.

For large mammals, approaches such as aerial or ground transects over a proportion of the protected area are commonly used to predict population sizes. For detection of disease outbreaks, which are often spatially and temporally localized, the precision of these estimates is usually only sufficient to detect major population changes. To estimate relative abundances of domestic and wild carnivore species and to investigate trends over time, night transects have been conducted on roads inside and outside the Serengeti National Park at monthly intervals (see figs. 15.3 and 15.4 in Craft et al., chapter 15). The count data generated by these transects provided useful indicators of abundance for small carnivore species such as bat-eared fox and white-tailed mongoose, which are not reliably detected by other survey techniques. However, the trade-off between accuracy and time/cost means that estimates of sufficient precision to detect mortality remain difficult for less common species.

Total counts, which provide the gold standard measure of abundance or density estimation, can be achieved by systematically counting every individual. These methods have provided valuable information for certain species; for example, total counts of buffalo allowed detailed analyses of the relative effects of rinderpest on these populations (Sinclair 1973; Sinclair 1977). Total counts can also be achieved through long-term behavioral studies, where individuals can be recognized (e.g., lion, cheetah, and wild dog studies). These studies are labor intensive and costly, and some species are difficult to observe, limiting the extent to which they can be studied (e.g., nocturnal species). However, intensive population monitoring can provide extremely valuable information. For instance, during an outbreak of canine distemper in 1994, lion mortality and morbidity was detected in large part due to the existence of a long-term lion behavioral study, which also enabled estimates of mortality through the disappearance of known individuals (Roelke-Parker et al. 1996; Munson et al. 2008). An intensive study on bat-eared foxes found that 60% of adult females ($n = 40$) were affected by rabies outbreaks in 1987 and 1988, the only information available on rabies mortality in this species (Maas 1993). When known individuals are followed longitudinally, it is even possible to measure key epidemiological parameters, particularly for chronic infections such as tuberculosis (Cross et al. 2009). Likewise sample banks of sera from known individuals also increase in value over time, and provide a window into past patterns of exposure and infection (Packer et al. 1999), as described in more detail below.

Laboratory-Guided Disease Detection

Presumptive diagnoses can be used for surveillance of some distinctive diseases, but for most pathogens, laboratory confirmation is required to confirm the presence of infection and disease. Methods for diagnosis of disease or confirmation of previous exposure to a given pathogen fall into several categories: (a) applying diagnostic pathology to characterize lesions associated with disease (e.g., histological and histochemical methods); (b) detecting the presence of the viable organism (e.g., culture, isolation, microscopy); (c) detecting the antigenic or genetic presence of a pathogen (immunohistochemical and molecular diagnostic methods); and (d) testing for antibodies to specific pathogens to detect previous infection (serological methods).

Biological samples may be obtained from live animals (using both invasive and noninvasive approaches) or from carcasses. Techniques for the collection and laboratory analysis of noninvasive samples (e.g. feces, saliva, urine, blood in excreted material) have generated infection data from great ape populations (Ashford, Reid, and Wrangham 2000; Murray et al. 2000; Lilly, Mehlman, and Doran 2002; Van Heuverswyn et al. 2006), and from Serengeti lions, as through coprological surveillance of lions' helminth status (Müller-Graf 1995; Müller-Graf, Woolhouse, and Packer 1999; Bjork, Averbeck, and Stromberg 2000). Despite advances in genetic technologies and the application of methods such as metabolomics to analysis of noninvasive samples for ecological studies, these techniques are still relatively underutilized for disease investigation in the Serengeti. Adapting and validating laboratory protocols for testing these types of samples from a range of species (required for much of the work in Serengeti) remains a major challenge. As a result, most data have relied on restraint and sampling of live animals. Domestic animal populations are generally accessible for safe and easy handling, but the need for immobilization makes wildlife sampling more difficult logistically, financially, and ethically. However, live sampling is often the sole means by which meaningful samples can be obtained from wildlife, and studies from the Serengeti have generated some of the most detailed information available on viral infection dynamics in wildlife (Packer et al. 1999; Munson et al. 2008).

Diagnostic pathology. Postmortem and histopathological (the study of the microscopic anatomy of tissues) examinations comprise a key element of all disease investigations. In the Serengeti, insights from pathology have been derived largely from materials generated through opportunistic carcass collection and passive surveillance. However, interpretation of results is often hampered by potential biases and data limitations. Bias may be introduced by nonrandom sampling of road kills or predated carcasses, which may represent a less healthy/vigorous section of the population (e.g., Thomson's gazelles affected by mange [*Sarcoptes scabiei*] appear to have a suppressed flight response and may be more vulnerable to predation or road-accidents). The poor quality and quantity of material from partly-scavenged and/or semidecomposed carcasses also hinders investigations. Invariably, opportunistic sampling results in patchy and incomplete data sets, though information generated is still valuable as in demonstrating the range of species infected with *Mycobacterium bovis* (the cause of bovine tuberculosis, bTB) (Cleaveland et al. 2005), seasonal and spatial patterns of anthrax outbreaks (Lembo et al. 2011), and for identifying new pathogens/diseases such as *Hepatozoon* infection in hyenas (East et al. 2008).

Clinical observations and gross pathology provide vital information, hence keeping accurate written and photographic records of clinical signs, postmortem presentation, and gross pathological lesions as part of routine carcass examination or targeted abattoir surveys is becoming extremely important. The presence of some pathogens, such as tuberculosis or hydatidosis, may be clear from postmortem examination alone. However, histopathological confirmation of lesions associated with disease is generally required for a definitive diagnosis. To distinguish "normal" from "abnormal" tissues in wildlife species, collection of standard specimens is particularly valuable. Although necropsies and sample collection should be ideally conducted by veterinarians or qualified personnel, appropriate samples can be obtained by field personnel with basic training, with the advantage that histopathology samples can be stored and preserved in formalin for subsequent laboratory analyses without the need for refrigeration.

Once in the laboratory, the histological anatomy of tissue samples and pathological changes within can be examined microscopically to confirm the presence and cause(s) of certain diseases, and staining can be used to enhance the differentiation of microscopic structures. Antigens (e.g., proteins) for a range of pathogens can be demonstrated using immunohistochemical identification by specific antibodies conjugated to an enzyme or tagged to a fluorescent compound to visualize the antibody-antigen interaction (see later). Specialized procedures are required for preparation of tissues (i.e., paraffin embedding) and slides, and accurate interpretation of histopathological changes is generally only possible when carried out by qualified and experienced histopathologists. On the contrary, simple microscopic techniques exist that enable the rapid diagnosis of certain diseases through the visualization of the pathogen or parts of it in blood (e.g., anthrax), tissue samples, or feces (e.g., helminths and intestinal protozoa).

Microbiological methods. While the causal agent of most diseases under investigation in the Serengeti is known, detection and culture of viable organisms remain an important tool to confirm the presence of pathogen in tissues and to obtain isolates for further analysis, such as whole genome sequencing (see later). For some diseases (e.g., tuberculosis and brucellosis), culture and isolation provide the principal means of definitive diagnosis. In vitro culture of parasites such as protozoa also allows their growth in sufficient quantities for immunodiagnosis and molecular genetics. In the Serengeti, culture of postmortem tissues has yielded valuable information on the presence and prevalence of infection of *M. bovis* in a range of Serengeti wildlife (Cleaveland et al. 2005) and highlighted difficulties of in vivo diagnosis of tuberculosis (T. Lembo, unpublished data). Furthermore, microbiological methods applied to the study of tuberculosis in Serengeti have demonstrated that "atypical" *Mycobacteria* species may be a much more important cause of disease in human and animal populations than previously recognized (Mfinanga et al. 2004; Cleaveland, Shaw et al. 2007).

Detecting the antigenic or genetic presence of a pathogen. Antigenic methods, such as immunohistochemistry and immunosorbent assays, and genetic molecular diagnostic techniques, such as DNA probes or polymerase chain reaction amplification (PCR), provide potential solutions for some of the problems associated with wildlife and field diagnosis.

Immunohistochemistry (the identification of antigens in diagnostic samples) has the advantage of being used on formalin-fixed tissues which can be easily preserved from field samples, as well as on tissue impressions. In the case of rabies, for example, a rapid technique (direct rapid immunohistochemical test, dRIT) for the application of immunohistochemistry to brain impressions and visualization of the antigen-antibody reaction by light microscopy has been developed (Niezgoda and Rupprecht 2006). The most important improvement of this technique, compared to the gold standard test for rabies diagnosis (fluorescent antibody test), is that it removes the need for specialized fluorescence microscope equipment, which is costly and necessitates high standards of maintenance, limiting the use in field conditions or laboratories with poor diagnostic infrastructure. Preliminary evaluation of the dRIT in the Serengeti highlighted its potential (Lembo et al. 2006), and local capacity for rabies diagnosis using the dRIT is currently being established in Tanzania. Immunodiagnosis can also be very effective in detecting *Echinococcus*-specific antigens directly in fecal samples of domestic dogs (Acosta-Jamett et al. 2010) as well as Serengeti wild carnivores, such as lions and cheetahs (E. Ernest, unpublished data).

As molecular techniques such as PCR identify pathogen genetic material rather than viable organisms, results can be obtained on sample material of lower quality or quantity and tests are generally much more sensitive than microbiological methods such as culture. With real-time PCR, quantitative information about the pathogen can also be obtained. These research techniques require investment in expensive laboratory equipment, but efforts are ongoing to develop field tests. For example, loop-mediated isothermal PCR (LAMP) for the detection of pathogens such as trypanosomes (Njiru et al. 2008) as well as viruses (Chen et al. 2010; Zhao et al. 2011), and helminths, as in the *Taenia* species (Nkouawa et al. 2009) are being considered. PCR techniques have provided a powerful tool for detecting and identifying known or new pathogens affecting wildlife and domestic animals in the Serengeti, including herpesviruses in wild carnivores (Ehlers et al. 2008) and zebra (Borchers et al. 2008); coronaviruses in hyenas (East et al. 2004);

and hemoparasites in domestic dogs (Barker et al. 2010), wild herbivores, lions (Fyumagwa et al. 2007), and their tick vectors (Fyumagwa et al. 2008; Fyumagwa et al. 2009). For trypanosomes in particular, the development of primers which can differentiate *Trypanosoma brucei rhodesiense* (which causes sleeping sickness in man) from its morphologically identical subspecies *Trypanosoma brucei brucei* (commonly found in wildlife and livestock but nonpathogenic in man) has been a big step in identifying potential reservoirs of sleeping sickness (see case study).

Serological assays. Serological assays detect antibodies to the pathogen of interest, indicating the presence of an immune response. Serology therefore provides a key method to assess exposure of wildlife populations to pathogens and can be of great value in: (1) understanding patterns of circulation retrospectively (Kock et al. 1998; Packer et al. 1999); (2) investigating the timing of infection if age-stratified seroprevalence data are available (Drakeley et al. 2005); (3) estimating the extent of outbreaks; (4) quantifying epidemiological parameters (Grenfell and Anderson 1985); (5) understanding factors influencing pathogenicity (Munson et al. 2008); and (6) measuring the effectiveness of control programs (Kock et al. 1998; Feliciangeli et al. 2003).

Long-term serological data sets available for the Serengeti lion population have been extensively used to evaluate temporal changes in exposure to a wide range of viruses and their effects on host fitness (Packer et al. 1999; Cleaveland, Mlengeya, et al. 2007; Munson et al. 2008). Serological profiles are uniquely linked to information on individual lions. The most important linked attribute is age, available as a result of intensive population monitoring or from observation of natural markings indicative of age, such as nose coloration (Whitman et al. 2004), which allows the timing of episodes of exposure to be determined with relative accuracy. Analyses of age-seroprevalence data in domestic dog populations (relying on age information provided by dog owners) have been equally useful and, combined with lion age-seroprevalence data, have enabled inferences about the spread of infection into and from wildlife (Cleaveland et al. 2000; Lembo 2007). Because snapshot studies of exposure can be misleading, these data are most valuable when collected over prolonged periods (Cleaveland, Mlengeya, et al. 2007).

Serology has also been valuable in highlighting variable degrees of morbidity and mortality following infection, such as those described earlier for CDV associated with *Babesia* coinfections (Packer et al. 1999; Munson et al. 2008). Similarly, wide variations in patterns of exposure to anthrax have been reported in Serengeti ungulates likely due to species-specific differences in susceptibility: zebra appear to be highly susceptible, whereas the presence of antibodies in 46% and 19% of Serengeti buffalo and wildebeest, respectively, suggests that nonfatal infection is not uncommon (Lembo et al. 2011). Such data also have potential to shed light on sublethal effects of pathogens (Jolles, Cooper, and Levin 2005), and elucidate interactions with the host immune system (Jolles et al. 2008).

Developing Effective Systems to Optimize the Use of Surveillance Data

Challenges presented by the limited infrastructure and availability of resources for effective disease surveillance in the Serengeti have stimulated interest in exploring innovative methods for improved data gathering and disease surveillance. These approaches include: (1) risk-based surveillance using sentinel/proxy populations; (2) syndromic surveillance using "sentinel" events; and (3) mobile computing technologies.

Sentinel populations often describe highly-susceptible populations that provide an early warning of infection or disease (the "canary in a coal mine" analogy). Proxy populations comprise host species that may not necessarily be of primary concern with respect to disease risk or epidemiology, but provide a useful indicator of whether infection or disease is present in an area. To be useful, these populations must be capable of developing a consistently detectable and specific response to the pathogen of interest, accessible, relatively inexpensive to monitor, and ideally sufficiently abundant for quantitative inference. Predators may be sensitive indicators, as infection prevalence is often higher than in prey because of consumption of infected carcasses, that is a suggested "bioaccumulation" effect (Cleaveland, Meslin, and Breiman 2006; Halliday et al. 2007). In addition, some predator and scavenger species are more cost-effective to sample. Among Serengeti carnivores, domestic dogs have proved to be useful proxies for detecting anthrax in areas where the disease had not been previously reported, determining the timing of outbreaks, and identifying high-risk areas for humans and livestock (Hampson et al. 2011; Lembo et al. 2011). Because of their close proximity to human, livestock and, in some areas, wildlife populations, and their abundance and accessibility for handling/sampling (large sample sizes of age-stratified data can be obtained more easily compared to wild carnivores), domestic dogs could be used as indicators of a range of other pathogens in the Serengeti and elsewhere. Among herbivores, buffalo show potential as useful wildlife indicators of the presence and prevalence of anthrax because they have high rates of seroconversion (Lembo et al. 2011) and, unlike wildebeest, their home ranges are relatively localized, potentially allowing high-risk areas to be identified (e.g., in combination with analysis of environmental determinants) (Hampson et al. 2011). Surveillance in buffalo can also be of value in monitoring a range of other pathogens, including bTB, foot-and-mouth disease (FMD), and rinderpest.

For livestock disease surveillance, interest is growing in the use of key syndromes, which can be easily reported to detect disease events. Around Serengeti, trials are under way to investigate surveillance approaches using abortion reporting (through interview data or mobile phone reporting systems) for early detection of important zoonoses such as Rift Valley fever, brucellosis, and Q fever. The use of pattern-recognition software also has considerable potential for detecting new constellations of signs/symptoms and the early detection of new emerging diseases (Perkins and Cousins 2007), but requires evaluation and validation for disease surveillance in Africa.

Over the last decade mobile phone use has been rising exponentially in Africa and the explosion in mobile computing technologies has great potential to improve disease surveillance and ecological monitoring infrastructure (Donner 2008; Mechael 2008; Srivastava 2008; Vital Wave Consulting 2009). Up until 2004, mobile phone coverage was nonexistent in the Serengeti, but now extends across most of the ecosystem and surrounding districts. Most simple mobiles are java-enabled so they can be easily programmed using open-source softwares for collection of questionnaire data. Using the standard mobile network, data can be transmitted directly to a centralized web-accessible server (Klungsöyr et al. 2008). General packet radio service (GPRS) technology means that large amounts of data can be collected and sent extremely cheaply, eliminating the requirement for paper-based forms and reducing the costs of data collection and transcription. The application of these technologies in the Serengeti is still at a very early stage, but pilot studies using mobile phones for data collection are under way both for rabies and for syndromic surveillance of livestock abortions and sudden death. The motivation behind these studies is the potential for generation of data in real time and for a more rapid and integrated surveillance and response system. Mobile phones are easy to use, cheap, and accessible to all members of the community. Therefore, livestock field officers, community-based animal health workers, trained community members, and even ecologists can play a role in collecting field data in otherwise remote locations. Uptake of these technologies, at least within local communities and government rather than purely within research projects, is only likely if their effectiveness can be proven. The novelty of technology can easily wear off without incentives for engagement. For infectious disease surveillance this most often means an active response to the reported disease problem. While mobile computing technologies can speed up the reporting process, lack of an active response is a more enduring problem associated with poor capacity and infrastructure (Hoinville et al. 2009). Nonetheless, these technologies provide a further opportunity to enhance passive surveillance and participatory techniques for data gathering.

Generating Ecological Data

Although disease surveillance data is obviously an integral part of understanding pathogen epidemiology, host-pathogen relationships must be considered within the wider ecological context. Disease circulation, persistence, and spread cannot be understood without also taking into account the relationships with (a) host ecology and demography, (b) vector species and their own often complex ecology, and (c) environmental and climatic factors.

Hosts. Abundance, distribution, demography, behavior, social structure, contact rates within and between species, and movement patterns are critical parameters for understanding disease dynamics in natural populations.

Data on the distribution and abundance of populations is clearly important for predicting where specific diseases are likely to occur and potential patterns of transmission and circulation. Presence/absence data can often provide fine-grained maps of host range (e.g., Tanzania bird atlas: http://tanzaniabirdatlas.com/) or indicators of host habitat selectivity (Durant et al. 2010). A drawback to presence data obtained through observations is that records are correlated with observer effort and absence data is difficult to confirm, although statistical techniques are available that help overcome this. Systematic use of motion-activated camera traps is a technique that avoids these biases (Pettorelli et al. 2009). Methods for abundance or density estimation are also available, as previously described (see section on monitoring of population abundance and density). The distribution of hosts can subsequently be inferred from these methods.

Parameters such as birth and death rates, age, and social structure have important effects on disease transmission. Demographic data are often obtained from behavioral studies of wild animals (Dobson 1988; Dobson 1995; Craft et al. 2008), and longitudinal and cross-sectional studies of domestic animals (Hampson et al. 2009). These parameters are also important when considering disease control since demographics influence the type or frequency of interventions. For example, knowledge of vital rates is useful for setting vaccination campaign intervals, as they determine the rate at which vaccination coverage declines.

When considering pathogen transmission, estimates for contact rates within species (for single host pathogens) and between species (for multi-host pathogens) are particularly valuable since they allow quantification of transmission potential. Although contact rates are notoriously difficult to measure, studies from the Serengeti have demonstrated the value of observational contact data, which are generally rare due to the intensive population monitoring that they require. Specifically, between-pride contact rates estimated from direct observations provided insight into the transmission dynamics of CDV in Serengeti lions (Craft et al. 2009). Focal follows of pride females and pride observations (number of pride-pride interactions per pride sighting) were used to estimate pride-pride and pride-nomad contact rates (nomads are nonterritorial lions). These data combined with other behavioral observations and lion movement data led to the conclusion that during the 1993–94 CDV outbreak (Roelke-Parker et al. 1996), prides were relatively well-mixed within the two-week viral infectious period. From the resulting network model it was possible to infer that, although lion prides are sufficiently well-connected to sustain chains of lion-to-lion transmission, in the fatal 1993–94 CDV outbreak, multiple hosts likely fueled the outbreak (Craft et al. 2009).

Movement and ranging data collected using VHF and GPS collars can be helpful to show potential for contact, as for animals ranging in and out of the park. But the presence of animals in the same area does not necessarily mean that contact occurs, as behavioral avoidance can happen at smaller spatiotemporal scales. Nonetheless, GPS collars can be useful for determining movements of healthy animals, as quantified in a contact network model (Craft et al. 2011), and allow individually known animals to be located, potentially generating information on other demographic parameters such as fecundity and longevity. Furthermore, if a collared animal dies (from disease or otherwise) there is a good chance of retrieving the carcass for postmortem and sample collection.

Vectors. For some pathogens an essential stage of development takes place within a vector, as in *Trypanosoma* spp. (tsetse—*Glossina* spp.), haemoparasites (Ixodid ticks), and Rift Valley fever virus (a number of mosquito species). In addition, some species act as mechanical vectors, carrying pathogens on their legs or mouthparts, such as blowflies and vultures which can disseminate *Bacillus anthracis*, the cause of anthrax, over a wide area (Braack and Devos 1990; Turnbull et al. 2008). The addition of a vector into a disease system introduces a new level of complexity. Parameters such as vector density, distribution, demography, behavior, and feeding preferences are as important in vector-borne disease epidemiology as the prevalence of the pathogen within vector or host populations. This is clearly illustrated for trypanosomiasis: the age structure of a tsetse population is a particularly important factor in human disease risk. It takes on average 18 days for *Trypanosoma brucei rhodesiense* to develop into a mature infection in tsetse that could be transmitted to a susceptible host (Dale et al. 1995). Mean tsetse life span is only a few weeks so only old flies present a risk in transmitting disease to man (Rogers 1988; Welburn and Maudlin 1999). Any intervention which affects vector survival can therefore have a substantial impact on human disease risk.

Abundance, density, and distribution of vector species can be measured by direct counts. Behavioral differences mean that techniques are usually specific to vector species, or even life stage: for example blue and black cloth traps are used to catch adult tsetse, and questing nymphal ticks can be counted by dragging a blanket over the ground. Host choice of vectors is important in vector-borne disease ecology since this determines onward transmission. For ticks, feeding preferences can be assessed by counting numbers present on individual animals of each species. For more mobile hematophagous vectors, a number of serological and genetic techniques are available to identify the host species present in blood meals. The principles of pathogen identification within a vector are similar to those in host species, with methods including direct visualization of pathogens by microscopic examination of vector organs, and molecular techniques such as PCR for identifying pathogen genetic material. Prevalence of transmissible infections can be used in conjunction with data on vector abundance and bite rates to give indicators of challenge, or disease risk.

Environment. Many Serengeti pathogens show spatial or temporal trends which can be related to environmental factors, including anthrax (rainfall, soil type), Rift Valley fever (rainfall), trypanosomiasis (vegetation, climate) and babesiosis (rainfall). Environmental influences are particularly evident for pathogens transmitted via a vector, where vector abundance and activity are related to factors such as vegetation and climate, or pathogens transmitted through environmental contamination (e.g., anthrax). However, environmental factors can also impact directly transmitted pathogens if they affect behavior and contact rates in ways which influence transmission, such as drought leading to animals clustering around water sources, increasing the risk of pathogen transmission, or alternatively geographic features like rivers and mountain ranges acting as barriers and reducing spread rates. Collection or assimilation of data such as rainfall, vegetation, and soil type allows these associations to be studied. This is particularly valuable for predicting times of high risk, or developing disease risk maps, which allows resources for surveillance and control to be targeted accordingly.

Approaches to Analysis

With continual improvements in computing power and statistical innovations, a wealth of sophisticated computational methodologies is becoming increasingly important and applicable in infectious disease ecology. Here we describe complementary analytical approaches, including phylogenetics, statistical inference, and mathematical modeling, which can be effectively integrated to address complex epidemiological questions. We provide examples of how the application of these approaches (alone or in combination) to the analysis of disease surveillance and ecology data from the Serengeti have led to insights into the dynamics of a range of pathogens.

Approaches to analysis of genetic and epidemiological data. Pathogens tend to have large population sizes, short generation times, and high mutation rates; hence their populations are characterized by relatively high levels of genetic diversity. The genetic changes resulting in this diversity serve as markers of population structure, enabling the relative relatedness of different pathogen genotypes to be determined through phylogenetic analysis. The very high mutation rates of viruses, particularly RNA viruses, result in measurable genetic diversity accumulating at single gene loci over just a few months or years. If the whole genome is sequenced, structure can be observed to arise in just days or weeks, and over the course of a handful of infections. Lower mutation rates of bacteria and protozoa have up to now limited insights to larger temporal and spatial scales, but with the advent of cheaper and faster full genome sequencing techniques, this is rapidly changing (Harris et al. 2010).

Early phylogenetic analyses of pathogens were used to determine their origin, as in Human Immunodeficiency Virus from chimpanzees infected with Simian Immunodeficiency Virus (Gao et al. 1992), and of hantavirus from rodents in the southwest United States (Nichol et al. 1993). Several technological step changes are revolutionizing the contributions that genetic data can make to our understanding of infectious disease ecology. These changes are the wider access and affordability of whole genome sequencing and the near ubiquitous access to high speed computing, coupled with the development of increasingly sophisticated models of evolution at the interface of phylogenetics and population genetics, along with the statistical methodology that enables these models to be fitted to data.

Consequently, modern phylogenetic methods can deliver a great deal more insights than a simple phylogeny. The development of models of evolution that include realistic "molecular clocks" (Drummond et al. 2006) makes it possible to date common ancestors of different pathogen genotypes, and the "root" of a phylogeny, thereby enabling estimation of the times at which particular variants arose (de Oliveira et al. 2006). Sophisticated methods of modeling the features of unsampled ancestral states (i.e., character states of determined clades which are also present in their ancestors) allow us to estimate the rate and direction of host switching in a phylogeny of a multihost pathogen, and thereby infer the most likely reservoir (Biek, Drummond, and Poss 2006), or even the spatial location of unsampled genotypes. The increasingly close integration of phylogenetics with population genetic and coalescent theory has enabled population level inferences to be made from purely genetic data. For example, it is possible to estimate the effective number of infections from sequence data, and even the reproductive number of the pathogen (Biek et al. 2007).

To date, insights about infectious diseases in the Serengeti obtained through phylogenetic analysis have been limited to only a few viral pathogens. Lembo et al. (2007) used rabies virus sequence data to show that circulating rabies genotypes in the greater Serengeti ecosystem are all of the Africa 1b group, and that the patterns of relatedness are consistent with high rates of between-species transmission. Similarly, phylogenetic analyses have confirmed that CDV variants from Serengeti carnivores were closely related and likely shared readily between species (Carpenter et al. 1998).

There is clearly a great deal more that can be done, especially with viral diseases like FMD, feline immunodeficiency virus, and MCF, as well as helminth infections, which is an area of research that still remains little explored in the Serengeti, limiting our understanding of the dynamics of macroparasitic diseases. Increasingly sophisticated sequencing approaches, coupled with high-throughput computational techniques, open the doors to a tremendous amount of scientific exploration, especially in relation to enhancing our understanding of biological diversity in complex ecosystems without prior information on what to look for. For instance, metagenomics combines the power of molecular biology and genetic analyses to assess the genetic diversity of entire communities of microbes/pathogens that exist in a given environment (e.g., to reveal which pathogens are present or most prevalent) without the need to isolate and culture individual species of pathogens (National Research Council 2007; Tang and Chiu 2010; Relman 2011). This is important given that the vast majority of microorganisms cannot be grown in the laboratory and therefore cannot be studied using classical microbiological methods, although approaches for interpretation of this kind of data, particularly with regard to the clinical significance of any microorganisms identified, are still under development.

When samples are collected at the finest spatial and temporal scales, their genetic relatedness can be used to infer the precise order of infections—

who infected whom, or a "transmission tree." An example of this in practice is the inference drawn on the spread of FMD infection during outbreaks in the United Kingdom (Cottam, Thebaud et al. 2008; Cottam, Wadsworth et al. 2008). Phylogenetic and transmission trees may be essentially identical, but because pathogens continue to evolve in an individual over the course of an infection (and usually prior to sampling), ambiguities may be introduced, and a single phylogeny can be consistent with several (often many) transmission trees.

Transmission trees can also be estimated from other types of data. If highly resolved spatiotemporal data can be acquired on the occurrence of cases, and knowledge of epidemiological parameters such as the generation time and the spatial infection kernel (time and distance between source cases and their resulting infection, respectively) is available, then statistical reconstruction of transmission trees can provide insights on the dynamics of infection. For pathogens of wildlife this technique is only practical when clinical signs are overt, coincide with transmission, and where there is little or no asymptomatic infection. In the Serengeti, epidemic reconstruction was used in animal populations in conjunction with epidemiological data obtained from contact tracing and provided support for the interpretation that rabies infection patterns are driven by dog-to-dog transmission, with no evidence that wildlife are capable of maintaining infection independently (Lembo et al. 2008; Hampson et al. 2009). Ultimately, transmission trees will be most effectively reconstructed by the combined use of data on the genotypes, location, timing, and epidemiology of the pathogen in question, but methods for the formal statistical integration of these different data types are still undergoing development.

Statistical inference using temporal and spatial data. The timing of disease occurrence can reveal a great deal about the dynamics of an infection including seasonal, demographic, stochastic, and immunological drivers. Methodologies for time series analyses of disease data range from the very simple (timing of exposure and identification of recurrent epidemics) to the extremely sophisticated (wavelet analyses investigating periodicity, and Bayesian and time series exposed-infectious-recover approaches for parameter fitting), but the utility of different approaches is largely dependent upon the quality and quantity of available data. Examples of simple but revealing analyses include the findings from archived serum samples that Serengeti lions had been repeatedly exposed to CDV long before the devastating epidemic in 1994 (Packer et al. 1999), and that domestic dogs in Serengeti district were the only population that had been exposed to CDV prior to the 1994 epidemic and were hence the most likely source of infection in this epidemic (Cleaveland et al. 2000). Similarly, relatively simple analyses of presumptive cases of anthrax in veterinary casebooks indicated an association of outbreaks with cumulative extremes in weather conditions (Hampson et al. 2011). The classic depiction of the release in ungulate populations from rinderpest through the vaccination of cattle, and consequent "knock-on" growth of carnivore populations, was also revealed as a result of time-series modeling using long-term population and disease monitoring (Holdo et al. 2009). But more generally, such complex time series methods have not been widely applied, because the resolution of data on disease in wildlife populations, as well as from animal and human populations in developing countries as a whole, is generally unsatisfactory. However, the wildlife in the Serengeti includes some of the most well-studied animal populations in the world, suggesting that the more sophisticated time series approaches could have exciting application.

Factors affecting disease transmission are frequently spatially localized. This is particularly true for vector-borne diseases where vector distribution depends on environmental attributes such as vegetation and climate. However, spatial heterogeneity is also evident in other disease systems and assessing the factors driving geographical variations is important both to understand disease epidemiology and to quantify disease risk and focus control. High-resolution satellite imagery and the development of powerful and accessible geographical information systems have allowed increasingly sophisticated analyses of spatial patterns. It has been possible, for instance, to demonstrate spatial heterogeneity for anthrax in Serengeti with suspected cases in wildlife and livestock occurring predominantly in pastoralist Maasai areas, which is also reflected in seroprevalence in domestic dogs (Lembo et al. 2011). Examination of the risk factors contributing to this pattern revealed a significant association between seropositivity in domestic dogs and buffalo and soil alkalinity (Hampson et al. 2011), likely because certain soil conditions provide a favorable environment for the survival of infective spores. Understanding these patterns may help in allocation of limited resources, such as guiding prophylactic livestock vaccination and public health education of human populations at risk, as well as informing wildlife management strategies.

Epidemiological modeling. Mathematical models can be used to gain insight into patterns of disease circulation, persistence, and spread, and have potential to inform control measures and predict future outbreaks, while shedding light on otherwise intractable hypotheses.

Traditional susceptible-infected-recovered (SIR) models often provide a surprisingly good approximation for disease dynamics despite their assumption of homogeneous mixing patterns. Their well-characterized structure considers rates of flow of individuals from susceptible, through infectious to recovered and potentially immune states, that can often be estimated from the known etiological characteristics of the pathogen (Anderson and May 1992). Heterogeneities in contact structure resulting from social structure, territoriality, age, geographical distance, and behavior, which are often more realistic assumptions for wildlife populations, can be incorporated into models in ways that can allow their relative importance to be examined. Network models, which explicitly incorporate heterogeneity in contact rate between individuals, have been successfully used for modeling disease dynamics in humans, and have provided the most insight into diseases where "super-spreading" is an important feature (Lloyd-Smith et al. 2005; Bansal, Grenfell, and Meyers 2007). Difficulties in quantifying contacts between individuals have however limited the use of these models in the study of infectious diseases of wildlife (McCallum 2009; Craft and Caillaud 2011). The Serengeti lion population provides an exception because of decades of behavioral observations of individually identified lions and their interactions (Craft et al. 2009). The application of network models to the study of CDV in Serengeti lions has allowed us to test the vulnerability of the network to pathogen spread, and infer best ways to target disease control (Craft et al. 2011).

A common misconception is that models are not useful without accurate parameter estimations. However, "best guesses" can be made for uncertain parameters, and sensitivity analyses can be used to test whether parameter uncertainty has any meaningful impact on model results. For example, in the network model of disease spread in lions, estimates of nomadic lion migration rates were based on small sample sizes. To test if migration rates were a crucial parameter, daily displacements were changed to unrealistically long distances and nomads were removed altogether from the model. Regardless of the nomad migration rate or the presence of nomads, disease dynamics in the lion system remained roughly the same, and hence the model was robust to this uncertainty (Craft 2010).

THE INTEGRATION OF TOOLS TO UNDERSTAND INFECTIOUS DISEASE DYNAMICS

Using rabies as an example, we show how major insights into pathogen dynamics in a range of populations can be achieved through effective integration of multiple methods for data collection and analysis spanning a range of disciplines and sectors. The application of this model to the study of other diseases, such as trypanosomiasis, will undoubtedly advance our understanding of other complex disease systems.

Rabies—an Example of Successful Integration of Approaches

Rabies is probably the best studied pathogen in the Serengeti ecosystem, as a consequence of its impact on public health and on the viability of endangered African wild dog populations. Almost all the tools discussed in this chapter have been applied to the study of rabies over the last 15 years (fig. 19.4a), from validating field diagnostics (Lembo et al. 2006), applying new methods of data collection (e.g., mobile computing technologies), engaging communities in participatory techniques, active surveillance, serological surveys (Cleaveland et al. 1999; Cleaveland et al. 2003; Lembo et al. 2008), phylogenetic analyses (Lembo et al. 2007), contact tracing (Hampson et al. 2009), epidemic reconstruction (Lembo et al. 2008; Hampson et al. 2009), metapopulation modeling, and new methods of statistical inference (Beyer et al. 2011). As a result, a great deal is known about the dynamics of infection and this has been translated directly into policy, with large scale rabies elimination demonstration projects now under way in Southern Tanzania, KwaZulu Natal, and the Philippines (WHO 2007). The study of rabies has many advantages over other pathogens; clinical signs are characteristic and the pathology is well understood and relatively straightforward. These attributes made it possible to investigate the challenging issue of disease reservoirs and conclude that domestic dogs are the key maintenance populations for rabies (Lembo et al. 2008). Nonetheless, retrieving samples from wildlife remained difficult, and even with detailed epidemiological data, statistical inference was complex. However, as a result there is now major potential to use rabies as a model system for developing a more mechanistic understanding of epidemiological processes such as transmission, and for bringing together detailed genetic and epidemiological data in new and insightful ways.

Exploring the Ecology of Trypanosomiasis in the Serengeti

Trypanosomes and their tsetse vectors have had a profound role in the history and ecology of the Serengeti (Ford 1971). Human African trypanosomiasis (HAT or sleeping sickness) is a debilitating disease which is fatal without treatment and imposes a major burden on the rural poor (WHO 2004). In addition, animal trypanosomiasis is the one of biggest economic threats to livestock-keeping in sub-Saharan Africa, with mortality losses and reductions in milk yield estimated at US $7.98 million annually in Tanzania alone (Daffa et al. 2005). Given the public health and economic

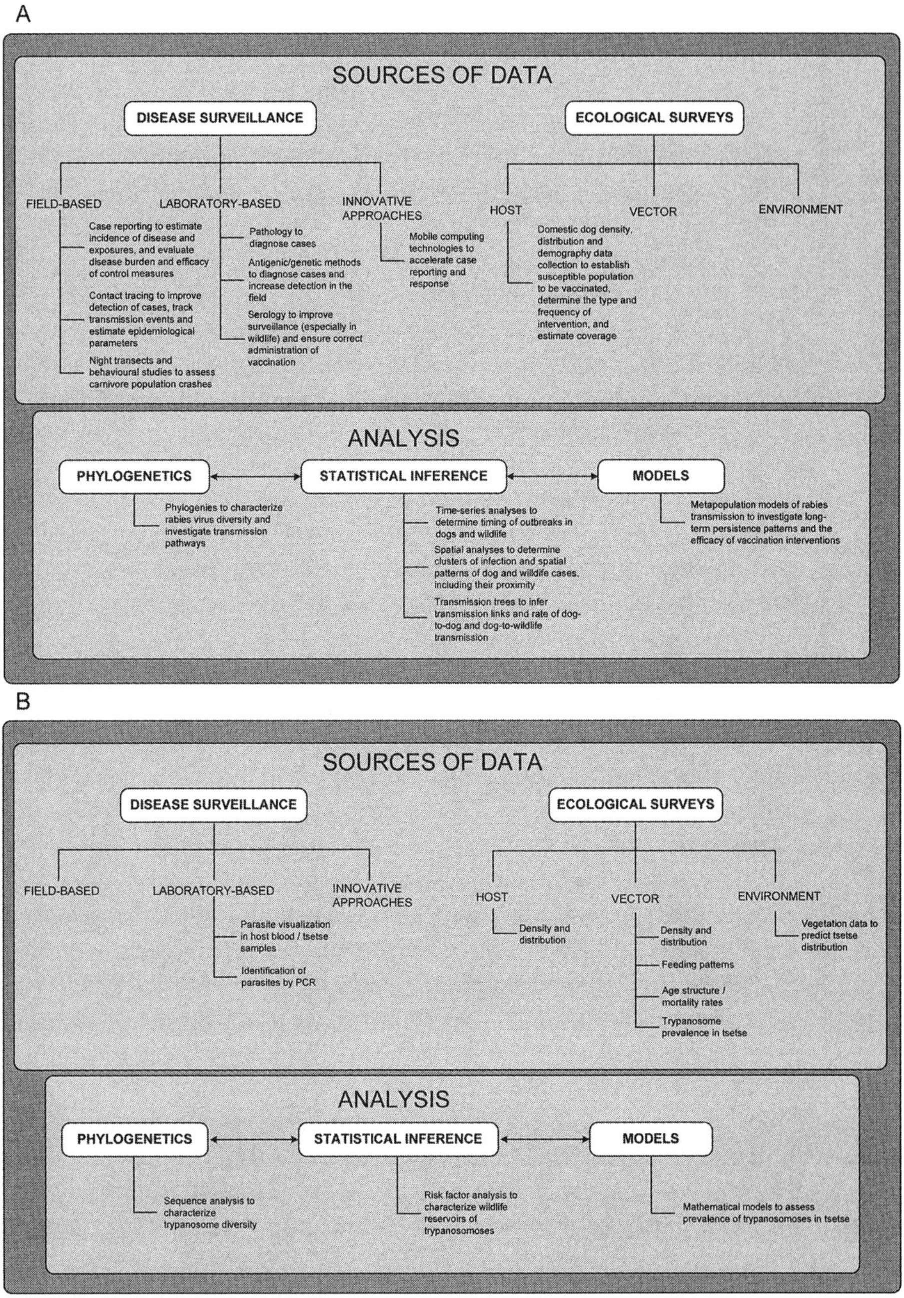

Fig. 19.4 Schema showing the tools used for rabies (A) and trypanosomiasis (B) research in the Serengeti.

importance of trypanosomiasis, and the century's worth of research conducted since trypanosomes were first identified in the blood stream of cattle and wildlife (Bruce 1895), it is perhaps surprising that we do not fully understand trypanosome epidemiology. However the disease is complex, with interactions of multiple host and vector species influencing transmission through their behavior and feeding patterns. Serengeti is an HAT focus (foci are specific geographical areas where HAT is endemic, with the potential for occasional epidemic activity, highlighted by sporadic cases in tourists and park staff over the last decade [Sinha et al. 1999; Jelinek et al. 2002]). Priority questions for understanding disease transmission and exploring options for control relate to the way that trypanosomes can persist in a focus over time; both livestock and wildlife can carry trypanosome infections but the relative roles of wildlife versus cattle, or of different wildlife species, in the maintenance and transmission of human trypanosome infections remain uncertain.

In contrast to rabies, trypanosome-infected wildlife hosts are difficult to detect as they often do not show clinical signs. Historically, identification of trypanosomes in animals has been conducted by microscopic examination of blood smears. While microscopy is cheap and easy to perform, it requires a parasitemia of 500 to 10^4 trypanosomes per ml of blood for parasites to be detected (Uilenberg 1998), which is a particular problem in wildlife species, which often show very low parasitemia. To improve sensitivity, concurrent subinoculation (inoculation of blood into other species, such as rodents) or xenodiagnosis (allowing susceptible tsetse to feed) was often used. These techniques were not only logistically and ethically difficult, but did not resolve a major constraint of microscopy: some trypanosomes are morphologically identical. In particular, *Trypanosoma brucei rhodesiense* (the causative agent of HAT) cannot be differentiated by microscopy from its subspecies *Trypanosoma brucei brucei* (which is found in wildlife and livestock, but does not cause disease in humans). This has limited investigations into the role of animals as reservoir hosts. The first studies to confirm the zoonotic nature of the disease relied on parasites isolated from animals causing sleeping sickness in human "volunteers," and showed that both wildlife and domestic cattle could carry *T. b. rhodesiense* (Heisch, McMahon, and Mansonbahr 1958; Onyango, Van Hoeve, and De Raadt 1966). The blood incubation infectivity test resolved the obvious ethical concerns of human infection, relying instead on the ability of *T. b. rhodesiense* to survive the trypanocidal effects of human serum in vitro (Rickman and Robson 1970), but often gave inconsistent results in field studies (Geigy, Mwambu, and Kauffmann 1971) because serum resistance demonstrated by *T. b. rhodesiense* was affected by passage through rodent hosts, an integral part of the technique (Targett and Wilson 1973). All these techniques have been used in Serengeti.

Both the constraints of low sensitivity and difficulties of trypanosome morphological identification have to an extent been addressed by the advent of molecular technology. Species-specific PCR primers have been developed for the main African trypanosome species and have a diagnostic sensitivity two to three times higher than microscopy (Solano et al. 1999; Picozzi et al. 2002). The discovery of the serum resistance associated (SRA) gene, responsible for the resistance of *T. b. rhodesiense* to human serum (Xong et al. 1998), finally allowed reliable differentiation between *T. b. rhodesiense* and *T. b. brucei* through the development of SRA PCR protocols (Welburn et al. 2001). While PCR-based techniques have not yet been developed into field tests and remain research tools, the sensitive and specific information they provide and the ease of processing large numbers of samples has led to progress in understanding the role that wildlife play in disease maintenance and transmission. PCR-based studies have confirmed the identification of *T. brucei* sensu lato (including both *T. brucei brucei* and *T. brucei rhodesiense*) trypanosomes in a wide range of wildlife species, with *T. b. rhodesiense* found in lion, hyena, bohor reedbuck, and warthog (Auty 2009; Kaare et al. 2007). Finding trypanosomes in a number of species suggests that rather than single reservoir species, a raft of species is making up a source of infection. The prevalence of *T. brucei* s.l. and *T. brucei rhodesiense* is unusually high in lions, possibly reflecting a different route of transmission, with infection accumulating via consumption of infected prey. PCR analysis has also revealed that coinfections with more than one trypanosome species are common, both in wildlife and in tsetse. Trypanosome sequence analysis confirmed a large number of trypanosome species circulating in wildlife, with *T. congolense*, *T. brucei* s.l., *T. vivax*, *T. simiae*, *T. simiae Tsavo*, and *T. godfreyi* all identified in Serengeti wildlife (Auty, Anderson et al. 2012). The proportion of tsetse carrying *T. brucei rhodesiense* is so low that it is difficult to measure accurately. Combining traditional dissection and microscopy approaches with molecular diagnostics, and using mathematical models to confirm that the low prevalence was consistent with the prevalence observed in wildlife gave a prevalence of less than 0.01% for *Glossina swynnertoni*, the main tsetse species in Serengeti (Auty, Picozzi et al. 2012). The nature of how *T. brucei* s.l. is maintained at such low prevalences is still unclear.

Most research so far has focused on establishing host range and generating estimates for prevalence in wildlife and tsetse populations (fig. 19.4B). We now have the tools to measure most of these parameters with reasonable certainty, but understanding how they fit together within this complex ecosystem requires expansion of our analytical toolbox, including (a) using sequence analysis and strain typing to look at the movement of trypanosomes between tsetse and wildlife, and see, for example, whether particular strains are adapted to different species; (b) developing mathematical models to study the dynamics of transmission, test the robustness of parameter estimates, and explore the effectiveness of potential control strategies; (c) conducting integrated analyses of prevalence of trypanosomes, tsetse feeding preferences, and the response of wildlife species to infection to assess the reservoir potential of individual species; and (d) analyses of spatial heterogeneity of human disease risk to help in allocation of limited resources for control.

CONCLUSIONS

Long-term studies in the Serengeti have provided important insights into how multiple approaches can be effectively deployed for generating and analyzing essential data on the dynamics and impacts of pathogens on different hosts and the concomitant effects on ecosystem function. This research has been used to inform conservation and public health policies and has served as a model for the development of ecosystem health approaches. The integrated efforts required for improved disease surveillance and ecological data gathering and analysis have linked a range of disciplines and sectors, which is an important achievement. Nonetheless, many practical and financial constraints still exist, highlighting the need for further investigating or developing creative, innovative approaches to disease surveillance and control, which make best use of the often-limited resources available. To ensure the design, implementation, and evaluation of cost-effective strategies, continuous synergistic interactions will be needed that integrate local communities, human and animal health workers, and scientists from various disciplines including epidemiologists, ecologists, social scientists, and economists. Adoption of any given strategy and allocation of sufficient resources will be ultimately dependent upon effective communication with and involvement of decision makers.

Box 19.1

Glossary of Technical Terms Used throughout the Chapter (Excluding Those Specifically Defined in the Main Text)

Age-prevalence (curve): *Prevalence* is the number of cases of a specific disease existing in a given population at a specific point in time divided by the number of individuals in the population at risk at the same point in time. An *age-prevalence curve* is a plot of prevalence against age.
Age-seroprevalence (curve): *Seroprevalence* is the number of individuals in a given population with antibodies (detected through serological testing) to the antigens of a particular pathogen or immunogen. Seroprevalence is often presented as a percent of the total samples tested. An *age-seroprevalence curve* is a plot of seroprevalence against age.
Antibody: Protein produced by the immune system to neutralize given *antigens* (e.g., microorganisms or chemicals) that are recognized as foreign and potentially harmful invaders.
Basic reproductive number: Average number of secondary cases which one case produces in a population consisting only of individuals susceptible to infection; that is, it determines whether or not a pathogen has the potential to spread in a host population.
Bayesian analysis: Method of statistical inference that incorporates previous information/knowledge ("prior" distribution) in statistical inference of a data set and uses the data at hand to update this distribution, converting it to a "posterior" distribution.
Case-fatality rate: Proportion of individuals with a given disease that die from it within a specified time period.
Coalescent theory: Retrospective model of population genetics that employs a sample of individuals from a population to trace all alleles of a gene shared by all members of a population to a single ancestral copy, the *most recent common ancestor* or coancestor.
Cross-sectional study: Collection of data on a population or sample of a population at one point in time.
Culture: Method to diagnose infectious disease by letting specific agents multiply in a predetermined growth medium.
Diagnostic pathology: Application of pathological methods to the examination of tissues or body fluids to determine causes of disease and structural and functional changes in abnormal conditions.
DNA probe: A DNA sequence designed to be complementary to a target se-

quence, so that it binds to the target sequence if it is present; for example, can be used to indicate the presence of a pathogen.

Gene locus: The specific place on a chromosome where a gene or DNA sequence is located.

Genome: Carrier of an organism's hereditary information, which in most organisms is encoded in DNA or, in many types of virus, in RNA.

Genotype: Genetic makeup of a cell, an organism, or an individual, usually with reference to a single trait, set of traits, or an entire complex of traits.

Gross pathology: Macroscopic manifestations of disease in organs, tissues, and body cavities observed during postmortem examination.

Immunodiagnosis: Diagnosis of disease based on antigen-antibody reactions.

Incidence: Rate of new cases of disease in a susceptible population over a defined period of time.

Isolation: Amplification of infectious disease agents in cultured cells, eggs, and laboratory animals.

Longitudinal study: Collection of data from repeated observations on the same individuals over a period of time.

Loop-mediated isothermal PCR (LAMP): Novel technique for amplification of genetic material for pathogen identification. As reactions occur at a single temperature (in contrast to PCR which requires temperature cycles), and amplification can be observed through changes in turgidity or colour, LAMP has the potential to become a tool for field diagnosis.

Maintenance host/population: Hosts/populations exceeding the minimum population size required for *disease persistence*, the *critical community size* (CCS). Populations smaller than the CCS (nonmaintenance populations) cannot maintain a pathogen independently, but together with other maintenance or non-maintenance populations, can constitute part of a reservoir system.

Mathematical modeling of disease: The process of constructing a mathematical representation of a disease system. These models can represent the transmission of disease by assigning values to the parameters which control the disease process, such as the rate at which new animals become infected or recover, or vector birth and death rates. It can be used to increase understanding of transmission dynamics, predict disease spread, or compare relative impacts of different control strategies.

Metagenomics: This term includes cultivation-independent genome-level characterization of communities or their members, high-throughput gene-level studies of communities with methods borrowed from genomics, and other "omics" studies (e.g., proteomics, transcriptomics, etc.), which are aimed at understanding transorganismal behaviors and the biosphere at the genomic level.

Metapopulation: "Population of populations" in which distinct subpopulations occupy spatially separated patches of habitat and interactions between patches occur at some level (i.e., dispersal as individuals move among patches).

Molecular clock: Method used to estimate divergence dates (of species or taxa) based on the assumption that sequence divergence accumulates at a roughly constant rate over time (in a clock-like fashion).

Mortality rate: Incidence rate of mortality (number of animals that die from all causes in a defined time period).

Mutation rate: Rate at which mutations arise at the DNA level, usually expressed as the number of nucleotide (or amino acid) changes per site, per replication cycle.

Network model: Type of mathematical model where individuals or groups can vary in the number of contacts that they have with other individuals or groups (heterogeneous contact structure), in contrast to compartment models where it is assumed that each individual has the same probability of contact (homogeneous contact structure).

Parasitaemia (bacteraemia, viraemia): The number of parasites (or bacteria or viruses) in the bloodstream, measured in parasites/ml.

Phylogenetics: The study of evolutionary relatedness between organisms through analysis of genetic sequence data and other characteristics.

Polymerase Chain Reaction (PCR): Technique by which a length of genetic sequence is amplified many times using specially designed primers. It is used in many molecular techniques, as in those that identify the presence of genetic material indicating pathogen presence, or to generate sufficient genetic material for further analysis such as identifying the genetic sequence.

Primer: A strand of nucleic acids designed to attach to a complementary sequence adjacent to a target sequence and used to start synthesis of genetic material in PCR.

Real-time PCR: Modification of PCR where fluorescent markers are used to monitor amplification in real time, allowing not only detection of genetic material but also quantification.

Reservoir: Set of epidemiologically connected populations that permanently maintain a pathogen and transmit infection to particular target populations considered to require protection (e.g., humans or endangered species).

Reservoir potential: For vector-borne diseases, the average number of infected vectors produced by an individual of a given species, used to estimate the relative contribution of different species to a reservoir system.
Sensitivity/specificity: The ability of a diagnostic test to correctly identify disease. Sensitivity is the proportion of true positives which are detected as positive. Specificity is the proportion of true negatives which are detected as negative.
Stochastic: A process where the outcome is determined both by predictable actions and by random or uncertain elements, so that the outcome is not known, but the probability of a particular outcome can be estimated.
Syndromic surveillance: Disease surveillance based on detection of patterns of clinical signs, even without a confirmed diagnosis. It can be used to detect known diseases, as in recording livestock abortions to detect potential Rift Valley fever activity, and may also detect unknown or emerging diseases if unusual patterns of clinical signs are detected.
Time-series analysis: Methods for analysis of data collected at intervals over a period of time. Analysis often has to allow for the fact that data points close together in time are likely to be more closely related than observations further apart, and be able to detect patterns at different time scales, such as seasonal cycles and/or overall trends over time.
Zoonosis: An infectious disease which is transmissible between animals and humans.

ACKNOWLEDGMENTS

We are grateful to the Tanzania Commission for Science and Technology, Tanzania Wildlife Research Institute, Tanzania National Parks, Ngorongoro Conservation Area Authority, and the National Institute for Medical Research for permission to undertake research; and the Viral Transmission Dynamics Project, Serengeti Lion and Cheetah Projects, Frankfurt Zoological Society, workers of the Tanzania Ministry of Livestock and Fisheries Development, and Ministry of Health and Social Welfare for their continuous assistance with field activities in the Serengeti and surrounding areas. Research findings presented herewith are based upon work supported by the joint National Institutes of Health (NIH)/ National Science Foundation (NSF) Ecology of Infectious Diseases Program under grant no. NSF/DEB0225453, Lincoln Park Zoo, the Wellcome Trust (including the Afrique One consortium), the Department for International Development Animal Health Programme, the Biotechnology and Biological Sciences Research Council (BBSRC), Google.org, and the Messerli Foundation. Lincoln Park Zoo (Conservation and Science Department) is also acknowledged for its scientific contributions to Serengeti research. Any opinions, findings, conclusions, or recommendations expressed in this material are those of the authors and do not necessarily reflect the views of the institutions with which the authors are affiliated or the funding bodies. We would like to extend our gratitude to Magai Kaare for his great contributions to Serengeti infectious disease research. His work on rabies in particular has shaped public health policies and has resulted in major international investment in canine rabies elimination in Tanzania. Magai's untimely death in a car accident has left a void in our team, but his research continues to inspire our work.

REFERENCES

Acosta-Jamett, G., S. Cleaveland, B. M. Bronsvoort, A. A. Cunningham, H. Bradshaw, and P. S. Craig. 2010. *Echinococcus granulosus* infection in domestic dogs in urban and rural areas of the Coquimbo region, north-central Chile. *Veterinary Parasitology* 169: 117–22.

Anderson, R. M., and R. M. May. 1992. *Infectious diseases of humans: Dynamics and control.* New York: Oxford University Press.

Ashford, R. W., G. D. F. Reid, and R. W. Wrangham. 2000. Intestinal parasites of the chimpanzee *Pan troglodytes* in Kibale forest, Uganda. *Annals of Tropical Medicine and Parasitology* 94:173–79.

Auty, H. 2009. Ecology of a vector-borne zoonosis in a complex ecosystem: Trypansomiasis in Serengeti, Tanzania. PhD diss., University of Edinburgh.

Auty, H., N. E. Anderson, K. Picozzi, T. Lembo, J. Mubanga, R. Hoare, R. D. Fyumagwa, B. Mable, L. Hamill, and S. Cleaveland, et al. 2012. Trypanosome diversity in wildlife species from the Serengeti and Luangwa valley ecosystems. *PLoS Neglected Tropical Diseases* 6:e1828.

Auty, H. K., K. Picozzi, I. Malele, S. J. Torr, S. Cleaveland, and S. Welburn. 2012. Using molecular data for epidemiological inference: Assessing the prevalence of *Trypanosoma brucei rhodesiense* in tsetse in Serengeti, Tanzania. *PLoS Neglected Tropical Diseases* 6 (1): e1501.

Bansal, S., B. T. Grenfell, and L. A. Meyers. 2007. When individual behaviour matters: Homogeneous and network models in epidemiology. *Journal of the Royal Society Interface* 4:879–91.

Barker, E. N., S. Tasker, M. J. Day, S. M. Warman, K. Woolley, R. Birtles, K. C. Georges, C. D. Ezeokoli, A. Newaj-Fyzul, and M. D. Campbell, et al. 2010. Development and use of real-time PCR to detect and quantify *Mycoplasma haemocanis* and "Candidatus *Mycoplasma haematoparvum*" in dogs. *Veterinary Microbiology* 140:167–70.

Bedelian, C., D. Nkedianye, and M. Herrero. 2007. Maasai perception of the impact and incidence of malignant catarrhal fever (MCF) in southern Kenya. *Preventive Veterinary Medicine* 78:296–316.

Beyer, H. L., K. Hampson, T. Lembo, S. Cleaveland, M. Kaare, and D. T. Haydon. 2011. Metapopulation dynamics of rabies, and the efficacy of vaccination. *Proceedings of the Royal Society B* 278:2182–90.

Biek, R., A. J. Drummond, and M. Poss. 2006. A virus reveals population structure and recent demographic history of its carnivore host. *Science* 311:538–41.

Biek, R., J. C. Henderson, L. A. Waller, C. E. Rupprecht, and L. A. Real. 2007. A high-resolution genetic signature of demographic and spatial expansion in epizootic rabies virus. *Proceedings of the National Academy of Sciences USA* 104:7993–98.

Bjork, K. E., G. A. Averbeck, and B. E. Stromberg. 2000. Parasites and parasite stages of free-ranging wild lions (*Panthera leo*) of northern Tanzania. *Journal of Zoo and Wildlife Medicine* 31:56–61.

Borchers, K., D. Lieckfeldt, A. Ludwig, H. Fukushi, G. Allen, R. Fyumagwa, and R. Hoare. 2008. Detection of equid herpesvirus 9 DNA in the trigeminal ganglia of a Burchell's zebra from the Serengeti ecosystem. *The Journal of Veterinary Medical Science* 70: 1377–81.

Braack, L. E. O., and V. Devos. 1990. Feeding habits and flight range of blow-flies (*Chrysomyia* Spp) in relation to Anthrax transmission in the Kruger National Park, South Africa. *Onderstepoort Journal of Veterinary Research* 57:141–42.

Breed, A. C., R. K. Plowright, D. T. S. Hayman, D. L. Knobel, F. M. Molenaar, D. Gardner-Roberts, S. Cleaveland, D. T. Haydon, R. A. Kock, and A. A. Cunningham, et al. 2009. Disease management in endangered mammals. In *Management of Disease in Wild Mammals*, ed. R. J. Delahay, G. C. Smith, and M. R. Hutchings, 215–39. Tokyo: Springer.

Bruce, D. 1895. *Preliminary report on the tsetse fly disease or nagana in Zululand*. Durban, South Afrrica: Bennett and Davis.

Carpenter, M. A., M. J. G. Appel, M. E. Roelke-Parker, L. Munson, H. Hofer, M. East, and S. J. O'Brien. 1998. Genetic characterization of canine distemper virus in Serengeti carnivores. *Veterinary Immunology and Immunopathology* 65:259–66.

Chen, Q., J. Li, X. E. Fang, and W. Xiong. 2010. Detection of swine transmissible gastroenteritis coronavirus using loop-mediated isothermal amplification. *Virology Journal* 7:206.

Cleaveland, S., M. G. J. Appel, W. S. K. Chalmers, C. Chillingworth, M. Kaare, and C. Dye. 2000. Serological and demographic evidence for domestic dogs as a source of canine distemper virus infection for Serengeti wildlife. *Veterinary Microbiology* 72: 217–27.

Cleaveland, S., J. Barrat, M. J. Barrat, M. Selve, M. Kaare, and J. Esterhuysen. 1999. A rabies serosurvey of domestic dogs in rural Tanzania: Results of a rapid fluorescent focus inhibition test (RFFIT) and a liquid-phase blocking ELISA used in parallel. *Epidemiology and Infection* 123:157–64.

Cleaveland, S., and C. Dye. 1995. Maintenance of a microparasite infecting several host species: Rabies in the Serengeti. *Parasitology* 111:S33–S47.

Cleaveland, S., E. M. Fevre, M. Kaare, and P. G. Coleman. 2002. Estimating human rabies mortality in the United Republic of Tanzania from dog bite injuries. *Bulletin of the World Health Organization* 80:304–10.

Cleaveland, S., M. Kaare, P. Tiringa, T. Mlengeya, and J. Barrat. 2003. A dog rabies vaccination campaign in rural Africa: Impact on the incidence of dog rabies and human dog-bite injuries. *Vaccine* 21:1965–73.

Cleaveland, S., L. Kusiluka, J. ole Kuwai, C. Bell, and R. Kazwala. 2001. *Assessing the impact of malignant catarrhal fever in Ngorongoro district, Tanzania*: Report for the Department for International Development, Animal Health Programme.

Cleaveland, S., F. X. Meslin, and R. Breiman. 2006. Dogs can play useful role as sentinel hosts for disease. *Nature* 440:605.

Cleaveland, S., T. Mlengeya, M. Kaare, D. Haydon, T. Lembo, M. K. Laurenson, and C. Packer. 2007. The conservation relevance of epidemiological research into carnivore viral diseases in the Serengeti. *Conservation Biology* 21:612–22.

Cleaveland, S., T. Mlengeya, R. R. Kazwala, A. Michel, M. T. Kaare, S. L. Jones, E. Eblate, G. M. Shirima, and C. Packer. 2005. Tuberculosis in Tanzanian wildlife. *Journal of Wildlife Diseases* 41:446–53.

Cleaveland, S., C. Packer, K. Hampson, M. Kaare, R. Kock, M. Craft, T. Lembo, T. Mlengeya, and A. Dobson. 2008. The multiple roles of infectious diseases in the Serengeti ecosystem. In *Serengeti III: Human impacts on ecosystem dynamics*, ed. A. R. E. Sinclair, C. Packer, S. A. R. Mduma, and J. M. Fryxell, 209–39. Chicago: University of Chicago Press.

Cleaveland, S., D. J. Shaw, S. G. Mfinanga, G. Shirima, R. R. Kazwala, E. Eblate, and M. Sharp. 2007. *Mycobacterium bovis* in rural Tanzania: Risk factors for infection in human and cattle populations. *Tuberculosis* 87:30–43.

Cottam, E. M., G. Thebaud, J. Wadsworth, J. Gloster, L. Mansley, D. J. Paton, D. P. King, and D. T. Haydon. 2008. Integrating genetic and epidemiological data to determine transmission pathways of foot-and-mouth disease virus. *Proceedings of the Royal Society, London, B* 275:887–95.

Cottam, E. M., J. Wadsworth, A. E. Shaw, R. J. Rowlands, L. Goatley, S. Maan, N. S. Maan, P. P. C. Mertens, K. Ebert, and Y. Li, et al. 2008. Transmission pathways of foot-and-mouth disease virus in the United Kingdom in 2007. *PloS Pathogens* 4:e1000050.

Craft, M. E. 2010. Ecology of infectious diseases in Serengeti lions. In *The biology and conservation of wild felids*, ed. D. Macdonald and A. Loveridge, 263–81. Oxford: Oxford University Press.

Craft, M. E., and D. Caillaud. 2011. Network models: An underutilized tool in wildlife epidemiology? *Interdisciplinary Perspectives on Infectious Diseases* 2011:676949.

Craft, M. E., P. L. Hawthorne, C. Packer, and A. P. Dobson. 2008. Dynamics of a multihost pathogen in a carnivore community. *Journal of Animal Ecology* 77:1257–64.

Craft, M. E., E. Volz, C. Packer, and L. A. Meyers. 2009. Distinguishing epidemic waves from disease spillover in a wildlife population. *Proceedings of the Royal Society London, B* 276:1777–85.

Craft, M. E., E. Volz, C. Packer, and L. A. Meyers. 2011. Disease transmission in territorial populations: The small-world network of Serengeti lions. *Journal of the Royal Society Interface* 8:776–86.

Cross, P. C., D. M. Heisey, J. A. Bowers, C. T. Hay, J. Wolhuter, P. Buss, M. Hofmeyr, A. L. Michel, R. G. Bengis, and T. L. F. Bird, et al. 2009. Disease, predation and demography: Assessing the impacts of bovine tuberculosis on African buffalo by monitoring at individual and population levels. *Journal of Applied Ecology* 46:467–75.

Daffa, J. W., P. Z. Njau, H. Mbwambo, and M. Byamungu. 2005. An overview of tsetse and trypanosomosis control in Tanzania 2003–2005. International Scientific Council for Trypanosomiasis Research and Control, 28th meeting, Addis Ababa, Ethiopia.

Dale, C., S. C. Welburn, I. Maudlin, and P. J. M. Milligan. 1995. The kinetics of maturation of trypanosome infections in tsetse. *Parasitology* 111:187–91.

de Oliveira, T., O. G. Pybus, A. Rambaut, M. Salemi, S. Cassol, M. Ciccozzi, G. Rezza, G. C. Gattinara, R. D'Arrigo, and M. Amicosante, et al. 2006. Molecular epidemiology—HIV-1 and HCV sequences from Libyan outbreak. *Nature* 444:836–37.

Dobson, A. P. 1988. The population biology of parasite-induced changes in host behavior. *The Quarterly Review of Biology* 63:139–65.

———. 1995. The ecology and epidemiology of rinderpest virus in Serengeti and Ngorongoro conservation area. In *Serengeti II: Dynamics, management and conservation of an ecosystem*, ed. A. R. E. Sinclair and P. Arcese, 485–505. Chicago: University of Chicago Press.

Donner, J. 2008. Research approaches to mobile use in the developing world: A review of the literature. *Information Society* 24:140–59.

Drakeley, C. J., P. H. Corran, P. G. Coleman, J. E. Tongren, S. L. R. McDonald, I. Carneiro, R. Malima, J. Lusingu, A. Manjurano, and W. M. M. Nkya, et al. 2005. Estimating medium- and long-term trends in malaria transmission by using serological markers of malaria exposure. *Proceedings of the National Academy of Sciences, USA* 102:5108–13.

Drummond, A. J., S. Y. W. Ho, M. J. Phillips, and A. Rambaut. 2006. Relaxed phylogenetics and dating with confidence. *PloS Biology* 4:e88.

Durant, S., M. E. Craft, C. Foley, K. Hampson, A. Lobora, M. Msuha, E. Eblate, J. Bukombe, J. Mchetto, and N. Pettorelli. 2010. Does size matter? An investigation of habitat use across a carnivore assemblage in the Serengeti. *Journal of Animal Ecology* 79:1012–22.

East, M. L., K. Moestl, V. Benetka, C. Pitra, O. P. Höner, B. Wachter, and H. Hofer. 2004. Coronavirus infection of spotted hyenas in the Serengeti ecosystem. *Veterinary Microbiology* 102:1–9.

East, M. L., G. Wibbelt, D. Lieckfeldt, A. Ludwig, K. Goller, K. Wilhelm, G. Schares, D. Thierer, and H. Hofer. 2008. A Hepatozoon species genetically distinct from *H. canis* infecting spotted hyenas in the Serengeti ecosystem, Tanzania. *Journal of Wildlife Diseases* 44:45–52.

Ehlers, B., G. Dural, N. Yasmum, T. Lembo, B. de Thoisy, M. P. Ryser-Degiorgis, R. G. Ulrich, and D. J. McGeoch. 2008. Novel mammalian *herpesviruses* and lineages within the *Gammaherpesvilinae*: Cospeciation and interspecies transfer. *Journal of Virology* 82:3509–16.

Feliciangeli, M. D., D. Campbell-Lendrum, C. Martinez, D. Gonzalez, P. Coleman, and C. Davies. 2003. Chagas disease control in Venezuela: Lessons for the Andean region and beyond. *Trends in Parasitology* 19:44–49.

Ford, J. 1971. *The role of the trypanosomiases in African ecology:A study of the tsetse fly problem*. Oxford: Oxford University Press.

Fraser, C., S. Riley, R. M. Anderson, and N. M. Ferguson. 2004. Factors that make an infectious disease outbreak controllable. *Proceedings of the National Academy of Sciences, USA* 101:6146–51.

Fyumagwa, R., V. Runyoro, I. G. Horak, and R. Hoare. 2007. Ecology and control of ticks as disease vectors in wildlife of the Ngorongoro crater, Tanzania. *South African Journal of Wildlife Research* 37:79–90.

Fyumagwa, R., P. Simmler, M. L. Meli, R. Hoare, R. Hofmann-Lehmann, and H. Lutz. 2009. Prevalence of anaplasma marginale in different tick species from Ngorongoro crater, Tanzania. *Veterinary Parasitology* 161:154–57.

Fyumagwa, R., P. Simmler, B. Willi, M. Meli, A. Sutter, R. Hoare, G. Dasen, R. Hofmann-Lehmann, and H. Lutz. 2008. Molecular detection of haemotropic Mycoplasma species in *Rhipicephalus sanguineus* tick species collected on lions (*Panthera leo*) from Ngorongoro crater, Tanzania. *South African Journal of Wildlife Research* 38:117–22.

Gao, F., L. Yue, A. T. White, P. G. Pappas, J. Barchue, A. P. Hanson, B. M. Greene, P. M. Sharp, G. M. Shaw, and B. H. Hahn. 1992. Human infection by genetically diverse SIVSM-related HIV-2 in West Africa. *Nature* 358:495–99.

Gascoyne, S. C., M. K. Laurenson, S. Lelo, and M. Borner. 1993. Rabies in African wild dogs (*Lycaon pictus*) in the Serengeti region, Tanzania. *Journal of Wildlife Diseases* 29: 396–402.

Geigy, R., P. M. Mwambu, and M. Kauffmann. 1971. Sleeping sickness survey in Musoma district, Tanzania. IV. Examination of wild mammals as a potential reservoir for *T. rhodesiense*. *Acta Tropica* 28:211–20.

Grenfell, B. T., and R. M. Anderson. 1985. The estimation of age-related rates of infection from case notifications and serological data. *Journal of Hygiene* 95:419–36.

Grenfell, B. T., and A. P. Dobson. 1995. *Ecology of infectious diseases in natural populations*. Cambridge: Cambridge University Press.

Halliday, J., C. Daborn, H. Auty, Z. Mtema, T. Lembo, B. M. Bronsvoort, I. Handel, D. Knobel, K. Hampson, and S. Cleaveland. 2012. Bringing together emerging and endemic zoonoses surveillance: Shared challenges and a common solution. *Philosophical Transactions of the Royal Society of London. Series B, Biological Sciences* 367:2872–80.

Halliday, J. E. B., A. L. Meredith, D. L. Knobel, D. J. Shaw, B. M. Bronsvoort, and S. Cleaveland. 2007. A framework for evaluating animals as sentinels for infectious disease surveillance. *Journal of the Royal Society Interface* 4:973–84.

Hampson, K., A. Dobson, M. Kaare, J. Dushoff, M. Magoto, E. Sindoya, and S. Cleaveland. 2008. Rabies exposures, post-exposure prophylaxis and deaths in a region of endemic canine rabies. *PloS Neglected Tropical Diseases* 2:e339.

Hampson, K., J. Dushoff, S. Cleaveland, D. T. Haydon, M. Kaare, C. Packer, and A. Dobson. 2009. Transmission dynamics and prospects for the elimination of canine rabies. *PLoS Biology* 7 (3): e53.

Hampson, K., T. Lembo, P. Bessell, H. Auty, C. Packer, J. Halliday, C. A. Beesley, R. Fyumagwa, R. Hoare, and E. Ernest, et al. 2011. Predictability of anthrax infection in the Serengeti, Tanzania. *Journal of Applied Ecology* 48:1333–44.

Harris, S. R., E. J. Feil, M. T. G. Holden, M. A. Quail, E. K. Nickerson, N. Chantratita, S. Gardete, A. Tavares, N. Day, and J. A. Lindsay, et al. 2010. Evolution of MRSA during hospital transmission and intercontinental spread. *Science* 327:469–74.

Heisch, R. B., J. P. McMahon, and P. E. C. Mansonbahr. 1958. The isolation of *Trypanosoma rhodesiense* from a bushbuck. *British Medical Journal* 2:1203–04.

Hoinville, L., F. Tomley, L. Wootton, and A. Cook. 2009. New approaches to surveillance. *Veterinary Record* 164:413–14.

Holdo, R. M., A. R. E. Sinclair, A. P. Dobson, K. L. Metzger, B. M. Bolker, M. E. Ritchie, and R. D. Holt. 2009. A disease-mediated trophic cascade in the Serengeti and its implications for ecosystem C. *PLoS Biol* 7 (9): e1000210.

Hudson, P. J., A. Rizzoli, B. T. Grenfell, H. Heesterbeek, and A. P. Dobson. 2002. *The ecology of wildlife diseases*. Oxford: Oxford University Press.

Jelinek, T., Z. Bisoffi, L. Bonazzi, P. van Thiel, U. Bronner, A. de Frey, S. G. Gundersen, P. McWhinney, and D. Ripamonti. 2002. Cluster of African trypanosomiasis in travellers to Tanzanian national parks. *Emerging Infectious Diseases* 8:634–35.

Jolles, A. E., D. Cooper, and S. A. Levin. 2005. Hidden effects of chronic tuberculosis in African buffalo. *Ecology* 86:2358–64.

Jolles, A. E., V. O. Ezenwa, R. S. Etienne, W. C. Turner, and H. Olff. 2008. Interactions between macroparasites and microparasites drive infection patterns in free-ranging African buffalo. *Ecology* 89:2239–50.

Kaare, M., K. Picozzi, T. Mlengeya, E. Fevre, L. Mellau, M. Mtambo, S. Cleaveland, and S. Welburn. 2007. Sleeping sickness—a re-emerging disease in Serengeti? *Travel Medicine and Infectious Disease* 5:117–24.

Kissui, B. M., and C. Packer. 2004. Top-down population regulation of a top predator: Lions in the Ngorongoro crater. *Proceedings of the Royal Society, B* 271:1867–74.

Klungsöyr, J., P. Wakholi, B. Macleod, A. Escudero-Pascual, and N. Lesh. 2008. OpenROSA, JavaROSA, GloballyMobile—Collaborations around open standards for mobile applications. Proceedings of the 1st international conference on M4D mobile communication technology for development (M4D 2008, General Tracks), December 11–12, 2008, Karlstad University, Sweden.

Kock, R., W. S. K. Chalmers, J. Mwanzia, C. Chillingworth, J. Wambua, P. G. Coleman, and W. Baxendale. 1998. Canine distemper antibodies in lions of the Masai Mara. *Veterinary Record* 142:662–65.

Lembo, T. 2007. An investigation of disease reservoirs in complex ecosystems: Rabies and canine distemper in the Serengeti. PhD diss., University of Edinburgh.

Lembo, T., K. Hampson, H. Auty, C. A. Beesley, P. Bessell, C. Packer, J. Halliday, R. Fyumagwa, R. Hoare, and E. Ernest, et al. 2011. Serologic surveillance of anthrax in the Serengeti ecosystem, Tanzania, 1996–2009. *Emerging Infectious Diseases* 17:387–94.

Lembo, T., K. Hampson, D. T. Haydon, M. Craft, A. Dobson, J. Dushoff, E. Ernest, R. Hoare, M. Kaare, and T. Mlengeya, et al. 2008. Exploring reservoir dynamics: A case study of rabies in the Serengeti ecosystem. *Journal of Applied Ecology* 45:1246–57.

Lembo, T., D. T. Haydon, A. Velasco-Villa, C. E. Rupprecht, C. Packer, P. E. Brandao, I. V. Kuzmin, A. R. Fooks, J. Barrat, and S. Cleaveland. 2007. Molecular epidemiology identifies only a single rabies virus variant circulating in complex carnivore communities of the Serengeti. *Proceedings of the Royal Society of London Series B* 274:2123–30.

Lembo, T., M. Niezgoda, A. Velasco-Villa, S. Cleaveland, E. Ernest, and C. E. Rupprecht. 2006. Evaluation of a direct, rapid immunohistochemical test for rabies diagnosis. *Emerging Infectious Diseases* 12:310–13.

Lilly, A. A., P. T. Mehlman, and D. Doran. 2002. Intestinal parasites in gorillas, chimpanzees, and humans at Mondika research site, Dzanga-Ndoki National Park, Central African Republic. *International Journal of Primatology* 23:555–73.

Lloyd-Smith, J. O., S. J. Schreiber, P. E. Kopp, and W. M. Getz. 2005. Superspreading and the effect of individual variation on disease emergence. *Nature* 438:355–59.

Maas, B. 1993. Behavioural ecology and social organisation of the bat-eared fox in the Serengeti National Park, Tanzania, PhD diss., University of Cambridge.

McCallum, H. 2009. Six degrees of *Apodemus* separation. *Journal of Animal Ecology* 78: 891–93.

Mechael, P. 2008. Health services and mobiles: A case from Egypt. In *Handbook of mobile communication studies*, ed. J. Katz, 91–103. Cambridge, MA: The MIT Press.

Mfinanga, S. G. M., O. Morkve, R. R. Kazwala, S. Cleaveland, M. J. Sharp, J. Kunda, and R. Nilsen. 2004. Mycobacterial adenitis: Role of *Mycobacterium bovis*, non-tuberculous mycobacteria, HIV infection, and risk factors in Arusha, Tanzania. *East African Medical Journal* 81:171–78.

Müller-Graf, C. D. 1995. A coprological survey of intestinal parasites of wild lions (*Panthera leo*) in the Serengeti and the Ngorongoro crater, Tanzania, East Africa. *Journal of Parasitology* 81:812–14.

Müller-Graf, C. D., M. E. Woolhouse, and C. Packer. 1999. Epidemiology of an intestinal parasite (Spirometra spp.) in two populations of African lions (*Panthera leo*). *Parasitology* 118:407–15.

Munson, L., K. A. Terio, R. Kock, T. Mlengeya, M. E. Roelke, E. Dubovi, B. Summers, A. R. E. Sinclair, and C. Packer. 2008. Climate extremes promote fatal co-infections during canine distemper epidemics in African lions. *PloS One* 3:e2545.

Murray, S., C. Stem, B. Boudreau, and J. Goodall. 2000. Intestinal parasites of baboons (*Papio cynocephalus anubis*) and chimpanzees (*Pan troglodytes*) in Gombe National Park. *Journal of Zoo and Wildlife Medicine* 31:176–78.

National Research Council (US) Committee on Metagenomics: Challenges and Functional Applications. 2007. *The new science of metagenomics—Revealing the secrets of our microbial planet.* The National Academies Collection: Reports funded by National Institutes of Health. Washington, DC: The National Academies Press.

Nichol, S. T., C. F. Spiropoulou, S. Morzunov, P. E. Rollin, T. G. Ksiazek, H. Feldmann, A. Sanchez, J. Childs, S. Zaki, and C. J. Peters. 1993. Genetic identification of a hantavirus associated with an outbreak of acute respiratory illness. *Science* 262:914–17.

Niezgoda, M., and C. E. Rupprecht. 2006. Standard operating procedure for the direct rapid immunohistochemistry test for the detection of rabies virus antigen. In *National laboratory training network course.* Atlanta: US Department of Health and Human Services, Centers for Disease Control and Prevention.

Njiru, Z. K., A. S. J. Mikosza, E. Matovu, J. C. K. Enyaru, J. O. Ouma, S. N. Kibona, R. C. A. Thompson, and J. M. Ndung'u. 2008. African trypanosomiasis: Sensitive and rapid detection of the sub-genus Trypanozoon by loop-mediated isothermal amplification (LAMP) of parasite DNA. *International Journal for Parasitology* 38:589–99.

Nkouawa, A., Y. Sako, M. Nakao, K. Nakaya, and A. Ito. 2009. Loop-mediated isothermal amplification method for differentiation and rapid detection of Taenia species. *Journal of Clinical Microbiology* 47:168–74.

OIE. 2010. *Terrestrial animal health code, 19th ed.* World Organisation for Animal Health, Paris.

Onyango, R. J., K. Van Hoeve, and P. De Raadt. 1966. The epidemiology of Trypanosoma rhodesiense sleeping sickness in Alego location, Central Nyanza, Kenya. I. Evidence that cattle may act as reservoir hosts of trypanosomes infective to man. *Transactions of the Royal Society of Tropical Medicine and Hygiene* 60:175–82.

Packer, C., S. Altizer, M. Appel, E. Brown, J. Martenson, S. J. O'Brien, M. Roelke-Parker, R. Hofmann-Lehmann, and H. Lutz. 1999. Viruses of the Serengeti: Patterns of infection and mortality in African lions. *Journal of Animal Ecology* 68:1161–78.

Perkins, N., and D. Cousins. 2007. Remote area surveillance workshop: Stakeholder review of Bovine Syndromic Surveillance System (BOSS). Final report, Australian Biosecurity CRC. Bardon Centre, Queensland.

Pettorelli, N., A. L. Lobora, M. J. Msuha, C. Foley, and S. M. Durant. 2009. Carnivore biodiversity in Tanzania: Revealing the distribution patterns of secretive mammals using camera traps. *Animal Conservation* 13:131–39.

Picozzi, K., A. Tilley, E. M. Fevre, P. G. Coleman, J. W. Magona, M. Odiit, M. C. Eisler, and

S. C. Welburn. 2002. The diagnosis of trypanosome infections: Applications of novel technology for reducing disease risk. *African Journal of Biotechnology* 1:39–45.

Potterat, J. J., R. L. Dukes, and R. B. Rothenberg. 1987. Disease transmission by heterosexual men with gonorrhea: An empiric estimate. *Sexually Transmitted Diseases* 14: 107–10.

Relman, D. A. 2011. Microbial genomics and infectious diseases. *The New England Journal of Medicine* 365:347–57.

Rickman, L. R., and J. Robson. 1970. The testing of proven *Trypanosoma brucei* and *T. rhodesiense* strains by the blood incubation infectivity test. *Bulletin of the World Health Organization* 42:911–16.

Riley, S., C. Fraser, C. A. Donnelly, A. C. Ghani, L. J. Abu-Raddad, A. J. Hedley, G. M. Leung, L. M. Ho, T. H. Lam, and T. Q. Thach, et al. 2003. Transmission dynamics of the etiological agent of SARS in Hong Kong: Impact of public health interventions. *Science* 300:1961–66.

Roelke-Parker, M. E., L. Munson, C. Packer, R. Kock, S. Cleaveland, M. Carpenter, S. J. Obrien, A. Pospischil, R. Hofmann-Lehmann, and H. Lutz, et al. 1996. A canine distemper virus epidemic in Serengeti lions (*Panthera leo*). *Nature* 379:441–45.

Rogers, D. J. 1988. A general model for the African trypanosomiases. *Parasitology* 97: 193–212.

Sinclair, A. R. E. 1973. Regulation, and population models for a tropical ruminant. *East African Wildlife Journal* 11:307–16.

———. 1977. *The African Buffalo: A Study of Resource Limitation of Populations*. Chicago: University of Chicago Press.

Sinha, A., C. Grace, W. K. Alston, F. Westenfeld, and J. H. Maguire. 1999. African trypanosomiasis in two travellers from the United States. *Clinical Infectious Diseases* 29:840–44.

Solano, P., J. F. Michel, T. LeFrancois, S. de La Rocque, I. Sidibe, A. Zoungrana, and D. Cuisance. 1999. Polymerase chain reaction as a diagnosis tool for detecting trypanosomes in naturally infected cattle in Burkina Faso. *Veterinary Parasitology* 86:95–103.

Srivastava, L. 2008. The mobile makes its mark. In *The handbook of mobile communications studies*, ed. J. Katz, 15–27. Cambridge, MA: The MIT Press.

Tang, P., and C. Chiu. 2010. Metagenomics for the discovery of novel human viruses. *Future Microbiology* 5:177–89.

Targett, G. A., and V. C. Wilson. 1973. The blood incubation infectivity test as a means of distingushing between *Trypanosoma brucei brucei* and *T. brucei rhodesiense. International Journal for Parasitology* 3:5–11.

Turnbull, P. C. B., M. Diekmann, J. W. Kilian, W. Versfeld, V. De Vos, L. Arntzen, K. Wolter, P. Bartels, and A. Kotze. 2008. Naturally acquired antibodies to *Bacillus anthracis* protective antigen in vultures of southern Africa. *Onderstepoort Journal of Veterinary Research* 75:95–102.

Uilenberg, G. 1998. *A field guide for the diagnosis, treatment and prevention of African animal trypanosomiasis*. Rome: Food and Agriculture Organization of the United Nations.

Van Heuverswyn, F., Y. Li, C. Neel, E. Bailes, B. F. Keele, W. Liu, S. Loul, C. Butel, F. Liegeois, and Y. Bienvenue, et al. 2006. Human Immunodeficiency Viruses—SIV infection in wild gorillas. *Nature* 444:164.

Vital Wave Consulting. 2009. *mHealth for development: the opportunity of mobile technology for healthcare in the developing world*: Washington, DC: UN Foundation-Vodafone Foundation partnership.

Welburn, S. C., and I. Maudlin. 1999. Tsetse-typanosome interactions: Rites of passage. *Parasitology Today* 15:399–403.

Welburn, S. C., K. Picozzi, E. M. Fevre, P. G. Coleman, M. Odiit, M. Carrington, and I. Maudlin. 2001. Identification of human-infective trypanosomes in animal reservoir of sleeping sickness in Uganda by means of serum-resistance-associated (SRA) gene. *Lancet* 358:2017–19.

Whitman, K., A. M. Starfield, H. S. Quadling, and C. Packer. 2004. Sustainable trophy hunting of African lions. *Nature* 428:175–78.

WHO. 2004. *The world health report 2004—changing history*. Geneva: World Health Organization.

———. 2007. Bill & Melinda Gates Foundation fund WHO-coordinated project to control and eventually eliminate rabies in low-income countries. Accessed March 25, 2013. http://www.who.int/rabies/bmgf_who_project/en.

WHO/DFID. 2006. The control of neglected zoonotic diseases—a route to poverty alleviation. Report of a joint WHO/DFID-Animal Health Programme meeting with the participation of FAO and OIE, Geneva, September 20–21, 2005. Geneva: The World Health Organization. http://www.who.int/zoonoses/Report_Sept06.pdf.

Xong, H. V., L. Vanhamme, M. Chamekh, C. E. Chimfwembe, J. Van Den Abbeele, A. Pays, N. Van Meirvenne, R. Hamers, P. De Baetselier, and E. Pays. 1998. A VSG expression site-associated gene confers resistance to human serum in *Trypanosoma rhodesiense*. *Cell* 95:839–46.

Zhao, K., W. Shi, F. Han, Y. Xu, L. Zhu, Y. Zou, X. Wu, H. Zhu, F. Tan, and S. Tao, et al. 2011. Specific, simple and rapid detection of porcine circovirus type 2 using the loop-mediated isothermal amplification method. *Virology Journal* 8:126.

PART IV

Coupled Human-Natural Interactions

TWENTY

Socioecological Dynamics and Feedbacks in the Greater Serengeti Ecosystem

Ricardo M. Holdo and Robert D. Holt

Ecologists often consider the effects of human activity on national parks and other protected areas as a one-way process, with the flow of information directed primarily inward. As a result, habitat loss, hunting pressure, disease outbreaks, fires, and other factors are often treated as fixed exogenous effects immune to feedback from the ecological systems they affect (Campbell and Hofer 1995; Pascual and Hilborn 1995), either for simplicity or because data on relevant feedbacks are lacking. Similarly, socioeconomic models of local human economies often make simplistic assumptions about the population dynamics of wildlife species (e.g., assuming logistic growth with an invariant carrying capacity) and about ecological processes in general (Bulte and Horan 2003; Horan and Bulte 2004; Costello et al. 2008). These approaches are useful as a first approximation, but over the past decade or so, both ecologists and social scientists have increasingly taken into consideration the importance of feedbacks between human-dominated and protected areas in socioecological models, treating both ecological and social systems with the same degree of dynamic complexity (Janssen et al. 2000; Walker and Janssen 2002; Milner-Gulland et al. 2006). These feedbacks can be particularly strong in subsistence societies where activities such as rain-fed agriculture, livestock herding, and harvesting of natural products (bushmeat, honey, firewood, etc.) form the basis of the local economy, as is the case in many of the agropastoral societies of sub-Saharan Africa (Galvin et al. 2006; Holdo et al. 2010). In such systems, economic activity is more tightly linked to the underlying ecology of the landscape and therefore less buffered against climatic or ecological perturbations than might be the case in communities that can rely on irrigation, tourism, or manufacturing, for example.

The system of protected areas that comprise much of the Serengeti ecosystem is in a privileged position among its peers worldwide because its boundaries preserve a fundamental ecological process, the wildebeest migration (Thirgood et al. 2004). This has rarely been the case in other ecosystems where fencing and habitat loss have largely led to the demise of most large ungulate migrations (Thirgood et al. 2004; Bolger et al. 2008; Harris et al. 2009). To a greater extent than most comparable systems it is an ecologically coherent, self-contained ecosystem, less reliant on intensive management for the maintenance of key processes than might be the case for protected areas elsewhere, where mass animal movements have become impeded. At the same time, the historical record shows that human activity has had strong detrimental effects on large animal populations in the Serengeti (Arcese, Hando, and Campbell 1995; Campbell and Hofer 1995; Hilborn et al. 2006), and the potential for such effects is likely to increase as the human population to the west of the Serengeti continues to grow and put pressure on the park ecosystem (Campbell and Hofer 1995).

THE SERENGETI AS A SOCIOECOLOGICAL SYSTEM (SES)

From an ecological standpoint, the Serengeti can be modeled either as an ecological system subject to constant exogenous influences (for example, a nonzero but temporally fixed offtake from hunting; fig. 20.1A), or as an SES with protected and unprotected areas that, in addition to having their own internal dynamics (the circular arrows in fig. 20.1), also exert feedbacks across the protected area boundary (fig. 20.1B). Underpinning the dynamics of the whole system is the influence of rainfall, which acts as the key forcing function for both ecological and socioeconomic processes in East Africa (Homewood et al. 2001; Barrett et al. 2003; Boone, Thirgood, and Hopcraft 2006). Rainfall drives primary productivity, and directly and indirectly affects herbivore populations and movement patterns and fire occurrence (Boone, Thirgood, and Hopcraft 2006; Holdo, Holt, and Fryxell 2009a, 2009b). On the human side, the lack of irrigation and dams makes crop production systems and livestock populations strongly susceptible to drought (Dercon 1996).

As is the case with most protected systems, the ecology of the Serengeti ecosystem is vulnerable to human impacts from the settled areas that surround the park (Campbell and Hofer 1995). In the past, most of the attention on these effects has focused on wildlife hunting (Pascual and Hilborn

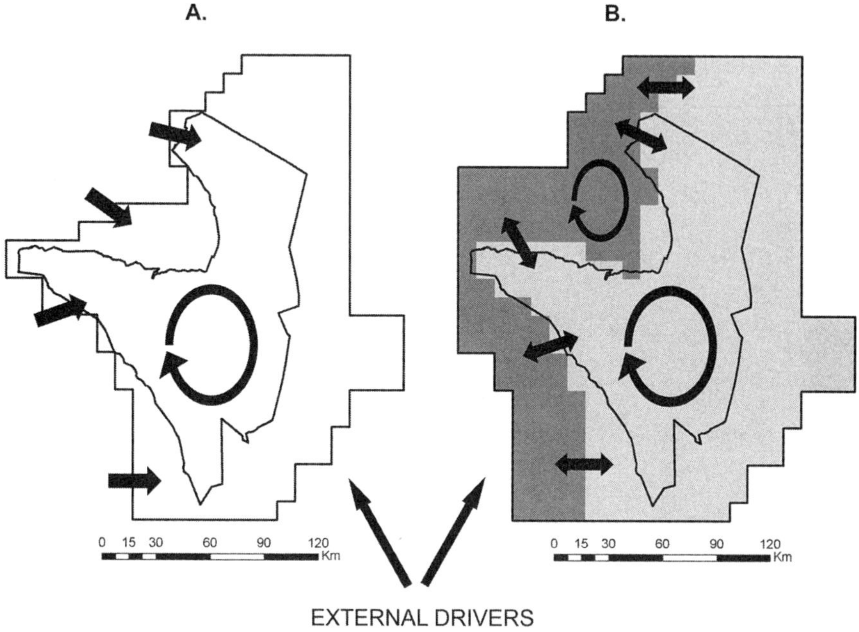

Fig. 20.1 The greater Serengeti ecosystem (Serengeti National Park, inset) as a self-contained ecological system (A), and a socioecological system (B). Model (A) depicts a closed system, and the dynamics of ecological variables interact with each other and respond to external forcing functions such as climate. It is assumed that human impacts are not dependent on ecological conditions. In model (B), both the socioeconomic (dark shaded areas, where humans and wildlife coexist) and ecological realms (light shaded areas, protected) have internal dynamics, but they are also linked with each other through a number of feedbacks.

1995; Barrett and Arcese 1998; Mduma, Sinclair, and Hilborn 1999), but human influences on the park also extend to the use of protected areas for livestock grazing, firewood extraction and tree cutting, the setting of fires, and transmission of diseases from domestic animals to wildlife (Dobson 1995; Holmern et al. 2004; Cleaveland et al. 2008; Knapp 2009). Many of these activities have knock-on indirect effects on the ecosystem. For example, intense fire use can lead to a reduction in soil fertility (Holdo et al. 2007) and a decline in woody biomass (Sinclair et al. 2007; Sinclair et al. 2010), whereas poaching can have downstream effects on vegetation and fire frequency (Holdo et al. 2009). Rabies and canine distemper, which become more prevalent as the human population increases, affect lion, wild dog, and hyena populations, and potentially the ability of these predators to regulate their prey populations (Packer et al. 1999; Cleaveland et al. 2008). The livestock vaccination campaign that led to the eradication of rinderpest in the Serengeti led directly to the explosion in the wildebeest population, with cascading effects of grass cover, fire extent, and tree population dynamics (Sinclair et al. 2007; Holdo et al. 2009). The park ecosystem can in turn exert significant effects on human livelihoods: wild ungulates provide a source of protein and income in areas adjacent to the park, but also may compete with livestock for forage and affect crop yields as a result of crop raiding (Knapp 2009).

MODELING THE GREATER SERENGETI ECOSYSTEM

Past studies of human impacts on wildlife populations in the Serengeti have explored the effects that changes in the human population, hunting pressure, and antipoaching enforcement are likely to have on wildlife populations (Pascual and Hilborn 1995; Pascual, Kareiva, and Hilborn 1997; Barrett and Arcese 1998). Other work has explored how proximity to the park boundary impacts human livelihoods outside the park (Costello et al. 2008; Knapp 2009). Comprehensive socioecological models of the system, that is, models that integrate a wide gamut of key ecological and socioeconomic processes and activities and interactions, have been lacking. Barrett and Arcese (1998) explored the interaction between humans and wildlife via a single interaction, namely, hunting. Costello et al. (2008) developed a household-level socioeconomic model in which wildlife dynamics occurred in spatially implicit space and responded only to hunting. In this model, humans devote labor resources to three competing activities: agriculture, livestock, and hunting. This greater economic scope sets the stage for greater dynamic complexity than might be the case with single-activity models, because the socioeconomic realm can interact with the ecological domain in multiple ways; for example, wildlife can affect crop yields through crop raiding, livestock through competition for grazing, and revenue from hunting.

The HUMENTS Model

More recently, Holdo et al. (2010) developed the HUMENTS model, which expanded the work of Costello et al. (2008) in two respects. First, HUMENTS adopts a spatially realistic framework to take into account the effects of complex environmental gradients (in rainfall, habitat type, and soil fertility) existing across the Serengeti and the shape of the protected area in relation to the human-dominated areas. Second, the HUMENTS model couples a modified form of the Costello et al. (2008) household model with a preexisting spatially explicit model of vegetation, fire, and herbivore dynamics in the Serengeti (Holdo, Holt, and Fryxell 2009a). The key players in the

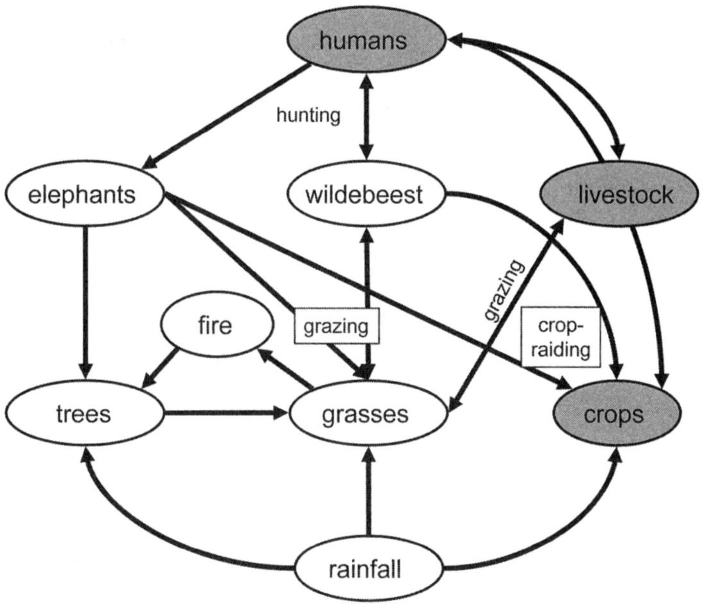

Fig. 20.2 Schematic representation of the Serengeti socioecological system variables and their key interactions as described in the HUMENTS model. The ecological components are represented by open symbols, and the components of the socioeconomic model are depicted as shaded symbols. Additional links (such as direct effects of humans on trees through tree cutting) are not presently incorporated in the model, but could also be considered.

HUMENTS model are depicted in fig. 20.2. From the ecological perspective, they include the following: herbivores (both grazers and browsers, with wildebeest and elephants being the dominant players in each guild, respectively), trees, grasses, and fire. The ultimate engine for this system is rainfall, which in the HUMENTS model is input as spatial layers of monthly rainfall. Rainfall drives grass dynamics, tree growth, and crop yields. From the socioeconomic perspective, the key variables are humans, livestock, and crops. At the root of the socioeconomic submodel is a household labor allocation algorithm that splits labor into components used for agriculture, livestock husbandry, and hunting. In the model, the human population occupies areas to the west of Serengeti National Park and its adjacent protected areas (fig. 20.1B). It is assumed that agropastoral ethnic groups, who engage both in agriculture and livestock husbandry, are dominant. Nomadic pastoral groups, more dominant east of the Serengeti, are not included in the current version of the model.

All of the key players in HUMENTS interact dynamically through a series of coupled difference equations, iterated over time steps appropriate to population dynamics and seasonal drivers, and are linked by interactions (Holdo et al. 2010). For example, crop yields depend jointly on rainfall, labor investment by humans, and crop raiding by wildebeest and elephants. As in other examples, increasing human populations can alter land use by resulting in habitat loss for crop production, and revenue from agriculture can result in livestock purchases as a form of savings, for example. These local interactions take place within individual 10 × 10 km cells in a spatial lattice, with the entire lattice comprising the greater Serengeti ecosystem as defined in fig. 20.1. In addition to these local interactions, however, the system is spatially coupled by the wildebeest migration. The wildebeest move across the system following a strategy that maximizes their intake of energy- and protein-rich forage (Holdo, Holt, and Fryxell 2009b), and respond numerically to resource availability and hunting. The wildebeest migration is a key aspect of this system, because it affects the distribution of fire and the spatiotemporal distribution of key processes in the socioeconomic model, such as competition for grass with livestock, the availability of prey for hunters, and crop raiding (Holdo, Holt, and Fryxell 2009a). Hunting intensity is in turn spatial in nature, because humans are assumed to travel over limited distances to hunt and their allocation of time for hunting over time for alternative activities depends in part on the proximity of wildebeest (Campbell and Hofer 1995).

It is worth noting here that, even with its added complexity, the HUMENTS model is not necessarily designed to capture the full range of behavior that the system is capable of. Human agents in the model make decisions based on only a limited set of variables, for example, compared to the actual complexity of human decision-making processes. Here, the model is used not with the objective of capturing the full range of behavioral complexity of the system, but rather as a case study designed to show that simply incorporating socioecological interactions can alter the trajectory of the system, even when the elements of the human-ecological interactions are quite simple.

A THEORETICAL EXAMPLE OF SOCIOECOLOGICAL COUPLING: INDIRECT EFFECTS OF ELEPHANTS ON WILDEBEEST

Holdo et al. (2010) used the HUMENTS model to explore the consequences of changing climatic regimes and alternative antipoaching enforcement scenarios in the Serengeti for the wildebeest population, and their downstream effects on fire and tree cover. This exercise permitted an investigation of the relative importance of a key external driver influencing the system (rainfall) and one of the fundamental links coupling the protected and human-dominated portions of the ecosystem. The main goal of this chapter is to explore how a perturbation in one of the key players in the system can propagate through the various links in the Serengeti SES. Specifically, one of the key objectives is to understand how the behavior of the Serengeti ecological system differs when treated as a self-contained ecological system versus a coupled natural-human system ([CNH]; fig. 20.1). As a case study, the HUMENTS model is used to explore how changes in the elephant population of the Serengeti can have downstream effects on wildebeest, fire, and trees in the system. Elephants were chosen as an independent variable for several reasons. As ecosystem engineers with a marked ability to shape ecosystem structure, elephants have long interested managers concerned about habitat change, particularly in the form of loss of tree cover, both in the Serengeti and elsewhere (Laws 1970; Croze 1974; Ruess and Halter 1990). Simultaneously, elephants are the primary source of elephant-human conflict in western Serengeti, where they are the dominant crop raiders on agricultural land (Knapp 2009). Human management could, for instance, aim to constrain elephant density to certain target levels (so it is reasonable to assume that elephant numbers are a constant), or instead let them freely vary in accord with their own inherent dynamics, poaching, and the like.

The role of elephants as agents of tree cover change in the Serengeti was previously explored with the SD model by Holdo, Holt, and Fryxell (2009a). Elephants directly reduce tree cover, but their impact is not only a function of this direct effect, as it is also modulated by feedbacks between tree cover, grass biomass, fire, and wildebeest grazing (Holdo, Holt, and Fryxell 2009a; fig. 20.2). For example, by reducing tree cover and thus the strength of tree-grass competition, elephants can indirectly promote grass biomass and thus enhance the carrying capacity of the system for wildebeest (Holdo, Holt, and Fryxell 2009a). This positive indirect link between elephants and wildebeest is here referred to as an "ecological pathway," via resource availability, that can link the two species (fig. 20.3A), and is the only one that might be considered in an isolated ecological model. When the socioeconomic component is taken into account, however, an additional pathway emerges: increases in the elephant population might affect crop yields, the relative allocation of household labor to hunting might then increase to compensate for the reduced viability of agriculture, and the wildebeest population might decline as a result (fig. 20.3B). This is one "socioeconomic pathway" linking elephants and wildebeest. Note that whereas the first interaction is positive, the second is negative. These two pathways are shown to illustrate some of the complexity of the system. Other indirect interactions between these components could (for example) be mediated through a numerical re-

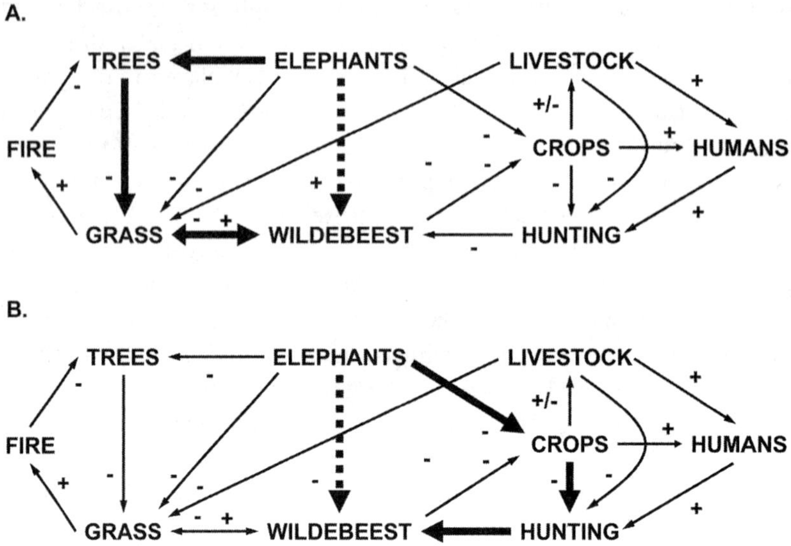

Fig. 20.3 Two hypothetical feedback pathways linking the elephant and wildebeest populations, one driven primarily through a purely ecological mechanism (A), and one driven by a socioeconomic mechanism (B). In (A), elephants reduce tree density, promoting grass growth and a higher carrying capacity for wildebeest; the net effect of elephants on wildebeest is positive. In (B), increases in elephant population size lead to a reduction in the viability of agriculture due to crop raiding, an increased allocation to hunting as an alternative source of income, and a consequent reduction in the wildebeest population; the net effect of elephants on wildebeest is negative. Thicker arrows denote dominant effects, and the dashed arrows indicate the net indirect effect of elephants on wildebeest. Note that other interactions (for example, mediated through fire, livestock, or numeric changes in the human population) are possible, but not depicted.

sponse of humans to crop losses, and increased labor allocations to livestock (with negative consequences for wildebeest due to competition for forage during the critical dry season) could potentially also play a role (fig. 20.3).

Model Scenarios

In the simulations, four scenarios are outlined for elephant population trajectories over the next 100 years, and these act as "treatments" in the simulations (table 20.1). To do this, alternative assumptions are made about elephant hunting and population growth. First, a default scenario with no hunting and no population growth is presented (labeled the CONST scenario), so that the elephant population remains at its present-day population size of 3,000. For the other three scenarios, it is assumed that the elephant population is dynamic, with logistic growth and a carrying capacity

Table 20.1 Elephant population change scenarios

Scenario	Population change	Hunting	Description
CONST	None	No	Constant population of 3,000
LOG	Increase	No	Logistic growth to population size of 10,000
PC	Decline	Yes	Fixed proportion of elephants hunted per year
FIX	Decline	Yes	Fixed number of elephants hunted per year

of 10,000. This value was chosen because it represents a population size that (a) has been predicted (Holdo, Holt, and Fryxell 2009a) to lead to significant changes in tree cover in the Serengeti (the ramifications of which can then be tested in this simulation), and (b) is ecologically plausible, given that it still represents a population density that is low compared to other savanna systems in sub-Saharan Africa, such as Hwange National Park (Chamaille-Jammes et al. 2008). The value is by necessity rather arbitrary, given the lack of data on elephant population regulation. In the LOG scenario, where no hunting occurs, the elephant population therefore grows logistically. Two hunting scenarios are also incorporated. In both of these, it is assumed that elephants are hunted with the same intensity recorded during their population collapse in the 1970s and 1980s, but in one case there is a constant per capita offtake (PC) where a constant fraction of the elephant population is removed per year, and in the other case there is a fixed absolute annual offtake (FIX). In all four scenarios it is assumed that hunting of wildebeest occurs, and a parameter (ϕ) controls the level of antipoaching enforcement to a value of 0.05. This value yields wildebeest poaching levels comparable to those that occurred during the period of the elephant population collapse. Therefore, even in scenarios when wildebeest are hunted, elephants may not be, so wildebeest and elephant poaching are somewhat decoupled. This is based on the observation that wildebeest poaching is used as a fairly generalized source of protein and family income, and is dictated by economic considerations at the household level (Campbell and Hofer 1995; Holmern et al. 2004). In contrast, elephant poaching is undertaken by commercial poachers, and it is assumed that the incentives that drive it are global demand for ivory. These four elephant population trajectory scenarios allow the independent ecological and hunting-mediated (socioeconomic) influences of elephants on wildebeest to be evaluated in the system. In all of the simulations, population and harvest parameters are used for elephants that were derived empirically from the fitting of a state-space model of elephant population dynamics to population census data from 1960 to the present (Holdo et al. 2009). Each scenario was run 100 times using random draws to generate rainfall surfaces from the historical record.

Model Results

Fig. 20.4A shows the four trajectories of the elephant population predicted by the model; with no hunting, the CONST and LOG scenarios lead (trivially) to no change and logistic growth, respectively. The two hunting scenarios, however, lead to very different outcomes. With hunting as a proportion of the total population (PC), the elephant population collapses over time (matching the historical pattern), whereas with a fixed offtake (FIX) it rises (fig. 20.4A) somewhat linearly (in the latter case, one could also observe an elephant collapse, if we were to start elephant numbers at a sufficiently low initial level). The alternative scenarios also lead to contrasting population trajectories for wildebeest and humans (fig. 20.4B, C). Wildebeest are generally projected to decline in population size from their initial state, given that the model assumes reduced levels of antipoaching enforcement than at present. Beyond this trend, however, the wildebeest population seems to be correlated with the elephant population size over time (fig. 20.4B). For humans, the opposite is true: the human population is predicted to asymptote at lower levels with increased elephant population density (fig. 20.4C). This is because the dominant effect of elephants in the model is to reduce agricultural output through crop raiding, so that elephant population increases lead to lower economic returns. The intensity with which wildebeest are hunted shows a less clear pattern, but generally, lower elephant numbers seem to reduce the annual wildebeest offtake (fig. 20.4D), most likely because of a reduction in the total number of hunters present (since the human population is negatively correlated with the elephant population), even though the *proportion* of hunters in the population may have increased (as shown in the next section, this is, in fact, the case). More elephants also lead to an increase in fire frequency (fig. 20.4E), despite the fact that wildebeest, the primary consumers of grass (which fuel fires), are more abundant. This suggests that the change in fire frequency is being controlled less by grazing than by a reduction in tree cover with increasing elephant population (fig. 20.4F). As tree cover declines due to increased elephant browsing, reduced tree-grass competition promotes the expansion of grasslands and thus the spread of fire, while simultaneously enhancing the carrying capacity of the system for wildebeest. It is also likely that as elephants reduce tree cover and indirectly promote fire, tree cover declines

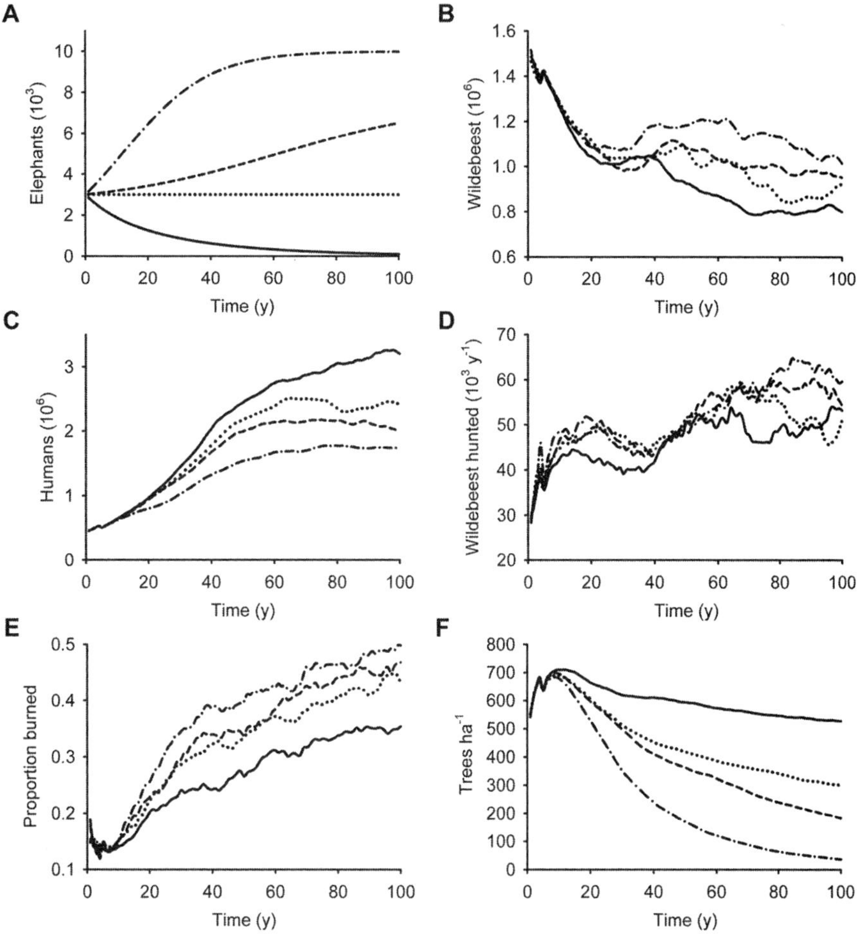

Fig. 20.4 Simulation results for the four "elephant scenarios," shown in order of their long-term impact on the elephant population over the next century (A) and their effects on the wildebeest (B) and human populations (C), the annual wildebeest offtake (D), the extent of fire in the ecosystem (E), and tree density (F). The results are weighted moving averages (with a five-year window) based on mean values for 100 model runs in each case. The scenarios are: elephant hunting with a fixed per-capita harvest coefficient (PC); no elephant hunting and no population growth (CONST); elephant hunting with a fixed annual offtake (FIX); and no hunting and logistic population growth assuming a carrying capacity of 10,000 (LOG).

even further, so that a positive feedback loop is established. These results essentially support the ecological pathway in fig. 20.3.

One can better tease apart these various effects through an examination of the bivariate relationships between some of the key variables (fig. 20.5). The final values (i.e., after 100 years of simulation) of selected variables are plotted against each other for each of the four scenarios. This approach

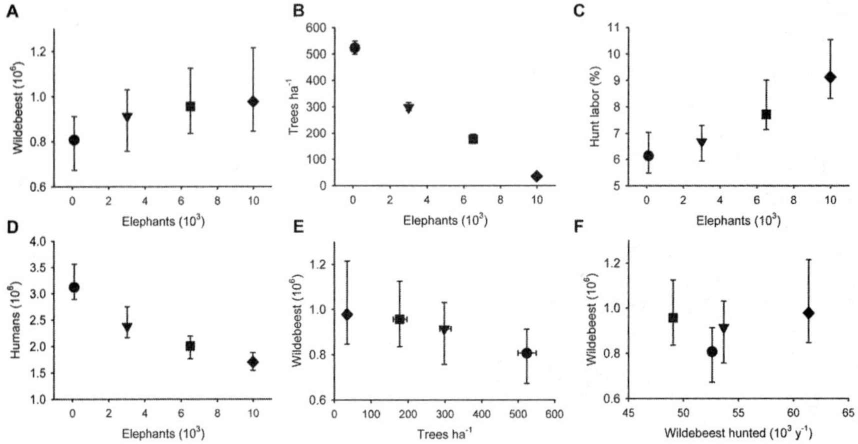

Fig. 20.5 Biplots showing bivariate relationships between model variables at an approximate steady state (100 years from now): wildebeest *W* versus elephants *E* (A); hunting labor allocation (*LH*) versus *E* (B); tree density *T* versus *E* (C); human population *H* versus *E* (D); *W* versus *T* (E); and *W* versus wildebeest offtake (F). Each point corresponds to a simulation scenario at the end of 100 years: PC (●), CONST (▼), FIX (■), and LOG (♦). The error bars (not shown in all cases) indicate the interquartile range across 100 runs.

shows more clearly the positive indirect effect of elephant population size on wildebeest (fig. 20.5A) and the negative effect of elephants on tree cover (fig. 20.5B). As elephants become more numerous, labor allocation also shifts increasingly toward hunting (fig. 20.5C) to the detriment of agriculture as a result of crop-raiding losses, even as these losses slow human population growth (fig. 20.5D). The final two panels of fig. 20.5 support the ecological pathway linking elephants and wildebeest: the wildebeest population declines (fig. 20.5E) as tree cover increases (and grassland retreats), but is relatively unaffected (fig. 20.5F) by changes in wildebeest offtake (the socioeconomic pathway depicted in fig. 20.3B).

These average results are somewhat misleading, however, because the strength of each of these effects varies considerably across space. Human settlements close to the protected area boundary are more exposed to the effects of crop raiding and also better positioned to take advantage of the wildlife resource within the park boundary as a source of revenue. Fig. 20.6 shows the spatial impact (in terms of deviations from the predictions of the CONST scenario) of each elephant scenario. Generally, human population size increases as a function of distance from the park boundary (fig. 20.6A), in agreement with empirical observations (Campbell and Hofer 1995), and there is a trade-off between agriculture and hunting (fig. 20.6E and I; note that labor allocation to livestock, which is fairly constant, is ignored

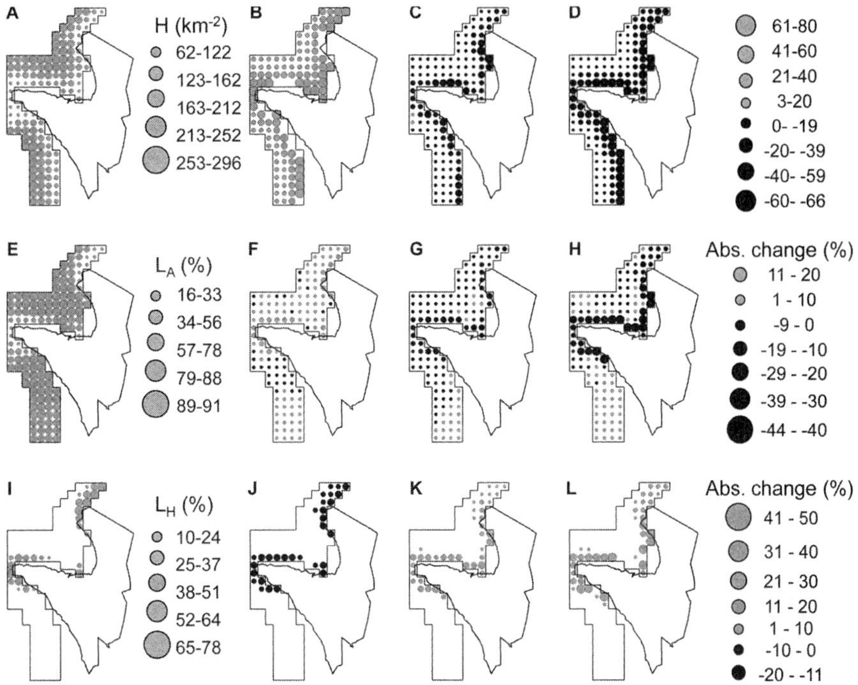

Fig. 20.6 Simulation results across space. The panels depict simulated values of human population density *H* (A), crop labor allocation *LA* (E), and hunting labor allocation *LH* (I) in western Serengeti 100 years from now under the CONST scenario. Also shown are: predicted percent changes in *H* relative to (A) under the PC (B), FIX (C), and LOG (D) scenarios; absolute changes in *LA* relative to (E) under the PC (F), FIX (G), and LOG (H) scenarios; and absolute changes in *LH* relative to (I) under the PC (J), FIX (K), and LOG (L) scenarios. In each case, values represent means for 100 runs (blanks are shown where default labor allocations or changes <1%).

here), with areas close to the Western Corridor and northern Serengeti that come into contact with the wildebeest during the dry season allocating more effort to hunting. Under the PC scenario, the elephant population collapses, crop raiding diminishes, the human population grows near the park boundary (fig. 20.6B), and there tends to be a small shift from hunting to crops throughout the western Serengeti in response to the enhanced viability of agriculture (fig. 20.6F and J). Under the FIX scenario, the elephant population grows and the human population declines near the park boundary (fig. 20.6C), with a corresponding increase in hunting to the detriment of agriculture (fig. 20.6G and K). This effect is even more pronounced under the LOG scenario (fig. 20.6D, H, and L), which allows elephants to reach carrying capacity and have an even stronger effect on agricultural viability adjacent to the park boundary. In areas that are not in close proximity to the park boundary and close to the wildebeest migration routes, however, the trade-off between growing crops and hunting is far weaker, so that there is a great amount of spatial variation in these effects (fig. 20.6).

In quantitative terms, the HUMENTS model may overestimate both the negative impact of elephants on human economic well-being outside the protected area and on tree cover within the Serengeti. Nevertheless, the qualitative results of this theoretical exercise are illustrative. The simulation results suggest a clear ecological link between elephants and wildebeest, mediated by tree cover, fire, and grass biomass, that trumps the pathway mediated by agriculture and hunting. Paradoxically, the socioeconomic pathway depicted in fig. 20.3B is also present, except that it is partially rerouted though humans in the form of a numeric response; this numeric response counters the negative correlation between agriculture and hunting. Offtake from hunting is a product of two components: fractional labor allocation to hunting and the total number of humans (the potential hunters; fig. 20.3). As currently parameterized, the HUMENTS model predicts a negative relationship between these two quantities, with the result that increases in crop raiding have only a minor effect on offtake. This prediction is strongly dependent on the assumptions of the model about the human demographic response to changing economic conditions, but it does show that when some of the feedbacks between the ecological and social components of an SES are considered, even in a simplified system, the behavior of the system can rapidly rise in complexity.

THE IMPORTANCE OF SPACE

One of the most striking aspects of the HUMENTS model results is how important the spatial extent and configuration of the ecosystem affect its dynamics. It is clear that proximity to the protected area boundary magnifies the intensity of the interactions between humans and the wildlife populations protected within Serengeti NP and adjacent game reserves. In the model (and this is strongly backed by empirical observations), this occurs because hunters have spatial "hunting kernels" within which the benefits of hunting decay with distance traveled. Traveling into the park imposes an opportunity cost on the one hand, and increases the likelihood of arrest on the other, so communities close to the park are more likely to hunt than those that are farther away. Similarly, the negative impacts of wildlife (notable from crop raiding) similarly decay in intensity with distance from the park boundary. In addition to the simple distance effect, human contact with the migratory wildebeest herds is highly variable along any given stretch of the park boundary. The wildebeest migrate roughly along a SE-NW gradient, and come into contact with humans to a greater degree in the Western Corridor and northern boundary than they do in the south, along the Maswa GR boundary. Other (resident) species also show marked pockets of abundance in distinct areas of the park, presumably with similar consequences. Therefore, even though the communities bordering Maswa probably engage in hunting to a far greater degree than shown in fig. 20.6, this hunting likely relies to a greater extent on resident herbivores rather than on migratory species. Our focus here has been on the spatial context of hunting pressure, but other factors, such as the distribution of water supplies, likely impose further constraints on the spatial distribution of both wildebeest and livestock (Ogutu et al. 2010).

In addition to affecting the spatial pattern of human-wildlife conflict, the size and spatial configuration of the Serengeti affects the mean trajectory of the system. Whereas a mean field model (i.e., one without explicit space) might predict a collapse of the wildebeest population at offtake rates predicted here (fig. 20.3D), in a spatially explicit model the wildebeest are able to avoid areas of high human population density, and much of the protected area acts as a spatial refuge within which hunting is not economically viable. A corollary of this is that the spatial pattern of human population change can be far more important than its average: a doubling of human population density in northern and western Serengeti, which borders the dry-season range of the wildebeest, is bound to have a greater impact on the wildebeest population than a similar increase in the south.

UNDERSTANDING THE LONG-TERM DYNAMICS AND RESILIENCE OF THE SERENGETI SOCIOECOLOGICAL SYSTEM

Modeling and Empirical Challenges

The case study presented here illustrates how, in complex dynamical systems with large numbers of variables, integrated socioecological modeling can help identify the dominant direct and indirect pathways that link external perturbations (in this case alternative elephant population scenarios) with changes in the state of the system, and differentiate among alternative hypotheses about how variables interact. An examination of HUMENTS model output in this and other simulations suggests that the Serengeti system responds in a graded fashion to perturbations, and there is little evidence for transitions among discrete alternative states or "regime shifts." Although this is not evident from the summary statistics plotted in figs. 20.4 and 20.5, an examination of the model output confirms that the distributions of values around the mean results are not bimodal (thus suggesting bifurcations or alternative stable states). To what extent this is true for the real (as opposed to the simulated) system remains unknown. If the Serengeti is capable of undergoing regime shifts and these are not captured by the model, it could be because certain key variables or processes have been omitted, or because the links between variables and their temporal dynamics have been mischaracterized or oversimplified in the modeling environment. Given the great deal of attention that regime shifts have received both in the ecological and socioecological literature (Scheffer et al. 2001; Westley et al. 2002; Cumming et al. 2005; Scheffer et al. 2009), the possibility of such shifts and the mechanisms responsible for them in the Serengeti are worth exploring further.

The existence of alternative stable or pseudostable (i.e., a not necessarily stable state in the purely mathematical sense, but rather slow to return to its previous state) states in a system requires either threshold effects of slow-changing variables or positive feedback loops. An example of the former might be a sudden shift in labor allocation strategies in subsistence communities as a result of declining soil fertility. If significant soil nutrient depletion occurs in human-dominated areas of the greater Serengeti ecosystem (suggested by M. E. Ritchie, unpublished data, but further research is required on this topic), areas where crops are a key source of income could potentially shift to hunting as an alternative economic activity. Soil nutrients may recover slowly, and even then, once shifts are made from crops to hunting, cultural inertia may make it difficult to shift back. HUMENTS and other socioecological models of the Serengeti system both ignore soil nutrient dynamics (a missing variable) and assume that human decision making is based on finding an optimal solution to a labor allocation problem (Barrett and Arcese 1998; Costello et al. 2008; Holdo et al. 2010), meaning that humans efficiently track changing environmental and economic conditions and are less likely to become "locked" in behavioral modes that might lead to regime shifts (assuming such optimality may be a possible mischaracterization of human behavior in the model; in reality cultural factors, game-like strategies, imperfect information, and suboptimal decision rules could play an important role). Other work has shown that agent-based formulations (for example) lead to different dynamics from classic optimization assumptions about human behavior (Walker and Janssen 2002). A critical area for further work is to develop a more complex and sophisticated portrayal of the human decision-making process, and how it impacts economic activity in both real and simulated environments.

A second example of a potential driver of regime shifts in the Serengeti is tree cover. Trees and grasses in savanna ecosystems compete and vary both spatially and temporally in relative cover (Scholes and Archer 1997). Fire is a dominant agent of tree cover change in this system, and declines in fire extent as a result of wildebeest grazing has led to an increase in tree cover over the past few decades (Holdo et al. 2009). Enhanced tree cover may in turn displace grasses and make less frequent fires even less frequent, leading to a positive feedback loop of progressively greater woody cover dominated by large trees that are relatively insensitive to fire. Under this scenario, the system shifts from a savanna with a relatively dynamic tree-grass balance to a relatively stable and persistent woodland (Eby et al., this volume). These changes do indeed occur at local scales in the HUMENTS model, but tend to become blurred on a global scale. Changes in tree cover also have the potential to lead to a number of important downstream ecological and economic effects that are presently not modeled by HUMENTS and other models. Increases in tree cover driven by the introduction of rinderpest may have caused regime shifts in human communities in western Serengeti as a result of massive eruptions of tsetse flies and the associated effects of trypanosomiasis (Sinclair 1979); loss of tree cover can affect regional rainfall patterns (Hoffmann, Schroeder, and Jackson 2002), leading to more drought and crop and livestock losses, changes in soil fertility (Belsky et al. 1989; Belsky 1994), and the loss of an important economic resource (Knapp 2009). Several of these changes could potentially lead to either ecological or socioeconomic regime shifts.

CONCLUSIONS

Complex ecological systems are practically defined by the wide array of direct and indirect links among species, functional groups, and abiotic factors in ecosystems. This is captured, for example, in the enormous complex array of links that describe the Serengeti ecosystem food web structure (Dobson 2009). Depending on the relative importance of individual links, changes in the abundance of any given node (e.g., elephants in fig. 20.3) can have positive, negative, or neutral effects on other elements in the network.

Inserting ecosystems into the broader context of socioecological systems enhances this complexity and imposes new challenges. One key challenge for any modeling approach is to capture the minimum level of model complexity that adequately addresses the question of interest, and whether modeling a system as a socioecological system (versus purely as an ecological system) is warranted. The above rinderpest example is a case in point. When rinderpest was endemic in the Serengeti, wildebeest numbers were kept relatively low, and the reduced grazing pressure resulted in widespread fires, which in turn led to a reduction in tree cover (Sinclair et al. 2007; Holdo et al. 2009). Rinderpest infection in wildebeest yearlings originated from livestock in areas surrounding the protected area (Dobson 1995). Once rinderpest was eradicated in livestock herds, seroprevalence for the disease plummeted in wild ungulates, and the wildebeest population shifted from being regulated by disease to being food (grass) limited (Dobson 1995; Mduma, Sinclair, and Hilborn 1999). The postrinderpest wildebeest population explosion dramatically altered the fire regime, and tree cover increased as a result (Holdo et al. 2009). It could therefore be argued that tree cover dynamics shifted from being best described by a socioecological system, where the force of rinderpest infection (a function of livestock management) was the ultimate driver, to being best described as a purely ecological system, where rainfall is the ultimate driver of tree dynamics (by affecting grass production, wildebeest population dynamics, and so on). The elephant example used in this chapter suggests that the scale at which the problem is being treated is also important: at the whole-ecosystem scale, the ecological pathways alone are dominant drivers of model outcome, but at local scales, socioecological elements can be of greater importance.

REFERENCES

Arcese, P., J. Hando, and K. Campbell. 1995. Historical and present-day anti-poaching efforts. In *Serengeti II: Dynamics, management, and conservation of an ecosystem*, ed. A. R. E. Sinclair and P. Arcese, 506–33. Chicago: Chicago University Press.

Barrett, C. B., and P. Arcese. 1998. Wildlife harvest in integrated conservation and development projects: Linking harvest to household demand, agricultural production, and environmental shocks in the Serengeti. *Land Economics* 74:449–65.

Barrett, C. B., F. Chabari, D. Bailey, P. D. Little, and D. L. Coppock. 2003. Livestock pricing in the northern Kenyan rangelands. *Journal of African Economies* 12:127–55.

Belsky, A. J. 1994. Influences of trees on savanna productivity—Tests of shade, nutrients, and tree-grass competition. *Ecology* 75:922–32.

Belsky, A. J., R. G. Amundson, J. M. Duxbury, S. J. Riha, A. R. Ali, and S. M. Mwonga. 1989. The effects of trees on their physical, chemical, and biological environments in a semi-arid savanna in Kenya. *Journal of Applied Ecology* 26:1005–24.

Bolger, D. T., W. D. Newmark, T. A. Morrison, and D. F. Doak. 2008. The need for integrative approaches to understand and conserve migratory ungulates. *Ecology Letters* 11: 63–77.

Boone, R. B., S. J. Thirgood, and J. G. C. Hopcraft. 2006. Serengeti wildebeest migratory patterns modeled from rainfall and new vegetation growth. *Ecology* 87:1987–94.

Bulte, E. H., and R. D. Horan. 2003. Habitat conservation, wildlife extraction and agricultural expansion. *Journal of Environmental Economics and Management* 45:109–27.

Campbell, B. M., and H. Hofer. 1995. People and wildlife: Spatial dynamics and zones of interaction. In *Serengeti II: Dynamics, management, and conservation of an ecosystem*, ed. A. R. E. Sinclair and P. Arcese, 534–70. Chicago: Chicago University Press.

Chamaille-Jammes, S., H. Fritz, M. Valeix, F. Murindagomo, and J. Clobert. 2008. Resource variability, aggregation and direct density dependence in an open context: The local regulation of an African elephant population. *Journal of Animal Ecology* 77: 135–44.

Cleaveland, S., C. Packer, K. Hampson, M. Kaare, R. Kock, M. Craft, T. Lembo, T. Mlengeya, and A. Dobson. 2008. The multiple roles of infectious disease in the Serengeti ecosystem. In *Serengeti III: Human impacts on ecosystem dynamics*, ed. A. R. E. Sinclair, C. Packer, S. A. R. Mduma, and J. M. Fryxell, 209–39. Chicago: Chicago University Press.

Costello, C., N. Burger, K. A. Galvin, R. Hilborn, and S. Polasky. 2008. Dynamic consequences of human behavior in the Serengeti ecosystem. In *Serengeti III: Human impacts on ecosystem dynamics*, ed. A. R. E. Sinclair, C. Packer, S. A. R. Mduma, and J. M. Fryxell, 301–24. Chicago: Chicago University Press.

Croze, H. 1974. The Seronera bull problem II: The trees. *East African Wildlife Journal* 12: 29–47.

Cumming, G. S., G. Barnes, S. Perz, M. Schmink, K. E. Sieving, J. Southworth, M. Binford, R. D. Holt, C. Stickler, and T. Van Holt. 2005. An exploratory framework for the empirical measurement of resilience. *Ecosystems* 8:975–87.

Dercon, S. 1996. Risk, crop choice, and savings: Evidence from Tanzania. *Economic Development and Cultural Change* 44:485–513.

Dobson, A. 1995. The ecology and epidemiology of rinderpest virus in Serengeti and Ngorongoro Conservation Area. In *Serengeti II: Dynamics, management, and conservation of an ecosystem*, ed. A. R. E. Sinclair and P. Arcese, 485–505. Chicago: Chicago University Press.

———. 2009. Food-web structure and ecosystem services: Insights from the Serengeti. *Philosophical Transactions of the Royal Society, B, Biological Sciences* 364:1665–82.

Galvin, K. A., P. K. Thornton, J. R. de Pinho, J. Sunderland, and R. B. Boone. 2006. Integrated modeling and its potential for resolving conflicts between conservation and people in the rangelands of East Africa. *Human Ecology* 34:155–83.

Harris, S., S. Thirgood, J. G. C. Hopcraft, J. Cromsigt, and J. Berger. 2009. Global decline in aggregated migrations of large terrestrial mammals. *Endangered Species Research* 7: 55–76.

Hilborn, R., P. Arcese, M. Borner, J. Hando, G. Hopcraft, M. Loibooki, S. Mduma, and A. R. E. Sinclair. 2006. Effective enforcement in a conservation area. *Science* 314:1266.

Hoffmann, W. A., W. Schroeder, and R. B. Jackson. 2002. Positive feedbacks of fire, climate, and vegetation and the conversion of tropical savanna. *Geophysical Research Letters* 29:2052.

Holdo, R. M., R. D. Holt, M. B. Coughenour, and M. E. Ritchie. 2007. Plant productivity and soil nitrogen as a function of grazing, migration and fire in an African savanna. *Journal of Ecology* 95:115–28.

Holdo, R. M., R. D. Holt, and J. M. Fryxell. 2009a. Grazers, browsers, and fire influence the extent and spatial pattern of tree cover in the Serengeti. *Ecological Applications* 19: 95–109.

———. 2009b. Opposing rainfall and nutrient gradients best explain the wildebeest migration in the Serengeti. *American Naturalist* 173:431–45.

Holdo, R. M., R. D. Holt, K. A. Galvin, S. Polasky, and E. Knapp. 2010. Responses to alternative rainfall regimes and antipoaching enforcement in a migratory system. *Ecological Applications* 20:381–97.

Holdo, R. M., A. R. E. Sinclair, K. L. Metzger, B. M. Bolker, A. P. Dobson, M. E. Ritchie, and R. D. Holt. 2009. A disease-mediated trophic cascade in the Serengeti and its implications for ecosystem C. *PLOS Biology* 7:e1000210.

Holmern, T., A. B. Johannesen, J. Mbaruka, S. Mkama, J. Muya, and E. Roskaft. 2004. Human-wildlife conflicts and hunting in the western Serengeti, Tanzania. *NINA Project Report* 026:1–26.

Homewood, K., E. F. Lambin, E. Coast, A. Kariuki, I. Kikula, J. Kivelia, M. Said, S. Serneels, and M. Thompson. 2001. Long-term changes in Serengeti-Mara wildebeest and land cover: Pastoralism, population, or policies? *Proceedings of the National Academy of Sciences* 98:12544–549.

Horan, R. D., and E. H. Bulte. 2004. Optimal and open access harvesting of multi-use species in a second-best world. *Environmental and Resource Economics* 28:251–72.

Janssen, M. A., B. H. Walker, J. Langridge, and N. Abel. 2000. An adaptive agent model for analysing co-evolution of management and policies in a complex rangeland system. *Ecological Modelling* 131:249–68.

Knapp, E. 2009. Western people shall not die: Examining the effects of household wealth on the resilience of a coupled human-natural system in Tanzania. PhD diss., Colorado State University.

Laws, R. M. 1970. Elephants as agents of habitat and landscape change in East Africa. *Oikos* 21:1–15.

Mduma, S. A. R., A. R. E. Sinclair, and R. Hilborn. 1999. Food regulates the Serengeti wildebeest: A 40-year record. *Journal of Animal Ecology* 68:1101–22.

Milner-Gulland, E. J., C. Kerven, R. Behnke, I. A. Wright, and A. Smailov. 2006. A multi-agent system model of pastoralist behaviour in Kazakhstan. *Ecological Complexity* 3: 23–36.

Ogutu, J. O., H.-P. Piepho, R. S. Reid, M. E. Rainy, R. L. Kruska, J. S. Worden, M. Nyabenge, and N. T. Hobbs. 2010. Large herbivore responses to water and settlements in savannas. *Ecological Monographs* 80:241–66.

Packer, C., S. Altizer, M. Appel, E. Brown, J. Martenson, S. J. O'Brien, M. Roelke-Parker, R. Hofmann-Lehmann, and H. Lutz. 1999. Viruses of the Serengeti: Patterns of infection and mortality in African lions. *Journal of Animal Ecology* 68:1161–78.

Pascual, M. A., and R. Hilborn, R. 1995. Conservation of harvested populations in fluctuating environments—The case of the Serengeti wildebeest. *Journal of Applied Ecology* 32: 468–80.

Pascual, M. A., P. Kareiva, and R. Hilborn. 1997. The influence of model structure on conclusions about the viability and harvesting of Serengeti wildebeest. *Conservation Biology* 11:966–76.

Ruess, R. W., and F. L. Halter. 1990. The impact of large herbivores on the Seronera woodlands, Serengeti National Park, Tanzania. *African Journal of Ecology* 28:259–75.

Scheffer, M., J. Bascompte, W. A. Brock, V. Brovkin, S. R. Carpenter, V. Dakos, H. Held, E. H. van Nes, M. Rietkerk, and G. Sugihara. 2009. Early-warning signals for critical transitions. *Nature* 461:53–59.

Scheffer, M., S. Carpenter, J. A. Foley, C. Folke, and B. Walker. 2001. Catastrophic shifts in ecosystems. *Nature* 413:591–96.

Scholes, R. J., and S. R. Archer. 1997. Tree-grass interactions in savannas. *Annual Review of Ecology and Systematics* 28:517–44.

Sinclair, A. R. E. 1979. Dynamics of the Serengeti ecosystem: Process and pattern. In *Serengeti: Dynamics of an ecosystem*, ed. A. R. E. Sinclair and M. Norton-Griffiths, 1–20. Chicago: Chicago University Press.

Sinclair, A. R. E., S. A. R. Mduma, J. G. C. Hopcraft, J. M. Fryxell, R. Hilborn, and S. Thirgood. 2007. Long-term ecosystem dynamics in the Serengeti: Lessons for conservation. *Conservation Biology* 21:580–90.

Sinclair, A. R. E., K. L. Metzger, J. S. Brashares, A. Nkwabi, G. Sharam, and J. M. Fryxell. 2010. Trophic cascades in African savanna: Serengeti as a case study. In *Trophic cascades: Predators, prey, and the changing dynamics of nature*, ed. J. Terborgh and J. A. Estes, 255–74. Washington, DC: Island Press.

Thirgood, S., A. Mosser, S. Tham, J. G. C. Hopcraft, E. Mwangomo, T. Mlengeya, M. Kilewo, J. M. Fryxell, A. R. E. Sinclair, and M. Borner. 2004. Can parks protect migratory ungulates? The case of the Serengeti wildebeest. *Animal Conservation* 7:113–20.

Walker, B. H., and M. A. Janssen. 2002. Rangelands, pastoralists and governments: Interlinked systems of people and nature. *Philosophical Transactions of the Royal Society of London, Series B, Biological Sciences* 357:719–25.

Westley, F., S. R. Carpenter, W. A. Brock, C. S. Holling, and L. H. Gunderson. 2002. Why systems of people and nature are not just social and ecological systems. In *Panarchy: Understanding transformations in human and natural systems*, ed. L. H. Gunderson and C. S. Holling, 103–20. Washington, DC: Island Press.

TWENTY-ONE

Living in the Greater Serengeti Ecosystem: Human-Wildlife Conflict and Coexistence

Katie Hampson, J. Terrence McCabe, Anna B. Estes, Joseph O. Ogutu, Dennis Rentsch, Meggan E. Craft, Cuthbert B. Hemed, Eblate Ernest, Richard Hoare, Bernard Kissui, Lucas Malugu, Emmanuel Masenga, and Sarah Cleaveland

Humans and wildlife have coexisted for millennia, but conflicts often occur and are intensifying with the rapid growth of human populations and concomitant shrinkage and isolation of natural habitats (Woodroffe 2000; Conover 2002; Graham, Beckerman, and Thirgood 2005). Human-wildlife conflicts are manifested through direct encounters that result in significant economic losses and can be injurious or fatal to either party, through competition for resources (water, pasture, and protein) or indirectly due to disease transmission or diversion of movement and/or activities. Human-wildlife conflict usually involves large carnivores, megaherbivores, and keystone species that shape ecosystems. Because it is these species that are also most prone to extinction (Woodroffe and Ginsberg 1998), human-wildlife conflicts pose a major threat to biodiversity conservation with long-term and disproportionate impacts. If, as a result of conflicts, wildlife is perceived as a liability rather than an asset, prevailing negative attitudes may influence land-use decisions and undermine conservation measures. The effectiveness of conservation efforts therefore hinges upon the relationship between people and wildlife (O'Connell-Rodwell et al. 2000). Yet most people living near protected areas in Africa bear the costs without receiving the concomitant benefits (Sibanda and Omwega 1996; Naughton-Treves 1998; Patterson et al. 2004) and the costs can be considerable in relation to standards of living (Adams and Hutton 2007). In particular, human-wildlife conflicts tend to be most intense and the ramifications most severe where livestock holdings and agriculture are an important part of rural livelihoods, as is the case in the greater Serengeti ecosystem.

The Serengeti is one of the few remaining intact ecosystems that supports mass ungulate migrations, megaherbivores, and large carnivore populations. Nonetheless, the ecological impacts of people living beside this vast and seemingly impenetrable ecosystem are pervasive. The wide-ranging and migratory patterns of key wildlife species in the ecosystem bring them into frequent contact with human populations (Holdo et al. 2010) and the Serengeti's human residents have played an important role in shaping the ecosystem (Homewood and Rodgers 1991). Crucially, most of the local human populations are dependent on the land through subsistence rather than through larger-scale commercial rangelands that exist elsewhere in sub-Saharan Africa (except for group ranches adjacent to the Maasai Mara National Reserve in Kenya; Homewood et al. [2001]). As a result, sizable and growing human populations exist largely in poverty (see E. Knapp et al., chapter 16; L. Knapp et al., chapter 23) and often in conflict with their wildlife neighbors. These human-wildlife conflicts are multifaceted, complex, can be severe (even fatal, both for humans and wildlife), and pose a major threat to ecosystem integrity by perpetuating antagonism among local people, and in future generations toward wildlife and the environment.

In this chapter we characterize the negative interactions between human and wildlife populations in the Serengeti focusing on their nature, intensity, and distribution (hunting and bushmeat demand and consumption are detailed in Rentsch et al., chapter 22). This synthesis highlights the complexity and urgency of the challenges arising along the human-wildlife interface that must be addressed.

BACKGROUND

The greater Serengeti ecosystem (GSE) comprises a variety of protected areas including: In Tanzania, the Serengeti National Park (SNP), Ngorongoro Conservation Area (NCA), and adjacent game reserves (Maswa, Ikorongo, and Grumeti), Loliondo Game Controlled Area (LGCA), two recently gazetted community-managed wildlife management areas (WMAs) Ikona and Makao, and communal village lands; and in Kenya, the Maasai Mara National Reserve (MMNR) and Maasai group ranches and conservancies. These land uses differ in the level of protection afforded to wildlife and the environment, as well as regulations regarding human activities. None of the protected areas are fenced. Human populations in the Serengeti include migratory/nomadic systems and considerable ethnic diversity (E. Knapp et al., chapter 16). Agropastoralist communities mainly occupy areas to the west of SNP, living at high densities with higher rates of population growth than the national average and particularly high immigration (Estes et al., chapter 18). Considerably lower density pastoralist populations (generally Maasai) dominate land to the north and east and have production systems based on traditional grazing, with limited, but increasing cultivation. A small number of agriculturist villages (Sonjo) are also interspersed to the east of SNP (Galvin et al., chapter 17).

The composition of wildlife populations across the region, especially outside the protected area complex, is largely influenced by interactions with humans. While wild ungulate and carnivore communities are reasonably abundant and diverse to the east in the Ngorongoro district, in the human-inhabited areas to the west, hunting and land-use changes have been less hospitable to wildlife. Opportunistic small carnivores apparently thrive in these human-inhabited areas (see Craft et al., chapter 15), but megafauna and ungulates are either absent, occur at extremely low densities, or only occasionally roam into human-settled areas, particularly during the migration. One result of the diversity and distribution of cultural practices, livelihoods, and environments is that quite different types, frequencies, and intensities of conflicts occur across the region (summarized in table 21.1).

DIRECT ENCOUNTERS WITH WILDLIFE

Encounters with wildlife are a part of life for those who live around the Serengeti. Consequences vary from minor nuisances to the loss of human life (table 21.1) and generate notoriety and negative press for the species involved, as well as retaliations that can be fatal for both humans and wildlife. Few studies systematically quantify fatalities and injuries caused by wildlife, with most examining specific conflicts in the context of livestock depredation, crop raiding, or threats to focal species of conservation concern, such as lions or elephants. We present preliminary data on the incidence and nature of conflicts, including risks to humans, from a range of human-wildlife interactions within the Serengeti ecosystem based on analysis of hospital records.

Hospital records provide a potentially valuable source of data on wildlife-related attacks and injuries, but have several shortcomings: fatal interactions, that are common with species like elephants and certain snakes, may not be recorded; patients frequently lie about the cause of injuries resulting from participation in illegal activities (poaching or ritual lion

Table 21.1 Summary of current human-wildlife conflicts in the GSE, their distribution, frequency, intensity, drivers, and potential consequences

Conflict	Context	Species	Risk to wildlife due to conflict	Impact of conflict on local communities	Frequency/ intensity & timing	Distribution (range & locations)	Risk factors
Encounters with humans	Accidental	Elephants	MEDIUM: if serious incident, lethal control measures often used	VERY HIGH: Potentially fatal, severe disruption (e.g., school closures)	Infrequent	Localized in western Serengeti, north Loliondo, and border of MMNR	Proximity to park and wooded refugia. Attempts to drive animals away
		Buffalo	LOW: occasional retaliation	HIGH: risk of serious injury, very occasionally fatal	Infrequent	Woodlands around NCA and Loliondo	Walking at night, particularly in woodland
		Hyenas	LOW: occasional retaliation	LOW: usually without incident	Infrequent	Eastern Serengeti and villages along western boundary	Falling asleep outside
		Lions	LOW	LOW: usually without incident	Infrequent	Eastern Serengeti, virtually absent in western Serengeti	Falling asleep outside
		Crocodiles, hippos	LOW	MEDIUM: risk of serious injury	Very rare	Mara River	

		Snakes	LOW	HIGH: risk of serious injury	Frequent	Throughout ecosystem	
	Rabies	Hyenas, mongoose, wild cats, honey badgers, jackals	LOW: occasional spillover from dog population	HIGH: fatal if untreated, and very serious injuries inflicted by hyenas	Rare	In areas where rabies circulates (mass dog vaccination has controlled rabies in some areas)	Low vaccination coverage in dog populations
	Ritual killing	Lions	LOW: little offtake in comparison to other threats	MEDIUM: risk of serious injury	Common following initiation of Morani cohorts	Localized: southern plains of NCA	New Morani cohorts, recent depredation
Depredation	Diurnal attacks on grazing cattle	Lions	LOW: large, relatively resilient populations and retaliatory killing in direct proportion to livestock predated (low offtake compared to other threats), but recent reports suggest increasing risks due to retaliatory poisoning	MEDIUM: small numbers predated, but losses are serious to owners with few animals. Risk of serious injury if attempting to intervene	Intermittent: incidents occurring most months	Prevalent in eastern Serengeti, virtually absent in western Serengeti	Isolated areas (low density human populations), child herders, and lack of Morani.
	Diurnal attacks on grazing small stock	African wild dogs	HIGH: small vulnerable populations and risk of retaliation; risk of losing entire packs through attacks while in den	MEDIUM: small numbers predated, but losses are serious to owners with few animals	Intermittent: more common when denning (May-July)	Mostly eastern Serengeti, localized conflict near dens	Isolated areas (low-density human populations)

continued

Table 21.1 continued

Conflict	Context	Species	Risk to wildlife due to conflict	Impact of conflict on local communities	Frequency/ intensity & timing	Distribution (range & locations)	Risk factors
	Nocturnal attacks on kraaled livestock	Lions, hyenas, leopards	LOW: large resilient populations (lions, hyenas), and evasive behavior. Retaliatory killing in proportion to livestock taken. But retaliatory poisoning may be increasing (also threatening vultures)	MEDIUM: small numbers predated, but serious to owners with few animals. Risk of serious injury if attempting to intervene	Frequent: particularly by hyenas. More common during rains and when migratory prey absent	Eastern Serengeti and villages along western boundary	Poor-quality kraals and lack of effective fencing. Absence of migratory prey
	Predation on guarding domestic dogs	Hyenas, leopards	LOW: large resilient populations (hyenas), little retaliatory killing	MEDIUM: commonly reported, but dogs are relatively easily replaced	Frequent	Widespread around the ecosystem	
	Predation on fowl and produce	White-tailed mongoose, genets, jackals	LOW: large resilient populations, retaliatory killing is opportunistic	MEDIUM: large numbers taken	Frequent	Widespread in western Serengeti and Sonjo villages in eastern Serengeti	Poultry keeping and agriculture
Crop damage	Trampling and consumption during raids	Elephants	MEDIUM: if mitigation measures fail, often results in lethal control	VERY HIGH: individually catastrophic crop losses, potentially fatal encounters while defending crops	Frequent	Localized in western Serengeti and north of MMNR	Proximity to park and wooded refugia, agriculture

	Defending crop from raiding	Warthogs, monkeys, baboons, mongoose	LOW	MEDIUM: small but consistent cumulative losses	Frequent	Widespread in western Serengeti and Sonjo villages in eastern Serengeti	Proximity to park, agriculture
	Trampling during migration	Wildebeest	LOW	MEDIUM: occasional but catastrophic crop trampling (cotton)	Seasonal and infrequent	Western Serengeti, migration route	Proximity to park
Competition for land (pasture and water)	Shared grazing areas	Wild herbivores	HIGH: Disruption of migration	HIGH: risk of disease transmission (MCF from wildebeest, FMD from buffalo, and trypanosomiasis from other species), forced to graze livestock on marginal lands, loss of productivity	Frequent and intensified during migration	Throughout ecosystem; most serious in eastern Serengeti, and migration routes in the Mara and western Serengeti	Proximity to park
	Land conversion to crops/ settlements	All	VERY HIGH: Serious loss and degradation of wildlife habitats, alteration of hydrology from irrigated agriculture	MEDIUM: conflicts with authorities	Frequent and intensifying	All areas surrounding the SNP	Proximity to park and irrigation upstream of ecosystem
	Cutting trees for charcoal	All	VERY HIGH: Serious loss and degradation of habitat	MEDIUM: conflicts with authorities	Frequent and intensifying	Western Serengeti	Proximity to park
	Grazing on protected area land	All	HIGH: risks of disease transmission and disturbance	MEDIUM: conflicts with authorities	Frequent and possibly intensifying	Villages bounding protected areas	Proximity to park

hunting); and often the detail is insufficient to understand the nature of the incident without comprehensive follow up. Nonetheless, these records indicate that clashes with large carnivores that require hospitalization are common in the eastern pastoralist communities, occurring almost every year (figs. 21.1, 21.2, and 21.3). In contrast, clashes with large carnivores are exceptional in western agriculturalist communities (table 21.2; figs. 21.3, 21.4), where small- to medium-sized carnivores are involved in more

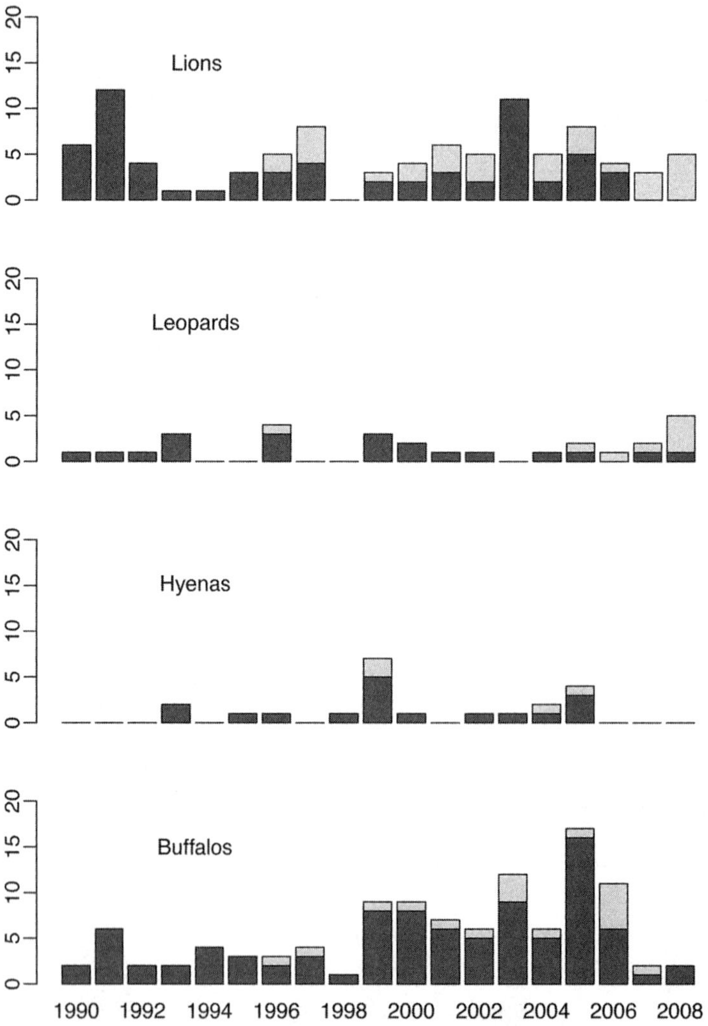

Fig. 21.1 Time series of hospitalizations in Ngorongoro district due to attacks by wildlife. Records from Endulen (1990–2008, dark gray) and Wasso (1997–2008, light gray) mission hospitals.

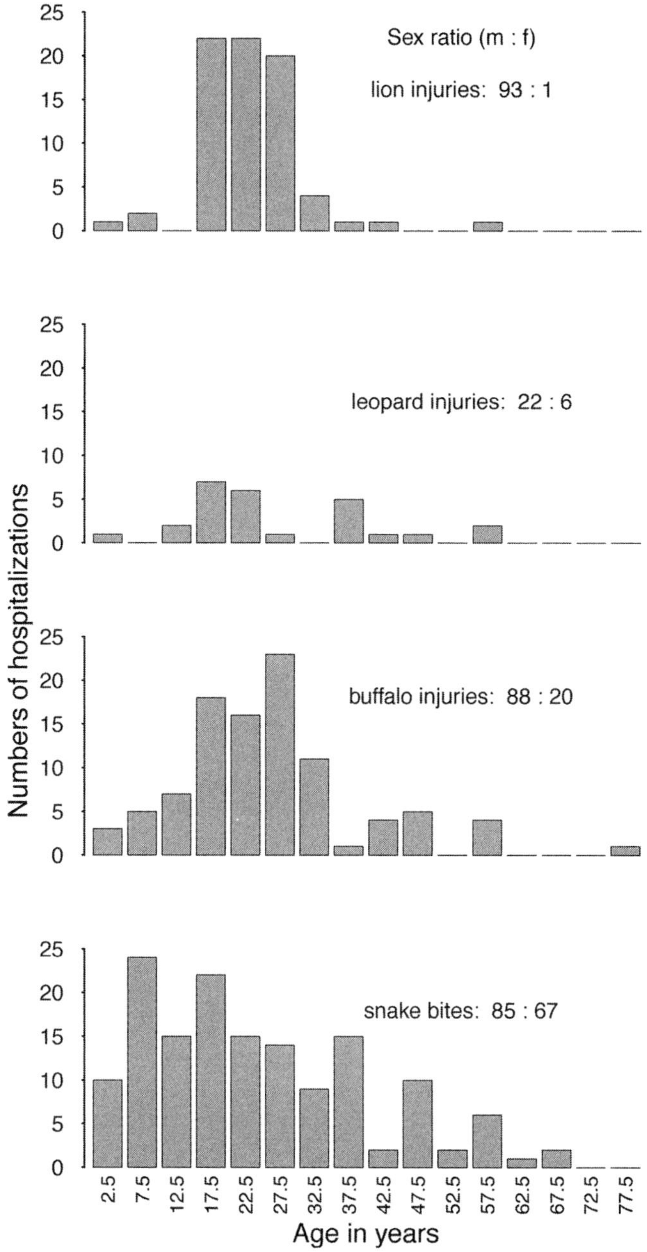

Fig. 21.2 Age and sex distribution of patients injured by lions, leopards, and buffalo or bitten by snakes in Ngorongoro district. Records from Endulen (1990–2008, dark gray) and Wasso (1997–2008, light gray) mission hospitals.

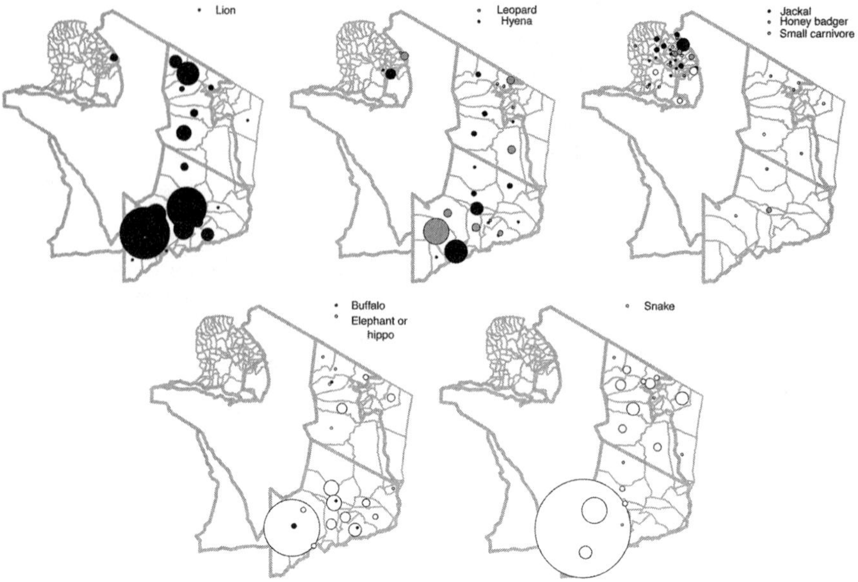

Fig. 21.3 Injuries caused by wild animals from 1999 to 2009 in the Serengeti and Ngorongoro districts. The radius of each circle is proportional to the number of victims that reported to hospital from each village. Circles are shaded according to the species responsible: other small carnivores include wildcat, bat-eared fox, genet, white-tailed mongoose, and civet. Information on injuries caused by buffalo, elephant, hippo, and snakes were only available for Ngorongoro district.

incidents. However, these incidents were due to abnormally behaving animals infected with rabies (fig. 21.3). Generally, human deaths due to rabies transmitted during attacks by wild animals were more commonly recorded than fatal attacks by wildlife, although both were underreported.

Patient demography provides further clues about the causality of conflicts with wildlife. The vast majority of Maasai hospitalizations from attacks by carnivores occurred in the context of *Ola-mayio*, or ritual lion killing, a rite of passage indicating the warrior progression to adulthood. *Ola-mayio* is an organized hunt, whereby Morani (Maasai warriors) compete against one another to spear a lion, which can be a purely cultural act or in retaliation to a lion attack on livestock (Ikanda and Packer 2008). Hospital records clearly show that *Ola-mayio* is not without risk. More than 95% of hospital records of injuries attributed to lions were of men aged 15–30, typical Morani age. Because the ritual is illegal, injuries are oftentimes not reported, or only partially described, but several young men coincidentally reporting "animal-injuries" usually indicates a lion hunt. The age and sex distributions of patients injured by leopards and hyenas are less pronounced, but still similar to those injured by lions, and probably reflect the greater role of

Table 21.2 Hospitalizations due to attacks by wild animals in Ngorongoro district (from Wasso and Endulen mission hospitals) and Serengeti district (from Nyerere Designated District Hospital, Mugumu) over a 10-year period from 2000 to 2009

Species	**Ngorongoro**	**Serengeti**
Snake	108	NA
Buffalo	73	NA
Lion	63	3
Hyena (spotted and striped)	29	5
Leopard	17	4
Honey badger	6	12
Elephant	5	NA
Baboon/ monkey	4	8
Wildcat	2	1
Genet	1	2
White-tailed mongoose	1	4
Hippo	1	0
Bat-eared fox	0	2
Civet	0	1
Jackal	0	26
Wildebeest	0	1
Other	27	11
Total	337	80

Note: NA = not available.

men in responding to nocturnal attacks on livestock kraaled (enclosed with a thornbush fence) in bomas (fig. 21.2).

Injuries caused by wild animals were also associated with substantial costs. Prolonged hospital stays were necessary for most injuries from encounters with large carnivores and buffalo. In some cases the NCA Authority subsidized treatments, but expenses were frequently in excess of 100,000 Tsh (~100 USD), especially if antirabies vaccine was considered necessary, while Tanzanians living around the GSE generally live on less than $1 per day (Schmitt 2010).

Although large carnivores typically evoke the most fearful images of dangerous encounters with wildlife, hospital data suggests that attacks by herbivores are roughly as common, and that snake envenoming is a frequent problem. In fact, buffalo and snakes present greater risks to the whole population rather than to specific high-risk groups (like young men; fig. 21.2) or as consequences of provocative practices (*Ola-mayio*; see table 21.1). Patient's origins are also likely to be risk factors (fig. 21.3); for example,

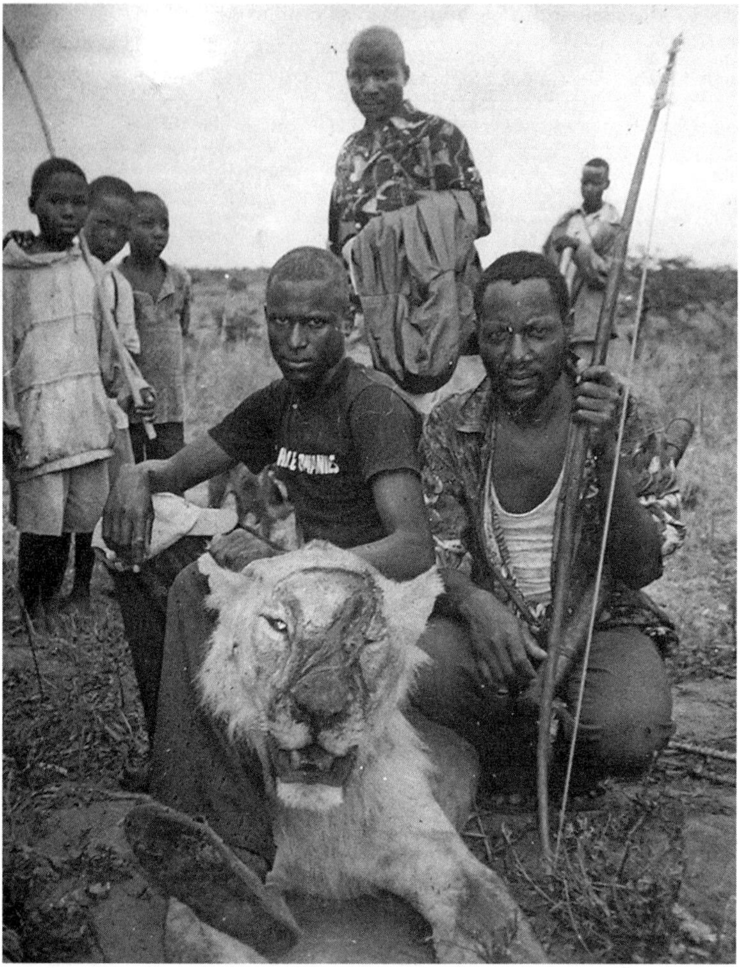

Fig. 21.4 A group of three lions strayed into Nyamburi village, western Serengeti in 2003. Villagers were quick to chase the group, killing one lion and leaving three men injured, but such incursions are rarely reported.

many injuries occur in wooded areas, which are preferred by buffalo, and anecdotal information suggests that women collecting firewood are often attacked by buffalo.

Encounters with elephants tend to be fatal, and so only a small number of incidents involving elephants were recovered from hospital records (five over a ten-year period; table 21.2). However, Wildlife Division reports and interviews with community members suggest that such encounters are much more common, with fatal incidents occurring each year. For example, in each of the districts bordering the north and west of the protected area complex (e.g., the Serengeti district to the west of SNP in Tanzania, and the Transmara and Narok districts bordering the MMNR in Kenya) fatalities were recorded in most years over the last three decades (Sitati et al. 2003; Kaelo 2007). Some incidents involve provocation; for instance, when local people attempt to drive elephants away from their crops, and these incidents typically occur close to protected area boundaries (particular conflict zones are reported in more detail below). Many incidents are accidental, catching both elephants and villagers unaware. When elephant groups do stray onto village land they typically wreak havoc. For example, in late 2007, a group of over 20 elephants were reported along the Mara River in villages more than 50 km from the park boundary, and in mid-2008 a group traversed a similar route and trampled six herdsmen while being chased by helicopter back toward the park. Repeat encounters also forced temporary school closures both to the east and west of the Serengeti (Olosokwan and Fort Ikoma, respectively, in 2007) when elephant bulls settled in these areas. The consequences of these conflicts with elephants are far reaching. In the Transmara district, school attendance and performance were poorer among children whose school routes overlapped with elephant ranges (Walpole et al. 2003).

While the numbers of people killed or injured by wildlife are relatively small in comparison to entrenched problems of poverty and disease, the perceived risks are large and the perpetrator is easily identified. The short-term and practical implications of these direct conflicts are usually by way of retaliatory actions that have longer-term repercussions for conservation and are discussed in detail below.

DEPREDATION AND CROP DESTRUCTION

Depredation of domestic animals by wild carnivores and raiding and destruction of crops by other wildlife are a common complaint of communities living in close proximity to SNP.

Livestock losses from predation have been shown to be significant in villages on the Serengeti's western border (over 25% of households reported livestock losses, amounting to 4.5% of livestock per affected household, per annum, or almost 20% of cash incomes equivalent to $27; Holmern, Nyahongo, and Roskaft [2007]). Spotted hyenas were overwhelmingly responsible for these losses (>98%), with only 12 cases (1 cow and 11 goats) due to lions and leopards reported from almost 500 affected households in 2003 (Holmern, Nyahongo, and Roskaft 2007). In pastoralist communities to the east and north of the Serengeti a greater proportion of households/bomas suffered predatory attacks on kraaled livestock during the night (37% in the NCA and >60% to the northeast of the MMNR) and attacks on grazing livestock, particularly by lions, were considerably more frequent than in agropastoralist areas to the west of the Serengeti (Kolowski and Holekamp 2006; Ikanda and Packer 2008). A direct quantitative comparison of the impacts of livestock depredation in these communities is not possible from these published studies, but proportional losses appear to be on a similar scale (~1–5% of overall livestock holdings [Karani 1994; Kolowski and Holekamp 2006; Holmern, Nyahongo, and Roskaft 2007; Ikanda and Packer 2008]). In general, Maasai keep more livestock per homestead than agropastoralists and are less reliant on a cash economy; so while perceived (and occupational) risks may be higher, relative losses for those affected may be lower. However, the Maasai population in the NCA has grown substantially over the last decades, while cattle numbers have remained stable; therefore, the Maasai now survive on reduced per capita numbers of cattle (McCabe, Leslie, and DeLuca 2010), making additional losses from depredation more significant and less tolerable.

The differences in patterns of depredation around the Serengeti relate primarily to predator composition and their hunting behavior and preferences in prey choice. To the west of the Serengeti, the natural prey base has been locally depleted and the extensive human disturbance is undesirable for large carnivores. Transgressions by lions into these areas are therefore very rare and not tolerated when they occur (see fig. 21.3), with only occasional reports of attacks on livestock (and people) by lions, in contrast to more frequent fatal attacks reported elsewhere in Tanzania (Packer, Ikanda et al. 2005). Hyenas, the predatory species most frequently involved in conflicts with humans, appear to be the only large carnivore capable of surviving in agropastoralist communities (and in the group ranches to the north and west of MMNR; Ogutu, Bhola, and Reid [2005]), with some resident individuals and others commuting from the park. Reported patterns of predation fit with the findings of Kissui (2008) from the Maasai steppe; hyenas and leopards are most likely to attack livestock, particularly small stock, at night while enclosed in a boma, whereas lions tend to take grazing cattle during the day (Kolowski and Holekamp 2006; Ikanda and Packer 2008). In the NCA, over one-third of cattle herds were tended solely by children, which was a significant risk factor for livestock depredation. Furthermore, the likelihood of attacks increased with the average number of cattle tended by each herder (Ikanda and Packer 2008).

African wild dogs became locally extinct in the Serengeti National Park in 1992 but have recently reestablished in parts of Ngorongoro district. The increase in their numbers has coincided with a concomitant increase in reports of depredation, particularly of small stock (Masenga and Mentzel 2005). Most of these reported incidents are from areas where packs have successfully denned (Sonjo and Maasai villages in Loliondo including Ololosokwan, Piaya, Arash, and Oldonyosambu), but also include surrounding areas, and as far away as Mara Somoche in Serengeti district to the west of SNP, which may have been due to dispersing groups.

Small mammals can also cause a myriad of problems. Jackals, honey badgers, mongoose, wildcats, and genet are all reported as pests, preying on small stock and fowl (chicken, ducks) and damaging crops such as maize and cotton. Besides elephants (discussed below), monkeys and baboons are considered crop pests and wildebeest and zebra are reported to cause crop damage during their seasonal migration. As a result, security of livestock (and crops) from wild animals is a major concern, and the primary reason for local people keeping domestic dogs (as reported by 90% of dog-owning respondents from over 500 household surveys in villages within 10 km of the western SNP boundary and in Ngorongoro district). Over half of the dog owners claimed that their dog had successfully defended livestock against predators, either by barking and alerting their owners when faced with larger carnivores, or killing or chasing away smaller carnivores. Ironically, wild animals regularly attack dogs in this role of security guard. In the past five years, almost 20% of all respondents had a domestic dog killed/eaten by a hyena (31% to the northwest of SNP and 16% to the east) and 4% reported a dog killed by a leopard (mostly from Ngorongoro district). Almost 60% of respondents claimed this to be another reason for wanting more dogs, but there was no evidence that losses of dogs led to retaliation from owners, as is seen with losses of livestock or crops.

Human-elephant conflict (HEC) has emerged as a major challenge to wildlife and protected area management in Africa (Sitati et al. 2003), and is recognized as one of the most serious threats to elephant conservation (Hoare 2000). The problem has intensified as human populations continue to increase in former elephant habitat, and as recovering elephant populations expand into parts of their former range now occupied by humans (Hoare and Du Toit 1999; Walpole et al. 2004). The HEC often undermines community support for the conservation goals of the protected areas they border (Gadd 2005; Naughton-Treves and Treves 2005; Sitati, Walpole, and Leader-Williams 2005). Indeed, in the Serengeti, it is elephants that have received the most attention with respect to damage of crops in cultivated areas surrounding the protected areas. The Wildlife Division in Tanzania, which is responsible for dealing with these issues on village land, generally receives reports of HEC from along two belts of land to the northwest

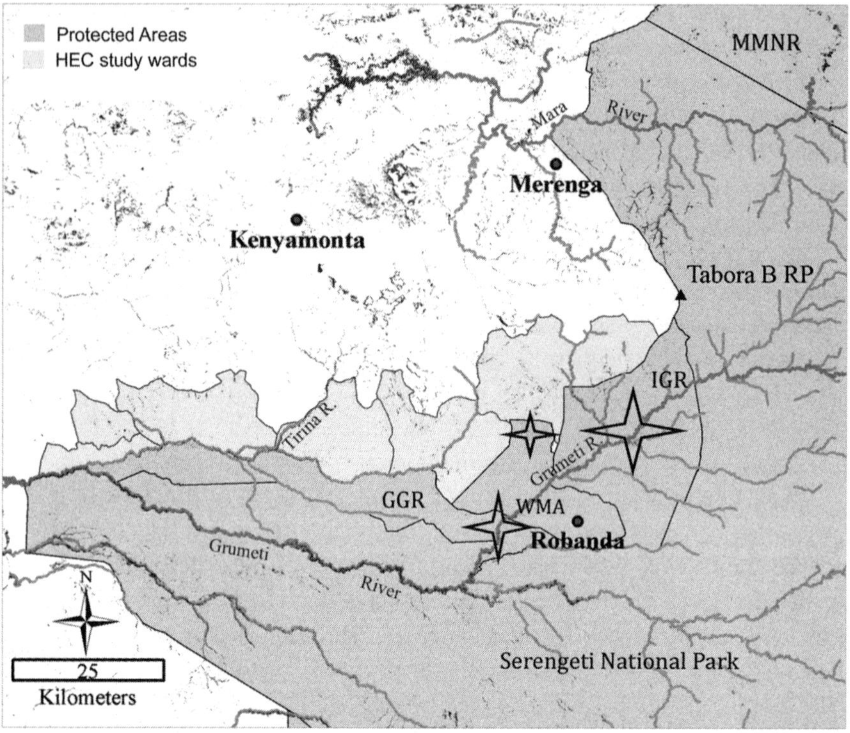

Fig. 21.5 Important features in western Serengeti that may coincide with increased incidents of human-elephant conflict (HEC). Starred areas indicate known intensive elephant occupation in the dry season coincident with HEC reports. IGR = Ikorongo Game Reserve; GGR = Grumeti Game Reserve; WMA = Ikona wildlife management area. Ranger reports also indicate intense HEC around Tabora B ranger post (RP). Dark gray indicates riverine forest and forest patches through which elephants often move when traversing human-dominated areas.

of SNP; the interface of villages bordering the protected areas (Merenga through to Robanda in Serengeti and Tirina in Bunda), and less frequently along the Mara river from Merenga to Kenyamonta (see fig. 21.5). The HEC appears to have intensified in recent years, both in western and northern Serengeti (around the MMNR) due to a combination of rapidly increasing human populations along the protected area boundaries (National Bureau of Statistics, Tanzania 2002), greater elephant numbers in the ecosystem (TAWIRI 2003; Walpole et al. 2003), and improved levels of security for game protection in the reserves bordering the SNP.

A study in eight wards (including 18 villages with about 90 subvillages) in a high HEC zone in western Serengeti, bordering Ikorongo and Grumeti GRs and the SNP, found almost 750 elephant crop-raiding events between March 2005 and February 2006 alone. In those 11 months, 10 categories of food and cash crops were damaged during forays by elephants in 2,258 crop fields (an average of 2–3 cultivated plots per incident). An activity peak of crop raiding corresponded with the harvest season (May to July) as elephants destroyed mature crops, mostly sorghum (20.8%) and maize (18.9%), while cassava, a perennial crop, was extensively damaged in the dry season (May to October). Relatively small groups of 6–10 elephants accounted for just over 25% of the reported incidents (Malugu and Hoare 2006). Elephants mainly leave the riverine and forested refugia to raid crops during the night, but there have also been reports of daytime raids by entire breeding herds. Although elephant crop damage is less frequent than raiding by smaller species, and may account for less economic loss in a given area, localized impacts of HEC are often catastrophic for individual landowners. As a result, elephants usually rank first in people's perceptions of the worst wildlife offenders (Naughton-Treves 1998).

Wards bordering the Ikorongo GR and encompassed within the WMA (fig. 21.5) had the highest number of HEC incidents (25.4% and 22.1%, respectively). During the dry season high concentrations of elephants are found around the rivers in Ikorongo and Grumeti GR and in the WMA (fig. 21.5). Historically, elephants resided in these areas before widespread poaching depleted their numbers (Sinclair and Arcese 1995; Walpole et al. 2003). Since the cessation of poaching in the late 1980s, the elephant population has expanded westward. The high incidence of raiding in these zones is therefore likely due to their proximity to these favored elephant refuges. In recent years, these areas have also been subject to the highest rates of land conversion (Estes et al., chapter 18), which must exacerbate the problem.

People with different livelihood strategies living within the ecosystem are differentially affected by depredation, crop damage, and dangerous encounters with wildlife. Agriculturalists suffer most HEC, whereas pastoralists suffer most from predators (table 21.1). These differences reflect the greater suitability of habitat and availability of wildlife prey on rangeland versus cultivated land, as well as cultural differences in hunting practices. Agriculturalists such as Kuria and Sukuma have a history of hunting and wildlife populations in these areas are depauperate, whereas Maasai pastoralists rarely hunt for meat. Conversion of land to agriculture and settlements further reduces the availability of wildlife habitats, yet at the same time attracts crop-raiding species such as elephants from neighboring habitats. Increasing cultivation in pastoralist areas is therefore also likely to result in an increase in HEC. Indeed in group ranches bordering the MMNR, intensifying HEC may directly reflect changes in land use, particularly increased sedentarization and agriculture (Kaelo 2007).

Cumulative losses of livestock and crop damage for impoverished communities living around the Serengeti have substantial impacts on livelihoods. The levels of depredation reported by Holmern, Nyahongo, and Roskaft (2007) are among the highest reported anywhere in Africa. Depredation, though relatively minor compared to other sources of livestock mortality, including disease (mortality due to disease is more than 10 times higher than due to depredation in the Maasai Steppe; Ogutu, Bhola, and Reid 2005; Kissui 2008) and drought-induced starvation (as observed in late 2009), is a threat for which the source can be easily identified. It is estimated that up to one-third of households in the western Serengeti regularly lose around a quarter of their harvest to wildlife, despite efforts to minimize crop damage (Emerton and Mfunda 1999). Extrapolations from household incomes suggest that losses amount to as much as US $500,000 a year in the western Serengeti or around US $155 for each of the 3,000 households who regularly lose crops to wildlife (Emerton and Mfunda 1999). Negative attitudes toward wildlife, with detrimental ramifications such as retaliatory killing, should therefore not be surprising.

Yet, despite these conflicts, a remarkable diversity of wildlife persists across the ecosystem. Few places in the world support large carnivores in such numbers, which would not be tolerated in North America and Western Europe (Breitenmoser 1998; Bangs et al. 2005; Houston, Bruskotter, and Fan 2010). In the predominantly pastoralist region to the east of the Serengeti, African wild dog populations have been growing. Despite heavy livestock losses, local people show remarkable tolerance, with >30% respondents living in proximity to African wild dogs in the eastern Serengeti ecosystem expressing positive attitudes toward the species. Indeed, the culture and practices of pastoralism are more compatible with wildlife conservation than agriculture, with studies from elsewhere in East Africa suggesting that there are fewer human-wildlife conflicts in areas with intact wild prey populations, and where traditional herding practices are retained (Woodroffe et al. 2005). This appears to be true in the GSE and, notwithstanding issues of poverty, the Maasai in Ngorongoro are proud of their pastoralist livelihood, which is increasingly threatened elsewhere. The NCA has therefore proven more successful than most other land-use policies in promoting coexistence of wildlife and people. Thus, although emphasis has been placed on conflict problems, these data suggest there are also positive lessons to be learned from the continued coexistence of human and large predators in the GSE. Nonetheless, rapid expansion of human populations in pastoral areas of East Africa and the concomitant drop in livestock per capita are progressively driving economic and livelihood diversification, including increasing adoption of subsistence and commercial agriculture. Hence, although coexistence has been more harmonious in pastoral than agricultural areas in the past, the accelerating transition to agropastoralism makes future prospects of continued peaceful coexistence less likely.

DISEASE

Living alongside wildlife poses many risks, but these are usually considered in terms of losses from depredation, crop damage, and dangerous encounters. Losses due to diseases spread by wildlife are also recognized as important in many parts of the world. However, living within the Serengeti ecosystem has a particularly complex range of repercussions regarding disease transmission. For example, from the 1950s, bush was cleared and woodland burned to reduce tsetse abundance and open areas for grazing. This created a lasting legacy, defining the protected area boundaries and facilitating the expansion of human settlements, which are now posing increasing threats for wildlife conservation (Lamprey and Reid 2004).

Wildlife are frequently thought of as a conduit for transmitting infection to humans and domestic animals, but in reality the reverse is often true (Haydon et al. 2002). This is exemplified by rinderpest, a devastating pandemic that swept across the African subcontinent at the end of the nineteenth century. Rinderpest had the single-greatest impact on shaping the GSE over the course of the last century (see Serengeti I, II, and III). The disease killed over 80% of the wildebeest population, reducing them to around 200,000 individuals for more than 50 years. Wild ungulates were believed to harbor the disease, but only when vaccination of livestock began in a cordon sanitaire surrounding the Serengeti did wildebeest numbers rebound. Successful elimination of rinderpest through livestock vaccination, therefore, identified cattle as the reservoir for infection, rather than wildlife, contrary to the original assumption (Dobson 1995). Similarly, it has now become clear that the large populations of domestic dogs living around the Serengeti are essential for maintaining rabies, a multihost pathogen, which is both a major public health concern and a threat to endangered carnivores such as African wild dogs (Lembo et al. 2008).

Large host populations are required for acute fatal infections (such as rabies) or highly immunizing pathogens (such as rinderpest) to persist. This is why multihost pathogens pose a particular threat for wildlife, because they can spill over from livestock and domestic animals, which live at sufficiently high densities and have rapid rates of population turnover. Such pathogens would be unlikely to persist in isolated wildlife populations and frequent spillover can prevent the recovery of small wildlife populations.

The GSE is large enough to support resilient populations of some species. For example, Serengeti lions recovered rapidly from the 1994 outbreak of canine distemper virus, which is estimated to have killed more than a third (~1,000 lions) of the population. But for endangered species, such as African wild dogs, which are now reestablishing a decade after major disease-induced population declines, the threats from disease are once again imminent. Wildlife populations in Ngorongoro Crater are also more vulnerable to disease due to their isolation and smaller populations, which are subject to greater demographic stochasticity and reduced genetic diversity.

For many shared pathogens in the Serengeti the directionality of transmission is unresolved (Lembo et al., chapter 19), despite the substantial impacts on public health, livestock production, and livelihoods. Trypanosomiasis, for example, is thought to be the biggest economic threat to livestock keeping in sub-Saharan Africa, and the Serengeti ecosystem is one of the historical foci of Trypanosomiasis in Tanzania (Fairbairn 1948). In Ngorongoro district, *Brucella* is the suspected cause of thousands of hospitalizations each year (fig. 21.6), while *Bacillus anthracis, Echinococcus*, and *Mycobacteria* spp. all cause very serious disease (anthrax, hydatid, and typical and atypical tuberculosis). Yet there is only relatively anecdotal information on the distributions of these pathogens and the burden that they cause. Hospital records, which are a crude indicator, show striking differences in their distributions (fig. 21.6), which likely relate to environmental and population risk factors. Misdiagnosis of these diseases is common, so hospital records should be treated with caution. Even when domestic animals are the known reservoir for disease, wildlife can often be perceived as the culprit. For example, many more people are bitten by domestic dogs than wildlife during outbreaks of rabies, but wildlife are often blamed for the spread of infection, even though domestic dogs are now known to be the population critical for maintaining the disease (Lembo et al. 2008). However, there is no evidence for current persecution of wildlife due to the perception of zoonotic disease risks. This contrasts with the 1960s, where vast numbers of bushbuck were culled due to fears that they were reservoirs of sleeping sickness.

One important disease with a known wildlife reservoir that greatly impacts local people in the Serengeti ecosystem is malignant catarrhal fever (MCF). Almost all wildebeest calves are born with the virus, *alcelaphine herpesvirus 1*, which is benign to wildebeest, but causes MCF among cattle. There is currently no available cure for MCF (although a vaccine is now being tested) and it is nearly 100% lethal to cattle that become infected. Wildebeest calves shed the virus in ocular and nasal secretions and the virus can survive outside the host from one to eighteen hours depending on the intensity and duration of exposure to sunlight (Rwambo et al. 1999). Wilde-

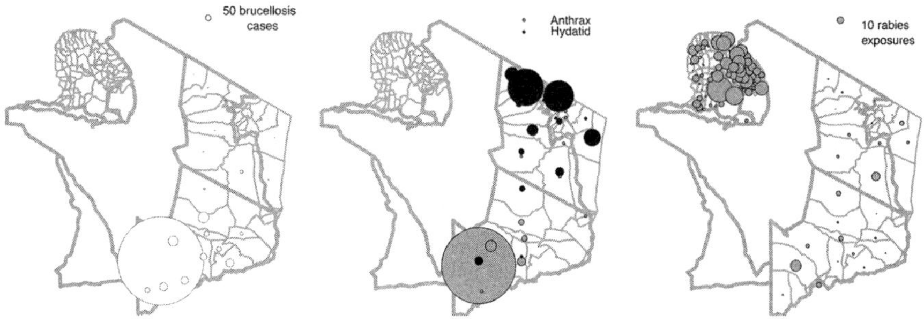

Fig. 21.6 Incidence of zoonotic pathogens in the Ngorongoro and Serengeti districts from 1999 to 2009. The radius of each circle is proportional to the number of victims that reported to hospital from each village. Brucellosis and bites by rabid animals have been scaled because of the high numbers of reports. Brucellosis data were only available for Ngorongoro.

beest calves remain infectious from birth until they are two to four months old. MCF is well known to the Maasai, who distinguish a time when cattle can mix with wildebeest and a time when they cannot, based on the turning point of when the color of calves' coats changes from tan to brown, when the calves are three to four months old.

The extensive wildebeest migration distinguishes the Serengeti ecosystem from almost all other national parks and protected areas in Africa, but it comes at a cost. Because of the risks of MCF transmission, cattle need to avoid wildebeest during the wildebeest calving season (with calving synchronized over a short two-week period during the wet season). Thus, the nutritious short-grass plains of the eastern Serengeti and western NCA become unavailable to cattle during this period. Lack of access to highly nutritious grazing is a major constraint for cattle, preventing animals from regaining condition at a critical time of year after the dry season.

Prior to the 1960s when wildebeest numbers were low (~200,000 animals), the Maasai were able to block off water holes and some migratory routes (such as the entrance to the valley, Angata Kiti) during the calving season. This secured pastures for their cattle far from the migrating wildebeest. From the mid-1960s to the late 1970s wildebeest numbers rebounded due to livestock vaccination programs (Sinclair and Arcese 1995). As wildebeest numbers grew it was no longer possible to fence off water holes or to block migratory routes. By the 1960s wildebeest began migrating into areas such as the Salei plains, where they had not been seen in living memory. The result was that Maasai were forced to abandon traditional movement patterns. Unable to utilize the short-grass plains during the wet season, herders increasingly confined cattle to unproductive highland and woodland pastures. This not only deprived the cattle of the nutritious grasses of the plains, but also increased exposure to ticks, which are vectors for east coast fever, among other diseases, and exacerbated land-use pressure on fragile highland ecosystems.

Although the Maasai rank MCF as one of the most important diseases that affect cattle in the NCA and LGCA, the incidence of cattle dying from MCF is relatively low (Cleaveland et al. 2000), but variable. In 2000, MCF was reported to have killed just over 5% in risk areas of the Ngorongoro district (Cleaveland et al. 2000), and in 1992 to 1993 the total estimated herd loss in the NCA was 20%, but in 1994 losses of less than 1% were reported (McCabe 1994). Climate is a major driver of this variability. With unpredictable rainfall patterns, wildebeest movements are also less predictable and avoidance by Maasai may be more difficult. Furthermore, in times of drought, many Maasai feel that they have no option but to take the chance and move onto the plains, feeling that without access to these grasses their cattle will not survive the dry season or reproduce. Many of those who take this risk seek blessings from Maasai laibons (ritual/spiritual leaders) to protect their cattle. Estimates suggest that the Maasai in the NCA could manage an additional 20,000 head of cattle if they could graze on the plains in February and March (Boone and Coughenour 2001). An old Maasai herd owner looking out over the plains during late February said to McCabe: "the wildebeest come and bring the 'disease of the wildebeest' with them, they have their babies, then they leave and take all the grass." This subtle but hidden cost has had a major impact on the Maasai and their livestock living on the eastern border of Serengeti.

RETALIATION

Many of the conflicts described inexorably lead to retaliation, usually in attempts to kill the animal(s) believed to be responsible. Historically, lethal control has been the most common method for resolving conflicts between wildlife and livestock (Holmern, Muya, and Roskaft 2007), with eradication campaigns carried out for many carnivores (Rasmussen 1999; Woodroffe and Frank 2005) leading to the near extinction of some species (like African wild dogs; Fanshawe et al. [1997]). Perceptions and practice have now changed, especially among conservation managers, with many species protected by law. However, options available to local people as a means of coping and responding to attacks by wildlife are limited and given the perceived threats that wild animals pose, the realized losses incurred and lack of compensation, retaliation by local people should perhaps be expected.

Of the large predators in the Serengeti, lions appear to be the most vulnerable species to direct retributive killing, at least in Maasai areas. This may be due to behavioral differences between predators, with lions being most aggressive and more likely to defend a livestock carcass against humans, thus exposing themselves to confrontations (Ogutu, Bhola, and Reid 2005; Kissui 2008). In comparison, hyenas are shier and quickly disappear after an attack event, while leopards are stealthy and secretive. Lions also more frequently predate cattle, which are of greater value both monetarily and culturally to Maasai pastoralists than the small stock typically attacked by hyenas and leopards, thus stimulating more resentment.

Species deemed to be either dangerous or pests are also killed proactively, even if not in immediate response to a depredation or crop-raiding event. Differentiating retaliation from unprovoked killings, however, is not easy, and retribution may be disproportionate to the scale of provocation, making this a tricky issue to both investigate and mitigate. Legally, wildlife can be killed in defense of human life and livestock in the pastoralist areas of the NCA (Homewood and Rodgers 1987). According to reports from Maasai elders, cattle depredation often merely provides an excuse for young Moranis to carry out ritual lion killings, with cultural practices remaining an important motivation in at least one part of the NCA (Ikanda and Packer 2008). Sometimes apparent retribution killings raise suspicion; villagers in the Serengeti district killed a cheetah in early 2010 in response to an alleged attack on livestock, although this would be unusual for such a typically shy species. The Wildlife Division also allows retaliation if it is deemed to be in response to an attack and will carry out problem-animal control in response to requests by villagers. For example, from 2005 to 2006, four people were killed during encounters with elephants in western Serengeti and wildlife officers killed four elephants in response (Malugu and Hoare 2006).

It is not clear how much of a problem retaliation poses to wildlife populations in the Serengeti. Studies in East Africa have found that pastoralists kill lions in direct proportion to the number of livestock killed by lions (Ogada et al. 2003; Holmern, Nyahongo, and Roskaft 2007; Kissui 2008). This also appears to be largely true in the NCA, except in the open plains where the rate of lion killing was nearly as high as in the woodlands, but depredation by lions was relatively rare (Ikanda and Packer 2008). Numbers killed in the plains (~2 lions per year from 1985–2005) did not seem to vary with number of livestock lost, but reflected the presence of recently initiated Morani cohorts. This suggests that in the pastoralist-occupied areas of the ecosystem lions are killed largely in retaliation, except in the short-grass plains, where Morani are reported to travel from other parts of Maasailand to participate in *Olamayio*. The impact of ritual lion killings is negligible on the overall Serengeti lion population, which has been consistently growing over the latter part of the twentieth century (Packer, Hilborn et al. 2005). Losses from trophy hunting in neighboring concessions (quotas set at 24 males per year) and snares are likely to outweigh the impact of the Maasai hunting practices, especially since nomadic-dispersing males are the lions that are expected to suffer from elevated mortality by entering these hunting areas (Ikanda and Packer 2008). In contrast, long-term data from north of the Mara shows that the proportion of hyena deaths caused by humans has been increasing, particularly over the last decade (Pangle and Holekamp 2010).

One form of retaliation that is likely to have substantial population impacts is the use of poison. Anecdotal reports from western and eastern Serengeti suggest that villagers sometimes lace livestock carcasses with poisonous substances to bait predators. These carcasses indiscriminately attract a range of predators and scavengers. Endangered species such as African wild dogs are particularly at risk, and there is still some debate as to the role that organophosphate poisoning may have played in a die-off of African wild dogs in Ololosokwan, Ngoronogoro district in 2007 (Fyumagwa et al. 2008). Increasing reports of predator and vulture poisoning from throughout East Africa (Frank, Woodroffe, and Ogada 2005), primarily involving the cheap and widely available organophosphate Furadan (carboran), are worrisome (Hazzah, Borgerhoff Mulder, and Frank 2009). Some conservationists fear that in parts of Maasailand lions are being poisoned and speared at an unsustainable rate (Hazzah 2006), while poisoning is likely to be responsible for dramatic declines in vulture populations in the MMNR and surrounding areas over the last 30 years (Virani et al. 2011).

Elephants are perceived particularly badly, because of their large size, physical threat, nocturnal raiding habits, and catholic diet. Lethal elephant control is therefore widely used and also has detrimental population impacts. For example, in 2006 a Serengeti elephant "offensive" was launched following high-level political pressure in Tanzania. By August of that year the Wildlife Division had killed 14 elephants, with ten animals killed in June alone. In the western Serengeti it is unlikely that the relatively small and currently recovering elephant population can sustain this level of offtake. Nor is this kind of retaliation likely to effectively target the "problem" animals.

Retaliation is an overt and direct threat to wildlife populations, and in some areas of the world, conflict-related mortality is thought to be so high that the borders of protected areas represent population sinks through an "edge effect" (Woodroffe and Ginsberg 1998; Kolowski and Holekamp 2006). In the Serengeti ecosystem it is clear that retribution killings of wildlife are common and probably increasing, but their population impacts have largely not been ascertained. This is clearly a priority if evidence-based policies for wildlife conservation and alleviation of human-wildlife conflicts are to be developed. Moreover, how attitudes toward wildlife influence the magnitude of retaliatory or retributive action is unclear, and the extent to which conservation education and awareness programs can change such attitudes and behaviors to ultimately improve conservation prospects is a relatively underresearched area in the GSE and requires attention.

COMPETITION FOR LAND

The diverse socioeconomic and political landscape means that the impacts from land competition vary around the ecosystem. While land policies and land-use change are beyond the scope of this chapter, we briefly discuss some of the repercussions to highlight the wider complexity of human-wildlife conflicts and the need for a holistic view in terms of mitigating these conflicts.

Human population growth and expansion of settlements around the park, including the sedenterization of formerly seminomadic pastoralists is probably the most persistent threat to the ecosystem (Western, Groom, and Worden 2009). This is the root cause for which many of the human wildlife conflicts described are symptomatic and has multiple negative and cascading consequences for the ecosystem. How the changes in human populations, settlements, and agricultural activities continue to affect the wildebeest migration, which historically moved in a wider circuit crossing many increasingly populated areas, is a consistent worry.

There are now concerted efforts to develop transboundary management plans, but the Kenya/Tanzania border bisecting the ecosystem has served as an experiment in land-use policies and management strategies. The main conclusions have been that land-use policies, including privatization of land tenure and land subdivision epitomized in the group ranches of the Mara (Lamprey and Reid 2004), have dramatically affected the integrity of parts of the ecosystem and the viability of its wildlife populations. The spread of mechanized agriculture on the Kenyan side of the ecosystem has been driven by increasing market opportunities, and is responsible for the greatest changes in land cover and depletion of wildlife populations in pastoralist areas (Homewood et al. 2001; Serneels and Lambin 2001). Marked declines in a number of ungulate populations in this part of the ecosystem were correlated with habitat deterioration due to changing land use in pastoral ranches driven by human population growth and increased settlements, while declines within the MMNR were most severe in the areas subject to most livestock incursions and poaching (Ogutu et al. 2009; Ogutu et al. 2011).

On the western border of the ecosystem most agriculture is still small scale for subsistence purposes. While the negative impacts of subsistence agriculture are less than large-scale mechanized agriculture, the resulting habitats are generally not hospitable to wildlife, nor are agriculturist livelihoods compatible with conservation. The population densities and rates of growth in the zone between Lake Victoria and SNP are among the highest in Tanzania. Human encroachment along the western boundary is intensifying the conflict between wildlife and human populations (Holmern, Nyahongo, and Roskaft 2007) and hardening the boundary of the protected areas, reducing any buffering effect of uncultivated land. A recent shift in management of hunting blocks adjacent to the national park has also led to more intensive wildlife management in the area, coupled with a decrease in consumptive wildlife utilization (trophy hunting). As a result, local wildlife populations show a significant increase in this buffer zone area, exacerbating the hard boundary between protected area and community land.

Growth of livestock holdings has not kept pace with the growth of pastoralist populations to the north and east of the Serengeti, with a range of repercussions. Fewer livestock per household means that Maasai are less buffered from stock losses caused by environmental variability and may be driven further into protected areas in search of quality forage during times of drought (Butt, Shortridge, and WinklerPrins 2009). Climate change could heighten these conflicts between people and wildlife over diminishing water and grazing resources. A more immediate consequence is that pastoralists have been forced to diversify their livelihoods. While more limited cultivation has spread through the Tanzanian pastoralist communities (McCabe, Leslie, and DeLuca 2010), the rate has been slower than on the Kenyan side due to regulations in the NCA. Because of difficulties surviving on fewer livestock per capita, Maasai requested the Tanzanian government to relax the ban on cultivation inside the NCA in 1992 (Thompson 1997). But, the increases of agriculture contravened the principles of sustainable coexistence agreed to by the Ngorongoro Maasai, and new legislation was effected in 2007 expressly forbidding all forms of agriculture in the NCA. These issues are understandably aggravating relationships between local Maasai and conservation stakeholders. While wildlife tourism has been seen as a potential alternative source of income for pastoralists, the distribution of benefits from conservation schemes in the Mara and the NCA have been far from equitable (Thompson and Homewood 2002). Particularly in the Mara, policies both empathetic with pastoralist culture and compatible with wildlife conservation have been lacking.

More generally, land-use change and increased agriculture upstream of the Mara River has been causing serious hydrological ramifications throughout the ecosystem. Agricultural encroachment and deforestation have caused soil erosion in the upper catchments within the Mara river basin and sediment build up downstream, leading to the expansion of wetlands in the lower reaches (Mati et al. 2008). As a result, river flows have become more extreme (lower flows and higher flood peaks), increasing the vulnerability of both dependent wildlife populations and people (Mati et al. 2008). The wider ecological impacts of the changing hydrological cycle is relatively understudied but will also be critical for long-term conservation planning and ecosystem management.

Despite the risks of living on the protected area boundaries, the rate of in-migration toward the Serengeti is still above the national average. This is consistent with the higher rates of population growth reported near protected areas throughout Africa and Latin America (Wittemyer et al. 2008). It seems that migrants generally suffer most from these conflicts with wildlife, as they are forced to occupy the most marginal land (Reid et al. 2008). Thus, there must still be perceived benefits to moving into the area surrounding the Serengeti. These benefits may perhaps be due to the availability of bushmeat (as a subsistence commodity or source of income; see Rentsch et al., chapter 22) or may perversely be generated by community-based conservation programs. Understanding these perceptions and taking measures to reduce the attractiveness of the park boundaries to potential immigrants should therefore be considered a crucial step for long-term ecosystem management.

CONFLICT MITIGATION

Potential methods for mitigating human-wildlife conflicts include statutory protection, shooting of problem animals, fencing of populations, aversion techniques, wildlife relocation, improved livestock husbandry, participatory management, monetary compensation, and sustainable utilization (e.g., wildlife tourism or sport hunting) to generate revenue to offset losses as well as long-term land-use planning. The effectiveness of legislation to protect wildlife from direct persecution is discussed elsewhere (Rentsch et al., chapter 22); fencing is generally not considered a viable option for Serengeti's migration, and wildlife relocations have only recently entered the agenda in Serengeti. Therefore, our discussions focus on conflict resolution efforts that have so far been employed in the ecosystem.

Compensation schemes for households suffering from depredation and crop destruction are generally difficult to implement and rarely successful in alleviating problems (Ogada et al. 2003; Treves and Karanth 2003; Hazzah, Borgerhoff Mulder, and Frank 2009). Until recently there were no compensation schemes in the Serengeti for damages caused by wildlife, but recent regulatory changes now permit payouts ("consolation," as described in the Tanzanian Wildlife Conservation Act of 2009). However, these are largely regarded as ineffectual, as the Wildlife Division must validate evidence of depredation or crop damage, yet they have little resources with which to do this, and the payout to victims is only issued annually, which is considered too late by impoverished villagers. In contrast, better-resourced pilot compensation schemes in Maasailand show promise (Maclennan et al. 2009), but are also not without problems and are unlikely to be adopted by government and replicated on a large scale in the near future.

Retaliatory killing, although permitted by law in the defense of human life or livestock (Tanzanian Wildlife Conservation Act of 2009), raises serious conservation challenges and may have only short-term impacts, if any at all. For species like lions that have cultural significance among certain tribes, retaliatory killing provides an incentive for people to participate in hunts, making lions unusually vulnerable to conflict with humans (Kissui 2008). For Sukuma, retaliatory hunts are used to disguise lion hunts for economic benefits (Borgerhoff Mulder et al. 2009). Oftentimes the wrong animals are targeted, especially if techniques such as poisoning are used and the responsible animal(s) are, in most cases, unlikely to be accurately identified. Elephants rapidly habituate to any scaring by humans and killing selected crop raiders merely results in their replacement by other individuals (Hoare 2001a, 2001b). Shooting elephants, therefore, has no long-term deterrent effect but only acts as a temporary palliative to those affected, by perceived retaliation and indirect compensation through the provision of some welcome free meat in village communities. Thus, conflict levels rapidly reoccur once new problem elephants replace those that were killed.

Human wildlife conflicts can be minimized through good management practices and approaches involving low-cost technologies. Research effort is going into improving livestock husbandry techniques to ameliorate depredation, with a variety of practical approaches being tested such as reinforcing bomas (kraals) where livestock are kept at night (Ogada et al. 2003; Kissui 2008). Improving livestock security during the daytime through better herding practices can potentially reduce incidences of livestock predation in pastoralist communities, where this is a more significant problem.

In many parts of Europe and North America, guarding animals have been used successfully to protect livestock against predation (Breitenmoser et al. 2005; Gehring, VerCauteren, and Landry 2010), and in Namibia guarding dogs are used to protect sheep and goats against cheetah predation (Marker, Dickman, and Macdonald 2005). Evidence as to whether dogs significantly reduce livestock depredation around the Serengeti has been conflicting (Ogada et al. 2003; Kolowski and Holekamp 2006; Woodroffe et al. 2007; Ikanda and Packer 2008), but most herders already use dogs and depredation remains a problem.

Methods that have proven successful for mitigating HEC elsewhere in Africa (Hoare 1999; Osborn and Parker 2002), including the Transmara region in Kenya (Sitati and Walpole 2006), are currently being trialed and adapted for local use in western Serengeti. This approach moves away from conventional measures of scaring and killing elephants, except in certain individual cases. Fences or barriers that are affordable for most local people cannot confine elephants, and so an opposite strategy of protecting the targets (crops) that elephants destroy is being pursued. It is recognized that one deterrent alone will not work, and therefore the approach involves the simultaneous use of several low-cost, low-tech deterrents based on upgrading and improving "traditional" self-defense already used by farmers. This "package" integrates improved vigilance, passive deterrents and active deterrents, and one of its main innovations is taking advantage of the olfactory deterrents *capsaicin*, the active ingredient of chilies (Osborn and Parker 2002), or even tobacco. Although HEC incidents can never be eliminated entirely with these methods they can be reduced to a level that can be tolerated locally. The success of trials in western Serengeti has led to the adoption of these methods by some farmers on the northwest boundary of SNP. Developing such a strategy at the district level in Tanzania, rather than relying only on central government policy, shows definite promise (Mpanduji and Malima 2006).

Community conservation work is considered critical to the mitigation of conflicts, but as already mentioned, can also contribute to the problem. Key issues include determining the responsibility for problems in partnership with the authorities, receptivity to implementing improved self-defense mechanisms—for example, the implementation of packages to reduce HEC (described above) if education about techniques and limited financial assistance is provided—and participation in local land-planning exercises with the objective of reducing conflict. There is considerable evidence that people's attitudes toward wildlife conservation are negatively affected by costs incurred from human-wildlife conflicts (summarized in Kideghesho, Roskaft, and Kaltenborn 2007) and positive attitudes supportive of conservation efforts have been reported from villages enrolled in the Serengeti's community conservation programs (Kideghesho, Roskaft, and Kaltenborn 2007). Tangible benefits from conservation without doubt motivate local people to initiate changes in attitudes and action, but these benefits need to be sufficiently large and equitably distributed for improvements to be sustained, and it is not yet clear whether conservation programs in the Serengeti have achieved this. Innovative management approaches are required to reconcile these issues, protecting the welfare and health of affected communities while conserving biodiversity, wildlife populations, and environmental integrity. Wildlife management areas (WMA) are a step toward empowering local communities who are on the front lines of wildlife conflict to be responsible for wildlife management and conflict resolution, as they in turn receive direct benefits from wildlife in the form of tourism investment. However these strategies are still at an early stage in the Serengeti, and their effectiveness in altering behavior of those most at risk from wildlife conflict remains to be seen.

In some parts of East Africa, for example the group ranches of Laikipia, the development of wildlife tourism operations is being seen as a potentially profitable way to offset the costs of wildlife damage (Frank, Woodroffe, and Ogada 2005; Walpole and Thouless 2005). Studies from other ecosystems show that local tolerance is often greater where revenue benefits are received, for example, through tourism (Naughton-Treves 1998). Tour operators have negotiated contracts with community groups in several of the areas where wild dog populations have been recovering and these may be playing a positive role in improving local tolerance in spite of livestock losses to dogs. Conservancies adjacent to the MMNR hold the promise of reducing conflicts and helping wildlife populations rebuild. They are popular and already operational in some areas, although their long-term success remains to be seen (Reid et al., chapter 25). However, tourism is a fragile business and challenges remain with building technical capacity for running local tourism operations, which mostly still require external subsidies, and appropriate distribution of benefits. In theory, trophy hunting may prove a more profitable approach, but sport-hunting operations are also difficult to administer and the reality of achieving sensitive and sustainable utilization of the wildlife resource is invariably beset by political and ethical challenges. In the long run, more efficient coping strategies for rural farmers must continue to be sought that explore alternative crops, agricultural practices, and income-generating schemes (O'Connell-Rodwell et al. 2000; Osborn and Hill 2005).

A complex problem like human-wildlife conflict cannot be effectively addressed unless the causes are understood. Systematic data collection and economic quantification of the problem is therefore the first and most important step (De Boer and Baquete 1998; Hill 1998), but with a few exceptions this has not been extensively carried out in the Serengeti. Monitoring wildlife populations and assessments of the population impacts of conflicts are needed to quantitatively evaluate mitigation strategies. There is also a need to monitor the changes in the perceptions and behavior of local people in response to conservation education and awareness programs, which will be integral to long-term conservation planning. Most critical is perhaps addressing the issues of human immigration and land-use change and their local effects on human-wildlife conflict. The consequences of continued human population growth adjacent to large wildlife populations could overwhelm even the most effective mitigation strategies.

Stakeholders in any proposed strategy to reduce wildlife conflict in the Serengeti ecosystem include the Wildlife Division, TANAPA, TAWIRI, district councils, WMAs, local NGOs, and a variety of tourism groups and privately managed wildlife concessions. Such potentially strong combined resources should make the area one of Africa's best potential testing grounds for conflict research and mitigation, but this is currently not the case. The Wildlife Division, which is immediately responsible for resolving most of these issues in Tanzania, is chronically underfunded and understaffed. Unless the magnitude of the problem is recognized by local and national government institutions, the local capacity for response and mitigation will remain weak. These are long standing and well-known problems, but besides staffing and funding, there is a need for policies that more directly promote wildlife conservation. For example in Kenya, in the Mara, Amboseli, and Athi-Kaputiei, official policies promote cultivation to reduce food insecurity, but policies are also promulgated that promote conservation and wildlife-based tourism on the same parcels of land—the pastoral lands. These incompatible land uses cannot persist for long, thus policies are needed that can truly enable coexistence. Often wildlife conservation statutes and efforts focus on species protection but ignore the destruction of the ecosystems in which the species live (e.g., through land conversion to agriculture, harvesting of fire wood, construction materials, felling trees for charcoal burning, etc.). A major problem in Kenya is that the government owns the wildlife and individuals privately own land, but without user rights for wildlife they have no motivation for protecting wildlife on their land. An approach to resolving human-wildlife conflicts that considers both the wildlife species involved and protection and restoration of their habitats is needed.

CONCLUSION

For communities living around the Serengeti, wildlife is often regarded as a burden causing significant losses of crops and livestock at the household level, large opportunity costs of lost grazing and farmland held in protected areas, and immediate dangers to human life and well-being. As agricultural land becomes more scarce and local sources of income and employment hard to access, community members are often unwilling and economically unable to bear the costs associated with conserving wildlife on and around their lands (Emerton and Mfunda 1999). Approaches to wildlife management imposed over the course of the last century have exacerbated, rather than improved this situation, with local communities' rights to natural resources expropriated by protected areas. The Serengeti situation mirrors that of many other parts of Africa, where communities living within the vicinity of protected areas that earn a large proportion of national wildlife tourism revenue receive very little (as little as 1%) of foreign exchange earnings derived from this source (Obunde, Ormiti, and Sirengo 2005).

Negative perceptions from human-wildlife interactions can be amplified because isolated and rare events cause catastrophic losses for individual households and the risks are ubiquitous, with a clear culprit. This view is conflated by poor relations between park staff and local residents including poaching retributions, wealth disparities, and the lack of economic benefits returning to local communities. Many residents live within 15 km of the park boundary but do not report a single benefit from wildlife and have never had the opportunity to view wildlife outside of a conflict scenario (Schmitt 2010), decreasing the likelihood of their appreciating wildlife.

In developed countries, human-wildlife conflicts also stir intense emotions, with classic examples including bovine tuberculosis and badgers in United Kingdom, beaver introductions in Scotland, habitats for wolves and spotted owl, and brucellosis and bison in North America. Resulting policies may not appear to be rational or to consider long-term conservation implications, while confusing personal interests and wider community benefits with those of powerful lobby groups and individuals pushing agendas to the detriment of the environment. The contrast in Serengeti is that (a) the people who have the most to lose from human-wildlife conflicts have the least power, and (b) the losses to individuals are often much more serious and imminent in comparison to those in developed countries. Hence, arguing for biodiversity conservation with respect to human wildlife conflicts in the Serengeti requires a more upfront assertion of what is at stake.

The long-term consequences of human-wildlife conflict for the ecosystem are serious, with human immigration, population growth, and expansion of settlements underlying most of the problems. For some species (like elephants and African wild dogs), the threats are particularly severe and pose a major challenge to conservation. For others, such as lions, their populations may be more resilient to the amount of retaliatory and ritual killing that takes place, at least in some parts of the ecosystem. However, the quantitative information currently available to evaluate the situation is limited, and so these have not been prioritized.

Remarkably, despite the losses that are incurred from wildlife, positive attitudes toward wildlife are still present among communities around the Serengeti, and the pride shown by local people in their wildlife heritage provides optimism that solutions can be found. However, there remains a serious risk that failure to address human-wildlife conflict will continue to generate resentment toward wildlife and undermine conservation efforts. Involvement of local communities is increasingly recognized as key to the successful management and conservation of wildlife populations (Newmark et al. 1994). The conflicts are complex, and solutions must address issues of historical land tenure, human settlements, local and national politics, and ecology.

There is a need to protect rural livelihoods, reduce their vulnerability, counterbalance losses from wildlife with benefits, and foster community-based conservation with tangible profits to communities unlike existing models of community conservation, otherwise antagonism toward conservation objectives are likely to increase. The nature of the problem requires both short-term mitigation tools and longer-term preventive strategies and compatible land-use policies. Although methods are being tested and policies developed, it is essential that further effort be put into comprehensive mitigation measures that will be effective and sustainable for local communities and the ecosystem as a whole.

ACKNOWLEDGMENTS

Thanks to Emmanuel Sindoya, Zilpah Kaare, and Dana Weldon for help with data compilation, and Andrew Ferdinands for collecting village questionnaire data. We are grateful to Tanzanian ministries, TANAPA, TAWIRI, NCA Authority, Tanzania Commission for Science and Technology, and the National Institute for Medical Research for permissions; TANAPA and TAWIRI Veterinary Units, research projects from the Serengeti, Frankfurt Zoological Society, Lincoln Park Zoo, Grumeti Reserves, the Wildlife Division, livestock field officers and health workers from the Serengeti and Ngorongoro districts and local communities for collaborative assistance. This work was supported by the Wellcome Trust, the Department for International Development Animal Health Programme, and Google.org.

REFERENCES

Adams, W. M., and J. Hutton. 2007. People, parks and poverty: Political ecology and biodiversity conservation. *Conservation and Society* 5:147–83.

Bangs, E. E., J. A. Fontaine, M. D. Jimenez, T. J. Meier, E. H. Bradley, C. C. Niemeyer, D. W. Smith, C. M. Mack, V. Asher, and J. K. Oakleaf. 2005. Managing wolf-human conflict in the northwestern United States. In *People and wildlife, conflict or co-existence*, ed. R. Woodroffe and S. Thirgood, 340–56. Cambridge: Cambridge University Press.

Boone, R. B., and M. B. Coughenour, eds. 2001. A system for integrated management and assessment of East African pastoral land: Balancing food security, wildlife conservation, and ecosystem integrity. Final Report to the Global Livestock Collaborative Research Support Program, University of California, Davis.

Borgerhoff Mulder, M., E. B. Fitzherbert, J. Mwalyoyo, and J. Mahenge. 2009. The national Sukuma expansion and Sukuma-lion conflict. Unpublished report, Panthera, Partners in Wild Cat Conservation.

Breitenmoser, U. 1998. Large predators in the Alps: The rise and fall of man's competitors. *Biological Conservation* 83:279–89.

Breitenmoser, U., C. Angst, J. M. Landry, C. Breitenmoser-Würsten, J. D. C. Linnell, and J. M. Weber. 2005. Non-lethal techniques for reducing depredation. In *People and wildlife: Conflict or coexistence?* ed. R. Woodroffe, S. Thirgood, and A. Rabinowitz, 49–71. Cambridge: Cambridge University Press.

Butt, B., A. Shortridge, and A. M. G. A. WinklerPrins. 2009. Pastoral herd management, drought coping strategies, and cattle mobility in southern Kenya. *Annals of the Association of American Geographers* 99:309–34.

Cleaveland, S., L. Kusiluka, J. Ole Kuwai, C. Bell, and R. R. Kazwala. 2000. Assessing the impact of malignant catarrhal fever in Ngorongoro district, Tanzania. A study commissioned by the animal health program, Department for International Development, University of Edinburgh.

Conover, M. 2002. *Resolving human-wildlife conflicts: The science of wildlife damage management.* Boca Raton, FL: CRC Press.

De Boer, W. F., and D. S. Baquete. 1998. Natural resource use, crop damage and attitudes of rural people in the vicinity of the Maputo Elephant Reserve, Mozambique. *Environmental Conservation* 25:208–18.

Dobson, A. 1995. The ecology and epidemiology of rinderpest virus in Serengeti and Ngorongoro Conservation Area. In *Serengeti II: Dynamics, management, and conservation of an ecosystem*, ed. A. R. E. Sinclair and P. Arcese, 485–505. Chicago: University of Chicago Press.

Emerton, L., and I. Mfunda. 1999. Making wildlife economically viable for communities living around the western Serengeti, Tanzania. Geneva: International Union for Conservation of Nature.

Fairbairn, H. 1948. Sleeping sickness in Tanganyika Territory, 1922–1946. *Tropical Diseases Bulletin* 45:1–17.

Fanshawe, J. H., J. R. Ginsberg, C. Sillero-Zubiri, and R. Woodroffe. 1997. The status and distribution of remaining wild dog populations. In *The African wild dog: Status survey and conservation action plan*, ed. R. Woodroffe, J. R. Ginsberg, D. W. Macdonald, and the IUCN/SSC Canid Specialist Group, 11–57. Gland, Switzerland: International Union for Conservation of Nature.

Frank, G., R. Woodroffe, and M. O. Ogada. 2005. People and predators in Laikipia District, Kenya. In *People and wildlife: Conflict or coexistence?*, ed. R. Woodroffe, S. Thirgood, and A. Rabinowitz, 286–304. Cambridge: Cambridge University Press.

Fyumagwa, R., E. Masenga, E. Eblate, and M. Kilewo. 2008. Report on wild dog mortality in Loliondo Game Controlled Area in the Serengeti ecosystem, northern Tanzania. Unpublished report to TAWIRI, Tanzania Wildlife Research Institute, Arusha.

Gadd, M. E. 2005. Conservation outside of parks: Attitudes of local people in Laikipia, Kenya. *Environmental Conservation* 32:50–63.

Gehring, T. M., K. C. VerCauteren, and J.-M. Landry. 2010. Livestock protection dogs in the 21st Century: Is an ancient tool relevant to modern conservation challenges? *Bioscience* 66:299–308.

Graham, K., A. P. Beckerman, and S. Thirgood. 2005. Human-predator-prey conflicts: Ecological correlates, prey losses and patterns of management. *Biological Conservation* 122:159–71.

Haydon, D. T., S. Cleaveland, L. H. Taylor, and M. K. Laurenson. 2002. Identifying reservoirs of infection: A conceptual and practical challenge. *Emerging Infectious Diseases* 8:1468–73.

Hazzah, L. 2006. Living among lions (*panthera leo*): Coexistence or killing? Community attitudes towards conservation initiatives and the motivations behind lion killing in Kenyan Maasailand. PhD diss., University of Wisconsin.

Hazzah, L., M. Borgerhoff Mulder, and L. G. Frank. 2009. Lions and warriors: Social factors underlying declining African lion populations and the effect of incentive based management in Kenya. *Biological Conservation* 142:2428–37.

Hill, C. M. 1998. Conflicting attitudes towards elephants around the Budongo Forest Reserve, Uganda. *Environmental Conservation* 25:244–50.

Hoare, R. 1999. Training package for enumerators of elephant damage. International Union for Conservation of Nature, African Elephant Specialist Group, Nairobi, Kenya. www.african-elephant.org/hec.

———. 2000. African elephants and humans in conflict: The outlook for co-existence. *Oryx* 34:34–38.

———. 2001a. A decision support system for managing human-elephant conflict situations in Africa. International Union for Conservation in Nature, African Elephant Specialist Group, Nairobi, Kenya. www.african-elephant.org/hec.

———. 2001b. Management implications of new research on problem elephants. *Pachyderm* 30:44–48.

Hoare, R. E., and J. T. du Toit. 1999. Coexistence between people and elephants in African savannas. *Conservation Biology* 13:633–39.

Holdo, R. M., R. D. Holt, K. Galvin, S. Polasky, E. Knapp, and R. Hilborn. 2010. Responses to alternative rainfall regimes and antipoaching enforcement in a migratory system. *Ecological Applications* 20:381–97.

Holmern, T., J. Muya, and E. Roskaft. 2007. Local law enforcement and illegal bushmeat hunting outside the Serengeti National Park, Tanzania. *Environmental Conservation* 34:55–63.

Holmern, T., J. Nyahongo, and E. Roskaft. 2007. Livestock loss caused by predators outside the Serengeti National Park, Tanzania. *Biological Conservation* 135:518–26.

Homewood, K. 1987. Pastoralism, conservation and the overgrazing controversy. In *Conservation in Africa: People, policies and practice*, ed. D. Anderson and R. Grove, 111–28. Cambridge: Cambridge University Press.

Homewood, K., E. F. Lambin, E. Coast, A. Kariuki, I. Kikula, J. Kivelia, M. Said, S. Serneels, and M. Thompson. 2001. Long-term changes in Serengeti-Mara wildebeest and land cover: Pastoralism, population, or policies? *Proceedings of the National Academy of Sciences* 98:12544–49.

Homewood, K. M., and W. A. Rodgers. 1991. *Maasailand ecology: Pastoralist development and wildlife conservation in Ngorongoro, Tanzania.* Cambridge: Cambridge University Press.

Houston, M. J., J. T. Bruskotter, and D. Fan. 2010. Attitudes toward wolves in the United States and Canada: A content analysis of the print news media, 1999–2008. *Human Dimensions of Wildlife* 15:389–403.

Ikanda, D., and C. Packer. 2008. Ritual vs. retaliatory killing of African lions in the Ngorongoro Conservation Area, Tanzania. *Endangered Species Research* 6:67–74.

Kaelo, D. 2007. Human-elephant conflict in pastoral areas north of Maasai Mara National Reserve, Kenya. MS. thesis, Moi University, Kenya.

Karani, I. W. 1994. An assessment of depredation by lions and other predators in the group ranches adjacent to Masai Mara National Reserve. PhM thesis, Moi University, Kenya.

Kideghesho, J. R., E. Roskaft, and B. P. Kaltenborn. 2007. Factors influencing conservation attitudes of local people in western Serengeti, Tanzania. *Biodiversity and Conservation* 16:2213–30.

Kissui, B. M. 2008. Livestock predation by lions, leopards, spotted hyenas, and their vulnerability to retaliatory killing in the Maasai steppe, Tanzania. *Animal Conservation* 11:422–32.

Kolowski, J. M., and K. E. Holekamp. 2006. Spatial and temporal variation in livestock depredation by large carnivores along a Kenyan reserve border. *Biological Conservation* 128:529–41.

Lamprey, R. H., and R. Reid. 2004. Expansion of human settlement in Kenya's Maasai Mara: What future for pastoralism and wildlife? *Journal of Biogeography* 31: 997–1032.

Lembo, T., K. Hampson, D. T. Haydon, M. Craft, A. P. Dobson, J. Dushoff, E. Ernest, R. Hoare, M. Kaare, T. Mlengeya, et al. 2008. Exploring reservoir dynamics: A case study of rabies in the Serengeti ecosystem. *Journal of Applied Ecology*: 45:1246–57.

Maclennan, S. D., R. J. Groom, D. W. Macdonald, and L. G. Frank. 2009. Evaluation of a compensation scheme to bring about pastoralist tolerance of lions. *Biological Conservation* 142:2419–27.

Malugu, L. T., and R. Hoare. 2006. Systematic recording and assessment of human-elephant conflict in western Serengeti, Tanzania. Unpublished report, Tanzania Wildlife Research Institute, Arusha.

Marker, L. L., A. J. Dickman, and D. W. Macdonald. 2005. Perceived effectiveness of livestock-guarding dogs placed on Namibian farms. *Rangeland Ecology and Management* 58:329–36.

Masenga, H. E., and C. Mentzel. 2005. Preliminary results from a newly established population of African wild dog (*Lyacon pictus*) in Serengeti–Ngorongoro ecosystem, northern Tanzania. Unpublished paper presented at the 5th TAWIRI Scientific Conference, December 1–3, 2005, Arusha, Tanzania.

Mati, B. M., S. Mutie, H. Gadain, P. Home, and F. Mtalo. 2008. Impacts of land-use/ cover changes on the hydrology of the transboundary Mara River, Kenya/Tanzania. *Lakes and Reservoirs: Research and Management* 13:169–77.

McCabe, J. T. 1994.Wildebeest/Maasai interactions in the Ngorongoro Conservation Area of Tanzania. Unpublished report submitted to the National Geographic Society, Washington, DC.

McCabe, J. T., P. Leslie, and L. DeLuca. 2010. Adopting cultivation to remain pastoralists: The diversification of pastoral livelihoods in northern Tanzania. *Human Ecology* 38: 321–34.

Mpanduji, D. G., and C. Malima. 2006.Workshop to improve wildlife conservation and human-elephant conflict mitigation in Kilwa District, Tanzania. Unpublished report for WWF/TPO Tanzania, Dar es Salaam, Tanzania.

National Bureau of Statistics, Tanzania. 2002. *Household budget survey 2000/2001*. Dar es Salaam: President's Office, Planning and Privatization.

Naughton-Treves, L. 1998. Predicting patterns of crop damage by wildlife around Kibale National Park, Uganda. *Conservation Biology* 12:156–68.

Naughton-Treves, L., and A. Treves. 2005. Socioecological factors shaping local support for wildlife: Crop-raiding by elephants and other wildlife in Africa. In *People and wildlife: Conflict or coexistence?* ed. R. Woodroffe, S. Thirgood, and A. Rabinowitz, 252–77. Cambridge: Cambridge University Press.

Newmark, W. D., D. M. Manyanza, D. M. Gamassa, and H. I. Sariko. 1994. The conflict between wildlife and local people living adjacent to protected areas in Tanzania: Human density as a predictor. *Conservation Biology* 8:249–55.

Obunde, P., J. M. Ormiti, and A. N. Sirengo. 2005. Policy dimensions in human-wildlife conflicts in Kenya: Evidence from Laikipia and Nyandarua districts. *Institute of Policy Analysis and Research, Policy Brief* 11:1.

O'Connell-Rodwell, C. E., T. Rodwell, M. Rice, and L. A. Hart. 2000. Living with the modern conservation paradigm: Can agricultural communities co-exist with elephants? A five-year case study in East Caprivi, Namibia. *Biological Conservation* 93: 381–91.

Ogada, M. O., R. Woodroffe, N. O. Oguge, and L. G. Frank. 2003. Limiting depredation by African carnivores: The role of livestock husbandry. *Conservation Biology* 17:1521–30.

Ogutu, J. O., N. Bhola, and R. Reid. 2005. The effects of pastoralism and protection on the density and distribution of carnivores and their prey in the Mara ecosystem of Kenya. *Journal of Zoology* 265:281–93.

Ogutu, J., N. Owen-Smith, H. P. Piepo, and M. Y. Said. 2011. Continuing wildlife population declines and range contraction in the Mara region of Kenya during 1977–2009. *Journal of Zoology*. 285:99–109.

Ogutu, J. O., H. P. Piepho, H. T. Dublin, N. Bhola, and R. S. Reid. 2009. Dynamics of Mara-Serengeti ungulates in relation to land use changes. *Journal of Zoology* 278:1–14.

Osborn, F. V., and C. M. Hill. 2005. Techniques to reduce crop loss: Human and technical dimensions in Africa. *People and wildlife: Conflict or co-existence?* ed. R. Woodroffe, S. Thirgood, and A. Rabinowitz, 72–85. Cambridge: Cambridge University Press.

Osborn, F. V., and G. E. Parker. 2002. Living with elephants II: A manual for implementing an integrated programme to reduce crop loss to elephants and improve livelihood security of small-scale farmers. Mid-Zambezi Elephant Project, Harare, Zimbabwe. www.elephantpepper.org.

Packer, C., R. Hilborn, A. Mosser, B. Kissui, M. Borner, G. Hopcraft, J. Wilmshurst, S. Mduma, and A. R. E. Sinclair. 2005. Ecological change, group territoriality, and population dynamics in Serengeti lions. *Science* 307:390–93.

Packer, C., D. Ikanda, B. Kissui, and H. Kushnir. 2005. Lion attacks on humans in Tanzania—Understanding the timing and distribution of attacks on rural communities will help to prevent them. *Nature* 436:927–28.

Pangle, W. M., and K. E. Holekamp. 2010. Lethal and nonlethal anthropogenic effects on spotted hyenas in the Masai Mara National Reserve. *Journal of Mammalogy* 91:154–64.

Patterson, B. D., S. M. Kasiki, E. Selempo, and R. W. Kays. 2004. Livestock predation by lions (*Panthera leo*) and other carnivores on ranches neighboring Tsavo National Parks, Kenya. *Biological Conservation* 119:507–16.

Rasmussen, G. S. A. 1999. Livestock predation by the painted hunting dog *Lycaon pictus* in a cattle ranching region of Zimbabwe: A case study. *Biological Conservation* 88: 133–39.

Reid, R., H. Gichohi, M. Said, D. Nkedianye, J. Ogutu, M. Kshatriya, P. Kristjanson, S. Kifugo, J. Agatsiva, S. Adanje, et al. 2008. Fragmentation of a peri-urban savanna, Athi-Kaputiei plains, Kenya. In *Fragmentation in semi-arid and arid landscapes*, ed. K. A. Galvin, R. S. Reid, R. H. Behnke, and N. T. Hobbs, 195–224. Dordrecht, Netherlands: Springer.

Rwambo, P., J. G. Grootenhuis, J. Martiini, and S. Mkumbo. 1999. Animal disease risk in the wildlife/livestock interface in the Ngorongoro Conservation Area of Tanzania. Ngorongoro Conservation and Development Project, CSU, CRSP, USAID. Colorado State University.

Schmitt, J. A. 2010. Improving conservation efforts in the Serengeti ecosystem, Tanzania: An examination of knowledge, benefits, costs and attitudes. PhD diss., University of Minnesota.

Serneels, S., and E. F. Lambin. 2001. Impact of land-use changes on the wildebeest migration in the northern part of the Serengeti-Mara ecosystem. *Journal of Biogeography* 28: 391–407.

Sibanda, B. M. C., and A. K. Omwega. 1996. Some reflections on conservation, sustainable development and equitable sharing of benefits from wildlife in Africa: The case of Kenya and Zimbabwe. *South African Journal of Wildlife Research* 26:175–81.

Sinclair, A. R. E., and P. Arcese, eds. 1995. *Serengeti II: Dynamics, management and conservation of an ecosystem*. Chicago: University of Chicago Press.

Sitati, N. W., and M. J. Walpole. 2006. Assessing farm-based measures for mitigating human-elephant conflict in Transmara District, Kenya. *Oryx* 40:279–86.

Sitati, N. W., M. J. Walpole, and N. Leader-Williams. 2005. Factors affecting susceptibility of farms to crop raiding by African elephants: Using a predictive model to mitigate conflict. *Journal of Applied Ecology* 42:1175–82.

Sitati, N. W., M. J. Walpole, R. J. Smith, and N. Leader-Williams. 2003. Predicting spatial aspects of human-elephant conflict. *Journal of Applied Ecology* 40:667–77.

Tanzania Wildlife Research Institute (TAWIRI). 2003. Total count of elephant and buffalo in the Serengeti ecosystem, wet season 2003. Unpublished report, Tanzania Wildlife Research Institute, Arusha, Tanzania.

Thompson, D. M. 1997. *Multiple land use: The experience of the Ngorongoro Conservation Area, Tanzania*. Gland, Switzerland: International Union for the Conservation of Nature.

Thompson, M., and K. Homewood. 2002. Entrepreneurs, elites, and exclusion in Maasailand: Trends in wildlife conservation and pastoralist development. *Human Ecology* 30: 107–38.

Treves, A., and K. U. Karanth. 2003. Human-carnivore conflict and perspectives on carnivore management worldwide. *Conservation Biology* 17:1491–99.

Virani, M. Z., C. Kendall, P. Njoroge, and S. Thomsett. 2011. Major declines in the abundance of vultures and other scavenging raptors in and around the Masai Mara ecosystem, Kenya. *Biological Conservation* 144:746–52.

Walpole, M. J., G. G. Karanja, N. W. Sitati, and N. Leader-Williams. 2003.Wildlife and people: Conflict and conservation in the Masai Mara, Kenya. In *Wildlife and development*, series no. 14. London: International Institute for Environment and Development.

Walpole, M. J., Y. Ndoinyo, R. Kibasa, C. Masanja, M. Somba, and B. Sungura. 2004. Assessment of HEC in western Serengeti, Tanzania. Consultant's report for FZS, TANAPA & Wildlife Division, Arusha, Tanzania.

Walpole, M. J., and C. R. Thouless. 2005. Increasing the value of wildlife through non-consumptive use? Deconstructing the myths of ecotourism and community-based tourism in the tropics. In *People and wildlife: Conflict or coexistence?*, ed. R. Woodroffe, S. Thirgood, and A. Rabinowitz, 122–39. Cambridge: Cambridge University Press.

Western, D., R. Groom, and J. Worden. 2009. The impact of subdivision and sedentarization of pastoral lands on wildlife in an African savanna ecosystem. *Biological Conservation* 142:2538–46.

Wittemyer, G., J. S. Brashares, P. Elsen, W. T. Bean, and A. C. O. Burton. 2008. Accelerated human population growth at protected area edges. *Science* 321:123–26.

Woodroffe, R. 2000. Predators and people: Using human densities to interpret declines of large carnivores. *Animal Conservation* 3:165–73.

Woodroffe, R., and L. G. Frank. 2005. Lethal control of African lions (*Panthera leo*): Local and regional population impacts. *Animal Conservation* 8:91–98.

Woodroffe, R., L. G. Frank, P. A. Lindsey, S. M. K. Ole Ranah, and S. Romañach. 2007. Livestock husbandry as a tool for carnivore conservation in Africa's community rangelands: A case-control study. *Biological Conservation* 16:1245–60.

Woodroffe, R., and J. R. Ginsberg. 1998. Edge effects and the extinction of populations inside protected areas. *Science* 280:2126–28.

Woodroffe, R., P. Lindsey, S. Romañach, A. Stein, and S. M. K. Ole Ranah. 2005. Livestock predation by endangered African wild dogs (*Lycaon pictus*) in northern Kenya. *Biological Conservation* 124: 225–34.

PART V

Consequences of Disturbance for Policy, Management, and Conservation

TWENTY-TWO

Bushmeat Hunting in the Serengeti Ecosystem: An Assessment of Drivers and Impact on Migratory and Nonmigratory Wildlife

Dennis Rentsch, Ray Hilborn, Eli J. Knapp, Kristine L. Metzger, and Martin Loibooki

The Greater Serengeti ecosystem (GSE) consists of a diverse mix of land uses, management areas, and local ethnic groups, each with a unique relationship to natural resources. Among the characteristics that make this ecosystem unique is the lack of fences demarcating protected areas. As a result of this policy, wildlife is not bound to the confines of legally protected areas. This is especially relevant in the Serengeti, where more than two million wildebeest (*Connochaetes taurinus*) and zebra (*Equus quagga burchelli*) pass through up to five different types of land in their migration throughout the ecosystem. These multiple land-use types are intended to help better conserve wildlife, while allowing various stakeholders to reap the benefits of conservation. However, as the annual migration of wildebeest and zebra passes through the ecosystem and the various land-use types, the herds are exposed to risk from human utilization, particularly bushmeat hunting. Bushmeat hunting, as defined here, is the illegal hunting of wildlife (both inside and outside of protected areas) for the purpose of consumption or sale of meat. Bushmeat hunting is an important economic activity for communities in the GSE, but is also a potential threat to wildlife populations. Due to the expanding human pressure within the ecosystem and the continued utilization of bushmeat, offtake from hunting has been implicated among the most serious threats to wildlife in the Serengeti ecosystem (Sinclair 1995). This chapter describes the current status of bushmeat hunting in the Serengeti, examining the drivers of hunting by local communities and the role of protected area managers in addressing this threat to the ecosystem.

Poaching has long been cited as a serious threat to wildlife populations throughout the ecosystem, largely due to the abundance of communities relying on bushmeat situated immediately adjacent the boundaries of unfenced protected areas (Sinclair 1995; Loibooki et al. 2002). However, it is extremely difficult to quantify the impact of illegal hunting on wildlife populations (Knapp et al. 2010). Wildlife in the GSE is constantly on the move, and the migration is reliant upon rainfall and climate, which are seldom consistent. Ground and aerial surveys provide the best source of data on trends in populations throughout the ecosystem, but it is difficult to pinpoint the effect of illegal hunting on a decreasing animal population. Selective hunting, in which a specific species is targeted to supply a niche market, is easier to observe in wildlife trends. For example, illegal ivory poaching, prevalent in the 1970s through the late 1980s was observed to have a negative impact on elephant populations (Sinclair, 1995). The majority of bushmeat hunting is conducted with wire snares which can indiscriminately catch a variety of animal species. Because migratory wildlife moves, vanguard herds leading the way may be disproportionately affected by bushmeat hunting due to the relative ease of snaring and greater accessibility by poachers to the remote areas of the park. Migratory wildlife is especially at risk, as predictable seasonal movement patterns make them targets of hunters' snares to supply local consumption. With an abundance of migratory wildlife in the Serengeti ecosystem, commonly used wire snares prove an efficient and indiscriminate tool for hunting (Hofer et al. 1996; Campbell, Nelson, and Loibooki 2001; Kaltenborn, Nyahongo, and Tingstad 2005; Holmern et al. 2006). Yet hunting continues even when the migratory herds are well away in a different part of the ecosystem. Since bushmeat hunting is distributed across a suite of species, the rate of offtake may be based on the relative abundance of various species. As a result, fluctuations in a certain species' population in the ecosystem often cannot be attributed solely to poaching. However, with a more thorough understanding of human activities, we now have a better idea where the threat of poaching originates within the ecosystem.

The question of the impact of offtake due to illegal bushmeat hunting is controversial, and assessments of the impact of poaching on wildlife populations have yielded different conclusions over time (Hofer et al. 1996; Barrett and Arcese 1998; Mduma, Sinclair, and Hilborn 1999; Campbell, Nelson, and Loibooki 2001). On the one hand, expenditure on enforcement within the national park commensurate with revenue generated through tourism is at a record high (Hilborn et al. 2006; Thirgood et al. 2008), and improved antipoaching efforts by investors in game reserves surrounding the national park help increase the probability of detection. Yet human populations within the Serengeti ecosystem are increasing at a rate of greater than 3% per annum (Tanzanian National Census 2002). Most of these communities rely heavily on bushmeat as a primary source of protein, with bushmeat considered one of the most important protein sources by 95% of households living in western Serengeti (Barnett 2000). Cross-sectional surveys of hunting practices and consumption patterns in high and low bushmeat communities found that bushmeat was consumed regularly by 45% to 60% of households in the northwest Serengeti (Emerton and Mfunda 1999; Barnett 2000; Campbell, Nelson, and Loibooki 2001), with migratory ungulates (wildebeest and zebra) accounting for an estimated two-thirds of total offtake (Campbell and Borner 1995).

Here we use a multifaceted approach to examine the current level of the threat of bushmeat hunting on wildlife in the greater Serengeti ecosystem. We employ a variety of data sources and methods including spatial dynamics of wildlife populations, aerial surveys, rainfall trends, enforcement data, nutritional surveys, and socioeconomic methods. By using spatially-explicit estimates of poaching mortality rates for buffalo, wildebeest, and other species, we are able to better understand the impact of bushmeat hunting on wildlife. Finally, we examine the results of these methods in the context of conservation management. Such information may be used to better understand—and potentially reduce—poaching in the GSE.

IMPACT OF POACHING ON WILDLIFE POPULATIONS

One way to determine the impact of poaching on wildlife in an ecosystem as large as the Serengeti is through utilizing existing ecological data on wildlife population sizes. By looking at the trends of abundance of individual species, we can explore the impact of poaching on populations. The four species for which we have the best estimates of abundance are elephant, rhino, wildebeest, and buffalo (fig. 22.1). Prior to 1978, populations of wildebeest and buffalo were growing exponentially after the elimination of rinderpest (Sinclair 1995). Elephants (*Loxodonta africana*) were also growing rapidly as they recovered from intense hunting pressure (Sinclair et al. 2007). At that time, black rhinos (*Diceros bicornis*) were common within the ecosystem (Metzger et al. 2007).

In 1978 there was an economic and political crisis in Tanzania, which led to a collapse of antipoaching resources in the Serengeti (Packer and Polasky 2008). In the midst of the national chaos brought on by a countrywide forced villagization scheme, both bushmeat and illegal trophy hunting, fed by the ivory trade, increased unabated within the park. At the same time,

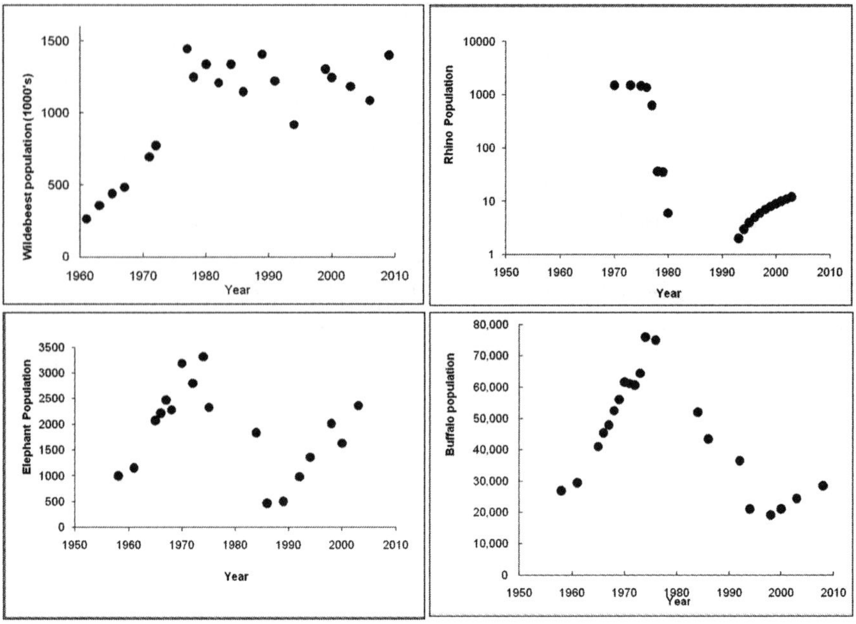

Fig. 22.1 Species abundance trends of wildebeest, rhino, elephant, and buffalo for the periods when census estimates are available.

there was increased global demand for ivory and rhino horn. The net result was a dramatic decline in the abundance of elephant, rhino, and buffalo. By 1990, park resources had increased, and regular antipoaching patrols had become more frequent. By 1995 (after a severe drought in 1993) these three species began increasing again, although rhino had been reduced to a handful of individuals, and buffalo were increasing much more slowly than they had in the 1960s and 1970s.

The trend in wildebeest is quite different, however. The exponential increase of the population in 1970s and 1980s has been followed by a period of stability, with a significant decline resulting from a 1993 drought, which was the driest year since 1978. The leveling off of the wildebeest population has been linked to a combination of food limitation and increased poaching (Mduma, Sinclair, and Hilborn 1999).

It is likely that between 1978 and 1990 poaching caused much of the declines in elephant, rhino, and buffalo. However, for buffalo, abundance data from different regions of the Serengeti ecosystem shows quite different trends to suggest that poaching is still high enough in the northwest portion of the park to prevent recovery to former levels (fig. 22.2).

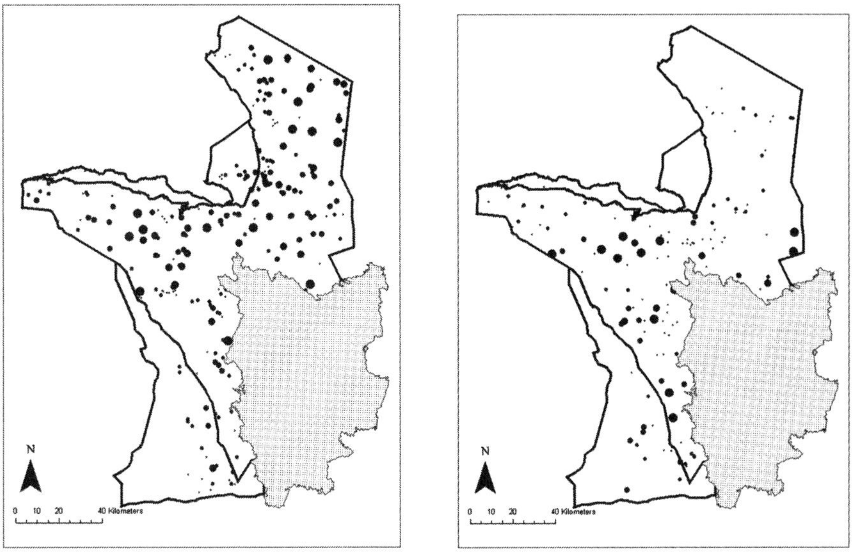

Fig. 22.2 Buffalo distribution in the Serengeti National Park based on 1970 (left) and 2003 (right) aerial census buffalo count. Administrative boundaries are identified in detail in figure. 2.1 (chapter 2).

Poaching for elephant and rhino has historically been a specific, targeted trade in trophy products. In contrast, buffalo and wildebeest are typically harvested for food. Due to its lucrative nature, individuals in the trophy trade used vehicles and firearms, making it a fundamentally different activity than the bushmeat harvest. The trophy trade appears to have diminished in the GSE in recent decades due in part to the influence of international policies such as the Convention on International Trade in Endangered Species (CITES) (Hilborn et al. 2006), whereas the bushmeat trade is ongoing and significant, relying primarily on the use of snares to capture animals.

Detailed Analysis of Buffalo

Buffalo are a useful species for examining the impact of poaching because they generally are sedentary and relatively easy to count. Therefore we have quality spatial and temporal information on their population in the GSE. We can use the population dynamics of buffalo to estimate the amount of the harvest and the temporal trends in harvest. Many factors can contribute to variation in animal population change including disease, food supply, drought, and natural predation. Using a spatially structured population

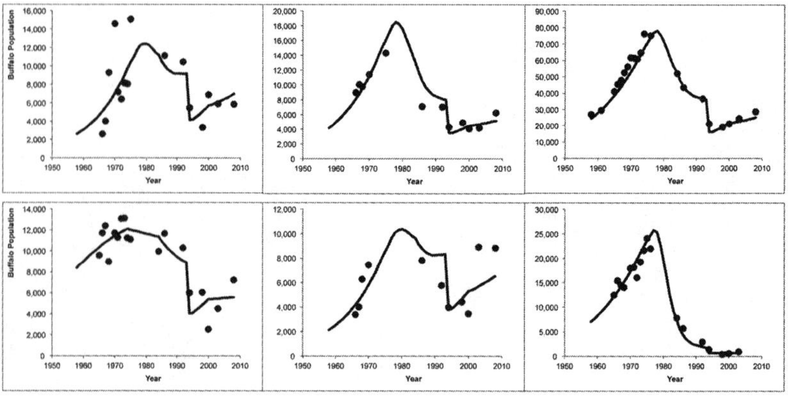

Fig. 22.3 Spatial trends in buffalo population within Serengeti National Park. Top graphs: Serengeti National Park center (left), far west (middle), total census (right); bottom graphs: far east (left), south (middle), north (right).

model, the potential contribution of food supply, lion predation rates, and poaching were considered to explain the spatial and temporal population dynamics of buffalo (Metzger et al. 2010). Disease has been well monitored in the buffalo population and we have no indication that the observed changes in population have been caused by epizootics (Rossiter et al. 1983; Dublin et al. 1990; Dobson 1995; Sinclair et al. 2008). The buffalo data were analyzed in detail to estimate the temporal trend in offtake in five different regions of the park. Poaching was the only variable able to explain the change in the buffalo population (Metzger et al. 2010). Figure. 22.3 shows the census estimates and model fits to the abundance for these five regions and the total buffalo population. The estimates indicate that the poaching pressure was highest in the north and lowest in the east and south (Metzger et al. 2010). The failure of the north population to show any significant recovery in the last two decades is ascribed to continuing poaching pressure.

From this analysis of population dynamics we can also calculate the number of individuals harvested per year and the annual exploitation fraction, while using rainfall as a covariate to account for natural fluctuations (table 22.1). Between 1979 and 1984 almost 13% of the population per year was being taken by poachers; the population went from a 6–7% per year increase to a 6–7% per year annual decline. The numbers harvested were very large initially, especially when the large population in the north was depleted, but in the last decade the total numbers harvested are estimated to be a few hundred animals per year, mainly from the north and west regions.

Table 22.1 The mean number of buffalo harvested and the mean proportion harvested per year (exploitation rate) for different time periods since 1979

Time period	Mean number harvested per year	Mean exploitation rate
1979–1984	8,131	0.127
1985–1991	2,621	0.062
1992–2000	598	0.025
2001–2008	315	0.014

The buffalo data show that buffalo numbers increased and then decreased, presumably due to poaching. Hilborn et al. (2006) used a single-area model, whereas Metzger et al. (2010) used a spatially explicit model. The single-area model was essentially an exponential growth model, whereas the spatial model does estimate some density dependence in the eastern area. Both analyses estimate significant changes in poaching pressure over time.

Analyzing long-term buffalo population data spatially can provide valuable information on the drivers and influences of poaching. Using long-term data on the buffalo population, we examine the spatial patterns of poaching through time and relate these patterns to proximity and changes in the human population residing adjacent to the park. Proximity of the human population to the national park boundary predict initial declines in the buffalo, and increases in human population density over time may influence the buffalo's recovery from poaching (fig. 22.4).

Impact on Other Species

We know that many other species are taken in the bushmeat harvest (Hofer et al. 1996; Campbell, Nelson, and Loibooki 2001; Holmern et al. 2006; Ndibalema and Songorwa 2008). For our purposes here, the major ungulates in the ecosystem can generally be divided into three groups: migratory zebra and wildebeest, resident gazelles of the plains, and varied resident species of the woodlands. The migratory wildebeest and zebra move throughout the park in an annual cycle, which means that their availability to poachers and snares, located primarily near human population centers in the north, west, and south is seasonal. When these migratory animals are on the plains, they are not subject to poaching because the local people in that area, the

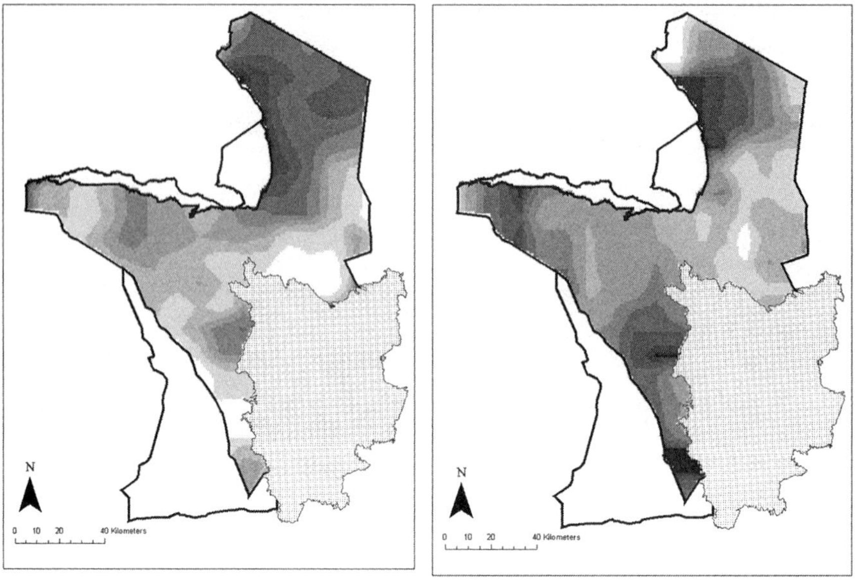

Fig. 22.4 Rate of increase of buffalo population in Serengeti National Park from 1970 to 1992 (left), and 1998 to 2003 (right).

Maasai, are not significant consumers of bushmeat (Campbell, Nelson, and Loibooki 2001). Even if they were, there are few trees on the plains for snare attachment. Thus we would expect poaching to be a seasonal phenomenon for the migratory species. The second group of animals are the Thomson's and Grant's gazelles, which are dominantly plains animals, and do not occur in large abundance in the woodlands where snaring takes place. For similar reasons mentioned previously, we would not expect significant poaching pressure on these two species. The third group is the species that are resident in the woodlands, predominantly impala, topi, buffalo, kongoni, warthog, and giraffe. Their habitat is subject to poaching year-round. We would expect these species to have the highest poaching rates because of their high abundance in these areas. We can examine the relative impact of poaching on the resident species.

The wildebeest data are more difficult to interpret than the buffalo, elephant, and rhino data because of several confounding factors. The major difficulty in assessing the overall impact of poaching is the interaction between density-dependent food limitation and removals by poaching. One can explain the leveling off post-1978 as caused by either factor. However, the relationship between per capita food availability in the dry season and survival rate has been demonstrated by Mduma, Sinclair, and Hilborn

(1999). Thus when one attempts to estimate the offtake from the wildebeest population dynamics data the model is considerably more complex and is fitted not only to the census data, but also to estimated natural mortality rates of wildebeest.

The second difficulty in estimating offtake of wildebeest is that the rate of removal by poachers has never been large enough to cause significant changes in the population. This means that the overall precision of the estimate is not high, and we cannot detect any trends. The estimated offtake between 1978 and 2008 is 50,000 animals per year with 95% confidence intervals of 20,000–85,000 individuals per year. Overall the estimated offtake appears to be about 3% of the total population.

The abundance of resident species is estimated periodically by aerial strip-transect counts. The data for the major resident species (except buffalo) are shown in (table 22.2). There are no statistically relevant trends in these data. Resident animals appear to be more abundant after 1978 than before, despite the acknowledged increase in bushmeat harvest. Using the abundance trend data we found no evidence for an impact of bushmeat harvest, or at least no change in impact before and after 1978.

Using another approach we can estimate the relative magnitude of bushmeat hunting on various species. In the process of antipoaching patrols, park rangers often encounter animals in snares or carcasses in poacher camps. At various times over the last 30 years, Serengeti National Park (SNP) rangers and scientists have recorded the number of carcasses recovered of each species in poacher camps and snares. Assuming that the probability of carcass discovery is the same for each species, carcass recovery data provides an estimate of the relative hunting impact by poachers for each species in question (fig. 22.5).

We assume that rangers find carcasses in proportion to the frequency with which the species is killed. If we know how many buffalo or wildebeest are killed, we can scale up the estimated kill of other species. Table 22.3 shows the number of carcasses found, and the estimated total number taken per year, assuming 50,000 wildebeest per year were harvested based on the ecological model predicting trends in wildebeest population due to dry season rainfall and annual offtake.

Looking at the vulnerability of each species relative to wildebeest, we see that in general, the resident species are considerably more vulnerable to poaching than wildebeest. Impala, topi, buffalo, warthog, kongoni, eland, waterbuck, and giraffe were caught two to nine times more frequently than wildebeest in relation to their abundance. The plains species, Thomson's and Grant's gazelle, are caught much less frequently than wildebeest per unit abundance. The implied fraction harvested of the total population is

Table 22.2 Abundance trends: Estimated abundance of resident herbivores in the Serengeti ecosystem

Year	Impala	Topi	Warthog	Kongoni	Eland	Waterbuck	Giraffe	Total
1971	57,000	27,000		9,000	10,000	1,000	7,000	111,000
1976	72,000	67,000		10,000		2,000	7,000	158,000
1988	67,000	55,000	7,000	8,000	8,000	2,000	6,000	153,000
1989	79,000	70,000		15,000	14,000	1,000	7,000	186,000
1991	62,000	71,000	5,000	10,000	10,000	2,000	7,000	167,000
1996	111,000	50,000	3,000	21,000	30,000	1,000	12,000	228,000
2001	82,000	47,000	2,000	14,000	18,000		13,000	176,000
2003	91,000	39,000	2,000	16,000	16,000	1,000	11,000	176,000
2006	149,000	52,000	4,000	27,000	27,000	2,000	9,000	270,000

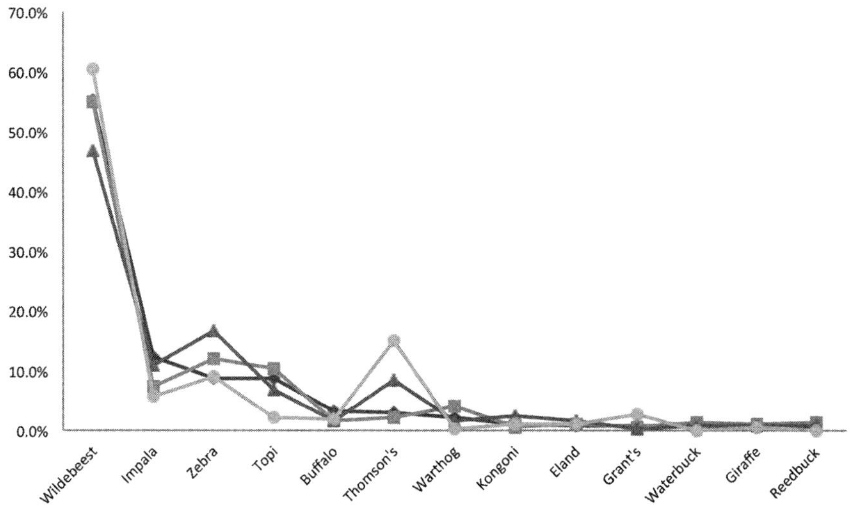

Fig. 22.5 Relative proportion of carcasses recovered by TANAPA rangers by three sources (1990s–2000s): Arcese and Mduma, unpublished data (diamond); Hofer and Campbell (1996) (square); TANAPA antipoaching records (triangle); relative average abundance from 2002 to 2006 based on aerial census counts conducted by the Tanzania Wildlife Research Institute (TAWIRI) (circle).

Table 22.3 Estimated offtake of key species based on number of carcasses found

Species	Carcasses	Total taken	Abundance	Harvest rate	Vulnerability
Wildebeest	3,377	50,000	1,278,000	4%	1
Impala	795	11,771	79,000	15%	3.81
Zebra	1,204	17,826	146,000	12%	3.12
Topi	486	7,196	95,000	8%	1.94
Buffalo	112	1,658	22,156	7%	1.91
T. gazelle	620	9,180	325,000	3%	0.72
Warthog	119	1,762	7,000	25%	6.43
Kongoni	167	2,473	12,000	21%	5.27
Eland	114	1,688	9,000	19%	4.79
G. gazelle	15	222	25,000	1%	0.23
Waterbuck	59	874	2,500	35%	8.93
Giraffe	52	770	8,000	10%	1.46

Notes: Vulnerability calculated relative to wildebeest harvest rate. T. gazelle = Thomson's gazelle, G. gazelle = Grant's gazelle.

Table 22.4 Offtake based on buffalo population estimates

Species	Carcasses found	Total offtake	Abundance	Harvest rate
Wildebeest	3,377	10,674	1,278,000	1%
Impala	795	2,513	79,000	3%
Zebra	1,204	3,806	146,000	3%
Topi	486	1,536	95,000	2%
Buffalo	112	354	22,156	2%
T. gazelle	620	1,960	325,000	1%
Warthog	119	376	7,000	5%
Kongoni	167	528	12,000	4%
Eland	114	360	9,000	4%
Grants	15	47	25,000	0%
Waterbuck	59	186	2,500	7%
Giraffe	52	164	8,000	2%

Note: T. gazelle = Thomson's gazelle.

high for some species. We know from the basic life history of having a single calf at 2 or 3 years of age that the maximum rate of increase of all of these species is in the range of 10–15%, depending on calf and adult survival. If a species (such as hartebeest or waterbuck) had been consistently harvested at 20–35% since 1978 they would be nearly extinct, and we would have expected to seen declines in the census estimates. However, any offtake for rarer species such as hartebeest and waterbuck is less certain, as indicated by higher error bars around such estimates.

Similar calculations can be done using the estimated buffalo offtake (table 22.4). Here the number of wildebeest carcasses found is much lower. The estimated number of wildebeest lost to poaching is less than the lower 95% confidence limit. It may be that buffalo carcasses are less likely to be found, although given their size, this seems unlikely.

Impact of Wildlife Based on Population Dynamics

From the population data we can detect historical changes in poaching pressure that caused significant declines in elephants, rhino, and buffalo in the 1980s. Each of these species is now recovering although this finding depends on the area involved. The data suggest that buffalo are still strongly affected by poaching in the north of the park, but recovering well in other areas. The total count data for other species including wildebeest do not provide any direct evidence for poaching impacts, either historically or at present. However using population models and estimates of natality and mortality from other factors, we can estimate average impacts on wildebeest, but these have wide confidence intervals (Pascual and Hilborn 1995; Mduma, Sinclair, and Hilborn 1999).

HUMAN IMPACT ON MIGRATORY POPULATIONS

The human population growth rate for Mara and Shinyanga Regions is 2.5% and 3.3% per annum, respectively (Tanzania National Bureau of Statistics 2002). In 1957, there were an estimated 118,000 people living in the villages or wards immediately adjacent to the Serengeti protected areas (the current Serengeti boundaries were not demarcated until 1959; see Sinclair et al. chapter 2). By the 2002 Tanzanian national census, 507,000 people were living in villages immediately adjacent to a protected area along the western boundary of Serengeti National Park. This increase led to the regions surrounding Lake Victoria, including the Western GSE, to rank among the highest in population density for the whole of Tanzania (figs. 22.5, 22.6).

Figure 22.7 demonstrates the ratio of people to wildebeest in the western Serengeti from the time of the earliest census to the most recent in 2006. Human population was estimated based on the intercensal population growth rates for Mara and Shinyanga Regions. Wildebeest numbers were based upon the aerial censuses conducted from 1957 to 2006.

Nonwildlife-Based Data Sets: Community Surveys

While many studies have suggested that poaching is a relatively common and widespread activity in the western Serengeti ecosystem (Campbell and Hofer 1995; Sinclair 1995; Hofer et al. 1996; Campbell, Nelson, and Loibooki 2001; Loibooki et al. 2002; Holmern et al. 2004; Kaltenborn, Nyahongo, and Tingstad 2005; Knapp 2007; Knapp et al. 2010), more detailed information about where it occurs and what villages are most actively involved has been less certain. Previous studies have clearly articulated the methods various authors have used to support the claim of widespread poaching, but few have thoroughly examined the methods themselves. However a recent study contrasted the most commonly used methods for assessing poaching activities to see if each method yielded similar results (Knapp et al. 2010).

The methods selected by Knapp et al. (2010) included three primary

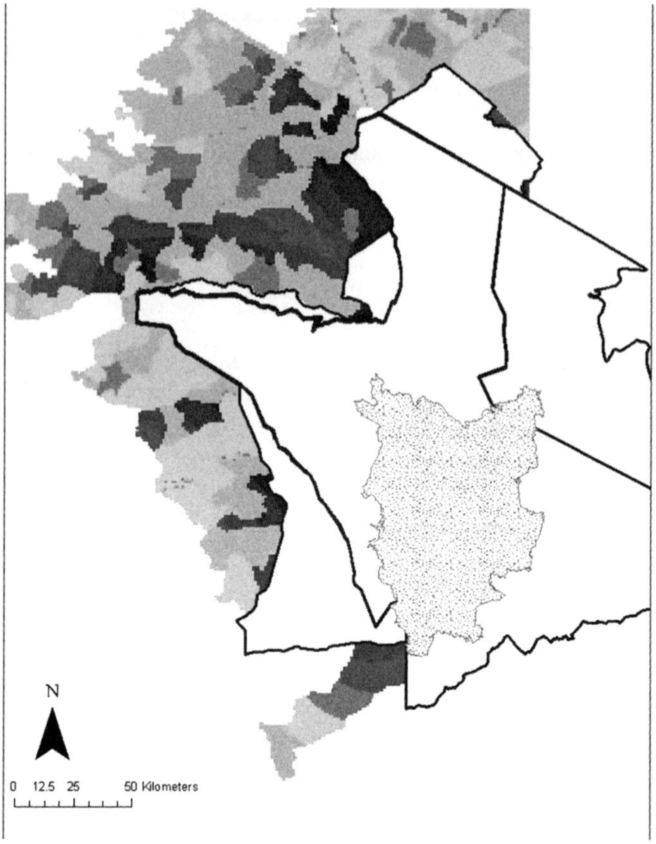

Fig. 22.6 Increase in the number of potential hunters adjacent to Serengeti National Park. Potential hunters are calculated as a function of density of people and the distance people are to the park boundary (Campbell & Hofer, 1995; Campbell, Nelson and Loibooki 2001). Dark areas represent high population growth of potential hunters, whereas light areas represent low population growth ($r = -0.6$ to $+0.59$). Location of fastest increase is adjacent to areas of slowest increase in buffalo numbers (fig. 22.3).

approaches—two of which were conducted using household interviews while another stemmed from data collected by antipoaching personnel. One method of household interviews used a self-assessment of poaching activity from the respondent. The respondent was directly asked to admit to involvement—or noninvolvement—in poaching activities. The other method of household interviews used dietary recall of bushmeat consumption where respondents were revisited monthly and asked to report on their consumption patterns for the previous seven days. In this approach, the respondent was asked how many times they had consumed bushmeat in the last day, week, month, and year. Results were contrasted with those from

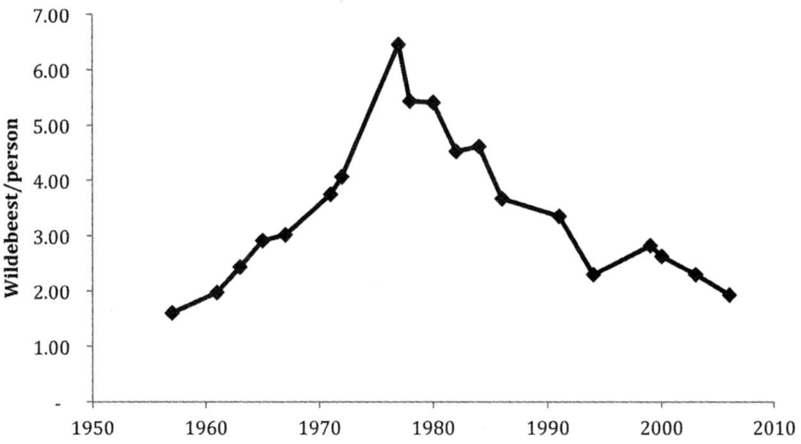

Fig. 22.7 Wildebeest and human population ratio in villages immediately adjacent to key protected areas in Serengeti ecosystem. Based on data from wildebeest aerial censuses from 1960 to 2006, and human population growth based on human census from 1957, 1967, 1970, 1972, 1975, 1978, 1985, 1988, 1994, and 2002.

an "enforcement method," in which arrests were tallied from antipoaching personnel. Each of these methods was examined in the context of three villages lying adjacent to the northwestern boundary of Serengeti National Park.

Each of the methods supported the claim that poaching is occurring in and around Serengeti National Park. When compared, however, the methods did not reveal a consistent pattern of poaching activity across the three villages. In fact, each method suggested that a different village was more implicated in poaching than the other two. Despite these nuanced results, two of the methods—dietary recall of bushmeat consumption and enforcement records—aligned more closely with each other than did the self-assessment method. The central question that naturally arises from these results concerns the accuracy, or validity, of each method.

Collecting representative data to measure local poaching levels is difficult due to the sensitive nature of the data itself. Since hunting without a permit inside or outside Serengeti National Park—and its affiliated protected areas—is illegal, respondents fear that acknowledgment of involvement will be incriminating and render themselves subject to fines, imprisonment, and/or mandated public service (Holmern, Muya, and Røskaft 2007). This is an overt weakness of the self-assessment method that directly asks respondents to report their involvement in poaching. It has also acted as an impetus for wildlife managers to rely more heavily on data recovered from antipoaching personnel (Hilborn et al. 2006). While data collected through enforcement avoids the sensitivity issue, it may be likewise subject to bias—and more expensive to implement. Antipoaching patrols are often dependent upon passable roads and established routes which poachers can quickly learn to avoid. Villages lying closest to the most heavily patrolled areas, therefore, may be more implicated in poaching merely because of proximity. Enforcement records can also bias results seasonally if patrols are not viable in the wet season. The strength of the dietary recall method is that it appears to diminish the fear of the respondent (especially when bushmeat consumption is queried within a list of other common protein sources) and can avoid the seasonality of poaching activity by asking consumption within a variety of time frames. Potential biases of dietary recall, however, involve possible respondent memory error and the indirect nature of the questions themselves. Just because a respondent acknowledges bushmeat consumption, the individual may not actually be a hunter.

Identifying the strengths and weaknesses of each method is fundamental to gaining a proper understanding of poaching levels. Moreover, the nuanced—and even contradicting—results of the three methods suggest that conclusions drawn from any one method in isolation should be done cautiously and cross-referenced whenever possible. Optimally, the study by Knapp et al. (2010) shows that the enforcement and dietary recall methods should be used in tandem when measuring local levels of poaching involvement. In situations where enforcement records are unavailable due to limited resources, dietary recall of bushmeat consumption appears to be a more cost-effective alternative to estimating poaching activities than self-assessment through household interviews. Although not as useful as a quantifiable tool at the village level, self-assessment may have a distinct utility in its ability to uncover key characteristics of a poaching household and what strategies are used to procure illegal bushmeat.

Quantifying Offtake through Consumption of Bushmeat

Another method for estimating the impact of poaching on wildlife is based on assessing the frequency and extent of bushmeat use by communities living adjacent to wildlife areas. Using monthly dietary recall surveys that include consumption of bushmeat from 2007 to 2008, we were able to estimate annual offtake of wildlife due solely to local consumption (i.e., how many wildebeest are lost to human consumption in the study area). These data provide a means of assessing the offtake of bushmeat based on local demand for protein consumption. As demonstrated in several studies (Wilkie et al. 2005; Knapp et al. 2010) dietary recall consumption data was found to be a cost-effective and reliable assessment of community involvement in bushmeat hunting, and less biased than random-sampled household interviews aimed at assessing involvement in illegal activity. By understanding the impact of bushmeat consumption on wildlife populations, wildlife managers are better able to assess the importance of addressing demand for wildlife as means to curb poaching. Based on a longitudinal panel study of household self-reported consumption data, as well as key informant interviews with former poachers in the area, the rate of bushmeat consumption per household per week was determined for a randomly selected sample of villages along the northwest boundary of protected areas in the Serengeti ecosystem (Rentsch 2012).

Bushmeat was by far the most commonly consumed meat-based protein source in the study area, with the exception of dagaa (small fish, which were virtually ubiquitous among the sample and are considered to be an inferior good by Tanzanians within the study area). Seventy-seven percent of the respondents who reportedly consumed bushmeat acquired the meat by purchasing it from within the village, while only 12% responded that they hunted the meat themselves. Consumption patterns showed high seasonal variability based on the proximity of the annual wildlife migration (fig. 22.8). Consumption remained high in the subsequent months after the migration had passed, as meat is most often dried to preserve after hunting and typically lasts for several months in this manner. The seasonality of consumption was commensurate with the types of species most commonly consumed during this same period (fig. 22.9), indicating that the consumption of wild meat (primarily wildebeest) was driven by availability and ac-

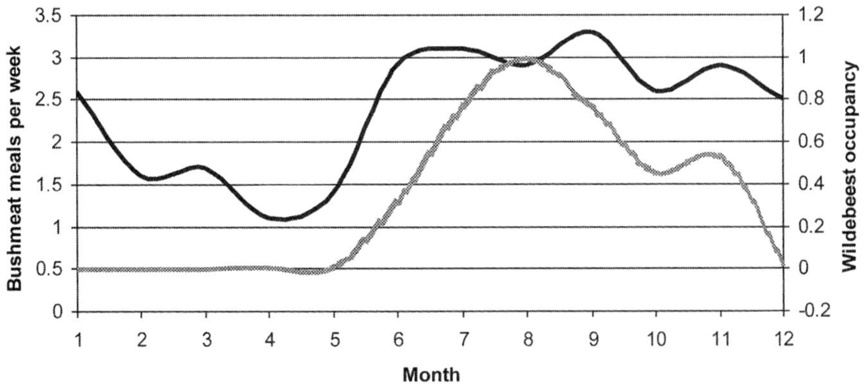

Fig. 22.8 Mean number of bushmeat meals consumed per household per week from October 2007 to November 2008 (*N* = 118) compared to time wildebeest spent per month in the study area. Wildebeest occupancy was calculated from monthly aerial survey data collected over three years (M. Norton Griffiths 1991, unpublished).

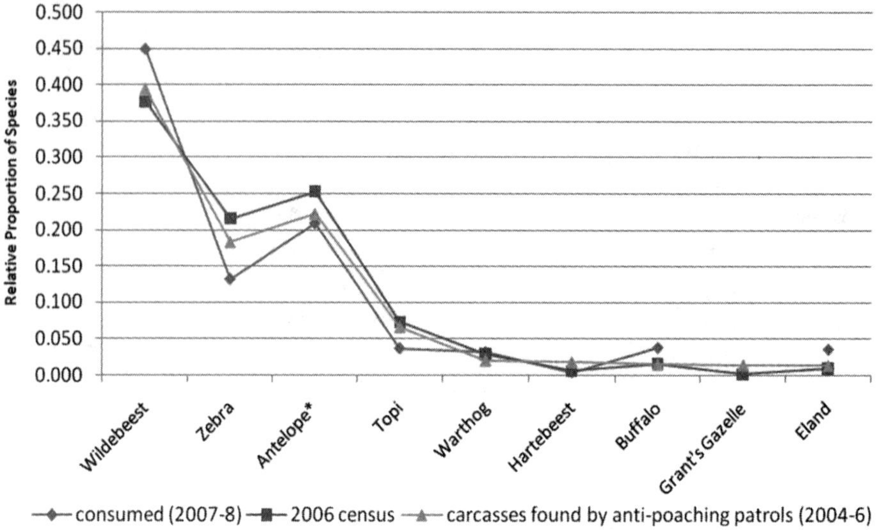

Fig. 22.9 Relative abundance of species in NW Serengeti ecosystem based on (1) consumption data from 2007 to 2008 (diamond), (2) aerial census of Grumeti and Ikorongo game reserves in 2006 (square), (3) carcasses found by antipoaching units in Ikorongo and Grumeti Game Reserves from 2004 to 2006 (triangle). (Courtesy of Grumeti Fund).

cess rather than a strong preference for a particular species, as is often the case in other bushmeat-exploited wildlife areas in Africa (Wilkie and Carpenter 1999; Wilkie and Godoy 2001; Milner-Gulland and Bennett 2003; Robinson and Bennett 2004; Cowlishaw, Mendelson, and Rowcliffe 2005).

The sample area for data collection was based on villages in the northwest area of the Serengeti ecosystem, where wildebeest and zebra are the most abundant animals. In order to determine if this pattern of consumption was based on abundance of available wildlife or on preference by the consumers, we compared the numbers of relative consumption with the 2006 aerial census of the same area by Grumeti Fund, a wildlife conservation and tourism company operating in Grumeti and Ikorongo Game Reserves. We also compared these relative numbers with the relative abundance of carcasses found during routine antipoaching patrols in the same area from 2004 to 2006. Species of antelope were lumped because they were not distinguished in the data, but they were mostly impala and Thomson's gazelle (fig. 22.9).

Using these data based on bushmeat meals consumed per household per week, we were able to extrapolate the total offtake per year by households living along the northwest boundaries of protected areas (game reserves and the national park) within the Serengeti ecosystem. By assuming that adjacent villages to those sampled consume bushmeat at a similar rate, we know that in 2008, households in northwest Serengeti consumed approximately 3.7 million meals of bushmeat. Of that, 45% was wildebeest. Based on key informant interviews, it was determined that poachers consistently harvested approximately 30 kg of meat from a single wildebeest to carry back to their village for sale.

Although we are able to estimate the offtake of a variety of species due to local consumption through the relative breakdown of species consumed, this model focuses primarily on wildebeest. This was done for several reasons. First, because wildebeest make up the largest biomass of mammals in the ecosystem and the wildebeest migration of more than one million animals is considered to be the major defining attribute of the ecosystem, examining the threat to wildebeest is a critical conservation issue. Second, because wildebeest were the most reported species consumed, the data are more robust within the sample. Species which were more rarely consumed may be skewed by single households consuming them multiple times per week, which may not be representative of their relative consumption overall.

All households within the protein survey sample were within 15 km of the closest protected area boundary. The bushmeat consumption patterns were extrapolated from sample households to those in immediately adjacent communities. Several studies examining involvement in poaching suggest that hunting is virtually nonexistent in communities more than 30–35 km away from the boundary of a protected area (Campbell and Hofer 1995; Campbell, Nelson, and Loibooki 2001; Nyahongo et al. 2009). While data on bushmeat consumption has not been directly assesssed, data on arrests of poachers found within the local game reserves indicated that there is a marked decline of poaching involvement in villages beyond 30 km from protected areas (fig. 22.10).

Based on these data, local consumption by communities adjacent to protected areas of the northwest Serengeti ecosystem resulted in offtake of approximately 65,485 wildebeest per annum (95% confidence intervals: 57,590; 79,223). According to the Tanzanian National Census conducted in 2002, human population growth for this region was 2.5% per annum. The US Census Bureau (2010) has estimated population growth rates for Tanzania until the year 2050, incorporating expected changes in birth and death rates (especially those expected changes due to HIV/AIDs). These estimates were applied to northwestern Serengeti as a rough estimate of the expected future impact on bushmeat consumption. While it is likely that

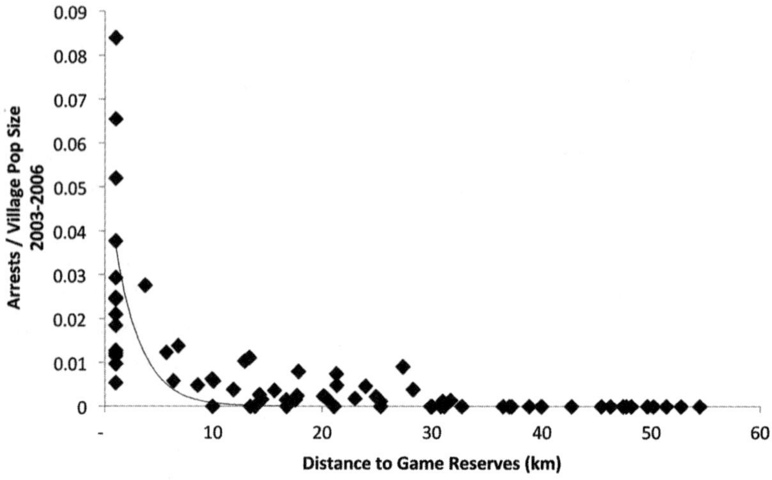

Fig. 22.10 Poacher arrests by village distance to Ikorongo and Grumeti Game Reserves 2004–2006 (data courtesy of Grumeti Fund).

consumption may vary slightly from year to year, it is unlikely that demand for wild meat would change much over such an interval. We can extrapolate the offtake of wildebeest for the near future based on consumption by households within 30 km of the national park or game reserves, assuming constant consumption rates at the 2008 level (fig. 22.11). Based on this projection of the northwestern region alone, bushmeat exploitation in the Serengeti ecosystem may soon surpass the rate of recruitment of wildebeest calves, and therefore reach unsustainable levels (Pascual and Hilborn 1995; Mduma, Sinclair, and Hilborn 1999). Although poaching offtake tracks abundance of wildlife in the area, migratory ungulates such as wildebeest and zebra are likely more susceptible as they travel through the high-poaching areas of the park en masse, providing easy targets for snare lines even at lower densities.

Further Impact of Local Consumption of Bushmeat

While the magnitude of bushmeat consumption by local communities has direct implications on the sustainability of migratory and nonmigratory ungulates throughout the system, examining trade-offs and patterns of consumption can also help to direct conservation interventions. Based upon the household consumption panel data set, weekly consumption of bush-

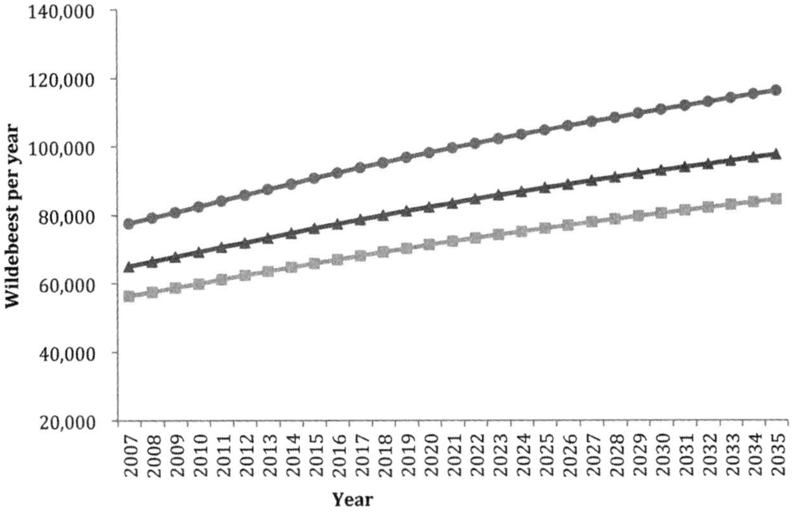

Fig. 22.11 Estimates of offtake of wildebeest based on local consumption patterns in NW Serengeti. Extrapolation based on census projections by the US Census Bureau. Assumption that local demand remains at current levels. Triangle line: offtake based on mean bushmeat meals consumed per household; square line = lower 95% confidence interval; circle line = upper 95% confidence interval.

meat by month showed a slight negative correlation to consumption of fish (fig. 22.12) but did not strongly correlate to consumption of other protein sources (excluding dagaa, a small river fish, previously mentioned as being considered an inferior food; consumption of this fish is negatively correlated with that of nearly all other protein sources). This trend was supported by a recent study examining the relationship between fish consumption and meat consumption by households of varying distances to Lake Victoria. It was also found that meat consumption increases with the abundance of wildlife and farther distance from the freshwater lake (Nyahongo et al. 2009). In northwest Serengeti, fish are available year-round due to the close proximity to Lake Victoria, the primary source of freshwater fish locally. The seasonal trade-off between consumption of bushmeat and fish (fig. 22.12) implies that (1) households in the survey do compensate somewhat for the seasonal inaccessibility of bushmeat as migratory animals move to distant reaches of the ecosystem, and (2) households do not increase consumption of domestic livestock during this time, but instead turn to another naturally occurring protein source. The interrelationship between bushmeat and fish has been demonstrated elsewhere in Africa (Brashares et al. 2004; Rowcliffe, Milner-Gulland, and Cowlishaw 2005; Nyahongo et al. 2009). This reliance

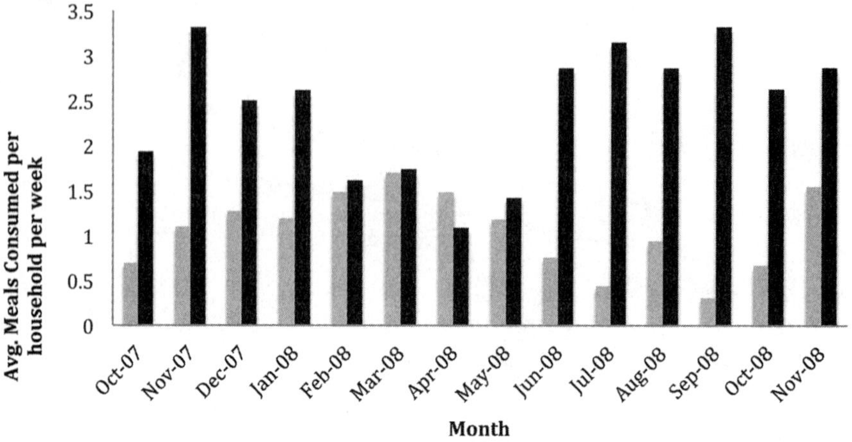

Fig. 22.12 Average weekly consumption of bushmeat and fish by month. White bars: fish; black histograms: bushmeat.

on wild-caught fish and bushmeat may be attributed to the cultural importance of livestock as a symbol of wealth and class, and a "bank account" only to be accessed in times of economic strain (Fleisher 2000). Cattle are unlikely to be slaughtered for direct consumption on a regular basis as long as cheaper protein sources are available locally. As access to bushmeat remains high, especially as migratory ungulates move closer to markets able to process them, wild meat becomes a more attractive alternative than dipping into long-term livestock savings for short-term nourishment.

Although more work is needed to understand the dynamics between fisheries in Lake Victoria and access to bushmeat from the GSE, studies show that harvest rates of fish and bushmeat are inextricably linked, and may be severely diminished if current practices continue (Brashares et al. 2004; Nyahongo et al. 2009). Should either resource of this balance become overharvested to the point of significantly reduced production, or closed in the form of more stringent regulation and enforcement, increased demand for the other may result. Lake Victoria fisheries are at risk from commercial fishing as well as increasing demands on water resources from the lake. Should the fish stock fail, it remains to be seen what the impact would be on wildlife from Serengeti, as demand for protein needs to be met elsewhere. Understanding these dynamics including market access and the sustainability of fisheries management is crucial. As well, wildlife management should be linked to management of other natural resources including fisheries and livestock grazing to better address the issues of demand for protein by a growing human population.

CONSEQUENCES OF OFFTAKE ON WILDLIFE POPULATIONS

From the earlier analysis of total offtake, and from census data for each species, we can calculate the annual harvest rate for each species. Based on the basic life history of each species (age at first births, number of offspring), we can determine for which species and time periods poaching constituted a serious conservation threat. At present, it has not been a conservation threat for wildebeest or zebra or the plains-oriented species. During the periods of intense poaching, there was a serious conservation threat for the resident species in the woodlands.

Given the large discrepancy between the estimates of wildebeest offtake between the ecologically-based model and the community-based consumption data, it is likely that there are other important factors involved. Dry season rainfall indicates migratory wildebeest survival rate in the ecosystem, as the limitation of food during the drying months cannot support a large population during times of extreme drought. Communities which rely on subsistence agriculture are also climate-dependent as the dry season rains often determine the crop yield for the primary subsistence crops. Farmers growing more labor-intensive crops such as cotton and maize were less likely to be involved in illegal hunting activities (Holmern et al. 2004; Johannesen and Skonhoft 2005). Overall, households with more labor-intensive income-generating activities are less inclined to poach. Involvement in poaching may be largely motivated as much by opportunity time to hunt as by need for income (Knapp 2007). Therefore in severe drought years when crops fail and farmers are not busy during harvest time, individuals may be more likely to engage in hunting to supplement income. Years with drought can influence the survival rate of the wildebeest population but also increase offtake pressure on wildlife populations (Barrett and Arcese 1998). As a result, the correlation between low dry season rainfall and declines in the migratory wildebeest population demonstrated by the ecological population model may not be caused solely by food regulation, but also partly by variation in hunting pressure by local communities—especially in years of severe drought such as 1993. However, the extent of this impact is not known at this time.

The ecologically based model of wildebeest regulation is based on the assumption of an equal wildebeest sex ratio, and that no poaching occurred prior to 1977 (Pascual and Hilborn 1995; Mduma, Sinclair, and Hilborn 1999). However, it is known that poaching in Serengeti was quite extensive, with estimates of offtake of migratory animals as high as 15,000–40,000 individuals per annum. (Schaller 1972; Turner 1987). Additionally, nonspecies specific hunting techniques, particularly snaring, may cause a skewed sex ratio in offtake due to poaching (Hofer et al. 1996). These techniques were shown to have a very strong male bias (14:4 ratio) in offtake in the Serengeti (Holmern et al. 2006). Studies of migratory and nonmigratory wildebeest populations in the western corridor of the national park showed a female-biased sex ratio which was also attributed to the proportionally greater loss of males through bushmeat hunting (Ndibalema and Songorwa 2008). The behavioral patterns of the Serengeti wildebeest support the likelihood of male-biased poaching harvest, even using indiscriminate methods such as snares placed in the wooded areas (Estes, Raghunathanb, and van Vleck 2008). Wildebeest bachelor herds tend to form large aggregations and occupy less desirable habitats than calving females, making them more likely to wander into wooded vegetation where they would be susceptible to snaring. There may also be financial gains to targeting male wildebeest by poachers, as they will yield more meat per animal than females or juveniles (Holmern et al. 2006). While this is based on a rather small sample size of hunter responses (151 total animals reported, based on hunter interviews), it is indicative of a male-biased hunting offtake which could lead to an underestimate in the observed impact on the wildebeest population overall (e.g., through aerial censuses). Mduma, Sinclair, and Hilborn (1999) do not report an observed sex-bias based on transects conducted in the plains in 1993; the model of offtake based on rainfall patterns and projected recruitment assume an even sex ratio of offtake through poaching (Pascual and Hilborn 1995). However a skewed sex ratio would be expected among migratory wildebeest as it is with nearly all Bovidae ungulates given their social structure in which bachelor herds of males tend to inhabit less desirable habitat and are thus more at risk of food scarcity (Estes 1966). It is possible that poaching done primarily through snaring would exacerbate this bias, which would be difficult to detect in the general population. Therefore, a male-biased hunting offtake of 3:1 would account for both the total offtake estimated through consumption surveys, while still validating the offtake predicted by Mduma, Sinclair, and Hilborn (1999) as the population recruitment is largely based on the number of female wildebeest.

However, despite the discrepancy in the estimates of total wildebeest offtake due to illegal poaching by the differing methods, what is known is that based on invaluable aerial census data dating back to the 1950s, the wildebeest population numbers appear to have stabilized in recent years, at least without a drastic decline for the past 30 years or more. Therefore the current levels of offtake have not reached unsustainable levels. However poaching pressure clearly remains high, primarily due to hunting by local communities to supply bushmeat markets locally and elsewhere. As human populations continue to expand and access to protected areas increases, the costs of abating poaching pressure through community outreach initiative as well as enforcement tactics will continue to climb.

MANAGEMENT IMPLICATIONS: WHAT THE DATA TELL US ABOUT POACHING, AND HOW TO ADDRESS IT?

There is a three-fold approach to mitigate poaching in Serengeti.

Increase the Negative Incentive for Hunting

Increasing the price of bushmeat is the most effective way to reduce bushmeat consumption. As little as a 10% increase in bushmeat price is predicted to lead to a reduction of 0.32 (±0.01) kilograms of bushmeat consumed per week per household in western Serengeti (Rentsch and Damon 2013). Likewise, ramping up enforcement provides a stronger negative incentive for hunters who supply this meat (Knapp 2012). This is done both by increasing penalties for hunting (fines or prison time) but also by increasing risk of detection. Even with increased patrol effort, the numbers of poachers are increasing commensurate with human population growth. This means that the risk of an individual being caught is unlikely to increase with enforcement effort. Law enforcement activities need to be looked at critically so that patrols become more effective at detecting illegal hunters. Training and retraining is of key importance, and steps need to taken seriously and evaluated and reviewed every year. Police and courts also need be sensitized to take poaching cases seriously and assess the correct legal penalties to deter habitual poachers. The recent reintroduction of black rhino to the Serengeti from a stock in South Africa has led to increased effort in selecting and training rangers as well as a shift in techniques. While it is too soon to assess the impact of these interventions, the goals remain to improve detection, actual arrests, and penalties for illegal hunters within the national park.

Improve Alternative Livelihood Opportunities

Data show that bushmeat hunters in Western Serengeti are hunting primarily as a source of income (Campbell, Nelson, and Loibooki 2001; Loibooki et al. 2002; Galvin et al. 2008; Holdo et al. 2010). While bushmeat is consumed at a high rate by local communities to the west of the national park, a system of supply and demand exists as well, with bushmeat hunters supplying the meat for sale both for local consumption and trade to external markets. Individuals living in western Serengeti have a variety of livelihood options, as discussed earlier. Bushmeat hunting will persist if it generates a higher profit than alternative income-earning activities. Poaching is a high-risk livelihood, but historically, the rewards have outweighed the costs (Knapp 2012). The other side of this involves improving incentives for nonpoachers by ensuring better reliability of alternative income opportunities. For example, if farming productivity on a per-hectare basis were to increase to the point where farming was consistently providing higher returns for effort than poaching, there would be little incentive for hunters to continue the high-risk, relatively low-reward activity of going into the national park to obtain bushmeat to sell. Tanzania National Parks (TANAPA) and the surrounding game reserves under the wildlife division have invested in community conservation programs since the early 1990s, but this has not resulted in a measurable decrease in bushmeat hunting, nor an improvement in interactions between people and protected areas. There is a need now to revisit the objective and activities in such programs and tailor them to address the underlying causes of bushmeat hunting where possible.

One important factor in the success of community-based conservation programs is understanding the relationship between household income and conservation outcomes. For example, it has been shown that as household income increases around western Serengeti, consumption of all meat will significantly increase as well (Rentsch and Damon 2013). Alarmingly, among all of the most commonly consumed meat types (beef, fish, chicken, bushmeat, goat, dagaa), bushmeat is the most responsive to increases in income. If household expenditure increases by 30% the quantity of bushmeat demanded would increase by 40%, or an increase of nearly 1.07 kilograms of bushmeat per household per week on average.

Increasing income is key to achieving development goals from both a poverty reduction and food security perspective. Total consumption of meat (including bushmeat) will increase as households accumulate more wealth through increased agricultural production and wage employment opportunities (Rentsch and Damon 2013). Thus, increasing income will also place more pressure on wildlife populations, fisheries, and likely increase demand for grazing land for beef. This suggests that development and conservation agencies working throughout the ecosystem to achieve Tanzania's rural development goals of reducing poverty and maintaining biodiversity would be best served to strategically collaborate to target interventions aimed at the producers of bushmeat, illegal hunters. Targeted alternative income-generating opportunities to producers of bushmeat coupled with antipoaching measures may curb production of bushmeat and significantly increase the price of bushmeat relative to alternatives, shifting consumption away from wildlife.

Reduce Local Demand for Bushmeat

It has been shown that local communities alone consume many thousands of wildebeest per year. Bushmeat is among the cheapest and most readily available protein sources for many impoverished agropastoralist communities living adjacent to GSE. As a naturally available protein source, bushmeat consumption also means not having to dip into the "bank account" of livestock holdings, which are considered a main economic asset for communities as well as an alternative source of protein. Shifting the economic incentives away from consumption of bushmeat to another replenishable protein source could reduce demand for bushmeat, and thereby decrease the extent of bushmeat hunting. However, this would likely require large-scale production of beef, fish, or chicken, each of which has its own challenges and potential negative externalities that impact the ecosystem functioning as well.

In reality, no single method can be effective in significantly reducing bushmeat hunting on its own. But if a concerted effort is made to address all three areas, it is possible that bushmeat hunting can be reduced to manageable levels in the ecosystem. It will be a challenge to overcome the significant logistical, financial, political, and social obstacles currently in place, especially the current human population growth over 3% in communities adjacent to SNP.

ACKNOWLEDGMENTS

We would like to thank the Tanzanian National Parks, Tanzanian Wildlife Research Institute, Singita Grumeti Fund, and Frankfurt Zoological Society for research support, permissions, and valuable data on bushmeat hunting and enforcement. Thanks to A. R. E Sinclair, Craig Packer, Jennifer Schmitt, Steven Polasky, Grant Hopcraft, Nelly Boyer, and Emile Smidt for input and insights into the drafting of this chapter.

REFERENCES

Barnett, R. 2000. Food for thought: The utilization of wild meat in eastern and southern Africa. TRAFFIC Report, 264–310. Nairobi, Kenya, TRAFFIC East/southern Africa.

Barrett, C. B., and P. Arcese. 1998. Wildlife harvest in integrated conservation and development projects: Linking harvest to household demand, agricultural production, and environmental shocks in the Serengeti. *Land Economics* 74:449–65.

Brashares, J. S., P. Arcese, M. K. Sam, P. B. Coppolillo, A. R. E. Sinclair, and A. Balmford. 2004. Bushmeat hunting, wildlife declines, and fish supply in West Africa. *Science* 306:1180–83.

Campbell, K., & M. Borner. 1995. Population trends and distribution of Serengeti herbivores: Implications for management. In *Serengeti II: Dynamics, management, and conservation of an ecosystem*, ed. A. R. E. Sinclair and P. Arcese, 117–45. Chicago: University of Chicago Press.

Campbell, K., and H. Hofer. 1995. People and wildlife: Spatial dynamics and zones of interaction. In *Serengeti II: Dynamics, management, and conservation of an ecosystem*, ed. A. R. E. Sinclair and P. Arcese, 534–70. Chicago: University of Chicago Press.

Campbell, K., V. Nelson, and M. Loibooki. 2001. Sustainable use of wildland resources: Ecological, economic and social interactions: An analysis of illegal hunting of wildlife in Serengeti National Park, Tanzania. Final technical report to the Department for International Development (DFID), Animal Health and Livestock Production Programmes, Project R7050. London: Department for International Development.

Cowlishaw, G., S. Mendelson and J. M. Rowcliffe. 2005. Evidence for post-depletion sustainability in a mature bushmeat market. *The Journal of Applied Ecology* 42:460–68.

Dobson, A. 1995. The Ecology and epidemiology of rinderpest virus in Serengeti and Ngorongoro Conservation Area. In *Serengeti II: Dynamics, management, and conservation of an ecosystem*, ed. A. R. E. Sinclair and P. Arcese, 485–505. Chicago: University of Chicago Press.

Dublin, H. T., A. R. E. Sinclair, S. Boutin, E. Anderson, M. Jago, and P. Arcese. 1990. Does competition regulate ungulate populations? Further evidence from Serengeti, Tanzania. *Oecologia* 82:283–88.

Emerton, L., and I. Mfunda. 1999. *Making wildlife economically viable for communities living around the western Serengeti, Tanzania.* Evaluating Eden Project, Working Paper no. 1. London: International Institute for Environment and Development.

Estes, R. 1966. Behaviour and life history of the wildebeest (*Connochaetes taurinus* Burchell). *Nature* 212:999–1000.

Estes, R. D., T. E. Raghunathanb, and D. van Vleck. 2008. The impact of horning by wildebeest on woody vegetation of the Serengeti ecosystem. *Journal of Wildlife Management* 72:1572–78.

Fleisher, M. 2000. *Kuria cattle raiders: Violence and vigilantism on the Tanzania/Kenya frontier.* Ann Arbor: University of Michigan Press.

Galvin, K. A., S. Polasky, C. Costello, and M. Loibooki. 2008. Human response to change: Modeling household decision making in western Serengeti. In *Serengeti III: Human impacts on ecosystem dynamics*, ed. A. R. E. Sinclair, C. Packer, S. A. R. Mduma, and J. M. Fryxell, 325–46. Chicago: University of Chicago Press.

Hilborn, R., P. Arcese, M. Borner, J. Hando, J. G. Hopcraft, M. Loibooki, S. A. R. Mduma, and A. R. E. Sinclair. 2006. Effective enforcement in a conservation area. *Science* 314:1266.

Hofer, H., K. L. I. Campbell, M. L. East, and S. A. Huish. 1996. The impact of game meat hunting on target and nontarget species in the Serengeti. In *The exploitation of mammal populations*, ed V. J. Taylor and N. Dunstone, 117–46. London: Chapman and Hall.

Holdo, R. M., K. A. Galvin, E. Knapp, S. Polasky, R. Hilborn, and R. D. Holt. 2010.

Responses to alternative rainfall regimes and antipoaching in a migratory system. *Ecological Applications* 20:381–97.

Holmern, T., A. B. Johannesen, J. Mbaruka, S. Mkama, and E. Røskaft. 2004. Human-wildlife conflicts and hunting in the western Serengeti, Tanzania. Norwegian Institute for Nature Research Project Report. Trondheim, Norway: (NINA).

Holmern, T., S. Mkama, J. Muya, and E. Røskaft. 2006. Intraspecific prey choice of bushmeat hunters outside the Serengeti National Park, Tanzania: A preliminary analysis. *African Zoology* 41:81–87.

Holmern, T., J. Muya, and E. Røskaft. 2007. Local law enforcement and illegal bushmeat hunting outside the Serengeti National Park, Tanzania. *Environmental Conservation* 34:55–63.

Johannesen, A. B., and A. Skonhoft. 2005. Tourism, poaching and wildlife conservation: What can integrated conservation and development projects accomplish? *Resource and Energy Economics* 27 (3): 208.

Kaltenborn, B. P., J. W. Nyahongo, and K. M. Tingstad. 2005. The nature of hunting around the Western Corridor of Serengeti National Park, Tanzania. *European Journal of Wildlife Research* 51:213–22.

Knapp, E. J. 2007. Who poaches? Household economies of illegal hunters in western Serengeti, Tanzania. *Human Dimensions of Wildlife* 12:195–96.

———. 2012. Why poaching pays: A summary of risks and benefits illegal hunters face in western Serengeti, Tanzania. *Tropical Conservation Science* 5:434–45.

Knapp, E. J., D. Rentsch, J. Schmitt, S. Polasky, and C. Lewis. 2010. A tale of three villages: Choosing an effective method for assessing poaching levels in western Serengeti, Tanzania. *Oryx* 44:178–84.

Loibooki, M., H. Hofer, K. L. I. Campbell, and M. L. East. 2002. Bushmeat hunting by communities adjacent to the Serengeti National Park, Tanzania: The importance of livestock ownership and alternative sources of protein and income. *Environmental Conservation* 29:391–98.

Mduma, S., A. R. E. Sinclair, and R. Hilborn. 1999. Food regulates the Serengeti wildebeest: A 40 year record. *Journal of Animal Ecology* 68:1101–22.

Metzger, K. L., A. R. E. Sinclair, K. L. I. Campbell, R. Hilborn, J. G. C. Hopcraft, S. A. R. Mduma, and R. M. Reich. 2007. Using historical data to establish baselines for conservation: The black rhinoceros (*Diceros bicornis*) of the Serengeti as a case study. *Biological Conservation* 139:358–74.

Metzger, K. L., A. R. E. Sinclair, R. Hilborn, J. G. C. Hopcraft, and S. A. R. Mduma. 2010. Evaluating the protection of wildlife in parks: The case of the African buffalo in Serengeti. *Biodiversity and Conservation* 19:3431–44.

Milner-Gulland, E. J., and E. L. Bennett. 2003. Wild meat: The bigger picture. *Trends in Ecology & Evolution* 18 (7): 351.

Ndibalema, V. G., and A. Songorwa. 2008. Illegal meat hunting in Serengeti: Dynamics in consumption and preferences. *African Journal of Ecology* 46:311–19.

Nyahongo, J. W., T. Holmern, B. P. Kaltenborn, and E. Røskaft. 2009. Spatial and temporal variation in meat and fish consumption among people in the western Serengeti, Tanzania: The importance of migratory herbivores. *Oryx* 43:258–66.

Packer, C., and S. Polasky. 2008. Introduction: Understanding the greater Serengeti ecosystem. In *Serengeti III: Human impacts on ecosystem dynamics*, ed. A. R. E. Sinclair,

C. Packer, S. A. R. Mduma, and J. M. Fryxell, 1–5. Chicago: University of Chicago Press.

Pascual, M. A., and R. Hiborn. 1995. Conservation of harvested populations in fluctuating environments: The case of the Serengeti wildebeest. *Journal of Applied Ecology* 32: 468–80.

Rentsch, D. 2012. The nature of bushmeat hunting in the Serengeti ecosystem, Tanzania: Socioeconomic drivers of consumption of migratory wildlife. PhD diss., University of Minnesota.

Rentsch, D., and A. Damon. 2013. Prices, poaching, and protein alternatives: An analysis of bushmeat consumption around Serengeti National Park, Tanzania. *Ecological Economics* 91 (C): 1–9.

Robinson, J. G., and E. L. Bennett. 2004. Having your wildlife and eating it too: An analysis of hunting sustainability across tropical ecosystems. *Animal Conservation* 7:397–408.

Rossiter, P. B., D. M. Jessett, J. S. Wafula, L. Karstad, S. Chema, P. Taylor, L. Rowe, J. C. Nyange, M. Otaru, and M. Mumbala. 1983. Re-emergence of rinderpest as a threat in East Africa since 1979. *Veterinary Record* 113:459–61.

Rowcliffe, J. M., E. J. Milner-Gulland, and G. Cowlishaw. 2005. Do bushmeat consumers have other fish to fry? *Trends in Ecology and Evolution* 20:274–76.

Schaller, G. B. 1972. *The Serengeti lion: A study of predator-prey relations*. Chicago: University of Chicago Press.

Sinclair, A. R. E. 1995. Serengeti past and present. In *Serengeti II: Dynamics, management, and conservation of an ecosystem*, ed. A. R. E. Sinclair and P. Arcese, 3–30. Chicago: University of Chicago Press.

Sinclair, A. R. E., S. A. R. Mduma, J. G. C. Hopcraft, J. G. M. Fryxell, R. Hilborn, and S. Thirgood. 2007. Long-term ecosystem dynamics in the Serengeti: Lessons for conservation. *Conservation Biology* 21:580–90.

Sinclair, A. R. E., J. G. Hopcraft, H. Olff, S. A. R. Mduma, K. A. Galvin, and G. J. Sharam. 2008. Historical and future changes to the Serengeti ecosystem. In *Serengeti III: Human impacts on ecosystem dynamics*, ed. A. R. E. Sinclair, C. Packer, S. A. R. Mduma, and J. M. Fryxell, 7. Chicago: University of Chicago Press.

Tanzania National Bureau of Statistics. 2002. Tanzania 2002 population and housing census. Dar Es Salaam, Tanzania: Tanzania National Bureau of Statistics.

Thirgood, S., C. Mlingwa, E. Gereta, V. Runyoro, R. Malpas, K. Laurenson, and M. Borner. 2008. Who pays for conservation? Current and future financing scenarios for the Serengeti ecosystem. In *Serengeti III: Human impacts on ecosystem dynamics*, ed. A. R. E. Sinclair, C. Packer, S. A. R. Mduma, and J. M. Fryxell, 443–70. Chicago: University of Chicago Press.

Turner, M. 1987. *My Serengeti years: The memoirs of an African game warden*. New York: W. W. Norton & Company.

US Census Bureau. 2010. International Database. (www.census.gov).

Wilkie, D. S., and J. F. Carpenter. 1999. Bushmeat hunting in the Congo basin: An assessment of impacts and options for mitigation. *Biodiversity and Conservation* 8:927–55.

Wilkie, D. S., and R. A. Godoy. 2001. Income and price elasticities of bushmeat demand in lowland Amerindian societies. *Conservation Biology* 15:761–69.

Wilkie, D. S., M. Starkey, K. Abernethy, E. N. Effa, P. Telfer, and R. A. Godoy. 2005. Role of prices and wealth in consumer demand for bushmeat in Gabon, Central Africa. *Conservation Biology* 19 (1): 268.

TWENTY-THREE

Human Health in the Greater Serengeti Ecosystem

Linda M. Knapp, Eli J. Knapp, Kristine L. Metzger, Dennis Rentsch, Rene Beyers, Katie Hampson, Jennifer Schmitt, Sarah Cleaveland, and Kathleen A. Galvin

The story of the health of Serengeti peoples is situated in a broader debate regarding the values of conservation versus development. Central to the polemics is the question of whether or not biodiversity conservation alleviates or exacerbates human poverty levels. Roe (2008) traces the history of this debate to an older discussion surrounding the links of environment and development. He writes, "As early as the 1950s a polarized debate about the purpose of conservation whether to establish national parks to protect species, or to benefit people, had begun to emerge" (Roe 2008, 493). Wilkie et al. (2006) distilled the social advocates' position to three main points: (1) successful biodiversity conservation can only occur if there is first poverty reduction; (2) protected areas take away the property and rights of local people, thereby affecting the welfare of these people; and (3) benefits from protected areas are distributed in an imbalanced manner with local people receiving the least benefits. Yet the conservationists' side is equally persuasive. They argue that benefits from intact ecosystems provide services both locally and globally (Upton et al. 2008) and that "environmental regulations are essential to ensure the sustainability of the planet's biological systems and the health and welfare of people" (Wilkie et al. 2006, 247).

Protected areas are not "islands," impervious to the whims of social, cultural, or economic contexts (Scherl 2004), nor can the "resilience of the poor" be strengthened without stewardship of natural resources (Scherl 2004). Our research fits into this larger discussion of the dynamics between conservation and poverty. However, we do not have sufficient data to clearly untangle the links between human health and biodiversity conservation; instead, we seek to get a better understanding of the overall health status of the human population within the GSE. Thus, we are asking the following questions: (1) What is the health status of the human populations living in the GSE compared to the health of people from across rural Tanzania? and (2) Is there geographical variation within the ecosystem in terms of human health? For example, are the pastoralists on the eastern side of the protected area better off than the densely populated agriculturalists, horticulturalists, and agropastoralists on the western edge? To answer the questions above, we will analyze data from two disparate research projects: (1) archival records of morbidity and mortality rates gathered from biomedical centers in western Serengeti, and (2) anthropometric measurements of populations across the entire GSE to assess nutritional status.

A RECENT HISTORY OF HUMANS IN THE GREATER SERENGETI ECOSYSTEM

The GSE contains a rich cultural heritage with thousands of years—even millennia—of sustainable human-ecological interactions occurring (Homewood and Rodgers 1991). Only in the past century have these dynamics become more strained as human population growth explodes leading to increased hunting of wildlife (Campbell and Hofer 1995), and as conservation policy affects the well-being of local people living within the system (Homewood and Rodgers 1991). In the following paragraphs we will briefly outline who these people are and what their history within the GSE has been.

The GSE inhabitants today are grouped into two main cultural categories: Nilotic Maasai pastoralists on the east, and Bantu horticulturalists or agropastoralists on the west. According to their cultural traditions, the Maasai center their lives around their cattle, live in small transhumant groups, and do not hunt wildlife for food. As will be described in greater detail below, they were the last inhabitants of the GSE when most of the conservation areas were established. On the western boundaries of the GSE live Bantu horticulturalists and agropastoralists. These people groups live in more densely populated villages, tend to have small farm plots (average of 2.5 ha), and grow corn, millet, sorghum, cassava, or cotton (Loibooki 1997; Emerton and Mfunda 1999). Contrary to the Maasai and in keeping with their cultural heritage, they are the source of the majority of bushmeat hunting in the GSE (Arcese, Hando, and Campbell 1995; Campbell and Hofer 1995). They too have been forced to alter their livelihoods and relocate as protected areas are added as buffer zones around the core of SNP. The story of both these groups—on the east and west—has been intertwined for hundreds of years.

Linguistic evidence shows that long after early hominids walked the Laetoli plain or carved tools in Olduvai Gorge, residents of the central GSE were southern Cushitic hunters, ancestors of the present day Iraqw people (Shetler 1998). Around 500 AD, Bantu-speaking farmers (ancestors of the Kuria, Ikoma, Natta, and Sukuma ethnic groups) as well as southern Nilotic herders (known as the Datog/Tatog or Barabaig) migrated into the region and began to peacefully intermarry with one another (Shetler 1998). Although individual tribes and languages remained, these early cross-cultural interactions resulted in diversified livelihood strategies of hunting wildlife, farming, and some animal husbandry (Shetler 2002). By 1,000 AD the marginalized Cushites were pushed out of the GSE and the dominant Bantu tribes controlled western GSE, while the Tatog settled present-day NCA (Shetler 1998). This domination continued until the rise of the powerful Maasai in the late 1800s across East Africa's Rift Valley.

In the GSE, Maasai took control of the Ngorongoro highlands during 1836–1850 by driving out Tatog pastoralists (Arhem 1985). By 1850, Maasai began pushing Bantu farmer-hunters westward out of present-day Serengeti National Park, raiding the few livestock herds they owned. Furthermore, the penetration of Maasai into the Serengeti caused the Bantu groups to flee their homes, which led to bush encroachment, an ultimate driver of the ensuing outbreak of African sleeping sickness (Shetler 1998). At this, the height of Maasai dominance across the GSE, exogenous forces of change led to great tumult and upheaval for all GSE people—whether pastoral or not. Smallpox, sleeping sickness, famine, and rinderpest crushed the resilience of the GSE people against the socioeconomic changes brought on by German (1885–1916) and English (1918–1961) colonialism (Shetler 1998). Under new taxation by these foreign governments and amid enforced economic changes (such as growing corn and cotton), colonial regimes brought with them new land-use management systems, including conservation areas.

The first GSE game reserve was established in what is now central SNP in 1929, although the national park itself was not founded until 1951 (Perkin 1997). Maasai first willingly relinquished their formal land rights to SNP under the Game Ordinance of 1940 with the understanding that they would be allowed to carry on with their normal livelihood strategies—except for hunting (Shivji and Kapinga 1998). When these promises were not kept by the colonial government, Maasai protests (over the loss of cultivation rights) led to the eventual formation of the Ngorongoro Conservation Area Authority (NCAA) in 1959, which would empty SNP of all human occupants but allow Maasai to have permanent land rights and occupancy in the Ngorongoro Conservation Area (NCA) (Arhem 1985). Despite maintaining these land rights, restrictions on Maasai livelihoods increased, including new sanctions on where to graze, live, access water, and the most devastating—cultivate small-scale farms. Although Maasai traditionally subsist on foods acquired directly from their livestock (milk, meat, and blood), they also practice small-scale agriculture in order to offset their nutritional needs as livestock-to-human rations decrease (McCabe 2003). Other scholars (Perkin, and Schofield 1992; McCabe 2002; Galvin et al. 2002; Galvin et al. 2008; Potkanski 1997) have chronicled in detail the nutritional and socioeconomic impacts of the NCA cultivation ban and its subsequent lifting.

Overall, the tensions between people and park in the GSE hinge on the fact that the establishment of conservation areas limits peoples' choices in responding to change by diminishing their rights to the land and the natural resources therein. At the same time, the biodiversity and ecological integrity of the system is also vulnerable to the overharvesting of these resources despite enforced disincentives. While these tensions do exist, the GSE holds a unique position in the global conservation arena for addressing these challenges and embracing creative solutions to reduce human and ecological vulnerability. For example, the GSE includes three different types of land-use systems: (1) formal conservation areas (such as SNP, the Maasai Mara National Reserve, and other game reserves where no people are allowed to live, hunt, farm, graze, or extract natural resources); (2) multiple land-use areas that seek to meet the needs of people and the conservation area (these include NCA, Loliondo Game Controlled Area, and Ikona Open Area); and (3) rangeland and agricultural areas under village jurisdiction where human interest overrides all else.

HUMAN HEALTH IN THE GSE: MORBIDITY AND MORTALITY RATES

Within this historical context of the GSE, we strive to understand more of the current patterns of human health in the ecosystem. To this end, archival data of morbidity and mortality rates were gathered from biomedical and community health centers within western Serengeti to address the question: What is the health status of western GSE people and how does this compare with the general population of rural Tanzania? Data were collected between 2004 and 2007 from the Serengeti District Council Health Sector, the Serengeti and Bunda Designated District Hospitals (DDH) located in Mugumu and Bunda towns, as well as from the Community-Based Health Promotion Program (CBHPP) HIV/AIDS clinic in Mugumu. The data presented in this chapter are grouped into three categories: (1) HIV/AIDS, (2) malaria and infectious disease, and (3) childhood health.

HIV/AIDS in Western Serengeti

Prevalence of HIV among the human population in the Serengeti district was estimated from blood donors who did not exhibit symptoms of clinical AIDS.

Over 8% of blood tests on nonsymptomatic individuals tested positive for HIV from 2001–2004 (table 23.1). Similarly, data gathered from Bunda DDH show that 9.7% of blood donors (combined male and female) in 2006 were HIV-positive. Infection rates for women in Serengeti district were consistently higher than those for men (table 23.1), and highest among married women (particularly those in polygamous marriages) rather than nonmarried women (Knapp 2010). The national infection rate for HIV in Tanzania at the time of this study was 7% (Tanzania Commission for AIDS 2005; National Bureau of Statistics [NBS]), which suggests that prevalence in western Serengeti is slightly higher. However, the data may not be fully comparable because sampling methods were different. In Serengeti district, only nonsymptomatic blood donors who visited the CBHPP clinic were sampled, while in the national survey a representative sample of 6,900 households was drawn from the total mainland population.

Malaria and Infectious Disease

In western Serengeti, infectious disease still dominates the lives of villagers. The top ten diagnosed illnesses of all patients who visited the district

Table 23.1 Percentiles of positive results from HIV blood screening from nonsymptomatic blood donors at CBHPP clinic in Mugumu, Serengeti district, Tanzania

	2001 (n = 195) (%)	2002 (n = 572) (%)	2003 (n = 1,419) (%)	2004 (n = 3,347) (%)
Female	7.4	6.8	15.0	10.8
Male	7.0	4.9	6.3	5.5
Combined	7.2	5.8	10.8	8.0

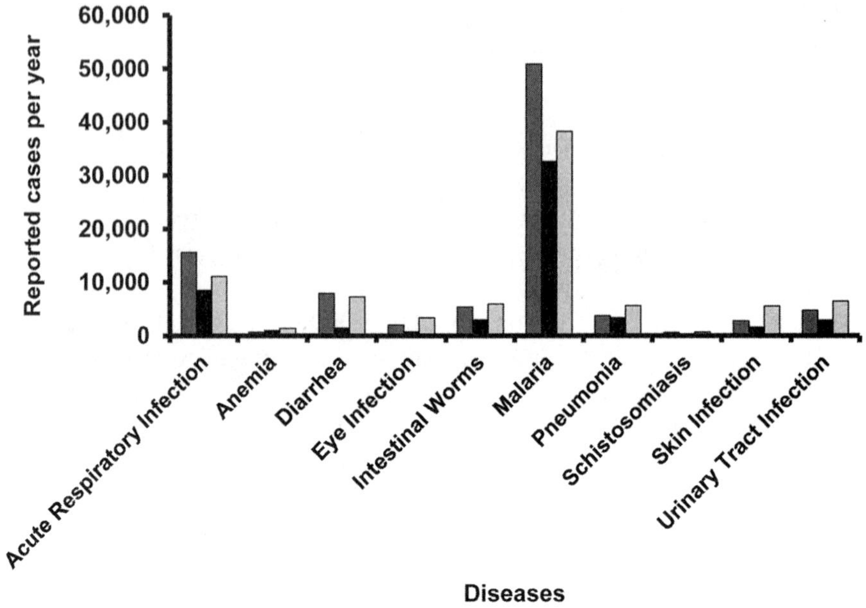

Fig. 23.1 Leading causes of morbidity for patients of Serengeti Designated District Hospital and its subsidiary health stations in 2004 (dark gray bars), 2005 (black bars), and 2006 (light gray bars).

hospital in Mugumu or any of the district health sector's subsidiary health stations in 2004–2006 were (in descending order): malaria, acute respiratory illness, intestinal worms, urinary tract infections, pneumonia, diarrhea, skin infections, eye infections, anemia, and schistosomiasis (fig. 23.1). Malaria alone comprised almost 60% of all cases seen in Serengeti district health centers during 2005, and was associated with high mortality (155 reported deaths/32,727 cases).

Human malaria is caused by one of four species of protozoan parasites carried by their hosts, the *Anopheles* mosquitoes. Symptoms include spiking fevers, chills, shakes, body and muscles aches, headache, diarrhea, vomiting, and a cough. The *Plasmodium falciparum* parasite is the most common source of malaria and also the most dangerous. It can lead to cerebral malaria, coma, and death. The most severe cases of malaria occur in individuals who have a compromised immune system or who have not yet developed immunity to the disease through exposure. Those most vulnerable include children under five and pregnant women (Holtz and Kachur 2004; TDHS 2005). Malaria is not only a problem in Tanzania but around the world and particularly across sub-Saharan Africa. Every 30 seconds, one child dies from malaria in sub-Saharan Africa (Holtz and Kachur 2004). Other scholars

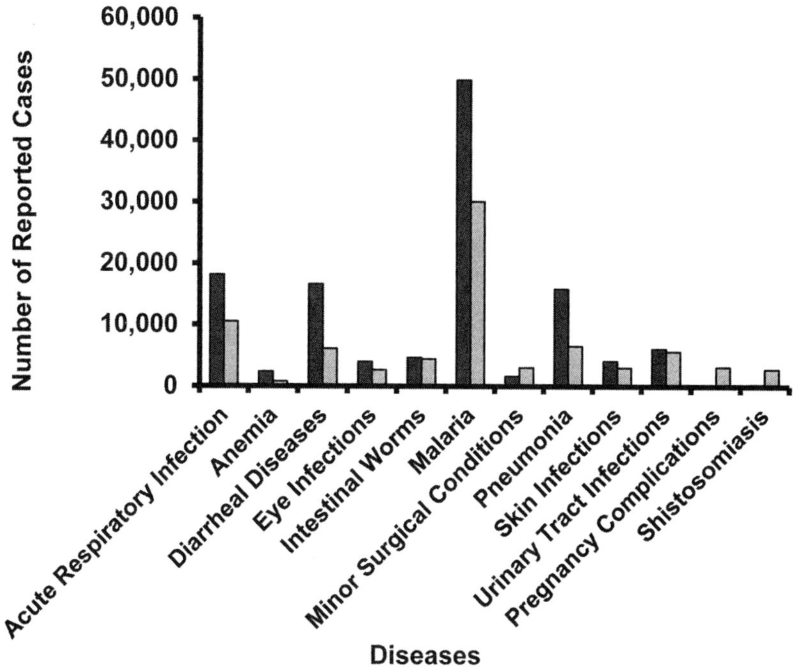

Fig. 23.2 Leading causes of morbidity in outpatient diagnoses from Bunda Designated District Hospital in 2006. Black bars represent morbidity rates for patients under age five, gray bars represent morbidity rates for patients age five and older.

report that of all the known diseases in the world, malaria has killed more people than any other (Inhorn and Brown 1997).

Archival data from Bunda Designated District Hospital (DDH) corroborate the findings from the Serengeti with infectious diseases dominating morbidity rates (see fig. 23.2). Malaria comprised over 40% of all diagnosed outpatient illnesses in Bunda and was the leading cause of death based on mortality data from that same hospital. Overall, these patterns of disease are similar to those across rural Tanzania. The Tanzania Demographic Health Survey (2005) reports that malaria is the leading cause of morbidity and mortality in Tanzania for both inpatient and outpatient attendance. Malaria accounts for about 40% of overall outpatient attendances in Tanzania (TDHS 2005).

Child Health

Children often bear the brunt of poverty and are the most likely to suffer in terms of health. In western Serengeti, small children are particularly vulner-

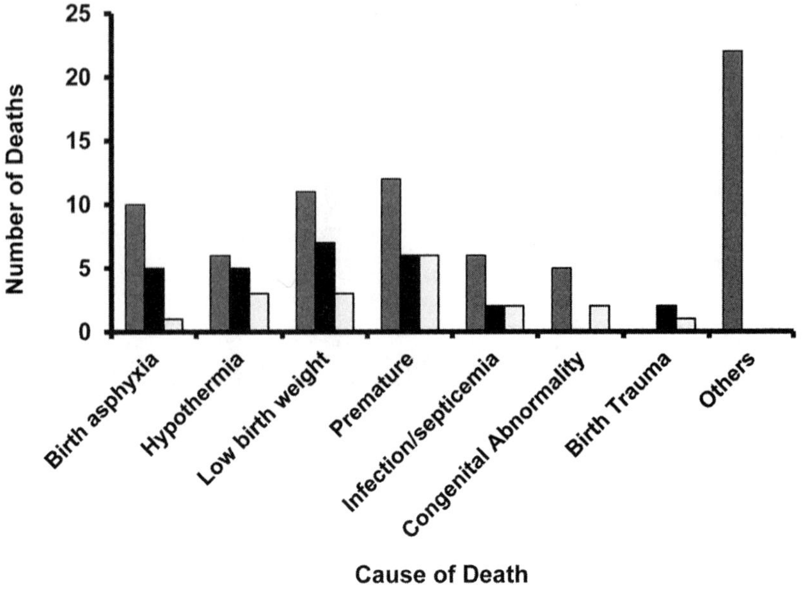

Fig. 23.3 Causes of neonatal mortality in Serengeti district, Tanzania. Dark gray bars represent data ($n = 72$ total deaths) from 2002, black bars represent data ($n = 28$ total deaths) from 2003, and light gray bars represent data ($n = 22$ total deaths) from 2004. All data gathered from Serengeti District Designated Hospital archives.

able to infectious disease. In 2006, Bunda Hospital recorded the deaths of 142 patients—all small children (under five)—from malaria while only 72 deaths from malaria occurred among patients age five and older. Moreover, most of these deaths from infectious diseases are easily preventable if proper sanitation and nutrition are available.

To assess children's health status in western Serengeti, we examined three important health indicators for early childhood: neonatal (NN) mortality (death of infants in the first 28 days of life); infant mortality (the probability of dying before the first birthday); and under-five mortality (death of children under five years old). The NN mortality is an important health indicator directly related to poverty (Adam et al. 2005). According to Serengeti District Health archives, NN mortality has decreased steadily in western Serengeti from 72 reported NN deaths in 2002 to 22 reported NN deaths in 2004. Causes of NN mortality during these years are listed in fig. 23.3. Similar to NN mortality rates, child deaths (under five) have been decreasing in the Serengeti district (fig. 23.3). The most common causes of under-five mortality are anemia and malaria (fig. 23.4)

In western Serengeti both infant mortality and under-five mortality are

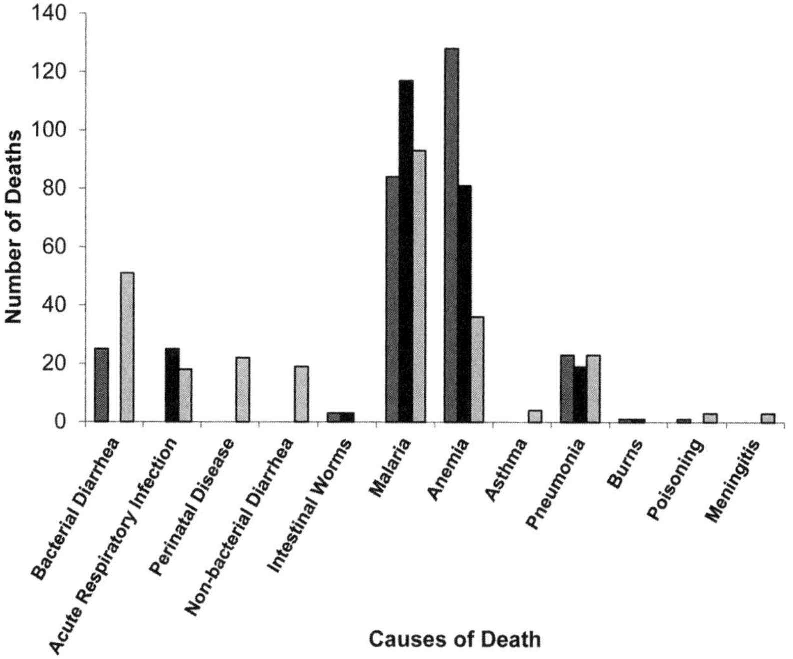

Fig. 23.4 Causes of under-five mortality in Serengeti district, Tanzania. Dark gray bars represent data (*n* = 396 deaths) from 2002, black bars represent data (*n* = 279 deaths) from 2003, and light gray bars represent data (*n* = 268 deaths) from 2004. All data gathered from Serengeti District Designated Hospital archives.

much worse than for rural Tanzania during the same time period (2001–2006). During 2000–2004, infant mortality rates (see fig. 23.5) recorded in Bunda district were approximately double that of the rate from across rural Tanzania (Tanzanian Demographic Health Survey [2005] recorded 68 deaths/1000 live births in rural Tanzania versus Bunda's records of 120–140 deaths/1,000 live births in western Serengeti). Similarly, under-five mortality rates were also higher in western Serengeti (120–150 deaths in Bunda district versus 112 deaths/1,000 live births in rural Tanzania). These high rates of early childhood mortality in western Serengeti—especially as compared to the rest of Tanzania—demonstrate the low quality of human health within or near the GSE.

PATTERNS OF HUMAN HEALTH WITHIN THE GSE: NUTRITIONAL STATUS

Anthropometry is the measurement of human bodies to study and understand biological, nutritional, and socioeconomic variation among and

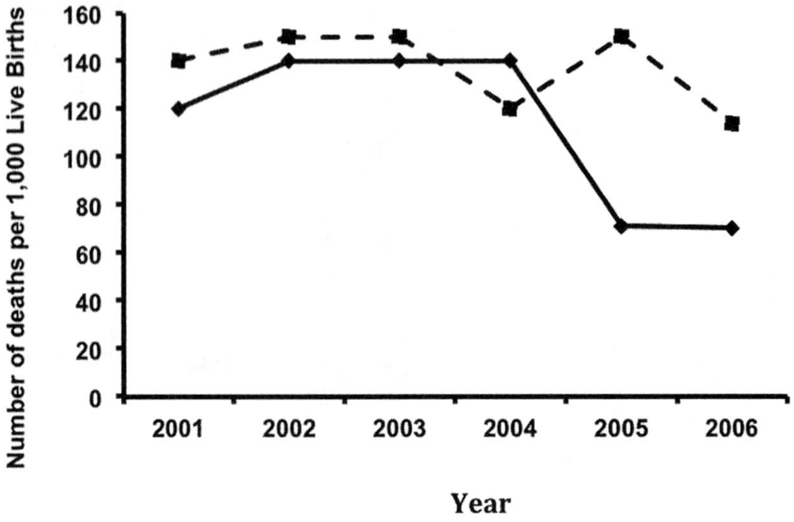

Fig. 23.5 Infant and under-five mortality rates from Bunda Designated District Hospital archives. Dashed lines represent under-five rates and the solid line represents infant mortality rates for Bunda district.

within populations or individuals. We used two of the most common anthropometric indices for children: height-for-age and weight-for-age (WHO 1995) to assess their nutritional status and, ultimately, their overall health.

Height-for-age reflects long-term nutritional status and low height-for-age is called stunting. Low weight-for-age, or wasting, can be a result of either long-term or short-term undernutrition. Undernutrition is defined as the outcome of insufficient food intake. Low weight-for-age is the best predictor of high mortality rates in children followed by height-for-age (WHO 1995). These two indices can reveal both current or past health problems as well as socioeconomic problems. Children's growth is positively related to socioeconomic conditions (WHO 1995), and anthropometric findings reflect the interconnectedness of nutrition, socioeconomic status, and infectious disease (WHO 1995).

Anthropometry Methods in the GSE

Anthropometric data from 20 villages representing four districts (Bunda, Loliondo, Meatu, and Serengeti; fig. 23.6) around the GSE were collected from August 2006 through February of 2007. Each of the sampled villages' boundaries are located within 5 km of a conservation area, whether it be a game reserve (GR), national park (NP), game controlled area (GCA), open

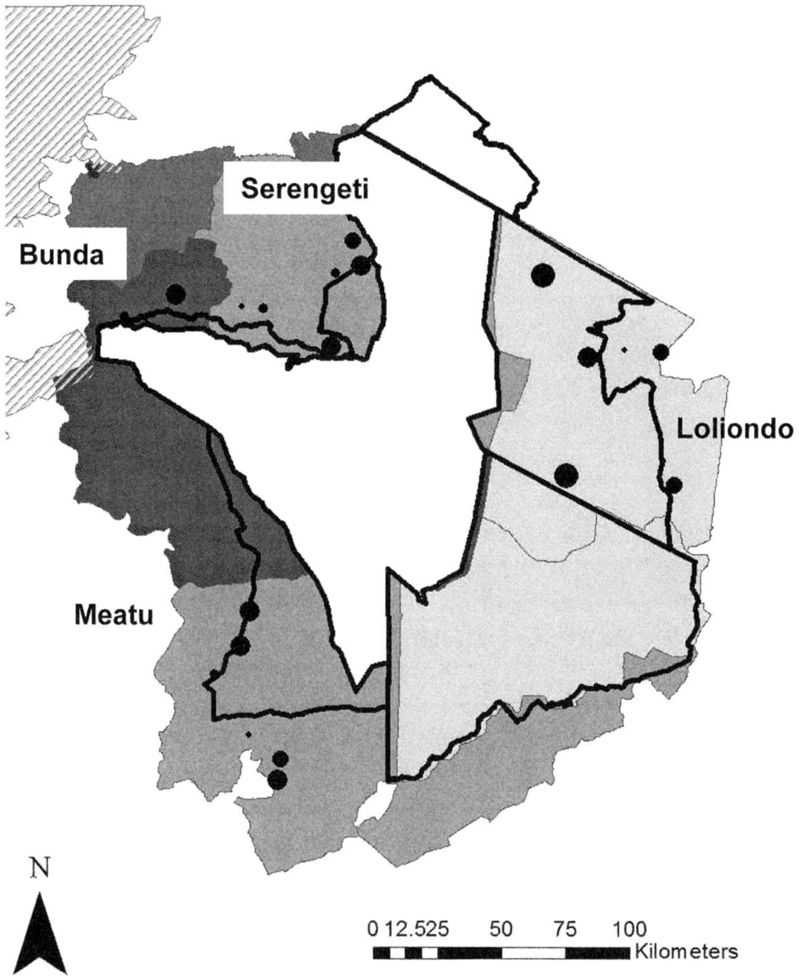

Fig. 23.6 Map of nutritional study sites (circles represent actual village locations where anthropometric data were gathered and size of circles represent variation in village population size). Small circles range from approximately 1,000–3,200 people and larger circles range from approximately 3,500–8,500 people. Gray areas represent variation in population density at the regional level; areas west of the Serengeti National Park boundary (dark areas) range from 93–500 people per km^2, while lighter shaded areas east of the park (i.e., Loliondo) range between 5–50 people per km^2. Population estimates are based on the 2002 census.

area (OA), or wildlife management area (WMA) (see fig. 23.6 for a map of the study sites). Stratified random sampling was used to ensure an even distribution of villages around the ecosystem. Field teams spent one day in the communal center of each village measuring heights and weights from as many children under age five within a village as possible. Scales and measuring instruments (both standing and recumbent) were provided by the

United Nations Children's Fund (UNICEF). Most mothers brought a health card annotated with their child's date of birth; those that did not have this document simply recalled the month and day of their child's birth. For the few women who only knew the month of their child's birth, the fifteenth day of that month was used. Field team members (trained by the WHO in Dar es Salaam, Tanzania) were able to measure 2,966 children under age five in the villages around the GSE.

Anthropometric Findings

The height and weight of the children being studied are compared to those from the same age and sex in the reference population (Alderman, Hoogeveen, and Rossi 2006). This international reference population, recommended by the WHO, is kept by the US National Center for Health Statistics (TDHS 2005). Z-scores are used to measure "the deviation of the value for an individual from the median value of the reference population, divided by the standard deviation (SD) for the reference population" (WHO 1995, 7). The two main ways to interpret the anthropometric data are prevalence-based reporting and summary statistics of Z-scores. We will present the findings from each of these methods separately (though both are summarized in table 23.2). First we will analyze the prevalence-based reporting and summary statistics of the height-for-age data from our sample, followed by the analyses of weight-for-age data.

Prevalence-based reporting. In prevalence-based reporting, *abnormal anthropometry* is defined as any anthropometric index, which is below–2 SD or above +2 SD from the mean Z-score since "these cut-offs define the central 95% of the reference distribution as the normality range" (WHO 1995, 219). Any individual or group that falls below the–2 Z-score can be considered *malnourished* (WHO 1995). As will be explained in greater detail below, many of the GSE villages sampled had high percentiles of malnourishment based on this definition (see fig. 23.7).

In prevalence-based reporting we analyze the percentage of a specific population that is below this–2 Z-score cut-off for height-for-age (stunting) and weight-for-age (wasting) to assess the extent of malnourishment in that population. Whenever more than 40% of a population is below the–2 SD of the Z-score for height-for-age, then the prevalence group for malnourishment (specifically stunting) is considered very high (WHO 1995). Table 23.3 (taken directly from the WHO [1995]) lists the other prevalence ranges for percentages of children below the normal Z-scores for height-for-age and weight-for-age. These groupings are important for our analyses.

Table 23.2 Anthropometric indices (height-for-age and weight-for-age) of children under age five in the greater Serengeti ecosystem. Data collected in Bunda, Loliondo, Meatu, and Serengeti district villages by Savannas Forever Tanzania (2006–2007)

	Height-for-age			**Weight-for-age**		
Village	**Percentage below −2 SD**	**Mean Z-score**	**Number of children**	**Percentage below −2 SD**	**Mean Z-score**	**Number of children**
Bunda village 1	29.9	−1.37	201	8.7	−0.7	206
Bunda village 2	25.6	−1.16	442	9.1	−0.57	453
Loliondo village 1	42.6	−1.85	54	29.6	−1.47	54
Loliondo village 2	29.9	−1.14	278	17.6	−1.04	284
Loliondo village 3	42.9	−1.59	70	22.7	−1.26	75
Loliondo village 4	31	−1.13	58	15.5	−0.6	58
Loliondo village 5	15.5	−0.84	220	8.8	−0.54	227
Loliondo village 6	31.6	−1.11	117	18.9	−0.87	122
Meatu village 1	40.8	−1.66	206	17.1	−0.99	216
Meatu village 2	42.2	−1.74	249	11.6	−0.81	251
Meatu village3	36.1	−1.55	277	15.2	−0.85	282
Meatu village 4	34.7	−1.48	98	13.9	−0.81	101
Meatu village 5	27.9	−1.43	165	12	−0.85	166
Meatu village 6	35.5	−1.71	155	13.6	−0.98	154
Serengeti village 1	27.2	−1.24	276	12	−0.81	276
Serengeti village 2	33.5	−1.42	275	12.8	−0.81	273
Serengeti village 3	13.4	−0.85	97	3.1	−0.34	97
Serengeti village 4	36.8	−1.45	163	19.5	−1.11	169
Serengeti village 5	25.8	−1.39	66	10.1	−0.67	69
Serengeti village 6	26	−1.27	180	8.3	−0.58	180

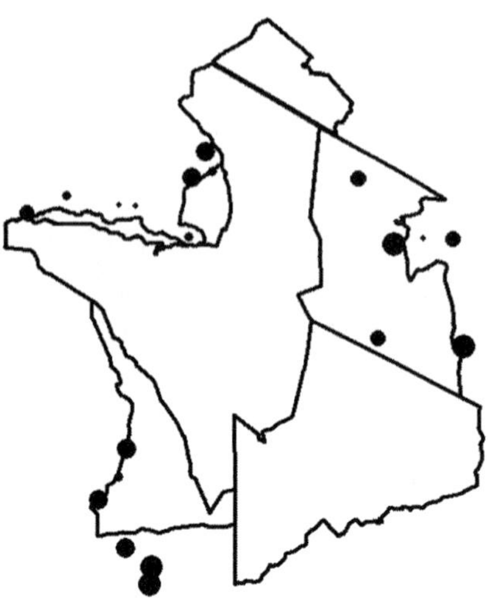

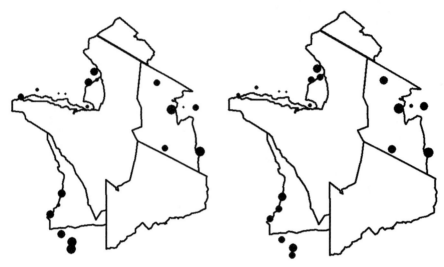

Fig. 23.7 Left, prevalence rates among children under age five for low height-for-age (stunting) and, *right*, low weight-for-age (wasting). Circle sizes correspond to prevalence rates in villages around the GSE. Circles range from small to large and represent low to very high prevalence rates, respectively. See table 23.3 for exact percentages for each prevalence group.

Using these WHO prevalence group categories (as listed in table 23.3) and combining all the districts around the GSE that we sampled, more than half of villages (11 out of the 20) have at least a high prevalence (30–39%) of stunting and a third of villages had a very high prevalence (40% and greater) of malnutrition. Fig. 23.7 visually demonstrates these findings as each village's prevalence rates are represented by a corresponding circle. If we examine the height-for-age data according to each district, we find that Loliondo and Meatu districts are the most malnourished. For both Loliondo and Meatu districts, one-third of villages maintain very high percentiles of stunting while 50% of the Meatu and Loliondo villages have high rates of stunting. In contrast, most of the Bunda and Serengeti district villages were in the medium prevalence group, meaning 20–29% of the population are stunted. Despite these better trends in the northwest region of the GSE, a third of the villages sampled in Serengeti district still were in the high prevalence group.

The prevalence-based reporting results for height-for-age are particularly interesting when compared to those from the rest of rural mainland Tanzania (this excludes the rural population from Zanzibar). Based on data from the 2004–2005 Tanzania Demographic and Health Survey (TDHS 2005), 40.9% of rural mainland Tanzanian children under age five are below–2 SD. This puts all of rural mainland Tanzania in the very high prevalence group

Table 23.3 Anthropometric prevalence-based reporting: Worldwide prevalence ranges (percent of children under age five below –2 Z-scores of low height-for-age and low weight-for-age)

Prevalence group	Low height-for-age (stunting) (%)	Low weight-for-age (wasting) (%)
Low	<20	<10
Medium	20–29	10–19
High	30–39	20–29
Very high	≥40	≥30

Source: Table taken directly from WHO (1995).

for stunting. On the other hand, only 4 out of the 20 sampled villages from around the GSE were in the very high group (table 23.3). This means that when analyzed as a whole (i.e., all districts combined), 80% of GSE villages have lower percentages of malnutrition than the national average for rural Tanzania. Of those villages that do have higher percentiles of malnutrition than the 40.9% across rural Tanzania, two are from Loliondo and two are from Meatu. The DHS research also includes percentages of the children in each region below the malnutrition "cut-off" of–2 SD for each index. Serengeti district and Bunda district are both in the Mara Region. Thirty-nine percent of the children in this region are malnourished based on the–2 Z-score for height-for-age (DHS 2005). All of Serengeti and Bunda district's villages that were sampled are below this percentage. This implies they are better nourished than the other villages located in the Mara Region.

Summary Statistics. The second method used to interpret anthropometric data is summary statistics of Z-scores. This method of analysis looks at the whole population (not just those above or below the–2 SD from prevalence-based reporting). Using this type of analysis, the WHO (1995) recommends that whenever a mean Z-score for a population is anywhere below the mean for the reference population, then the entire population being studied is affected. When a study population is experiencing extreme famine, the mean Z-score will be two or three units below the reference (WHO 1995).

In terms of results of the summary statistics of Z-scores for height-for-age, all of the GSE villages are below the mean for the reference population and 90% (or 18 out of 20) of the GSE villages sampled maintain a mean Z-score that is at least one unit below the mean for the reference population's mean Z-scores. According to the WHO standards, this means that the entire sample from the GSE population is stunted and experiencing chronic undernutrition. This finding is in keeping with the Tanzanian DHS (2005) data, which shows a mean Z-score of–1.7 for height-for-age for all of rural mainland Tanzania. Most of the villages (17 out of 20) sampled in the GSE did not have a mean Z-score that was as low as this national average. However, of those that were worse off (below–1.7), two came from Meatu district, and 1 from Loliondo. Thus, in keeping with the prevalence-based reporting of height-for-age, the northwestern villages of the GSE (in Bunda and Serengeti districts) are less malnourished than the eastern and southwestern villages.

The weight-for-age index in prevalence-based reporting showed fewer underweight children than stunted children in the GSE villages we sampled. This pattern reflects anthropometric indices gathered across rural Tanzania. According to the TDHS (2005), 40.5% of Tanzania's rural mainland population is considered stunted compared to 22.9% being underweight. In our GSE sample, the majority (65%) of villages were in the medium prevalence group for lightness. This means that 19 out of 20 sampled villages have a lower prevalence of malnutrition (low weight-for-age) than the rest of rural mainland Tanzanians. Villages on the eastern side of the GSE (in Loliondo district) are more malnourished (based on being underweight) than those on the western side of the ecosystem (see fig 23.7). Also, a greater proportion of villages in Meatu district were underweight than those in Serengeti or Bunda districts.

Analyses of the mean Z-scores for the weight-for-age index show that 16 out of 20 villages in the GSE sample were between 0.34 and 0.99 SD below the mean for the reference population. The remaining nine sampled villages fared worse since they were more than 1.0 SD below the reference population's mean. Once again, according to the methods of analysis recommended by the WHO (1995), these data demonstrate that the entire population sampled in the GSE is underweight for their age since all villages' means were below the reference population mean. The summary statistics of Z-scores for the weight-for-age index for all of rural mainland Tanzania is–1.1 units below that of the reference population (TDHS 2005). Only 3 out of 20 (15%) villages sampled in the GSE were worse off than this average. Of those that were worse off, two came from Loliondo district and one from Serengeti district.

CONCLUSION

What is the health status of Serengeti human populations compared to the health of people from rural Tanzania? What variation is there within the GSE in terms of human health?

The archival morbidity and mortality data suggest that HIV/AIDS prevalence in western Serengeti is higher than the average prevalence for HIV/AIDS in rural Tanzania. In addition, children bear the brunt of health problems in the GSE, with infant and under-five mortality rates worse in western Serengeti than the rest of Tanzania for the same period. Such discrepancies beg the question, Is it simply the remote location of western Serengeti that makes these health indicators worse than the rest of Tanzania? Or are these patterns at all related to the presence of conservation areas (which might limit villagers' decision-making rights and alter livelihood strategies) in the region?

While some human health indicators (as seen above) are worse in the GSE, others (particularly high rates of infectious disease) are similar to the rest of the country. Malaria is the leading cause of morbidity and mortality in both western Serengeti districts and rural Tanzania. This national pattern of disease in Tanzania and the data we gathered in western Serengeti provides evidence that these areas have not undergone the *epidemiological transition*, a shift that occurs in the causes of morbidity and mortality from infectious to chronic disease (Holtz and Kachur 2004). During the process of industrialization, people in developed countries become wealthier and more sedentary in their lifestyles and therefore become more prone to diabetes, cancer, and cardiovascular disease. On the other hand (and as is the case in the GSE), before the epidemiological transition occurs, high rates of infectious disease plague people due to their low socioeconomic status (Inhorn and Brown 1997).

Unfortunately, some medical practitioners within western Serengeti do not recognize the role that socioeconomics plays in human disease patterns and instead blame local people and their culture for the spread of infectious disease (Knapp 2010). For example, one medical professional working in the GSE claimed that the unwillingness of local residents to sell their cattle in order to make certain household-level improvements (such as buying mosquito nets) is to blame for the high incidence of malaria. Yet qualitative interviews among women demonstrate that it is not ignorance or unwillingness to accept biomedical advice; rather, it is purely economic constraints faced by households (it costs nearly half of an entire month's salary for a mosquito net) that hinder them from being able to prevent the spread of malaria or other infectious diseases (Knapp 2010). Essentially, what these patterns of infectious disease in western Serengeti and around rural Tanzania demonstrate is the need for improved access to adequate housing, clean water, sufficient clothing, and decent nutrition (Inhorn and Brown 1997), as well as improved (both in terms of supplies, staff, facilities, and cultural sensitivity) health care (Knapp 2010).

Summary statistics of mean Z-scores for height-for-age and weight-for-age reveal that the entire population of GSE is stunted (due to chronic undernutrition) and underweight (which could be a sign of long- or short-term malnutrition). In keeping with the WHO (1995) recommendations for interpreting summary statistics, we base the previous statement on the fact that all GSE villages have mean Z-scores that are below those of the reference population. Thus, we learn that the villagers in the GSE are living in a constant state of poor nutrition and compromised health. When compared to the mean Z-scores for all of rural mainland Tanzania, we see that these health problems are not unique to the GSE. In fact, only 15% of GSE villages sampled had mean Z-scores (for both height-for-age and weight-for-age) that were below the mean Z-scores for all of rural mainland Tanzania. Such a finding suggests that conservation agendas alone are not the cause of malnutrition in the GSE, but poor nutrition is more a sign of the ongoing political and economic struggle of all rural Tanzanians.

What variation is there within the GSE in terms of human health? Combining all indices of the prevalence-based reporting, villages in the east of the GSE (specifically in Loliondo district) seem to have the worst levels of malnutrition, followed closely by villages in the southwest of the GSE (in Meatu district). The malnutrition levels of eastern villages in the GSE were particularly noticeable when analyzing the weight-for-age index, though both Loliondo and Meatu districts reveal high malnutrition rates as measured by height-for-age. These uneven patterns of malnutrition could reflect the sampling of some northwestern districts (Serengeti and Bunda) during the dry season when more food is available (Frisancho 1993), or the ecological heterogeneity of the ecosystem (Meatu and Loliondo districts are more arid and less arable than Serengeti and Bunda districts). With greater difficulty in growing crops due to less rainfall and poorer soils, some villagers have more limited access to markets and affordable food. Finally, the poorer nutrition of Maasai in the eastern GSE could also be evidence of the cultural differences in the pastoral system—both in terms of their preferred livelihood strategy (as cattle keepers) and in terms of their loss of power (or rights) in land-use decision making and land tenure (Galvin et al. 2008).

To further understand these patterns of heterogeneity in human health across the GSE, more research is needed to uncover the ultimate drivers of malnutrition and poor health.

ACKNOWLEDGMENTS

We especially want to thank the thousands of villagers who allowed us to enter their homes, interview them, and measure their children. In addition, we are grateful to the Tanzanian ministries of Tanzania Wildlife Research Institute (TAWIRI), Tanzania National Parks (TANAPA), Ngorongoro Conservation Area Authority (NCAA), and the Commission of Science and Technology (COSTECH) for allowing this research to be conducted within the greater Serengeti ecosystem, Tanzania. This chapter would not be complete without the efforts of the Savannas Forever Tanzania field team who gathered all the anthropometric data. Finally, we must thank the hard-working health practitioners and statisticians from the Bunda and Serengeti district hospitals, as well as those at the CBHPP clinic in Mugumu for keeping records of morbidity and mortality rates around the GSE. Funding for this research was provided through the National Science Foundation (grant no. DEB-0308486).

REFERENCES

Adam, T., S. S. Lim, S. Mehta, Z. A. Bhutta, H. Fogstad, M. Mathai, J. Zupan, and G. L. Darmstadt. 2005. Cost effectiveness analysis of strategies for maternal and neonatal health in developing countries. *British Medical Journal* 331:1107–10.

Alderman, H., H. Hoogeveen, and M. Rossi. 2006. Reducing child malnutrition in Tanzania: Combined effects of income growth and program interventions. *Economics and Human Biology* 4:1–23.

Arcese, P., J. Hando, and K. Campbell. 1995. Historical and present day anti-poaching efforts in Serengeti. In *Serengeti II: Dynamics, management, and conservation of an ecosystem*, ed. A. R. E. Sinclair and P. Arcese, 506–33. Chicago: University of Chicago Press.

Arhem, K. 1985. *Pastoral man in the Garden of Eden: The Maasai of the Ngorongoro Conservation Area, Tanzania*. Uppsala, Sweden: University of Uppsala.

Campbell, K., and H. Hofer. 1995. People and wildlife: Spatial dynamics and zones of interaction. In *Serengeti II: Dynamics, management, and conservation of an ecosystem*, ed. A. R. E. Sinclair and P. Arcese, 534–70. Chicago: University of Chicago Press.

Emerton, L., and I. Mfunda. 1999. Making wildlife economically viable for communities living around the western Serengeti. Evaluating Eden series: Working Paper no. 1, University of Manchester, Manchester, England. http://pubs.iied.org/pdfs/7794IIED.pdf.

Frisancho, A. R. 1993. *Human adaptation and accommodation*. Ann Arbor: University of Michigan Press.

Galvin, K. A., J. Ellis, R. B. Boone, A. L. Magennis, N. M. Smith, S. J. Lynn, and P. Thornton. 2002. Compatibility of pastoralism and conservation? A test case using integrated assessment in the Ngorongoro Conservation Area, Tanzania. In *Displacement forced settlement and conservation*, ed. D. Chatty and M. Colester, 36–60. Oxford: Berghahn.

Galvin, K. A., P. K. Thornton, R. B. Boone, and L. M. Knapp. 2008. Ngorongoro Conservation Area: Fragmentation of a unique region of the greater Serengeti ecosystem. In *Fragmentation in semi-arid and arid landscapes: Consequences for human and natural sys-*

tems, ed. K. A. Galvin, R. S. Reid, R. H. Behnke, and N. T. Hobbs, 255–80. Dordrecht, Netherlands: Springer.

Holtz, T., and S. P. Kachur. 2004. The reglobalization of malaria. In *The corporate assault on global health*, ed. M. Fort, M. A. Mercer, and O. Gish, 131–43. Cambridge, MA: South End Press.

Homewood, K. M., and W. A. Rodgers. 1991. *Maasailand ecology: Pastoral development and wildlife conservation in Ngorongoro, Tanzania.* Cambridge: Cambridge University Press.

Inhorn, M. C., and P. J. Brown. 1997. The anthropology of infectious disease. In *The anthropology of infectious disease*, ed. M. C. Inhorn and P. J. Brown, 31–67. Amsterdam: Gordon and Breach Publishers.

Knapp, L. M. 2010. Human health in western Serengeti: Using three methodologies to better understand the interactions and impacts of conservation, culture, and poverty. MA thesis, Colorado State University.

Loibooki, M. 1997. People and poaching: The interactions between people and wildlife in and around Serengeti National Park, Tanzania. MS thesis, Reading University, UK.

McCabe, J. T. 2002. Giving conservation a human face? In *Conservation and mobile indigenous people: Displacement, forced settlement and sustainable*, ed. D. Chatty and M. Colester, 61–76. New York: Berghahn Books.

———. 2003. Sustainability and livelihood diversification among the Maasai of northern Tanzania. *Human Organization* 62:100–11.

McCabe, J. T., S. Perkin, and C. Schofield. 1992. Can conservation and development be coupled among pastoral people? An examination of the Maasai of the Ngorongoro Conservation Area, Tanzania. *Human Organization* 51:353–66.

Perkin, S. L. 1997. The Ngorongoro Conservation Area: Values, history and land-use conflicts. In *Multiple land use: The experience of the Ngogorngoro Conervation Area, Tanzania*, ed. D. M. Thompson, 7–18. Gland, Switzerland: International Union for Conservation of Nature.

Potkanski, T. 1997. *Pastoral economy, Property rights and traditional mutual assistance mechanisms among the Ngorongoro and Salei Maasai of Tanzania.* London: International Institute for Environment and Development (IIED) Drylands Programme.

Roe, D. 2008. The origins and evolution of the conservation-poverty debate: A review of key literature, events and policy processes. *Oryx* 42:491–503.

Scherl, L. M. 2004. *Can protected areas contribute to poverty reduction?: Opportunities and limitations*. Gland, Switzerland: IUCN/The World Conservation Union.

Shetler, J. B. 1998. The landscapes of memory: A history of social identity in western Serengeti, Tanzania. PhD diss., University of Florida, Gainsville.

———. 2002. The politics of publishing oral sources from the Mara Region, Tanzania. *History in Africa* 29:413–26.

Shivji, I. G., and W. B. Kapinga. 1998. *Maasai rights in Ngorongoro, Tanzania.* London: International Institute for Environment and Development (IIED) and Land Rights Research and Resources Institute.

Tanzania Commission for AIDS, National Bureau of Statistics, and ORC Macro. 2005. Tanzania HIV/AIDS indicator survey 2003–2004. Calverton, Maryland: USA TACAIDS, NBS, and ORC Macro.

Tanzania Demographic and Health Survey (TDHS). 2005. National Bureau of Statistics, Dar es Salaam, Tanzania.

Upton, C., R. Ladle, D. Hulme, T. Jiang, D. Brockington, and W. M. Adams. 2008. Are poverty and protected area establishment linked at a national scale? *Oryx* 42:19–25.

Wilkie, D. S., G. A. Morelli, J. Demmer, M. Starkey, P. Telfer, and M. Steil. 2006. Parks and people: Assessing the human welfare effects of establishing protected areas for biodiversity conservation. *Conservation Biology* 20:247–49.

World Health Organization Expert Committee on Physical Status. 1995. *Physical status: The use and interpretation of anthropometry; report of a WHO Expert Committee*. Geneva: World Health Organization.

TWENTY-FOUR

Multiple Functions and Institutions: Management Complexity in the Serengeti Ecosystem

Deborah Randall, Anke Fischer, Alastair Nelson, Maurus Msuha, Asanterabi Lowassa, and Camilla Sandström

The global importance of the Serengeti ecosystem for biodiversity conservation, cultural heritage, and economic development is recognized by designation of two UNESCO World Heritage Sites and a Biosphere Reserve. Lying at the heart of the ecosystem, the Serengeti National Park (SENAPA) is one of the best known wildlife parks in the world and one of Tanzania's most valuable foreign currency earners whose revenue is invested in Tanzania's other protected areas, with additional benefits flowing to local communities and the regional economy. Other protected areas surrounding the park provide a buffer zone where wildlife and human use overlap. With roughly 2.3 million people in the seven districts that abut the park and a population growth rate of approximately 3% annually (United Republic of Tanzania [URT] 2002), the potential for conflict between wildlife and communities adjacent to the park is substantial and growing. Thus, discussions on the management of the Serengeti ecosystem are often framed as a debate on management for livelihoods versus conservation (Prins 1992; Sinclair 2008). However, this oversimplifies the complexity of the goods and services (or functions) provided by the ecosystem, which can be roughly classified as ecological (sustaining key species and habitats), economic (enhancing national economic development, local poverty alleviation, and sustainable livelihoods), and sociocultural (safeguarding traditions and cultural identity). Some of these functions will inevitably be mutually exclusive, creating potential conflicts, but others will be compatible or even mutually dependent. In this chapter, we use an institutional perspective to illustrate the conflicts and synergies in the management of the Serengeti ecosystem as a result of this multifunctionality.

Institutions can be defined as the formal and informal norms, rules, values, and customs that regulate society, as well as the corresponding incentives and sanctions used to enforce compliance (North 1990; Brett 2000). Institutions provide the frameworks within which individuals and organizations operate on both a national level (e.g., national law, country-specific customs) and local level (e.g., district-related bylaws, village-specific norms). We focus in particular on the institutions governing the management of the ecosystem's protected area landscape (i.e., the Serengeti National Park, neighboring game reserves, Loliondo Game Controlled Area, Ngorongoro Conservation Area and the wildlife management areas) as well as national laws and policies concerning wildlife, tourism, and land management. We consider to what degree existing institutions and management approaches recognize and reconcile the ecological, economic, and sociocultural functions of the Serengeti ecosystem. We outline some of the challenges to date and make recommendations toward more sustainable management, emphasizing in particular the need for governance approaches that address the multifunctionality and institutional complexity across the ecosystem.

MULTIFUNCTIONALITY OF THE SERENGETI ECOSYSTEM

The Serengeti ecosystem is defined by the annual migration of two million wildebeest, zebra, and gazelle from the short-grass plains of Ngorongoro and Serengeti in Tanzania to the savanna woodlands of the Maasai Mara in Kenya (Sinclair 1995; Thirgood et al. 2004). The area of interest is more than 33,000 km^2 (table 24.1) and in Tanzania alone, includes seven districts and conservation areas with varying levels of "protection" (fig. 24.1). Given its size and mosaic of protected and unprotected areas with different land uses, resource users, and management priorities, the Serengeti ecosystem provides a range of ecological, economic, and sociocultural functions briefly described below.

Ecological Functions

The grasslands and woodlands of the Serengeti ecosystem support some of the greatest concentrations of large mammals in the world, including one of the world's largest herds of migrating ungulates (Sinclair 1995). The migration alone consists of over 1.3 million wildebeest and 200,000 zebra.

Table 24.1 Protected areas in the Serengeti ecosystem

Protected area	**Management authority**	**Management objective**	**Area (km^2)**
Serengeti National Park	Tanzania National Parks Authority	Protection with nonconsumptive use only	14,763 km^2
Ngorongoro Conservation Area	Ngorongoro Conservation Area Authority	Protection with multiple use but no consumptive use	8,285 km^2
Loliondo Game Controlled Area	Wildlife Division and district councils	Consumptive and nonconsumptive use	4,500 km^2
Maswa Game Reserve	Wildlife Division	Protection with consumptive and nonconsumptive use	2,897 km^2
Ikorongo Game Reserve			605 km^2
Grumeti Game Reserve			420 km^2
Ikona Wildlife Management Area	Authorized associations (with local government authority and Wildlife Division input)	Consumptive and nonconsumptive use	242 km^2
Makao Wildlife Management Area			480 km^2
Maasai Mara National Reserve	Narok County council and Trans Mara County Council (with the Mara Conservancy)	Protection with nonconsumptive use only	1,510 km^2

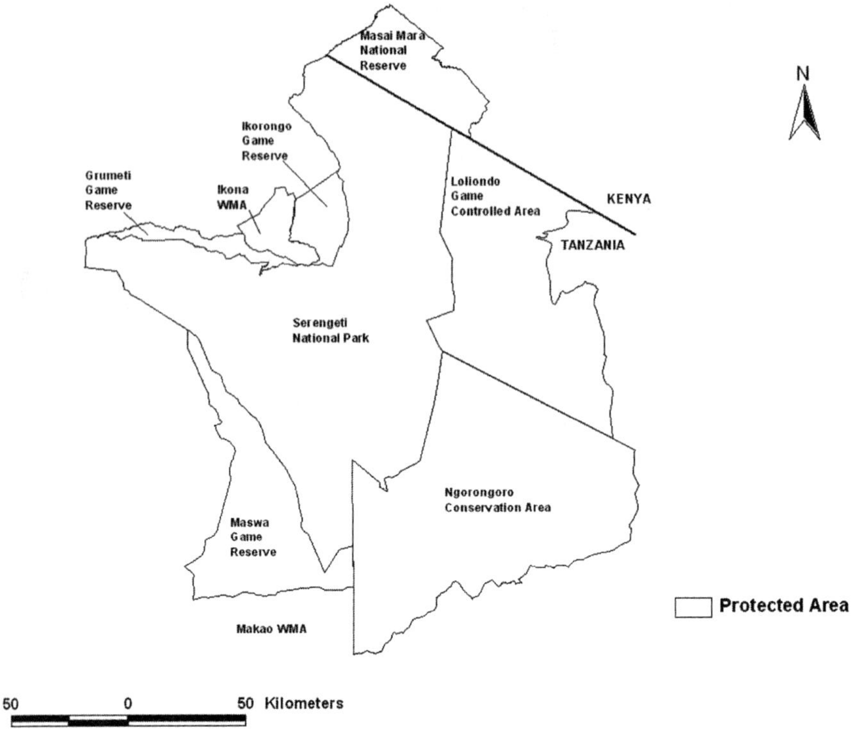

Fig. 24.1 Protected areas in the Serengeti ecosystem.

There are also important populations of resident ungulates, east African black rhino (*Diceros biconis michaeli*), 14 species of predators, over 500 species of birds, and many other taxa. The ecosystem is composed of a mosaic of habitat types broadly classified as southern grassland plains, northern woodlands, and western mixed woodland-grassland within which there is substantial heterogeneity in floral diversity and vegetation assemblages. Seasonality in rainfall and its effects on nutrient availability is thought to be the driving force behind the migration (Fryxell 1995; Murray 1995) with wildebeest concentrated in the southern grasslands during the wet season (December to May), and thereafter moving northwest to spend the dry season (August to November) in the northern woodlands (Thirgood et al. 2004). This seasonality, as well as other biotic and abiotic components and processes (e.g., fire, hydrology, predation) that sustain such tremendous biological diversity, underscore the important ecological functions of the Serengeti ecosystem.

Economic Functions

Poverty is prevalent in the Serengeti ecosystem with the per capita income below that for the country as a whole (Borge 2003), and approximately 75% of households below the basic needs poverty line for Tanzania of US $0.76/day (Schmitt 2010). As in the rest of rural Tanzania, subsistence agriculture in the form of farming and livestock keeping is the predominant livelihood activity (Schmitt 2010) and the majority of households own livestock as a form of household savings. Thus, the ecosystem provides an important economic function by providing land and ecosystem services that sustain local livelihoods. For instance, to the west of the park the ecosystem protects the Simiyu and Duma watersheds which are critical for agriculture and livestock, and it is estimated that the Mara River alone supports agricultural yields worth US $12 million (Gereta, Wolanski, and Chiombola 2003). Furthermore, opportunities for wage labor or other forms of paid employment are limited and bushmeat hunting is an important source of cash income for many households (Schmitt 2010). Other forms of natural resource extraction and use (e.g., charcoal making, beekeeping, fuelwood collection) also contribute to household needs and income.

On a wider scale, the economic potential of the Serengeti ecosystem and its ability to generate revenue, primarily through wildlife tourism, is substantial. Serengeti National Park is the second highest earning national park in Tanzania (behind Kilimanjaro National Park) with revenues of US $21 million in 2006 (URT 2006), and the Ngorongoro Conservation Area makes approximately US $30 million primarily from visitor and concession fees (United Nations Educational, Scientific, and Cultural Organization/International Union for Conservation of Nature [UNESCO/IUCN] 2007). Other areas in the ecosystem retain their own revenue from tourism (e.g., villages in Loliondo collected over US $300,000 in 2007 from private tourism enterprises, Tanzania Natural Resource Forum [TNRF] 2008). The potential for wildlife to contribute to poverty alleviation and rural development in the Serengeti ecosystem through employment in the tourism sector and sustainable natural resource use is enormous, and local communities on both sides of the international border derive some benefits from tourism revenue. Nationally, wildlife underpins the tourism industry in Tanzania which generated an estimated US $1.3 billion in 2008 (United Nations World Tourism Organization [UNWTO] 2008). Tanzania's tourist hunting industry contributed an estimated US $30 million to the Wildlife Division with an additional US $9 million going to the leasing companies (Baldus and Cauldwell 2004). As one of the most visited areas in Tanzania, wildlife-based tourism in the Serengeti ecosystem is a key driver of macroeconomic growth in Tanzania and, if managed effectively, will contribute significantly toward Tanzania meeting its national goals for poverty reduction as laid out in the National Strategy for Growth and Reduction of Poverty (developed in 2005 and more widely known by its Swahili acronym MKUKUTA).

Sociocultural Functions

The Serengeti ecosystem is comprised of a diverse mix of people, including more traditional pastoralists (primarily Maasai and Sonjo) to the east and southeast and agropastoralist hunters (primarily Sukuma, Kuryia, Ikizu, Taturu, Natta, Issenye, Ikoma) to the west and southwest (Kaltenborn et al. 2008; Schmitt 2010). Each of these groups use (or aspire to use) the wider Serengeti in their own ways. The Maasai are thought to have inhabited the area since the seventeenth century and there are a number of rock paintings, other artifacts of Maasai culture, and traditional worship sites scattered across the ecosystem (Tanzania National Parks Authority [TANAPA] 2006). Interwoven with this cultural diversity and history is a wealth of traditional knowledge (cultural and ecological) acquired in some cases over hundreds of years. Paleontological and archaeological records are found in four major sites (Olduvai Gorge, Laetoli, Lake Ndutu, and Nasera rock shelter in the Gol Mountains). The discovery of fossil remains of early hominids, such as *Homo habilis*, 3.5 million-year-old human footprints, and other artifacts, make it one of the world's most important research sites on human evolution (Leakey and Hay 1979; Johanson et al. 1987). Natural resources also constitute a part of the local culture, as in wildlife products being used for traditional medicine, clothing, decorations, and traditional ceremonies and marriages (Kaltenborn, Nyahongo, and Tingstad 2005). There are also social and cultural functions associated with trophy hunting.

INSTITUTIONS—THE NATIONAL CONTEXT

The principles of wildlife conservation in Tanzania were first laid out in President Nyerere's famous Arusha Declaration in 1967. Now, the contribution of wildlife and other natural resources to poverty reduction and economic growth are recognized in Tanzania's National Strategy for Growth and Reduction of Poverty (MKUKUTA), which is also in line with the millennium development goals of addressing poverty, hunger, disease, illiteracy, environmental degradation, and discrimination against women. In addition to these international and national development goals, Tanzania's formal institutions that relate to the functions and management of the Serengeti ecosystem include a number of policies and laws related to wildlife and protected area management, natural resource management, tourism, and land use (see also Polasky et al. 2008). We overview these institutions and evaluate whether they promote resilient local livelihoods and contribute to economic development without threatening wildlife populations, ecosystem health, or cultural integrity.

Wildlife

The 1974 Wildlife Conservation Act No. 12, revised most recently in 2009, lays out the national system of state protected areas comprised of national parks, game reserves and game controlled areas, and regulates all consumptive use of wildlife through allocation of hunting permits. The Wildlife Policy (1998, revised in 2007) defines the vision for the protected areas network in Tanzania, and clearly sets out to integrate conservation with rural development and broader socioeconomic development in Tanzania (National Development Vision 2025). It emphasizes the contribution of protected areas to the local and national economy through tourism (photographic and hunting) and resident hunting in game controlled areas or areas outside but adjacent to protected areas. And whereas previously, wildlife management was entirely controlled by the government, the 1998 Wildlife Policy embraced community-based conservation while recognizing community user rights and the importance of traditional knowledge and practices for the management of wildlife in Tanzania. In doing so, it called for the legislation that would enable the establishment of a new type of protected area where communities would have legal rights to manage and benefit from wildlife on village land. The 2002 WMA Regulations (revised in 2005) under the Wildlife Conservation Act provide for the creation of wildlife management areas (WMAs) as areas set aside by village councils for wildlife conservation and sustainable use on village land. The Wildlife Policy also promotes (a) the delivery of extension services and "good neighborliness" by protected areas; (b) communication, public education, and awareness building; (c) equitable distribution of costs and benefits; (d) gender issues, and (e) international collaboration and cooperation with neighboring countries with regard to transboundary issues as well as other government sectors and management authorities.

In one way or another, wildlife conservation falls under the jurisdiction of the Wildlife Division (WD) of the Ministry of Natural Resources and Tourism (MNRT) or two parastatal organizations: the Tanzania National Parks Authority (TANAPA) and the Ngorongoro Conservation Area Authority (NCAA). WD has oversight of game reserves, game controlled areas, wildlife management areas, and open areas, while the management, conservation, and use of all national parks in Tanzania falls under TANAPA. NCAA oversees wildlife conservation in the Ngorongoro Conservation Area (NCA). WMAs are managed by legally recognized community-based authorized associations (AAs) under the jurisdiction of the WD. Although WMAs delegate wildlife use and management rights to the village, the WD regulates gazettement of areas, safari and hunting quotas, and prosecution of offenders under the Wildlife Act. WMAs obtain a certain number of wildlife hunting permits according to the WD quota system, which they can either use or sell to a private tourist hunting company.

The Ngorongoro Conservation Area Ordinance 413 (1959) set up the earliest nationally recognized form of community-based conservation in Tanzania by designating the NCA as a multiple land use area for the integration of human development and conservation. The NCAA was established under a separate ordinance to manage the NCA for indigenous Maasai residents, including regulating cultivation, grazing, tourism, and natural resource use.

Tourism

Like the Wildlife Policy, the Tourism Policy (1991, revised in 1999) also sets out to enhance ecological, economic, and sociocultural functions by promoting "the development of sustainable and quality tourism that is culturally and socially acceptable, ecologically friendly, environmentally sustainable and economically viable." Involvement of and benefit-sharing with communities within and adjacent to protected areas is specifically highlighted in the policy, as is adherence to wildlife conservation and sustainable use strategies. Implementation of the Tourism Master Plan (1996), the implementation strategy for the Tourism Policy, is the responsibility of the Tourism Division of the MNRT in coordination with other government ministries. The Tanzania Tourism Board (TTB) was established under the MNRT to promote Tanzania as a tourist destination.

Land, Settlement, and Local Governance

Recognizing and securing traditional rights to land is one of the main objectives of the National Land Policy (1997), which also recognizes the linkages between conservation, development, and sustainable use by promoting a land tenure system "to encourage optimal use of land resources and to facilitate broad-based social and economic development without upsetting or endangering the ecological balance of the environment." The 1999 Land Act (No. 4) categorized public land into three categories—general land, village land, and reserved land. Reserved land includes all land set aside for special purposes to be managed by the government, including those protected areas established under the Wildlife Conservation Act of 1974. Management of village land was made the responsibility of the village council under the 1999 Village Land Act (No. 5). Conflicts do arise, for instance where game controlled areas (GCAs) and the NCA demarcated as reserved land under the Land Act have been established on areas traditionally occupied by communities, and thus considered village land where customary rights over tenure and management prevail (TNRF 2008). In addition, the 1999 revision of the 1982 Local Government Act (No. 6) stated that "the Minister shall endeavour to ensure that the local government authorities are strong and effective institutions that are more and more autonomous in managing their own affairs and they operate in a more transparent and democratic manner." Theoretically, the policy and legal framework serves to strengthen decentralization of decision making to local institutions while recognizing community rights to manage their own land and resources, and thus derive financial benefits from wildlife resources, these being key economic and sociocultural functions of the Serengeti ecosystem.

INSTITUTIONS—THE SERENGETI CONTEXT

Policies and regulations set at the national level determine permissible land uses and other activities in different land categories, including protected and unprotected areas. The Tanzanian sector of the Serengeti ecosystem includes five types of protected areas (fig. 24.1). In addition, the Maasai Mara National Reserve on the Kenyan side of the ecosystem has its own set of institutions which are covered only briefly in this chapter.

The Serengeti National Park (SENAPA) is managed for conservation and tourism and, in recognition of its outstanding biodiversity value, was inscribed as a Natural World Heritage Site in 1981. As a national park it has the highest level of protection, and thus all forms of consumptive use are prohibited (including settlement, grazing, hunting, and resource extraction) other than that needed for park, tourism, or research staff. The SENAPA management plan (2006–2016), which lays out the vision and activities for the management of the park, was developed by a series of working groups and interdisciplinary planning teams representing a range of stakeholders involved in the planning process (TANAPA 2006). The plan is divided into four programs: ecological management, tourism management, park outreach, and park operations. In this sense, the general management plan (GMP) broadly recognizes the multifunctionality of the ecosystem and lays out a number of objectives and actions outlining the park's role in enhancing these functions.

The Ngorongoro Conservation Area (NCA) is a multiple-use area that, in theory, recognizes and accommodates ecological, economic, and sociocultural functions. Management objectives aim to promote natural resource conservation in conjunction with pastoralism and permanent settlement by indigenous Maasai. Hunting is not permitted but substantial revenue is earned from photographic tourism. The area was inscribed as a Natural World Heritage Site in 1979 in recognition of its global importance for biodiversity conservation. In recognition of the area's archaeological importance, it was also inscribed as a Cultural World Heritage Site in 2010. A 10-year GMP for the area was approved in 2006.

Maswa, Grumeti, and Ikorongo Game Reserves (GR) are managed primarily for conservation through the consumptive use of wildlife through hunting tourism (Maswa GR) and nonconsumptive photographic tourism (Ikorongo GR and Grumeti GR). Maswa GR has three hunting blocks operated by different companies (Big Game Safaris, Tanzania Game Tracker Safaris, and Robin Hurt Safaris) on five-year leases, while Grumeti and Ikorongo GRs each have one hunting block operated by Singita Grumeti Reserves for photographic tourism. Settlement, pastoralism, cultivation, and other forms of resource extraction are prohibited within the GRs.

Loliondo Game Controlled Hunting Area (GCA) is a multiple-use area where settlement, pastoralism, and cultivation by Maasai are allowed, as are consumptive uses of wildlife through hunting and nonconsumptive wildlife or cultural tourism. Tourism is run by a handful of top-end private operators (largely photographic) who have entered into local contracts with individual Maasai villages for lease and exclusive use of land in exchange for fixed payments.

Three wildlife management areas (WMAs) were initiated in the Serengeti ecosystem: Loliondo to the east, Ikona to the west, and Makao to the south.

WMAs are multiple-use areas that can be managed for consumptive and nonconsumptive use of wildlife with associated revenue shared between the communities and government. Following the gazettement of Ikona WMA in Serengeti district and Makao WMA in Meatu district, user rights were given to the authorized association (AA) by the Ministry of Natural Resources and Tourism (MNRT) which enabled bylaw and constitution formation in both areas. Management plans have been prepared for Ikona and Makao WMAs, and Ikona WMA has steady revenue from private tourism investors and appears to be capable of maintaining its operational and management costs through this revenue. Makao WMA is still in the process of setting up contracts with investors (primarily hunting companies) to earn revenue from wildlife. The WMA process has not proceeded in Loliondo, but villages in the area are earning revenue from tourism by entering into legal agreements with private investors in ways that are, arguably, less complex or costly than undergoing WMA gazettement (TNRF 2008).

In Kenya, management of the Maasai Mara National Reserve (MMNR) is mandated to the Narok County council (east of the Mara River) and the Trans Mara County council (west of the Mara River). Day-to-day management of the Trans Mara part of the MMNR is contracted to the Mara conservancy (a local nonprofit organization representing the Maasai). Grazing and other forms of consumptive use are not allowed inside the reserve but revenue generated from tourism aims to benefit both local and national economies. A draft management plan (2009–19) was developed, involving a range of stakeholders, to address threats to the reserve and guide ecological management, tourism, community outreach, and protected area operations. At the time of writing it was undergoing revision before approval.

Informal institutions also exist including the customs and traditions related to bushmeat hunting, livestock herding, and other forms of land and resource use. There is a considerable amount of knowledge on traditional institutions of land tenure and access to land in areas inhabited by the Maasai (Pander 1995; Seno and Shaw 2002; Fratkin and Mearns 2003). However, there is comparatively little information about the informal institutions that regulate bushmeat hunting, trophy hunting, or wildlife tourism (for a historical perspective see Steinhard 1989) or their interaction with national policies and legislation. Bushmeat hunting is not believed to be a strong part of the Maasai's pastoral culture, although this may not be the case today in the face of changing livelihoods and modernization. Initial qualitative research suggests that while bushmeat hunting on the western side of the Serengeti National Park does not seem to have strong cultural functions, it is clearly organized through informal rules and associated incentives, and hunting is currently gaining in social status due to the increasing need for cash in the market economy (Lowassa, Tadie, and Fischer 2012). Until recently, clans and tribes tended to have taboos relating to specific species (elephants or zebra) or hunting methods—for example, the use of pitfall traps could potentially cause misfortune, as could bushmeat hunting in general for cattle owners. We can only speculate about the ecological impact of these taboos and social norms—most likely, they provided migrating wildlife populations with a quasi-closed season during the migration, and a mosaic of "safe" areas for nonmigrating species. In recent times, many of these informal rules may be decreasing in importance. In addition, institutions that seem at first rather unrelated to the ecological functions of the Serengeti ecosystem, draw on the economic functions of wildlife populations. For example, the need for cash or cattle—governed by formal (tax, school fees) and informal (e.g., customs related to dowry) institutions—are strong drivers of hunting as a livelihood activity. There may be other informal institutions that, although unwritten, are socially embedded in local society, and thus sustaining shared norms and prevalent practices.

INTERACTIONS BETWEEN FUNCTIONS AND INSTITUTIONS

Although the protected area landscape is designed to reflect the multiple land uses and users in the ecosystem, this multifunctionality is not easily divisible geographically or institutionally. Where overlap occurs, interactions are inevitable and, more often than not, lead to conflict. For example, wildebeest spend about a third of their time in less-protected (e.g., GRs, GCA, WMAs) or unprotected open areas of the ecosystem (Thirgood et al. 2004) where they are exposed to agricultural intensification (Homewood et al. 2001) and high levels of (illegal) hunting (Arcese, Hando, and Campbell 1995; Campbell and Hofer 1995; Mduma 1996). Historically wildlife hunting was not a major conservation concern in Tanzania, but increasing rural populations, expanding agriculture, food insecurity, and the demand for increased household income pose ever greater threats to wildlife inside and outside protected areas. Previous studies suggest that poverty and lack of alternative sources of protein are the primary drivers of local bushmeat consumption in the Serengeti ecosystem (Loibooki et al. 2002), making bushmeat hunting a major component of rural livelihoods (sociocultural and economic function) and, at the same time, a serious threat to the viability of migratory and resident wildlife populations (ecological function) and thus potentially also to the attractiveness of the Serengeti for tourists (economic function). This leads to mounting conflict between economic functions for different groups (e.g., local communities and national economy) as well as between institutions that aim to protect ecological functions and communities with few alternative livelihood options. Other conflicts ensue from the interactions between functions, including:

- Agricultural intensification, deforestation, and unsustainable water use in the headwaters and catchment on the Kenyan side are causing decreased water levels and increasing seasonality of flow in the ecosystem's only perennial river, the Mara River (Gereta, Wolanski, and Chiombola 2003). This has the potential to seriously affect the stability of the entire ecosystem and all its functions.
- Disturbance to habitat and wildlife from tourist vehicles are evident, particularly in the NCA and on the Mara side where severe overcrowding and overuse are widespread (UNESCO/IUCN 2007, 2008; MMNR, 2009). Other negative impacts from tourism are degradation of local culture and traditional sites, as well as unsustainable water extraction and poor waste management at lodges and other tourist facilities.
- Wildlife are negatively impacted by diseases from livestock (Dobson 1995) and domestic dogs (Roelke-Parker et al. 1996; Lembo et al. 2008) and vice versa (Kaarea et al. 2007).

While management debates tend to focus on the conflicts and alleged trade-offs between land uses and users, the ecological, economic, and sociocultural functions of the ecosystem are, for the most part, highly interdependent. Thus, activities that aim to promote one function to the detriment of another are unsustainable in the long run. For example, biodiversity loss and natural resource degradation disproportionately affect the poorest of society who greatly depend on natural resources for their livelihoods (Millennium Ecosystem Assessment [MEA] 2005). Even where poverty is a major driver of illegal activities such as bushmeat hunting, community development can increase well-being in ways that are counterproductive to sustainable development if increased wealth leads to consequent degradation of the resources on which people intimately depend (Walpole 2006). Sustainable development ultimately depends on the mutual enhancement of ecological, economic, and social goals (World Commission on Environment and Development [WCED] 1987), and this in turn depends on institutions that reflect the Serengeti's multifunctionality and promote good governance.

A key question then is, do existing institutions recognize and reconcile the multifunctionality of the Serengeti ecosystem? For the most part the multiple functions of the Serengeti ecosystem and their linkages are enshrined in Tanzania's policies, and broadly speaking, these put forth strategies for sustainable wildlife utilization that have the potential to contribute to poverty alleviation and rural economic development (World Bank 2005). They are also in line with broader development strategies such as Tanzania's National Strategy for Growth and Reduction of Poverty (or MKUKUTA), which call for greater revenue generation from tourism and wildlife for local and national economies. Nevertheless, there are at times contradictions between institutions, leading to conflicts that are made worse by inadequate adherence to and poor implementation of policies and laws that should otherwise promote sustainability. We use examples from community-based wildlife management and tourism to highlight this.

Community-based wildlife management has become the political mantra for integrating rural development with conservation (Hulme and Murphree 1999), the underlying rationale being that if wildlife provides sufficient economic value to communities, then community-based conservation can compete with agriculture, pastoralism, and hunting. This not only provides a basis for resilient livelihoods by alleviating poverty but also reduces illegal poaching. Thus, in theory, community-based conservation and wildlife management areas (WMAs) should go some way toward reconciling wildlife conservation and local livelihoods, particularly for communities that incur most of the costs but receive few of the benefits of living adjacent to wildlife and protected areas. Unfortunately, progress to date in the development of the WMAs has been mixed and challenges remain. For one thing, the formation of authorized associations constitutes the creation of a new local institution distinct from the village council and other preexisting, elected governing bodies at the local level, thus complicating governance (Nelson 2007). Tanzania's policies and laws have also been criticized for not enabling suitable levels of community ownership and empowerment to adequately ensure the dual goals of wildlife conservation and rural development (TNRF 2008). For example, the Village Land Act No. 5 (1999) and the Local Government Act No. 7 and 8 (1982) allow for responsibility for the conservation, management, and development of wildlife to be devolved to the village level, but the state retains ownership of wildlife and other regulatory functions (e.g., authority for granting hunting block concessions in the WMA, approval of investment agreements with private operators, and determining how revenue from WMAs will be shared between the community and government) that limit the economic benefits that communities can actually accrue from wildlife management in WMAs (Nelson 2007). The apparent reluctance on the part of the Wildlife Division (and private sector) to grant full mandate for wildlife management as per the Wildlife Policy means that the benefits that actually go to rural communities are minimal (World Bank 2005). Contrast this with Namibian conservancies which are given exclusive user rights and retain all revenue earned from wildlife, making Namibia's model one of the most successful for community-wildlife management in east and southern Africa (Nelson 2007).

Tanzania's Tourism Master Plan calls for "low volume, high yield" tourism as a means of promoting sustainable development. Yet high tourist numbers and vehicles in the NCA, one of Tanzania's most popular destinations, is leading to overcrowding of wildlife, degradation of roads and other infrastructure, soil erosion and runoff in some areas, and unsustainable extraction of water from the crater for lodges on the crater rim (UNESCO/IUCN 2007, 2008). Environmental degradation as a result of tourism is also occurring on the Kenyan side of the ecosystem as a result of high tourist numbers (MMNR 2009). This growing conflict between economic and ecological functions undermines conservation goals as well as livelihoods that depend on wildlife and other natural resources. Furthermore, while MKUKUTA calls for greater revenue generation from tourism and wildlife, a World Bank report (2005) suggested that the Wildlife Division is not adequately capturing the market value of Tanzania's wildlife resources (e.g., trophy fees are low compared to other African countries, Baldus and Cauldwell 2004). This reflects a lack of implementation of the Policy and Management Plan for Tourist Hunting developed in 1995. Lastly, expanding cultural tourism is a main objective of Tanzania's tourism policy, but restrictions imposed by the NCAA on land and natural resources use are adversely affecting Maasai cultural identity (DeLuca 2002) as more households turn to agriculture as a means of supplementing traditional pastoral livelihoods (UNESCO/IUCN 2007).

MANAGEMENT APPROACHES TO MULTIFUNCTIONALITY

Approaches for managing multifunctionality and institutional interplay in the Serengeti ecosystem have evolved over time. Historically, SENAPA and NCAA have contended with illegal hunting primarily through strong law enforcement consisting of heavily armed antipoaching patrols and arrests (Arcese, Hando, and Campbell 1995). While law enforcement is an effective wildlife protection strategy (Hilborn et al. 2006), it has long been recognized that a combination of incentives (community development) and law enforcement (wildlife protection) may better reconcile the multiple functions of the Serengeti ecosystem. That said, the trade-offs between incentives and protection in reducing conflicts are tricky to disentangle (Adams et al. 2004).

The Serengeti Regional Conservation Strategy (SRCS), initiated in 1985, was one of the first attempts to enhance community development as a means of reconciling the ecological, economic, and sociocultural functions of the Serengeti ecosystem (Mbano et al. 1995). This was a government project under the wildlife sector of the MNRT with funding from the Norwegian Agency for Development Cooperation (NORAD) and other donors (e.g., Frankfurt Zoological Society). The SRCS recognized the need for reconciling wildlife conservation and human development and, in doing so, tackling growing resource conflicts associated with poaching and encroachment through an integrated approach that addressed human needs and livelihoods. The project was implemented in three phases: phase one focused on identifying problems and opportunities for community participation in conservation and benefit sharing; phase two focused on building awareness and coordination among stakeholders, and phase three focused on implementation of community natural resource management and benefit sharing (Mbano et al. 1995). The Serengeti Regional Conservation Project (SRCP) had some positive results in terms of empowering communities, delivering social services, training game scouts, and mitigating poaching to some degree, but the overall sustainability of the program was called into question (Bryceson et al. 2005).

The Ngorongoro Conservation and Development project set up in 1985 was also a response to growing conflict between the NCAA and Maasai communities. Its objectives were to incorporate land use and development needs of resident pastoralists into management strategies (IUCN 1987). Since then the NCAA has also established a community development department and supported the establishment of the Maasai pastoral council to involve the Maasai community in management decisions. Nevertheless, the effectiveness of the Maasai pastoral council has been criticized for its lack of decision making power, weak mechanisms for dissemination of information, consultations with the wider Maasai population, and the absence of transparency in how it uses its share of the revenue from tourism in the NCA (Odhiambo 2003; UNESCO/IUCN 2007). Subsequently, Ereto was set up as a local nongovernmental organization (NGO) to provide services to Maasai communities.

Both TANAPA's community conservation services (CCS) and SENAPA's support to community initiated projects (SCIP) schemes also aims to distribute benefits from conservation to local communities around the Serengeti National Park, thereby improving relations with neighboring stakeholders (Bergin 1996). Since 2000, SENAPA has contributed about US $100,000 per year to community development projects (Thirgood et al. 2008). Direct economic benefits are provided to communities through the revenue sharing program, which provides a percentage of park fees for community development projects largely aimed at improving school, health, and road infrastructure in villages adjacent to parks (TANAPA 2006). Given the size of the target population (approximately 2.3 million people in the seven districts that abut the park) and an under-resourced/under-staffed outreach department (four staff and only 7.5% of the park budget according to the SENAPA GMP), the overall impact of the park outreach program is questionable. Until recently, there was also no official forum for coordinating activities or facilitating communication with communities and other stakeholders in the ecosystem. As a result, the views and needs of communities have typically not been taken into account, marginalized groups have often been excluded, and benefits have not been equitably distributed (Schmitt 2010). The SENAPA GMP (2006–16) identifies three areas for current and future outreach activities: (1) the identification and establishment of conservation friendly income generating activities; (2) the mitigation of human wildlife conflicts, and (3) support for community-based natural resources management (TANAPA 2006).

Frankfurt Zoological Society (FZS) and TANAPA recently implemented the Serengeti Ecosystem Management Project (SEMP), a five year (2005–2010) project aimed at piloting the Convention on Biological Diversity (CBD)'s ecosystem approach in the Serengeti ecosystem. The ecosystem approach recognizes that since local stakeholders are important beneficiaries of healthy ecosystems as well as a major threat to the maintenance of ecosystem functions, it is vital that they are at the forefront of conservation and sustainable use. Strengthening local institutions is also fundamental to the success of community-based wildlife management (Nelson 2007). SEMP has gone beyond other projects in trying to reconcile the multiple functions and institutions in the ecosystem by focusing project activities on the approach's five operational guidelines:

- establishing intersectoral ecosystem cooperation mechanisms
- improving understanding of ecosystem processes and functions
- decentralizing management to local institutions
- improving benefits and incentives for local stakeholders
- introducing adaptive management systems

Among other things, SEMP has supported the Serengeti ecosystem community conservation forum (SECCF). SECCF is a local NGO established to improve intersectoral ecosystem cooperation and mobilize stakeholders to collaborate on key issues "in order to strengthen sustainable conservation utilization and social economic development" (SECCF 2010). In doing so, the SECCF also aims to increase awareness of the unsustainable nature of resource use and encourage participative approaches that mutually enhance sustainable livelihoods and conservation across the ecosystem. A number of values are outlined that underpin the overall governance strategy, including trust, commitment, transparency, accountability, sustainability, fair and equitable benefits, cooperation, partnership, and empowerment. A wide range of stakeholders are represented, a memorandum of understanding (MoU) was signed in 2008 (albeit not by all members), and a draft strategic plan (2010–15) has been developed. Although challenges remain (e.g., community interests are underrepresented at present, sustainable funding is lacking, transparency and accountability need strengthening, Borner and Mwageni 2010), the SECCF has the potential to be an effective platform for stakeholder participation in the governance and, ultimately, management of the Serengeti ecosystem.

CURRENT MANAGEMENT CHALLENGES

It is apparent that most management initiatives to date have encountered a range of problems that have undermined their effectiveness or limited their contribution to a multifunctional approach to management. As a result many of the ecosystem's functions remain threatened. In particular, high levels of illegal use persist (see Rentsch et al., chapter 22); hunting accounts for nearly two-thirds of all incidences of illegal activity picked up by the SENAPA law enforcement department (TANAPA 2006). Furthermore, the majority of people report no benefits from conservation (Schmitt 2010) or insufficient benefits to offset costs from wildlife (e.g., crop damage) or foregoing other activities (e.g., hunting, Holmern et al. 2002). In fact, despite previous outreach/development activities in the Serengeti ecosystem, less than one-third of the local population is even aware of the existence of the park or other protected areas (Schmitt 2010). Lack of benefits is also cited as an underlying cause of negative attitudes among communities adjacent to the Maasai Mara (Sitati 2003). It is clear that these circumstances undermine wildlife conservation, poverty reduction, and sustainable resource use goals laid out in many of the national development strategies. But what are some of the underlying reasons for the current unsustainability of management approaches, particularly those aimed at curbing illegal behavior and promoting local development and conservation-compatible livelihoods in park-adjacent communities?

First, poverty and illegal behavior do not interact straightforwardly. For instance, it is generally assumed that bushmeat hunting in developing countries is poverty driven and that hunters can be transformed into conservationists given the right economic opportunities or incentives. This may seriously underestimate the cultural, social, and political benefits of hunting (Gibson 1999) and the range of attitudes and values to wildlife which influence hunting or other illegal behavior (Fulton, Manfredo, and Lipscomb 1996; Kaltenborn, Bjerke, and Vittersø 1999). A number of other obstacles are evident (Kaltenborn et al. 2008; Schmitt 2010). First, there are still few, if any, alternatives for poor, resource-dependent households in many villages or conservation-friendly alternatives are not economically competitive with other resource uses. Second, where benefits accrue from wildlife tourism, there is a lack of transparency, accountability, and equitability with regards to revenue distribution and benefit sharing. Third, such benefits are not tangible at the individual or household level nor are they conditional on compliance. Fourth, law enforcement does not render the costs of illegal activity high enough to outweigh the benefits from wildlife.

Also, until recently, there were no clear or transparent mechanisms for establishing shared goals, strengthening coordination among stakeholders, or resolving conflict among stakeholders. The creation of the SECCF is a step in the right direction in this respect. For one thing, it established an MoU out of what were previously informal social links between actors. However, limited donor funding has restricted representation on the forum to only three of the seven districts on the Tanzanian side of the Serengeti ecosystem, and key members, including TANAPA and WD, have not signed the MoU because of institutional obstacles preventing them from doing so. Poor representation of communities as well as lack of accountability and transparency are obstacles to effective governance (Borner and Mwageni 2010). Similarly, in the Ngorongoro Conservation Area there seems to be little genuine representation of Maasai interests and rights in the management of the NCAA (Lane 1996; Odhiambo 2003; UNESCO/IUCN 2007). Within WMAs power sharing and revenue distribution between AAs, village governments, and districts are often not clear or transparent (Walsh 2000). These local governance challenges are exacerbated by the need for policies and institutional structures for transboundary cooperation in the ecosystem.

THE WAY FORWARD FOR MANAGEMENT OF THE SERENGETI ECOSYSTEM

Despite several attempts to establish institutional arrangements that reconcile multiple functions and institutions, the sustainability of the Serengeti ecosystem remains in question. In theory most institutions reflect the multiple claims towards the ecosystem, yet in practice these institutions are either in conflict or poorly adhered to. As the gaps between institutional goals and reality continue to widen, the need for institutional reform becomes increasingly apparent. The main challenge is how to manage the inevitable complexity (and tensions) arising out of diverse interests and power relations across the landscape. We put forward the following recommendations taking into account international best practice:

- Strengthen institutions for protected area management
- Enhance comanagement approaches to governance
- Promote institutional compliance
- Evaluate management effectiveness

Strengthen Institutions for Protected Area Management

As part of the process of institutional reform, Tanzania could make wider use of the IUCN protected areas management categories and guidelines for best practice in protected area management. The IUCN categories were established with the aim of developing a common (albeit flexible) understanding of the aims of different types of protected areas (Bishop et al. 2004). Broadly speaking, a protected area is designated as one of the six categories based on its primary management objective and the level of human modification or intervention (table 24.2). Categories V and VI were established to explicitly recognize the compatibility between the rights of traditional people, sustainable use, and conservation (IUCN 1994), and the IUCN guidelines outline mechanisms for reconciling conflicts between competing objectives and priorities as part of sustainable development (IUCN 2002). These guidelines provide internationally recognized norms and standards that can be used to guide institutional issues with respect to land tenure, user rights, participation, partnership building, governance, and other issues to do with protected area management in the Serengeti ecosystem.

The Convention on Biological Diversity (CBD), to which Tanzania is a signatory, endorses the categories system in their Programme of Work on Protected Areas (CBD/COP7) adopted in 2004 and "encourages Parties, other Governments and relevant organisations to assign protected area management categories to their protected areas, and provide information consistent with the refined IUCN categories for reporting purposes." Although Tanzania's obligations under international commitment such as the CBD and Convention on International Trade in Endangered Species of Wild Fauna and Flora (CITES) are mentioned, no reference is made to the IUCN categories system in either the wildlife policy or the Wildlife Conservation Act. Nevertheless, IUCN categories have been designated for Serengeti National Park (Category II), Maasai Mara National Reserve (Category II) and Ngorongoro Conservation Area (Category VI) although designations are lacking for game reserves, game controlled areas, and wildlife manage-

Table 24.2 The IUCN protected area management categories (IUCN 1994)

Areas managed mainly for:	
I.	Strict protection (a) strict nature reserve and (b) wilderness area)
II.	Ecosystem conservation and protection (national park)
III.	Conservation of natural features (natural monument)
IV.	Conservation through active management (habitat/species management area)
V.	Landscape/seascape conservation and recreation (protected landscape/seascape)
VI.	Sustainable use of natural resources (managed resource protected area)

ment areas. Assigning IUCN categories through a participatory approach that involves both national agencies and local stakeholders could help clarify the objectives of these areas and using the relevant guidelines wherever possible would help ensure management approaches are in line with best practice internationally.

Enhance Comanagement Approaches to Governance

While institutions are the rules by which activities are regulated (North 1990), governance covers the structures, processes and traditions by which responsibilities are allocated, decisions are made, and authority is exerted (Graham, Amos, and Plumptre 2003). In a protected area landscape with multiple functions and institutions, management will involve a wide range of stakeholders on different societal levels making governance particularly challenging. Governance structures and processes will need to incorporate complex relationships, objectives, and power struggles (Lockwood 2010) and, given the transboundary nature of the Serengeti ecosystem, national sovereignty, legislation, and country-specific interests will also play a role.

There are, in general, four protected area governance models that are recognized by the IUCN: governance by government, shared governance, private governance, and governance by indigenous people and local communities (Dudley 2008), each of which could be applied to any of the IUCN protected area categories. No single governance structure will be sufficient to address the multifunctionality of the Serengeti ecosystem, especially as there are multiple centers of power and layers of authority with respect to the management of geographic areas and sectors. This is a case of polycentric governance, that is, systems in which "political authority is dispersed to separately constituted bodies with overlapping jurisdictions that do not stand in hierarchical relationship to each other" (Skelcher 2005, 89). Polycentrism or any governance structure that involves a plurality of institutions can be advantageous when complex problems need to be addressed (Imperial 1999) since different geographical scopes can be managed at different scales, and where governance systems overlap, resources can be shared and social learning can take place (Ostrom 2005). However, a certain level of coordination is needed for pluralistic governance approaches to be effective at distributing authority across multiple institutions. This is one of the greatest challenges currently facing the Serengeti ecosystem.

In Tanzania, the state became the direct coordinator and main implementer of government services after the Arusha Declaration in 1967. However, decentralization of decision making to local actors gave rise to comanagement approaches to governance. Comanagement is more or less interchangeable with terms such as cooperative management, collaborative management, joint management, or participatory management, and generally indicates some form of cooperation between the state and local actors (Berkes, George, and Preston 1991). The IUCN (1996, sec. 1.42) defines it as "a partnership in which government agencies, local communities and resource users, nongovernmental organizations and other stakeholders negotiate, as appropriate to each context, the authority and responsibility for the management of a specific area or set of resources."

Comanagement is often understood as a governance *structure* that formalizes power sharing between stakeholders (Jentoft 1985), but it may also comprise a set of loosely connected actors coordinating management of complex issues as is the case, for instance, in polycentric governance (Imperial 1999). Importantly, it also describes a *process* for collecting and exchanging information, coordinating activities, and solving problems in an iterative way, as in adaptive management (Carlsson and Berkes 2005). More specifically, adaptive comanagement is "a process by which institutional arrangements and ecological knowledge are tested and revised in a dynamic, ongoing, self-organized process of learning-by-doing" (Folke et al. 2002). Adaptive comanagement is particularly suited to multifunctional landscapes where there are diverse values, interests, and levels of power since it emphasizes dialogue, deliberation, and negotiation toward joint problem solving (Olsson, Folke, and Berkes 2004; Carlsson and Berkes 2005). The dynamic and iterative nature of adaptive comanagement not only enable decision-making feedback loops conducive to solving complex problems through trial and error, but also incorporate the uncertainty and unpredictability that stem from incomplete knowledge of the ecosystem as well as environmental and social change (Olsson, Folke, and Berkes 2004).

There is no panacea for how to operationalize adaptive comanagement as a governance approach, but some essential features include: (a) local decision-making power and financial support; (b) information exchange and knowledge sharing toward collaborative learning; (c) linkages between organizations and institutions at different levels and scales; (d) analytical and strategic decision making based on multiple sources of information; and (e) monitoring and evaluation feedback systems for adaptive management (adapted from Olsson, Folke, and Berkes 2004). Importantly, local actors need to feel there is a genuine commitment to cooperation and collaboration on the part of the state (rather than mere consultation or coercion). This not only requires that the rights and experiences of local actors are recognized but that local institutions have the capacity and incentives to participate in problem solving and decision making (see Carlsson and Berkes (2005) for practical steps that can be taken to build stakeholder capacity for genuine participation in comanagement arrangements).

Promote Institutional Compliance

Institutions are enforced and entrenched through incentives and sanctions (or disincentives). That being said, compliance is also dependent on a number of political, sociocultural, and economic factors. First and foremost is the need for consistency and coherence between institutions relating to wildlife management, land tenure and rights, tourism development, and poverty reduction in Tanzania. Coherence should exist horizontally (across sectors) and vertically (across local, national, and international institutions) to facilitate compliance with policy and legal frameworks. Compliance also depends on legal clarity and a mutual understanding of the institutional framework among stakeholders. Clear, unambiguous laws and policies reduce the possibility of arbitrary interpretation by government officials and facilitate the work of the police, judiciary, and others with discretionary power. Stakeholder participation in the negotiation of institutions will not only improve consistency and coherence, but also promote legitimacy, transparency, and awareness of the policies and laws they will be expected to enforce and/or adhere to.

Awareness alone, however, will not deter illegal behavior unless a number of other conditions are also met. First, wildlife must provide economic gain at the household level for there to be adequate incentives for conservation-compatible activities. Furthermore, the benefits of conservation must be sufficient to offset the costs of foregoing other activities that are seen as more economically profitable (Ottichilo et al. 2000; Emerton 2001; Homewood et al. 2001). Second, whether a "fences and fines" or benefits-based approach is taken, law enforcement authorities (including the police and judiciary) must have sufficient capacity to ensure detection and conviction of offenders. Unfortunately, protected areas are often ill-equipped and under-resourced and thus, the risk of illegal activities being detected and punished is too low to deter potential offenders. Law enforcement activities need to be strengthened in protected areas and community-managed areas which also need sufficient human, financial, and institutional capacity to enforce rules over common property management. Third, ensuring land tenure and user rights would improve compliance in community-managed areas by promoting a local stake in sustainable management. This would not only increase the incentives for local people to comply with the law, but also insist on the compliance (or exclusion) of outsiders.

Evaluate Management Effectiveness

Ecological monitoring has been a key activity in the Serengeti National Park and Ngorongoro Conservation Area for decades, including censuses of buffalo and wildebeest since 1958 (Sinclair 1973; Sinclair et al. 2007). A number of studies have also assessed the socioeconomic status of communities (e.g., Campbell, Nelson, and Loibooki 2001; Schmitt 2010) which provides a foundation for long-term monitoring of community development activities in the ecosystem. A number of studies have used monitoring data to examine long-term trends and assess changes in the ecosystem (Sinclair and Arcese 1995; Sinclair et al. 2007), including the impact of policies and institutional issues such as land tenure (Hilborn et al. 1995; Homewood et al. 2001), but much more could be done to examine the institutional and governance dimensions of effective (and adaptive) management.

Collecting and evaluating information on how well a protected area is being managed in terms of the extent to which resources, plans, and action are being generated, and in turn objectives and goals are being met (Hockings et al. 2006) is a key part of the adaptive management cycle (Conservation Measures Partnership [CMP] 2007). The CBD also emphasizes setting concrete performance targets for assessing management effectiveness as part of the program of work on protected areas (seventh meeting of the Conference on the Parties to the Convention on Biological Diversity [CBD/COP7]). The IUCN guidelines provide an internationally recognized framework for assessing management effectiveness with respect to the appropriateness of the design of individual areas or protected area systems (context and planning), the adequacy of management approaches (inputs and processes), and progress towards protected area goals and objectives (outputs and outcomes) (Hockings et al. 2006, fig. 24.2). Governance can also be evaluated by indicators

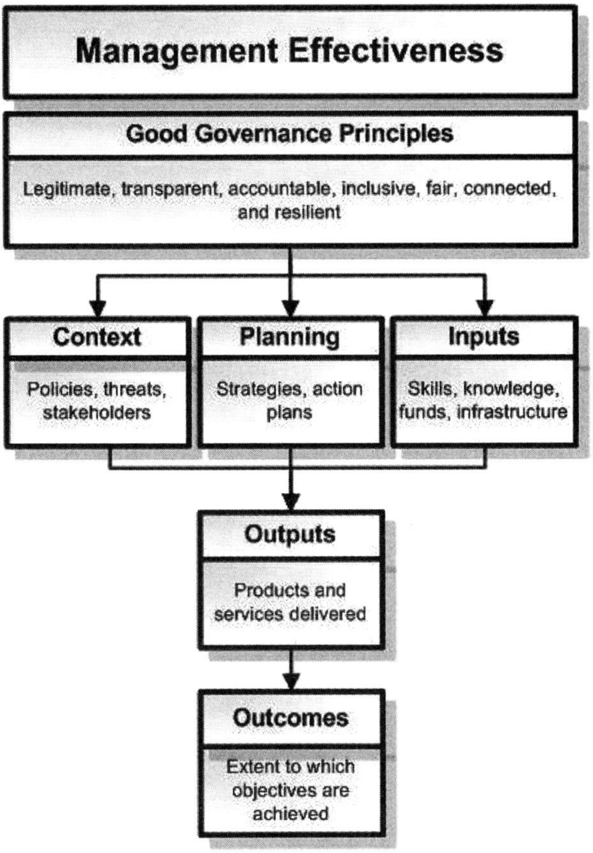

Fig. 24.2 Framework for management effectiveness (adapted from Hockings et al. 2006 and Lockwood 2010).

that define standards for good protected area governance using, for example, a framework such as Lockwood's (2010), which characterizes governance quality according to seven principles: legitimacy, transparency, accountability, inclusiveness, fairness, connectivity, and resilience (fig. 24.2). Indicators could be developed to specifically assess essential elements of adaptive co-management, including multiscale characteristics and the capacity of the system to react to feedback and adapt accordingly. Table 24.3 lists key variables and indicators from the literature (see Cundill and Fabricius 2010) that could be used as the basis for monitoring collaborative governance in the Serengeti ecosystem. Developing a comprehensive monitoring program for evaluating management effectiveness in the Serengeti ecosystem is an area for future collaboration between organizations and institutions.

Table 24.3 Key variables and indicators for collaborative governance monitoring

Attribute	Key variable	Indicator
Social capital (Pretty and Ward 2001; Pretty 2003)	Trust building	Trust building takes place among the groups involved in collaborative decision making. Decision making is perceived as open and fair. Information is shared and understood by all participants.
	Common rules and norms	See "rule compliance" under "preconditions for adaptive governance."
	There are common-interest groups	There is a common interest and a shared vision. Participants jointly identify and agree on the problems to be solved and what the future should look like. It is clear to all participants why a decision-making body is needed. Participants agree on what the major problems are and what the benefits might be of resolving these problems.
	Financial and capacity support from higher levels of organization	A long-term investment has been made. The state or its partners are committed to making a substantial and long-term financial investment in the project. Long-term skills and leadership development programs are in place. Planning and decision-making support is offered.
	Security of tenure over the resources of concern	There is long-term security of access to resources. The decision-making body is confident that they are or will be able to prevent outsiders from using the resources.
	Economic or other incentives for collective action	People who contribute more are rewarded. People who lose ways of earning a living because of the project are compensated.
Adaptive capacity (Armitage 2005)	Willingness to learn from mistakes	All actors, within and outside the community, listen to each other and are willing to change what they are doing in response. The organization or committee involved in the initiative is made up of people from the community and from outside the community. These actors respect one another and listen to each others' points of view.
	Willingness to engage in collaborative decision making	All participants are willing to engage in collaborative learning and decision making. Participants recognize the value of sharing information among actors. Experts are willing to learn from resource users, and resource users are open to alternative ways of doing things. The project is viewed as a learning process by everyone involved.

	Willingness to accept a diversity of institutions	Participants understand that it is unlikely that one institution will be able to manage the entire ecosystem. Although a broad institution should be established to provide vision and overall coordination, members of the institution are aware that smaller groups may be formed to deal with specific issues.
	Maintaining options for adaptation (e.g., diversity of ecosystems, livelihoods and institutions)	Projects can bring many benefits, but they cannot solve all the problems. For example, it should be understood that not everyone can be employed on the project. People understand this and continue to do their work as usual. Over time, the projects provide some new opportunities.
Self-organization (Olsson, Folke, and Berkes 2004)	Enabling legislation is in place, is accessible, and is understood	Legislation is in place that allows people to form legal entities to manage natural resources. Project participants have access to and an understanding of the legislation.
	Funds are available for adaptive management	See "a long-term investment has been made" under Social Capital.
	Information flow and social networks	Networks are established that connect the local decision-making body with other institutions. Outside partners such as government officials, researchers, and nongovernmental organizations are involved and willing to devolve decision-making powers. Other relevant, local decision-making bodies are consulted and included in decision making. The roles of these different actors are clearly defined.
	Various sources of information are combined for sense making	Information flow. There is good communication among everyone involved. People are informed about what is happening, and their views and opinions are heard.
	Arenas of collaborative learning	See "all actors, from within and outside the community, listen to each other and are willing to change what they are doing in response" and "all participants are willing to engage in collaborative learning and decision making" under adaptive capacity.
	Leadership	Leadership is effective and recognized. The leaders of the initiatives care about more than just their own interests. The leaders are trusted and acknowledged by all actors.

continued

Table 24.3 continued

Attribute	Key variable	Indicator
Preconditions for adaptive governance (Dietz, Ostrom, and Stern 2003)	Access to accurate and relevant knowledge and information	Combination of "enabling legislation" and "networks are established that connect the local decision-making body with other institutions" under self-organization.
	Conflict resolution mechanisms in place	Participants are aware that there will be conflict. The decision-making body is prepared for conflict and solves problems before they become serious. People are kept informed and their complaints and problems are heard.
	Compliance with rules and regulations	There are a management plan and rules for the use of natural resources, especially those that people depend on for their livelihoods. Resource users respect and adhere to the rules.
	Being prepared for change	A combination of "all actors, from within and outside the community, listen to each other and are willing to change what they are doing in response" under adaptive capacity and "conflict resolution mechanisms are in place" under adaptive governance.

Source: Taken from Cundill and Fabricius 2010.

CONCLUSION

The multifunctionality of the Serengeti ecosystem is evident in the range of ecological, economic, and sociocultural goods and services across the landscape and the many institutions designed to manage these functions. Navigating the institutional complexity with respect to wildlife, land, and local community rights in Tanzania remains a huge challenge. Most of the relevant policies recognize the linkages between functions and institutions, but they are still, for the most part, sector-oriented with no clearly defined mechanisms for collaboration with other sectors or stakeholders. As a result the existing institutions tend to guard the ecosystem's functions only partially or selectively. In some cases, local institutions—both formal (e.g., police, judiciary, rural development authorities) and informal (e.g., social norms and customs)—exist but are not being used effectively and, in the case of local informal rules, may be eroding over time. This institutional interplay is increasingly giving rise to conflicts between actors and the functions they seek to enhance.

Given all this, "strengthening governance systems at relevant scales is perhaps the most important challenge of the next century for biodiversity conservation" (Agrawal and Ostrom 2006, 682). In this chapter, we have described the concept and principles of adaptive comanagement as a framework for a more dynamic approach to governance in the Serengeti ecosystem that explicitly recognizes the multifunctionality of the landscape. The state may continue to play a strong coordinating role in comanagement arrangements, but reforms should establish clearer mechanisms for collaboration between stakeholders and, in particular, communication with and participation of communities. Strengthening local institutions will be important in order to empower local stakeholders to participate in decision making in ways that support their interests. In this sense, the role of the SECCF could be strengthened given its potential to be a platform for exchanging information, encouraging dialogue, and resolving conflict at the local level. Tensions arising from the competing needs and demands of stakeholders will persist, but a commitment to building political will, requisite management capacity, and problem-solving mechanisms through comanagement approaches to governance will go a long way to resolving conflict and improving management effectiveness. Enhancing adaptive comanagement can not only help reconcile the multifunctionality of the Serengeti ecosystem but also improve its resilience to environmental and social change, and ultimately help ensure sustainability.

ACKNOWLEDGMENTS

This work was conducted as part of the project "HUNT" (Hunting for Sustainability; http://fp7hunt.net/) funded by the European Union's Framework Programme 7. Support was also provided by Frankfurt Zoological Society, the Tanzania Wildlife Research Institute, the James Hutton Institute, and Umeå University. Thanks to Honori T. Maliti for producing the map in figure 24.1.

REFERENCES

Adams, W. M., R. Aveling, D. Brockington, B. Dickson, J. Elliot, J. Hutting, D. Roe, B. Vira, and W. Wolmer. 2004. Biodiversity conservation and the eradication of poverty. *Science* 306:1146–49.

Agrawal, A., and E. Ostrom. 2006. Political science and conservation biology: A dialogue of the deaf. *Conservation Biology* 20:681–82.

Arcese, P., J. Hando, and K. Campbell. 1995. Historical and present-day antipoaching efforts in Serengeti. In *Serengeti II: Dynamics, managament, and conservation of an ecosystem*, ed. A. R. E. Sinclair and P. Arcese, 506–33. Chicago: University of Chicago Press.

Armitage, D. 2005. Adaptive capacity and community-based natural resource management. *Environmental Management* 35:703–15.

Baldus, R. D., and A. E. Cauldwell. 2004. Tourist hunting and its role in development of wildlife management areas in Tanzania. Dar es Salaam, Tanzania. http://www.wildlife-programme.gtz.de/.

Bergin, P. 1996. Tanzania national parks community conservation service. In *Community-based conservation in Tanzania*, ed. N. Leader-Williams, J. Kayera, and G. Overton, 67–70. Gland, Switzerland: IUCN.

Berkes, F., P. J. George, and R. J. Preston. 1991. Comanagement: The evolution in theory and practice of the joint administration of living resources. *Alternatives* 18:12–18.

Bishop, K., N. Dudley, A. Phillips, and S. Stolten. 2004. Speaking a common language: The uses and performance of the IUCN system of management categories for protected areas. Cardiff University, IUCN, and UNEP-WCMC.

Borge, A. 2003. Essays on the economics of African wildlife and utilization and management. PhD diss., University of Trondheim, NTNU.

Borner, M. and H. Mwageni. 2010. Serengeti-Luangwa ecosystem management project final evaluation. Frankfurt Zoological Society, Tanzania National Parks, Tanzania Wildlife Division, Zambia Wildlife Authority.

Brett, T. 2000. Understanding organizations and institutions. In *Managing development: Understanding interorganizational relationships*, ed. D. Robinson, T. Hewitt, and T. Harriss, 17–48. London: Sage Publications in association with The Open University.

Bryceson, I., K. Havnevik, A. Isinika, I. Jørgensen, L. Melamari, and S. Sønvinsen. 2005. Management of natural resources programme, mid-term review of Tan–092, phase III (2002–2006). Ministry of Natural Resources and Tourism, Tanzania.

Campbell, K., and H. Hofer. 1995. People and wildlife: Spatial dynamics and zones of interaction. In *Serengeti II: Dynamics, management and conservation of an ecosystem*, ed. A. R. E. Sinclair and P. Arcese, 534–70. Chicago: University of Chicago Press.

Campbell, K., V. Nelson, and M. Loibooki. 2001. An analysis of illegal hunting of wildlife in Serengeti National Park, Tanzania. Sustainable use of wildland resources: Ecological, economic, and social interactions. Chatham, Kent, United Kingdom, Department for International Development (DFID) Animal Health Programme and Livestock Production Programmes, Final Technical Report, Project R7050. Natural Resources Institute (NRI).

Carlsson, L., and F. Berkes. 2005. Comanagement: Concepts and methodological implications. *Journal of Environmental Management* 7:65–76.

CBD/COP7. www.biodiv.org/doc/meetings/cop/cop-07/official/cop-07-I-32-en.pdf.

Conservation Measures Partnership (CMP). 2007. Open standards for the practice of conservation, version 2.0. www.conservationmeasures.org.

Cundill, G., and C. Fabricius. 2010. Monitoring the governance dimension of natural resource comanagement. *Ecology and Society* 15 (1): 15 [online].

DeLuca, L. M. 2002. Tourism, conservation, and development among the Maasai of Ngorongoro district, Tanzania: Implications for political ecology and sustainable livelihoods. PhD diss., University of Colorado.

Dietz, T., E. Ostrom, and P. C. Stern. 2003. The struggle to govern the commons. *Science* 302:1907–12.

Dobson, A. 1995. The ecology and epidemiology of rinderpest virus in Serengeti and Ngorongoro Conservation Area. In *Serengeti II: Dynamics, management, and conservation of an ecosystem*, ed. A. R. E. Sinclair and P. Arcese, 485–505. Chicago: University of Chicago Press.

Dudley, N. 2008. *Guidelines for applying protected area management categories*. Gland, Switzerland: IUCN.

Emerton, L. 2001. Why wildlife conservation has not economically benefited communities in Africa. In *African wildlife and livelihoods: The promise and performance of community conservation*, ed. D. Hulme and M. Murphree, 208–26. Oxford: James Currey.

Folke, C., S. Carpenter, T. Elmqvist, L. Gunderson, C. S. Holling, and B. Walker. 2002. Resilience and sustainable development: Building adaptive capacity in a world of transformations. *Ambio* 31:437–40.

Fratkin, E., and R. Mearns. 2003. Sustainability and pastoral livelihoods: Lessons from East African Maasai and Mongolia. *Human Organization* 62:112–22.

Fryxell, J. M. 1995. Aggregation and migration by grazing ungulates in relation to resources and predators. In *Serengeti II: Dynamics, management and conservation of an ecosystem*, ed. A. R. E. Sinclair and P. Arcese, 257–73. Chicago: University of Chicago Press.

Fulton, D. C., M. J. Manfredo, and J. Lipscomb. 1996. Wildlife value orientations: A conceptual and measurement approach. *Human Dimensions of Wildlife* 1:24–27.

Gereta, E. J., E. Wolanski, and E. A. T. Chiombola. 2003. Assessment of the environmental, social and economic impacts on the Serengeti ecosystem of the developments in the Mara River catchment in Kenya. Unpublished report, Tanzanian National Parks, Arusha.

Gibson, C. C. 1999. *Politicians and poachers. The political economy of wildlife policy in Africa*. Cambridge: Cambridge University Press.

Graham, J., B. Amos, and T. Plumptre. 2003. Governance principles for protected areas in the 21st Century. Ottawa: Institute on Governance.

Hilborn, R., P. Arcese, M. Borner, J. Hando, G. Hopcraft, M. Loibooki, S. Mduma, and A. R. E. Sinclair. 2006. Effective enforcement in a conservation area. *Science* 314:1266.

Hilborn, R., N. Georgiadis, J. J. Lazarus, J. M. Fryxell, M. D. Broten, et al. 1995. A model to evaluate alternative management policies for the Serengeti-Mara ecosystem. In *Serengeti II: Dynamics, management, and conservation of an ecosystem*, ed. A. R. E. Sinclair and P. Arcese, 617–37. Chicago: University of Chicago Press.

Hockings, M., S. Stolten, F. Leverington, N. Dudley, and J. Courrau. 2006. Evaluating effectiveness: A framework for assessing the effectiveness of protected areas, 2nd ed. Best practice protected area guidelines series no. 14, ed. P. Valentine. Gland, Switzerland: IUCN.

Holmern, T., E. Røskaft, J. Mbaruka, S. Y. Mkama, and J. Muya. 2002. Uneconomical game cropping in a community-based conservation project outside the Serengeti National Park, Tanzania. *Oryx* 36:364–72.

Homewood, K., E. Lambin, E. Coast, A. Kariuki, J. Kivelia, M. Said, S. Serneels, and M. Thompson. 2001. Long-term changes in Serengeti-Mara wildebeest and land cover: Pastoralism, population, or policies? *Proceedings of the National Academy of Sciences* 98:12544–49.

Hulme, D., and M. Murphree. 1999. Communities, wildlife and "the new conservation in Africa." *Journal of International Development* 11:277–85.

Imperial, M. T. 1999. Institutional analysis and ecosystem-based management: The institutional analysis and development framework. *Environmental Management* 24: 449–65.

IUCN. 1987. Ngorongoro conservation and development project: Work plan of activities. Unpublished report, Gland, Switzerland.

———. 1994. Guidelines for protected area management categories. IUCN, Gland, Switzerland and Cambridge, UK, CNPPA with the assistance of WCMC.

———. 1996. Resolutions and recommendations. Gland, Switzerland, World Conservation Congress.

———. 2002. Management guidelines for IUCN Category V protected areas. Gland, Switzerland: IUCN.

Jentoft, S. 1985. Models of fishery development: The cooperative approach. *Marine Policy* 9:322–31.

Johanson, D. C., F. T. Masao, G. G. Eck, T. D. White, R. C. Walter, W. H. Kimbel, B. Asfaw, P. Manega, P. Ndessokia, and G. Suwa. 1987. New partial skeleton of *Homo habilis* from Olduvai Gorge, Tanzania. *Nature* 327:205–09.

Kaarea, M. T., K. Picozzi, T. Mlengeya, E. M. Fèvre, L. S. Mellau, M. M. Mtambo, S. Cleaveland, and S. C. Welburn. 2007. Sleeping sickness—A reemerging disease in the Serengeti? *Travel Medicine and Infectious Disease* 5:117–24.

Kaltenborn, B. P., T. Bjerke, and J. Vittersø. 1999. Attitudes toward large carnivores among sheep farmers, wildlife managers, and research biologists in Norway. *Human Dimensions of Wildlife* 4:57–73.

Kaltenborn, B. P., J. W. Nyahongo, J. R. Kidegesho, and H. Haalanda. 2008. Serengeti National Park and its neighbours—Do they interact? *Journal for Nature Conservation* 16:96–108.

Kaltenborn, B. P., J. W. Nyahongo, and K. M. Tingstad. 2005. The nature of hunting around the Western Corridor of Serengeti National Park, Tanzania. *European Journal of Wildlife Research* 51:213–22.

Lane, C. 1996. *Ngorongoro voices: Indigenous Maasai residents of the Ngorongoro conservation area in Tanzania give their views on the proposed general management plan*. Rome: Food and Agriculture Organization of the United Nations.

Leakey, M. D., and R. L. Hay. 1979. Pliocene footprints in the Laetoli beds at Laetoli, northern Tanzania. *Nature* 278:317–23.

Lembo, T., K. Hampson, D. T. Haydon, M. Craft, A. Dobson, J. Dushoff, E. Ernest, R. Hoare, M. Kaare, and T. Mlengeya, et al. 2008. Exploring reservoir dynamics: A case study of rabies in the Serengeti ecosystem. *Journal of Applied Ecology* 45: 1246–57.

Lockwood, M. 2010. Good governance for terrestrial protected areas: A framework, principles and performance outcomes. *Journal of Environmental Management* 91:754–66.

Loibooki, M., H. Hofer, K. L. I. Campbell, and M. East. 2002. Bushmeat hunting by communities adjacent to Serengeti National Park, Tanzania: The importance of livestock ownership and alternative sources of protein and income. *Environmental Conservation* 29:391–98.

Lowassa, A., D. Tadie, and A. Fischer. 2012. On the role of women in bushmeat hunting—Insights from Tanzania and Ethiopia. *Journal for Rural Studies* 28:622–30.

Maasai Mara National Reserve (MMNR). 2009. Maasai Mara National Reserve general management plan (2009–2019). Narok and Trans Mara County Councils, Narok, Kenya.

Mbano, B. N. N., R. C. Malpas, M. K. E. Maige, P. A. K. Symonds, and D. M. Thompson. 1995. The Serengeti regional conservation strategy. In *Serengeti II: Dynamics, management, and conservation of an ecosystem*, ed. A. R. E. Sinclair and P. Arcese, 605–16. Chicago: University of Chicago Press.

Mduma, S. A. R. 1996. Serengeti wildebeest population dynamics: Regulation, limitation and implications for harvesting. PhD diss., University of British Columbia.

Millennium Ecosystem Assessment (MEA). 2005. *Ecosystems and human well-being: Synthesis*. Washington, DC: Island Press.

Murray, M. G. 1995. Specific nutrient requirements and migration of wildebeest. In *Serengeti II: Dynamics, management and conservation of an ecosystem*, ed. A. R. E. Sinclair and P. Arcese, 231–56. Chicago: University of Chicago Press.

Nelson, F. 2007. Emergent or illusory? Community wildlife management in Tanzania. Pastoral civil society in East Africa. Issue Paper no. 146. IIED and Pastoral civil society. Nottingham, United Kingdom: Russell Press.

North, D. C. 1990. *Institutions, institutional change and economic performance*. Cambridge: Cambridge University Press.

Odhiambo, M. O. 2003. Ereto-Ngorongoro pastoralist project draft 1, potential institutional placement and links to phase II. An appraisal of the institutional framework for sustaining the development interventions of ERETO-NPP. West Sussex, United Kingdom: Lion House.

Olsson, P., C. Folke, and F. Berkes. 2004. Adaptive comanagement for building resilience in social-ecological systems. *Environmental Management* 34:75–90.

Ostrom, E. 2005. *Understanding institutional diversity*. Princeton, NJ: Princeton University Press.

Ottichilo, W. K., J. de Leeuw, A. K. Skidmore, H. H. T. Prins, and M. Y. Said. 2000. Population trends of large nonmigratory wild herbivores and livestock in the Maasai Mara ecosystem, Kenya, between 1977 and 1997. *African Journal of Ecology* 38:202–16.

Pander, H. 1995. Land tenure and land policy in the Transmara district, Kenya: Situations and conflicts, GTZ. http://www.mekonginfo.org/mrc/html/pander/pan_inh.htm.

Polasky, S., J. Schmitt, C. Costello, and L. Tajibaeva. 2008. Larger-scale influences on the Serengeti ecosystem: National and international policy, economics, and human demography. In *Serengeti III: Human impacts on ecosystem dynamics*, ed. A. R. E. Sinclair, C. Packer, S. A. R. Mduma, and J. M. Fryxell, 347–77. Chicago: University of Chicago Press.

Pretty, J. 2003. Social capital and the collective management of resources. *Science* 302: 1912–14.

Pretty, J., and H. Ward. 2001. Social capital and the environment. *World Development* 29: 209–27.

Prins, H. 1992. The pastoral road to extinction: Competition between wildlife and traditional pastoralism in East Africa. *Environmental Conservation* 19:117–23.

Roelke-Parker, M. E., L. Munson, C. Packer, R. Kock, S. Cleaveland, M. Carpenter, S. J. O'Brien, A. Pospischil, R. Hofmann-Lehmann, and H. Lutz, et al. 1996. A canine distemper virus epidemic in Serengeti lions (*Panthera leo*). *Nature* 379:441–45.

Schmitt, J. A. 2010. Improving conservation efforts in the Serengeti ecosystem: An examination of knowledge, benefits, costs and attitudes. PhD diss., University of Minnesota.

SECCF 2010. The Serengeti ecosystem community conservation forum draft strategic plan 2010–2015. The Serengeti ecosystem community conservation forum, Tanzania.

Seno, S. K., and W. W. Shaw. 2002. Land tenure policies, Maasai traditions, and wildlife conservation in Kenya. *Society and Natural Resources* 15:79–88.

Sinclair, A. R. E. 1973. Population increases of buffalo and wildebeest in the Serengeti. *East African Wildlife Journal* 11:93–107.

———. 1995. Serengeti past and present. In *Serengeti II: Dynamics, management and conservation of an ecosystem*, ed. A. R. E. Sinclair and P. Arcese, 3–30. Chicago: University of Chicago Press.

———. 2008. Integrating conservation in human and natural ecosystems. In *Serengeti III: Human impacts on ecosystem dynamics*, ed. A. R. E. Sinclair, C. Packer, S. Mduma, and J. Fryxell, 443–69. Chicago: University of Chicago Press.

Sinclair, A. R. E., and P. Arcese, eds. 1995. *Serengeti II: Dynamics, management and conservation of an ecosystem.* Chicago: University of Chicago Press.

Sinclair, A. R. E., S. A. R. Mduma, J. G. C. Hopcraft, J. M. Fryxell, R. Hilborn, and S. Thirgood. 2007. Long-term ecosystem dynamics in the Serengeti: Lessons for conservation. *Conservation Biology* 21:580–90.

Sitati, N. W. 2003. Human-elephant conflict in Transmara district, Kenya. In *Wildlife and people: Conflict and conservation in Maasai Mara, Kenya*, ed. M. Walpole, G. Karanja, N. Sitati, and N. Leader-Williams, 27–33. London: International Institute for Environment and Development (IIED).

Skelcher, C. 2005. Jurisdictional integrity, polycentrism, and the design of democratic governance. *Governance* 18:89–110.

Steinhard, E. I. 1989. Hunters, poachers and gamekeepers: Towards a social history of hunting in colonial Kenya. *Journal of African History* 30:247–64.

TANAPA 2006. Serengeti National Park general management plan (2006–2016). Arusha: Tanzania National Parks.

Tanzania Natural Resource Forum (TNRF). 2008. Wildlife for all Tanzanian: Stopping the loss, nurturing the resource and widening the benefits. An information pack and policy recommendations, unpublished report. Tanzania Natural Resource Forum, Arusha, Tanzania.

Thirgood, S., C. Mlingwa, E. Gereta, V. Runyoro, R. Malpas, K. Laurenson, and M. Borner. 2008. Who pays for conservation? Current and future financing scenarios for the Serengeti ecosystem. In *Serengeti III: Human impacts on ecosystem dynamics*, ed. A. R. E. Sinclair, C. Packer, S. Mduma, and J. Fryxell, 443–69. Chicago: University of Chicago Press.

Thirgood, S., A. Mosser, S. Tham, G. Hopcraft, E. Mwangomo, T. Mlengeya, M. Kilewo, J. Fryxell, A. R. E. Sinclair, and M. Borner. 2004. Can parks protect migratory ungulates? The case of the Serengeti wildebeest. *Animal Conservation* 7:113–20.

UNESCO/IUCN 2007. Ngorongoro conservation area (United Republic of Tanzania). Report of the reactive monitoring mission. April 29, 2007–May 5, 2007. World Heritage Committee, thirty-first session. Paris: UNESCO.

———. 2008. Ngorongoro conservation area (United Republic of Tanzania). Report of the reactive monitoring mission. December 1–6, 2008. World Heritage Committee, thirty-third session. Sevilla: UNESCO.

United Nations World Tourism Organization (UNWTO). 2008. Tourism highlights 2008 edition. Madrid: UNWTO.

United Republic of Tanzania (URT). 2002. Population census and housing census. Dar es Salaam: Bureau of statistics, president's office, planning commission, United Republic of Tanzania.

———. 2006. The economic survey 2006. Dar es Salaam: The government printer, the United Republic of Tanzania.

Walpole, M. J. 2006. Partnerships for conservation and poverty reduction. *Oryx* 40: 245–46.

Walsh, M. T. 2000. The development of community wildlife management in Tanzania, lessons learned from the Ruaha ecosystem. Conference on African wildlife management in the new millenium, Mweka, Moshi, Tanzania, College of African Wildlife Management.

World Bank. 2005. Study on growth and environmental link for preparation of country economic memorandum (CEM), part 2: Uncaptured growth potential—Foresty, wildlife and marine fisheries, final report. COWI, Tanzania.

World Commission on Environment and Development (WCED). 1987. *Our Common Future*. Oxford: Oxford University Press.

TWENTY-FIVE

Sustainability of the Serengeti-Mara Ecosystem for Wildlife and People

Robin S. Reid, Kathleen A. Galvin, Eli J. Knapp, Joseph O. Ogutu, and Dickson S. Kaelo

KEY ISSUES AFFECTING THE SUSTAINABILITY OF THE SERENGETI-MARA ECOSYSTEM

The word "sustainability" evokes multiple meanings to different people and this is the essential complexity of attaining sustainability: we need to address multiple goals and understand the trade-offs among them. In this chapter, we use sustainability to mean the long-term maintenance (and hopefully improvement) of linked ecological, social, and economic systems. In the context of the Serengeti-Mara ecosystem (fig. 25.1), conservationists from Tanzania, Kenya, and elsewhere have the goal to maintain the grand wildlife populations, including the wildebeest migration and the savanna ecosystem that supports those populations. To this end, the governments of Kenya and Tanzania created a set of core protected areas including Kenya's Maasai Mara National Reserve and Tanzania's Serengeti National Park, Loliondo Game Controlled Area, Ngorongoro Conservation Area, as well as Grumeti, Ikorongo, and Maswa Game Reserves. It would simplify our discussion of sustainability if we focused only on the part of the ecosystem that resides within the boundaries of these core protected areas; the goals of those protected areas would then take precedence. However, the ecosystem, defined by the movements of the wildebeest, goes beyond this core to some of the farming land to the west and the grazing lands to the north, east, and south of the protected areas (Sinclair and Norton-Griffiths 1979; Thirgood

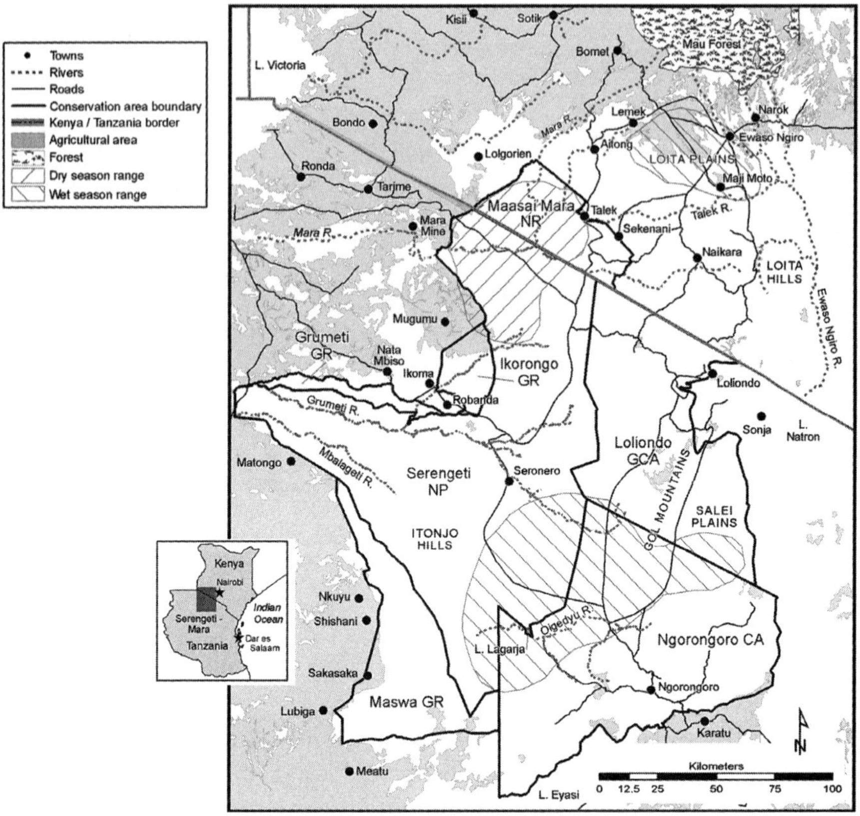

Fig. 25.1 Map of Serengeti-Mara ecosystem, straddling the Kenya-Tanzania border. Wet and dry season ranges show approximate areas where the wildebeest migration grazes in these two seasons.

et al. 2004). The ecosystem also extends far upstream to the headwaters of the Mara River watershed in Kenya, which starts at the top of the Mau Forest Complex, extends down through the pastoral ranches and the Maasai Mara National Reserve, then crosses the border into Serengeti National Park and out to the west through the Mara Wetlands and into Lake Victoria. In this entire ecosystem, we find a diversity of peoples who have a stake in this land from the Okiek hunters in the Mau Forest to the pastoral Maasai and Tatoga to a diversity of farming peoples like the Asi, Nata, Ishenya, Ikizu, Ngoreme, Ikoma, Sukuma, Kuria, Kipsigis, and Sikazi. Even beyond these immediately adjacent lands and peoples, we must account for the goals of people far away who have a "stake" in the system, like the Kenyan and Tanzanian national governments and their citizens, tourism businesses, and the foreign conservation community.

We will focus our attention on three main goals for the sustainability of the Serengeti-Mara ecosystem (fig. 25. 2). The first is that of the conservationists, which include local tourism businesses, local and foreign conservationists, national governments, and some of the local people. From their perspective, the Serengeti-Mara represents an all-too-rare standard of an ecosystem "undisturbed" by the human hand (Sinclair, Mduma, and Arcese 2002), a benchmark that is crucial to our efforts to understand how humans are modifying the earth. The system also represents a gold mine of profits for governments (Honey 2008; Norton-Griffiths et al. 2008; Thirgood et al. 2008), the tourism industry (Norton-Griffiths 1995; Norton-Griffiths and Southey 1995; Norton-Griffiths 2003), and some local elites (Thompson and Homewood 2002). A second perspective is that of many of the Maasai pastoralists living in Kenya and Tanzania. Their livelihoods are sometimes threatened by wildlife directly (Sitati et al. 2003; Kolowski and Holekamp 2006), but even more so by their loss of access to the land they have grazed for centuries as more and more of it has been converted to protected areas (Neumann 1995; Brockington 2002). A third perspective is that of the farmers and hunters west of the Serengeti and Mara, many of whose cultures predate the Maasai (Shetler 2007). From their perspective, access to wild meat may be important to the sustainability of their households, and thus they may include this access as part of their goals to sustain their livelihoods. The wildlife are not heard in this discussion, but presumably are represented by the conservationist views.

The Serengeti-Mara is also unusual for its ancient nature of human use and the possibility that people had a hand in creating the ecosystem. If people created the Serengeti-Mara ecosystem in the first place, it could be suggested that the best way to sustain the ecosystem is for people to continue to use and maintain it. Our best evidence suggests that herders, hunters, and farmers have affected the ecosystem by building water points (which elephants use; Shetler [2007]), by burning woodlands and creating and maintaining the Mara grasslands (Lamprey 1983), by creating nutrient hotspots in their old settlement sites, and potentially by protecting grazing wildlife from predators (Reid et al. 2003; Reid 2012). But there is no general evidence that people created the ecosystem; rather, by using it only moderately, they intentionally or unintentionally sustained it over millennia. In particular, the Maasai habit of only hunting wildlife in hard times gives support to the claim that they are the guardians of wildlife (Parkipuny 1991). This suggests that local peoples' goals for sustainability are important and should be considered alongside those of conservationists (and the wildlife).

Despite this ancient and likely "co-existence" between people and wildlife in this ecosystem, recent changes in human populations and land use bring in new forces that are incompatible with all three goals for the ecosystem (abundant wildlife populations, herder and hunter access) in some places. In the part of this ecosystem in Kenya, hereafter called the Mara (which includes the Maasai Mara National Reserve and the surrounding pastoral lands), expansion of wheat farming, poaching, and pastoral human population and settlement growth may be extinguishing wildlife outside the Mara Reserve (Homewood et al. 2001; Ottichilo, de Leeuw, and Prins 2001; Serneels and Lambin 2001; Ogutu et al. 2009). These effects also reach inside the reserve, where most resident wildlife species declined by more than half from 1977 to 2003 (Ottichilo et al. 2000; Ogutu et al. 2009). Strong efforts have been made in the western part of the Mara Reserve (also called the Mara Conservancy) to halt poaching but, so far, this does not seem to be stopping the overall decline in wildlife in this large ecosystem (Ogutu et al. 2011).

Maasai access to pastures is also changing. After agreeing to move from Tanzania's Serengeti in 1959 into the Ngorongoro Conservation Area (NCA), government policy increasingly restricted the ways Maasai could use the land. Human populations also grew. Unfortunately for the Maasai, the benefits they receive in exchange for leaving the Serengeti are not an equal trade. For example, Ngorongoro Maasai are poorer in terms of their own nutrition and the number of livestock they own and acres they cultivate than those living to the north in the Loliondo Game Controlled Area, where conservation policy is less stringent (Galvin and Magennis 1999; Galvin et al. 2002). And, if current trends continue, these restrictions will likely become tighter and tighter as human populations grow. In Tanzania's Loliondo, some Maasai villages granted ecotourism operators exclusive access to their lands for walking and camping safaris in exchange for a significant share of tourism profits (Nelson et al. 2009). This is also a way for villages to defend their land rights (Nelson et al. 2009). The creation of several wildlife management areas (WMAs) on local village lands (on both Maasai village lands and those of other groups) around the Serengeti and NCA (Ikoma, Eramatare, Lake Natron, and Makao WMAs) is meant to devolve the management of wildlife to the local level (Polasky et al. 2008). In the Kenya's Mara, recent privatization and subdivision of land gave some Maasai families more secure access to grazing land through private land holdings, but this process also forced others to move from the area (Reid, pers. obs).

Finally, the western Serengeti peoples have lost most or all of their access to their ancestral lands inside Serengeti Park with little prospect of regaining it (Shetler [2007]; see also Sinclair et al., chapter 2 for more recent information). They were largely pushed out of the current park area by the Maasai in the last two centuries, but some still lived within the boundaries of the current Serengeti Park before it was formed.

This chapter is about how to reconcile the apparently conflicting sustainability goals of these different actors. Some of the goals are irreconcilable, like maintaining wildlife populations and cultivating the land intensively for crops. This has happened already on the western border and 20 km north of the Mara Reserve and in the western Serengeti, and thus farming increasingly replaced mixed livestock-wildlife grazing over the last half century. In farmlands, wildlife and people do not mix well and must be separated to avoid major conflicts. But where access to forage and water for both livestock and wildlife exists, there is good prospect that local peoples can benefit strongly from the profits from wildlife, and wildlife can benefit strongly from community protection. Here we focus on what we view as the "big six" sustainability issues for this ecosystem now and in the near future. Key to all of these issues, in our view, is how governance can be improved so that wildlife populations are healthy and local communities have access to land and share the benefits from wildlife. There are other issues affecting sustainability of the wildlife and people in this system like diseases, grass poaching, and other issues, but we focus here on those we think are currently most important, realizing that many of these are interrelated. The six are:

Changing human populations, land access, and use. Pastoralism forms an important wildlife-compatible land use around the core protected areas in this ecosystem. How can this grazing land use be sustained in the face of human population growth, growing per capita needs, privatization of land, and expansion of settlements and farming?

Construction of the Serengeti highway. Can grazing populations be sustained if the Tanzanian government constructs a highway bisecting the northern Serengeti?

Drying of the Mara River. What are the options to sustain the flow of the Mara River, which provides critical support for people and wildlife in both the Kenyan and Tanzanian part of this ecosystem?

Changing environmental governance. Environmental governance determines who has the responsibility and power to manage, use, and profit from land, water, and wildlife in the ecosystem through policies, institutions, rules, and practices. What is a sustainable mix of local and national governance over wildlife, and how can all forms of governance support both social and ecological goals?

Heavy hunting pressure. How can wildlife populations be sustained in the face of strong pressure from hunters? Can poverty be reduced that seems to be at the root of the heavy hunting and in a way that is compatible with conservation?

Climate change, globalization. How can the ecosystem be maintained in the face of climate change, globalization, and other changes in the future?

We will explore these questions by first describing prehistorical, historical, and current land and resource use in the Serengeti-Mara ecosystem. We will then look at the rapid and complex changes occurring in the ecosystem today with a view to the future. Next, we will describe the short- and long-term consequences, as far as we know, of these changes for species and ecosystems (or biodiversity) and human livelihoods across the ecosystem. The chapter finishes with a look into the future prospects for sustainability for both wildlife and people.

HISTORICAL AND CURRENT LAND AND RESOURCE USE IN THE SERENGETI-MARA ECOSYSTEM

The Serengeti-Mara ecosystem has arguably some of the most ancient evidence of human use of any ecosystem on the planet. It is clear that ancient hominins walked here at least 3.7 million years ago (Leakey and Hay 1979) and since then, have developed more and more proficient ways to use the land and its resources. For the vast majority of this time, the area was probably thinly populated and human ancestors (and subsequently, modern humans) probably had little influence on the ecosystem. Major leaps in human influence probably occurred as hominins developed more complex tools (Schick and Toth 1993), as they relied more and more on hunting and less on scavenging (e.g., Blumenschine and Peters 1998), and when they started using fire to burn large areas of vegetation (e.g., Brain and Sillen 1988). Up until about 5,000 years ago (Newman 1995, Marshall 1998), it appears that the region was used only by click-speaking hunter-gatherers. Southern Cushitic herders slowly migrated south into the region and absorbed the hunter-gatherers (Ambrose 1984), reaching the Serengeti-Mara Region about 2,000–3,000 years ago (Homewood and Rodgers 1991). A bit later, Bantu farmers migrated into the region from the west (Newman 1995). Since then, waves of herders (eastern Cushites, southern and western Nilotes, and then eastern Nilotes) and Bantu farmers overwhelmed the original hunter-gatherers and became the dominant users of the land (Newman 1995). Only 200–300 years ago (Homewood and Rodgers 1991), the Maasai moved into the Serengeti-Mara ecosystem from the north, replacing the earlier pastoral peoples (the Tatoga or Datog).

In the twentieth century, European colonial governments imported the notion of protecting wildlife in the absence of people (Neumann 1995). In the process, the Asi, Nata, Ishenya, Ikizu, Ngoreme, Ikoma, Sukuma, Kuria, Tatoga, Sikazi, and Maasai peoples of this region lost land and ways to make a living when the colonial and postcolonial governments established the Tanzanian parks and reserves, which was done in consultation with the Sukuma and Maasai but without consultation of the western Serengeti peoples (Shetler 2007). By 1959, the Serengeti National Park (SNP) was established under the control of the central government and both the western Serengeti peoples and the Maasai could no longer use this part of the landscape. Moringe ole Parkipuny, former Tanzania minister of parliament from Ngorongoro, sums up the creation of Serengeti National Park this way, "with the creation of the national park in 1959, the Maasai of western Serengeti and Loliondo lost vast grazing areas, salt lick grounds, and permanent sources of water which were critical to the viability of their pastoral economy. These rangelands, although partially infested with tsetse flies, provided an important livestock refuge in times of drought" (Parkipuny 1991, 8). To compensate the Maasai for their part in the loss of the Serengeti, Ngorongoro was set aside for joint use by Maasai and wildlife (Shetler 2007). The Maasai leadership agreed, once promised water development and veterinary care in Ngorongoro (Thompson 1997), although one of the 12 Maasai traditional leaders who signed the agreement says they were forced out of the Serengeti (Lissu 2000).

In Tanzania today, a diverse mixture of government ministries and offices govern the lands around SNP, which itself is governed by the Tanzania National Parks Association (TANAPA). Tanzania's Ministry of Natural Resources governs the Maswa, Grumeti, and Ikorongo Game Reserves and the Loliondo Game Controlled Area (GCA). The relatively new wildlife management areas (WMAs) are governed by the Ministry of Wildlife. The NCA is governed by the Ngorongoro Conservation Area Authority (NCAA). The conservation area is the only one of its kind in Tanzania, with all of the country's other protected areas designated as game reserves and national parks. Village lands are public lands held in trust by the office of the president (Shivji 1998). All these organizations and institutions associated with them affect conservation and human land use, and add complexity to the sustainability of this ecosystem.

In Kenya, the Maasai Mara National Reserve was set up as a different model of conservation. As Maasai lost access to the Serengeti, the colonial government in Kenya allowed district councils to adopt game reserves as "African District Council Game Reserves" in 1956 (Lamprey and Reid 2004).

After independence in 1963, the Narok District Council, run by Maasai, took over the management of the Mara Reserve and started collecting the profits from gate and camping fees and lodge concessions, which its successor, the Narok County Council, still does today (Lamprey and Reid 2004). Thus, this reserve is run by the local country council, much like more than half of Kenya's protected areas, and not by the central government hundreds of kilometers away like the Serengeti is (Parkipuny 1991).

Today, diverse peoples surround these two core protected areas, and how they use the land, wildlife, and water is crucial to the sustainability of the entire ecosystem. Directly west and 20 km north of the core protected areas, farmers cultivate the land for crops (gray shading, fig. 25.1) and some hunt wildlife. North, east, and south of the core protected areas, herders use the land in a way that is largely compatible with wildlife (white areas outside protected areas on fig. 25.1). Pastoral lands in Kenya, which used to be managed as large group ranches, are now being subdivided into individual parcels, each owned by individual families (Thompson et al. 2009). On pastoral land in both Kenya and Tanzania, new policies and local initiatives are now changing how wildlife are managed in these lands, who benefits from the profits from wildlife, and where Maasai can graze their livestock.

CAUSES OF CHANGE IN THE SERENGETI-MARA ECOSYSTEM

Some drivers (or causes) of change in this large ecosystem affect all parts of the system and originate either globally (climate change, globalization), nationally (policy), or within the general region (deforestation of the Mau Forest), and some are caused from within the ecosystem itself (fig. 25.2). To understand the complexity of change in this large ecosystem, it is important to distinguish those changes that involve farming peoples (western Serengeti in Tanzania, west and far north of the Mara Reserve in Kenya) from those that involve herders (Ngorongoro and Loliondo in Tanzania and northern and eastern borders next to the Mara Reserve in Kenya) and those that involve both groups (fig. 25.1).

Changing Human Populations, Land Access and Use

Human population change. With the exception of the protected areas, human populations are growing in all areas of this ecosystem. There are not only more people, but the basic needs for each person are also growing, as people access better human and livestock health care and need to pay for it, as more

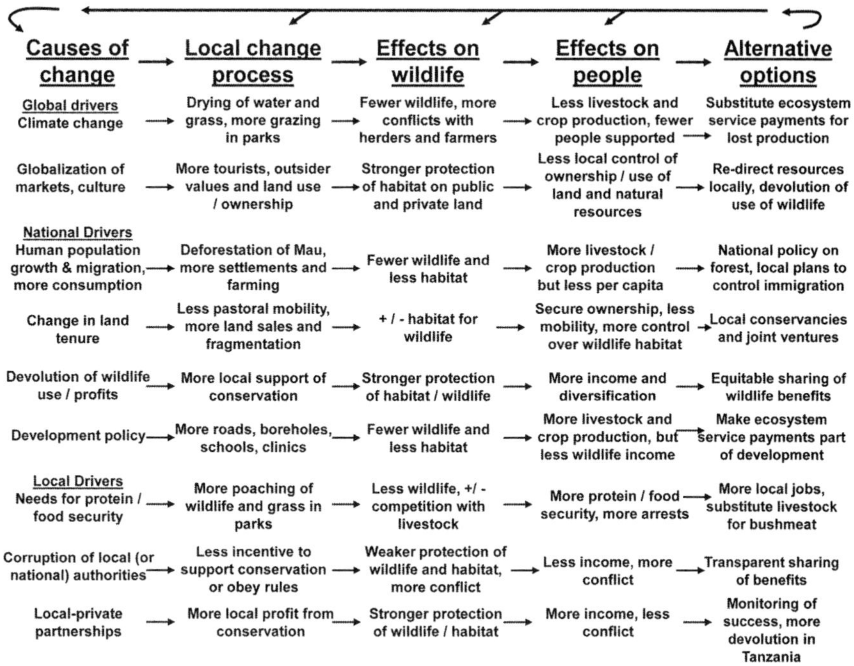

Causes of change
Local change process
Effects on wildlife
Effects on people
Alternative options
Global drivers
Climate change
Drying of water and grass, more grazing in parks
Fewer wildlife, more conflicts with herders and farmers
Less livestock and crop production, fewer people supported
Substitute ecosystem service payments for lost production
Globalization of markets, culture
More tourists, outsider values and land use / ownership
Stronger protection of habitat on public and private land
Less local control of ownership / use of land and natural resources
Re-direct resources locally, devolution of use of wildlife
National Drivers
Human population growth & migration, more consumption
Deforestation of Mau, more settlements and farming
Fewer wildlife and less habitat
More livestock / crop production but less per capita
National policy on forest, local plans to control immigration
Change in land tenure
Less pastoral mobility, more land sales and fragmentation
+ / - habitat for wildlife
Secure ownership, less mobility, more control over wildlife habitat
Local conservancies and joint ventures
Devolution of wildlife use / profits
More local support of conservation
Stronger protection of habitat / wildlife
More income and diversification
Equitable sharing of wildlife benefits
Development policy
More roads, boreholes, schools, clinics
Fewer wildlife and less habitat
More livestock and crop production, but less wildlife income
Make ecosystem service payments part of development
Local Drivers
Needs for protein / food security
More poaching of wildlife and grass in parks
Less wildlife, +/- competition with livestock
More protein / food security, more arrests
More local jobs, substitute livestock for bushmeat
Corruption of local (or national) authorities
Less incentive to support conservation or obey rules
Weaker protection of wildlife and habitat, more conflict
Less income, more conflict
Transparent sharing of benefits
Local-private partnerships
More local profit from conservation
Stronger protection of wildlife / habitat
More income, less conflict
Monitoring of success, more devolution in Tanzania

Fig. 25.2 Causes, processes, and effects of change in the Serengeti-Mara ecosystem and possible alternative options to improve sustainability.

children go to school and often need to pay school fees, and as families pay for improved communication. By 2002, for example, the Tanzanian census reported that two million people lived in districts west of the Serengeti (URT 2002) and about 0.4 million people lived within 12 km of the edge of Serengeti National Park (Ndibalema and Songorwa 2008). Most of these people live on the western side, although populations have also grown in the still thinly populated lands to the east in Loliondo and in Ngorongoro to the southeast. In Kenya's Mara Region, situated in the southern parts of Narok and Transmara districts, in-migration and average population growth has been lower (an estimated 3.4% per year in the 1990s; Coast [2001]) than in Narok district as a whole (6% per year in the 1980s, our calculations). Within the Mara Region, however, the density of pastoral settlements grew, on average, an astounding 9% per year within 10 km of the park between 1967 and 2003, which was more than double the national average (Norton-Griffiths et al. 2008). Growth of settlements far from the park was only 2.8%. Much of the growth near the Mara Reserve is of small villages near the park where settlers reap the benefits of tourism jobs or wildlife payments from the park (Lamprey and Reid 2004), an example of the "honey pot effect" of parks, where tourism opportunities attract local settlers to the edges of parks.

Conversion of savanna to agriculture. The incentives to convert open savanna pastures into subsistence crop farming are high in areas with high rainfall and lower elsewhere in this ecosystem (Norton-Griffiths et al. 2008), partly because production is higher where there is more rainfall, but also because the net returns (or "cash in the pocket") from crops grows faster as rainfall increases than net returns from livestock. Except where it is too dry to grow crops, livestock never return the profits that crops do. Clearly, the high population, adequate rainfall, and the history of farming people who live west of the Serengeti has expanded the land under cultivation in this area over time (Campbell and Hofer 1995). Maasai also will cultivate crops to supplement household food security in places where they are allowed to do so, like in Ngorongoro (Runyoro 2007). In the past, it is likely that the Maasai culture of livestock keeping has prevented large-scale conversion to farming, but many Maasai have cultivated small plots of land for many years (McCabe, Molle, and Tumaini 1997). More and more Maasai are taking up farming as their populations grow and because they receive only a very small part of their income from wildlife (Runyoro 2007; Homewood, Trench, and Kristjanson 2009b).

Wheat farming. In Kenya, members of some of the Maasai group ranches leased a portion of the northern Loita Plains to commercial wheat farmers in the 1980s causing major conversion of land to crops (Serneels, Said, and Lambin 2001). This leasing declined by 73% from 1998 to 2004 in areas far from the Mara Reserve in Lemek and Nkorinkori (Thompson et al. 2009). Wheat farmers said they are shifting away from farming in the Mara area because of subdivision of land. When the land was governed by group ranches, the farmers negotiated leases with one or two parties representing many families, but with subdivision of land into private family parcels, farmers had many new landowners to negotiate with and many of those landowners wanted short-term contracts (Thompson et al. 2009).

Private ownership. The shift from common to private landownership started in the early 1970s in Kenya's Mara, with wealthy Maasai elites receiving larger landholdings in this process (Thompson and Homewood 2002, Thompson et al. 2009). Privatization started first in ranches far from the Mara Reserve (with some exceptions), but now most of those group ranches close to the reserve are also subdivided, with the exception of Siana, which is still trust land (Thompson et al. 2009). The former group ranches were large contiguous blocks of land managed by all the group ranch members (often 200–400 or more) as a single unit. Subdivision and privatization means each of those group ranch members get their own demarcated piece of land within the former group ranch. This change drives two new processes: land sales (often out of Maasai ownership) and less agreement about how different landowners use their land (Reid et al. 2008), sometimes resulting in the introduction of more intensive land uses and fences (intensive livestock keeping, crop farming) or entirely new land uses (wealthy residential housing, churches, businesses).

Settlement and intensification. All the above changes caused pastoralists to settle and both farmers and pastoralists to intensify how they use the land (Homewood, Trench, and Kristjanson 2009a). This means that their imprint on the landscape becomes more concrete and permanent, compared to more fluid use in the past.

Disease transmission. As human and wildebeest populations grow, there is more contact between domestic livestock, dogs, and wildlife, and thus increased disease transmission. Control of rinderpest in the 1960s caused the wildebeest population to quadruple; as the larger population of wildebeest spread out to find more pasture and larger calving grounds, their calves increasingly mingled with Maasai cattle. Wildebeest calves transmit malignant catarryl fever (MCF) for the first three months after birth and this disease kills cattle that graze near the calves (Cleaveland et al. 2008). Domestic dogs transmit rabies to wild dogs and canine distemper to lions, spotted hyenas, and bat-eared foxes, which killed a huge portion of the Serengeti lion population in 1994 (Packer et al. 1999). Canine distemper originated with the high human and dog populations in the western Serengeti and swept north to Kenya, east to Loliondo, and south to Ngorongoro (Cleaveland et al. 2008).

Growing Road Infrastructure to Serve Growing Populations

The districts west of the Serengeti are isolated from the rest of Tanzania, partly because of the poor road infrastructure serving the area. Improved roads would allow for more people and goods to flow into and out of this region, which would presumably promote economic development. For two decades, the Tanzanian government has considered building a better road to serve the area (Dobson et al. 2010). The favored location of the road in 2010 was across the northern part of Serengeti National Park, which cut across the northern path of the wildebeest migration before the migration reaches the Kenyan border. The road has now been put on hold while the alternative southern road is considered (Sinclair et al., chapter 26).

Drying of the Mara River

The northern part of the Serengeti-Mara ecosystem is part of the Mara River watershed, which starts at the top of the Mau Forest Complex in Kenya, flows 50 miles downstream passing through the pastoral ranches to the Mara Reserve, then crosses into Tanzania's Serengeti and then out to the west through the Mara Wetlands and into Lake Victoria. This watershed is important to people and wildlife alike: its banks support loggers, hunters, and farmers in the Mau Forest, farmers below the forest, the Maasai and their livestock, millions of migrating wildlife in the Serengeti, and farmers downstream northwest of the Serengeti (Reid 2012). People cleared about 34% of the forest between 1973 and 2000 (Metzger et al., chapter 3) with about a 1% loss every year since then (Akotsi, Gachanga, and Ndirangu 2006). They cut trees for charcoal, farms, tea estates, schools, churches, and roads. Beginning around 2001 with the introduction of irrigation pumps to produce high-value horticultural crops, Kenyan farmers were able to withdraw larger amounts of water (Thompson et al. 2009). As a consequence of deforestation and water withdrawals, the Mara River now floods more quickly in the wet season than in the past, and then dries out faster in the dry season (Gereta et al. 2002). As a result, between the 1970s and 2005, the dry season flow of the Mara River has dropped by 68% (Gereta, Mwangomo, and Wolanski 2009).

Changing Environmental Governance (Sharing of Wildlife Profits and Power)

Environmental governance determines who has the responsibility and power to manage, use, and profit from land, water, and wildlife in the ecosystem through policies, institutions, rules, and practices. As described in the previous section, the core protected areas are structured differently, with local, county-level control of conservation in Kenya and national-level control in Tanzania. In the 1980s, with the advent of community-based conservation, authorities controlling the two core protected areas, Tanzania's Serengeti and Kenya's Mara, started to share about 10–20% of their annual budgets with surrounding communities (Lamprey and Reid 2004; Thirgood et al. 2008).

Privatization of land has recently created a window of opportunity to create new land-use institutions in the form of wildlife conservancies in the Mara (Reid 2012). New Maasai landowners are creating agreements with private tourism companies to gain a larger share in the profits from tourism in return to access to their land. In most cases, landowners agree to move off their land parcels to create contiguous areas mainly for wildlife use and tourism (and some livestock grazing in the dry season). It is unknown at this point if these new institutions will persist into the future. They can only be maintained if the benefits outweigh the costs, there is strong and honest leadership, and if government regulations are harmonious with local regulations and conditions (Ostrom 2009).

Outside the Serengeti in Tanzania, government and local villages are evolving new institutions and organizations to share management of and profits from wildlife. Wildlife management areas (WMAs), developed in 1998 and legally established in 2002, are supposed to allow rural communities and private land holders to manage wildlife on their lands and benefit from it. In reality, WMAs continue to embody a high degree of central government control (MNRT 1998; Nelson et al. 2009). Maasai and western Serengeti peoples are often suspicious of the motives of the government in these ventures, partly because of their long history of little consultation and land loss to central government (Shetler 2007). But some villages have taken some control of the wildlife profits from their lands by making agreements with private tourism companies, not unlike their brethren over the border in Kenya. In Loliondo Game Controlled Area, directly east of the Serengeti, Maasai and private tourism companies have developed a community-based wildlife management (CWM) plan that returns significant profits from tourism to the villages involved (Nelson et al. 2009). This partnership, however, has been illegal since the WMA statue was issued in 2002, so its existence is somewhat tenuous (Nelson et al. 2009).

Heavy Hunting Pressure

There is a huge offtake of wildlife by local hunters living in farming communities of the western Serengeti and west of the Mara (in Transmara district) in Kenya. Some hunt for subsistence while others for the market, and some hunt legally while many hunt illegally. In Tanzania, many of the peoples west of the Serengeti hunted for millennia in what is now Serengeti National Park before the colonial government began restricting hunting by natives in 1921 (Shetler 2007). Today, some legal and illegal hunting occurs outside the Serengeti as the wildebeest migration spills out of the park into the villages and on farmland each year (Thirgood et al. 2004; Nyahongo et al. 2009). However, illegal hunting inside the protected areas (Serengeti and reserves) in Tanzania is widespread: more than 50,000 people (Loibooki et al. 2002) may be involved in snaring and transporting up to 160,000 animals of all species (Hofer et al. 1996), including 40,000 wildebeest (Mduma, Sinclair, and Hilborn 1999) each year. Western Serengeti peoples hunt illegally in the park for food or cash income (Loibooki et al. 2002), and when crops fail during drought (Barrett and Arcese 1998) families that hunt receive about a third to a half of their income from bushmeat (Loibooki et al. 2002; Knapp 2009). Knapp (2009) found that 93% of the 104 illegal hunters that he interviewed sell their bushmeat, compared to just 7% who directly consume it. About 75% of the meat sold stays local in the hunter's own village or another nearby, but the other quarter of the bushmeat goes out of the district to distant consumers (Knapp 2009).

Illegal hunting is widespread in Kenya also, where hunting is illegal both inside and outside protected areas. The staff of the Mara Conservancy, who manage the western third of Kenya's Maasai Mara National Reserve, arrested 278 poachers and recovered 1,201 snares and 1,200 kg of meat from 9 wildlife species over 35 months from 2001–2004 (Ogutu et al. 2009). By late 2011, the Mara Conservancy arrested a total of 1,511 poachers and has collected more than 14,000 snares over a decade (Mara Conservancy 2011).

We do not yet know how legal hunting by locals, safari hunters, and others affects wildlife in the Serengeti. Maasai informants have complained that hunting and live animal capture by hunters from the United Arab Emirates associated with the Ortello Business Corporation on Loliondo village lands impeded Maasai livestock movements and caused excessive loss of wildlife and tree cutting (Ojalammi 2006).

Climate Change and Globalization

Climate change. This process affects the entire ecosystem. Over the long term, CO_2 will rise, wet season rainfall may fall, and the variability in rainfall will likely increase (Ritchie 2008). In addition, in Kenya's Mara, mean daily temperature has risen, especially the minimum daily temperature, over the course of the twentieth century (Ogutu et al. 2007) and this may continue into the future. Diminishing rainfall coupled with the rising temperatures through the 1990s and the early years of the twenty-first century appeared responsible for progressive habitat desiccation and reduced vegetation production in the Mara (Ogutu et al. 2007). However, just south in Tanzania in the central Serengeti woodlands, rainfall did not change between 1938 and 2002 (Sinclair et al. 2007, Ritchie 2008).

Increased CO_2 and decreased rainfall will likely have opposing effects on the amount and quality of vegetation for wildlife and livestock in the ecosystem. Increasing CO_2 concentrations may increase vegetation productivity but lower plant nutrient contents for herbivores (Ritchie 2008). However, lower rainfall may reduce production but increase nutrient contents (Ritchie 2008). It is thus difficult to predict how vegetation will change with climate change.

Globalization. The ecosystem is becoming more connected to and influenced by peoples and economies elsewhere around the globe, bringing western culture and other sources of capital. Generally, globalization brings more tourists to the rural ecosystems, attracts nonlocals to buy land and develop commercial enterprises, and brings in more migrant labor (Woods 2007). To cater for people from around the world there is likely more pressure to build more tourist amenities, which disrupts the ecosystem that tourists come to see. Globalization is also bringing faster communication to local communities with better access to the Internet and mobile phones. Some Maasai families now have better access to opportunities to profit from livestock and wildlife, but much of these profits go to the rich (Thompson and Homewood 2002), while women, the poor, sick, and young are likely being largely left out of the profits (Woods 2007). In Botswana, globalization is increasing the inequalities of livestock holdings and income between rich and poor livestock keepers and gives incentives to livestock keepers to overstock rangelands to meet international market demand (Darkoh and Mbaiwa 2002). Since the beginning of the last century, outside conservation values have replaced more local values about sustainable use of savannas and wildlife (Woods 2007). As elsewhere, with the expansion of the influence of global nongovernmental organizations (NGOs) and international agreements, communities in this ecosystem may feel that political authority is "scaled up" out of their reach (Woods 2007).

IMPLICATIONS FOR BIODIVERSITY IN THE SERENGETI-MARA ECOSYSTEM

Human Population Growth and Increasing Human Needs

Growing human populations can remove and/or fragment wildlife habitat and increase human wildlife conflict by expanding the number of pastoral settlements. For example, between 1989 and 2003, the area north of the Mara Reserve lost about 25% of its resident wildlife (like topi, impala, warthog), and this coincided with the growth in the number of settlements (Ogutu et al. 2009), although it is not clear if the settlements themselves caused the wildlife loss (poaching is another likely cause). More recent analyses suggest that this trend is only continuing; over the entire period from 1977 to 2009, the average decline in numbers of resident wildlife in the pastoral ranches was 82% and 74% in the reserve (Ogutu et al. 2011).

But there is some contradictory evidence that pastoral settlements and land use can attract certain species of wildlife in the Mara of Kenya (Reid et al. 2003; Reid 2012) and across the border in Loliondo (Maddox 2003). In the case of the Mara, most species of grazers and browsers prefer to forage in the pastoral lands north of the park in the wet season, where fast-growing grass is kept short and nutritious by nearby herds of livestock. Pastoralists also chase predatory wildlife away from their settlements at night, which may provide some protection for grazing wildlife. However this only applies to some grazers, other species of wildlife like rhino, elephant, and lion are more abundant where there are no people inside the Mara Reserve.

East of Tanzania's Serengeti in Loliondo, herders may also "protect" mesocarnivores like cheetah and wild dog. Here, wild dog and cheetah apparently survive better on pastoral lands than in the nearby park because they have less competition from lions; in Loliondo pastoralists actively chase lions away from their settlements (Maddox 2003; Durant et al. 2007), as they do in Kenya's Mara (Ogutu and Dublin 1998). This "safe haven" is under question, however, with an increase in poisoning (presumably by pastoralists, but this is not entirely clear) of lions in the Mara and the death of 25 wild dogs in Loliondo in 2007 (Nkwame 2009).

Effects of Settlements, Expansion of Farming, and Privatization of Land

Pastoral areas. In pastoral areas, how the moderate expansion of crop farming or spread of pastoral settlements affect wildlife is an open question. Pastoral landscapes are mostly open grazing land with sparse settlements dotting the savanna, creating a landscape with few and soft (= porous) boundaries (Reid 2012). In Kenya's Mara, the maximum number of wildlife occurs where there are a moderate number of pastoralists, about seven people per km^2 or one settlement every 4 km^2. But above this there is a tipping point, where more and more people and settlements mean fewer and fewer wildlife (Reid et al. 2003; Reid 2012).

Settled farming. Above a certain level, expansion of settled farming is usually incompatible with wildlife. Farms create "hard boundaries" on the landscape that slow or stop migration of wildlife like wildebeest and elephants and create greater conflicts between people and wildlife (Reid 2012). In Tanzania's Ngorongoro, for example, cultivation covered 0.5% (Boone et al. 2006) of the land surface in 2000 and only marginally more a few years later (Runyoro 2007). Many of the smaller farm plots are cultivated by Maasai. Boone et al.'s computer modeling suggests that cultivation at this level has little impact on wildlife, but brings important improvements in household food security (Boone et al. 2006). However, they calculated what would happen if cultivation increased by five times to 2.5% coverage in Ngorongoro in the future, and the results are rather different: there is a loss of 33% of the resident zebra, 10% of the elephants, 8% of grazing antelopes, and 3% of buffalo (but a 15% increase in browsing antelopes). Expansion of farming (and settlement) can also greatly reduce the extent of many wildlife populations. For example, in the Mara, maps of the extent of warthog populations, based on aerial survey data, contracted by 80% from 1977 to 2009 (Ogutu, unpublished data). These changes suggest that cultivation can have outsized impacts on wildlife, even when the area of land cultivated is still rather small.

West of Kenya's Mara Reserve in the Transmara area, wildebeest and elephants once used the areas on top of the Siria Escarpment, but these movements are less common with the expansion of farms. In the Serengeti, bird populations are significantly lower in the farming areas of the western Serengeti than in the adjacent Serengeti National Park (Sinclair, Mduma, and Arcese 2002).

Wheat farming. Expansion of commercial wheat farming 40 km north of Kenya's Mara Reserve devastated the ecosystem's second, and smaller, Loita wildebeest migration. This migration used to prefer the nutritious grass of the Loita plains in the wet season for calving. In the 1980s, Maasai leaders started to lease this land to commercial wheat farmers. The 130,000-strong Loita migration lost 100,000 animals over the next 15 years, partly because the fences blocked routes to calving grounds (Serneels and Lambin 2001).

Effects of a Serengeti Highway

The planned road across the northern Serengeti would cut across the path of the Tanzanian wildebeest migration of about 1.4 million animals. In other parts of the world fences, roads, and other sources of habitat fragmentation have extinguished a full quarter of the remaining 24 land animal migrations (Harris et al. 2009). Computer models predict a loss of 37% of the wildebeest population if the road provides an impenetrable barrier to animal movement (Holdo et al. 2011). While it is unlikely that the road would be impenetrable at first, if vehicle collisions with migrating wildebeest cause human deaths along the highway, it is likely that there will be a push to fence the highway in the future (Dobson et al. 2010), which would create a strong barrier. Also, the highway would allow illegal hunters better access to migrating wildlife and an easy way to move carcasses quickly out of the area. The road would also cross areas where conservationists are attempting to restore both wild dog and rhinoceros populations (Dobson et al. 2010).

Drying of the Mara River

Hydrological models suggest that if the worst droughts of the last 50 years recurred (1948 or 1972), the Mara River would dry out for one to two months and the wildebeest migration would collapse (Gereta et al. 2002; Gereta, Mwangomo, and Wolanski 2009). Perhaps 30% of the population would die in the first two weeks, with a similar proportion dying in succeeding weeks. If 80% of the wildebeest population were lost, it might take decades to recover (Gereta, Mwangomo, and Wolanski 2009). Drying of the river would affect not only wildebeest but a whole suite of water-dependent grazers and large mammals living in the river, like hippopotamuses and crocodiles. The Mara River is also home to many rare fish species, like *Oreochromis* (a cichlid related to tilapia) and *Esculenta* (B. Mnaya et al., unpublished manuscript), which have disappeared from Lake Victoria. The river is also an important breeding ground for fish in Lake Victoria.

Evolution of New Governance Structures: Is it Making a Difference for Wildlife?

It is most certainly true that the very large Serengeti National Park has been generally good for wildlife (Hilborn et al. 2006), even if it has not been entirely good for local people. Kenya's Mara Reserve has not fared as well; resident wildlife populations have declined dramatically in the part of the reserve east of the Mara River. Recent antipoaching operations in the Mara Conservancy (Ogutu et al. 2009), the part of the reserve west of the Mara River, appear to be promoting a resurgence of wildlife populations. It is not clear yet if the pastoral conservancies north of the Kenya's Mara, the community-based wildlife management in Tanzania's Loliondo, or the wildlife management areas (WMAs) are promoting healthy wildlife populations, partly because these efforts are relatively new. Up until 2009, there has been no evidence in the aerial wildlife survey data that the conservancies are increasing wildlife populations, but it may be too soon to tell (J. Ogutu, unpublished data). These conservancies find that fewer settlements and cattle are good for carnivores, but, if grass is not actively managed (mowed or grazed short), many grazing wildlife follow the people to the safety of shorter grass grazed by livestock (D. Kaelo, pers. obs.). This supports the idea that pastoral settlements, when there are not too many of them, can attract grazing wildlife, but repel predatory wildlife (Reid et al. 2003, Reid 2012). This is a classic demonstration of the complexity of the interactions between people and wildlife in these savannas; the answer is in the middle ground, neither wholly positive nor wholly negative for both parties.

Heavy Hunting Pressure

Population growth west of the Serengeti and Mara in both countries likely means more illegal hunters closer to the park and heavy poaching of both resident and migratory wildlife (e.g., Campbell and Hofer 1995). In the Serengeti, Simon Mduma and colleagues (Mduma, Sinclair, and Hilborn 1999) suggest that a hunting offtake of 40,000 animals is sustainable, but if doubled to 80,000, the wildebeest population will decline to low levels after 25 years. In 2003, Tanzania's Grumeti and Ikorongo Game Reserves and the Ikoma Open Area supported 70% fewer impala, and they ran farther when approached on foot than in the nearby park (Setsaas et al. 2007), presumably from the effects of illegal hunting. However, more recent law enforcement in these game reserves may have changed this situation. Over the border in Kenya, unchecked illegal hunting contributed to the decline in wildlife populations in the Maasai Mara National Reserve between 1989 and 2003 (Ogutu et al. 2009). After 2001, strong efforts to arrest illegal hunters and remove their traps has likely led to increases in wildlife populations (B. Heath, pers. comm.), but this has not been clearly documented yet.

Climate Change and Globalization

Climate change is affecting wildlife and pastoral livestock populations in this ecosystem through its influence on both temperature and rainfall. As temperature rises, evaporation increases, and thus more rainfall is required to maintain the same amount of food for grazing wildlife. If rainfall also declines, this ecosystem may support much fewer wildlife and livestock. Rainfall also affects the retention of surface water for water-dependent wildlife and livestock. Not only is rainfall and temperature changing, but the timing, location, and size of each rainfall event may become more unpredictable. This may increase the frequency of extreme rainfall events like floods and droughts.

Droughts affect wildlife directly through their forage and indirectly by magnifying the effects of changes in land use. For example, extreme droughts not only kill livestock because of lack of forage, but can cause severe wildlife mortality through starvation (Hillman and Hillman 1977; Mduma, Sinclair, and Hilborn 1999). In the Mara, the Cape buffalo population crashed after the 1993 drought (Sinclair et al. 2007) and is only now beginning to recover (J. Ogutu, unpublished data). Some species, like topi and warthog, respond to drought by changing and synchronizing the timing of the birth of young in the following year (Ogutu et al. 2010). However, large herbivores like elephant (*Loxodonta africana*) and hippo (*Hippopotamus amphibius*), have shown little response to annual rainfall variation in the Serengeti-Mara ecosystem (Ottichilo et al. 2000; Kanga et al. 2011). We also expect that fragmentation and loss of savannas through increased numbers of settlements, expansion of cropped farming and fencing, will cause more mortality during major droughts, more competition with livestock for forage, and more contact between wildlife and people.

It is not clear how globalization will affect wildlife in this ecosystem. If consumers around the globe demand livestock products from the pastoralists living around the parks (Darkoh and Mbaiwa 2002), they could have either negative or positive effects on wildlife. On one hand, global demand may give herders the incentive to bring more livestock into the system. If this happens, more separation of livestock from wildlife to prevent disease transmission will be needed and may require building veterinary cordon fences, as was done in southern Africa (Darkoh and Mbaiwa 2002). This may remove large parts of the pastoral lands from use by wildlife; if all the pastoral lands in the Kenya's Mara were inaccessible to the migration, perhaps a third of the wildebeest migration would be lost (Norton-Griffiths 1995). There may be potential for increased profits from "conservation beef" where high-end consumers pay more for beef produced in land conserved for wildlife (like "Serengeti Beef"; Reid 2012). If Maasai brand their livestock as conservation beef (or lamb, goat meat), the higher prices this could bring may increase local incomes and encourage families to hold fewer, higher-quality livestock, thus reducing competition for forage with wildlife.

Another issue of globalization is the political power of the global conservation community. For more than a century, the political influence of the global conservation community has strongly affected this ecosystem through pressure to remove local residents to establish parks to protect wildlife (Neumann 1995). Conservation interests are stronger in Tanzania than Kenya, partly because of Kenya's devolution of power over conservation management to the county level in the Mara (Parkipuny 1997).

THE BIG SIX: HOW CAN WILDLIFE AND HUMAN LIVELIHOODS BE SUSTAINED IN THE SERENGETI-MARA ECOSYSTEM IN THE FUTURE?

Sustainability Goals and the Crucial Role of Protected Areas

To sustain wildlife populations, local livelihoods and ecosystem structure and function, it is important to achieve the following outcomes:

* Less death and injury of wildlife caused by people (less illegal hunting)
* Less death and injury of people and livestock caused by wildlife (less predation, less conflict with large dangerous animals)
* Maintenance and restoration of land use compatible with wildlife around core protected areas (sustainable pastoralism)
* Secure access of local people and wildlife to land and water (tenure)
* More benefits flowing from wildlife to local communities (sharing, devolution of governance)
* Active experimentation by all stakeholders to develop sustainable ways to support livestock and wildlife at the same time
* Strong local institutions, supported by national institutions, that share benefits equitably (more transparency and accountability, stronger local institutions)

Before addressing critical sustainability issues, it is important to recognize the crucial role of the protected areas for sustaining wildlife populations in this ecosystem (Hilborn et al. 2006). The Serengeti National Park is built around a traditional protectionist model with control by the central government. This model been successful in protecting wildlife through enforcement (Hilborn et al. 2006) and, as long as resources and political will for that enforcement exists, we expect this park will continue to protect wildlife.

Issue 1: Increasing the Incentives for Wildlife-Compatible Pastoral Land Use

Wherever wildlife conservation is a goal within this ecosystem, crop farming has to be kept to minimal levels. Currently, however, crop farming is more profitable than pastoralism and/or wildlife conservation in the northern part of this ecosystem (Norton-Griffiths et al. 2008). How can profits from wildlife and livestock be improved to provide economic incentives for wildlife-compatible land uses? First, there may be more scope for sharing the profits from wildlife with local communities; for example, Serengeti National Park shares less than 10% of their annual budget with local communities. Private sector companies may also be able to share more profits, but this has not been studied in any transparent way. Another possibility is to build up several layers of profits from tourism and other ecosystem services, like carbon, water, or biodiversity (Daily et al. 2009), so that the sum creates an attractive incentive. Here, government regulations and private citizens can create a demand for wildlife-compatible land management by paying land users to improve their management or avoid adopting crop farming or other incompatible activities. These payments for ecosystem services are "direct, contractual and conditional payments to local landholders and users in return for adopting practices that secure ecosystem conservation and restoration" (Wunder 2005, 1).

Another layer of income on top of wildlife profits are livestock development activities and conservation employment to make wildlife-compatible pastoralism more profitable. For example, efforts to reduce livestock diseases like trypanosomosis in cattle and Newcastle's disease in chickens will improve livestock enterprises (Cleaveland et al. 2008).

Issue 2: Avoiding Fragmentation of the Serengeti by a Highway

There is the possibility of a "win-win" for this issue: construction of a road that keeps the Serengeti intact. This would be accomplished by building the road south of the Serengeti, which would require it to be 50 km longer (and thus more expensive). In the long run, this option may be more economical, given the importance of tourism and the wildebeest migration to the Tanzanian economy, and it appears to be the one favored by the Tanzanian government (Dobson et al. 2010; Sinclair et al., chapter 26).

Issue 3: Maintaining and Restoring the Mara River Flow

The competing needs for the Mara River water are complex and difficult to manage. Ultimately, all downstream water users (herders, farmers, wildlife) are entirely dependent on what happens upstream in relation to deforestation of the Mau, potential diversion of water for hydroelectric generation, and diversion of water for irrigation and other uses. Deforestation is a national issue and its resolution depends on a highly contentious and long-delayed process with a wide range of stakeholders. Some of this process may lead to a trade-off between the land rights of indigenous forest dwellers in the Mau and the rights of downstream users to water. Some of the solution here is stronger and implemented government regulation. The

Kenyan government is currently making progress in addressing these issues in the Mau Forest.

Gereta (2004) suggests that a Mara River transboundary management plan needs to be drawn up, in which the governments of Kenya and Tanzania have equal voices in planning for the water needs of a wide range of powerful and less powerful stakeholders. In addition to regulation and coordinated planning, stakeholders in this watershed could establish a water services payment program where downstream users (tourism, irrigated farming) pay upstream Mau foresters to reforest catchments areas to restore more reliable flows (Reid 2012).

It is not realistic to suggest that reservoirs be built in the Serengeti to maintain enough water for the wildebeest population if the Mara River were to dry up in a drought. The extension of the park from the western end of the corridor to Speke Gulf in Lake Victoria could allow wildlife access to the water in Lake Victoria in the dry season or drought (Sinclair et al., chapter 26).

Issue 4: Improving Governance at the Local to the National Level and Beyond

Better environmental governance. For many of the sustainability issues (land and resource use, sharing of wildlife profits and power, Mara River flows), better environmental governance is a critical part of sustainability, and thus we think this issue may be most important. From our perspective, new institutions, rules, or policies must be: (1) adaptive and diverse to promote resiliency, (2) collaborative to promote whole system governance, (3) equitable to strengthen local involvement and control, and (4) transparent and accountable.

Adaptive and diverse institutions. These are desirable because the ecosystem is under a complex array of pressures such that one approach to conservation (such as only community-controlled conservancies or only central government-controlled parks) limits the options available to adapt to these changes to maintain sustainability (Reid 2012). Recent studies of institutions suggest that diverse, flexible, and decentralized organizations operating at different autonomous levels can be more successful (Mwangi and Ostrom 2009). This suggests that it is important to maintain a diverse mix of ways to manage land and wildlife in this ecosystem. Political systems, with policy input at various levels from different sources, may have the potential to be more responsive to change than are traditional institutions or bureaucracies (Pahl-Wostl 2009).

Collaboration. In an ecosystem as large and connected as the Serengeti-Mara, collaboration is also essential to knit together the policies, actions, and practices of the diverse organizations and peoples who manage land and water for conservation and agriculture (Reid 2012). Thus, as suggested above for water, a broader, whole-ecosystem, coordinated, internationally recognized and implemented management plan is needed, with local landowners and villages as equal partners with government in both Kenya and Tanzania. Beyond water, the plan needs to address all resources and actions that need coordinated management in this ecosystem (like fire, wildlife, disease, trade). Some progress has already been made to do this, but more needs to be made. Joint Kenyan-Tanzanian actions and monitoring of the consequences of those actions is often elusive, but badly needed. Just as elusive is inclusion of local (indigenous) knowledge in the development of indicators to measure the success of the process and outcomes of collaborative management.

Equitable responsibility for governance. Another desirable characteristic is equitable responsibility for governance and the sharing of its benefits. Here, the objective is to create and strengthen diverse institutions at multiple scales with sufficient flexibility and autonomy to make and change rules as needed to share responsibility for resource management. Mwangi and Ostrom (2009, 43) state that "this type of regime champions the principle of subsidiarity, the notion that a central authority should have an auxiliary function, performing only those tasks that cannot be performed effectively at a more immediate or local level." In the Serengeti-Mara ecosystem there are examples of an entire gamut of power sharing between local communities and government from very little in Tanzania's Serengeti National Park, to local county control of Kenya's Mara Reserve (but national control of wildlife), to new partnerships between private or village landowners and the private sector over management of government-owned wildlife in the Mara conservancies, and the Loliondo community-based wildlife management (CWM) plans between villages and the private sector. In this range of arrangements there is no classic "comanagement" where local communities and government manage natural resources, sharing responsibility and authority (Carlsson and Berkes 2005; Armitage et al. 2009). If local communities gain more power and responsibility over conservation management, there will also need to be a revolution in the capacity of local institutions to manage wildlife across the entire ecosystem in a collaborative manner and implement transparent and accountable revenue capture and distribution systems (Reid 2012).

Devolution of power and responsibility. While local devolution of power and responsibility clearly meet social "fairness" goals, we think it is unlikely that national parks, like the Serengeti, will start to comanage wildlife with local communities in the near term. But steps could be made toward more inclusion of local voices with real influence over the way the park is managed (Reid 2012). In Tanzania, there may be an opportunity to do this through village natural resource committees around the Serengeti. This would not only bring good will, but incorporation of local knowledge would undoubtedly improve park management and sustainability.

Accountable institutions. Effective decentralization requires the construction of accountable institutions at all levels of government and a secure domain of autonomous decision making at the local level (Ribot, Agrawal, and Larson 2006). Around the world, many central governments obstruct attempts to decentralize power over natural resources to the local level. In Tanzania, for example, decisions on who hunts where on village lands are made by central government with little participation of local villages, and thus these decisions have little downward accountability (Nelson and Agrawal 2008). A hunting concession in Loliondo was granted by the central government to the Ortello Business Company, which was created by a brigadier from the United Arab Emirates. The media accused Ortello of obtaining the right to hunt through high-level corruption (see citations in Nelson et al. [2009]). In Kenya, returns from conservation of the Mara Conservancy part of the Mara Reserve quintupled when a local and accountable system was put into place to manage the profits from conservation (Walpole and Leader-Williams 2001). Thus, sustainability of this system depends partly on more support and expansion of local and accountable institutions that share the profits from and management of wildlife (Reid 2012).

Strong local capacity. Efforts to further devolve the rights to manage and profit from wildlife will likely fail unless there is strong local capacity to responsibly manage wildlife and equitably share those profits. Thus, building the ability of local institutions and individuals to do this is completely essential to maintaining the bank account (the wildlife) and the interest (the profits from wildlife) well into the future (Reid 2012).

International accountability. There also needs to be better accountability in wildlife conservation at the international level. Powerful international NGOs, the private sector, and international agencies have an inordinate influence on how sustainability will play out in this system and who wins and who loses at the local level (Goldman 2003; Igoe 2003). As Homewood (2009, 362) says, it is time for these organizations to be much more aware of who they work with and why, "to make choices that stabilize and foster inclusive representation, even if such choices appear to run counter to their own ends in the short term."

International linkages. In addition, linkages among different institutions managed by government, communities, and the private sector are also particularly important. Institutions at broad levels are often slow to learn and respond to problems; most responses are reactive, not proactive. However, by linking groups at various levels, including the local, there is the potential for innovative approaches to problem solving, making the most of local knowledge linked with other perspectives from outside the ecosystem.

Regardless of type, linkages, and goals of various governance structures, the nature of conservation comes with a significant lag time before results are seen for locals, the wildlife, and the ecosystem (Mee, Dublin, and Eberhard 2008). In addition, until distrust among government, private society, and NGOs is overcome, it is unlikely that long-term sustainability is possible. We must keep in mind that all institutions serve needs; what we must ask continually in the Serengeti-Mara ecosystem is whose needs are being served and met and whose are not under each governance structure. We can no longer consider the ecosystem solely as a wildlife system but must ask whose needs are being met for wildlife AND people in this social-ecological system.

Issue 5: Heavy Hunting Pressure, Sustaining Wildlife, and Reducing Poverty

There is a real danger that heavy hunting pressure on wildlife in the western Serengeti, which spills over the border into Kenya (Ogutu et al. 2009), could damage wildlife populations. What could reduce that hunting? On the surface, more enforcement of antipoaching will get to the symptom (Hilborn et al. 2006), but will not provide a cure. To improve efficiency, sentencing of illegal hunters could occur locally with the proceeds from fines retained at the village level (Holmern, Muya, and Roskaft 2007). This could provide a faster and more cost-effective processing system and eliminate the highly ineffectual transportation and drawn-out prosecution that occurs at the district level.

Digging deeper, there is a need for more protein for local populations, so livestock development may help (Loibooki et al. 2002). Greater availability of less expensive protein, from livestock or wildlife, may reduce poaching. However, a greater availability of less expensive meat through wildlife cropping has not been a viable alternative in western Serengeti (Holmern et al. 2002).

But more deeply, it is likely that culture (Shetler 2007) and poverty are driving hunting, both of which are harder to address (E. Knapp et al., chapter 16). Poverty can be reduced by direct financial assistance, along with development and/or expansion of income-generating activities. It could also be done by redirecting more of the profits from hunting and tourism to local communities. The utility of direct financial assistance is limited because it creates short-term dependency and is subject to long-range changes in donor policies and strategies. More opportunities for income-generating activities and redirecting profits may be the only viable long-term strategies for reducing hunting pressure. Although potentially viable, increasing employment opportunities and redirecting profits should be done judiciously. For example, seasonal employment may have little effect on reducing hunting because it fails to account for enough of a hunter's time (Knapp 2007). Another potential pitfall is that an increase in employment opportunities or redirecting profits will likely draw more people into an area, thereby exacerbating the original dilemma (Estes et al., chapter 18). Despite these possible drawbacks, long-term employment and redirecting profits may be the only workable options for reducing hunting pressure. In western Serengeti, for example, over two-thirds (69%) of respondents reported that they would permanently cease hunting activities if they procured year-round employment (Knapp 2009; E. Knapp et al., chapter 16).

Park-related employment is one way to provide long-term employment to a small part of the population of illegal hunters. Local persons—and households—with this form of employment should have greater incentive to protect the system that provides their livelihood. A sample of households with park-related employment ($n = 50$) earned US $1,722, which was more than double the amount of households without this form of employment (Knapp 2009). Households with park-related employment had better-constructed houses (e.g., concrete foundation and walls, tin roof), more assets, and higher protein intake than households without (Knapp 2009). Such households were also able to financially assist other households, thereby decreasing poverty throughout the area.

In a recent survey of 722 households across three districts adjacent to western SNP, however, just 6% currently had park-related employment (Knapp 2009). There are several reasons for this, which appear to be linked. On the eastern side of SNP, for example, the primary obstacle is education and skills (Charnley 2005). Here, many people lack sufficient education required to compete for jobs that require higher skills (e.g., English proficiency, auto mechanics, accounting, etc.). As a result, park-related employers have targeted areas far from protected areas, like urban locales, where there are larger pools of educated potential employees. For example, only 28% of the people employed by the park were from a district that directly borders the Serengeti (Knapp 2009). Park employees came from 23 districts and represented 20 ethnic groups (Knapp 2009). The ethnic group with the greatest representation—the Chagga—were largely from the area around Mount Kilimanjaro, several hundred kilometers away from the Serengeti (Knapp 2009; E. Knapp et al., chapter 16). Park employees made more than six times as much (US $1,722) than households living near but outside the park (US $371; Knapp 2009). This results in continued poverty for people who live with wildlife and employment of people from outside the region.

Issue 6: Future Climate Change and Globalization

As described above, climate change will only create more stress on this ecosystem and the peoples it supports. If rainfall becomes more unpredictable, it will be critical to sustain opportunities for livestock and wildlife to move so that they can respond in a flexible and opportunistic way to changes in forage and water availability.

To sustain this linked system, Kenyans and Tanzanians must recognize and plan for the stresses that globalization is bringing. Global markets may increase the disparity in income between the rich and poor (Omotola 2008), both in relation to livestock (Darkoh and Mbaiwa 2002) and tourism. More tourists mean that tourism planning will be increasingly important. Profits will continue to be difficult to keep at the local level as people in businesses around the world profit from the riches this ecosystem provides. It will continue to become more difficult for pastoral communities to assert their local values concerning this ecosystem, if the influence of global NGOs and international agreements grows (Woods 2007).

Most important, building resilience and social capital of the linked human and ecological community is the foundation for future sustainability of both wildlife and people in the face of climate change, globalization, and the other changes described above. Wildlife will be more resilient to illegal hunting, changes in land use, climate change, and drying of the Mara River if they have sufficient high-quality habitat (land) and water to survive. Livestock and pastoral peoples need similar access to land, forage, and water. For people, better education, stronger leaders, more transparent and accountable institutions, and diverse sources of income will all contribute to resilience in the face of future challenges to sustainability. And slower (even negative) human population growth and slowing the spread of wildlife-incompatible land use will give all people and wildlife in the system more "breathing room" to access land and other resources to grow stronger and healthier human and natural communities.

ACKNOWLEDGMENTS

Any insights we have to these issues were strongly informed by working with Maasai families in the Serengeti and the Mara, the management of the Serengeti National Park and Mara National Reserve, and tourism operators in both areas, and others with a stake in this socioecosystem. We appreciate the thought-provoking work of our scientific colleagues who we cite here and the comments of two anonymous reviewers. Russ Kruska kindly created the map in fig. 25.1.

REFERENCES

Akotsi, E. F. N., M. Gachanja, and J. K. Ndirangu. 2006. Changes in forest cover in Kenya's five "water towers" 2003–2005. Department of Resource Surveys and Remote Sensing and Kenya Forest Working Group, Nairobi, Kenya.

Ambrose, S. H. 1984. The introduction of pastoral adaptations to the highlands of East Africa. In *From hunters to farmers: The causes and consequences of food production in Africa*, ed. J. D. Clark and S. A. Brandt, 212–39. Berkeley: University of California Press.

Armitage, D. R., R. Plummer, F. Berkes, R. I. Arthur, A. T. Charles, I. J. Davidson-Hunt, A. P. Diduck, N. C. Doubleday, D. S. Johnson, M. Marschke, et al. 2009. Adaptive co-management for social-ecological complexity. *Frontiers in Ecology and the Environment* 7:95–102.

Barrett, C. B., and P. Arcese. 1998. Wildlife harvest in integrated conservation and development projects: Linking harvest to household demand, agricultural production, and environmental shocks in the Serengeti. *Land Economics* 74:449–65.

Blumenschine, R. J., and C. R. Peters. 1998. Archaeological predictions for hominid land use in the paleo-Olduvai Basin, Tanzania, during lowermost Bed II times. *Journal of Human Evolution* 34:565–607.

Boone, R. B., K. A. Galvin, P. K. Thornton, D. M. Swift, and M. Coughenour. 2006. Cultivation and conservation in Ngorongoro Conservation Area, Tanzania. *Human Ecology* 34:809–28.

Brain, C. K., and A. Sillen. 1988. Evidence from the Swartkrans cave for the earliest use of fire. *Nature* 336:464–66.

Brockington, D. 2002. *Fortress conservation: The preservation of the Mkomazi Game Reserve, Tanzania*. Oxford: International African Institute.

Campbell, K., and H. Hofer. 1995. People and wildlife: Spatial dynamics and zones of interaction. In *Serengeti: Dynamics, management and conservation of an ecosystem*, ed. A. R. E. Sinclair and P. Arcese, 534–70. Chicago: University of Chicago Press.

Carlsson, L., and F. Berkes. 2005. Co-management: Concepts and methodological implications. *Journal of Environmental Management* 75:65–76.

Charnley, S. 2005. From nature tourism to ecotourism? The case of the Ngorongoro Conservation Area, Tanzania. *Human Organization* 64:75–88.

Cleaveland, S., C. Packer, K. Hampson, M. Kaare, R. Kock, M. Craft, T. Lembo, T. Mlengeya, and A. Dobson. 2008. The multiple roles of infectious diseases in the Serengeti ecosystem. In *Serengeti III: Human impacts on ecosystem dynamics*, ed. A. R. E. Sinclair,

C. Packer, S. A. R. Mduma, and J. M. Fryxell, 209–39. Chicago: University of Chicago Press.

Coast, E. 2001. *Demography of the Maasai*. PhD diss., Department of Anthropology, University College London.

Daily, G. C., S. Polasky, J. Goldstein, P. M. Kareiva, H. A. Mooney, L. Pejchar, T. H. Ricketts, J. Salzman, and R. Shallenberger. 2009. Ecosystem services in decision making: Time to deliver. *Frontiers in Ecology and the Environment* 7:21–28.

Darkoh, M. B. K., and J. E. Mbaiwa. 2002. Globalisation and the livestock industry in Botswana. *Singapore Journal of Tropical Geography* 23:149–66.

Dobson, A. P., M. Borner, A. R. E. Sinclair, P. J. Hudson, T. M. Anderson, G. Bigurube, T. B. B. Davenport, J. Deutsch, S. M. Durant, R. D. Estes, et al. 2010. Road will ruin Serengeti. *Nature* 467:272–73.

Durant, S. M., S. Bashir, T. Maddox, and M. K. Laurenson. 2007. Relating long-term studies to conservation practice: The case of the Serengeti Cheetah Project. *Conservation Biology* 21:602–11.

Galvin, K., J. Ellis, R. B. Boone, A. L. Magennis, N. M. Smith, S. J. Lynn, and P. K. Thornton. 2002. Compatibility of pastoralism and conservation? A test case using integrated assessment in the Ngorongoro Conservation Area, Tanzania. In *Conservation and mobile indigenous peoples: Displacement, forced settlement and sustainable development,* ed. D. Chatty and M. Colchester, 36–60. New York: Berghan Books.

Galvin, K. A., and A. L. Magennis. 1999. Compatibility of pastoralism and conservation? A test case comparing nutritional status among two Maasai populations in Tanzania. *American Journal of Human Biology* 11:111–12.

Gereta, E. J. 2004. Transboundary water issues threaten the Serengeti ecosystem. *Oryx* 38: 14–15.

Gereta, E., E. Mwangomo, and E. Wolanski. 2009. Ecohydrology as a tool for ensuring the survival of the threatened Serengeti ecosystem. *Ecohydrology and Hydrobiology* 9: 115–24.

Gereta, E., E. Wolanski, M. Borner, and S. Serneels. 2002. Use of an ecohydrology model to predict the impact on the Serengeti ecosystem of deforestation, irrigation and the proposed Amala Weir Water Diversion Project in Kenya. *Ecohydrology Hydrobiology* 2: 135–42.

Goldman, M. 2003. Partitioned nature, privileged knowledge: Community-based conservation in Tanzania. *Development and Change* 34:833–62.

Harris, G., S. Thirgood, G. C. Hopcraft, J. P. G. M. Cromsigt, and J. Berger. 2009. Global decline in aggregated migrations of large terrestrial mammals. *Endangered Species Research* 7:55–76.

Hilborn, R., P. Arcese, M. Borner, J. Hando, G. Hopcraft, M. Loibooki, S. Mduma, and A. R. E. Sinclair. 2006. Effective enforcement in a conservation area. *Science* 314:1266.

Hillman, J. C., and A. K. K. Hillman. 1977. Mortality of wildlife in Nairobi National Park during the drought of 1973–74. *East African Wildlife Journal* 15:1–18.

Hofer, H., K. L. I. Campbell, L. M. East, and S. A. Huish. 1996. The impact of game meat hunting on target and non-target species in the Serengeti. In *The exploitation of mammal populations*, ed. V. J. Taylor and N. Dunstone, 117–46. London: Chapman & Hall.

Holdo, R. M., J. M. Fryxell, A. R. E. Sinclair, A. Dobson, and R. D. Holt. 2011. Predicted impact of barriers to migration on the Serengeti wildebeest population. *PloS ONE* 6: e16370. doi:10.1371/journal.pone.0016370.

Holmern, T., J. Muya, and E. Roskaft. 2007. Local law enforcement and illegal bushmeat hunting outside the Serengeti National Park, Tanzania. *Environmental Conservation* 34:55–63.

Holmern, T., E. Roskaft, J. Mbaruka, S. Y. Mkama, and J. Muya. 2002. Uneconomical game cropping in a community-based conservation project outside the Serengeti National Park, Tanzania. *Oryx* 36:364–72.

Homewood, K. 2009. Policy and practice in Kenya rangelands: Impacts on livelihoods and wildlife. In *Staying Maasai? Livelihoods, conservation and development in East African rangelands*, ed. K. Homewood, P. Kristjanson, and P. C. Trench, 335–67. New York: Springer.

Homewood, K., E. F. Lambin, E. Coast, A. Kariuki, I. Kikula, J. Kivelia, M. Said, S. Serneels, and M. Thompson. 2001. Long-term changes in Serengeti-Mara wildebeest and land cover: Pastoralism, population, or policies? *Proceedings of the National Academy of Sciences U. S. A.* 98:12544–49.

Homewood, K. M., and W. A. Rodgers. 1991. *Maasailand ecology: Pastoralist development and wildlife conservation in Ngorongoro, Tanzania*. Cambridge: Cambridge University Press.

Homewood, K., P. C. Trench, and P. Kristjanson, eds. 2009a. *Staying Maasai? Livelihoods, conservation and development in East African rangelands*. London: Springer.

———. 2009b. Staying Maasai? Pastoral livelihoods, diversification and the role of wildlife in development. In *Staying Maasai? Livelihoods, conservation and development in East African rangelands*, ed. K. Homewood, P. Kristjanson, and P. C. Trench, 369–408. New York: Springer.

Honey, M. 2008. *Ecotourism and sustainable development*. Washington, DC: Island Press.

Igoe, J. 2003. Scaling up civil society: Donor money, NGOs and the pastoralist land rights movement in Tanzania. *Development and Change* 34:863–85.

Kanga, E., J. O. Ogutu, H. Olff, and P. Santema. 2011. Population trend and distribution of the vulnerable common hippopotamus *Hippopotamus amphibius* in the Mara Region of Kenya. *Oryx* 45:20–27.

Knapp, E. J. 2007. Who poaches? Household economies of illegal hunters in western Serengeti, Tanzania. *Human Dimensions of Wildlife* 12:195–96.

———. 2009. *Western Serengeti people shall not die: The relationship between Serengeti National Park and rural household economies in Tanzania*. PhD diss., Colorado State University.

Kolowski, J. M., and K. E. Holekamp. 2006. Spatial, temporal, and physical characteristics of livestock depredations by large carnivores along a Kenyan reserve border. *Biological Conservation* 128:529–41.

Lamprey, H. F. 1983. Pastoralism yesterday and today: The overgrazing problem. In *Tropical savannas. Ecosystems of the world, vol. 13*, ed. F. Bourlière, 643–66. Amsterdam: Elsevier.

Lamprey, R. H., and R. S. Reid. 2004. Expansion of human settlement in Kenya's Maasai Mara: What future for pastoralism and wildlife? *Journal of Biogeography* 31:997–1032.

Leakey, M. D., and R. L. Hay. 1979. Pliocene footprints in the Laetolil beds at Laetoli, northern Tanzania. *Nature* 278:317–23.

Lissu, T. 2000. Policy and legal issues on wildlife management in Tanzania's pastoral lands: The case study of the Ngorongoro Conservation Area. *Law, Social Justice & Global Development* http://elj.warwick.ac.uk/global/issue/2000-1/lissu.html.

Loibooki, M., H. Hofer, K. L. I. Campbell, and M. L. East. 2002. Bushmeat hunting by communities adjacent to the Serengeti National Park, Tanzania: The importance of livestock ownership and alternative sources of protein and income. *Environmental Conservation* 29:391–98.

Maddox, T. M. 2003. The ecology of cheetahs and other large carnivores in a pastoralist-dominated buffer zone. PhD diss., Dept. of Anthropology, University College London.

Mara Conservancy. 2011. Monthly report, the Mara triangle. http://maratriangle.org.

Marshall, F. 1998. Early food production in Africa. *Review of Archaeology* 19:47–58.

McCabe, J. T., N. Molle, and A. Tumaini. 1997. Food security and cultivation. In *Multiple land-use: The experience of the Ngorongoro Conservation Area, Tanzania*, ed. D. M. Thompson, 397–416. Gland, Switzerland: International Union for Conservation of Nature.

Mduma, S. A. R., A. R. E. Sinclair, and R. Hilborn. 1999. Food regulates the Serengeti wildebeest: A 40-year record. *Journal of Animal Ecology* 68:1101–22.

Mee, L., H. Dublin, and A. Eberhard. 2008. Evaluating global environmental facility: A goodwill gesture or a serious attempt to deliver global benefits? *Global Environmental Change* 18:800–10.

MNRT. 1998. *The wildlife policy of Tanzania*. Ministry of Natural Resources and Tourism. Dar es Salaam, Tanzania: Government Printer.

Mwangi, E., and E. Ostrom. 2009. Top-down solutions: Looking up from East Africa's rangelands. *Environment* 51:34–44.

Ndibalema, V. G., and A. N. Songorwa. 2008. Illegal meat hunting in Serengeti: Dynamics in consumption and preferences. *African Journal of Ecology* 46:311–19.

Nelson, F. and Agrawal, A. 2008. Patronage or participation? Community-based natural resource management reform in sub-Saharan Africa. *Development and Change* 39: 557–85.

Nelson, F., B. Gardner, J. Igoe, and A. Williams. 2009. Community-based conservation and Maasai livelihoods in Tanzania. In *Staying Maasai? Livelihoods, conservation and development in East African rangelands*, ed. K. Homewood, P. C. Trench, and P. Kristjanson, 299–333. London: Springer.

Neumann, R. 1995. Ways of seeing Africa: Colonial recasting of African society and landscape in Serengeti National Park. *Ecumene* 2:149–69.

Newman, J. L. 1995. *The peopling of Africa*. New Haven: Yale University Press.

Nkwame, M. 2009. Wild dogs to be taken to Serengeti. *Tanzania Daily News*. http://www.dailynews.co.tz/home/?n=5861&cat=home, accessed 24 Dec 2009.

Norton-Griffiths, M. 1995. Economic incentives to develop the rangelands of the Serengeti: Implications for wildlife conservation. In *Serengeti: Dynamics, management and conservation of an ecosystem*, ed. A. R. E. Sinclair and P. Arcese, 588–604. Chicago: University of Chicago Press.

———. 2003. The case of private sector investment in conservation: An African perspective. In *World parks congress*. Durban, South Africa: World Parks Congress.

Norton-Griffiths, M., M. Y. Said, S. Serneels, D. S. Kaelo, M. B. Coughenour, R. Lamprey,

D. M. Thompson, and R. S. Reid. 2008. Land use economics in the Mara area of the Serengeti ecosystem. In *Serengeti III: Human impacts on ecosystem dynamics*, ed. A. R. E. Sinclair, C. Packer, S. A. R. Mduma, and J. M. Fryxell, 379–416. Chicago: Chicago University Press.

Norton-Griffiths, M., and C. Southey. 1995. The opportunity costs of biodiversity conservation in Kenya. *Ecological Economics* 12:125–39.

Nyahongo, J. W., T. Holmern, B. P. Kaltenborn, and E. Roskaft. 2009. Spatial and temporal variation in meat and fish consumption among people in the western Serengeti, Tanzania: The importance of migratory herbivores. *Oryx* 43:258–66.

Ogutu, J. O., and H. T. Dublin. 1998. The response of lions and spotted hyenas to sound playbacks as a technique for estimating population size. *African Journal of Ecology* 36: 83–95.

Ogutu, J. O., N. Owen-Smith, H.-P. Piepho, and M. Y. Said. 2011. Continuing wildlife population declines and range contraction in the Mara region of Kenya during 1977–2009. *Journal of Zoology* 285:99–109.

Ogutu, J. O., H.-P. Piepho, H. T. Dublin, N. Bhola, and R. S. Reid. 2007. El Nino-Southern Oscillation, rainfall, temperature and Normalized Difference Vegetation Index fluctuations in the Mara-Serengeti ecosystem. *African Journal of Ecology* 46:132–43.

———. 2009. Dynamics of Mara-Serengeti ungulates in relation to land use changes. *Journal of Zoology* 278:1–14.

———. 2010. Rainfall extremes explain interannual shifts in timing and synchrony of calving in topi and warthog. *Population Ecology* 52:89–102.

Ojalammi, S. 2006. *Contested lands: Land disputes in semi-arid parts of northern Tanzania.* PhD diss., University of Helsinki.

Omotola, J. S. 2008. Political globalization and citizenship: New sources of security threats in Africa. *Journal of African Law* 52:268–83.

Ostrom, E. 2009. A general framework for analyzing sustainability of social-ecological systems. *Science* 325:419–22.

Ottichilo, W. K., J. de Leeuw, and H. H. T. Prins. 2001. Population trends of resident wildebeest (*Connochaetes taurinus hecki* [Neumann]) and factors influencing them in the Maasai Mara ecosystem, Kenya. *Biological Conservation* 97:271–82.

Ottichilo, W. K., J. de Leeuw, A. K. Skidmore, H. H. T. Prins, and M. Y. Said. 2000. Population trends of large non-migratory wild herbivores and livestock in the Maasai Mara ecosystem, Kenya, between 1977 and 1997. *African Journal of Ecology* 38:202–16.

Packer, C., S. Altizer, M. Appel, E. Brown, J. Martenson, S. J. O'Brien, M. Roelke-Parker, R. Hofmann-Lehmann, and H. Lutz. 1999. Viruses of the Serengeti: Patterns of infection and mortality in African lions. *Journal of Animal Ecology* 68:1161–78.

Pahl-Wostl, C. 2009. A conceptual framework for analyzing adaptive capacity and multilevel learning processes in resource governing regimes. *Global Environmental Change* 19:354–65.

Parkipuny, M. S. 1991. *Pastoralism, conservation and development in the Greater Serengeti Region.* Drylands Network Programme, Paper no. 26, London, International Institute for Environment and Development.

———. 1997. Pastoralism, conservation and development in the Greater Serengeti region. In *Multiple land-use: The experience of the Ngorongoro Conservation Area, Tanzania*, ed. D. M. Thompson, 143–68. Gland, Switzerland: IUCN.

Polasky, S., J. Schmitt, C. Costello, and L. Tajibaeva. 2008. Larger-scale influences on the Serengeti ecosystem: National and international policy, economics, and human demography. In *Serengeti IV: Human impacts on ecosystem dynamics*, ed. A. R. E. Sinclair, C. Packer, S. A. R. Mduma, and J. Fryxell, 347–77. Chicago: University of Chicago Press.

Reid, R. S. 2012. *Savannas of our birth: People, wildlife and change in East Africa*. Berkeley: University of California Press.

Reid, R. S., H. Gichohi, M. Y. Said, D. Nkedianye, J. O. Ogutu, M. Kshatriya, P. Kristjanson, S. C. Kifugo, J. L. Agatsiva, S. A. Adanje, et al. 2008. Fragmentation of a peri-urban savanna, Athi-Kaputiei Plains, Kenya. In *Fragmentation in semi-arid and arid landscapes: Consequences for human and natural systems*, ed. K. A. Galvin, R. S. Reid, R. H. Behnke, and N. T. Hobbs, 195–224. Dordrecht: Springer.

Reid, R. S., M. Rainy, J. Ogutu, R. L. Kruska, Nyabenge, M. McCartney, K. Kimani, M. Kshatriya, J. Worden, L. N'gan'ga, et al. 2003. *Wildlife, people, and livestock in the Mara ecosystem, Kenya: the Mara count 2002*. Nairobi, Kenya: International Livestock Research Institute.

Ribot, J. C., A. Agrawal, and A. M. Larson. 2006. Recentralizing while decentralizing: How national governments reappropriate forest resources. *World Development* 34:1864–86.

Ritchie, M. E. 2008. Global environmental changes and their impact on the Serengeti. In *Serengeti III: Human impacts on ecosystem dynamics*, ed. A. R. E. Sinclair, C. Packer, S. A. R. Mduma, and J. M. Fryxell, 183–208. Chicago: University of Chicago Press.

Runyoro, V. A. 2007. Analysis of alternative livelihoods strategies for the pastoralists of Ngorongoro Conservation Area, Tanzania, PhD diss., Sokoine University of Agriculture, Morogoro, Tanzania.

Schick, K. D., and N. Toth. 1993. *Making silent stones speak: Human evolution and the dawn of technology*. New York: Simon and Schuster.

Serneels, S., and E. F. Lambin. 2001. Impact of land-use changes on the wildebeest migration in the northern part of the Serengeti-Mara ecosystem. *Journal of Biogeography* 28: 391–407.

Serneels, S., M. Y. Said, and E. F. Lambin. 2001. Land cover changes around a major east African wildlife reserve: The Mara ecosystem (Kenya). *International Journal of Remote Sensing* 22:3397–420.

Setsaas, T. H., T. Holmern, G. Mwakalebe, S. Stokke, and E. Roskaft. 2007. How does human exploitation affect impala populations in protected and partially protected areas?—A case study from the Serengeti ecosystem, Tanzania. *Biological Conservation* 136:563–70.

Shetler, J. B. 2007. *Imagining Serengeti: A history of landscape memory in Tanzania from earliest times to the present*. Athens, Ohio: Ohio University Press.

Shivji, I. G. 1998. *Not yet democracy: Reforming land tenure in Tanzania*. Dar es Salaam, Tanzania: IIED/HAKIARDHI/Faculty of Law, University of Dar es Salaam.

Sinclair, A. R. E., S. A. R. Mduma, and P. Arcese. 2002. Protected areas as biodiversity benchmarks for human impact: Agriculture and the Serengeti avifauna. *Proceedings of the Royal Society of London Series B*. 269:2401–05.

Sinclair, A. R. E., S. A. R. Mduma, J. G. C. Hopcraft, J. M. Fryxell, R. Hilborn, and S. Thirgood. 2007. Long-term ecosystem dynamics in the Serengeti: Lessons for conservation. *Conservation Biology* 21:580–90.

Sinclair, A. R. E., and M. Norton-Griffiths. 1979. *Serengeti: Dynamics of an ecosystem.* Chicago: University of Chicago Press.

Sitati, N. W., M. J. Walpole, R. J. Smith, and N. Leader-Williams. 2003. Predicting spatial aspects of human-elephant conflict. *Journal of Applied Ecology* 40:667–77.

Thirgood, S., C. Mlingwa, E. Gereta, V. Runyoro, R. Malpas, K. Laurenson, and M. Borner. 2008. Who pays for conservation? Current and future financing scenarios for the Serengeti ecosystem. In *Serengeti III: Human impacts on ecosystem dynamics*, ed. A. R. E. Sinclair, C. Packer, S. A. R. Mduma, and J. M. Fryxell, 443–70. Chicago: University of Chicago Press.

Thirgood, S., A. Mosser, S. Tham, G. Hopcraft, E. Mwangomo, T. Mlengeya, M. Kilewo, J. Fryxell, A. R. E. Sinclair, and M. Borner. 2004. Can parks protect migratory ungulates? The case of the Serengeti wildebeest. *Animal Conservation* 7:113–20.

Thompson, D. M., ed. 1997. *Multiple land-use: The experience of the Ngorongoro Conservation Area, Tanzania.* Gland, Switzerland: International Union for Conservation of Nature.

Thompson, D. M., and K. Homewood. 2002. Entrepreneurs, elites, and exclusion in Maasailand: Trends in wildlife conservation and pastoralist development. *Human Ecology* 30:107–38.

Thompson, D. M., S. Serneels, D. Kaelo, and P. C. Trench. 2009. Maasai Mara—Land privatization and wildlife decline: Can conservation pay its way? In *Staying Maasai: Livelihoods, conservation and development in East African rangelands*, ed. K. Homewood, P. Kristjanson, and P. Trench, 77–114. London: Springer.

United Republic of Tanzania. 2002. Population and housing census—Dar es Salaam, United Republic of Tanzania (URT). Bureau of Statistics, President's Office Planning Commission, Dar es Salaam, Tanzania.

Walpole, M. J., and N. Leader-Williams. 2001. Maasai Mara tourism reveals partnership benefits. *Nature* 413:771.

Woods, M. 2007. Engaging the global countryside: Globalization, hybridity and the reconstitution of rural place. *Progress in Human Geography* 31:485–507.

Wunder, S. 2005. *Payments for environmental services: Some nuts and bolts.* Jakarta, Indonesia: Centre for International Forestry Research.

PART VI

Synthesis

TWENTY-SIX

The Role of Research in Conservation and the Future of the Serengeti

Anthony R. E. Sinclair, Julius D. Keyyu, Simon A. R. Mduma, Mtango Mtahiko, Emily Kisamo, J. Grant C. Hopcraft, John M. Fryxell, Kristine L. Metzger, and Markus Borner

The first formal ecological research in the Serengeti began in 1952 with the study of Thomson's gazelles on the Serengeti plains by Alan Brooks (1961), and research has continued for the subsequent 60 years. The purpose of the series of research volumes has been to synthesize this research both to describe a unique ecosystem for posterity—a major objective for civilization—and to provide the data and understanding to advise the managers of Serengeti National Park. Nevertheless, the question has been asked by both the wardens of Serengeti and by politicians in parliament: What good has all this research been to conservation and management in Tanzania? Fundamentally the question arises because of a lack of appreciation that what is taken as common knowledge is, in fact, derived from research (Turner 1987).

It would, therefore, be appropriate to summarize the main problems facing the park wardens since the inception of the protected area and how research has helped solve those problems.

THE CONTRIBUTION OF SCIENTIFIC RESEARCH TO MANAGEMENT

The 1960s and Earlier

The Serengeti, as we describe in chapter 2, was never set up to protect the great migration. It was formed to control profligate shooting of lions at the end of the 1920s. The migration was not considered as a conservation objective. There was no formal knowledge of where the migrating animals went, merely wild speculation; it was thought the migrants came from the Athi plains near Nairobi in Kenya (Johnson 1929). The first warden had no knowledge of where the migrants disappeared to when they left the plains in May or came from in November (Moore 1937). Professor Grzimek conducted the first aerial surveys of the migration in 1958–59 at the request of the director of national parks, but even he thought the migration ended up near Musoma in the dry season, and he opposed the new boundaries in 1959 when the park was divided because they did not capture the routes of the migrants (Grzimek and Grzimek 1960a, 1960b).

In the early 1960s a major issue was the use of wildlife as a meat source for local peoples; the prevailing political view was that game cropping could save the wildlife by making it economically valuable and help people at the same time. However, so little was known about the ecology and populations that there was a real worry that wildlife could be eliminated by overharvesting and so destroy the Serengeti. Thus, John Owen, the director of national parks, raised the funds for the Serengeti Research Project and brought in scientists to begin the task of documenting and monitoring the Serengeti.

The most important conservation question at the start of the Serengeti Research Project in 1962 was: Where did the migrants go and where should the correct boundaries of the protected area lie? Murray Watson was the first to record correctly the areas used by the migrants. He recorded by aerial survey the distribution of the herds at monthly intervals and built up the familiar picture of the migration, a picture that is now taken for granted (Watson 1967).

Mike Norton-Griffiths organized the huge systematic aerial survey of the whole 25,000 km^2 ecosystem, employing 12 observers and 24 hours of flying each month for 30 months in 1969–1972. From these data a detailed picture of the migration of all species emerged along with information on resident species, water supplies, food, pastoralists, and agriculture (Pennycuick 1975; Maddock 1979). Forty years later these data continue to be of value for local issues of human-wildlife conflicts—for example, where elephants are likely to meet agriculture and conservation. The data were the basis for the analysis of the original rhino population of the park and, hence, where rhino reintroductions should be made with respect to the best habitat and safety from poachers (Metzger et al. 2007). Thus, by the early 1970s scientists had established the basic information on wildlife distribution in the park.

Carnivore research began in the 1960s with Hans Kruuk and George Schaller. Common perceptions by wardens of some predator species were that they were intolerable vermin and had to be exterminated. The perception of wild dogs epitomized this dogma; Martin Johnson (1929) refers to shooting these animals near Kuka. The first park warden, Monty Moore, found a pack of 13 near Banagi and killed nine of them in the 1930s (Moore 1937). Wardens in the 1950s shot both wild dogs and hyena within the Serengeti National Park. All were justified as the removal of vermin.

Lions, ironically, were also shot, indeed massacred, in the 1920s (see chapter 2) because they were seen as majestic animals that proved the bravery of the shooter. It was these massacres that led to the first conservation action, and led to the setting up of the Serengeti Game Reserve in 1929.

Such was the state of ignorance and bigotry when the first scientists started in the 1960s. Hans Kruuk established that hyenas were not only true predators with a complex social ecology but also the most important predator in the system, and that hyenas also moved considerable distances following the migrants (Kruuk 1972). George Schaller (1972) provided the first data on the lion population, showing that they depended on the migration for the success of their cubs. At the same time Schaller conducted the first wild dog studies and presented a different face on this species; they also had interesting social ecology and were very vulnerable to attack from the larger predators, especially hyena. Far from being common pests, they were rare and vulnerable. In summary, these studies changed the view of wardens about predators from being vermin to valuable components of the ecosystem.

The 1970s

Throughout savanna Africa in the 1960s elephants were branded as the destroyers of national parks, laying waste to the vegetation, and creating a desert (Pienaar, van Wyk, and Fairall 1966; Laws 1969). The reaction by management was to institute culling of elephant populations in South Africa, Uganda, and elsewhere, and this was proposed for Tsavo, Kenya, and the Serengeti (Laws, Parker, and Johnstone 1975; Pienaar 1983). In both Tsavo and the Serengeti, however, the proposals were rejected and research was instituted to examine whether elephants were the cause of the decline in tree populations. For the Serengeti it was found that elephants damaged or killed trees at a rate too slow to have caused the changes in vegetation (Croze 1974a, 1947b). Studies of fire and its effects on tree regeneration showed that this was the fundamental cause of the decline in mature trees—there was simply too much burning and it was preventing the regeneration of small trees (Norton-Griffiths 1979). In short, the elephant problem was in fact a fire problem. Management accepted these results and instituted an "early burning" program to reduce the effects of fire.

Research in the 1960s had established that wildebeest and buffalo populations were increasing at a fast rate (Watson 1967; Sinclair 1973). It was unknown why this increase was occurring, and managers were concerned that eventually there could be overstocking. Research combined the knowledge from veterinarians, virologists, and population ecologists to highlight the fundamental role that the viral disease rinderpest had on the ruminant populations of the Serengeti; the arrival of the disease had decimated the populations in the 1890s, and conversely, its disappearance in 1963–64 had allowed an outbreak of the same populations—this was the reason for the increase (Sinclair 1977; Plowright 1982; Dobson 1995). More importantly, this disease underlay most of the changes in the Serengeti ecosystem that researchers had documented in the 1970s (Sinclair and Norton-Griffiths 1979). In addition, this understanding supported other research that these two species increased because they were understocked; they had been kept down by the disease and so had abundant food. As numbers increased through the 1970s they progressively found less food and the death rate increased from undernutrition so that eventually both populations stabilized; it was established that this was a natural regulating process and managers did not have to control numbers (Sinclair 1977; Sinclair, Dublin, and Borner 1985; Mduma, Sinclair, and Hilborn 1999).

Studies of predator populations and their predation on resident and migrant wildebeest populations showed that predators could impose a heavy mortality on resident numbers in Ngorongoro Crater (Kruuk 1972) but only a minor mortality on the migrants (Schaller 1972; Sinclair 1979), supporting the conclusion that migrants were limited by their food supply.

The 1980s

With the closure of the border with Kenya in 1977, the collapse of tourism, and, hence, park revenues, protection from poachers and bandits effectively stopped. The major management question was: What effect was illegal hunting having on the wildlife of the Serengeti? Research based on censuses and studies of carcasses showed the complete collapse of rhino numbers and a drop of 80% of the elephant and buffalo numbers—until these results were obtained, managers were unaware that there was a problem (Dublin et al. 1990; Sinclair 1995b). However once the problem became known, a major investment in antipoaching was put in place by Tanzania National Parks with the help of the Frankfurt Zoological Society in the late 1980s. Although rhino have never been able to recover, both elephant and buffalo numbers have shown a remarkable recovery in the following decade;

without the original monitoring data showing the normal population size for this habitat and the subsequent collapse, management would not have had the information to save the remaining animals.

Further research on the role of elephants in the ecosystem produced a surprising result: the increase in the wildebeest population had, through their grazing, reduced the incidence of burning (Norton-Griffiths 1979). His results predicted an increase in the survival of small trees and photographic records confirmed this for the Tanzania part of the Serengeti (Sinclair 1995a). However, in the Mara Reserve no such increase was seen. Holly Dublin's research established that elephants could reduce tree recruitment and keep the vegetation in a grassland state (Dublin, Sinclair, and McGlade 1990; Dublin 1995). The important result for managers was that elephants can keep the vegetation in two different states, one with trees and one without, and both are normal ecological situations. Subsequently, it became clear that within the Serengeti the increase in tree recruitment was not just because of a reduction in burning—which was essential as confirmed by Stronach's (1989) work—but also because there was a reduction in elephant browsing from the poaching (Sinclair 1995a; Sinclair et al. 2007). These results confirmed the advice to managers that elephants were a normal component of savanna ecosystems rather than a management problem. Fire, on the other hand, was a major management problem if it was not controlled.

The 1990s

The effect of illegal hunting on elephant and buffalo in the 1980s had drawn attention to the possibility that the increasing numbers of people surrounding the Serengeti could be threatening both the movements and the numbers of the migrating species: management needed to know what the effect of poaching was on the wildebeest (Campbell and Hofer 1995; Arcese, Hando, and Campbell 1995). Simon Mduma's research on the wildebeest population established the scale of the illegal hunting. He confirmed first that the wildebeest population was limited by its food supply at around 1.3 million animals. Second, he measured the number of animals killed by hunters as between 20,000 and 40,000 per year. Although this seems a large number it is well within the capacity of the population to compensate by producing more juveniles; it is only if hunting reached 80,000 that the population would decline. There was, however, a warning for managers—the increasing numbers of people around the Serengeti would mean an increasing poaching rate with the possibility that this threshold of 80,000 could be reached (Mduma, Hilborn, and Sinclair 1998; Mduma, Sinclair, and Hilborn 1999). The advice was that poaching had to be brought under control and stabilized. At the present time (2015), we do not yet know if this has been achieved.

It was in the 1990s that managers became aware that disease was a threat to both wildlife and human ecosystems. In 1992 the last wild dog in the Serengeti National Park died. An outcry followed with different groups blaming each other for the demise of the wild dog in this ecosystem (Creel 1992). In 1994 large numbers of lions died following the most severe drought of the twentieth century. The management question was: What was killing the Serengeti predators? Research was begun on the diseases of carnivores and it was eventually established that rabies and canine distemper virus (CDV) were implicated in the deaths of the Serengeti's wild dogs and lions. These viral diseases have long been associated with domestic dogs, but their impact on wildlife had not been well understood. In particular, it had not previously been known that CDV could inflict such high mortality in large cats (Roelke-Parker et al. 1996).

The suspicion was that the burgeoning population of domestic dogs around the park—which is tied to the increasing human population—was transmitting the disease to wild predators. Virulent pathogens that have short infection cycles, such as rabies and CDV, cannot persist in small populations in the absence of a more abundant reservoir, in which the infection can be maintained. Research has shown that domestic dogs around the park act as such a reservoir and they harbor important pathogens that affect the health of wild carnivore populations (Cleaveland et al. 2000; Lembo et al. 2008).

In terms of CDV, long-term research on the Serengeti and Ngorongoro lions has provided critical information for understanding the impact of CDV and the factors that influence the severity of disease outbreaks. Over the past 30 years, monitoring of CDV infection in lions, coupled with detailed data on individual morbidity and survival, has revealed that CDV only inflicts substantial mortality if infection coincides with outbreaks of tick-borne disease. Severe droughts in 1993 and 2000 caused widespread tick infestations in wild herbivores, particularly buffalo, resulting in higher levels of tick-borne parasites, such as *Babesia* spp. in lions. Coinfections involving CDV and *Babesia* appear to exacerbate the severity of each of these infections, causing unusually high levels of mortality. Thus, coinfection of lions with high levels of *Babesia* resulted in high mortality from CDV in 1994, and coinfection of lions with CDV in the Ngorongoro Crater in 2001 was associated with unusually severe pathology from *Babesia* (Munson et al. 2008). Infection of lions with CDV in low-*Babesia* years has never been linked with signs of disease or identified as a cause of death. In combination with studies of CDV in domestic dogs, these findings have important implications for understanding whether and how to target control measures to reduce the impact of CDV on Serengeti carnivores.

One management result of this research has been the application of a vaccination campaign against rabies and CDV in domestic dogs surrounding the Serengeti National Park, a program that also benefits the human population. These benefits arise from a reduction in (a) human deaths from rabies, (b) the incidence of human bite injuries from rabid animals, and (c) the direct and indirect medical costs associated with postexposure rabies vaccination, which is required to prevent rabies after the bite of a rabid animal. Data from these campaigns has shown that rabies can be controlled by vaccinating a critical proportion of the domestic dog population, and that these vaccination programs can lead to cost-effective and immediate benefits for both human and animal health (Cleaveland et al. 2003; Hampson et al. 2008; Kaare et al. 2009; Beyer et al. 2010). The rapid recovery of the African wild dog population in pastoral communities adjacent to the park has also coincided with the onset of mass domestic dog vaccination campaigns, with elimination of canine rabies in some of these communities as well as in the Serengeti National Park (Lembo et al. 2010).

The Early Twenty-First Century

The increasing pressures from the expanding human population surrounding the Serengeti has raised concerns about the ability of migrant species to move to their essential areas of refuge in the dry season, in particular the areas around the Mara River in the Kenya Mara Reserve and northwest Serengeti (Serneels and Lambin 2001; Norton-Griffiths et al. 2008). This problem was investigated by placing radio collars on wildebeest and zebra. The results have confirmed that a major route from western Serengeti to northern Serengeti runs along the Grumeti River outside the protected area. If this route is closed by using fences or other forms of land transformation, then the migration could be impeded or even prevented from reaching the north (Thirgood et al. 2004; Hopcraft et al., chapter 6).

Wildlife populations extend outside the Serengeti National Park. Topi populations occur in the Grumeti Reserve; the only roan antelope remaining from the northern population are now based in the Ikorongo and Grumeti Reserves. Greater kudu are found only in Maswa Game Reserve. Wild dogs have been found in the Loliondo district east of Serengeti and remain outside the protected area. Management has focused on developing community-based conservation areas—called wildlife management areas—so as to provide protection for these species and benefit for local peoples (Thirgood et al. 2008; Randall et al., chapter 24).

Research has now highlighted a major problem with water resources in the Serengeti (Gereta et al. 2002, 2009). The Mara River is the only major permanently flowing water supply for the migrants in the dry season. Indeed, the migration moves to the north for this water. However, research has shown that water flow in the dry season has been declining and late in the first decade of the twenty-first century it stopped flowing. The management implication is that if water in the Mara River dries up then the migrating populations will be seriously affected, and so will the rest of the Serengeti ecosystem, which depends on the migration (Sinclair et al. 2007; Holdo et al. 2009).

RESEARCH BENEFITS FOR TANZANIA

This overview of research and its relevance to the conservation and management of Serengeti National Park does not cover all the issues; it merely illustrates that present common knowledge was originally unknown and came from research. Furthermore, this research has been of value to Tanzania as a whole; we summarize their relevance to the major national concerns of economics, health, social welfare, ecosystem services, and conservation.

Economics: Tourism and the Great Migration

As we mentioned above, the great Serengeti migration was only fully known by the end of the 1960s when researchers discovered that the wildebeest moved north to the Mara River and into Kenya, and that the Ngorongoro Crater animals were separate from those of the Serengeti. Publications from this research, particularly the books and film of the Grzimeks', created such international publicity that the Serengeti became the foremost destination for tourists. Since then tourism to Tanzania, drawn by the famous migration, has steadily increased; it has become one of the two greatest tourist attractions for Tanzania—Mt Kilmanjaro is the other one—and is becoming the most important foreign income earner for the country. This research, by highlighting these great events, has contributed significantly to the economy.

Health: Impacts and Implications of Wildlife Infections

Tanzanians are fully aware of the threats posed by diseases such as rabies, sleeping sickness, anthrax, malignant catarrhal fever, foot-and-mouth disease, and rinderpest. These diseases involve a wide range of different host species, including wildlife, and their effective management and control requires an understanding of the transmission dynamics in wildlife, domestic animal, and human populations. Serengeti research has shown not only that diseases in domestic animal reservoirs, such as rinderpest, rabies, and CDV can have major impacts on wildlife populations, but that infectious diseases transmitted from wild animals also have important impacts on livestock health and production, with implications for national economies, land-use policy, and rural livelihoods (e.g., foot-and-mouth disease, malignant catarrhal fever, trypanosomiasis) (Cleaveland et al. 2008). The findings from long-term research in the Serengeti have thus provided important insights that not only shape local conservation management actions and protect ecosystem health, but also inform the development of national and regional policies for control of transboundary human and animal diseases in and around protected wildlife ecosystems.

Rinderpest has for the whole of the twentieth century posed a major threat to the livelihoods of both pastoralists and agriculturalists in Tanzania. The decades of research have now resulted in the extinction of rinderpest from the world (Normile 2008), so providing a major benefit for the country.

Research from the Serengeti has also had a major influence in shaping national and international rabies policy in terms of demonstrating the epidemiological, economic, and logistic feasibility of eliminating canine rabies in rural Africa (Lembo et al. 2010). This has previously been considered implausible due to perceptions that wildlife might act as independent reservoirs of rabies, and that domestic dog populations would be difficult to vaccinate due to a large proportion of inaccessible or stray dogs. Partly as a result of the Serengeti research, which demonstrated that neither of these assumptions was true, large-scale rabies elimination demonstration projects are now being undertaken in Africa and Asia, with the World Health Organization (WHO) and World Organisation for Animal Health (OIE) declaring the global elimination of canine rabies a feasible objective. From the wildlife perspective, it is now well-recognized that disease can pose a threat to endangered species, even in well-protected ecosystems, and planning for disease management is now more routinely integrated into park management plans (TANAPA 2006).

Social Welfare: Lion Attacks

In recent decades lion attacks on humans have become more prominent as the human population has almost doubled. Areas that once supported wild ungulates have now been cleared for agriculture. Wild ungulates have been hunted for food until few remain, leaving little for lions. Instead, lions turned to feeding on cattle and humans. Predation on domestic animals occurs outside of protected areas such as Tarangire (Kissui 2008) and the Ngorongoro Conservation Area (Ikanda and Packer 2008). Lions tend to feed on cattle, while hyenas and leopards feed on sheep and goats.

Lion predation on humans is particularly prevalent in southern Tanzania. Craig Packer and his associates Dennis Ikanda, Bernard Kissui, and Hadas Kushnir have shown some of the causes for lions becoming maneaters (Kushnir 2009; Kushnir et al. 2010; Packer et al. 2005, 2008). Incidents of human predation by lions have increased dramatically since 1990, especially in southern Tanzania where wildlife has been reduced. The human population effectively doubled between 1990 and 2005 so that more land has been transformed into agriculture. The lion researchers found that in some areas of Tanzania the main problem for agriculturalists is the bush pig, a secretive animal that feeds nocturnally and is surprisingly common—it can live in heavily cultivated areas with dense human populations provided it has a refuge, usually a swamp or wet thicket. Bush pigs love to eat beans and other legumes. Peasant farmers protect their crops at night by building a small shelter with a bed raised off the ground—the shelter is open on all sides so the farmer can see his crop and chase off the bush pigs. Lions like to eat bush pigs and follow them out to the crops, and in this situation they come across farmers sleeping in their shelters. This presents a too tempting opportunity for the lions, and some have taken advantage of it, taking the sleeping man. Soon they become accustomed to eating humans. Packer has advised that the best solution for avoiding the human-lion conflict is to control bush pig numbers (Packer et al. 2005). Since Tanzania has one of the largest lion populations remaining in Africa, this advice has the added advantage of conserving lion numbers as well.

Ecosystem Services

Pests in agricultural areas affect the livelihoods of people. Rodents transmit diseases and insect pests eat crops. Research has shown that rodent populations are much higher in villages than in nearby natural woodland and insect outbreaks, which destroy crops, may be more frequent. Large birds of prey that eat rodents and smaller birds that eat insects are common in natural areas, but are almost absent in agricultural areas. This is because trees that the birds require to nest and roost have been cut down. Allowing more native large trees and bushes to grow around fields would provide habitat for birds and reduce pest populations (see Byrom et al., chapter 12).

Conservation: Park Management and Sustainable Ecosystems

Ecological baselines are essential to monitor changes imposed by humans on their ecosystems, and protected areas play this role (Arcese and Sinclair 1997; Sinclair 1998). They are the insurance policy for future generations. The reason for baselines is that ecosystems change without us noticing unless we have a way of measuring change, and we do this by comparing natural areas with human-use areas. It takes a long time to do this, usually many decades.

The long-term research in the Serengeti illustrates several aspects of this approach. The most important change has been the increase in the wildebeest population due to the removal of the rinderpest disease. This change has shown us how the different components of the ecosystem work together and so demonstrate how human ecosystems can be sustainable (Holdo et al. 2009).

THE FUTURE OF THE SERENGETI

The future of the Serengeti is threatened by three major developments outside the ecosystem that have impacts inside it—the drying of the Mara River, changing levels of Lake Victoria, and the construction of a trunk road across the migration routes in northern Serengeti.

The Drying of the Mara River

The Mara River is the most important water source for the migration in the dry season. The animals move north to the Mara River because it is the only flowing river of any size in the dry season and provides vital water for the millions of animals. However, the Mara has its origins in the Mau forests of the highlands of Kenya, far from the Serengeti ecosystem. These forests are being cut down at an accelerating pace and the flow of water has declined. There is now an initiative by the government of Kenya (2010) to develop a restoration and rehabilitation program for the Mau forests.

However, progress with the Mau Forest could be undermined by water offtake downstream. The Mara flows through agricultural land where unregulated irrigation is tapping off the water. As a result, the Mara has, in recent years, stopped flowing for the first time in a century; these impacts take place in Kenya, out of reach of controls from Tanzania. Unless agreements are drawn up to regulate the irrigation offtake, particularly in the dry season, the wildebeest could find there is no water for them when they arrive in the next few years (Gereta et al. 2002, 2009; Dybas 2011).

Lake Victoria

Lake Victoria is a vast but shallow lake immediately to the west of the Serengeti. It is so large that it creates its own weather system, producing rainstorms in the dry season in the west and northwest of the ecosystem. These storms are really an extension of the Congo rainforest weather systems, and it is these storms that provide the food for the wildebeest migration. The lake has only one major water inlet, the Kagera River on the west side, and some smaller rivers on the east side, the main one being the Mara River. If the Mara stops flowing there is a possibility that the lake, being only a few meters deep around its shores, could recede as water levels drop. On Speke Gulf the shore could recede several kilometres from the current shores. Millions of people dependent on the lake for fishing, irrigation, and drinking water would find themselves stranded. In addition, the weather systems of the lake could be changed, with dry season rain declining as the lake receded. The migration pattern of the wildebeest would also change. We should learn the lessons from the drying up of the Aral Sea in Asia due to uncontrolled offtake of water for irrigation.

The Road across Northern Serengeti and Infrastructure Development

Early in 2010 the Tanzanian government announced that it would construct a major international trunk road through the northern part of Serengeti National Park. Scientists put together the biological facts derived from the forty years of accumulated information (Dobson et al. 2010; Sinclair 2010; Holdo et al. 2010). The proposed road was to cut the northern extension of the park in two. This is the area that the wildebeest and zebra migrate to in the dry season as their refuge when food and water are in short supply. They head for the Mara River, the only permanent water supply in the ecosystem and the region of highest rainfall, where they can find green food. Migrants remain in this region from June to November, the duration depending on the vagaries of the rainfall; they usually move outside the park boundaries on both the west and east sides, just as they did a hundred years ago, although now not so far (see the map of S. E. White in 1913, fig. 2.4 in Sinclair et al., chapter 2). Numbers of animals are in the range of one million wildebeest, zebra, eland, and gazelle. In addition, large herds of buffalo, elephant, and topi are resident there. All of these animals could be feeding near the road, crossing it, and because it is open and flat, resting there to ruminate. As a result of the information from long-term research, the Tanzanian government announced on June 22, 2011, that it would not go ahead with the north road as previously planned. Instead, the government announced it would confine the tarmac to roads east and west of the park.

Although the tarmac road has currently been put on hold, it is worth reviewing the conservation, social, and economic issues that arise with such roads so as to avoid future problems. There are many roads in the Serengeti National Park but none of them produce the human-wildlife conflict that a tarmac road across the northern Serengeti is likely to create. The present roads in the park are made of gravel and are used by local traffic for tourism, a few buses, and local trucks. These trucks are small because all traffic currently passes up the tortuous, steep, and narrow Ngorongoro escarpment. This road is impassable for large 18-wheel trucks and semitrailers; indeed, they are not allowed to use the road. The traffic cannot travel at more than 50 kilometers per hour.

The problem that remains with the northern road, although currently planned as a gravel road, is that there would be a great temptation to make it part of a major trunk route across Africa used by very heavy multiwheelers because it would be so convenient. However, there is a limit to the amount of traffic that a gravel road can sustain, the estimates of traffic use are already beyond this limit, and traffic flow will only increase in the future. Thus, if the northern route does develop into a trunk road across the park it will have to be tarmac, suitable for very heavy traffic. The problem with such tarmac roads is not that they impede wildlife—they do not—but that they encourage fast driving. With such a vast number of animals on the road there are bound to be accidents, just as occurred at Banff National Park in Canada with moose and elk—and that was with only 800 elk. In Hluhluwe-iMfolosi Park, South Africa, a 10-mile road constructed in 2002 has resulted in numerous road kills of wildlife, especially involving wild dog, now a threatened species, but also lion, leopard, buffalo, rhino, and many antelope species. Elephants have been hit and injured. Human fatalities have occurred. The road was constructed with road humps to control vehicle speeds, but cars have been recorded at 120 kilometers per hour. There are no fences along the whole length, but barriers were put into place to reduce the risk of animals crossing in dangerous places. This served to impound animals on the road and actually increased collisions. In general the park manager, Sihle Nxulalo, has concluded that the road has become a serious problem (Sue van Rensburg, pers. comm.).

The effects of the tarmac road constructed in 1972 through Mikumi National Park, Tanzania, show that animals that are disturbed by the road suffer proportionately lower fatalities than species that take no notice of the road (Newmark et al. 1996). This scientific observation is relevant to the Serengeti because the huge migrating herds pay little attention to vehicle traffic. In an extensive review Benitez-Lopez, Alkemade, and Verweij (2010) report that impacts from roads and corridors include habitat loss, intrusion of edge effects in natural areas, isolation of populations, barrier effects, road mortality, and increased human access. In general, roads are a major driving factor for biodiversity loss.

Accidents cause human fatalities, as occurred at Banff, and as a result there is pressure to build fences along the road. This may not happen immediately, it could be twenty or even fifty years away, but eventually it will happen. More realistically, it will happen in the next ten years if the road were to be constructed. The important point is that once the process of building a road is engaged there is no turning back. The concern is that the road will never be taken out again; development will proceed inevitably toward a fence. The fence will prevent the migrants from reaching their dry season refuge containing their last water and food reserves, and so there will be a collapse in numbers (Holdo et al. 2010). For example, the migratory wildebeest in Botswana collapsed to 10% of their former population when they were cut off by fences from their dry season water and food supplies (Williamson and Mbano 1988). Gadd (2012) has shown from the wide experience of fences in southern Africa that there are negative effects on wild species, natural communities, and whole ecosystems; indeed, they even exacerbate human-wildlife conflict. Fences have already been shown to cause the end of migrating systems not only in Africa (Bolger et al 2008; Bartlam-Brooks, Bonyongo, and Harris 2011), but around the world (Harris et al. 2009). Holdo et al. (2010) have calculated that a minimum of 35% of the population of migrants will die if they fail to reach northern Serengeti; this is a conservative figure, for the area south of the proposed road is far less suitable as dry season habitat. The long-term consequence is that the Serengeti as we know it will change: the accumulated information over the past 50 years shows that the Serengeti ecosystem is structured by the huge populations of migrants. Without the migrants it will change into a different system, and the Serengeti as a great migration will be lost.

In addition, a major trunk road attracts settlement along it, thus bringing buildings to the very edge of the park boundary. As we have seen in earlier chapters, the migrants move outside the park boundary on either side and thus will come up against settlement. People will have hundreds of animals in their gardens and along the streets. Inevitably, fences will be constructed along the park boundaries. The problem with fences is that wildebeest and zebra do not know what they are. They have never met them in their lives and are not adapted to tolerate them. They usually do not see the wires, or they think they can push through the fence. Thus, the great herds run straight into the wires. Fences were built across the short-grass plains at Angata Kiti in the Gol Mountains in 1964 to keep the wildebeest out. When the wildebeest arrived they stampeded and the fences fell over within a few minutes (Turner 1987). Fences built to withstand the charging masses will result in a catastrophic mortality with bodies lined up along it. Lions will use the fence to trap prey, as they have with the fence around Keekerok airfield in the Mara Reserve. In summary, a major human-wildlife conflict would develop where none now exists if a tarmac road were to proceed. The environmental impacts are likely to be substantial.

Development of the people in the Mara Region west of the park can be better achieved by routing the road south of the Serengeti ecosystem. Although a road cutting through the Serengeti would be shorter than that passing south of it by about 80 km, the cost of building that road would be more expensive: a north road would have to climb 1,500 m up the Rift Valley escarpment along a new route whereas the route south is already established, and consequently 428 km of new pavement would be required for the north compared to 332 km for the south. Assuming vehicles were to travel at the speed limit, which is 50 km/hr inside protected areas and 80 km/hr outside, the travel times from Mwanza to Arusha would be 7.8 hours to go south of the park compared to 7.9 hours to go via the northern route.

In addition, the southern route provides more economic returns for Tanzania because it provides access to almost twice as many people as the northern route. There is almost three times as much agriculture and livestock along the southern route than the northern route. Currently, most of this agricultural produce in the south has little or no access to regional markets; export of the major cash crop in the region, cotton, is fraught with difficulties due to the poor state of gravel roads. As can be seen in other areas of Tanzania where roads have been paved (for instance, east of the Serengeti between Makayuni and Karatu), the development of a national highway facilitates small businesses, provides access to markets, schools, and hospitals, and offers a backbone for other infrastructure such as power lines, railways, pipelines, and eventually ethernet cables. Because most of the northern road traverses uninhabited areas, these economic spin-offs would not be realized for the Tanzanian economy, making a road through the north more expensive and with fewer economic gains than the southern alternative. In general, the southern route provides access to the Mara Region using many roads that are already built and that merely need upgrading (Hopcraft, Bigurube et al., forthcoming; Hopcraft, Mduma et al., forthcoming).

Currently, the government intends to confine the tarmac to roads east and west of the park. In between, across the park, there would remain the current gravel road to be used only for tourists. Meanwhile, the government has asked for help in finding funds for the southern route around the park. This is a positive step because if the southern route is built then there is no economic need for the northern road. Thus, the future of the Serengeti depends on funds being found for construction of the south road. Unless this occurs the Serengeti will be cut in half by a northern trunk road. It is up to the international community to find these funds (*Nature* editorial 2010).

The north road, however, is only one of several other developments in the Serengeti; the increasing tourism is creating demands for increased infrastructure such as camps, lodges, hotels, and tourist roads. These need careful study by Tanzanians themselves, and this requires capacity building. Tanzanians must conduct the high-quality environmental and social impact assessments both within the ecosystem and outside it to generate information for managers and politicians to make informed decisions. Such capacity building requires outside support.

CONCLUSION

The history of research in the Serengeti shows that with every decade different problems have appeared. These problems have been addressed by researchers with a view to providing information to be used by the park managers. The earliest problems related to the lack of basic information on how the ecosystem worked—Where did the migration go? What limited populations? Later decades produced problems related to the impact of increasing human populations on the habitats and animal numbers, and also how the protected area affects surrounding peoples—How was fire changing the savanna? How was poaching changing ungulate populations? How were animals damaging crops? Recent problems focus on the impacts of development—how does a road across northern Serengeti affect the migration? How would offtake of water from the Mara River affect migrant species? Research has provided the relevant information, and in most cases successful management decisions have been taken.

Tanzania as a whole has also benefited from this research. We show how it is directly relevant to the major national concerns of economics, health, social welfare, ecosystem services, and conservation.

Research can be divided into two categories. First is basic research to understand the system—largely the work in the 1960–1980 period—and second is monitoring, also a form of research, which documents long-term changes relevant to more immediate and focused problems. In essence, without monitoring information managers must act by guesswork.

ACKNOWLEDGMENTS

We thank Hans Kruuk and George Schaller for information on the 1960s, Sarah Cleaveland and Andy Dobson for help with aspects of disease research, Craig Packer for information on man-eating lions, and Stephanie Eby for help with the list of research priorities.

REFERENCES

Arcese, P., J. Hando, and K. Campbell. 1995. Historical and present-day anti-poaching efforts in Serengeti. In *Serengeti II: Dynamics, management and conservation of an ecosystem*, ed. A. R. E. Sinclair and P. Arcese, 506–33. Chicago: University of Chicago Press.

Arcese, P., and A. R. E. Sinclair. 1997. The role of protected areas as ecological baselines. *Journal of Wildlife Management* 61:587–602.

Bartlam-Brooks, H. L. A., M. C. Bonyongo, and S. Harris. 2011. Will reconnecting ecosystems allow long-distance mammal migrations to resume? A case study of a zebra *Equus burchelli* migration in Botswana. *Oryx* 45:210–16.

Benitez-Lopez, A., R. Alkemade, and P. A. Verweij. 2010. The impacts of roads and other infrastructure on mammal and bird populations: A meta-analysis. *Biological Conservation* 143:1307–16.

Beyer, H. L., K. Hampson, T. Lembo, M. Kaare, S. Cleaveland, and D. T. Haydon. 2010. Metapopulation dynamics of rabies occurrence and transmission, and the efficacy of vaccination. *Proceedings of the Royal Society B* 278:2182–90.

Bolger, D. T., W. D. Newmark, T. A. Morrison, and D. F. Doak. 2008. The need for integrative approaches to understand and conserve migratory ungulates. *Ecology Letters* 11: 63–77.

Brooks, A. C. 1961. *A study of the Thomson's gazelle (Gazella thomsonii* Gunther*) in Tanganyika.* London: Colonial Research Publication no. 25, Her Majesty's Stationary Office.

Campbell, K., and H. Hofer. 1995. People and wildlife: Spatial dynamics and zones of interaction. In *Serengeti II: Dynamics, management and conservation of an ecosystem*, ed. A. R. E. Sinclair and P. Arcese, 534–70. Chicago: University of Chicago Press.

Cleaveland, S., M. G. J. Appel, W. S. K. Chalmers, C. Chillingworth, M. Kaare, and C. Dye. 2000. Serological and demographic evidence for domestic dogs as a source of canine distemper virus infection for Serengeti wildlife. *Veterinary Microbiology* 72: 217–27.

Cleaveland, S., M. Kaare, P. Tiringa, T. Mlengeya, and J. Barrat. 2003. A dog rabies vaccination campaign in rural Africa: Impact on the incidence of dog rabies and human dog-bite injuries. *Vaccine* 21:1965–73.

Cleaveland, S., C. Packer, K. Hampson, M. Kaare, R. Kock, M. Craft, T. Lembo, T. Mlengeya, and A. Dobson. 2008. The multiple roles of infectious diseases in the Serengeti ecosystem. In: *Serengeti III: Human impacts on ecosystem dynamics*, ed. A. R. E. Sinclair, C. Packer, S. A. R. Mduma, and J. M. Fryxell, 209–40. Chicago: University of Chicago Press.

Creel, S. 1992. Cause of wild dog deaths. *Nature* 360:633.

Croze, H. 1974a. The Seronera bull problem, part 1. The elephants. *East African Wildlife Journal* 12:1–28.

———. 1974b. The Seronera bull problem, part 2. The trees. *East African Wildlife Journal* 12: 29–48.

Dobson, A. 1995. The ecology and epidemiology of rinderpest virus in Serengeti and Ngorongoro Conservation Area. In *Serengeti II: Dynamics, management and conservation of an ecosystem*, ed. A. R. E. Sinclair and P. Arcese, 485–505. Chicago: University of Chicago Press.

Dobson, A. P., M. Borner, A. R. E. Sinclair, P. J. Hudson, T. M. Anderson, G. Bigurube, T. B. B. Davenport, J. Deutsch, S. M. Durant, R. D. Estes, et al. 2010. Road will ruin Serengeti. *Nature* 467:272–74.

Dublin, H. T. 1995. Vegetation dynamics in the Serengeti-Mara ecosystem: The role of elephants, fire and other factors. In *Serengeti II: Dynamics, management and conservation of an ecosystem*, ed. A. R. E. Sinclair and P. Arcese, 71–90. Chicago: University of Chicago Press.

Dublin, H. T., A. R. E. Sinclair. S. Boutin, E. Anderson, M. Jago, and P. Arcese. 1990. Does competition regulate ungulate populations? Further evidence from Serengeti, Tanzania. *Oecologia* 82: 238–88.

Dublin, H. T., A. R. E. Sinclair, and J. McGlade. 1990. Elephants and fire as causes of multiple stable states for Serengeti-Mara woodlands. *Journal of Animal Ecology* 59: 1157–64.

Dybas, C. L. 2011. Saving the Serengeti-Masai Mara: Can ecohydrology rescue a key East African ecosystem? *BioScience* 61:850–55.

Gadd, M. E. 2012. Barriers, the beef industry and unnatural selection: A review of the impacts of veterinary fencing on mammals in southern Africa. In *Fencing for conservation: Restriction of evolutionary potential or a riposte to threatening processes?* ed. M. J. Somers and M. W. Hayward, 153–86. New York: Springer.

Gereta, E., E. Mwangomo, J. Wakibara, and E. Wolanski. 2009. Ecohydrology as a tool for the survival of the threatened Serengeti ecosystem. *Ecohydrology and Hydrobiology* 9: 115–24.

Gereta, E., E. Wolanski, M. Borner, and S. Serneels. 2002. Use of an ecohydrological model to predict the impact on the Serengeti ecosystem of deforestation, irrigation and the proposed Amala weir water diversion project in Kenya. *Ecohydrology and hydrobiology* 2:127–34.

Government of Kenya. 2010. Rehabilitation of the Mau Forest ecosystem. Prepared by the Interim Coordinating Secretariat, Office of the Prime Minister, on behalf of the Government of Kenya, with support from the United Nations Environment Programme.

Grzimek, B., and M. Grzimek. 1960a. *Serengeti shall not die.* London: Hamish Hamilton.

Grzimek, M., and B. Grzimek. 1960b. A study of the game of the Serengeti plains. *Zeitschrift fur Saugetierkunde* 25:1–61.

Hampson, K., A. Dobson, M. Kaare, J. Dushoff, M. Magoto, E. Sindoya, and S. Cleaveland. 2008. Rabies exposures, post-exposure prophylaxis and deaths in a region of endemic canine rabies. *PloS Neglected Tropical Diseases* 2 (11): e339.

Harris, G., S. Thirgood, J. G. C. Hopcraft, J. P. G. M. Cromsigt, and J. Berger. 2009. Global decline in aggregated migrations of large terrestrial mammals *Endangered Species Research* 7:55–76.

Holdo, R. M., J. M. Fryxell, A. R. E. Sinclair, A. Dobson, and R. D. Holt. 2010. Predicted impact of barriers to migration on the Serengeti wildebeest migration. *PLoS ONE* 6 (1): 1–7, e16370. doi:10.1371/journal.pone.0016370.

Holdo, R. M., A. R. E. Sinclair, K. L. Metzger, B. M. Bolker, A. P. Dobson, M. E. Ritchie, and R. D. Holt. 2009. A disease-mediated trophic cascade in the Serengeti and its implications for ecosystem C. *PLoS Biology* 7:e1000210.

Hopcraft, J. G. C., G. Bigurube, J. D. Lembeli, and M. Bornar. Forthcoming. Alternatives to the Serengeti highway: Finding solutions that benefit both socio-economic development and conservation. *PLoS ONE*.

Hopcraft, J. G. C., S. A. R. Mduma, M. Borner, G. Bigurube, A. Kijazi, D. T. Haydon, W. Wakilema, D. Rentsch, A. R. E. Sinclair, A. P. Dobson, and J. D. Lembeli. Forthcoming. Road around Serengeti provides greater economic development opportunities than a road through it. *Conservation Biology*.

Ikanda, D., and C. Packer. 2008. Ritual vs. retaliatory killing of African lions in the Ngorongoro Conservation Area, Tanzania. *Endangered Species Research* 6:67–74.

Johnson, M. 1929. *Lion*. New York: G. P. Putnam's & Sons.

Kaare, M., T. Lembo, K. Hampson, E. Eblate, A. Estes, C. Mentzel, and S. Cleaveland. 2009. Rabies control in rural Africa: Evaluating strategies for effective domestic dog vaccination. *Vaccine* 27:152–60.

Kissui, B. M. 2008. Livestock predation by lions, leopards, spotted hyenas, and their vulnerability to retaliatory killing in the Maasai steppe, Tanzania. *Animal Conservation* 11:422–32.

Kruuk, H. 1972. *The spotted hyena*. Chicago: University of Chicago Press.

Kushnir, H. 2009. Lion attacks on humans in southeastern Tanzania: Risk factors and perceptions. PhD diss., University of Minnesota.

Kushnir, H., H. Leitner, D. Ikanda, and C. Packer. 2010. Human and ecological risk factors for unprovoked lion attacks on humans in southeastern Tanzania. *Human Dimensions of Wildlife* 15:315–31.

Laws, R. M. 1969. The Tsavo research project. *Journal of Reproductive Fertility (Supplement)* 6:495–531.

Laws, R. M., I. S. C. Parker, and R. C. B. Johnstone. 1975. *Elephants and their habitats.* Oxford: Oxford University Press.

Lembo, T., K. Hampson, D. T. Haydon, M. Craft, A. Dobson, J. Dushoff, E. Ernest, R. Hoare, M. Kaare, T. Mlengeya, et al. 2008. Exploring reservoir dynamics: A case study of rabies in the Serengeti ecosystem. *Journal of Applied Ecology* 45:1246–57.

Lembo T., K. Hampson, M. Kaare, E. Ernest, D. Knobel, R. Kazwala, D. Haydon, and S. Cleaveland. 2010. The feasibility of eliminating canine rabies in Africa: Dispelling doubts with data. *PLoS Neglected Tropical Diseases* 4:e626.

Maddock, L. 1979. The "migration" and grazing succession. In *Serengeti: Dynamics of an ecosystem*, ed. A. R. E. Sinclair and M. Norton-Griffith, 104–33. Chicago: University of Chicago Press.

Mduma, S., R. Hilborn, and A. R. E. Sinclair. 1998. Limits to exploitation of Serengeti wildebeest and implications for its management. In *Dynamics of tropical communities*, British Ecological Society Symposium 37, ed. D. M. Newbery, N. Brown, and H. H. T. Prins, 243–65. Oxford: Blackwell Science.

Mduma, S. A. R., A. R. E. Sinclair, and R. Hilborn. 1999. Food regulates the Serengeti wildebeest population: A 40-year record. *Journal of Animal Ecology* 68:1101–22.

Metzger, K., A. R. E. Sinclair, K. Campbell, R. Hilborn, J. G. C. Hopcraft, S. A. R. Mduma, and R. Reich. 2007. Using historical data to establish baselines for conservation: The black rhinoceros (*Diceros bicornis*) of the Serengeti as a case study. *Biological Conservation* 139:358–74.

Moore, A. 1937. *Serengeti*. London: Country Life Ltd.

Munson, L., K. A. Terio, R. Kock, T. Mlengeya, M. E. Roelke, E. Dubovi, B. Summers, A. R. E. Sinclair, and C. Packer. 2008. Climate extremes promote fatal co-infections during canine distemper epidemics in African lions. *PLoS ONE* 3:1–6. e2545.

Nature. 2010. Editorial. An alternative route: A proposed road through the Serengeti can be halted only by providing a viable substitute, not by criticism. *Nature* 467: 251–52.

Newmark, W. D., J. I. Boshe, H. I. Sariko, and G. K. Makumbule. 1996. Effects of a highway on large mammals in Mikumi National Park, Tanzania. *African Journal of Ecology* 34:15–31.

Normile, D. 2008. Driven to extinction. *Science* 319:1606–09.

Norton-Griffiths, M. 1979. The influence of grazing, browsing, and fire on the vegetation dynamics of the Serengeti. In *Serengeti: Dynamics of an ecosystem*, ed. A. R. E. Sinclair and M. Norton-Griffiths, 310–52. Chicago: University of Chicago Press.

Norton-Griffiths, M., M. Y. Said, S. Serneels, D. S. Kaelo, M. Coughenour, R. H. Lamprey, D. M. Thompson, and R. S. Reid. 2008. Land use economics in the Mara area of the Serengeti ecosystem. In *Serengeti III: Human impacts on ecosystem dynamics*, ed. A. R. E. Sinclair, C. Packer, S. A. R. Mduma, and J. M. Fryxell, 379–416. Chicago: University of Chicago Press.

Packer, C., D. Ikanda, B. Kissui, and H. Kushnir. 2005. Lion attacks on humans in Tanzania. *Nature* 436:927–28.

———. 2008. The ecology of man-eating lions in Tanzania. *Nature and Faune* 21:10–15.

Pennycuick, L. 1975. Movements of the migratory wildebeest population in the Serengeti area between 1960 and 1973. *East African Wildlife Journal* 13:65–88.

Pienaar, U. de V. 1983. Management by intervention: The pragmatic option. In *Manage-*

ment of large mammals in African conservation areas, ed. N. R. Owen-Smith, 23–26. Pretoria: Haum Educational Publishers.

Pienaar, U. de V., P. W. van Wyk, and N. Fairall. 1966. An aerial census of elephant and buffalo in the Kruger National Park and the implications thereof on intended management schemes. *Koedoe* 9:40–108.

Plowright, W. 1982. The effects of rinderpest and rinderpest control on wildlife in Africa. *Symposium of the Zoological Society, London* 50:1–28.

Roelke-Parker, M. E., L. Munson, C. Packer, R. Kock, S. Cleaveland, M. Carpenter, S. J. O'Brien, A. Pospischil, R. Hofmann-Lehmann, H. Lutz, et al. 1996. A canine distemper virus epidemic in Serengeti lions (*Panthera leo*). *Nature* 379:441–45.

Schaller, G. B. 1972. *The Serengeti lion: A study of predator-prey relations*. Chicago: University of Chicago Press.

Serneels, S., and E. F. Lambin. 2001. Impact of land-use changes on the wildebeest migration in the northern part of the Serengeti-Mara ecosystem. *Journal of Biogeography* 28: 391–407.

Sinclair, A. R. E. 1973. Population increases of buffalo and wildebeest in the Serengeti. *East African Wildlife Journal* 11:93–107.

———. 1977. *The African buffalo. A study of resource limitation of populations*. Chicago: University of Chicago Press.

———. 1979. The eruption of the ruminants. In *Serengeti: Dynamics of an ecosystem*, ed., A. R. E. Sinclair and M. Norton-Griffith, 82–103. Chicago: University of Chicago Press.

———. 1995a. Equilibria in plant-herbivore interactions. In *Serengeti II: Dynamics, management and conservation of an ecosystem*, ed. A. R. E. Sinclair and P. Arcese, 91–114. Chicago: University of Chicago Press.

———. 1995b. Serengeti past and present. In *Serengeti II: Dynamics, management and conservation of an ecosystem*, ed. A. R. E. Sinclair and P. Arcese, 3–30. Chicago: University of Chicago Press.

———. 1998. Natural regulation of ecosystems in protected areas as ecological baselines. *Wildlife Society Bulletin* 26:399–409.

———. 2010. Road proposal threatens existence of Serengeti. *Oryx* 44:478–79.

Sinclair, A. R. E., H. Dublin, and M. Borner. 1985. Population regulation of Serengeti wildebeest: A test of the food hypothesis. *Oecologia* 65:266–68.

Sinclair, A. R. E., S. A. R. Mduma, J. G. C. Hopcraft, J. M. Fryxell, R. Hilborn, and S. Thirgood. 2007. Long-term ecosystem dynamics in the Serengeti: Lessons for conservation. *Conservation Biology* 21:580–90.

Sinclair, A. R. E., and M. Norton-Griffiths, eds. 1979. *Serengeti: Dynamics of an ecosystem*. Chicago: University of Chicago Press.

Stronach, N. 1989. Grass fires in Serengeti National Park, Tanzania: Characteristics, behaviour and some effects on young trees. PhD diss., University of Cambridge, Cambridge.

TANAPA (Tanzania National Parks). 2006. The Serengeti 10-year management plan, 2005–2015. Tanzania National Parks, Arusha.

Thirgood, S., C. Mlingwa, E. Gereta, V. Runyoro, R. Malpas, K. Laurenson, and M. Borner. 2008. Who pays for conservation? Current and future financing scenarios for the Serengeti ecosystem. In *Serengeti III: Human impacts on ecosystem dynamics*, ed. A. R. E.

Sinclair, C. Packer, S. A. R. Mduma, and J. M. Fryxell, 443–70. Chicago: University of Chicago Press.

Thirgood, S., A. Mosser, S. Tham, J. G. C. Hopcraft, E. Mwangomo, T. Mlengeya, M. Kilewo, J. M. Fryxell, A. R. E. Sinclair, and M. Borner. 2004. Can parks protect migratory ungulates? The case of the Serengeti wildebeest. *Animal Conservation* 7: 113–20.

Turner, M. 1987. *My Serengeti years*. London: Elm Tree Books/Hamish Hamilton Ltd.

Watson, R. M. 1967. The population ecology of the wildebeest (*Connochaetes taurinus albojubatus*) in the Serengeti. PhD diss., Cambridge University, Cambridge.

Williamson, D., and B. Mbano. 1988. Wildebeest mortality during 1983 at Lake Xau, Botswana. *African Journal of Ecology* 26:341–44.

TWENTY-SEVEN

The Future of Conservation: Lessons from the Serengeti

Anthony R. E. Sinclair, Andy Dobson, Kristine L. Metzger, John M. Fryxell, and Simon A. R. Mduma

> *In accepting the trusteeship of our wildlife we solemnly declare that we will do everything in our power to make sure that our children's grandchildren will be able to enjoy this rich and precious inheritance.*
>
> *The conservation of wildlife and wild places calls for specialist knowledge, trained manpower, and money, and we look to other nations to cooperate with us in this important task—the success or failure of which not only affects the continent of Africa but the rest of the world as well.*
>
> —Julius K. Nyerere, Arusha Manifesto, 1961
> First President of Tanzania and Father of the Nation

In chapter 1 we raised the issue of whether we can create sustainable human landscapes and whether protected areas are a necessary component to achieve this. Specifically we addressed three questions: Do protected areas play a role in conservation that is not achieved in human ecosystems? Do human-dominated systems contribute to conservation objectives? Do these two—protected areas and human-dominated areas—support each other so that human systems are sustainable? In essence we ask how the greater Serengeti ecosystem, including both the human and natural components, can be made sustainable. So in this volume we examine whether the Serengeti is maintaining its biota, and how it relies on outside influences. In other words to maintain a protected area indefinitely, we must maintain the greater ecosystem within which it is embedded.

We also emphasize the generalities of many of the insights from the present research for studies of other ecosystems with particular emphasis on the implications for long-term conservation and management. This involves a deeper understanding of the human population that increasingly surrounds the park. The philosophy of conservation in protected areas is that such areas should not significantly reduce the welfare of people who live around them; they should instead provide economic benefits to those who incur the costs from living near the park. Then protected areas have a value to those living near them. We emphasize that the primary purpose of national parks and World Heritage Sites is to preserve the wonders of nature intact for future generations and for the continued well-being of the multitude of nonvoting species that live in and around them.

Optimistically we believe that a sentient world will ultimately see an increased appreciation for the ethical rights of other species; in its simplest form this optimism stems from the simple observation that the last century has seen greater equality for individuals of both sexes and all races. From neither an evolutionary, nor any religious perspective, can we readily differentiate any logical grounds for assuming the superiority of one species over any other. From a purely evolutionary perspective it is a simple matter of chance that humans are the species in charge of managing the park and have only played this role for a tiny portion of its history. The present emphasis on modern humans (in the past few hundred years) as part of management is because they are a very recent and dominant threat to the existence of the park. In the absence of modern humans, the park has managed itself in a state of wondrous abundance for several million years. The ultimate goal of management, therefore, should be to minimize the influence of humans and ensure this continues so that Serengeti acts as a baseline for comparison with outside systems. Serengeti is one of very few parks in the world where it is still possible to appreciate what an unmanaged and naturally functioning ecosystem looks like. We can think of very few parks where intensive management has come anywhere near mimicking the natural processes that drive a large fully functioning ecosystem. Perhaps the ultimate thing we have learned from 50 years of research in Serengeti is that in the absence of humans, the park has always done a phenomenally good job of managing itself, continually recovering from droughts, volcanic eruptions (which ultimately nurture it), and disease outbreaks. The predominant emphasis of management in recent years is to minimize the damage induced by direct and indirect human actions: disease introduction, poaching, invasive species, and conversion of surrounding habitat.

We first address the three questions outlined above. As with earlier volumes we provide explicit recommendations for the ways in which we perceive that this research should influence management. We fully appreciate that it has not always been possible to act on this advice, but we take the liberty of emphasizing areas where in the past, research has been critical to the development of sound management policy; we also emphasize areas where ignoring insights gained from research has led to mismanagement.

IS THE PROTECTED AREA NECESSARY FOR CONSERVATION OF BIODIVERSITY?

Research within the protected area has increased the known species of several groups including microbes, nematodes (chapter 8), insects (chapters 10, 11), rodents (chapter 12), and birds (chapter 13). Synthesis of the half century of data on the wildebeest migration by Hopcraft et al. (chapter 6) establishes the changes in demography, the spatial extent of the population, and the impacts on the Serengeti ecosystem. They confirm the important lesson that the Serengeti would have completely different dynamics if not for wildebeest, and reciprocally, wildebeest would not be so abundant but for the unique features of the Serengeti; the two are inextricably connected.

What are these unique attributes? Fire and its impacts is one of them and these details are synthesized by Eby et al. (chapter 4): fire drives changes in tree populations, determines the movements of migrant ungulates, and changes grass structure for insects and birds. However, we still have a lot to learn about its impact on other animal groups, and also how fire is to be managed in the ecosystem. Spatial mosaics are another attribute: they provide the variability that determines the migration patterns as described by Metzger et al. (chapter 3). There are spatial changes in virtually every physical parameter including geology, soils, vegetation, climate, and human impacts of agriculture. Furthermore, there are different scales of change and these differ in how they structure the ecosystem. Managing the habitats to maintain the natural spatial mosaic (heterogeneity) is important to maintain biodiversity (Christensen 1997).

Comparison of biodiversity inside and outside the protected area shows a consistent pattern where restricted range species inside the park are lost outside—only the global or Africa-wide species are found there (Sinclair et al., chapter 11). Even microbe species assemblages show changes with the different grazing regimes inside and outside the park because inside grazing is seasonal, while outside it is more persistent through the year (Verchot et al., chapter 8). Top carnivore species are lost outside the park, particularly in agricultural areas, and less so in pastoralist areas (Craft et al., chapter 15). Loss of top predators such as raptors outside the park may be having impacts on rodent dynamics and possibly disease transmission in human ecosystems (Byrom et al., chapter 12).

Large mammals are not only safer inside the park but they are *aware* they are safer: this has now been documented for elephants. Our long-term records of recruitment of juvenile elephants to the population showed high recruitment from the start of records (1966) to 1980. There was a major drop in recruits during the 1980s and rates only improved after 1990. The 1980s was the period of severe elephant poaching but poaching did not target juveniles. The recruitment ratios should have increased during this period both because of the higher adult mortality and because lower numbers stimulate higher birth rates (as was indeed seen after 1990). We hypothesized that the lower recruitment was due to higher stress from human killing. This stress has now been confirmed by Tingvold et al. (2013); elephants have higher stress hormones not only when they leave the park but even when they approach the boundaries of the park prior to leaving.

Processes that determine the distribution and time trends in plants are summarized by Anderson et al. (chapter 5). Tree dynamics have been described here and elsewhere for riverine forests (Turkington et al., chapter 9) and savanna tree species (Sinclair et al. 2008). Nevertheless, underlying drivers for such change remain largely unexplored. As Anderson et al. (chapter 5) state "despite four decades of research on the topic, the direct and indirect effects of fire, soil moisture, grass competition and herbivory on the germination, recruitment, growth, and survival of trees in savannas is still largely unknown." They comment that there is relatively weak coupling between the processes that determine germination, establishment, and the growth of small trees and the processes that maintain the density of adult trees. They also give an important warning: invasive plants are threatening the Serengeti and must be combated immediately by the Tanzania national parks managers. Overall the dynamics of plant communities need more intensive study for their conservation.

Studies of processes such as grazing show major top-down impacts on soils, microbes, and nematodes (Verchot et al., chapter 8), and on birds (together with those of fire) (Nkwabi et al., chapter 14). Modeling of tree dynamics, fire, herbivory, and herbivore populations by Holdo and Holt (chapter 20) has confirmed top-down processes that emanated from the massive perturbations of rinderpest (see chapter 2). Holdo and Holt also show that both a pure ecological model and a coupled social-natural ecosystem model are necessary depending on the scale and the process under examination.

Climate change is now appearing as an important driver (see also chapter 25). Fryxell et al. (chapter 7) show that modeling of long-term data supports the hypothesis that global variation in climate, as measured by the Southern Oscillation Index (SOI), has substantial effects on juvenile recruitment of wildebeest, topi, and lions in the Serengeti ecosystem, whereas the majority of other herbivore species are apparently insensitive to SOI variation. Byrom et al. (chapter 12) also refer to how climate change could alter the trophic dynamics of small mammals, raptors, and small carnivores based on changes in the SOI. These environmental drivers are particularly relevant to climate change. Thus, dry season rainfall may be increasing in Serengeti and if this trend were to continue, our data imply that wildebeest and topi recruitment should benefit, as would that of lions. This suggests that deeper appreciation of global controls on climate change may have important long-term implications for population abundance and persistence in the Serengeti ecosystem.

Anderson et al. (chapter 5) point out that increased atmospheric CO_2 is having a fertilization effect in savannas that could be changing competition between trees and grasses by favoring the former (Bond 2008). Thus, if tree seeds can germinate successfully, perhaps from the increase in rainfall mentioned above, then elevated CO_2 levels could increase growth rates so that seedlings have a higher probability of escaping fire and herbivory during their first few seasons, which are crucial for establishment. Both more intensive research and modeling are required in the future.

DO HUMAN-DOMINATED SYSTEMS CONTRIBUTE TO CONSERVATION OBJECTIVES?

We ask whether the protected area is dependent on a sustainable human-dominated system. The Serengeti ecosystem is shaped by the wildebeest migration which in turn depends almost entirely on the flow of water in the Mara River during the dry season. So the Serengeti depends vitally on water flows from outside the system, specifically on the Mau highland forests of Kenya. Thus, the Serengeti requires a sustainable human-dominated system within which it lives—but there are indications the water supply is declining from unsustainable offtake (chapter 26).

The wildebeest migration uses areas outside the protected area in the far north of the greater Serengeti ecosystem (on the Loita plains and above the Isuria escarpment). Such areas are under increasing human use and becoming less available to wildlife. The migration also depends on the Maswa Reserve during the short dry season—that period January–March after the short rains or even more importantly when the short rains fail—because the animals move from the plains westward to Maswa, this being the closest water sources on the southern plains. In this period the migration does not reverse itself and move north toward Kenya. Thus, Maswa is essential during one period of the year. The Maswa Reserve has been continuously eroded over the past decades, the boundary being moved to accommodate encroaching settlement for political expediency. If the Maswa Reserve disappears then the Serengeti ecosystem will likely change.

Illegal hunting is a continuing pressure on wildlife populations and it is likely to increase as human numbers increase in surrounding areas. Bushmeat hunting is more a commercial enterprise than a subsistence activity. Perversely, alleviation of poverty (and these hunters are very poor) will increase hunting rather than reduce it. Thus, hunting has the potential to reach unsustainable levels as economic levels in the human-dominated areas increase in the future (chapter 22). There are more signs that ivory poaching is increasing due to economic demands from Asia. If history repeats the events of the 1980s then there will be few elephants left 20 years from now.

IS THE HUMAN-DOMINATED PART OF THE GREATER SERENGETI ECOSYSTEM SUSTAINABLE?

Conservation of the Serengeti will only succeed by understanding the human ecosystem surrounding it. Reid et al. (chapter 25) synthesize the six major issues that are currently determining whether the greater Serengeti ecosystem persists in the long term. These are: (1) human population increase and land sequestration around the protected area, and the two issues resulting from this being (2) offtake of water from the Mara River, and (3) illegal hunting. There are two socioeconomic issues being (4) the threat of a trunk road across northern Serengeti, and (5) changes to the governance of resources. Finally, (6) there are threats resulting from environmental climate change. We have addressed the issues of the Mara River and the northern road above. Here we consider the other issues.

Human impacts on the protected ecosystem are largely due to illegal hunting. In the most detailed study to date Rentsch et al. (chapter 22) document the drivers for and degree of bushmeat hunting by local communities on the edge of the protected area. They consider three essential policies for reducing these impacts on wildlife: (1) increase the negative incentive for hunting because at present they are ineffective; (2) improve alternative livelihood opportunities to replace bushmeat as a commodity. We address the issues concerning livelihoods below, and (3) shift the economic incentives away from consumption of bushmeat to another replenishable protein derived from domestic species. Their most important finding is that bushmeat hunting is not driven by poverty but by commercial economics—hunting will increase when wealth increases. So poverty is not necessarily a driver of threats to wildlife (Rentsch and Damon 2013).

The trick to reducing human impacts on wildlife lies in understanding social and economic drivers determining the livelihoods of the peoples within the greater Serengeti ecosystem and the causes of human-wildlife conflict. The origin of human-wildlife conflicts in the greater Serengeti ecosystem began not with the confiscation of lands and resources as is often supposed. Indeed humans were not practicing agriculture (nor even pastoralism in the majority of the system) due to the presence of tsetse fly (Sinclair et al., chapter 2); rather conflict began and then intensified by peoples arriving on the boundaries of what became the protected area having been pushed out of their traditional areas elsewhere; this occurred among pastoralists on the eastern side a century ago (chapter 2), but much more recently (after the boundaries were demarcated) among agriculturalists on the western boundary (Estes et al., chapter 18). Estes et al. show that people on the western boundaries of Serengeti National Park were not attracted by wildlife resources nor related benefits from the protected area but rather by new agricultural land. This is also the same process occurring in the Ngorongoro Conservation Area; agriculturalists are invading what was supposed to be a pastoralist-wildlife community-based conservation system in order to find new agricultural land. This process is a major cause of poverty and malnutrition among the pastoral Maasai (Galvin et al., chapter 17).

Those arriving on the boundaries looking for new land are among the poorest in the communities. They are least able to withstand the impacts of the adjacent wildlife populations. As Hampson et al. (chapter 21) emphasize "there is a need to protect rural livelihoods, reduce their vulnerability, counterbalance losses from wildlife with benefits, and foster community-based conservation with tangible profits to communities unlike existing models of community conservation, otherwise antagonism toward conservation objectives are likely to increase. The nature of the problem requires both short-term mitigation tools and longer-term preventive strategies and compatible land-use policies."

Transmission of disease between wildlife and human systems is a further burden on local peoples as well as a threat to conservation of rare wildlife species. Lembo et al. (chapter 19) describe the new approaches for improved disease surveillance and ecological data gathering, for analyzing impacts of pathogens on different hosts and for reciprocal effects on ecosystem function. Their long-term work is a model for conservation, public health policies, and for the development of ecosystem health approaches. The hope is that these efforts will also improve public relations between national parks authorities and local communities.

Linda Knapp et al. (chapter 23) highlight the chronic state of malnutrition in the communities living outside the Serengeti Park, and particularly in the eastern side; they are in a constant state of poor nutrition and compromised health. The authors note that it is not the protected area itself that is the cause of malnutrition; instead such poor nutrition is more a sign of the economic state of all rural Tanzanians. Potentially this is an issue where conservation could provide benefit. To mitigate these problems of health and well-being, both Linda Knapp et al. and Hampson et al. (chapter 21) emphasize two important issues that relate to governance. First, successful biodiversity conservation can only occur if there is poverty reduction and improved well-being; and secondly, benefits from protected areas have to be distributed equitably so that those suffering the costs also receive the benefits (see also Wilkie et al. 2006). So what are these costs and benefits from being close to protected areas?

Eli Knapp et al. (chapter 16) distill the four important human-wildlife interactions, two having negative consequences for local communities (crop destruction, livestock depredation), and two having positive consequences (park-related employment, and illegal hunting). The most important negative effects came from crop destruction (although depredation of livestock was regarded as important in some areas). Surprisingly, the overall effect of these impacts was a financial benefit, contrary to previous opinion (e.g., Colchester 2004). However, these benefits and losses are not evenly distributed (some households experience benefits while others losses) and it is the losers that create the negative attitudes in the community.

THE GREATER SERENGETI ECOSYSTEM: THE MANAGEMENT PRIORITIES

In 2010 the Tanzania National Parks drew up a list of priorities for future research to address management concerns in Serengeti based on the present research. This list is presented in table 27.1, ranked according to priority category but not ranked within each category. It is clear that management saw information on conservation as the predominant need; there are five conservation, four economic, two social, and one health project identified under the highest priority. Conservation covers water resources and protection of iconic species; economic issues are entirely concerned with problems of tourism; social issues identify human-wildlife conflict as a looming problem, and the transmission of disease from tsetse flies as a serious health problem.

Table 27.1 Research priorities for the Serengeti National Park prepared by Tanzania National Parks

Category	Issue	Topic	Justification	Priority
Conservation	Water resources conservation and sustainable utilization	Eco-hydrological studies of the (major) rivers, Mara, Grumeti, Mbalageti, Simiyu	Long-term data needed for informed decision making on water utilization	Very high
Conservation	Sustainable conservation of flagship wildlife species wildebeest and elephants	Assessment of the status of elephant and wildebeest outside protected areas	They spend considerable time outside protected areas where they are potentially threatened	Very high
Conservation	Rhino conservation	Assessment of the population genetics of Moru rhinos (Inbreeding?)	Small isolated rhino population may potentially result in inbreeding leading to genetic depression	Very high
Conservation	Rhino conservation	Survival of reintroduced rhino	Reintroduction program ongoing, and needs updated information	Very high
Conservation	Poaching	Research on effective poaching control options	Applied research needed to solve the problem of poaching	Very high
Economic	Water resources conservation and sustainable utilization	Surveying of underground water availability and quality for human use	Water scarcity for human consumption is a major management issue including improving tourism conditions	Very high
Economic	Tourism	Study on visitor satisfaction and attitudes	Enhancing quality of services to visitors	Very high
Economic	Tourism	Assessment of visitor capacity in Serengeti	Over use in the Mara Reserve has had negative effects on wildlife and wildlife experience for tourists; international concern (UNESCO/IUCN)	Very high
Economic	Tourism	The status and potential contribution of tourism to the economy of local communities	Needed to engage and link local communities to tourism benefits of the park	Very high
Health	Animal health and diseases	Evaluation of control options for tsetse flies	Adverse impact on human health (compromises tourism)	Very high

continued

Table 27.1 continued

Category	Issue	Topic	Justification	Priority
Social	Human-wildlife conflicts	Socioeconomic impacts of human-wildlife conflicts in adjacent villages	Increasing local community outcry on the problem	Very high
Social	Human-wildlife conflicts	Control options for human-wildlife conflicts	Innovative and effective control measures urgently needed	Very high
Conservation	Sustainable conservation of flagship wildlife species wildebeest and elephants	Monitoring the population dynamics and migration patterns of wildebeest and elephants	Flagship species are key indicators of ecosystem health; long-term (unbroken) studies are necessary	High
Conservation	Conservation of rare animal species	Assessment of the population status and dynamics of (selected) rare animal species	Status of rare animal species unknown: roan, oryx, greater kudu, mountain reedbuck, striped hyena, caracal, patas, colobus	High
Conservation	Poaching	Long-term monitoring of poaching dynamics and trend	Poaching major budgetary expenditure (>40%); poaching stable for 15 years, monitoring required for decisions	High
Conservation	Species specific studies	Carnivores (lions, cheetah, hyena, wild dogs); ungulates (giraffe, hippo, waterbuck, bohor reedbuck, Grant's gazelle)	Carnivores ongoing; ungulates new projects needed.	High
Ecological	Fire ecology	Impact of fire on vegetation (grassland/ woodland)	Regular prescribed and unprescribed (hot) burning may destroy the natural plant and animal biodiversity: riverine forests, hill thickets, swamps	High
Ecological	Biodiversity status of key taxa	Inventories for amphibians, fish, reptiles, and insects	Low taxa biodiversity status poorly understood; needed for conservation; used for taxa-specific tourism guide books	High
Economic	Tourism	Long-term monitoring of the impact of tourism facilities on park resource values	Tourism is growing; data needed for informed tourism management decisions	High

Economic	Socioeconomic issues	Long-term monitoring of the socioeconomic status and trends for local communities adjacent to the park	Needed to monitor adjacent human population pressure on park resources	High
Health	Animal health and diseases	Epidemiology of human trypanosomiasis	Long-term monitoring needed	High
Health	Animal health and diseases	Impact of selected diseases (anthrax, foot-and-mouth disease, Rift Valley fever, malignant cattarrhal fever, skin disease, rabies, genital disease, mange, etc.) to key species (wildebeest, zebra, giraffe, baboon)	Significant mortalities due in outbreaks, annual and seasonal records; short & long term monitoring needed	High
Social	Animal road kills	Monitoring of road kills on public roads traversing the Park	Data necessary for better management of public transportation through the park	High
Conservation	Sustainable conservation of flagship wildlife species wildebeest and elephants	Assessment of the population ecology of resident wildebeest and zebra in Western Corridor	Resident wildebeest in the park are poorly studied	Medium
Ecological	Climate change	Impacts of global warming on park resources and values	New area of global interest	Medium
Ecological	Climate change	Studies on possibilities of carbon exchange and trading	Possibilities for the role of the park as a major carbon-sinking platform; a new area of global interest and collaboration	Medium
Health	Animal health and diseases	Prevalence of zoonotic diseases in and around the park	Incidents of rabies and other diseases reoccur frequently	Medium
Health	Animal health and diseases	Prevalence of trypanosome rhodesiense in large ruminants	Visitors and park staff confirmed with Hathipaon disease; need to establish hosts and causative agent	Medium

In the next category, management saw a need for continued monitoring of wildlife to obtain data on a number species for ecological and conservation reasons. An important economic need is the welfare of local peoples. Health identifies a number of human and wildlife diseases for which data are needed. In the medium category, further information on wildlife species, climate change, and diseases are identified.

In general this list confirms the requirement for continued monitoring of the main biological species and environmental variables for water and climate monitoring. It also confirms that social issues need urgent attention; both the impact of wildlife on humans and the reverse effect of poaching on wildlife are seen as important areas to focus research for the future.

CONCLUSION

We began this book by considering the recent emphasis on community-based conservation, whether this approach can be biologically sustainable, and whether protected areas are necessary. We asked three questions. First, does the protected area provide sanctuary unavailable outside it? The studies of biodiversity comparing the natural system in Serengeti with that in the surrounding human-modified areas showed a consistent loss of endemic biota in human areas; the local and endemic species are less able to withstand human disturbance. This pattern appears in every group studied, and in particular, greater losses occur at higher trophic levels—carnivores, raptors, insectivorous birds. Thus, the park is essential for protecting species that cannot live in human-dominated landscapes—particularly carnivores, megaherbivores, and fragile species that require special habitats and ones with restricted ranges, as we have seen with birds and insects. This supports evidence from other large protected areas (Cantú-Salazar and Gaston 2010). Are these populations sustainable entirely within the protected area? We do not yet know the answer to this question. Some small species may have viable population sizes. But some such as forest birds have such fragmented habitats (chapter 9) that they may not be able to exist entirely within the Serengeti and may depend on supplements from other areas. They cannot depend on the greater Serengeti ecosystem because all such forest has been destroyed. If this is the case, then there is a need to make the human-dominated ecosystem more biologically sustainable.

Second, therefore, is the human-dominated system biologically sustainable? The distortion of the trophic cascade due to the loss of ecologically important species will have important impacts on the stability of human ecosystems—in the case of the Serengeti region through outbreaks of diseases, rodents, insects, and invasive plants. The protected area was a necessary instrument in detecting these changes in ecology.

Third, do human-dominated systems contribute to conservation objectives? The human ecosystem is a necessary component that allows the protected area to exist, and so we need to make that system self-sustaining. One way is to diversify habitats within agriculture—the loss of plant species and the homogenization of crops has led to the loss of sensitive (fragile) endemic species at higher trophic levels. This suggests an obvious experiment in conservation biology where sample sites within agriculture of regenerated native plants are compared with both the surrounding unaltered agriculture and the protected area, with respect to reinvading native species of insects, birds, reptiles, and small mammals.

Shellenberger and Nordhaus (2011) imply that it is up to conservationists to redirect efforts towards human-dominated ecosystems. But it should not fall on just the conservationists to do this; it is up to society as a whole to value its own environment and protect it as society's obligation to future generations. It is now emerging that species loss is driving global environmental and ecosystem change (Cardinale, et al. 2012; Sutherland et al. 2013). So the decline in global sustainability is a global problem. At the same time society has to value the presence of the protected area—peoples need to obtain economic benefits from these areas on a sustainable basis so that there is a stable local economy and improved livelihood. Conservation in protected areas must avoid the pitfalls documented in Kareiva, Lalasz, and Marvier (2011) where peoples reacted against these areas because they saw no benefit. One of the most instructive findings from the present research is the high percentage of households in areas surrounding the park that are unaware of even the existence of the national park and what it stands for (Eli Knapp et al., chapter 16). As these authors point out "local people with limited knowledge of the greater Serengeti ecosystem are unlikely to know how natural resource use is restricted or regulated, be it hunting, grazing, or fuelwood collection." The people will also not value the park.

This extraordinary finding has important implications for the role of governance. A consistent theme repeated in many chapters (Hampson et al., chapter 21; Linda Knapp et al., chapter 23; Randall et al., chapter 24; Reid et al., chapter 25) emphasizes that the strengthening of governance systems is one of the most important challenges of the next century for biodiversity conservation (Agrawal and Ostrom 2006). Governance requires the direction of funds and services to peoples on the boundaries of protected areas so that those who suffer the costs of conservation can obtain compensatory benefits. But in the light of the above finding we see that even if such communities receive benefits, they do not appreciate (and therefore value) the source of the funds, namely the protected area. In short, benefits cannot translate into care and protection through community-based conservation. Clearly education as well as governance is essential.

President Nyerere's statement made at the outset of independence in 1961 provides the rationale for the conservation of Serengeti for the benefit of both Tanzania and the world. The role of research is to inform the world about exceptional areas such as the Serengeti and to advise management on potential threats to the system. Information is the fundamental basis of this advice and it is obtained from careful, systematic data gathering over long periods of time. Little was known about the Serengeti in the 1950s but the decades of research has resulted in this system becoming internationally famous and a major economic asset for Tanzania. It was the accumulated data from the long-term monitoring that has allowed scientists to address and resolve the main management problems over the decades, and these data have allowed scientists to warn of the consequences of blocking the migration. At the same time, the supporting data from historical studies documented in chapter 2 were crucial to developing and protecting the Serengeti.

The present program of monitoring is coming to an end. If scientists are to continue providing the basic knowledge for conservation, then the work will have to be supported from a trust fund. This has yet to be set up, but if this is not forthcoming, there will be little or no further information.

REFERENCES

Agrawal, A., and E. Ostrom. 2006. Political science and conservation biology: A dialogue of the deaf. *Conservation Biology* 20:681–82.

Bond, W. J. 2008. What limits trees in C4 grasslands and savannas? *Annual Review of Ecology, Evolution and Systematics* 39:641–59.

Cantú-Salazar, L., and K. J. Gaston. 2010. Very large protected areas and their contribution to terrestrial biological conservation. *BioScience* 60:808–18.

Cardinale, B. J., J. E. Duffy, A. Gonzalez, D. U. Hooper, C. Perrings, P. Venail, A. Narwani, G. M. Mace, D, Tilman, and D. A. Wardle, et al. 2012. Biodiversity loss and its impact on humanity. *Nature* 486:59–67.

Christensen, N. L. 1997. Managing for heterogeneity and complexity on dynamic landscapes. In *The ecological basis for conservation: Heterogeneity, ecosystems, and biodiversity*, ed. S. T. A. Pickett, R. S. Ostfeld, M. Shachak, and G. E. Likens, 167–86. New York: Chapman and Hall.

Colchester, M. 2004. Conservation policy and indigenous peoples. *Environmental Science and Policy* 7:145–53

Kareiva, P., R. Lalasz, and M. Marvier. 2011. Conservation in the anthropocene. Beyond solitude and fragility. In *Love your monsters. Post-environmentalism and the Anthropocene*, ed. M. Shellenberger and T. Nordhaus, 26–36. Washington, DC: Breakthrough Institute.

Rentsch, D., and A. Damon. 2013. Prices, poaching, and protein alternatives: An analysis of bushmeat consumption around Serengeti National Park, Tanzania. *Ecological Economics* 91 (C): 1–9.

Shellenberger, M., and T. Nordhaus. 2011. Introduction. In *Love your monsters. Post-environmentalism and the Anthropocene*, ed. M. Shellenberger and T. Nordhaus, 5–7. Washington, DC: Breakthrough Institute.

Sinclair, A. R. E., J. G. C. Hopcraft, H. Olff, S. A. R. Mduma, K. A. Galvin, and G. J. Sharam. 2008. Historical and future changes to the Serengeti ecosystem. In *Serengeti III: Human impacts on ecosystem dynamics*, ed. A. R. E. Sinclair, C. Packer, S. A. R. Mduma, and J. M. Fryxell, 7–46. Chicago: University of Chicago Press.

Sutherland, W. J., S. Bardsley, M. Clout, M. H. Depledge, L. V. Dicks, L. Fellman, E. Fleishman, D. W. Gibbons, B. Keim, and F. Lickorish, et al. 2013. A horizon scan of global conservation issues for 2013. *Trends in Ecology and Evolution* 28:16–22.

Tingvold, H. G., R. Fyumagwa, C. Bech, L. F. Baardsen, H. Rosenlund, and E. Røskaft. 2013. Determining adrenocortical activity as a measure of stress in African elephants (*Loxodonta africana*) in relation to human activities in Serengeti ecosystem. *African Journal of Ecology* doi: 10.1111/aje.12069.

Wilkie, D. S., G. A. Morelli, J. Demmer, M. Starkey, P. Telfer, and M. Steil. 2006. Parks and people: Assessing the human welfare effects of establishing protected areas for biodiversity conservation. *Conservation Biology* 20:247–49.

CONTRIBUTORS

T. Michael Anderson
Department of Biology
206 Winston Hall
Wake Forest University
Winston-Salem, NC 27109, USA

Harriet Auty
Boyd Orr Centre for Population and Ecosystem Health
Institute of Biodiversity, Animal Health and Comparative Medicine
College of Medical, Veterinary and Life Sciences
University of Glasgow
Graham Kerr Bldg., Room 314
Glasgow G12 8QQ, UK

Tyler A. Beeton
Department of Anthropology
Colorado State University
Fort Collins, CO 80523, USA

Jayne Belnap
US Geological Survey
Southwest Biological Science Center
Canyonlands Research Station
2290 S. West Resource Blvd.
Moab, UT 84532, USA

Rene Beyers
Beaty Biodiversity Research Centre
6270 University Boulevard
University of British Columbia
Vancouver, BC, Canada V6T 1Z4

Randall B. Boone
Department of Ecosystem Science and Sustainability
and
Natural Resource Ecology Laboratory
Colorado State University
Fort Collins CO, 80523, USA

Markus Borner
Frankfurt Zoological Society
Box 14935
Arusha, Tanzania

Deborah Bossio
Director Soils Research
International Center for Tropical Agriculture (CIAT)
PO Box 823-00621
Nairobi, Kenya

John Bukombe
Serengeti Biodiversity Program
Tanzania Wildlife Research Institute
PO Box 661
Arusha, Tanzania

Andrea E. Byrom
Landcare Research
PO Box 40
Lincoln 7640, New Zealand

Sarah Cleaveland
Institute of Biodiversity, Animal Health and Comparative Medicine
College of Medical, Veterinary and Life Sciences
University of Glasgow
Glasgow G12 8QQ, UK

Michael Coughenour
NESB B221
Natural Resource Ecology Laboratory
Campus Mail 1499
Fort Collins, CO 80523-1499, USA

Meggan E. Craft
Division of Ecology and Evolutionary Biology
University of Glasgow
Glasgow G12 8QQ, UK

Jan Dempewolf
Department of Geographical Sciences
University of Maryland
4321 Hartwick Road, Suite 410
College Park, MD 20740, USA

Sara N. de Visser
Community and Conservation Ecology Group
Centre for Ecological and Evolutionary Studies
University of Groningen
PO Box 11103
9700 CC Groningen, The Netherlands

Junyan Ding
Beaty Biodiversity Research Centre
6270 University Boulevard
University of British Columbia
Vancouver, BC, Canada V6T 1Z4

Andy Dobson
Ecology and Evolutionary Biology
117 Eno Hall
Princeton University
Princeton, NJ 08544-1003, USA

Sarah M. Durant
Institute of Zoology
Zoological Society of London
Regent's Park, London, UK
and
Wildlife Conservation Society
2300 Southern Boulevard
Bronx, NY 10460, USA
and
TAWIRI
Box 661
Arusha, Tanzania

Stephanie Eby
Department of Biology
Syracuse University
107 College Place
Syracuse, NY 13210, USA

Eblate Ernest
Tanzania Wildlife Research Institute
PO Box 661
Arusha, Tanzania

Anna B. Estes
Department of Environmental Sciences
University of Virginia
Charlottesville, VA 22903, USA

Anke Fischer
Frankfurt Zoological Society
PO Box 14935
Arusha, Tanzania
and
The James Hutton Institute
Aberdeen AB15 8QH, UK

Guy J. Forrester
Landcare Research
PO Box 40
Lincoln 7640, New Zealand

Robert F. Foster
Northern Bioscience
363 Van Horne Street
Thunder Bay, Ontario, Canada P7A 3G3

Bernd P. Freymann
Community and Conservation Ecology Group
Centre for Ecological and Evolutionary Studies
University of Groningen
PO Box 11103
9700 CC Groningen, The Netherlands

John M. Fryxell
Department of Integrative Biology
University of Guelph
50 Stone Road E
Guelph, Ontario, Canada N1G 2W1

Robert Fyumagwa
Tanzania Wildlife Research Institute
PO Box 661
Arusha, Tanzania

Kathleen A. Galvin
Department of Anthropology
and
Natural Resource Ecology Laboratory
Colorado State University
Fort Collins, CO 80523, USA

John Gibson
The Centre for Genetic Analysis and Applications
C. J. Hawkins Homestead
University of New England
Armidale, NSW, Australia

Katie Hampson
Boyd Orr Centre for Population and Ecosystem Health
Institute of Biodiversity, Animal Health and Comparative Medicine
College of Medical, Veterinary and Life Sciences
University of Glasgow
Graham Kerr Bldg., Room 314
Glasgow G12 8QQ, UK

Olivier Hanotte
School of Biology
University of Nottingham
NG7 2RD, UK

Andrew W. Harvey
IDEA-NEW
North Central Region
Mazar-i-Sharif, Afghanistan

Dan Haydon
Boyd Orr Centre for Population and Ecosystem Health
Institute of Biodiversity, Animal Health and Comparative Medicine
College of Medical, Veterinary and Life Sciences
University of Glasgow
Graham Kerr Bldg., Room 314
Glasgow G12 8QQ, UK

Cuthbert B. Hemed
Wildlife Division
P.O. Box 176
Mugumu, Mara Region, Tanzania

Ray Hilborn
School of Aquatic and Fishery Sciences
Box 355020
University of Washington
Seattle, WA 98195, USA

Richard Hoare
Tanzania Wildlife Research Institute—Messerli Foundation
Wildlife Veterinary Programme
PO Box 707
Arusha, Tanzania

Ricardo M. Holdo
Division of Biological Sciences
University of Missouri
217 Tucker Hall
Columbia, MO, 65211-7400, USA

Robert D. Holt
Department of Biology
University of Florida
Gainesville, FL 32611, USA

J. Grant C. Hopcraft
Community and Conservation Ecology Group
Centre for Ecological and Evolutionary Studies
Groningen University
PO Box 11103
9700 CC Groningen, The Netherlands
and
Frankfurt Zoological Society
Box 14935
Arusha, Tanzania

Jill E. Jankowski
Beaty Biodiversity Research Centre
6270 University Boulevard
University of British Columbia
Vancouver, BC, Canada V6T 1Z4

Magai Kaare (Deceased)
Formerly Centre for Tropical Veterinary Medicine
Royal (Dick) School of Veterinary Studies
University of Edinburgh
Midlothian EH25 9RG, UK

Dickson S. Kaelo
Basecamp Foundation
PO Box 43369
Nairobi 00100, Kenya

Julius D. Keyyu
Tanzania Wildlife Research Institute
PO Box 661
Arusha, Tanzania

Emily Kisamo
Tanzania National Parks
PO Box 3134
Arusha, Tanzania

Bernard Kissui
African Wildlife Foundation
Tarangire Lion Project
PO Box 2658
Arusha, Tanzania

Eli J. Knapp
Graduate Degree Program in Ecology
Colorado State University
Fort Collins, CO 80523-1401, USA

Linda M. Knapp
Graduate Degree Program in Ecology
Colorado State University
Fort Collins, CO 80523-1401, USA

Tobias Kuemmerle
Geography Department
Humboldt-University Berlin
Unter den Linden 6
10099 Berlin, Germany

Hadas Kushnir
Department of Ecology, Evolution and Behavior
University of Minnesota
100 Ecology Building
1987 Upper Buford Circle
Saint Paul, MN 55108, USA

Felix Lankester
Lincoln Park Zoo
PO Box 395
Usa River, Arusha, Tanzania

Tiziana Lembo
Boyd Orr Centre for Population and Ecosystem Health
Institute of Biodiversity, Animal Health and Comparative Medicine
College of Medical, Veterinary and Life Sciences
University of Glasgow
Graham Kerr Bldg., Room 314
Glasgow G12 8QQ, UK

Martin Loibooki
Tanzania National Parks
PO Box 3134
Arusha, Tanzania

Asanterabi Lowassa
Tanzania Wildlife Research Institute
PO Box 661
Arusha, Tanzania

Alexander (Sandy) Macfarlane
Formerly Mineral Resources Division
Dodoma, Tanzania
and
The British Geological Survey
Keyworth, Nottingham NG12 5GG, UK

Ann L. Magennis
Department of Anthropology
Colorado State University
Fort Collins, CO 80523

Stephen Makacha
Serengeti Biodiversity Program
Tanzania Wildlife Research Institute
PO Box 661
Arusha, Tanzania

Lucas Malugu
Tanzania Wildlife Research Institute
PO Box 661
Arusha, Tanzania

Emmanuel Masenga
Tanzania Wildlife Research Institute
PO Box 661
Arusha, Tanzania

J. Terrence McCabe
Institute of Behavioral Science
and
Department of Anthropology
University of Colorado
Boulder 80309, USA

John Mchetto
Serengeti Biodiversity Program
Tanzania Wildlife Research Institute
PO Box 661
Arusha, Tanzania

Simon A. R. Mduma
Serengeti Biodiversity Program
and
Director-General
Tanzania Wildlife Research Institute
PO Box 661
Arusha, Tanzania

Kristine L. Metzger
Beaty Biodiversity Research Centre
6270 University Boulevard
University of British Columbia
Vancouver, BC, Canada V6T 1Z4

Titus Mlengeya
Tanzania National Parks
PO Box 3134
Arusha, Tanzania

Maurus Msuha
Tanzania Wildlife Research Institute
PO Box 661
Arusha, Tanzania

Mtango Mtahiko
Chief Park Warden
Serengeti National Park
PO Box 3134
Arusha, Tanzania

Andrew N. Muchiru
African Conservation Centre
PO Box 15289-00509
Nairobi, Kenya

Ephraim Mwangomo
Tanzania National Parks
PO Box 3134
Arusha, Tanzania

Alastair Nelson
Frankfurt Zoological Society
PO Box 14935
Arusha, Tanzania

Ally K. Nkwabi
Serengeti Biodiversity Program
Tanzania Wildlife Research Institute
PO Box 661
Arusha, Tanzania

Joseph O. Ogutu
International Livestock Research Institute
Nairobi 00100, Kenya
and
Universitaet Hohenheim
Institut für Pflanzenbau und Gruenland
Stuttgart, Germany

Han Olff
Community and Conservation Ecology Group
Centre for Ecological and Evolutionary Studies
Groningen University
PO Box 11103
9700 CC Groningen, The Netherlands
and
Frankfurt Zoological Society
Box 14935
Arusha, Tanzania

Craig Packer
Dept. of Ecology and Evolution
University of Minnesota
St. Paul, MN 55108, USA

Susan L. Phillips
US Geological Survey
Forest and Rangeland Ecosystem Science Center
777 NW 9th St. Suite 400
Corvallis, OR 97330, USA

V. C. Radeloff
Department of Forest and Wildlife Ecology
University of Wisconsin-Madison
1630 Linden Drive
Madison, WI 53706, USA

Deborah Randall
Frankfurt Zoological Society
PO Box 14935
Arusha, Tanzania

Denne N. Reed
Department of Anthropology
University of Texas at Austin
1 University Station C3200
Austin, TX 78712, USA

Robin S. Reid
Center for Collaborative Conservation
and
Natural Resource Ecology Laboratory
Colorado State University
Fort Collins, CO 80523, USA

Dennis Rentsch
Frankfurt Zoological Society
PO Box 14935
Arusha, Tanzania

Wendy A. Ruscoe
Landcare Research
PO Box 40
Lincoln 7640, New Zealand

Camilla Sandström
Department of Political Science
Umeå University
SE-901 87 Umeå, Sweden

Jennifer Schmitt
100 Ecology Building
University of Minnesota
1987 Upper Buford Circle
St. Paul, MN 55108, USA

Gregory Sharam
Beaty Biodiversity Research Centre
University of British Columbia
Vancouver, BC, Canada V6T 1Z4

H. H. Shugart
Department of Environmental Sciences
University of Virginia
Charlottesville, VA 22903, USA

Anthony R. E. Sinclair
Beaty Biodiversity Research Centre
6270 University Boulevard
University of British Columbia
Vancouver, BC, Canada V6T 1Z4

Blaire Steven
Department of Molecular Biology
University of Wyoming
Dept 3944
1000 E. University Ave.
Laramie, WY 82071, USA

Simon J. Thirgood (Deceased)
The James Hutton Institute
Craigiebuckler, Aberdeen AB15 8QH, UK

Dominic Travis
Veterinary Population Medicine
College of Veterinary Medicine
University of Minnesota
385 Animal Science/Veterinary Medicine
1988 Fitch Avenue
St. Paul MN 55108, USA

Roy Turkington
Beaty Biodiversity Research Centre
and
Department of Botany
University of British Columbia
Vancouver, BC, Canada, V6T 1Z4

Louis V. Verchot
Climate Change Mitigation Research
Center for International Forestry Research
PO Box 0113 BOCBD
Bogor 16000, Indonesia

Diana H. Wall
Department of Biology
and
Natural Resource Ecology Laboratory
Colorado State University
Fort Collins, CO 80523, USA

Naomi L. Ward
Department of Molecular Biology
University of Wyoming
Dept 3944
1000 E. University Ave.
Laramie, WY 82071, USA

INDEX

Page numbers in italics refer to figures and tables.

abundance of small mammals, 333–37; peaks in, 335–37, 344
Acacia polyacantha, 240–44, 256; establishment in grassland, 242; facilitate establishment, 240, 256; tool for restoration, 240
acacia rat, *326*, 328
Acacia tortilis, 245, 257; establishment in grassland, 245; germination of, 255; growth and survival of, 245; regeneration of, 245; seedlings, 249
Acacia woodland, 325, *326*, 329, *334*, *337*; small mammals in, 328; species richness of small mammals in, 328, *334*
Acidobacteria, 214–18
Actinobacteria, 214–18, 224
Actinomycete, 213, 218, 224–26
African dormouse, *326*; in montane habitat, 332; in riverine grassland, 330
African grass rat, *326*, 328, 338, 349; agriculture, 338, 345; kopjes, 332; long-grass plains, 331; outbreaks, 328, 340, 346, 349; riverine floodplain, 329; *Terminalia* woodland, 329
African groove-toothed rat, *327*; in agriculture, 332; *Terminalia* woodland, 329
African Humid Period, 49
African mole rat, *327*, 328; long-grass plains, 331; short-grass plains, 331
African pouched rat, *326*, 328
African swamp rat, *327*; riverine floodplain, 329
agent-based models, 600
agricultural expansion, and human populations, 513–31; future development, impacts on land-cover change, 527; human population trends, 516–17; human population trends and drivers, 523–26; immigration, causal factors, 513–14; immigration, management interventions, 528; interactions with human demography, 523; land-cover change, 516–23; livelihoods in the ecosystem, 514; park-people interactions, 513; population growth, impacts on the ecosystem, 526–27. *See also* land-cover change
agricultural land use, disturbance effects, 284, 285–89
agriculture, 57, 463–64, 492, 494, 503, 682; agropastoral, 489; butterfly abundance, 315, 317–18; conflicts with hunting, tourism, 713–14; conversion from savanna, 746; cotton, 463; crop damage, 463; crop raiding, 463; disturbance of avifauna,

agriculture (*continued*)
395; FAO Africover data, 57; impacts on wildebeest, 146–48; intensification, 457, 747; and Mara River, 705; opportunity costs, 464; and pastoralism, 803; rainfall patterns, 463; sales, 476; watersheds, 705; and wildlife, 712
agriculture, villages and cultivation: carnivores in, 343–44, 347; cassava, 332; disease and small mammals, 350; economic impacts of rodents in, 351; intensification of, 349; land-use change and small mammals, 349–50; maize, 332; millet, 332; outbreaks of small mammals in, 347; predator dynamics in, 344; as refuge habitat for small mammals, 348; rice, 332; rodent dynamics in, 344, 347; and rodents, 323, 325, *326*, *334*, 337; sisal, 332; small mammal dynamics in, 344, 347; small mammals in, 332, 337, 347, 350; species richness of small mammals, 332, *334*, 345; stored grains, 347
agropastoralism, 609, 620, 632; carnivore abundance, *424*; carnivore diversity, 437–38
AIDS. *See* HIV/AIDS
Albert rift, 38
alternative states, 256, 599–601
ancient drainage: Congo, 38; Miocene, 38
Anopheles mosquito, 684. *See also* malaria
anthrax, 535–36, 539, 540–41, 543, 551, 554–55, 557–58, 562, 783
anthropometry: abnormal, 690; definition, 687; findings, 690–94; and height-for-age, 688, 690, 692, 695; and malnourished, 690, 692; methods, 688–90; prevalence-based reporting, 690–94; stunting, 690, 692; summary statistics, 693–95; underweight, 694–95; wasting, 690, 694; and weight-for-age, 688, 690, 694–95; WHO prevalence groups, 693; Z-scores, 690, 694–95
antigenic and genetic detection, 552–53; immunohistochemistry, 552; polymerase chain reaction (PCR), 552–53
antipoaching, 452, 588, 590, 593, 650, 664; buffalo, 651–56; elephant, 651–53; impact of ivory poaching, 650; impact on wildlife populations, 651–61, 670–73; penalties for, 673; rhino, 651–53; snaring, 650, 656–57, 670–71; wildebeest, 651–53, 655–61, 665–73, 662–73; wildebeest offtake sex ratio, 671–72
arbuscular mycorrhizae fungi, 213
arthropods, 265–94; abundance, 268, 276–82, 289–94; collection methods, 277, 280–81; grazing disturbance on, 281–82, 285–89; morphospecies, 277; seasonal activity, 289–94. *See also* invertebrates
Arusha Manifesto, President Nyerere, 797
atmospheric CO_2, 801
avian community, effect of natural disturbance, 395–416; aims of study, 397–98; bird fauna, 398; bird species abundance, 401–2; bird species composition, 403–5; bird species richness, 403; censusing birds, 400–401; compensation, 405–13; data analysis, 401; discussion, 413–16; disturbance in savanna, 396; role of disturbance, 395–96; savannas, 398; sites and treatments, 398–400
avifauna, Serengeti: bird species abundance, 401–2; bird species composition, 403–5; bird species richness, 403; censusing birds, 400; compensation, 405–13; data analysis, 401; insectivores, 403–5; seed and fruit feeders, 405

Babesia, 780–81
Bacteroidetes, 215, 217
ban on cultivation, in Ngorongoro Conservation Area, 486–89, 491–92, 496
barn owl, 323
baseline, for small mammals, 352
bat-eared fox, 422, 432–33, 438; abundance, *424*, *431*; density, *428*, *441*; population decline, 422
belowground community structure, Acidobacteria, 214–18; Actinobacteria, 214–18, 224; actinomycete, 213, 218, 224–26; arbuscular mycorrhizae fungi, 213; Bacteroidetes, 215, 217; BRC1, 217; carbon turnover, 228; Chloroflexi, 215, 217, 224–25; cyanobacteria, 201, 217; Firmicutes, 215–18, 223–24, 226; fungi, 196, 211–14, 220; Gemmatimonadetes, 215, 217, 223–24; Gram-negative bacteria, 211–13, 224, 226; Gram-positive bacteria, 211–13, 224–26; herbivory, aboveground, 218–20; herbivory, belowground, 218–20; nematodes, 218–19, 220–21, 224, 226; Nitrospira, 217; OP10, 217; operational taxonomic unit (OUT), 204, 214–16; phospholipod fatty acid, 202–4, 211–13, 220, 222–29; Planctomycetes, 215, 217, 223; Proteobacteria, 215–18, 223,

225–26; protozoa, 213, 220, 225–26; 16S rRNA, 204, 214–17, 219, 222–25, 227–29; sulfate reducers, 211–13, 225–26; Thermomicrobia, 217, 224; Verrucomicrobia, 215–18, 224–25, 228; Xiphinematobacteriaceae, 216, 218–19
benefits to people, lack of, 718
beta diversity, 111
biodiversity, inside and outside park, 799
biomass, livestock, 492, 503
bird counts, 400–401
bird data analysis, 401
bird fauna, Serengeti, 398
birds of prey, 323; abundance, 323, 341; lagged response to prey, 342, 347; population dynamics, 342–43; in relation to rodent outbreaks, 341–43, 346
bird species: abundance, 401–2; composition, 403–5; richness, 403
black-backed jackal, 422; density, *428*, 429, *430*; population decline, 422; rodent outbreaks, 343; small mammals and disease, 350
black-chested snake eagle, rodent outbreaks, *339*, 342–43
black-headed heron, rodent outbreaks, *339*, 342–43
black rat, *327*; in agriculture, 332
black-shouldered kite, 323, 341–42; in agriculture, 344, 347; breeding response to prey, 342; lagged response to prey, *341*, 342; nomadic response to prey, 342; numerical response to prey, 347, 351; population dynamics, 341–43; productivity, 342; in relation to rodent outbreaks, *339*, 341–42
body mass index (BMI), nutrition measurement, 493, *494*
bottom-up drivers of small mammal dynamics, 346; with climate change, 350
bovine tuberculosis, 535–36, 538, 551–52
Bray-Curtis index of similarity, for small mammals, 333, *334*
broad-headed mouse, *326*; in long-grass plains, 331
brown snake eagle, rodent outbreaks, *339*, 342–43
browsing by elephants, 236; reduction in, 250; removal of, 247, *248*
bruchid beetles, 253, 254, 255
brush-furred mouse, *327*; in *Terminalia* woodland, 329
buffalo: collapse of, 778; exploitation rate, 655; poaching, 651–60; spatial distribution of, 656
Bunda Designated District Hospital records, 682–83; HIV/AIDS rates, 683; malaria and infections disease, 685–86
burned area, 79–80, 83, 90–93, 94
burning, 248; as disturbance to butterflies, 303; removal of, *248*. *See also* fire
bush baby, lesser, 328
bushmeat, 633, 750; alternative protein to, 669–70; consumption projections, 669; definition of, 649; and human population growth, 526; income effects on consumption, 674; local consumption of, 649–51; Maasai consumption of, 656; price, 673; profitability, 674; sales, 478; seasonality of consumption, 665; supply and demand, 673–74
bushmeat hunting, 649–75, 711–12; antipoaching, 650, 664; definition of, 649; human impact on wildlife populations, 661–70; ivory poaching, population impact, 650; local consumption of, 649–51; management implications, 673–75; poaching impact on wildlife populations, 651–61, 670–73; relative impact on wildlife species, 650; snaring, 650, 656–57, 670–71. *See also* poaching
bush rat, *326*, *327*, 328; in riverine forest, 330; in riverine grassland, 330; in *Terminalia* woodland, 329
butterflies
—abundance and biodiversity, 308–11; migrants, 309–11, 312; rare and endemic species, 308–9
—distribution and biodiversity, 304–8; food plant families, 304, 306–7; geographic range, 304, 308; habitats, 304–5
—disturbance and abundance, 311–16; agriculture, 315; counts and grass biomass, 314; fluctuations in rainfall, 311–15; grass biomass, 314; grazing and burning, 315–16; interannual peaks, 311, 313; seasonality, 311
—disturbance and biodiversity, 316–18; disturbance from agriculture, 317–18; habitat heterogeneity and biodiversity, 316–17
—environmental disturbance and biodiversity, 301–19; abundance and biodiversity, 308–11; agriculture, 315, 317–18; distribution and biodiversity, 304–8; disturbance and abundance, 311–16; disturbance and biodiversity,

butterflies (*continued*)
—environmental disturbance and biodiversity (*continued*)
316–18; fluctuations in rainfall, 311–15; grazing and burning, 315–16; habitat heterogeneity and biodiversity, 316–17; methods, 303–4; migrants, 309–11; rare and endemic species, 308–9; synthesis, 318–19
—methods, 303–4; grazing and burning disturbance, 303; habitats, 303; identification, 304; visual method, 303; walk method, 303

caldera, conservation, 489, 501
camera trapping, carnivores, 425
canine distemper, 533, 535–36, 539, 541–43, 545, 553, 557, 560–61, 626, 747, 780–81, 783
Cape hare, 331
caracal, rodent outbreaks, 343
carbon: analysis and ground biota, 228; labile, 211; turnover and ground biota, 228
carnivore(s): abundance, *341*, 343–44; in agriculture, 343–44, 347; conservation of, 347, 350–51; in-depth studies, 421–22; interspecific competition in, 432–33, 438; intraguild killings, 432–33; lagged response to prey, *341*, 343–44, 347; murid preference, 343; numerical response to prey, 347, 351; opportunistic species, 434–35, 438; population dynamics, 344, 347, 439–40; population trends, 439–40, 422; response to rodent outbreaks, *341*, 343–44, 346, 351; response to seasonal migration, 426–31, *428*, *430*, 437, *441*; small mammalian, *341*, 343–44, 350–51; species, 419, 425–26; species richness of, 426, 434–36, *436*
carnivore communities, 419–42; abundance, 419, *424*, *431*; abundance per land-use type, 434–35; assemblages, 437–39; camera trapping, 425; daytime transect surveys, 422–23; density in relation to humans, 436–37; diversity, 435–36, *436*; habitat, 431–32; human influence, 434–39; interspecific competition, 432–33; per land-use type, 434–35; methods, 421–26; nighttime transect surveys, 423–25; opportunistic sightings, 425; population trends, 439–40; richness and diversity, 435–36; by season, 426–31; seasonal migration, 426–31; species, 425–26; tools for research, 421–26; transect surveys, 422–25
carnivore density, 426, *428*, *430*, *441*; Maasai Mara National Reserve, 429–30; relation to humans, 436–37; seasonal, 426–31; Serengeti National Park, 429–30; time of day, 439
carnivore transect surveys, 422–25; daytime, 422–23; line- or distance-based, 422–23; nighttime, 423–25; strip or fixed-width, 422–23
Caspian Sea, 49
catena, 45; *A. drepanolobium*, 45; *A. gerarrdii*, 45; *A. kirkii*, 45; *A. mellifera*, 45; *A. polyacantha*, 42; *A. robusta*, 45; *A. senegal*, 45; *A. seyal*, 45; *A. tortilis*, 45; *A. xanthophloea*, 45; niche partitioning, 45; small mammals, 330
census: birds, 400–401; future research priorities, 166; wildebeest population trend, 145–47; zebra population trend, 146
cerebral bovine theileriarosis, 495
cheetah, 421–22, 431, 440; abundance, *424*; density, *428*; interspecific killing, 432–33; seasonal prey migration, 427–28
chemical defenses, 249; of *Euclea divinorum*, 249
child health, 685–87, 695; archival records, 682; indicators, 688; infant mortality, 686, 695; and malaria, 684; and mortality rates, 686–88; neonatal mortality, 686; stunting, 690; under-five mortality, 686–87, 695; and underweight, 694–95; wasting, 690
Chloroflexi, 215, 217, 224–25
CITES, 653, 720
climate, 46–53; broad spatial scale, 46–48; changes over short time scales, 52–53; precipitation, 46–48; short time scales, 52; variability and patterns of drought, 48–50; variability on flora and fauna, 51–52
climate change, 121, 750–51, 755–56, 764; and ecosystem stability, 800–801; future research priorities, 166; and small mammals, 349–51; threats to the migration, 167–68
climate effects: environmental variation, 176–77; herbivore recruitment, 177–80; interannual variability, 52; lion recruitment, 181–84; predator-prey interactions, 184, 186–88; wildebeest rate of population change, 180–81

CO_2 fertilization, 121
Coke's hartebeest, 126–40; breeding and rut, 134–35; calving, 136–37; calving synchrony, 129–31; diet, 127–28; gestation, 135–36; intake rates, 127; lactation, 137–38; lek behavior, 134; maturity in, 138; migration, 126; reproduction, 129–38; spatial distribution, 129; water requirements, 128–29
collaring programs: future research, 166; lions, 164; wildebeest, 155–57, 161; zebra, 155–57
comanagement between state and local, 722–23
common genet, rodent outbreaks, 343; small mammals and disease, 350
common land ownership, 746
communities and disturbance, Serengeti avifauna, 395–416
community: conservation services, 716–17; development, 715
community-based conservation, 2–5, 425, 633–36, 639; in agriculture, 3; novel ecosystems, 3; for sustainable communities, 2; unsustainability of, 4–5
community-based wildlife management, 714
community structure, microbes belowground, 16S rRNA, 204, 214–17, 219, 222–25, 227–29
compensation: bird species replacement, 405–13; schemes for wildlife, 628, 633
competition
—in animals: grazing competition and migration, 159–62; male competition and breeding synchrony in wildebeest, 131–32, 138, 144; wildebeest-zebra grazing competition, 155
—in plants, 236; with grasses, 236, 249
compositional turnover of plants, 112; rainfall, in relation to, 112; Serengeti plains vs. tall grasslands, 112
conflict(s): from agriculture, hunting, tourism, 713–14; mitigation, 633–35, 637–39
conservation, 453, 478–79
—education and awareness, 631, 637
—and ground biota, 195–96
—history: beginning in Serengeti, 21–25; Finch-Hatton, 21; Grzimek surveys, 24; Huxley, 21; Lamai wedge, 25; lion massacres, 21; Maasai and Ngorongoro Conservation Area, 24–25; northern extension, 25; redrawing boundaries, 23–25; Serengeti Closed Reserve, 22; Serengeti Game Sanctuary, 22; Serengeti National Park establishment, 22–25
—in a human-dominated world, 1–6
—in Ngorongoro Conservation Area: area, 502; caldera, 488–89, 501; compatibility with cultivation, 486; land use, 492; natural resources, 487; officials, 487; policy, 503; small-scale cultivation, 491, 494; soil and cultivation ban, 489, 491–92; vegetation degradation, 487; water resources, 488; wildlife, 486–87, 489, 502; World Heritage, 488–89, 503
—park management and sustainable ecosystems, 785
—and tourism, Ngorongoro Conservation Area, 500–502; environment, 501; tourism impacts, 501; tourism revenue, 501–3; wildlife population, 501
contagious bovine pleuropneumonia, 485
Convention on Biological Diversity, 720
Convention on International Trade in Endangered Species (CITES), 653, 720
coupled human-natural (CNH) systems, 591
Crawshays hare, 328
crop: raiding and damage, 452–53, 459, 475–78, 588, 594, 596, 609–13, 619, 621–25, 629, 634–36, 638; yields, 588–89, 591
Croton dichogamus, 245, 246
cultivation, 57; boundary, 497; environmental concerns, 497–99; food security, 497, 500; impact on livestock, 497; impact on wildlife, 497; Loliondo, 57; Ngorongoro Conservation Area, 57, 497–500; small-scale, in Ngorongoro Conservation Area, 488, 491, 494, 497–98, 500
cultivation ban in Ngorongoro Conservation Area, 486–89, 496; food security, 486–88, 496; impact on tourism; 501; large-scale, 498; small-scale, 497–98, 500
cyanobacteria, 201, 217

Danish International Development Agency (Danida), 488
deforestation, 744, 748, 758; Mara River and threats to the migration, 167–68
devolution of power, 756, 760–61
disease, 153; *Acacia* recruitment, 60; *A. drepanalobium*, 60; cerebral bovine theileriarosis, 495; contagious bovine pleuropneumonia, 485; east coast fever, 495–96; human population density, 60; livestock, 465–67; livestock interactions

disease (*continued*)
with wildlife, 495; malignant catarrhal fever, 495–96; rinderpest, 60, 145, 485; small mammals, 349–50; threats to the migration, 167–68; wildebeest, 496; wildebeest recovery, 145–47
disease detection: antigenic and genetic detection, 552–53; field-based approaches, 545–49; case reporting, 546–47; contact tracing, 547–48; diagnostic pathology, 550–51; laboratory-guided, 549–54; microbiological methods, 551–52; population abundance and density, 548–49; serology, 553–54
disease surveillance
—active surveillance, 544
—approaches and definitions, 544–56
—methods, 554–56; mobile-phone surveillance, 555–56; proxy populations, 554; risk-based surveillance, 554–55; sentinel populations, 554; syndromic surveillance, 555
—monitoring, 544
—passive surveillance, 544–45
—surveillance, 544
disease transmission, 610, 613, 616, 625–28, 747, 756, 803–4; health and malnutrition, 804; human-wildlife interactions, 804
disturbance in ecological communities: agriculture, 395; fire and grazing, 396; heterogeneity, 395; Serengeti avifauna, 395–416; Serengeti savanna, 396
diversity: carnivores, 435–36, *436*; insect, 266, 276–89; insect, disturbance effects on, 282–84; Pielou's index of species evenness, 281; Shannon-Wiener index, 281; of small mammals, 323–24, 333, 344–45, 351; small mammal species richness, 325, 329, 333, 344–45, 351
DNA analysis of small mammal species, 335
Dodoman Orogenic Belt, 35
domestic cat and small mammals, 343–44; in agriculture, 344
domestic dog: abundance, *424*, 435; in agriculture, 344; density, *430*, 438–39; and small mammals, 343–44; vaccination, 781
drought: environment, 490–91, 494–96, 499–500; seasonal and yearly cycles, 51–52
dry season, 52–53; carnivore densities, *428*, *430*; long, 52–53; short, 53
dung and ground biota, 201, 205–6
dung beetles, 269–71, 276; effect of fire, 282; effect of grazing, 284; functional role, 270; species, 271

early burning, 74
east coast fever (ECF), 495–96
Echinococcus, 535–36, 539, 552
ecological data for disease, 556–58; environmental and climatic factors, 558; host ecology, 556–57; vector ecology, 557–58
ecological functions, 702–4
ecological model, environment, 494
ecological perturbations, 586
economic drivers, 803
economic functions, 705–6
economic gains and incentives, 723
economics, tourism and the migration, research benefits, 782
ecosystem change: climate change, 755–56; comanagement, 754; conservation beef, 756; devolution of power, 756; disease transmission, 756; environmental governance, 754, 759; expansion of farming, 752–53, 756; expansion of settlement, 753, 756; globalization, 755–56; Grumeti Game Reserve, 755; habitat fragmentation, 753, 756; herders protecting wildlife, 752; hunting pressure, 755; Ikorongo Game Reserve, 755; illegal hunting, 753, 755; implications for biodiversity, 751–56; Loita Plains, 753; Maasai pastoralists, 755–56; Mara Conservancy, 754; Mara River drying, 754; Ngorongoro Conservation Area, 752–53; poaching, 754–55; privatization of land, 752; Serengeti highway, 753; wheat farming, 753; wildlife conservancies, 754; wildlife decline, 754–55; wildlife management area (WMA), 754; wildlife poisoning, 752
ecosystem functioning, carnivores, 419
ecosystem services, 784–85; conservation, park management and sustainable ecosystems, 785; insect outbreaks, 785; payments for, 758–59; rodent outbreaks, 784–85
eland, poaching, 657–60
elephant grass and small mammals, 330
elephants: conflict with humans, 609–10, 612, 616–19, 621–24, 629–30, 634–35; as ecosystem engineers, 591; impacts on tree cover, 591, 596, 777; indirect effects on wildebeest, 591, 596, 598; as

key players, 591; poaching of, 592–93, 650, 651–53, 656, 660, 800; population collapse of, 593–94, 778; population growth of, 592; stress hormones in, 800; and tree decline, 777
elephant shrew, long-eared, *327*, *338*
El Niño/Southern Oscillation (ENSO), 52
employment, 468; agricultural sales, 468; livelihood diversification, 456, 468, 476; livestock sales, 468; opportunities, 479; park-related, 453, 468, 470, 475–79; remittances, 469, 476; seasonal, 476; tourism, 501
endozoochory, 110
environment: drought, 495–96, 499; revegetation, 498–99; soil erosion, 497; stress, 501
environmental governance, 741, 748, 754, 759
environmental variation, 176–77
epidemiological transition, 695
erosion-deposition processes, 36–46
establishment of plants: facilitation of, 246–47, 249, 256; mass, 236; pulses of, 240; riparian forest, 240, 249; thickets, 245, 247; trees, 236; woodlands, 245
ethnic diversity, 453, 455; and colonialism, 455; and household characteristics, 455–56; and mixed settlements, 455. *See also* ethnic groups
ethnic groups: Ikoma, 455; Issenye, 455; Kuria, 455; Maasai, 455; Natta, 455; Ngurime, 455; Sukuma, 455
Euclea divinorum, 244–45, 249, 257; chemical defenses in, 249; establishment of, 248–49; facilitating establishment, 246; as nurse species, 256; as pioneer species, 245, 249; seedlings of, *248*

farming expansion, 740–41, 745, 752–53, 756
fat mouse, *326*, 328; in long-grass plains, 331; in short-grass plains, 331; in *Terminalia* woodland, 329
feedback loops, 594, 601
fences and fines, 723–24
fencing, threats to the migration, 167–68
fire, 61, 107, 109, 115–16, 119–20, 236, 242, 245, 247, 250, 256–57, 590
—biomass accumulation, 61
—disturbance: on arthropods, 276, 278–79, 282, 285–89; on avifauna, 396
—drivers and consequences: history, 74–76; influence on ecosystem structure and function, 85–91; intensity, 84–85; management plan, 92–96; in savannas and grasslands, 73–74; scale-dependent drivers of frequency, 76–83
—on ecosystem structure and function, 85–91; birds, 90; carnivores, 91; herbaceous communities and leaf nutrients, 88–89; invertebrates, 90; large herbivores, 90–91; soil nutrient dynamics, 89; spatial-temporal patterns of woody cover change, 85–86; stimulated effects on woody cover change, 86; woody cover change, 85–86
—effects on herbaceous communities and leaf nutrients, 88–89; fire frequency, 89; nitrogen, 88–89; phosphorus, 88–89; plant diversity, 88; sodium, 89; soil carbon, 89; soil nutrients, 89; species composition, 88
—effects on tree cover, 601
—frequency, 74–77, 82, 86, 89, 94, 96, 244, 249, 594; annual precipitation (AP), 80, 82; biomass, 76–78, 82–83; fire return interval (FRI), 83; grazing, 77–80, 82–83; northern Serengeti (NS), 79–80, 82; rainfall, 76, 78–79, 82–83; scale-dependent drivers of, 76–83; tree cover, 76, 82–83; Western Corridor (WC), 79–80, 82; wildebeest, 77–80, 82–83
—fuel load, 74, 82–84, 86, 89, 95
—history, 74–76; early burning, 75; human population, 75; wildebeest population, 75
—impacts, 799
—intensity, 84–85
—management plan, 92–96; adaptive ecosystem management, 95; cool fires, 92–93; disease vectors, 94; fire frequency, 94; forage, 93–94; herbivores, 92, 94; rare and sensitive species, 93; sensitive vegetation communities, 93; ticks, 94–95; tourism, 92; tsetse flies, 94; wildfires, 95
—in northern Serengeti, 79–80, 82, 85
—removal of, 247; resistance to, 249
—return interval, 83, 86, 93–94
—seasonal precipitation, 61
—in western corridor, 79–80, 82
—and woody cover change, 85–86; elephants, 85–86; fire return interval (FRI) 86; wildebeest, 85–86
—*See also* burning
Firmicutes, 215–18, 223–24, 226

floodplain, 325, *326*, *334*, *337*; as refuge habitat for small mammals, 329, 345; small mammals in, 329; species richness of small mammals, 329, *334*
food: availability, carnivores, 426; security, Maasai livelihood, 486–88, 496–97
foot-and-mouth disease, 783
forest birds, 25–52; decline of, 251
forests: decline of, 236; establishment, 245, 249; regeneration, 236, 255, 257; restoration of, 240
forest types: comparisons, 238; decline, 236; density of trees in, 252; descriptions of, 236–39; establishment, 245–46; gallery, 235; hilltop, *237*; lowland, 236, 239; montane, 238–39; regeneration, 236; riparian, 235, *237*, 240; riverine, 235
Frankfurt Zoological Society, 717
frugivorous birds, 251–55
fuel load for fires, 74, 82–84, 86, 89, 95
fungi and microbes, 196, 211–14, 220
future of conservation, lessons from Serengeti. *See* Serengeti: lessons for future of conservation
future of Serengeti, role of research. *See* research in conservation, future of Serengeti

Game Ordinance, 485
Game Parks Laws Act, 487
Game Preservation Act, 485
game reserves, 493, 485; tourism, 502–3
Gemmatimonadetes, 215, 217, 223–24
genet, large-spotted: response to rodent outbreaks, 343; small mammals and disease, 350
genetic and epidemiological data analysis, 559–61; phylogenetic analyses, 559–61; transmission trees, 561
geology, 34–36; Archean floor, 36; Archean foreland margin, 36; calcareous tufts, 35–36; candelabra trees, 36; carbonatitic volcanoes, 36; clastic deposits, 35; exfoliation, 36; fig trees, 36; Gondwanaland, Pan-African orogeny, 36; granite gneisses, 35; granites, 35; greenstone belts, 35; heterogeneity, 34–36; heterogeneity, causes of, 34–36; heterogeneity, origins of, 34–36; inselbergs, 36; Isuria escarpment, 36; Kavirondian systems, 35; kopjes, 36; Lake Victoria, 36; metasedimentary group, 35; minor politic beds, 35; Moru kopjes, 36; Mozambique Ocean, 35–36; Mozambique Orogenic Belt, 35–36; Neogene, 35–36; Nyanzian-Kavirondian Orogenic Belt, 35; Nyanzian system, metamorphosed lava, 35; Oldoinyo Lengai, 36; Pan-Africa deformation, 36; parent material, 34, 61; pebble grits, 35; phonolite lava, 36; processes, 34; quartzites, 35; recrystalized granitic rock, 36; river alluvium, 35; Serengeti plains, 36; shrubs, 36; spheroidal weathering, 36; synorogenic granites, 35; timescales, 34, 61; and vegetation, 36
geomorphology and landforms, 36–46; distribution of soil properties, 41–46; soil formation, 41–46
gerbil, large naked-soled, *326*, 328; in agriculture, 332, 345; in long-grass plains, 331; in short-grass plains, 331
gerbil, pygmy, *326*, *327*; in long-grass plains, 331; in short-grass plains, 331
German era, 18–21; absence of human settlement, 20–21; Anglo-German boundary, 18–19; Baumann's journey, 18, 20; Maasai limits, 19; rhino distribution, 20; roan antelope distribution, 21; White's journey, 19; wild dog distribution, 20; wildebeest distribution in 1913, 19–20
germination rate, *253*, 254
giraffe, poaching, 657–60
global change, response of small mammals, 349–52
globalization, 742, 750–51, 755–56, 764
golden jackal, 422, 431–32; density, *428*; population decline, 440, *441*; rodent outbreaks, 343; small mammals and disease, 350
Gondwanaland, Pan-African orogeny, 36
governance: enhance comanagement, 721–23; land, settlement, and local governance, 709; models, 721–22; monitoring key variables and indicators, 725–28; process, 722; structure, 722; transparency in, 719
government policy: DANIDA, 488; Game Ordinance, 485; Game Parks Laws Act, 487; Game Preservation Act, 485; Ngorongoro Conservation Ordinance, 486; Wildlife Conservation Act, 487
Gram-negative bacteria, 211–13, 224, 226
Gram-positive bacteria, 211–13, 224–26

Grant's gazelle, poaching, 656–60
grass, 244; biomass and ground biota, 201, 206; competition from, 249; density of, 244; removal of, *248*
grasshoppers, 274–76; diversity, 274–76; grass consumption, 274; habitat, 275; species composition, 274–76
grasslands, 240, 245; birds of, 251–52
grazing, 110–11, 119, 487, 492, 496; ban on, 486; exclosure experiments, 110
grazing disturbance: arthropods, 281–82, 285–89; avifauna, 396; butterflies, 303; soil biota, 800
greater galago, in riverine forest, 330
greenstone belt, 35
groove-toothed creek rat, *326*; in agriculture, 332; in riverine floodplain, 329; in riverine forest, 330
Grumeti Game Reserve, 737, 755

H evenness, for small mammals, 333, *334*, 345
habitat: fragmentation, 753, 756, 758; heterogeneity, potential niches, 40; loss, 585–86; selectivity, carnivores, 431–32, *433*; selectivity, carnivores and body size, 432
habitats for small mammals, 324–33, *325*, *337*, 349, 351; *Acacia* woodland, 325; agriculture, 325, 344; floodplain, *325*; heterogeneity, 345, 348–49; intermediate-grass plains, *325*; kopje, *325*; long-grass plains, *325*; montane, *325*; productivity, 348; as refuge for small mammals, 329, 330, 345, 348–49; riverine forest, 325; riverine grassland, *325*; rocky hills, *325*; short-grass plains, *325*; and small mammal diversity, 344–46, 351; *Terminalia* woodland, *325*; wetlands, *325*
handicraft, and tourism, 502
health: impacts and wildlife infections, research benefits, 783–84; Maasai, 485, 503; and malnutrition, 804; nourishment, 491–92, 495; survival rates, 489
health care: blaming local people, 695; lack of, 695
herbivores: attraction to burned areas, 90–91; recruitment, 177–80
herbivory and ground biota: aboveground, 218–20; belowground, 218–20
herders protecting wildlife, 752
heterogeneity, home range of resident wildebeest, 142
highland forest, 51; Crater Highlands, 51; expansion of, 51; Loita Forest, 51; across Maasai Mara National Reserve, 51; Mara River, 51. *See also* montane forest
hilltop forests, *237*, 245; thickets, 257. *See also* montane forest
history
—of humans in the GSE, 680–82; Bantu farmers, 680–81; Barabaig/Tatog, 681; colonialism, 681; Cushites, 681; Iraqw, 681; Maasai culture, 680, 682; Maasai dominance, 681; Nilotes, 680–81
—of land use, Ngorongoro Conservation Area, 484–89; conservation, 486–88; cultivation, 485–89; government policy, 485–87; land use and resource rights, 485–88; livestock diseases, 485; Maasai livelihood, 485–89; tourism, 487–89; wildlife populations, 485–87
—lessons for conservation, 25–27; boundaries and absence of humans, 27; distribution of wildebeest, 26; increase of trees, 26; ivory trade and elephant populations, 26; rinderpest and exodus of humans, 26; rinderpest and increase of ungulates, 26
HIV/AIDS, 682–83; Community-Based Health Promotion Program (CBHPP) in Mugumu, 682–83; Bunda Designated District Hospital records, 683; prevalence rates, 683
honey badger, rodent outbreaks, 343
households
—labor allocation, 588, 590–92, 596, 598; model, 588; shifts in, 597, 600; tradeoffs, 597
—surveys: consumption surveys, 662–68; dietary recall surveys, 662–68; household income, 674; income-generating projects, 674
human(s)
—activity, impact on small mammals, 345
—conflict and coexistence between wildlife and, 452–53, 478, 526, 588, 607–40, 804; background, 608–9; competition for land, 631–33; conflict mitigation, 633–37; depredation and crop destruction, 619–25; direct encounters with wildlife, 609–19; disease, 625–28; retaliation, 628–31

human(s) (*continued*)
—environment dominated by, conservation in, 1–6
—health in the GSE, 680–96; abnormal anthropometry, 690; anthropometric findings, 690–96; anthropometry methods, 688–90; child health, 685–87, 695; compared to all of rural Tanzania, 695–96; height-for-age, 688, 690, 692–94, 695–96; heterogeneity in health patterns, 696; HIV/AIDS, 683–85, 695; human history in the GSE, 680–82; indicators, 688; malaria and infectious disease, 683–85, 695; malnourished, 690, 692; malnutrition, 692, 694, 696; morbidity and mortality rates, 682–87; nutritional status, 687–95; patterns, 692–96; and poverty, 685–86; prevalence-based reporting, 690–96; stunting, 688, 690, 692–93, 695; summary of anthropometry, 693–96; weight-for-age, 694
—history in the GSE (*see* history: of humans in the GSE)
—immigration: causal factors, 513–14; impact on ecosystem, 526–27; land-cover change, 516–23; livelihoods in the ecosystem, 514; management interventions, 528; park-people interactions, 513; population growth, 526–27
—impact on wildlife populations, 661–70; antipoaching, 650–64; bushmeat, alternative protein, 669–70; bushmeat, consumption projections, 669; bushmeat, quantifying offtake through consumption, 664–68; consumption surveys, 662–68; dagaa, 665; demand for meat protein, 669–70, 673–75; dietary recall surveys, 662–68; human community surveys, 661–64; human population, growth rate, 661–63, 667–70, 673–75; poachers, density, 662; poaching, methods for community assessment, 661–64; poaching, relative offtake of wildlife species, 658–60, 666–68; seasonality of bushmeat consumption, 665
—influence on carnivores, *424*, *430*, *436*, 434–39, 440
—migration and ground biota, 195–96
—and paleolandscape, 53–61
—settlement: absence in Serengeti, 20–21, 27; boundaries, 27
—threats to the migration, 167–68
—*See also* human population; human systems

human population
—density: changes, 55–56; illegal hunting, 55; immigration, 55; between Lake Victoria and the western boundary, 55; Maasai Mara Nature Reserve boundary, 55; natural resource extraction, 55; Ngorongoro Conservation Area District, 55; population growth, 55; rate of increase, 55; rural population growth, 55; west of game reserves, 55
—growth, 452, 517–19, 523, 526–27, 586, 589, 599, 661–63, 667–70, 673–75, 740–41, 751; agricultural expansion, 523–26; bushmeat, 526; densities, 452, 518–19; human-wildlife conflict, impact on ecosystem, 526–27; and land-cover change, 523; and park boundary, 517–18; size, 451
—immigration, 489, 513–14, 524–28; drivers, influence on management, 527–28; drivers, in Serengeti, 524–25; human-wildlife conflict, 526; around protected areas, 513–14; push and pull factors, 514
—impact on tourism, 501
—land access and use, 741, 744–47
—livestock ratio, 491, 494–95
—Maasai, 488–89
—migration, 195–96, 490
—and well-being, Ngorongoro Conservation Area, 489–95; cultivation, 489–92, 494; diet and nutrition, 491–93; environment, 490–91, 494; food security, 491, 494; human health, 489, 491–93, 495; human population growth, 489–91, 494–95; land-use restrictions, 489, 491–92; livestock, 490–92, 494–95; pastoralism, 489–91

human systems
—and conservation objectives, 801–2; Maswa wildebeest refuge, 801–2; minimizing impacts, 798; outside water supplies, 801; poaching and poverty, 802
—sustainability, 802–4; agriculture and pastoralism, 803; health and malnutrition, 804; human-wildlife interactions, 804; illegal hunting, 802; major conservation issues, 802; rural livelihoods, 803; social and economic drivers, 803; transmission of disease, 803–4

HUMENTS model, 588–91, 599–600

hunting, 588–90, 598; blocks, 710; in game reserves, 485; pressure, 741, 749, 755, 762–63; subsistence, 485–86; and taboos, 712; travel cost of, 598; wildlife impact of, 485

hydrology, 59–60; Amala tributary, 59; clearing for agriculture and charcoal production, 59; deforestation, 59; forest and grassland clearing, 59; hydrological cycle, 613, 633; hydrological cycle, processes, 36–46; land-use change, 59–60; Mara River drainage, 59–60; Napuiyapi Swamp, 59; Nyangores tributary, 59; water diversion, 59
hyrax, 324; bush hyrax, 332; rock hyrax, 332; tree hyrax, 330

ice cores, Mt. Kilimanjaro, 49
Ikorongo Game Reserve, 710, 737, 755
illegal hunting, 452–53, 471–75, 718, 741, 749–50, 753, 755, 757, 762–63; agricultural sales, 474; beer sales, 473; bushmeat sales, 474; household economies, 474; income generation, 475; livestock sales, 473–74; primary school education, 472; seasonal employment, 473; secondary education, 472; wildebeest, 471; year-round employment, 473–74. *See also* poaching
immigration of humans. *See* human(s): immigration
impala, poaching, 656–60
incentives, for conservation, 715
infant mortality rates, 686–87, 695
infectious disease
—data analysis, 559–63; epidemiological modeling, 562–63; genetic and epidemiological data, 559–61; temporal and spatial data, 561–62
—in humans, 683, 686; blaming locals, 695; connections to poverty, 688, 695
—surveillance, 533–73; analysis of data, 559–63; antigenic and genetic detection, 552–53; case reporting, 546–47; contact tracing, 547–48; diagnostic pathology, 550–51; ecological data, 556–58; environmental and climatic factors, 558; epidemiological modeling, 562–63; field-based approaches to case detection, 545–49; general surveillance approaches and definitions, 544–56; host ecology, 556–57; innovative methods for surveillance, 554–56; integration of tools for infectious disease research, 563–68; laboratory-guided disease detection, 549–54; microbiological methods, 551–52; pathogens, 534–35; population abundance and density, 548–49; serology, 553–54; temporal and spatial data analysis, 561–62; tools for infectious disease research, 535–63; vector ecology, 557–58
insects, 265–94; outbreaks, 785. *See also* arthropods; invertebrates
institutions: compliance, 723–24; interactions with functions, 712–15; the national context, 706–9; the Serengeti context, 709–12; strengthen for management, 720–21
intermediate-grass plains, 325
International Union for Conservation of Nature (IUCN), 54; categories of protection, 720–21
interspecific competition, carnivores, 432–33, 438
intraguild killings, carnivores, 432–33
invasive plant species, 121–22; *Chromolaena odorata*, 122; monitoring, 122; Ngoronongoro Crater, 122; northern road development, 122; *Parthenium hysterophorus*, 122; threats to the migration, 167–68
invertebrates, 265–94; disturbance effects on, 276–89; diversity, 266, 276–89; ecosystem importance, 270–76; historical expeditions, 266–67; new records, 269–70; nutrient cycling, 270–71, 273; seasonal abundance, 289–94; spatial patterns, 289–94; spiders, 273–74; as vertebrate resource, 267–69
irrigation from Mara River, 786
IUCN. *See* International Union for Conservation of Nature
ivory poaching, population impact, 650–53
ivory trade, 14–15; elephants in Serengeti in 1800s, 14–15; industrial revolution in Britain, 14–15; slave trade, 14–15; Zanzibar Arab traders, 14–15

kongoni, poaching, 656–60
kopje, 323, 325, *326*, *334*; hyraxes in, 332; as refuge habitat for small mammals, 332, 345; small mammals in, 331–32; species richness of small mammals, 332, *334*
Kruger National Park, 2

lagomorph, 324
Lake Chad, 49
Lake Naivasha, 49
Lake Victoria: fisheries, 669–70; water levels, 786

land
—availability in western Serengeti, 456–58; agricultural intensification, 457; communally held lands, 458; human-wildlife conflicts, 457; immigrants, 457; land ownership, 457; land subdivision, 456, 458; livelihood strategies, 453, 458, 462–79; population trends, 456; protected areas, 456; reciprocity, 458; rented land, 457; tragedy of the commons, 458
—conversion, competition for land, 609, 613, 623–25, 627–28, 631–33, 637, 639
—ownership, 457
—rights, 740
—settlement, and local governance, 709
—subdivision of, 456, 458, 740, 746
—tenure, 745, 757
land and resource use, historical and current, 742–44; climate change, 742, 744; globalization, 742, 744; Grumeti Game Reserve, 743; hominins, 742; human population growth, 744; Ikorongo Game Reserve, 743; Loliondo Game Controlled Area, 743; Maasai Mara National Reserve, 743–44; Maasai pastoralists, 742–44; Maswa Game Reserve, 743; Mau Forest, 744; Ngorongoro Conservation Area, 743–44; Tatoga pastoralists, 742–43; western Serengeti peoples, 743; wildlife management area (WMA), 743
land-cover change, 516–28; change map, *519*; east-west comparison, 517–18, 523–24; impact of Serengeti highway, 527; impact on ecosystem, 526; mapping, 516–17; rates, over time, distance from park, 521–23; relation to population growth and density, 523; trends in relation to park boundary, 520–21
landform, 34, 36–46; alluvial floodplain, 38; Gol Mountains, 38; Kuka hill, 39; Lake Victoria, 38; Lobo kopje and hill, 39; nutrient availability, 38; Nyamuma hill, 39; Nyaraboro plateau, 39; seasonal rainfall gradient, 38
landscape
—heterogeneity in relation to tree cover, 115–16; petro calcic layer (caliche), 116; soil and water limits to seedling recruitment, 116; topographic moisture index (TMI), 116; upper boundary of tree cover, 115
—patterns of tree composition and diversity, 114; Fabaeceae, 114; rainfall, in relation to, 114
land use: changes in, 483–502; changes in, agricultural impacts on wildebeest, 146–48; changes in, deforestation threat to migration, 167–68; conservation policies, 492; cultivation, 488, 501; designations, 451; fragmentation, 483; livestock, 489, 501; and microbes, 196–97; policies, 635–37; privatization, 485; restrictions, 486, 489; rights, 485–87; in the Serengeti ecosystem, 53–54; systems, 682; tourism, 502
law enforcement, 715, 724; costs and benefits, 719
legal clarity, 723
leks, topi, 134
lemur, 324
lion(s), 421, 422, 431, 438; abundance, *424*; attacks, social welfare, 784; density, *428*, 429, *430*, 437; feeding, 184–85; interspecific killing, 432–33; nomadic lions, 429; pride takeovers, 182–83; rate of population change, 184–85; ritual killing of, *Ola-mayio*, 609, 611, 616–17, 629–30, 634, 639; seasonal prey migration, 427–28, 437
livelihood strategies, 453, 462–79, 681–82; cash crops, 462; and climate, 462–63; diversification, 456; and land availability, 458; land subdivision, 462; livestock holdings, 462, 470; poaching, 471; restrictions on, 682; subsistence farming, 462
livestock, 465–68, 589; annual sales, 465; biomass, 503; bomas, 60; cattle, 465–67; chickens, 465–67; contagious bovine pleuropneumonia, 485; corrals, 60; depredation, 475–78, 609–12, 616–17, 619–21, 625, 629, 634–35; *Dicrostachys*, 60; disease, 465–67; goats, 467; grazing, 587; grazing and microbes, 196–99; herding, 585; household economies and protein, 467; livestock-human ratios, 492, 494–95; loss of, 452–53; Maasai livelihood, 502–4; purchases, 590; rinderpest, 485; sales, 476; sheep, 466; tourism, impact, 501
livestock and wildlife, Ngorongoro Conservation Area, 495–97; cultivation, 496–97; environmental concerns, 495; livestock health, 495–96; Maasai livelihood, 495–96; resource competition, 495–96
local knowledge, attitudes, and awareness of protected areas, 452, 458–59, 461–62

lodges, tourism, 501–2
Loita Plains, 746, 753
Loliondo Game Controlled Area, 710, 737, 740, 747, 749
Loliondo revenues, 705
long-crested hawk-eagle, rodent outbreaks, *339*, 342–43
long-grass plains, 325, *326*, *334*, *337*; carnivore densities, *428*; small mammals in, 330–31, 337; species richness of small mammals, 331, *334*
lowland forest, 51, 236, 239, 244; conversion to grasslands, 240; dynamics of, 240–44; along Grumeti River, 51; around Lake Victoria, 51; on Mbalageti River, 51; regeneration of, 243–44; wet periods, 51

Maasai: consumption of bushmeat, 656; group ranches and conservancies, 608, 623, 631, 636; history in Ngorongoro Conservation Area, 24–25; pastoralists, 738–40, 742, 746–49, 750, 755–56. *See also* Maasai livelihood
Maasai livelihood: cultivation, 502; education, 503–4; food security, 503–4; health, 485; human population, 488–89, 503; livestock, 502–4; revenue, 488
Maasai Mara National Reserve, 711, 737–41, 744–46, 748, 750–51, 760–61; carnivore abundance, *424*; local migration in, 426–27; microbes in, 196–98, 200–201, 227–28
malaria, 683–85, 695; under-five mortality, 686
malignant catarrhal fever (MCF), 495–96, 501, 626–28, 747, 783
malnourished, 690, 692, 694; definition, 690
malnutrition, 692, 696; in Bunda District, 694; and culture, 696; and ecological heterogeneity in the GSE, 696; in Loliondo District, 694, 696; in Meatu District, 694, 696; in Serengeti District, 694; uneven patterns in the GSE, 696
management: approaches to multifunctionality, 715–18; current challenges, 718–19; evaluate effectiveness, 724–28; framework, 725; monitoring, 724; priorities, 804–8; the way forward, 719–28. *See also* management and research
management and research, 775–82; *Babesia*, 780–81; canine distemper, 780–81, 783; collapse of rhino, elephant, and buffalo; 778; domestic dog vaccination, 781; elephants and tree decline, 777; extermination of predators, 776–77; increase in trees, 779; Mara River drying, 782, 785–86; migration routes, 781; the 1960s and earlier, 775–77; the 1970s, 777–78; the 1980s, 778–79; the 1990s, 779–81; rinderpest, 778, 783; Serengeti Research Project, 776; the twenty-first century, 781–82; wildebeest and buffalo increase, 778; wildebeest poaching, 779–80; wildlife management areas, 782
Mara Conservancy, 711, 740, 750, 754, 761
Mara Region, human population, 661
Mara River, 737–38, 741, 758, 764; and agriculture, 705; drying of, 741, 748, 754, 764, 782, 785–86; hydrology, 59–60, 62; irrigation, 786; threats to the migration, 167–68
martial eagle, rodent outbreaks, *339*, 342–43
Maswa Game Reserve, 53, 710, 737, 743; as wildebeest refuge, 801–2
Mau Forest, 737–38, 748, 758–59; deforestation of, 167–68, 786
Mau highlands, deforestation threats to migration, 167–68
maximum daily temperature, 49–50; dry season, 49; wet season, 49
microbes: carbon, labile, 211; carbon analysis, 203; dung, 201, 205–6; grass biomass, 201, 206; human migration, 195–96; land use, 196–97; livestock grazing, 196–99; Maasai Mara National Reserve, 196–98, 200–201, 227–28; nematode analysis, 205; nitrogen analysis, 203; people and wildlife, 195–96; phospholipid fatty acids analysis, 203–4, 211; site description, 16S rRNA analysis, 204–5; vegetation cover, 207–8; vertisols, 197–98; wildlife conservation, 195–96; wildlife density, 199
migration, 702; abundance and, 159–62; advantages of, 162; Coke's hartebeest, 126; compensatory vegetation growth, 163; drivers of, 162–64; elemental composition of grass, 163; establishing protected area boundaries, 157; extent of the wildebeest, 159–62; grass protein, 163; human population, 490–91; impact on carnivores, 426–31; rainfall, 162; rotational grazing, 163; routes, 781; to avoid predation, 161–62; topi, 126; water availability, 163; wildebeest, 154–65, 586, 590, 599; wildlife, 487; zebra, 154–65

Miombo woodland, and small mammals, 328
Moderate Resolution Imaging Spectrometer (MODIS), 57
monitoring: governance, key variables and indicators, 725–28; management, 724; wildlife, 724
montane forest, 238, 244–56, 325, *326*, *334*; disturbances to, 244–56; dry periods, 51; dynamics of, 250; *Ilex*, 48; *Juniperus*, 48; northeastern Serengeti, 48; *Olea*, 48; *Podocarpus*, 48; as refuge habitat for small mammals, 345; regeneration of, 254–56; small mammals in, 332, *334*. *See also* hilltop forests
morbidity/mortality rates in humans, 682; from Bunda Designated District Hospital, 682–83; from Serengeti Designated District Hospital, 682
mouse, *326*, *327*, 328, 333; diversity, 333; in long-grass plains, 331
Mozambique Orogenic Belt (MOB), 35–36; Archean floor, 36; Archean foreland margin, 36; Gondwanaland, Mozambique Ocean, 35–36; metasedimentary group, 35; Pan-Africa deformation, 36; Pan-African orogeny, 36; recrystalized granitic rock, 36
multimammate rat, 323, *326*, 328, 346, 348; agriculture, 323, 338, 345; external drivers of outbreaks, 348, *349*; in kopjes, 332; in montane habitat, 332; outbreaks, 323, *327*, 328, 330, 338, 340, 346, 348, *349*; predator responses to outbreaks, 341–44, 346; refuge habitat, 330, 332, 346, 347; in riverine floodplain, 329; in riverine forest, 330; in *Terminalia* woodland, 329
multiple states, 256, 600, 601
multiple-use protected area, Ngorongoro Conservation Area, 502–5; conservation, 502–3; cultivation, 502–3; human health, 503; livestock, 502–4; Maasai livelihood, 502–5; tourism, 502–4
murid preference, of predators, 343

national park: conservation, 486–88; land use, 486–87
National Strategy for Growth and Reduction of Poverty, 706
Natron-Magadi basin, 49
nematodes, 205, 218–19, 220–21, 224, 226
Neogene, 35–36; calcareous tufts, 35–36; candelabra trees, 36; carbonatitic volcanoes, 36; exfoliation, 36; fig trees, 36; inselbergs, 36; Isuria escarpment, 36; kopjes, 36; Lake Victoria, 36; Moru kopjes, 36; Oldoinyo Lengai, 36; phonolite lava, 36; river alluvium, 35; Serengeti plains, 36; shrubs, 36; spheroidal weathering, 36; vegetation of, 36
neonatal mortality, 686
Ngorongoro Conservation Area, 708, 737, 740, 744–47, 752–53; conservation and tourism, 500–502; cultivation, 497–500; history of land use, 484–89; history of Maasai, 24–25; human population and well-being, 489–95; livestock and wildlife, 495–97; multiple-use protected area, 502–5; transitions, 483–505
Ngorongoro Conservation Development Project, 716
Ngorongoro Conservation Ordinance, 486
nitrogen analysis, ground biota, 203
Nitrospira, 217
Normalized Difference Vegetation Index (NDVI), 57, 107; vegetation production, 57
nourishment, health, 491–92, 495
nutrition, measurements, 492–93
nutritional status, 687–95; prevalence-based reporting, 690–96; stunting, 688, 690, 692–93, 695; summary statistics of anthropometry, 693–96; underweight, 694–95; wasting, 688, 690, 693–94; weight-for-age, 688, 690, 694–96; Z-scores, 690, 694–96
Nyanzian-Kavirondian Orogenic Belt (NKOB), 35; clastic deposits, 35; granites, 35; granitic gneisses, 35; greenstone belts, 35; Kavirondian systems, 35; minor politic beds, 35; Nyanzian system, metamorphosed lava, 35; pebble grits, 35; quartzites, 35; synorogenic granites, 35
Nyerere, and Arusha Manifesto, 797

Okiek hunters, 738
operational taxonomic unit (OUT), 204, 214–16
Orangi River, 38
outbreaks of small mammals, 323, 328, 335–37, 346, 348, *349*; of African grass rat, 328; in agriculture, 345, 347; of multimammate rat, 328, 330, 338; predator responses to, 341–44; in relation to rainfall, 338, 340, 346, 348, *349*
owl: response to rodent outbreaks, 342–43; small mammals in pellets, 333

paleolandscape and humans, 53–61
paleontological sites, 706
park effects
—through employment, 468–70; adult equivalent value, 469; livelihood strategy, 470; livestock holdings, 470; park-related employment, 453, 468, 470, 475–79; poverty line, 469; private safari companies, 468; remittances, 469; Tanzania National Parks, 468
—on household economies, 475–79; agricultural sales, 476; beer sales, 476; bushmeat sales, 478; conservation, 478–79; crop destruction, 475–78; human-wildlife conflict, 478; illegal hunting, 475; in-migration, 478; job opportunities, 479; livestock depredation, 475–78; livestock sales, 476; park-related employment, 475–79; remittances, 476; seasonal noncontract wage labor, 476; trade/small business, 476; year-round contractual employment, 476
—on livelihoods, 463–75; agriculture, 463–64; employment, 468–70; livestock, 465–68
pastoralist areas: carnivore abundance, *424*; carnivore diversity, 437–38
pastoralists and pastoralism, 609, 623–25, 629, 631–34
pedogenesis, volcanic soils, 41
people and wildlife: coexistence, 739; ground biota, 195–96
people-park tensions, 682
people's awareness of protected areas, 452, 458–59, 461–62
phospholipid fatty acids and microbes, 202–4, 211–13, 220, 222–29
Pielou's index, invertebrates, 281
Planctomycetes, 215, 217, 223
plant compositional turnover, 112; rainfall, in relation to, 112; Serengeti plains versus tall grasslands, 112
plant diversity
—patterns, in space and time, 106–14; drivers of plant species richness, 107; fire frequency, 107; map of plant species richness, 108; models, 109; normalized difference vegetation index (NDVI), 107; potential evapotranspiration, 107; rainfall, 107; topography, 107
—spatial and temporal drivers: caveats to model, 110–11; comparing Serengeti to other systems, 112–14; compositional turnover, 112; diversity in space, 106–9; future directions, 120–22; invasive species, 121–22; landscape heterogeneity in tree cover, 115–16; landscape patterns of tree diversity, 114; other systems, 113; patterns, 106–14; plant species number, 112; riverine forests, 120; space, 106–9; spatial patterns of tree density and basal area, 116–17; temporal dynamics in tree cover, 118–20; woody cover change in African savannas, 120–21; woody species density and diversity, 114–20
plant families, food of butterflies, 304, 306–7
plant species richness, total savanna, 112–13
Plasmodium falciparum, 684. *See also* malaria
Pleistocene, 38; Lake Victoria, 38; Nile River, 38
poaching, 453, 471–75, 740–41, 751, 754–55, 762, 802; annual harvest rates, 657–60, 671; buffalo, 651–56; buffalo, exploitation rate, 655; buffalo, spatial distribution of, 656; effects on wildebeest habitat use, 161; elephant, 651–53; future research priorities, 166; impact on wildlife populations, 651–61, 670–73; ivory poaching, 650–53; Maasai consumption of bushmeat, 656; methods for community assessment, 661–64; other species, 655–60; population trends in wildlife, 651–61; potential density of, 662; rainfall effects on, 671; rhino, 651–53; risk of detection, 673; sex-biased mortality in wildebeest, 143; snaring, 650, 656–57, 670–71; subsistence agriculture, 671; threats to the migration, 167–68; wildebeest, 651–53, 655–61, 665–73, 662–73; wildebeest and food regulation, 671; wildebeest offtake sex ratio, 671–72
poaching, management implications, 673–75; bushmeat, alternative protein, 674–75; bushmeat, price, 673; bushmeat, supply and demand, 673–74; demand for meat protein, 669–70, 673–75; household income, 674; human population, growth rate, 661–63, 667–70, 673–75; improving alternative livelihood options, 673–75; income effects on bushmeat consumption, 674; income-generating projects, 674; increasing negative incentive, 673; poachers, risk of detection, 673; poaching, penalties, 673; poaching, profitability, 674; reduce local demand for bushmeat, 675

poaching and antipoaching, 650, 664; buffalo, 651–56; elephant, 651–53; impact of ivory poaching, 650; impact on wildlife populations, 651–61, 670–73; penalties for, 673; rhino, 651–53; snaring, 650, 656–57, 670–71; wildebeest, 593, 651–53, 655–61, 665–73, 662–73; wildebeest offtake sex ratio, 671–72
poaching and humans: agricultural sales, 474; beer sales, 473; bushmeat sales, 474; household economies, 474; income generation, 475; livestock sales, 473–74; and poverty, 802; primary school education, 472; seasonal employment, 473; secondary education, 472; wildebeest, 471; year-round employment, 473–74
poisoning, 611–12, 630, 634
policy and legal frameworks, 723
pollen records, 51
population dynamics: in agriculture, 345; carnivores, 439–40; decline phase, 343; density-dependence, 340, 348; low phase, 347; in relation to rainfall, 348; small mammals, 337–40
population monitoring, 636–37
population regulation: future research priorities, 166; wildebeest abundance, 147–48; zebra abundance, 161–62
poverty, 452, 741; and bushmeat, 712; and child health, 685; and health indicators, 686; and hunting, 718–19; the line for, 469; reduction vs. conservation, 679
power sharing and revenue distribution, 719
precipitation, 46–48; gradient, 47; Lake Malawi, 48; Lake Victoria, 47–48; mean annual, 46; temperature, 47. *See also* rainfall
predation: dilution of risk, 162; impact on female wildebeest, 137; impact on wildebeest recruitment, 129, 136–37, 143, 148–50; impact on zebra, 161–62, 164; nonlethal effects of, 139–40; predator swamping, 129; regulation of migrant population, 153, 164; regulation of resident wildebeest populations, 148
predator-prey interactions, 184, 186–88, 587
predators, extermination of, 776–77
predators, responses to rodent outbreaks, 341–44; in agriculture, 344; lagged response to prey, 342, 347; numerical response to prey, 347, 351; population dynamics, 342, 347; responses to global change, 350; rodent specialists, 342
pre-Nyanzian, 35; crystal-containing gneisses, schists, amphibolites, 35
private land ownership, 746
privatization: of land, 740–41, 746–48, 752; land rights, 483, 485
protected areas, 2
—administrations, 703
—attrition in, 2
—function of, 2
—loss of biodiversity in, 2
—necessary for conservation, 799–801; atmospheric CO_2, 801; biodiversity inside and outside, 799; climate change and ecosystem stability, 800–801; elephant poaching, 800; elephant stress hormones, 800; fire impacts, 799; grazing impacts on soil biota, 800; rodent outbreaks, 800; soil biota and grazing, 800; top carnivores, 799; tree germination in savanna, 800; wildebeest migration, 799
—outside processes, 3
—perceived costs and benefits, 458–63; attitudes and behavior, 462; awareness of, by local peoples, 452, 458–59, 461–62; conservation initiatives, 460; crop damage, 459; income, 460; infrastructure, 460–62; knowledge and awareness, 458–59, 461–62; land-use designations, 458–59; legal uses, 459; nongovernmental organizations, 459; park-related employment, 460; resource utilization, 458, 460
—testing community-based conservation, 5–6
Proteobacteria, 215–18, 223, 225–26
protozoa, 213, 220, 225–26

rabies, 533, 535–36, 538, 541–43, 546–48, 552, 560, 564–65, 611, 616, 625–27, 747, 783; policy for, 783
rainfall, 107–8, 112, 115–17; as driver of rodent outbreaks, 337, 338, 340, 348, 351; as driver of small mammal abundance, 338, 340, 348, 351; gradient of, 345; as key driver, 589–91; Musoma, 49; Shirati, 49; trends with climate change, 350
rain shadow, Crater Highlands, 47–48
red rock hare, 332
refugia for small mammals, 345–46, 349
regeneration of forests, 236, 249–50, 255, 257; episodes of, 245
regime shifts, 256, 600, 601
research benefits for Tanzania, 782–85; anthrax, 783; canine distemper, 783;

economics, tourism and the migration, 782; ecosystem services, 784–85; foot-and-mouth disease, 783; health and wildlife infections, 783–84; insect outbreaks, 785; malignant catarrhal fever, 783; rabies, 783; rinderpest, 783; rodent outbreaks, 784–85; sleeping sickness, 783; social welfare and lion attacks, 784; World Health Organization, 783; World Organization for Animal Health, 783
research in conservation, future of Serengeti, 775–91; anthrax, 783; conservation, park management and sustainable ecosystems, 785; contribution of scientific research to management, 775–82; economics, tourism and the migration, 782; ecosystem services, 784–85; first research, 775; foot-and-mouth disease, 783; health, impacts and wildlife infections, 783–84; insect outbreaks, 785; irrigation from Mara River, 786; Lake Victoria water levels, 786; malignant catarrhal fever, 783; Mau Forest deforestation, 786; rabies, 783; research benefits for Tanzania, 782–85; rodent outbreaks, 784–85; Serengeti future, 785–90; Serengeti northern road, 786–90; sleeping sickness, 783; social welfare, lion attacks, 784; World Health Organization, 783; World Organization for Animal Health, 783. *See also* management and research
resilience, 764
resources, Ngorongoro Conservation Area: competition with wildlife, 496; conservation, 487; natural, 490; vegetation, 495; water, 487–88, 495
retaliation and lethal control, 611–13, 616–19, 624, 628–31, 633–34
revegetation of Ngorongoro Conservation Area, environment, 498–99
revenue: Maasai livelihood, 488; tourism, 501–3
rhino: collapse of, 778; poaching, 651–53, 660
Ricker logistic, 184, 186
rinderpest, 15–18, 60, 240, 257, 485, 501, 533, 535–36, 539, 562, 601, 625, 747, 778, 783; arrival from Asia, 15; buffalo collapse, 16; disappearance of, 17–18; eradication of, 587, 602; exodus of humans, 26; in giraffe, 16; increase of ungulates, 26; starvation of Maasai, 16; in warthog, 16; wildebeest die-off, 16; wildebeest increase, 18; zebra unaffected, 18
riverine forest, 120, 235, *237*, 241, 244–45, 257, 325, *326*, *334*, *337*; birds and tree seed dispersal, 120; establishment of, 240; fire effects on, 120; as refuge habitat for small mammals, 345; small mammals in, 329–30, 337; species richness of small mammals, 330, *334*
riverine grassland, 325, *326*, *334*; small mammals in, 330; species richness of small mammals, 330, *334*
road, northern Serengeti, 786–90; threats to the migration, 167–68. *See also* Serengeti highway
rodent, 324–25, 335–37, 349; abundance, 335–37, 341, 345, 349, 351; in agriculture, 344–45; breeding, 336; and climate change, 349–51; decline phase, 341; density-dependence, 340, 348; and disease, 349–50; diversity, 345, 351; extrinsic drivers of, 336–52; food availability, 346; and global change, 349–51; habitats, 323, 324–33, 325, 336–37; and land-use change, 349; low phase, 347; numerical response of predators to, 347, 351; outbreaks, 335–37, 341, 346, 347, 348, *349*, 784–85, 800; population dynamics, 323, 337–40, 346, 348, *349*; predator responses to outbreaks, 341–44; predators with murid preference, 343; productivity, 345; rainfall, driver of abundance, 324, 337, 338, 340, 346, 348, *349*, 351; refuge habitats, 345; in relation to shrew abundance, *338*; species richness, 345; trap catch, 335, 344; trapping, *337*, *338*; trophic role in ecosystem, 344–49, 351
rufous-nosed rat, *327*
rural livelihoods, 803

safaris, tourism, 502
savanna, converted to agriculture, 746
scales of change in the greater Serengeti ecosystem, 33–62
SD model, 591
seasonality, effects on migration, 155–57
secretary bird, rodent outbreaks, *339*, 342–43
seed: bank, 249; germination of, 252, 254; predators of, 253; viability of, 252
seedlings, 235, 246; of *Acacia*, 247; density of, *243*, 246, 247, 254, 255; of *Euclea divinorum*, *248*; influence of shrubs on, 241; recruitment of, 252, 254; survival of, 241, 243, 246

Serengeti
—as a coupled human-natural interaction: dynamics and resilience, 599–601; human impacts, 586–87; HUMENTS model, 588; hunting, 588; modeling and empirical challenges, 599–601; modeling the Serengeti, 588–90; model results, 594–98; model scenarios, 592–93; rainfall, key role of, 586, 589; role of space, 598–99; socioecological coupling, 590–98; socioecological feedbacks, 585–86; socioecological system, 586–88
—current management challenges, 718–19; illegal offtake, 718; lack of benefits to people, 718; law enforcement, costs and benefits, 719; poverty and hunting, 718–19; power sharing and revenue distribution, 719; tourism benefits, 719; transparency in governance, 719
—function of protected areas, 2; road development in, 1; sustainable over millennia, 1; threats to, 1
—functions: ecological, 702–4; economic, 705–6; Loliondo revenues, 705; Mara River and agriculture, 705; migration, 702; National Strategy for Growth and Reduction of Poverty, 706; paleontological sites, 706; protected area administrations, 703; resident ungulates, 704; sociocultural functions, 706; tourism revenues, 705; traditional medicine, 706; watersheds for agriculture, 705
—future, 785–90; irrigation from Mara River, 786; Lake Victoria water levels, 786; Mara River drying, 785–86; Mau Forest deforestation, 786; Serengeti northern road, 786–90
—institutions, national context, 706–9; land, settlement, and local governance, 709; Ngorongoro Conservation Area, 708; Tanzania National Parks, 708; tourism, 708; wildlife, 707–8; Wildlife Division, 708; wildlife management areas, 707, 711
—institutions, Serengeti context, 709–12; bushmeat hunting, 711–12; hunting blocks, 710; Loliondo Game Controlled Area, 710; Maasai Mara National Reserve, 711; Mara Conservancy, 711; Maswa, Grumeti, and Ikorongo, 710; taboo related to hunting, 712; World Heritage Site, 710
—interactions of functions and institutions, 712–15; community-based wildlife management, 714; conflicts from agriculture, hunting, tourism, 713–14; poverty and bushmeat, 712; Tourism Master Plan, 715; wildlife and agriculture, 712
—lessons for future of conservation, 797–810; agriculture and pastoralism, 803; atmospheric CO_2, 801; biodiversity inside and outside, 799; climate change and ecosystem stability, 800–801; elephant poaching, 800; elephant stress hormones, 800; fire impacts, 799; grazing impacts on soil biota, 800; health and malnutrition, 804; human systems and conservation objectives, 801–2; human system sustainability, 802–4; human-wildlife interactions, 804; illegal hunting, 802; major conservation issues, 802; Maswa wildebeest refuge, 801–2; minimizing influence of modern humans, 798; Nyerere, Arusha Manifesto, 797; outside water supplies, 801; poaching and poverty, 802; protected areas are necessary for conservation, 799–801; rodent outbreaks, 800; rural livelihoods, 803; Serengeti management priorities, 804–8; social and economic drivers, 803; soil biota and grazing, 800; top carnivores, 799; transmission of disease, 803–4; tree germination in savanna, 800; wildebeest migration, 799
—management: approaches to multifunctionality, 715–18; community conservation services, 716–17; community development, 715; Frankfurt Zoological Society, 717; incentives, 715; law enforcement, 715; Ngorongoro Conservation Development Project, 716; priorities, 804–8; Serengeti ecosystem community forum, 717–18; Serengeti Ecosystem Management Project, 717; Serengeti Regional Conservation Strategy, 715–16
—shaping the ecosystem, 11–27; beginning of conservation, 21–25; German era, 18–21; ivory trade, 14–15; lessons from history, 25–27; the nineteenth century, 14–15; rinderpest, 15–18; Serengeti ecosystem, 12–13
—way forward for management, 719–28; CITES, 720; comanagement, state and local, 722–23; Convention on Biological Diversity, 720; economic gains and incentives, 723; enhance comanagement to governance, 721–23; evaluate management effectiveness, 724–28; fences and

fines, 723–24; framework for management effectiveness, 725; governance models, 721–22; governance monitoring, key variables and indicators, 725–28; governance process, 722; governance structure, 722; IUCN categories, 720–21; law enforcement, 724; legal clarity, 723; management monitoring, 724; policy and legal frameworks, 723; promote institutional compliance, 723–24; stakeholders, 721; strengthen institutions for management, 720–21; wildlife monitoring, 724
Serengeti Biodiversity Program, 324
Serengeti Closed Reserve, 22
Serengeti ecosystem, 12–13
—boundaries and tribes, 12
—management complexity, multiple functions and institutions, 701–29; current management challenges, 718–19; functions, 702–6; institutions, the national context, 706–9; institutions, the Serengeti context, 709–12; interactions of functions and institutions, 712–15; management approaches to multifunctionality, 715–18; way forward for management, 719–28
—other administrations, 12–13
Serengeti Ecosystem Management Project, 717; Serengeti ecosystem community forum, 717–18
Serengeti Game Sanctuary, 22
Serengeti highway, 741, 753, 758, 786–90; land-cover change, 527
Serengeti-Mara ecosystem, causes of change, 744–51; bushmeat, 750; canine distemper, 747; climate change, 750–51; common land ownership, 746; conversion of savanna to agriculture, 746; deforestation, 744, 748; disease transmission, 747; environmental governance, 748; expansion of settlement, 745; globalization, 750–51; human population growth, 751; human populations, land access and use, 744–47; hunting pressure, 749; illegal hunting, 749–50; intensification of agriculture, 747; land tenure, 745; Loita Plains, 746; Loliondo Game Controlled Area, 747, 749; Maasai Mara National Reserve, 744–46, 748, 750–51; Maasai pastoralists, 746–49, 750; malignant catarrhal fever, 747; Mara Conservancy, 750; Mara River drying, 748; Mau Forest, 748; Ngorongoro Conservation Area, 744–47; private land ownership, 746; privatization of land, 746–48; poaching, 751; rabies, 747; rinderpest, 747; subdivision of land, 746; western Serengeti peoples, 750; wheat farming, 746; wildlife conservancies, 748; wildlife decline, 751; wildlife management area (WMA), 749; wildlife profits, 748–49, 751
Serengeti National Park: establishment, 22–25; Grzimek surveys, 24; Lamai wedge, 25; Maasai and Ngorongoro Conservation Area, 24–25; northern extension, 25; redrawing boundaries, 23–25
Serengeti northern road. *See* road, Serengeti northern; Serengeti highway
Serengeti Regional Conservation Strategy, 715–16
Serengeti Research Project, 776
serval, rodent outbreaks, 343
settlement and intensification, 631–33, 637–39
shaggy swamp rat, *326*; in riverine floodplain, 329; in riverine forest, 330
Shannon-Wiener index of diversity: invertebrates, 281; small mammals, 333, *334*, 345
Shinyanga Region, human population, 661
short-grass plains, 325, *326*; carnivore densities, *428*; small mammal species richness in, 331
short rains, 52
shrew, 324–25, *326*, *327*, 328, 335, 337; abundance, 337; extrinsic drivers of, 337; in long-grass plains, 331; in relation to rodent abundance, *338*; in riverine floodplain, 329; in riverine grassland, 330; in short-grass plains, 331; trapping, 335
side-striped jackal, 432; density, *428*; rodent outbreaks, 343; small mammals and disease, 350
sleeping sickness, 626, 783
small mammals, 323–57; acacia rat, *326*; in *Acacia* woodland, 325; African dormouse, *326*; African grass rat, *326*, 328, 338, 349; African groove-toothed rat, *327*; African mole rat, *327*, 328; African pouched rat, *326*; African swamp rat, *327*; in agriculture, 344–45; black rat, *327*; bottom-up drivers of, 346, 348, 349, 351; breeding, 336, 345; broad-headed mouse, *326*; brush-furred mouse, *327*;

small mammals (*continued*)
bush rat, *326*, *327*, 328; and climate change, 349–51; community composition, 333, 345, 351; conservation, 350; density-dependence, 340, 348; and disease, 349–50; diversity, 323–24, 333, 344–46, 351; DNA analysis, 335; extrinsic drivers of, 336–52; fat mouse, *326*; in floodplain, 325; food availability, 346; and global change, 349–51; groove-toothed creek rat, *326*; habitats, 323–33, 336–37; hyrax, 324; in intermediate-grass plains, 325; in kopjes, 323, 325; and land-use change, 349; large naked-soled gerbil, *326*; lemur, 324; long-eared elephant shrew, *327*; in long-grass plains, 325; low phase, 347; in montane habitat, 325; mouse, *326*, *327*; multimammate rat, 323, *326*, 328; numerical response of, predators to, 347, 351; outbreaks, 323, 338, 340, 346–49; population dynamics, 323–24, 337–40, 345–46, 348–49, 351; predator responses to outbreaks, 341–44, 346; predators of, 323, 341–44; predators with murid preference, 343; pygmy gerbil, *326*, *327*; rarity, 325; refuge habitats, 345, 348; in relation to rainfall, 337–38, 340, 346, 348, 349, 351; in relation to shrew abundance, *338*; in riverine forest, 325; in riverine grassland, 325; rodent, 324; rufous-nosed rat, *327*; Serengeti Biodiversity Program, 324; shaggy swamp rat, *326*; in short-grass plains, 325; shrew, 324; soft-furred rat, *327*; spatial distribution, 324–33, 346, 349; species richness, 325, 345, 351; spiny mouse, 326; spring hare, *327*; striped grass mouse, *326*, *327*; taxonomy, 325; in *Terminalia* woodland, 325; thicket rat, *327*; trap catch, 335, 344; trapping, 333–35, *337*, *338*; tree mouse, *326*; trophic role in ecosystem, 344–49, 351; in wetlands, 325
snaring, 650, 656–57, 670–71
social and economic drivers, 803
social capital, 764
social welfare, lion attacks, 784
sociocultural functions, 706
socioecological systems (SES), 586, 591; models of, 585, 588, 599
socioeconomic status and infectious disease, 695; links to human health, 688; malnutrition in GSE, 696
soft-furred rat, *327*; in kopjes, 332; in riverine forest, 330
soil: biota and grazing, 800; catena, and small mammals, 330 (*see also* catena); degradation, 487; distribution of soil properties, 41–46; erosion, 497; fertility, depletion of, 587, 600; formation, 41–46; moisture, 41–42, 236, 241, 243–44; pedogenesis, 41–46; volcanic, 489
soil chemistry: C:N ratio, 210–11, 222; carbon, soil organic, 210; carbon, total, 203, 210, 222; exchangeable bases, 203, 208, 222; extractable elements, 203, 208–9, 221–22; nitrogen, total, 210, 221; nutrients, 208, 220, 223
soil nutrients, 208, 220, 223; fire effects on, 89; nitrogen, 89; nutrient availability and water-holding capacity, 41–42; soil carbon, 89
solar minima, 49; Maunder, 49; Sporer, 49; Wolf, 49
Southern Oscillation Index (SOI), 176–83, 186, 188. *See also* El Niño/Southern Oscillation (ENSO)
spatial coupling, 590
spatial processes, importance of, 598–99
species richness, of small mammals, 325, 333, *334*, 344–46
spiny mouse, *326*; in agriculture, 332; in kopjes, 332; in montane habitat, 332; in *Terminalia* woodland, 329
spotted eagle owl, 323
spotted hyena, 421, 422, 431, 438; abundance, *424*, 429; commuting, 427–28; density, *428*, *430*; interspecific killing, 432–33; seasonal prey migration, 427
spring hare, 327; in short-grass plains, 331
stakeholders, 721
starvation: annual wildebeest energy demands, 133; interacting with other mortality, 151–53; lactation and starvation, 133, 137–38
striped grass mouse, *326*, *327*; in riverine forest, 330; in *Terminalia* woodland, 329
structural equation model, 176–77
stunting in children, 688, 690, 692–95; in Bunda District, 692; definition, 688; within the GSE human population, 696; in Loliondo District, 692; in Meatu District, 692; prevalence groups, 692–93; across rural Tanzania, 692–94; in Serengeti District, 692

sulfate reducers, microbes, 211–13, 225–26
surveillance, infectious disease, 533–73; analysis of infectious disease data, 559–63; antigenic and genetic detection, 552–53; case reporting, 546–47; contact tracing, 547–48; diagnostic pathology, 550–51; ecological data, 556–58; environmental and climatic factors, 558; epidemiological modeling, 562–63; field-based approaches to case detection, 545–49; general surveillance approaches and definitions, 544–56; host ecology, 556–57; innovative methods for disease surveillance, 554–56; integration of tools for infectious disease research, 563–68; laboratory-guided disease detection, 549–54; microbiological methods, 551–52; pathogens, 534–35; population abundance and density, 548–49; serology, 553–54; temporal and spatial data analysis, 561–62; tools for infectious disease research, 535–63; vector ecology, 557–58
survival dynamics, of forests, 236; tree seedlings, 243
survival rate, health in humans, 489
sustainability for wildlife and people, key issues, 737–42; coexistence, people and wildlife, 739; environmental governance, 741, 748, 754, 759; expansion of farming, 740–41; expansion of settlement, 740–41; globalization, 742; Grumeti Game Reserve, 737; guardians of wildlife, 739; human population growth, 740–41; human populations, land access and use, 741; hunting pressure, 741; Ikorongo Game Reserve, 737; land rights, 740; Loliondo Game Controlled Area, 737, 740; Maasai Mara National Reserve, 737–41; Maasai pastoralists, 738–40, 742; Mara Conservancy, 740; Mara River, 737–38, 741; Mara River drying, 741; Maswa Game Reserve, 737; Mau Forest, 737–38; Ngorongoro Conservation Area, 737, 740; Okiek hunters, 738; poaching, 740–41; poverty, 741; privatization of land, 740–41; Serengeti highway, 741; subdivision of land, 740; Tatoga pastoralists, 738, 742; western Serengeti peoples, 741; wheat farming, 740; wildlife decline, loss, 740; wildlife management area (WMA), 740; wildlife profits, 739, 741
sustaining the Serengeti, 757–64; accountable institutions, 759–61; climate change, 764; collaboration, 759–60; deforestation, 758; devolution of power, 760–61; environmental governance, 759; globalization, 764; habitat fragmentation, 758; hunting pressure, 762–63; illegal hunting, 757, 762–63; land tenure, 757; Maasai Mara National Reserve, 760–61; Mara Conservancy, 761; Mara River, 758, 764; Mara River drying, 764; Mau Forest, 758–59; payments for ecosystem services, 758–59; poaching, 762; poverty, 741; resilience, 764; Serengeti highway, 758; social capital, 764; transboundary management plan, 759; wildlife conservancies, 759–60; wildlife profits, 757–59, 761, 764
synchrony: advantages for predator swamping, 129, 134, 139; behavioral cues, 132; calving, 129–34, 142; environmental cues, 131, 134; positive feedback loops, 132; role of male competition, 131–32, 138, 144

taboo related to hunting, 712
Tanzania National Parks, 708
Tanzanian craton, 34
Tanzanian Geological Survey, 34
Tanzanian Wildlife Conservation Act, 634
Tatoga pastoralists, 738, 742
tawny eagle, rodent outbreaks, *339*, 342–43
taxonomy of small mammals, 325, 335
Teclea trichocarpa, 245–46
Terminalia woodland, 238, 325, *326*, *334*, *337*; small mammals in, 328–29, 337; species richness of small mammals, 329, *334*
termitaria, 45; agama lizards, 45; animal species, 45; anteater chat, 45; *Arvicanthis*, 45; *Cynodyn dactylon*, 45; fine scale spatial variation, 45; grass species, 45; snakes, 45; sooty chat, 45; vegetation, 45–46
termite mounds. *See* termitaria
termites, 270, 272–74, 289; decomposition, 273; dung removal, 272, 292–93; landscape topography, 273; mound density, 273; predators, 273–74; seasonal activity, 289–90, 292–93; termitophiles, 273–74
Thermomicrobia, 217, 224
thicket rat, *327*; in kopjes, 332; in riverine forest, 330; in *Terminalia* woodland, 329
thickets, 244, 249; establishment of, 245–46, 249; as refuges for animals, 245

Thompson's gazelle, poaching, 656–60
tick-borne diseases, 628
ticks and fire, 90, 94–95
TLU. *See* tropical livestock units
top-down drivers, small mammal dynamics, 347
topi, 126–40; breeding and rut, 134–35; calving, 136–37; calving synchrony, 129–31; diet, 127–28; gestation, 135–36; intake rates, 127; lactation, 137–38; leks, 134; maturity, 138; migration, 126; poaching, 656–60; recruitment, 177–79; reproduction, 129–38; spatial distribution, 129; water requirements, 128–29
topographic diversity, 40; evenness, 40; Loliondo, 40; Ngorongoro Conservation Area, 40; richness, 40; riverine forest, 40
total exchangeable bases, 41–42
tourism, 708; benefits, 719; cultivation, 501; employment, 501; game reserves, 503; handicraft, 502; human population, 501; livestock, 501; lodges, 501–2; revenues, 501–3, 705; safaris, 502; Tourism Master Plan, 715; water resources, 502; wildlife, 501
traditional medicine, 706
transboundary management plan, 759
trapping, of small mammals, 333, 335; in agriculture, 344
tree cover, 115–19; change in, 591, 593–94, 601; limitations on woody cover, 115–16; photographs and, 119; rinderpest, wildebeest, and fires, 119; temporal dynamics, 118–20
tree germination in savanna, 800
tree harvesting, 57–58; tree canopy, 58; vegetation continuous fields (VCF), 57–58
tree hyrax, 330
tree mouse, *326*; in long-grass plains, 331; in short-grass plains, 331
trees: basal area in relation to rainfall, 117; harvesting, 57–58; increase of, 779; size class density, 117; spatial patterns of, 116–17; species richness in, 114
triceps skinfold (TSF), nutrition measurement, 492, *493*
trophy hunting, 630, 632, 636
tropical livestock units (TLU), livestock biomass, *490*, 492, 495
trypanosomiasis, 535–36, 539, 540, 542, 553, 557–58, 564–68, 626. *See also* sleeping sickness
Tsavo National Park, 51
tsetse flies and fire, 90, 94
TSF. *See* triceps skinfold

UAC. *See* upper-arm circumference
under-five mortality rates, 686–87, 695
undernutrition in humans: definition, 688; in the GSE, 696. *See also* wasting
underweight, in humans, 694–96; wasting, 688, 690, 693–94; weight-for-age, 688, 690, 694–96; Z-scores, 690, 694–96. *See also* wasting
upper-arm circumference (UAC), nutrition measurement, 493

vaccination, domestic dogs, 781
vector ecology, 557–58
vegetation cover and ground biota, 207–8
vegetation resources, Ngorongoro Conservation Area, 495
Verrucomicrobia, 215–18, 224–25, 228
vertisols and ground biota, 197–98
villages and cultivated areas. *See* agriculture, villages and cultivation
volcanic soils, 41; alkaline, 41; ash deposits, 41; nutrient rich, 41

warthog, poaching, 656–60
wasting, in children, 688, 690, 693–94; in Bunda District, 694; definition, 688; in Loliondo District, 694; in Meatu District, 694; prevalence groups, 692–93; in Serengeti District, 694. *See also* underweight
water: environmental concerns, 488, 495; resource rights, 487–88; tourism, 502
waterbuck, poaching, 657–60
water-holding capacity of soils, 41–42
watersheds for agriculture, 705
western Serengeti peoples, 741, 750
wetlands, 325
wet season, carnivore densities, *428*, *430*
wheat farming, 740, 746, 753
white-tailed mongoose: rodent outbreaks, 343; small mammals and disease, 350
wildcat: abundance, 429, *431*; rodent outbreaks, 343
wild dog, 421, 439; seasonal prey migration, 427; small mammals and disease, 350
wildebeest, 589
—abundance, 140–54; and buffalo increase, 778; disease, 153; increase of, 778; instantaneous growth rate, 147; life tables, 149; mortality, 143, 151–53; popu-

lation abundance, 144–47; population density, 144–47; population dynamics, 144–53; population regulation, 147–48; recruitment, 148–51; resident, 140–42; resident-migrant comparison, 142–44; survival, 150–51
—annual harvest rates, 657–60, 671; food regulation, 671; impact of rainfall on population growth, 671; poaching, 593, 651–53, 655–73, 779–80; recruitment, 177–80; sex ratio, 671–72
—biology, 125–74; annual energy expenditure, 133, 138; annual reproductive cycle, 135; body size, 142; breeding and rut, 134–35; calf survival, 136, 148–51; calving, 136–37; calving grounds, 134, 137, 163; calving synchrony, 129–34, 142; diestrus cycle, 131; diet, 127–28; digestion, 127, 154–55; general biology, 126–40; genetics, 140–42; gestation, 135–36; habitat requirements, 142; intake rates, 127; lactation, 133, 137–38; maturity, 138, 143–44; recruitment, 177–80; reproduction, 129–38; rumination vs. hindgut fermenting, 154–55; sex ratio, 143; taxonomy and evolution, 126; territoriality, 134; water requirements, 128–29; weaning, 150–51
—crop-raiding, 594
—disease in, 496
—distribution, 19–20
—forage intake, 590
—migration, 154–65, 184, 586, 590, 599, 799; advantages of migration, 162; collaring program, 155–57; compensatory vegetation growth, 163; daily activity, 164; daily movement, 158–59; density-dependent movement, 159–65; distance of migration, 157–59; drivers of migration, 162–64; element composition of grass, 163; future research priorities, 166; grass protein, 163; historic migratory patterns, 159–62; poaching, 161, 167–68; predation, 163–64; rainfall, 162; range, 140–42, 155–57; rate of population change, 180–81; recruitment, 177–80; rotational grazing, 163; soil fertility, 162; Speke Gulf at, 51; threats, 167–68; water availability, 163
—poaching, 593, 651–53, 655–73, 779–80
—population: collapse from rinderpest, 16; decline, 594; increase, 18, 587, 602; rate of change in, 180–81
—refuge in Maswa, 801–2
wildlife, 707–8; and agriculture, 712; attacks and injuries related to, 609–11, 614–19, 623, 629; conservancies, 748, 754, 759–60; conservation and ground biota, 195–96; decline of, 740, 751, 754–55; density and ground biota, 199; depredation on livestock, 452–53; guardians, 739; monitoring of, 724; poisoning, 752; populations, perturbations in, 51–52; profits, 739, 741, 748–49, 751, 757–59, 761, 764; tourism, 632–33, 636–38
wildlife, in Ngorongoro Conservation Area: competition with livestock, 496; conservation, 486–87, 489; disease in, 501; migratory routes, 487; tourism, 501
Wildlife Conservation Act, 487
Wildlife Division, 708
wildlife management areas (WMAs), 608, 622–23, 636, 707, 711, 740, 749, 754, 782
woodlands, 245; *Acacia*, 24; establishment of, 245
woody cover change in African savannas, 85–86, 120–21; climate change as driver, 121; CO_2 fertilization effects, 121; cover change in Serengeti, 121; future predictions of, 121; spatial-temporal patterns of, 85–86; stimulated effects on, 86
World Health Organization, 783
World Heritage: conservation, 488–89, 503; World Heritage Site, 710
World Organization for Animal Health, 783

Xiphinematobacteriaceae, 216, 218–19

Yellowstone National Park, 2

zebra, 154–65; collaring program, 155–57; daily movement, 158–59; diet, 154; digestion, 154–55; drivers of migration, 162–64; grazing competition, 155; historic migratory patterns, 159–62; migration, 157–59; poaching, 657–60; population abundance, 146; population regulation, 161–62; predation, 161–62; range extent, 155–57; rumination vs. hindgut fermenting, 154–55
zoonoses, 626–27